Island Arcs
Deep Sea Trenches
and
Back-Arc Basins

Maurice Ewing Series 1

Manik Talwani and Walter C. Pitman III, Editors

Contributors

R. N. Anderson
M. Barazangi
P. Bird
R. L. Bruhn
P. Buhl
C. A. Burk
J. G. Caldwell
L. Chaqui
W. Connelly
A. K. Cooper
I. W. D. Dalziel
F. J. Davey
J. N. Davies
S. E. DeLong
W. R. Dickinson
W. L. Donn
E. R. Engdahl
T. J. Fitch
P. J. Fox
H. S. Gnibidenko
W. Hamilton
J. W. Hawkins, Jr.
W. F. Haxby
E. M. Herron
B. B. Hill
M. Hill
R. E. Houtz
B. L. Isacks
K. H. Jacob
H. Kanamori
D. E. Karig
R. W. Kay
J. Kelleher
J. P. Kennett
W. S. F. Kidd
L. D. Kulm
J. W. Ladd
M. G. Langseth
M. S. Marlow
A. Masias
W. McCann
D. P. McKenzie
G. F. Moore
J. C. Moore
K. Nakamura
D. Ninkovich
J. R. Ockendon
A. E. Ringwood
A. D. Saunders
D. W. Scholl
W. J. Schweller
D. R. Seely
R. H. Sillitoe
P. L. Stoffa
M. Talwani
J. Tarney
S. R. Taylor
M. J. Terman
R. C. Thunell
M. N. Toksöz
D. L. Turcotte
S. Uyeda
T. Watanabe
J. S. Watkins
A. B. Watts
S. D. Weaver
J. K. Weissel
C. C. Windisch
M. Winslow
J. L. Worzel

Maurice Ewing Series 1

Island Arcs
Deep Sea Trenches
and
Back-Arc Basins

Edited by
Manik Talwani and
Walter C. Pitman III

American Geophysical Union
Washington, D.C.

Island Arcs
Deep Sea Trenches
and
Back-Arc Basins

Standard Book Number: 0-87590-400-9
Library of Congress Catalog Card Number:
76-58102

Printed in the United States of America by
LithoCrafters, Inc.
Ann Arbor, Michigan

PREFACE

A three day symposium was held in honor of the late Maurice Ewing at Arden House, Harriman, New York on March 28-31, 1976. The symposium is the first of a planned Maurice Ewing series of symposiums to be held biennially. The American Geophysical Union has agreed to publish the proceedings of the symposiums in a special Maurice Ewing series. This volume represents the first of the series.

In planning the symposium we deliberately chose a subject that engaged Maurice Ewing's interest and efforts for many years, but one which still holds promise for much future research and exploration. The participants were researchers active in the study of Island Arcs, Deep Sea Trenches and Back-Arc Basins. We made a special effort to invite a number of students who were interested in the field. This blend of senior and junior scientists exemplifies the philosophy of Professor Ewing, who, throughout his academic life, did extraordinary amounts of original research and study which constantly involved young scientists and students in a wide area of oceanographic research.

The symposium was supported by the G. Unger Vetlesen Foundation, the U.S. Office of Naval Research, the U. S. National Science Foundation, and the Lamont-Doherty Geological Observatory of Columbia University.

Manik Talwani
Walter C. Pitman III

MAURICE EWING

We are here to remember Maurice Ewing who was a colleague and friend to nearly all of us. He was a fortunate man; he was born at the right time and in the right country. That it was the right time was far from obvious when he took his Ph.D. in 1931. The Depression was at its worst and it certainly did not look a good time to embark on a scientific career. However, Ewing got started sufficiently early to do very substantial original work, including important investigations at sea, before war came. His work in the 1930's led naturally to important tasks for the Navy and left him, at the end of the war, as one of the conspicuously effective members of the scientific community. The war had given him substantial experience not only of science at sea, but of how to get things done.

I do not know how Maurice first became concerned with the earth. He had done his Ph.D. in the Physics Department of the Rice Institute in Houston and to leave the main line of physics was not the obvious thing to do. It was probably the chance of summer employment while a student; he did seismic work for three summers in Louisiana, and I think it was this which turned his thoughts to a career in Earth Science. However it happened, it suited his personality. I cannot see Maurice spending his life in a laboratory; he needed a wider field and a closer relation with the outside world.

Maurice was much more than a good physicist. You can be a good physicist if you are interested in physics and have exceptional intelligence, ability and application. To do what Maurice did needed more, it needed what are now unfashionable gifts of character. It required a passionate belief in the worthwhileness of what he was doing and also great skill in the techniques of coercing bureaucracies and seizing opportunities. He had the ability to make people feel that what he wanted was not only what he wanted but what the Good Lord would have done. It was a useful back-up to this position that everyone realised how extremely troublesome he would be if thwarted. This self confidence combined with the powers of righteous indignation were part of the secret of his incredible achievement.

For forty years Maurice Ewing used his very great abilities to build up an organisation for studying the earth. It is not too much to say that he made respectable the study of the two-thirds of the earth's surface that is beneath the sea. When Ewing was young it was eccentric for a bright young physicist to become a geologist; bright young men stayed in physics. There were the great men of pre-war physics, Rutherford and the rest of them, whom the young Ewing described as 'hardly mortal;' they were discovering the inside of the atom; the study of the inside of the earth was something of a backwater.

Ewing was sent to sea by that eccentric genius Dick Field, who, although he did rather little original work himself, was one of the most influential geologists of this century. He persuaded (perhaps coerced is the better word) not only Ewing but also Harry Hess and myself to embark on the study of the ocean floor.

Ewing's first achievement was to show that seismic refraction shooting could be done at sea. I went with him from Woods Hole in the research ship ATLANTIS in 1937. He always said that I was sent to check that what he was doing was sound. It was not so; I went to learn and to go and do likewise. It was my first experience of a small oceanographic vessel, it was tremendously stimulating and also great fun. Among many memories, I recall Maurice saying, just as we were going to bed: "You know I think we really need cast TNT and not this flaked stuff." I said, "Yes Maurice, but there aren't any shops and we've only got flaked TNT." He replied, "Don't you think, perhaps, we could . . ." He produced an electric coffee pot, we melted the TNT in it and poured it into moulds made of folded paper. In the middle of the operation the skipper of the vessel, Captain McMurray, came in and started to knock out his pipe on the box of TNT. Ewing looked at him and, after what seemed a long time, said, "You know Captain McMurray, if I was you I wouldn't knock my pipe out on that there box." McMurray continued to knock out his pipe and said "And for why Dr. Ewing, if you was me, wouldn't you knock your pipe out on this box?" I could stand it no longer and said "That box is full of TNT."

McMurray didn't say a word. He just turned round and walked out.

Ewing was tremendously inventive, and ingenious, he could make anything. His first achievement was to show that, with relatively small facilities, he could build equipment which would work at sea, and that he could do a whole range of things that no-one had thought of doing before. The trouble had been not only the technical difficulty and the lack of physical and engineering expertise, but also, I think, a feeling on the part of geologists that the floor of the ocean was not their business, and even that it was useful to have the oceans unknown. The oceans were a sink into which you could cast your difficulties. In the literature of the 1920's people assumed anything they liked about the ocean floor, it could bob up and down, it could be permanently there, it could come and go. There is a famous Annual Address to the Geological Society of London, given by J. W. Gregory in 1929, in which he talks about the Geology of the Oceans. The odd thing about this very interesting discussion is that it contains no suggestion that something ought to be done about the uncertainties; the whole argument is pure speculation with no clue that one might have a programme for investigating it. This was the jam that Ewing broke. He fitted means to ends, he knew what he wanted to do, he found ways of doing it and persuaded other people to help him. The achievement was indeed remarkable: you could number on less than the fingers of one hand the other institutions that played any large part in the enterprise.

Ewing's prime interest in life was finding out about the oceans. The things he found led to the great synthesis which is now one of the standard examples of a scientific revolution. It has become part of the story of man's understanding of Nature. I think I can say without giving offence, that this synthesis was not quite Ewing's field. He was interested in all the details, in the techniques and in the gathering of the data. He had a certain hesitation about its interpretation in global terms. I always felt a good deal of sympathy with him over this after reading some of the papers of his competitors. It is quite clear, however, that the recent revolution in ideas about earth history has been based predominately on the work done at Lamont-Doherty by Maurice and his collaborators.

Finally, I would like to express my own personal debt to Maurice as an example and an encouragement all through my career. We didn't meet very often, we were both rather busy and I knew him well in a funny kind of way. I usually knew what he would do and what he would say, but I didn't spend very many days with him in my whole life. When we did meet, it was always fun and we both enjoyed it. We were much of an age, he was one year older than I, and there was no feeling of constraint between us. Maybe I had an easier relation with him than almost anyone; we were not competitors for funds or ships and neither of us was going to try to get the things the other wanted.

The great enterprise that Ewing started has been fun. We have found the nature and history of two thirds of the earth's surface and we have enjoyed ourselves while doing it. To take part in such an adventure is an exceptional good fortune which does not come in every generation; none of us will ever forget who was the prime mover in it.

Now, as a worthy memorial to Maurice, we have this fascinating meeting. I have sat all day listening to discussions of problems that could not even have been formulated when Ewing started his work. The only sad thing is that he is not here to enjoy it with us.

An Address Given at the Dinner Held in Memory of Maurice Ewing on March 29th, 1976 in the Low Library of Columbia University

by

Sir Edward Bullard
Institute of Geophysics and Planetary Physics,
University of California, San Diego

CONTENTS

SOME BASIC PROBLEMS IN THE TRENCH-ARC-BACK ARC SYSTEM

Seiya Uyeda

Earthquake Research Institute, University of Tokyo, Tokyo, Japan

Abstract. Some basic problems related to the thermo-mechanical aspects of trench-arc systems are discussed. The origin of high heat flow, volcanism and back arc basins, in particular, is considered from two basic models, i.e. the frictional heating model and the secondary mantle flow model. The possible effect of increasing subduction rate within the frictional model is considered also. The analysis reveals that the frictional model is not necessarily self-defeating. Finally, possible mechanisms of marginal basin formation are considered in the light of one reconstruction history of the western Pacific and its margins. Four mechanisms are suggested: 1) tensional rifting due to a subducted ridge, 2) entrapment of an old ocean basin, 3) back-arc opening and 4) "leaky" transform fault.

Introduction

The trench-arc back arc systems, now found mainly along the continental margins around the Pacific, are among the most spectacular tectonic features in the world. In the paradigm of plate tectonics, they are the sites where an oceanic plate subducts under another plate. Among the three major types of plate boundaries, these converging boundaries seem to be the most complex. In this paper, the author attempts to list the problems that he considers basic and unsolved. Some review and author's ideas will also be presented on these problems. It is hoped that many, if not all, of the problems cited below will find solutions in the aftermath of the first Ewing Symposium.

Basic Unsolved Problems

The trench-arc-back arc systems (T-A-BA systems hereafter) have characteristic features in common (e.g. Sugimura and Uyeda, 1973; Burk and Drake, ed., 1974). Theories must be able to explain the origin of these features. It has to be added, moreover, that at the present stage of the game, theories must also be able to explain the diversity of the systems, e.g. T-A-BA systems are generally arcuate in shape convex toward the ocean but not always. Suggested mechanisms for the "arc" (Frank, 1968; Matsuda and Uyeda, 1971) do not seem to explain this diversity.

In addition to that cited above, the following three problems seem to be the most basic. Of course, these problems are not independent but are closely inter-related.

1) Why is the upper mantle "hot" in the inner zone of the arc?
(origin of volcanism and high heat flow)
2) Why is the stress "tensional" in the back-arc area?
(origin of back-arc basins)
3) Why do the arcs rise?
(origin of arc mountain belts)

To understand these problems, the solutions of the following more specific problems may be helpful.

4) Origin of the trench outer swell and its diversity (Watts and Talwani, 1974).
5) Shallow outer-arc seismicity and its diversity (Kanamori, 1971; Kelleher et al., 1974).
6) Real nature of Wadati-Benioff zone (Umino and Hasegawa, 1975; Hasegawa, 1975; Hasegawa, Umino and Takagi, 1976).
7) Constance of dip angle of Wadati-Benioff zones and its variations (Jischke, 1976).
8) Time-space distribution of arc volcanic rocks (Miyashiro, 1974).
9) Source material and depth of magma production. Does descending oceanic crust melt (Ringwood, 1974; this Symposium)?
10) Possible role of water and CO_2 in magma production (Wyllie, 1971; Kushiro, 1972).
11) Is "frictional heating" important? (Turcotte and Oxburgh, 1969; Hasebe, Fujii and Uyeda, 1970).
12) Possible effects of higher subduction rate (Sugimura and Uyeda, 1973; Sugisaki, 1976).
13) Is there outer arc volcanism, or such a thing as initial "geosynclinal volcanism?

14) Is there such a thing as a "cannibalistic" trench? (Scholl et al., 1970; Karig and Sharman, 1975).
15) Processes in the trench-arc gap (Dickinson, 1974; Burk and Drake, 1974).
16) Is back-arc spreading the same as the conventional sea-floor spreading? (Sclater, 1972; Hawkins, 1976).
17) Why some arcs develop back-arc basins and others do not? (Wilson and Burke, 1972).
18) What does a collision of ridge and trench do? (Uyeda and Miyashiro, 1974; Delong, Fox and McDowell, 1976).

In the following, some of the problems listed above will be considered.

Features of Trench-arc-back Arc Systems as Present Day Manifestations of Pacific-Type Orogeny

A typical cross-section of shallower features of T-A-BA systems would be as shown in Fig. 1 (Karig, 1974). Physiographic features are variously named but may be cited as follows. From the oceanic side, ocean floor, trench outer swell, trench, subduction zone, trench-slope break, non-volcanic (frontal) arc, volcanic arc, and back-arc area and remnant arc. The arc-trench gap is defined as the zone between the trench-slope break and the volcanic arc (Dickinson, 1973). These features are not common to all the arcs. Zonal arrangement is also apparent in the more deeply rooted geophysical and geological activities, such as seismic, geodetic and thermal activities (Fig. 2) (Sugimura, 1960; Utsu, 1971; Sugimura and Uyeda, 1973; Yoshii, 1975). Seismicity oceanward of the trench is shallow and of tensional focus, reflecting the bending of the oceanic plate (Stauder, 1968). Closer to the land, high seismic activity related to the underthrusting of the oceanic slab takes place. All the great earthquakes are found in this zone. Most of the great earthquakes are low angle thrust type but occasionally normal fault type events, possibly related to the breakdown of the slab, have been reported (Kanamori, 1971; Abe, 1972). Mantle earthquakes in the continental or island-arc side of the trench are apparently confined in a small wedge above the slab as shown in Fig. 2. Yoshii (1975) noted that in the Japan Arc the landward limit of these events coincides with the oceanward limit of the low Q, low V zone under the arc (Utsu, 1971). Yoshii proposed to call this boundary the "aseismic front". The aseismic front is clearly separated from the volcanic front which also runs parallel to the trench, but farther back from the trench. Crustal earthquakes inland show that the stress field there is predominantly compressional in the east-west direction (Ichikawa, 1971). Seismic stress patterns are in good agreement with those inferred from active geologic faults

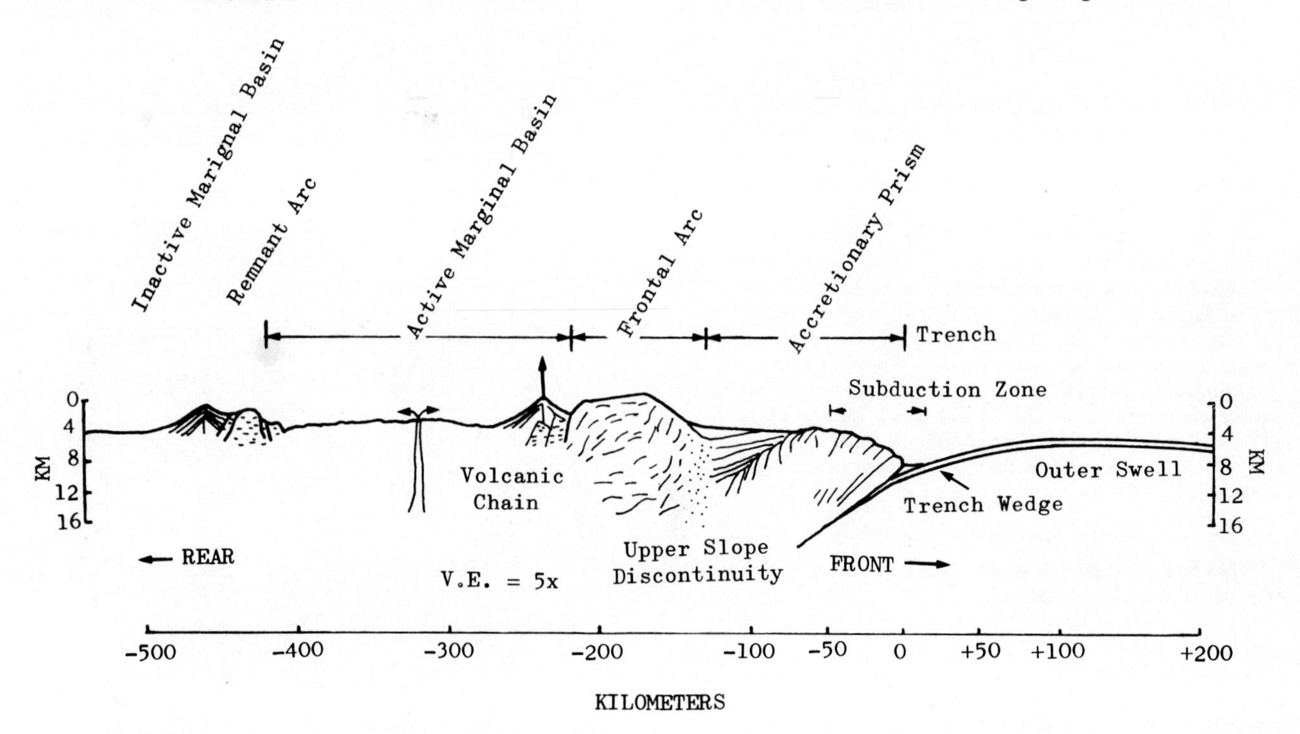

Figure 1. Generalized framework of a western Pacific island arc system (Karig, 1974)

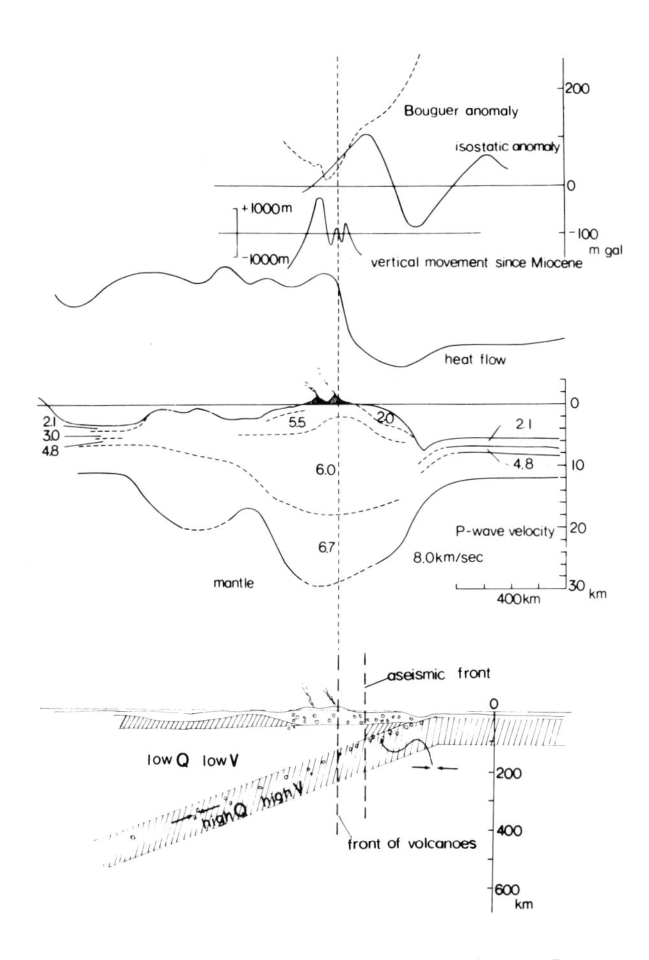

Figure 2. Cross-section of the Northeast Japan arc. (Combined from Sugimura and Uyeda, 1973; Utsu, 1971; Yoshii, 1975)

(Huzita, 1969), suggesting that today's activities have persisted at least for the period concerned with the active geologic faults (late Quaternary). The continuation of mantle earthquakes extends along the Wadati-Benioff zone. Intermediate and deep earthquakes have been interpreted, from their source mechanism (Isacks and Molnar, 1971) and from their locations relative to the slab (Engdahl, Sleep and Lin, 1976) as intra-plate events (McKenzie, 1969; Toksöz, Sleep and Smith, 1973). The general seismotectonic situations outlined above appear to fit with the steady state plate tectonic model. However, there are still problems if one looks into more details. First, general seismic features in the trench areas outlined above are not the same for all the arcs, but are different from one arc to another. As is well-known (Kanamori, 1971; Kelleher et al., 1974), the largest magnitude of great earthquakes varies from arc to arc. They are, for instance, great in Alaska and the Aleutians and become of smaller magnitude along the northwestern Pacific margin down to the Japan arc. In the Izu-Bonin-Marianas arc, no great earthquakes are known. In addition, there are few shallow inter-plate events in the latter case. Kanamori (1971) presents an interesting explanation for these features in terms of gradual decoupling at the interface zone between the landward and subducting plates. This view is consistent with the existence to non-existence of trench outer swells from north to south along the Pacific plate margin (Watts and Talwani, 1974). Why does this decoupling proceed, systematically along the Pacific margin, and how does this idea fit with other observations, such as heat flow, volcanism, presumed duration of the present mode of subduction? In other words, do the differences in seismicity correspond to the geologic evolution of the T-A-BA system?

Secondly, deep earthquakes of the Wadati-Benioff zones also present problems. What is the significance of the length of the Wadati-Benioff zones? Does it represent the length of subduction that started some 10 million years ago (Isacks, Oliver and Sykes, 1968), or the time-constant for the slab to become heated to such a degree that it loses its brittleness (McKenzie, 1969)? The former view does not seem to be compatible with the general ideas of the plate tectonic models of the Pacific that assume that the present mode of plate subduction started much longer ago (e.g. Uyeda and Miyashiro, 1974). Even if the latter idea were correct, it still remains somewhat mysterious that brittleness is preserved to the depth of many hundreds of kilometers when the inter-plate earthquakes under the San Andreas fault, for instance, are limited to the uppermost 10 to 20 km, "because" the earth material below that depth cannot cause earthquakes (Brace and Byerlee, 1970). Dehydration of hydrous minerals in the slab might cause restoration of brittleness at some depth (Raleigh, 1967); would this mechanism be applicable to depths of the order of the deepest earthquakes? Source mechanisms also vary from arc to arc (Isacks and Molnar, 1971). Toksöz, Sleep and Smith (1973) try to explain the observed variety of source mechanisms in terms of the balance between the gravitational body force and the mantle support and Forsyth and Uyeda (1975) explain this observation in terms of the differences in the rate of subduction. These ideas must be checked against further observations. Recently a distinctly two-layered Wadati-Benioff zone has been ascertained for the Northeast Japan arc (Umino and Hasegawa, 1975; Hasegawa, Umino and Takagi, 1976) (Fig. 3), substantiating an earlier observation of Tsumura (1973). It was further shown that the events in the upper layer are down-dip compressive whereas those on the lower layer are down-dip tensional. The source mechanism may have something to do with

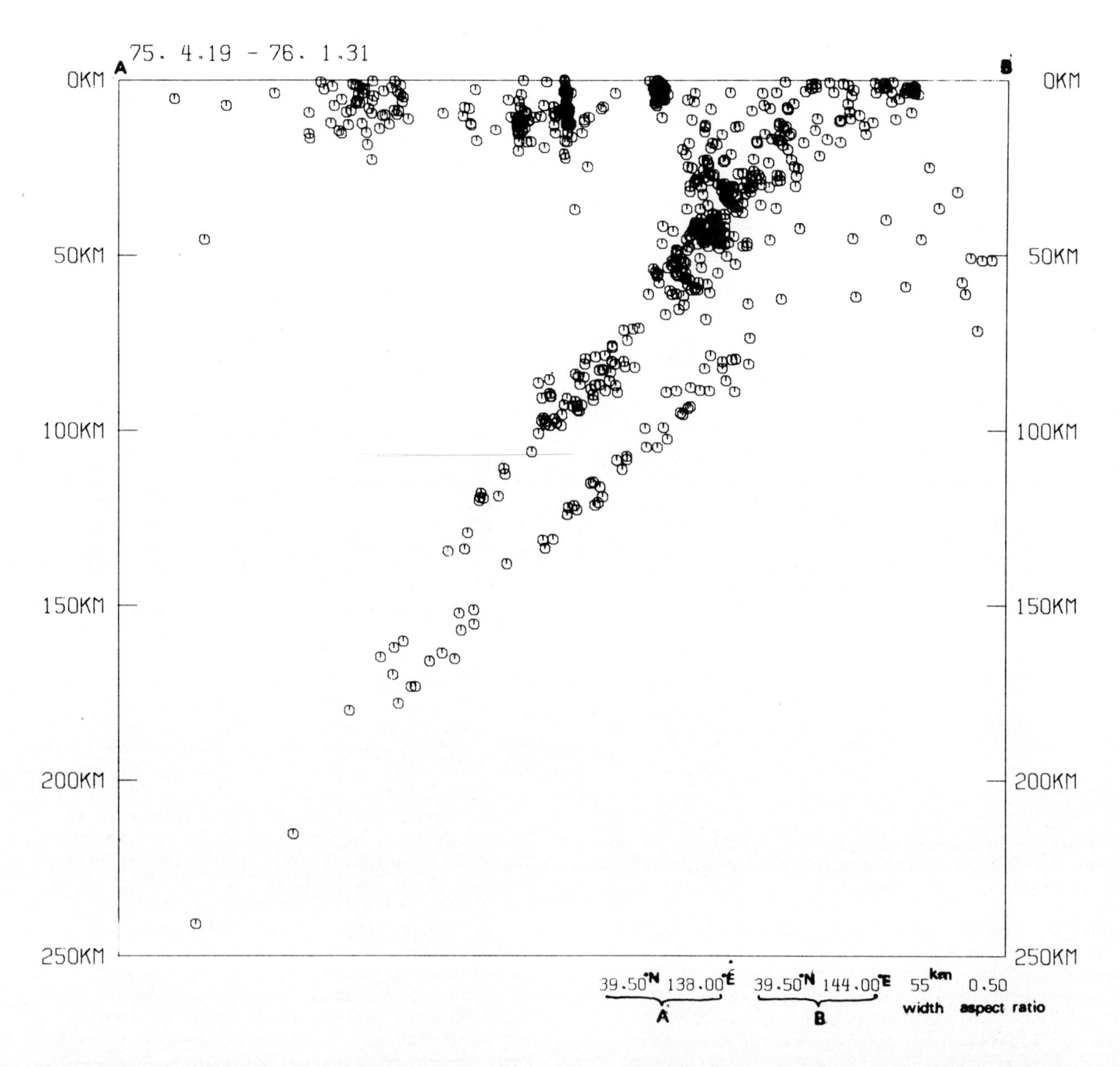

Figure 3. Two-layered deep seismic plane under the Northeast Japan arc (Hasegawa, Umino and Takagi, 1976)

the long-existing problem of re-stretching of the down-going slab (Lliboutry, 1969; Jischke, 1976). At any rate, we must realize there is much to be further studied regarding the true nature of the Wadati-Benioff zones.

Thirdly, if we look into the stress field on the landward side of the arc, a further problem arises. It has been generally accepted that back arc basins have been formed by extensions: some of them are actively opening at present (e.g. Karig, 1974). If extension is in progress today as postulated for the Andaman Sea (Lawver, Curry and Moore, 1976), Mariana Trough (Karig, 1974) or Lau Basin (Sclater et al., 1972), shallow normal fault type earthquakes may be expected. Strangely few observations have been made on this problem, although Sykes, Isacks and Oliver (1969) studied the Fiji-Tonga region and found some strike-slip and thrust events and Fitch (1972) found a few normal faulting events in the Andaman Sea and Gorontalo Basin (Celebes). A cursory look at the world's seismicity map (ESSA, 1970) shows relatively few shallow epicenters in these supposedly actively opening back arc basins compared with the zones of oceanic spreading centers. In the Japan Basin or the Philippine Sea Basin, practically no shallow earthquakes have been located. Almost no seismicity was detected by OBS observation in both Kita-Yamato Bank area in the Japan Sea (Nagumo, 1971) and the Central Basin fault area in the West Philippine Basin (Shimamura, Tomoda and Asada, 1975). The Japan Sea and the Central Basin fault may be no longer active even though high in heat flow (Yasui et al., 1968; Hilde et al., 1976; Watanabe, Langseth and Anderson, this volume). If so, the inactivity of the Japan Sea opening must be explained if one maintains that active subduction should cause back-arc extension. Fukao and Furumoto (1975), on the other hand, gave examples of shallow thrust events along the eastern margin of the Japan Sea suggesting that the eastern Japan Sea may be underthrusting the Honshu Island. Noritomi and Saeki (1970) even presented some evidence of northeast dipping seismic plane under the Japan Sea side of the Northeast Honshu. Although these activities do not necessarily attest to the active opening of the Japan Sea, it might also be conceivable that back-arc extensional opening is accompanied by little or no seismic activity. When we come to the origin of the extensional forces behind island arcs required for the tension tectonics, the problem becomes more serious. Why is the stress tensional behind an island arc which is converging with an oceanic plate? Artyushkov (1973) argues that inhomogeneity in the crustal thickness generates major tectonic stresses, the stress being tensional where the crust is thick. This would provide tensional forces for continental margins to thin themselves toward the ocean. But, the marginal sea opening takes place only in conjunction with oceanic subduction, so that some mechanism specifically connected with subduction would be needed to explain this problem. Several thoughts have been put forward. One is to assume that some diapiric rise in the back arc area causes seaward displacement of the arc (Karig, 1971; Matsuda and Uyeda, 1971). Alternatively, the tensional force is due to the secondary forced convective flow behind the arc (McKenzie, 1969; Sleep and Toksöz, 1973). In addition to this, tension caused by subduction of an active ridge (Uyeda and Miyashiro, 1973) and suction due to the retreat of the oceanic slab (Elsasser, 1971) may also be possibilities. Clearly the problem is not merely a mechanical one but is closely related to other aspects, notably the thermal regime of the system.

The heat flow distribution characteristic of the T-A-BA systems has been known for some time (e.g. Uyeda and Horai, 1964; Vacquier et al., 1967). Heat flow is generally subnormal in the outer zone and high in the inner zone, the outer boundary of the high heat flow zone being coincident with the "volcanic front" (Fig. 2). Such a contrasting thermal structure seems to have persisted in the past orogenic belts around the Pacific. During the late Cenozoic orogeny in Japan (Mizuho orogeny), volcanic activity has been confined in the western zone of Northeastern Honshu (Matsuda, Nakamura and Sugimura, 1967). For the Mesozoic and Paleozoic periods, evidence of contrasting thermal belts is now exposed in the paired belts of metamorphism in the circum-Pacific areas (Miyashiro, 1961, 1972, 1973). The pair consists of an outer, low temperature-high pressure type metamorphic belt and an inner high temperature-low pressure one. It was shown that the subterranean temperature field for these paired metamorphic belts is consistent with the thermal regime in the present day arcs (Takeuchi and Uyeda, 1965). Widespread occurrences of Mesozoic granitic rocks in the inner areas of arcs also support the inner high temperature zone. Matsuda and Uyeda (1971) postulated that the pair of contemporaneous outer and inner belts having contrasting characteristics is the basic element of orogeny and they called this the Pacific-type orogeny. Although it was called Pacific-type because of present day prevalence of T-A-BA systems in the Pacific margins, it has become increasingly clear that it is not a local variant of orogeny but is the most important type of orogeny (Dewey and Bird, 1970). The new concepts of orogeny seem to require major revisions of classical geosynclinal theories of orogeny (Mitchel and Reading, 1969). For instance, in the classical eugeosynclinal evolution, the "geosynclinal phase" of subsidence, sedimentation and basic volcanism is followed by the "orogenic phase" of uplift and granitic magmatism. In the new type of orogeny, the eugeosynclinal environment and the granitic magmatism are active almost simultaneously in the juxtaposed outer and inner belts.

In the Pacific-type scheme of orogeny, the outer belt is always "cold", so that no in-situ volcanism is likely to take place. The ultrabasics and basic volcanics found in eugeosynclinal formations, traditionally called geosynclinal volcanics or "ophiolites", must be pieces of ocean crust which were left behind during the subduction of the oceanic plate (Dewey and Bird, 1970). But, whether this view is really always right or not has still to be tested.

Although the new concept has greatly helped to improve our understanding of orogeny, it is still unclear how mountain ranges are formed. It is easier, perhaps, to see how mountain belts are formed when continent/continent collision takes place, than in the cases of Pacific-type or the Cordilleran mountain belts. Is it due to faulting and folding in the inner zone by the compression from the subducting plate or to the uprise of plutonic bodies from the mantle as in the Sierra Nevada? How can one make these possibilities compatible with the predominance of tensional forces behind arcs? Maybe one is looking at features at different stages of evolution. In the Japanese arc area, the Japan Sea opened presumably tens of millions of years ago but the present high mountains in Japan have arisen essentially in the Quaternary time (Sugimura and Uyeda, 1973). But, this would not present a possible solution to the problem unless a mechanism for a change in the stress pattern were delineated. Even in the case of the collision type orogeny, like the Himalayas, the mountains rose to the present height only in the last few million years or so (Gansser, 1964), whereas it apparently was many millions of years ago that continental collision started between India and Eurasia. In spite that theories are called mountain building theories, the very problem of mountain building has not yet been quite solved.

Thermal Regime of T-A-BA Systems

In order to better understand the thermal regime of the T-A-BA systems, two major thoughts have been developed. One is called the frictional heating model and the other is the secondary mantle flow model. The two major lines of thought mentioned earlier to explain the mechanism of back-arc opening stem from these two types of thermo-mechanical regime.

Frictional heating models (Turcotte and Oxburgh, 1969; Minear and Toksöz, 1970; Hasebe, Fujii and Uyeda (1970) (HFU hereafter), instead of solving the direct problem, took a different approach. They tried to solve the inverse problem by finding the best set of concerned parameters, such as rate of heat production and effective thermal conductivity, to explain the observed heat flow distribution. They found that the observed heat flow, in particular the high heat flow in the back-arc basin, such as the Japan Sea, can be explained only if stress for shear heating of the order of a few kilobars exists along the slip plane and the effective thermal conductivity at temperature above the wet-solidus of basic rocks amounts to some ten times as large as the ordinary thermal conductivity value. The latter condition implied that the effective heat transfer mechanism must operate, i.e. some massive transfer of heat, such as extensive magmatism or diapirism must be at work. This uprise of mass might account for the opening of marginal back-arc basins (Fig. 4a) or the upheaval of island arc mountain belts.

Heat flow profiles across arcs have another important characteristic feature: the low heat flow in the trench-arc gap zone (Fig. 2).

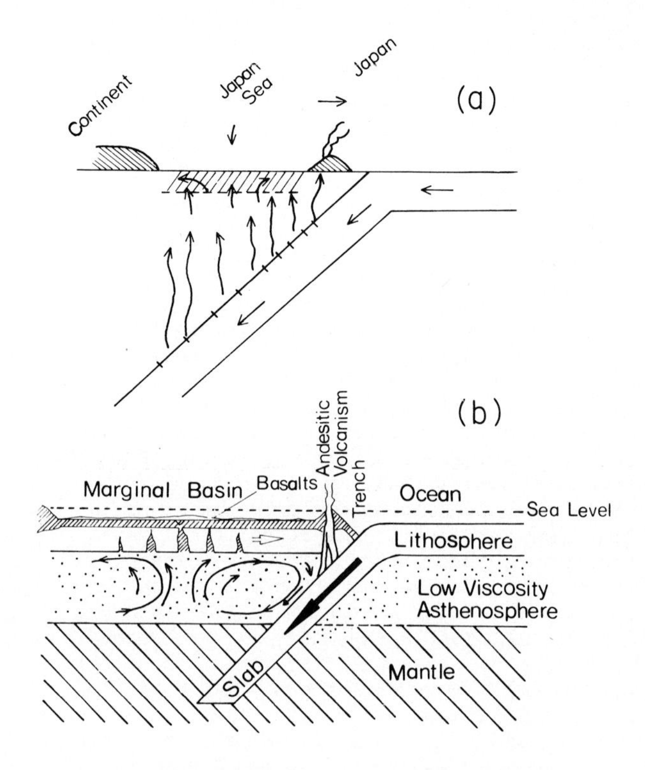

Figure 4. Two possible types of back-arc opening.
(a) Frictional model (Matsuda and Uyeda, 1971)
(b) Secondary flow model (Sleep and Toksöz, 1973)

In frictional heating models, it is difficult to explain this zone of low heat flow because large quantities of heat are being generated along the slip zone. In order to keep the heat flow in the trench-arc gap as low as observed, the HFU model found it necessary to assume that heating takes place only at depth greater than 60 km. HFU attributed this to possible cancellation of frictional heat by a negative heat due to dehydration of hydrous minerals in the subducting crust. Recently, Anderson, Uyeda and Miyashiro (1976) noticed the possible importance of this point and showed, through thermodynamic calculations, that the dehydration of ocean crust can indeed effectively cancel out the frictional heat. Dehydration of hydrous oceanic crust will provide water that may play an important role in the generation of magma (e.g. Wyllie, 1971; Kushiro, 1972; Ringwood, 1974). The details of the role of water in this respect, however, must be the subject of further study since it is not known whether the water immediately leaves the slab or is carried down with the slab to greater depth, or even driven into the slab.

The alternative way to explain the thermo-mechanical regime of the subduction system is, as mentioned before, to assume that the upper mantle wedge above the slab is a viscous fluid capable of sustaining convective flow forced by the descending slab as shown in Fig. 4b (McKenzie, 1969; Sleep and Toksöz, 1973; Andrews and Sleep, 1974; Toksöz, this symposium). In this way, heat from the lower part of the mantle can be brought to shallow depth so that the time constant for heat transfer becomes short enough and, at the same time, tensional force is exerted at the bottom of the crust in the back-arc area to open a basin.

In both arguments, the mechanism of massive transfer of heat is employed. The basic difference is, of course, in the source of heat and the force for the mass movement. As long as the interface between the upper surface of the slab and the mantle wedge remains sufficiently below the solidus temperature, it may be presumed that friction would be large and little flow would be induced. However, as soon as the temperature approaches the solidus, the material can no longer sustain high stress and will start to flow. It is this point that proponents of the flow model emphasize: the frictional model may be right to start with but as the temperature rises, no more heat will be produced and in this way it may be a self-defeating process.

Possible Effects of Subduction Rate

If the arc activities are caused by subduction of the oceanic slab, it may be anticipated that a higher rate of subduction would produce higher arc activities. In this section, a brief survey will be made of this point.

Comparative studies of activities of various subduction systems in relation to different subduction rates are limited because some systems are ill-defined in terms of both subduction rates and activities. It is not always clear, moreover, whether the subduction rate or the convergence rate is relevant, although in most cases these two rates are similar. Besides, some activities, especially thermal ones, are not expected to directly reflect the present day subduction but are "convoluted" effects of the past history. However, it has been noticed that some parameters, such as depth of the trench, mantle seismicity, depth of the deepest mantle earthquakes and silica index of the magma at the volcanic front are significantly correlated with the convergence rate (Sugimura and Uyeda, 1973). Miyashiro (1972) further noticed that tholeiitic magma is formed only where the convergence rate is 8 cm/year or more, whereas only the calc-alkaline and alkaline series magmas are formed where the rate is less than 8 cm/year. Sugisaki (1972, 1976) demonstrated that the frequency of occurrence of andesite and the chemical compositions of volcanics are closely related to the convergence rate given by Le Pichon (1968) as shown in Fig. 5a,b. The percent of andesite relative to the total occurrence of calc-alkaline volcanics increases and the alkalinity decreases with increasing convergence rate. Based upon the assumption that andesite is generated only in the presence of water (Kushiro, 1972), it may be reasonable to suggest that more andesite occurs when the rate of subduction of the oceanic plate is higher because more water is provided to the mantle. As to the chemistry of magma, the solidus temperature is possibly raised to a shallower depth with higher rates of subduction because of more active frictional heating (Sugimura and Uyeda, 1973; Sugisaki, 1976). Let us examine this point, because it is not evident that higher subduction rates raise the isotherms. As noted by HFU (1970), among others, subduction of cold oceanic slab is an effective cooling system by which isotherms are lowered more and more as the subduction rate increases were there no frictional heat production.

Viscous heating per unit volume ϕ is expressed as:

$$\phi = \eta(v/a)^2 \qquad (1)$$

if the thickness of the shear zone is a and the velocity difference across the shear zone is v, and η is the viscosity. The total heat production, then, is

$$W = \phi \cdot a = \eta \cdot v^2/a \ (= \sigma \cdot v) \qquad (2)$$

where σ is the shear stress. This shows that viscous heat generation may be proportional to v^2. Numerical model experiment by Minear and Toksöz (1970) showed that $W \propto v^2$ is not compatible with observations because the mantle temperatures become unrealistically high for high subduction rates. In reality, as the temperature rises, η decreases. If then we assume $W \propto v$, instead of v^2, the temperature rise of any part of the slab at any given depth would be independent of the subduction rate, provided heat conduction is small. But in fact the conduction is not negligible and our numerical experiments showed that even with the assumption of $W \propto v$, the heating effect is too great for higher subduction rates. This would mean that in the expression $W = \sigma \cdot v$, σ decreases with increasing temperature. If, on the other hand, it is assumed that $W \propto v^0$, the effect of increasing v is purely to depress the isotherms. Thus, if we express the relation as,

$$W \propto v^n \qquad (2)$$

the value of n should be $0<n<1$.

Ida, Fujii and Uyeda (1976) have recently examined this problem. Their simple model is a viscous medium, as shown in Fig. 6, which is

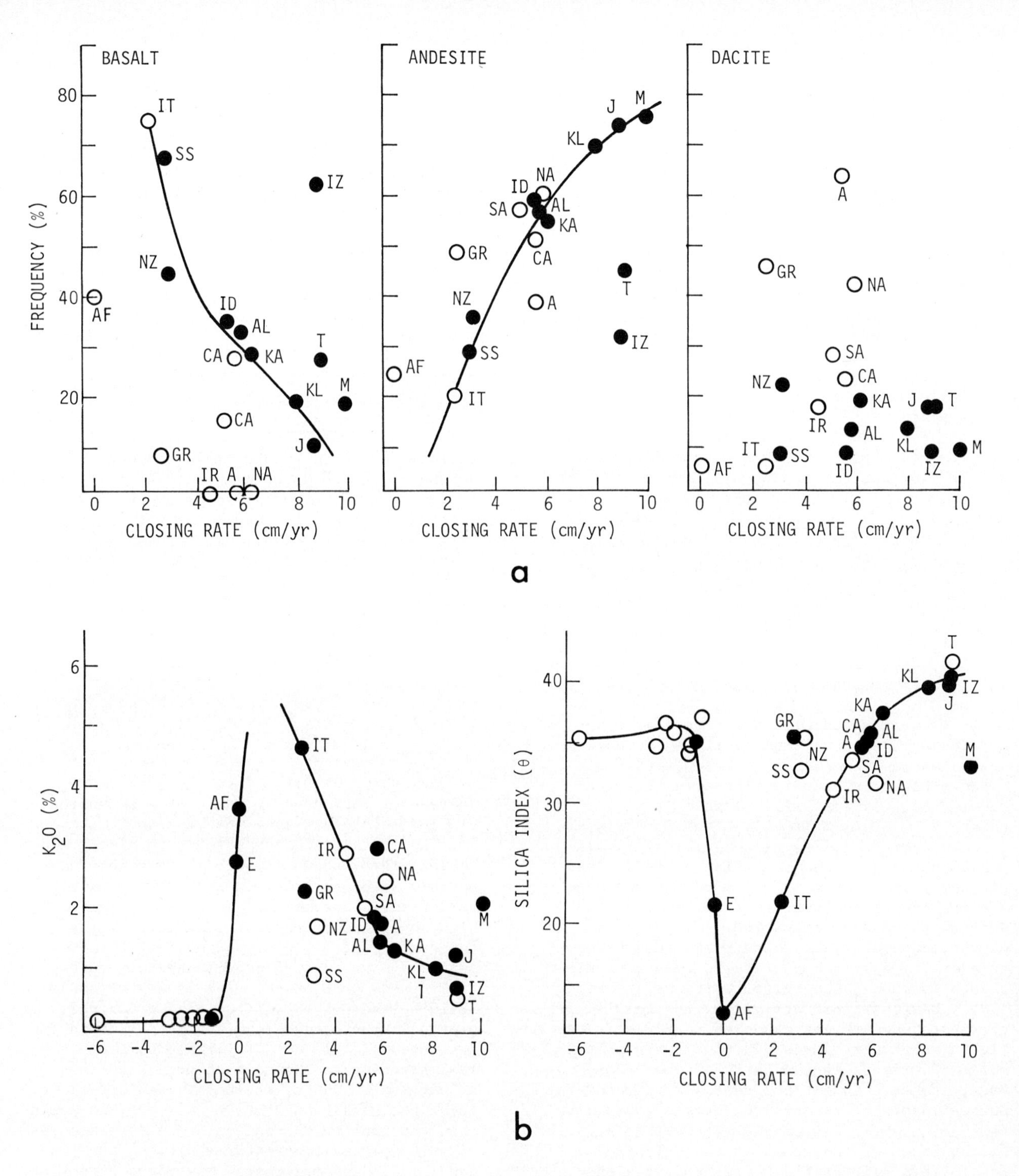

Figure 5. Correlation between characteristics of volcanic rocks and converging rates at plate boundaries (Sugisaki, 1976)
(a) Correlation between proportion of occurrence of various rock types and closing (converging) rate of plates.
(b) Relationship between chemical compositions of volcanic rocks and closing (converging) rate of plates. Negative rates are for diverging boundaries.
Symbols: A: Central Andes, AL: Aleutian and Alaska, AF: Kenya and Tanzania, CA: Central A-merica, GR: Greece, ID: Indonesia, IR: Iran, IT: Italy, IZ: Izu-Mariana, J: Japan, KA: Kamchatka, KL: Kurile, M: Melanesia, NA: North America, NZ: New Zealand, SA: South Andes, SS: South Sandwich Islands, T: Tonga

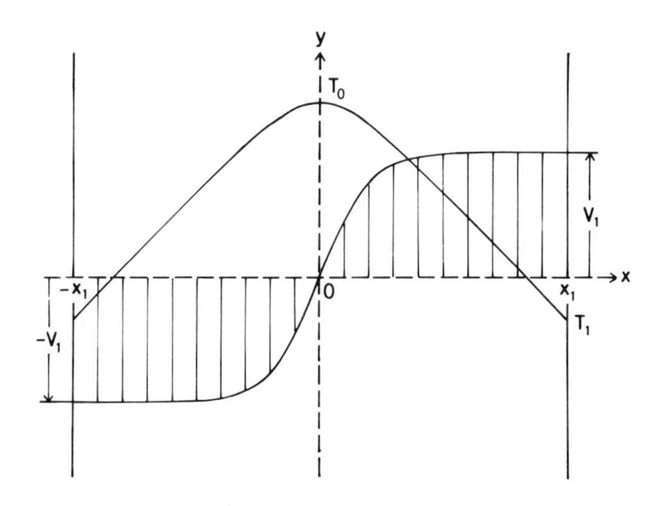

Figure 6. Model of the system considered. Viscous fluid between $-x_1$ and $+x_1$ is in steady laminar flow. Velocities at $-x_1$ and $+x_1$ are $-V_1$ and $+V_1$. Arrows indicate the velocity and the symmetric curve is the temperature distribution. (Ida, Fujii and Uyeda, 1976)

bounded by $-x_1$ and $+x_1$ and extends to infinity in the y direction. Assume steady state in which boundaries are moving in the y directions with $v = -v_1$ and $v = +V_1$, both kept at temperature $T = T_1$. Basic equations are:

σ = const: mechanical equilibrium

$W = \partial/\partial x(v\cdot\sigma)$: heat generation

$\partial v/\partial x = \sigma^m, f(T)\cdot\exp(-E/T)$: constitutive equation

$Q = -K\ \partial T/\partial x$: heat flow

......(3)

where m is 1 for a Newtonian fluid, E is the activation energy for the temperature dependence of viscosity which also contains f(T), K is the thermal conductivity. When K is constant, the energy balance equation leads to:

$$d^2T/dx^2 + (\sigma^{m+1}/K)f(T)\exp(-E/T) = 0 \quad (4)$$

which can be integrated analytically. The results are as follows (see Figs. 6 and 7). (The details are reported elsewhere.) The flow pattern is such that the boundary shear zone is confined to a narrow region near x = 0 where the temperature is maximum T_o. T_o and the shear stress σ depend on the velocity V_1 as shown in Fig. 7. Thus, in a stationary state, higher rates of movement result in higher temperature and heat flow. As was expected, σ decreases with V_1 or with T_o. Similar effects were obtained by Turcotte and Schubert (1971). Thus, n in Eq. (2) does appear to be $0 < n < 1$. Numerically, it is shown that the temperature rise and increases in heat flow due to increases in velocity are quite small; for instance, velocity increases of an order of magnitude would entail a temperature rise of a few hundred degrees and a heat flow increase of 0.1×10^{-6} cal/cm^{2}sec. These seem to be compatible with the observations that heat flow values do not correlate with the subduction rate explicitly but the chemistry of volcanics does. Thus, as far as this simple model is concerned, the frictional heating model does not seem to be entirely self-defeating.

Possible Origin of Marginal Seas and the Evolution of the Western Pacific

As mentioned repeatedly, the origin of marginal basins is one of the most important unsolved problems in plate tectonics today. Recently, magnetic lineation studies in various marginal seas have rendered considerable new information pertinent to the problem (Watts and Weissel, 1975; Weissel and Watts, 1975; Tomoda et al., 1975; Kobayashi and Isezaki, 1976; Watts, Weissel and Larson, 1976). From these studies, it may be said that sea-floor spreading occurred in such basins as Shikoku Basin and South Fiji Basin. In other basins, like the Japan Basin, the lineations are hard to correlate with magnetic time scales although the symmetric nature of the profiles has been suggested (Isezaki and Uyeda, 1973; Isezaki, 1975). Yet in some other basins such as the West Philippine Basin (Ben-

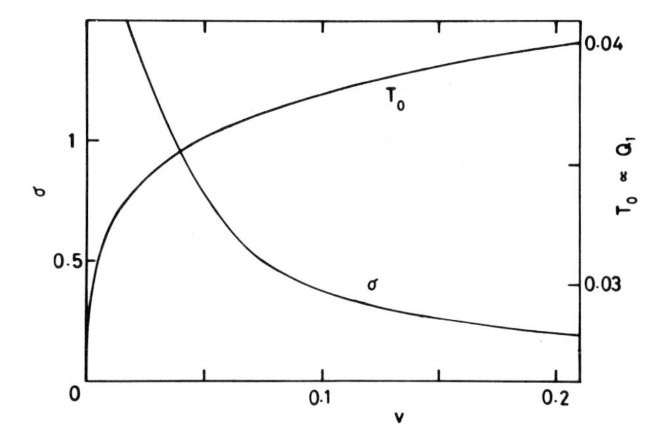

Figure 7. Relationship between the maximum temperature T_o at x = 0 and the velocity at the boundary, V_1 (curve T_o), and relationship between the stress σ and V_1 (curve σ). (Ida, Fujii and Uyeda, 1976)

Avraham, Bowin and Segawa, 1972; Louden, 1976; Watts, Weissel and Larson, 1976) and Bering Sea Basin (Cooper, Marlow and Scholl, 1976) magnetic lineations indicate that these basins might have been parts of old Pacific Ocean basins trapped by the formation of trench-arc systems. (Uyeda and Ben-Avraham, 1972; Cooper, Scholl and Marlow, 1976). Thus there have apparently been three kinds of mechanisms for generation of magnetic lineations in marginal seas. One shows back-arc basin spreading similar to ocean-floor spreading, the second is the case where lineations are not regular enough and not correlative with magnetic time scales, and the third is the case attesting to a trapped ocean basin. Here the important tectonic question about the back-arc spreading would be: "Is the back-arc opening the same process as the ordinary sea-floor spreading or not? What are the similarities and dissimilarities between these processes?" In addition to magnetic studies, more detailed petrologic and crustal as well as heat flow studies of the two types of the sea floor (trapped and back-arc extension) should be carried out to solve this problem. One aspect to be noted here is that the magnetic lineations of all the marginal basins investigated so far are distinctly less pronounced than those in the oceans. They suggest much lower magnetic intensity and poorer regularities. In this regard, there may be some important difference between ocean spreading and back-arc spreading. The marginal sea opening, whatever its causes may be, is generally smaller in scale and less orderly compared to the ocean-ridge spreading. But it should also be kept in mind that the less well-defined nature of magnetic lineations is true also for the basins which are supposedly trapped oceans. Perhaps processes related to or subsequent to the entrapment might have disturbed the original, well-defined lineations.

Recently, Hilde, Uyeda and Kroenke (1976) have attempted to present a comprehensive model of the development of the western Pacific and its margin, mainly based on the marine magnetic lineations, paleomagnetic data, DSDP results and the hot-spot hypothesis related to the Hawaiian-Emperor chain. Their reconstruction with special references to the possible modes of marginal sea formation, may be summarized as follows (see Fig. 8). (For full explanation of the reconstruction, see Hilde, Uyeda and Kroenke (1976).)

The present Pacific plate started to grow from almost a point at about -190 m.y. and generally has migrated north-westward (Larson and Chase, 1972; Hilde, Isezaki and Wageman, 1976). At about -135 m.y., the Pacific plate had grown as shown in Fig. 8a. There were two subparallel spreading ridges in the western part of the Pacific plate. These two ridges are needed to generate the Japanese and Phoenix lineations observed today. Further west, the ridge was probably formed of segments connected by a few north-south trending transform faults. This is an assumption employed to explain the later entrapment of the Western Philippine basin in a manner similar to the one suggested by Uyeda and Ben-Avraham (1972): (see later). In the north, the Kula-Farallon ridge was moving northeastward. An east-west trending transform fault is assumed here to locate the Mesozoic sequence of lineations in the right position in the Bering Sea region to be trapped later (see later). At about -100 m.y., the situation would have been as shown in Fig. 8b. The western part of the Kula-Pacific ridge was about to collide with or subduct under the eastern margin of Asia. Subduction of the actively spreading Kula-Pacific ridge might have caused the opening of the Japan Basin (Uyeda and Miyashiro, 1974).

Formation of the South China Sea probably was initiated at about this time. In Fig. 8b, a back-arc spreading is indicated for the South China Sea, but it could have been related to a ridge subduction, depending on the amount of the off-set of ridge segments (Ben-Avraham and Uyeda, 1973). Up in the northern rim of the Pacific, the Kula-Farallon ridge was subducted and the east-west trending long transform fault, assumed in Fig. 8a stage, was still left behind (Fig. 8b). At about -70 m.y., subduction started along the fault and the fault became a converging boundary. In Fig. 8c, (-65 m.y.), the Bering Sea and the Aleutian arc were already in existence. The manner of entrapment of the Aleutian Basin assumed here is similar to the entrapment of the western Philippine Basin in that a transform fault was transformed to an island arc. The entrapment of the western Philippine Basin happened at about -40 m.y. when the direction of the motion of the Pacific plate changed from north-northwest to west-northwest (Clague and Jarrard, 1973). Hilde, Uyeda and Kroenke believe that this change in the plate motion was caused by the subduction of the Kula-Pacific ridge under the Aleutians. Forsyth and Uyeda (1975) demonstrated the major driving force of an oceanic plate is the slab-pull. When a ridge comes to the subduction zone, the slab will have no more pulling force, especially when the overthrusting plate is also oceanic as the Bering Sea plate. Then the Pacific plate "feels" the pull from the western rim only so that the direction of the motion changes. For a reason not entirely clear, such a change in the plate motion apparently did not happen when the Tethys ridge and the ridge north of Australia collided with Asia. Instead, it appears that on these occasions new rifting started along the opposite side of the plate as shown in Figs. 8b and 8d and the direction of plate motion was apparently maintained.

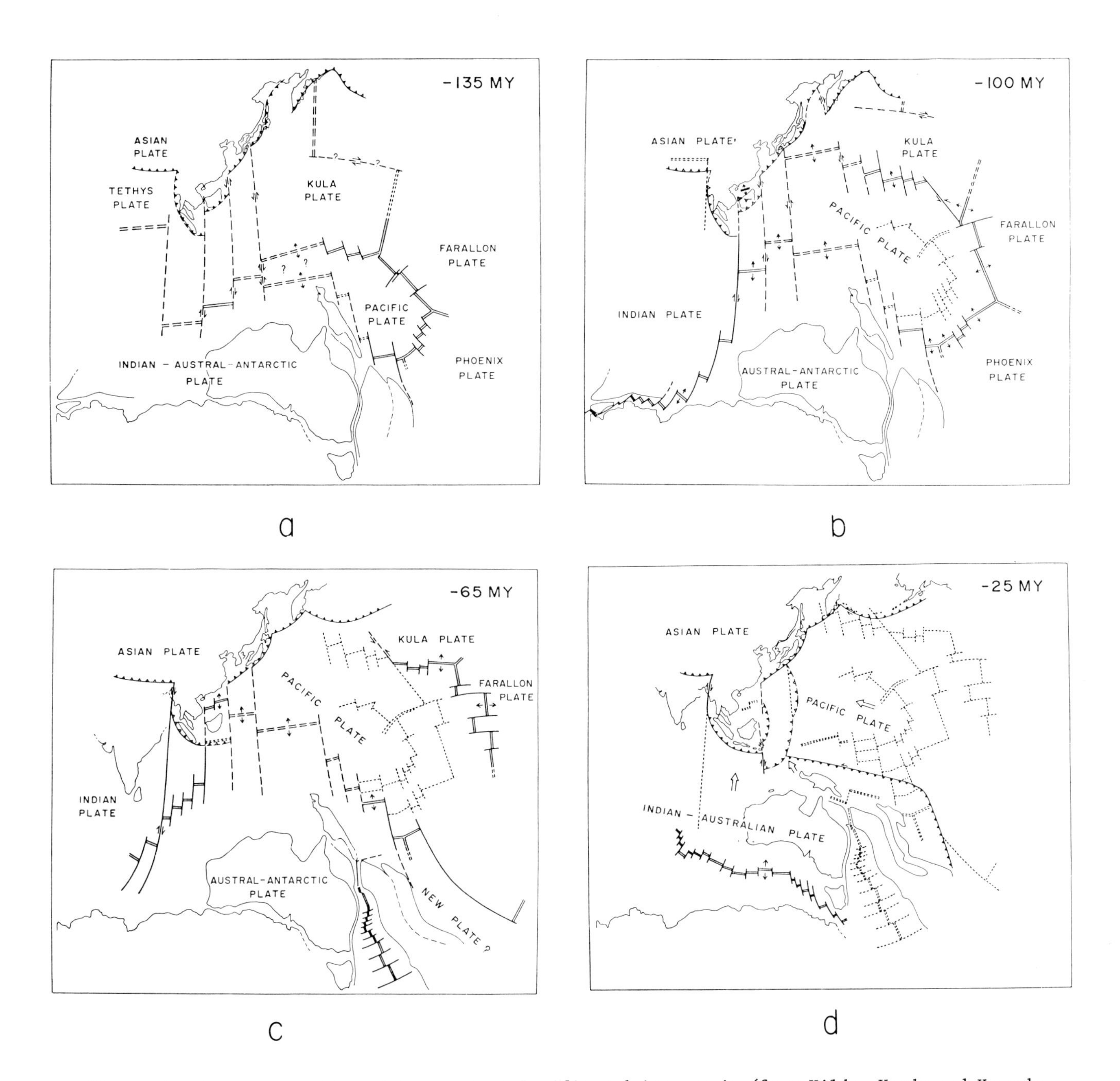

Figure 8. A reconstruction of the western Pacific and its margin (from Hilde, Uyeda and Kroenke, 1976)
double lines: ridges, single lines: transform faults, lines with ticks: subduction zones, full lines: data supported features, broken lines: assumed features, dotted lines: data supported features in previous Figures. (a) at -135 m.y., (b) at -100 m.y., (c) at -65 m.y. and (d) at -25 m.y.

(A change in the velocity, possibly a pause, could have happened at the time of ridge subduction.) These are the points, in the authors' opinion, of some basic importance in plate tectonics in general and worthy of further investigation. Since -53 m.y., Australia has been moving north with the new rifting between Australia and Antarctica and a system of subduction zones was formed in the north of the Australia-New Guinea-Solomon System (Fig. 8d). The system obliquely subducted the ridge system lying to the immediate north. Because of the westward movement of the Pacific plate, many of the young back-arc basins were formed, i.e. Yamato, Parece Vela and South Fiji Basins, Mariana Trough, Okinawa Trough and Law Basin, etc. The process concerned here seems to be the back-arc opening advocated by Karig (1971, 1974). One might notice that according to Van Andel (1974) the rate of the Pacific plate motion was tripled about this time. It may be that Karig's process requires a very high rate of subduction. Another possible case of small basin opening is being documented for the Andaman Sea (Lawyer, Curray and Moore, 1976). In this case, the plate motions involved do not seem to fit with Karig's process or other hitherto suggested processes. Here the relative plate motion seems to be a mega-shear between the Indian and Eurasian plates with large normal rifting component or a very "leaky" transform fault.

In summary, there may be at least four possible processes for forming marginal basins, i.e.

1) ridge subduction
2) entrapment of old ocean basin by transformation of a transform fault to a subduction system
3) back-arc opening (Karig's process)
4) very leaky transform fault

As to the first case, it might be pointed out that, although this hypothesis may appear unlikely to some, at least one good example can be cited: namely, the opening of the Gulf of California where the East Pacific Rise certainly has migrated from the Pacific side to the continental side of the Baja California.

References

Abe, K., Mechanisms and tectonic implications of the 1966 and 1970 Peru earthquakes, Phys. Earth Planet. Interiors, 5, 367-379, 1972.

Anderson, R.N., S. Uyeda, and A. Miyashiro, Geophysical and geochemical constraints at converging plate boundaries - Pt. I: Dehydration in the downgoing slab, Geophys. J. Roy. astr. Soc., 44, 333-357, 1976.

Andrews, D.J., and N.H. Sleep, Numerical modeling of tectonic flow behind island arcs, Geophys. J. Roy. Astr. Soc., 38, 237-251, 1974.

Artyushkov, E.C., Stresses in the lithosphere caused by crustal thickness inhomogeneities, J. Geophys. Res., 78, 7675-7708, 1973.

Ben-Avraham, Z., J. Segawa, and C. Bowin, An extinct spreading center in the Philippine Sea, Nature, 240, 453-455, 1972.

Ben-Avraham, Z., and S. Uyeda, The evolution of the China Basin and the Mesozoic Paleogeography of Borneo, Earth Planet. Sci. Lett., 18, 365-376, 1973.

Brace, W.F., and J.D. Byerlee, California earthquakes: why only shallow focus? Science, 168, 1573-1575, 1970.

Burk, C.A., and C.L. Drake, ed., The Geology of Continental Margins, Springer-Verlag, New York, 1009 pp., 1974.

Clague, D.A., and R.I. Jarrard, Tertiary Pacific plate motion deduced from Hawaiian-Emperor chain, Geol. Soc. Amer. Bull., 84, 1135-1154, 1973.

Cooper, A.K., M.S. Marlow, and D.W. Scholl, Mesozoic magnetic lineations in the Bering Sea marginal basin, J. Geophys. Res., (in press), 1976.

Cooper, A.K., D.W. Scholl, and M.S. Marlow, A plate tectonic model for the evolution of the Eastern Bering Sea Basin, Geol. Soc. Amer. Bull., (in press), 1976.

Dewey, J.F., and J.M. Bird, Mountain belts and the new global tectonics, J. Geophys. Res., 75, 2625-2647, 1970.

Delong, S.E., P.J. Fox and F.W. McDowell, Subduction of the Kula Ridge at the Aleutian Trench, (preprint), 1976.

Dickinson, W.R., Width of modern arc-trench-gaps proportional to past duration of igneous activity in associated magma arcs, J. Geophys. Res., 78, 3367-3389, 1973.

Dickinson, W.R., Plate tectonics and sedimentation, in Tectonics and Sedimentation, W.R. Dickinson, Ed., Soc. Economic Pal. and Min., Special Pub. No. 22, 1-27, 1974.

Elsasser, W.M., Sea-floor spreading as thermal convection, J. Geophys. Res., 76, 1101-1112, 1971.

Engdahl, E.R., N.H. Sleep, and Ming-Te Lin, Plate effects in north Pacific subduction zones, Tectonophysics, (in press), 1976.

Fitch, T.J., Plate convergence, transcurrent faults, and internal deformation adjacent to southeast Asia and western Pacific, J. Geophys. Res., 77, 4432-4461, 1972.

Forsyth, D., and S. Uyeda, On the relative importance of driving forces of plate motion, Geophys. J. Roy. astr. Soc., 43, 163-200, 1975.

Frank, F.C., Curvature of island arcs, Nature, 220, 363, 1968.

Fukao, Y., and M. Furumoto, Mechanism of large earthquakes along the eastern margin of the Japan Sea, Tectonophysics, 25, 247-266, 1975.

Gansser, A., Geology of the Himalayas, Inter-Science, 1964.

Hasebe, K., N. Fujii, and S. Uyeda, Thermal processes under island arcs, Tectonophysics, 10, 335-355, 1970.

Hasegawa, A., N. Umino, and S. Takagi, Fine structure of deep seismic plane in Northeast Japan, abstract to Spring Meeting of Seismological Society of Japan, 1976.

Hawkins, J.W., Jr., Petrology and geochemistry of basaltic rocks of the Lau Basin, Earth Planet. Sci. Lett., 28, 283-297, 1976.

Hilde, T.W.C., N. Isezaki, and J.N. Wageman, Mesozoic sea-floor spreading in the north

Pacific, Amer. Geophys. Un., Geophys. Monograph 19 (in press), 1976.
Hilde, T.W.C., C.S. Lee, T. Watanabe, F.C. Liu, N. Isezaki, and M. Ozima, Central Basin fault-Philippine Sea (in preparation).
Hilde, T.W.C., S. Uyeda, and L. Kroenke, Evolution of the western Pacific and its margin, Tectonophysics, (in press), 1976.
Huzita, K., Tectonic development of southwest Japan in the Quaternary Period, J. Geosci., Osaka Univ., 12, 53-70, 1969.
Ichikawa, M., Re-analysis of mechanism of earthquakes which occurred in and near Japan and statistical studies on the nodal plane solution obtained, 1926-1968, Geophys. Mag., 35, 207-274, 1971.
Ida, Y., N. Fujii and S. Uyeda, Relation between frictional heat and velocity of plate, J. Phys. Earth (in press), 1976.
Isacks, B., J. Oliver, and L.R. Sykes, Seismology and the new global tectonics, J. Geophys. Res., 73, 5855-5899, 1968.
Isacks, B., and P. Molnar, Distribution of stresses in the descending lithosphere from a global survey of focal mechanism solution of mantle earthquakes, Rev. Geophys. and Space Phys., 9, 103-174, 1971.
Isezaki, N., Possible spreading centers in the Japan Sea, J. Marine Geophys. Res., 2, 265-277, 1975.
Isezaki, N., and S. Uyeda, Geomagnetic anomaly of the Japan Sea, J. Marine Geophys. Res., 2, 51-59, 1973.
Jischke, M.C., On the dynamics of descending lithosphere plates and slip zones, J. Geophys. Res., 80, 4809-4814, 1976.
Kanamori, H., Great earthquakes at island arcs and the lithosphere, Tectonophysics, 12, 187-198, 1971.
Karig, D.E., Origin and development of marginal basins in the western Pacific, J. Geophys. Res., 76, 2542-2560, 1971.
Karig, D. E., Evolution of arc systems in the western Pacific, Ann. Rev. Earth Planet. Sci., 2, 51-75, 1974.
Karig, D.E., and G.F. Sharman, III, Subduction and accretion in trenches, Geol. Soc. Amer. Bull., 86, 377-389, 1975.
Kelleher, J., J. Savino, H. Rowlett, and W. McCann, Why and where great thrust earthquakes occur along island arcs, J. Geophys. Res., 79, 4889-4899, 1974.
Kobayashi, K., and N. Isezaki, Magnetic anomalies in Japan Sea and Shikoku Basin and their possible tectonic implications, Amer. Geophys. Union, Geophys. Monogr., 19, (in press), 1976.
Kushiro, I., Effect of water on the magmas at high pressures, J. Petrol., 13, 311-334, 1972.
Larson, R.L., and C.G. Chase, Late Mesozoic evolution of the western Pacific, Geol. Soc. Amer. Bull., 83, 3645-3662, 1972.
Lawver, L.A., J.R. Curray, and D.G. Moore, Tectonic evolution of the Andaman Sea, EOS, 57, 333, 1976.
Le Pichon, X., Sea-floor spreading and continental drift, J. Geophys. Res., 73, 3661-3697, 1968.
Lliboutry, L., Sea-floor spreading, continental drift and lithosphere sinking with an asthenosphere at melting point, J. Geophys. Res., 74, 6525-6540, 1969.
Louden, K., Magnetic anomalies in the west basin of the Philippine Sea, Amer. Geophys. Union Geophys. Monograph 19, (in press), 1976.
Matsuda, T., K. Nakamura, and A. Sugimura, Late Cenozoic orogeny in Japan, Tectonophysics, 4, 595-613, 1967.
Matsuda, T., and S. Uyeda, On the Pacific-type orogeny and its model-extension of the paired belts concept and possible origin of marginal seas, Tectonophysics, 11, 5-27, 1971.
McKenzie, D.P., Speculations on the consequences and causes of plate motions, Geophys. J. Roy. astr. Soc., 18, 1-32, 1969.
McKenzie, D.P., and J.C. Sclater, Heat flow inside the island arcs of the northwestern Pacific, J. Geophys. Res., 73, 3137-3179, 1968.
Mitchell, A.H., and H.G. Reading, Continental margins, geosynclines, and ocean-floor spreading, J. Geol., 77, 629-646, 1969.
Minear, J.W., and M.N. Toksöz, Thermal regime of a downgoing slab and new global tectonics, J. Geophys. Res., 75, 1397-1419, 1970.
Miyashiro, A. Evolution of metamorphic belts, J. Petrol., 2, 277-311, 1961.
Miyashiro, A., Metamorphism and related magmatism in plate tectonics, Amer. J. Sci., 272, 629-656, 1972.
Miyashiro, A., Paired and unpaired metamorphic belts, Tectonophysics, 17, 241-254, 1973.
Miyashiro, A., Volcanic rock series in island arcs and active continental margins, Amer. J. Sci., 274, 321-355, 1974.
Nagumo, S., Ocean bottom seismographic observation, Preliminary Report of the Hakuho Maru Cruise KH-69-2, 173-196, Ocean Res. Inst., Univ. Tokyo, 1971.
Noritomi, K., and Y. Saeki, On the earthquakes off Akita, Northern Japan, Tohoku Chiiki Saigai Kagaku Kenkyu, 99-111, (in Japanese), 1970.
Raleigh, C.B., Tectonic implications of serpentine weakening, Geophys. J. Roy. astr. Soc., 13, 113-118, 1967.
Ringwood, A.E., The petrological evolution of island arc systems, J. Geol. Soc. London, 130, 183-204, 1974.
Scholl, D.W., M.N. Christiansen, R. von Huene, and M.S. Marlow, Peru-Chile trench sediment and sea-floor spreading, Geol. Soc. Amer. Bull., 81, 1339-1360, 1970.
Sclater, J.G., Heat flow and elevation of marginal basins of the western Pacific, J. Geophys. Res., 77, 5705-5719, 1972.
Sclater, J.G., J.W. Hawkins, Jr., J. Mammerickx, and C.B. Chase, Crustal extension between the Tonga and Lau ridges: petrologic and geo-

physical evidence, Geol. Soc. Amer. Bull., 83, 505-518, 1972.
Shimamura, H., Y. Tomoda, and T. Asada, Seismographic observation at the bottom of the Central Basin fault of the Philippine Sea, Nature, 253, 177-179, 1975.
Sleep, N., and M.N. Toksöz, Evolution of marginal basins, Nature, 233, 548-550, 1971.
Stauder, W., Tensional character of earthquake foci beneath the Aleutian Trench with relation to sea-floor spreading, J. Geophys. Res., 73, 7693-7701, 1968.
Sugimura, A., Zonal arrangement of some geophysical and petrological features in Japan and its environs, J. Fac. Sci., Univ. Tokyo, XII, 133-153, 1960.
Sugimura, A., and S. Uyeda, Island Arcs: Japan and Its Environs, Elsevier, Amsterdam, 1973.
Sugisaki, R., Tectonic aspects of andesite line, Nature Phys. Sci., 240, 109-111, 1972.
Sugisaki, R., Chemical characteristics of volcanic rocks: relation to plate movements, Lithos, 9, 17-30, 1976.
Sykes, L.R., B.L. Isacks, and J. Oliver, Spatial distribution of deep and shallow earthquakes of small magnitudes in the Fiji-Tonga Region, Bull. Seism. Soc. Amer., 59, 1093-1113, 1969.
Takeuchi, H., and S. Uyeda, A possibility of present day regional metamorphism, Tectonophysics, 2, 59-68, 1965.
Toksöz, M.N., N.H. Sleep, and A.T. Smith, Evolution of the downgoing lithosphere and mechanisms of deep focus earthquakes, Geophys. J. Roy. astr. Soc., 35, 285-310, 1973.
Tomoda, Y., K. Kobayashi, J. Segawa, M. Nomura, K. Kumura, and T. Saki, Linear magnetic anomalies in the Shikoku Basin, J. Geomag. Geoelectr., 28, 47-56, 1975.
Tsumura, K., Microearthquake activity in the Kanto District, Publication for the 50th Anniversary of the Great Kanto Earthquake, 1923, Earthq. Res. Inst., Univ. Tokyo, (in Japanese with English abstract), 1973.
Turcotte, D.L., and E.R. Oxburgh, A fluid theory for the deep structure of dip-slip fault zones, Phys. Earth Planet. Interiors, 1, 381-386, 1969.
Turcotte, D.L., and G. Schubert, Structure of the olivine-spinel phase boundary in the descending lithosphere, J. Geophys. Res., 76, 7980-7987, 1971.
Turcotte, D.L., and G. Schubert, Frictional heating of the descending lithosphere, J. Geophys. Res., 78, 5876-5886, 1973.
Umino, N., and A. Hasegawa, On the two-layered structure of deep seismic plane in northeastern Japan Arc, Zisin, 27, 125-139, (in Japanese with English abstract), 1975.
Utsu, T., Seismological evidence for anomalous structure of island arcs with special reference to the Japanese region, Rev. Geophys. Space Phys., 9, 839-890, 1971.
Uyeda, S., and K. Horai, Terrestrial heat flow in Japan, J. Geophys. Res., 69, 2121-2141, 1964.
Uyeda, S., and Z. Ben-Avraham, Origin and development of the Philippine Sea, Nature, Phys. Sci., 240, 176-178, 1972.
Uyeda, S., and A. Miyashiro, Plate tectonics and Japanese Islands, Geol. Soc. Amer. Bull., 85, 1159-1170, 1974.
Vacquier, V., S. Uyeda, N. Yasui, J. Sclater, C. Corry, and T. Watanabe, Heat flow measurements in the northwestern Pacific, Bull. Earthq. Res. Inst., 44, 1519-1535, 1967.
Van Andel, T.H., Cenozoic migration of the Pacific plate, northward shift of the axis of deposition, and paleobathymetry of the central equatorial Pacific, Geology, 2, 507-510, 1974.
Watts, A.B., and M. Talwani, Gravity anomalies seaward of deep-sea trenches and their tectonic implications, Geophys. J. Roy. astr. Soc., 26, 57-90, 1974.
Watts, A.B., and J.K. Weissel, Tectonic history of the Shikoku marginal basin, Earth. Planet. Sci. Lett., 25, 239-250, 1975.
Watts, A.B., J.K. Weissel, and R.L. Larson, Sea-floor spreading in marginal basins of the western Pacific, Tectonophysics, (in press), 1976.
Weissel, J.K., and A.B. Watts, Tectonic complexities in the South Fiji marginal basin, Earth Planet. Sci. Lett., 28, 121-126, 1975.
Wilson, J.T., and K. Burke, Two types of mountain building, Nature, 239, 448-449, 1972.
Wyllie, P.J., Role of water in magma generation and initiation of diapiric uprise in the mantle, J. Geophys. Res., 76, 1328-1338, 1971.
Yasui, M., T. Kishii, T. Watanabe, and S. Uyeda, Heat flow in the Sea of Japan, in The Crust and Upper Mantle of the Pacific Area, Geophys. Monograph 12, ed. by L. Knopoff et al., 3-16, Am. Geophys. Union, Washington, D.C., 1968.
Yoshii, T., Proposal of the "aseismic front", Zisin, 28, 365-367, (in Japanese), 1975.

SUBDUCTION IN THE INDONESIAN REGION

Warren Hamilton

U.S. Geological Survey, Denver, Colorado 80225

Abstract. The outer-arc ridge between Java and Sumatra and the active Java Trench is the top of a wedge of melange and imbricated rocks whose steep to moderate dips are sharply disharmonic to the gently dipping, subducting oceanic plate beneath. The wedge has grown by scraping off of oceanic sediments and basement against and beneath its toe, and also by internal imbrication which is a gravitationally driven counter to subductive dragging at the base. The outer-arc basin behind the ridge originated from a Paleogene continental shelf-and-slope assemblage whose seaward side was raised by melange stuffed beneath it by Neogene subduction. The Banda Arc is now colliding with the Australian-New Guinea continent. As the frontal wedge of melange has been driven onto the continental shelf, abundant shallow-water shelf strata have been imbricated into the wedge. The Philippines are the product of a complex aggregation of segments of various island arcs. The extinct Palawan and Sabah-Sulu-Zamboanga arcs bridge between Borneo and the Philippines. The Sangihe and Halmahera arcs are now colliding in the south, but the collision is complete in the north, where the sutured components form central Mindanao. New subduction zones, including the Philippine Trench, are breaking through on both sides of the arc aggregate.

Introduction

This short report is concerned primarily with characteristics of active or recently deactivated subduction systems, and summarizes some of the findings of my 6-year study, now almost complete, of the onshore and offshore tectonics of Indonesia and surrounding areas. This study has involved the examination of most of the published primary geologic and geophysical papers on the region. Van Bemmelen (1949) listed most of the pre-1940 papers in his compendium. Space limitations permit me to cite only a few of the more recent papers here. Unpublished data, primarily geophysical, were provided by oil and mining companies and by oceanographic institutions, particularly Lamont-Doherty Geological Observatory of Columbia University. The late Maurice Ewing, then of Lamont-Doherty, gave me all of the seismic-reflection profiles obtained in the study region by the research vessels *Vema* and *Robert Conrad*, and a number of segments of these profiles, and of others given by oil companies, illustrate this paper. The major products published so far from my study are the 1:5,000,000 maps of bathymetry, sedimentary basins, and earthquakes of the Indonesian region (U.S. Geological Survey, 1974; Hamilton, 1974a, b). Figure 1 is an index map of part of the region.

Pre-Cenozoic tectonics

The effects of Cenozoic subduction, arc migration, rifting, oroclinal folding, and arc magmatism in the Indonesian region are superimposed on the products of similar and equally complex processes that operated during Paleozoic and Mesozoic time. The present locations and characteristics of many of the pre-Cenozoic tectonic elements are summarized on figure 2. Some general relationships are noted here to provide a background for the discussion of young tectonic systems.

The north-trending terrain of late Paleozoic and Triassic tin granites and comagmatic volcanic rocks of eastern Australia and New Guinea, and the craton flanking it on the west, are matched by analogous and correlative terrains in Sumatra and Malaya. (Whole-rock Rb/Sr data are interpreted by Bignell and Snelling, in press, in terms of assumed low and uniform initial ratios

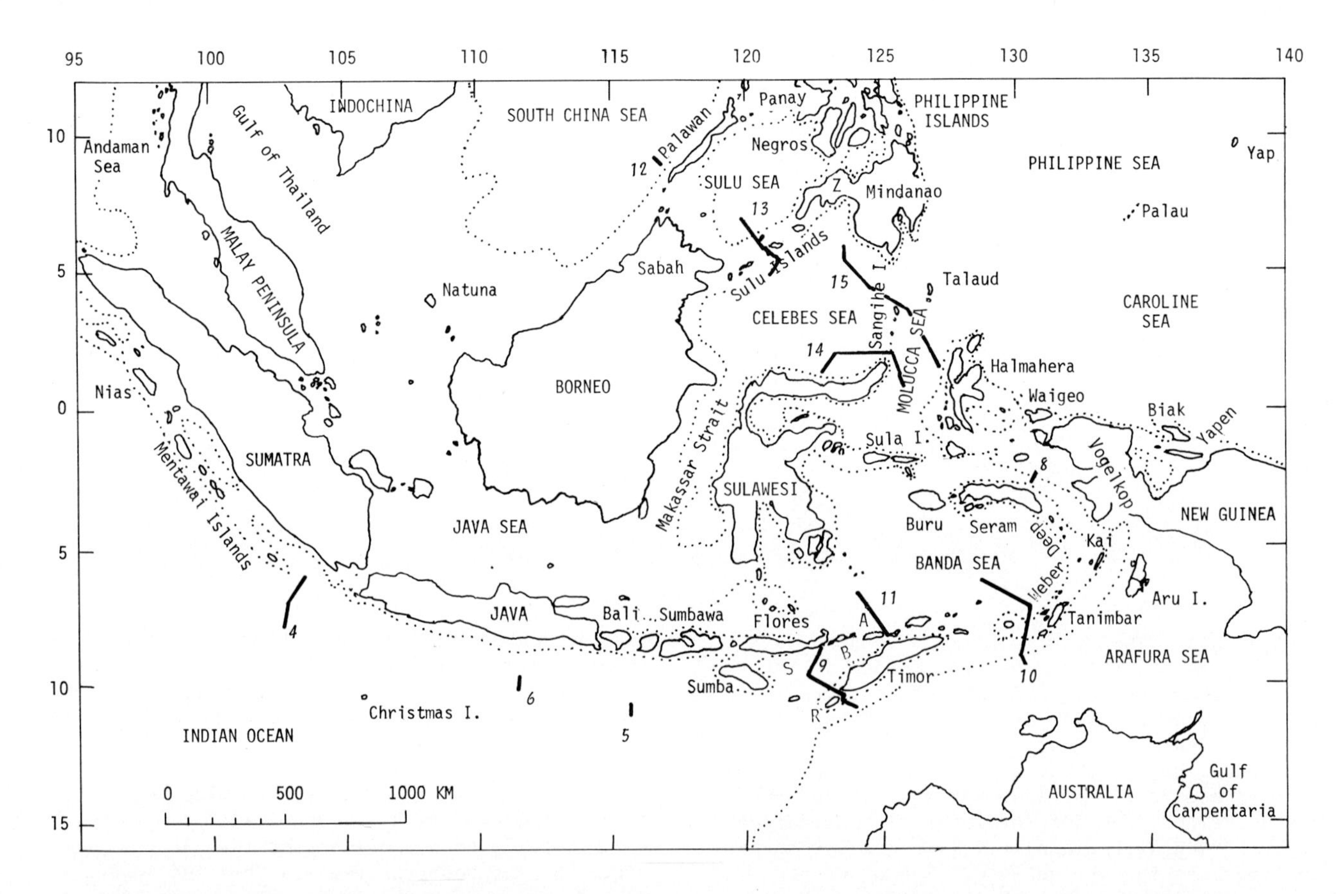

Fig. 1. Index map of the Indonesian region, showing locations of the seismic-refraction profiles (heavy lines, with figure numbers) accompanying this paper. Dotted line is generalized 1000-metre bathymetric contour; not shown around small islands and shoals. Abbreviations: A, Alor; R, Roti; S B, Savu Basin; Z, Zamboanga.

of $^{87}Sr/^{86}Sr$, to indicate both Paleozoic and Mesozoic granites to be present also in western Malaya. As granites of this type characteristically have high and erratic initial ratios--see Cooper, Webb, and Whitaker, 1975--I infer that the western Malayan granites are in fact all Mesozoic, as indeed is indicated by concordant-mineral K/Ar dates, by Rb/Sr mineral isochrons, and by regional stratigraphy.) Some landmass with this combination of eastern tin granites and western craton was rifted from medial New Guinea in Early Jurassic time, and Sumatra and Malaya provide the likely mass. (The contrary inference by McElhinny, Haile, and Crawford, 1974, that Malaya has undergone little change in latitude since Carboniferous time is weakly based on a paleomagnetic reconnaissance of uncertainly dated rocks formed at times of undefined geomagnetic polarity; great translations and rotations are permitted by the paleomagnetic data as actually constrained.) Later in Jurassic time, India broke away from Australia and migrated relatively northwestward, at twice the speed of the intervening northeast-trending oceanic spreading center, until a new rift system broke across the oceanic crust northwest of Australia in middle Late Cretaceous time (fig. 2). During the late Late Cretaceous and early Paleogene, the oceanic crust which now lies south of Sumatra formed at a spreading center that migrated northward and has since been mostly or wholly subducted beneath Sumatra and Java (Sclater and Fisher, 1974). Then, in the Eocene, the modern spreading center broke across between Australia and East Antarctica, and Australia and the Indian Ocean have since been moving northward. The Australia-New Guinea continent developed stable-margin continental shelves on its west and north sides, during the intervals between the continental riftings and late Cenozoic arc collisions.

Paired Cretaceous belts of subduction melange and of granitic and silicic-volcanic rocks trend southwestward along the Pacific side of Asia from Japan and Korea to Viet Nam, from which they can be tracked, in wells and islands, curving eastward across the shelf of the South China Sea to Borneo (where relationships are poorly understood), and thence southwestward again across the Java Sea shelf to Java and southern Sumatra (fig. 2). The melange terrain is present also in the microcontinent of Sulawesi, which was rifted from Borneo in Tertiary time, and perhaps in the microcontinent of Sumba, the Paleogene stratigraphy of which suggests an origin in the shelf of the Java Sea. These paired Cretaceous belts are not defined by published data within central or northern Sumatra, and may have ended against a transform fault, or have extended south beyond Sumatra but have been removed by Late Cretaceous or early Paleogene rifting. As is mentioned in a subsequent section, the arc defined by these and other complexes between Viet Nam and central Borneo appears to be an orocline, produced by Cenozoic counterclockwise rotation of Borneo.

Cenozoic tectonics

A number of Cenozoic tectonic features are located on figure 3. Some of these elements, and particularly those active or inactive ones which best illuminate the shallow processes of subduction, are interpreted briefly on the following pages. The order of presentation is a general progression from the simple to the complex.

Java Trench subduction system

A continuous subduction system along the Indian Ocean side of Java and Sumatra can be identified clearly first in the upper Oligocene record. Older Oligocene and Eocene rocks near this margin are nonvolcanic strata likely deposited on a stable continental shelf, to the south of which lay open water. (Several reconnaissance K/Ar Eocene age determinations for granitic rocks in Sumatra if correct however indicate otherwise.) In southwest Java, for example, the upper Eocene and lower Oligocene section consists of nonvolcanic quartzose sediments, and volcanic debris first appears in upper Oligocene or lower Miocene strata (Baumann and others, 1973). Magmatism has been generally continuous since the late Oligocene. During the late Oligocene and very early Miocene the main magmatic belt lay approximately along what is now the south coast of Java, magmatism having shifted to about its present axis later in the early Miocene.

Figure 4 illustrates the morphologic features that characterize the present subduction system along Java and Sumatra. Between the Java Trench and the major islands, with their active volcanic arcs, is a high submarine outer-arc ridge and a continuous outer-arc basin. The trench is about 6 km deep along Java, but progressively shallower, by an amount correlative with the amount of Bengal Fan turbidites on the ocean floor and in

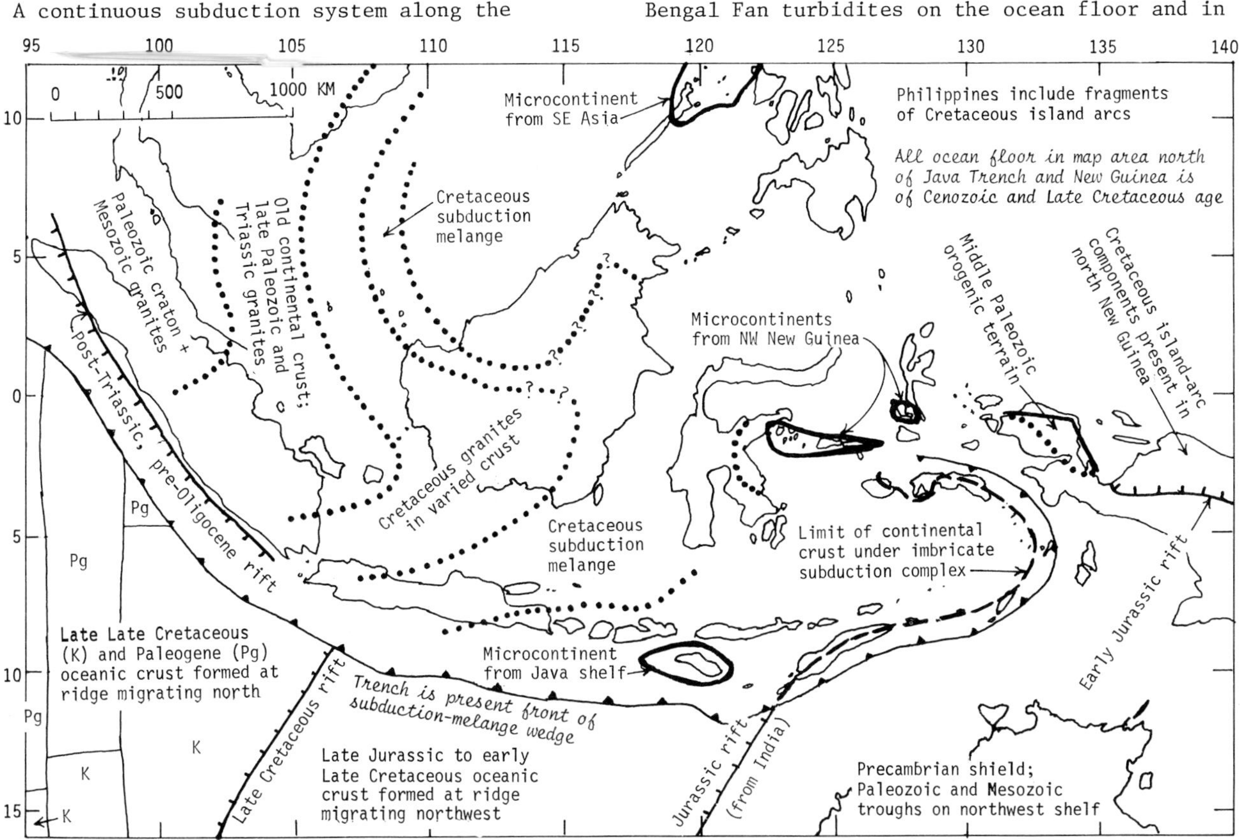

Fig. 2. Map showing pre-Cenozoic tectonic elements in the Indonesian region. Relative and absolute positions, shapes, and orientations of these elements have changed profoundly since they formed.

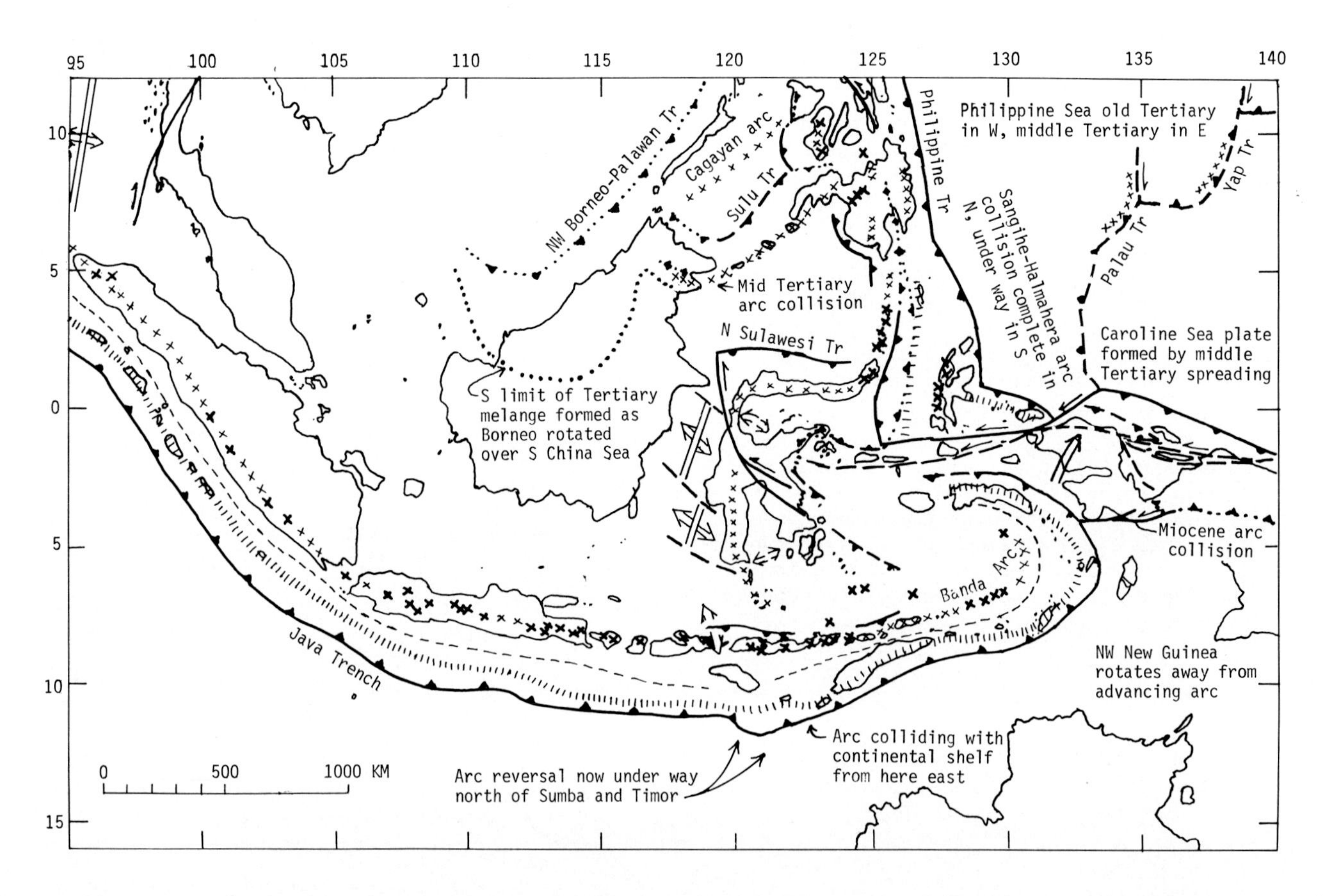

Fig. 3. Map showing some Cenozoic tectonic elements in the Indonesian region. Trace of subduction zone indicated by thrust-fault symbol, teeth on overriding plate; strike-slip fault indicated by arrow giving motion relative to Southeast Asian side. Fault shown as solid where active, dashed where inactive, dotted where covered by younger strata or melange wedge. Dotted thrust-fault symbol also used on land for suture marking are collision; barbs on both sides where two arcs collided. Middle Tertiary spreading center marked by double line and by double arrow giving direction and amount of spreading, or by double arrow alone. Active volcanoes marked by bold X's, and other Neogene volcanic axes marked by light x's. Crest of melange wedge (outer-arc ridge) shown by hachured line. Axis of outer-arc basin shown by light dashed line.

the trench (Hamilton, 1974a), going northwestward along Sumatra. The basement surface beneath the trench is gently convex upward, and the turbidite wedge deposited in it shows a landward fanning of dips, indicating that the oceanic crust is rolling through a gentle flexure, in conveyor-belt fashion, as sediments are deposited on it. The trench sediments become deformed at the toe of the continental slope, but most of the continental slope is acoustic basement as seen in low-energy, high-exaggeration records such as that of figure 4. The slope from the trench to the crest of the ridge on these records is relatively rough. On the opposite side of the outer-arc ridge, the thick strata of the outer-arc basin lap onto basement or older strata on the landward side, and become increasingly deformed toward the outer-arc ridge on the seaward side, where they merge in records such as that of figure 4 with the acoustic basement of the ridge. Although not obvious on this profile, the deep strata beneath much of the outer-arc basin thicken seaward past its axis, and integration of this feature with the onshore stratigraphic record suggests that a stable continental shelf and slope were present here in early Tertiary time, and were deformed subsequently into the present basin configuration.

High-energy, multichannel seismic-reflection records, such as those of figures 5 and 6, show the gentle slope between trench and outer-arc ridge to be the top of a wedge of highly deformed rocks, mostly low-velocity sediments, that lie on top of oceanic crust which dips gently landward. The strongly reflecting basement surface in figure 5 can be followed for 60 km, with a dip of only 7°, beneath the wedge; where last visible on the record, it lies at a depth of 15 km, beneath

11 km of wedge material. The internal structure of the wedge is dominated by imbrication that dips moderately landward and is strongly disharmonic to the gently dipping oceanic crust beneath, which may itself be sliced by faults (fig. 6).

The outer-arc ridge--the top of the wedge--is 1-3 km below sea level off Java and southernmost Sumatra, but it is shallower than 1 km along most of Sumatra, where it is exposed in the Mentawai and other islands. Presumably the height of the ridge correlates with the thickness of the underlying low-density wedge, which in turn is related to the amount of sediment that has been moved in conveyor-belt fashion into the subduction system. The islands expose complexes of subduction melange--a chaos, dipping steeply to moderately, of scaly clay, broken formations, serpentine and peridotite, basalt and spilite, greenschist, amphibolite, and low-grade metasedimentary rocks--plus interfolded and intersliced coherent Miocene strata.

The deformation apparent on reflection profiles decreases northward from the crest of the outer-arc ridge into the adjacent outer-arc basin (for example, fig. 4) and generally is directed northward. Farther east, between Sumba and Timor, a profile presented by Crostella and Powell (1976, fig. 6) shows the northern deformation front to be a thrust fault that dips gently southward, subparallel to the bedding in the underlying strata. I infer that such shallow thrusting typifies the behavior on the basin side of the wedge crest off Java and Sumatra also, but this conclusion may be invalid, for the underlying materials in the profile of Crostella and Powell are part of the aberrant Sumba microcontinent.

The deeper trajectory of the subducting oceanic plate is shown by the Benioff zone of mantle earthquakes (Hamilton, 1974b). The plate maintains a gentle dip to beneath the outer-arc basin, landward of which it steepens gradually to a depth of about 125 km beneath the main volcanic arc. Beyond this, the dip is mostly steeper than 60°, and the zone reaches depths of more than 600 km beneath the Java Sea, but generally less than 200 km beneath Sumatra.

The preceeding discussion and other data are integrated in the cross sections of figure 7.

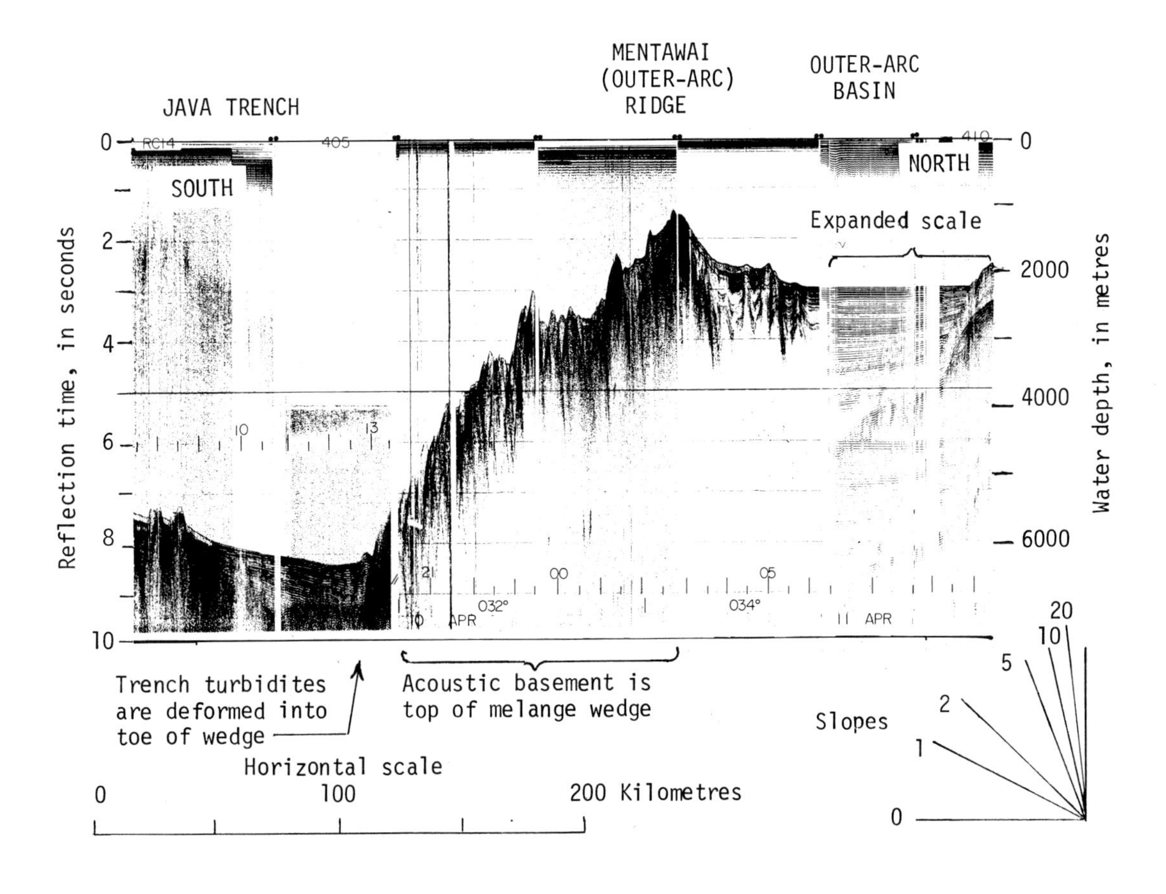

Fig. 4. Seismic-reflection profile across the Java Trench, and outer-arc ridge and basin, off southern Sumatra (fig. 1). Profile from Cruise 14 of R. V. Robert Conrad, Lamont-Doherty Geological Observatory of Columbia University.

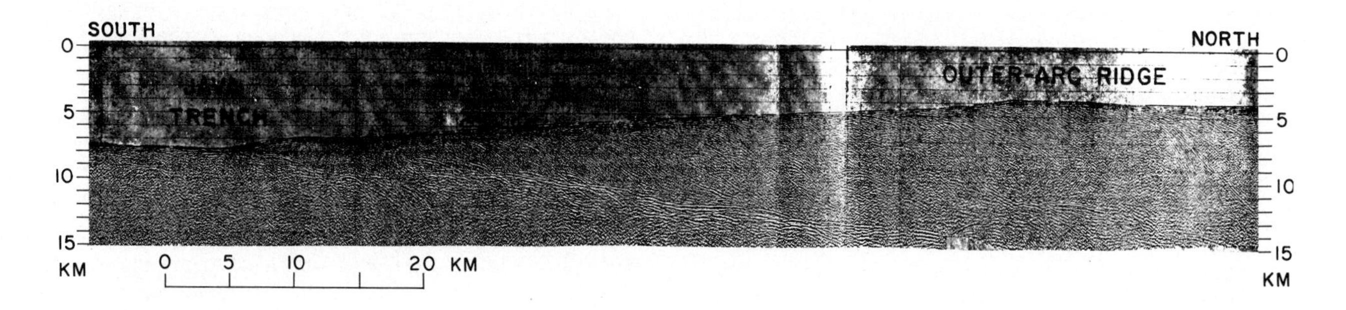

Fig. 5. Depth-converted seismic-reflection profile across the wedge of melange and imbricated sediments south of Bali (fig. 1). Horizontal and vertical scales equal. Profile provided by Shell Internationale Petroleum Maatschapij N. V.; see also Beck and Lehner (1974, fig. 20).

(Grow, 1973, presented a similar model.) Geometry and behavior of the system from trench to outer-arc ridge are reasonably well established. Sediment on the oceanic lithosphere, which is sliding northward beneath Java and Sumatra at a velocity of about 7 cm/yr (Minster and others, 1974), is scraped off against, and becomes part of, the accretionary wedge. This wedge, which is dominated by low-density, wet sediments, is dragged back and thickened by the underthrusting oceanic crust, but simultaneously is driven forward by gravitational flow (see Elliott, 1976, for discussion of mechanics) so that its shape is one of dynamic equilibrium, not static accretion. The opposed forces result in imbrication by internal shear which is discordant at a sharp angle to the dip of the oceanic plate beneath. Although deformation within the wedge is concentrated near its toe, where presumably most of the scaly clay is formed, lesser shearing is distributed throughout the wedge, so that strata deposited on top of it become downfolded or imbricated into it, and so that wedge components, including slices of oceanic crust and mantle, are mixed tectonically. Trajectories of masses within the wedge can be visualized as obliquely upward along thrust faults, slumping at the surface from fault scarps, and obliquely downward again beneath other thrusts. Possibly this gravitationally driven mixing process can explain the present shallow occurrence of glaucophane schist and other high-pressure metamorphic rocks in such melange wedges. The gentle-sided trench and the gentle rise (not shown on the figures) that marks its oceanward rim probably are eleastic responses of the lithosphere to the weight of the melange wedge. Elastic departures of bending from isostatic equilibrium are however minor; given the densities indicated on figure 7, the weight of columns to a depth of 50 km throughout the section is within 2 percent of $150 \times 10^9 g/m^2$. The often-cited extreme gravity "deficiency" deduced by Vening Meinesz (for example, 1954) is a measure of the thickness of the low-density melange wedge, and not, as he considered it to be, of gross disequilibrium in crust of uniform density.

The following explanation is suggested, with less confidence, for the evolution of the landward side of the outer-arc ridge and of the outer-arc basin. In middle Paleogene time, Sumatra and Java were coupled to the Indian Ocean plate, or else were separated from it by a left-lateral transform fault. Some of the convergence of Indian and east Asian plates was accommodated by rapid southward subduction of part of the South China Sea beneath Borneo, and some by rotation of Sumatra and Java. (Both counterclockwise rotation of Borneo, described in a subsequent section, and clockwise rotation of Sumatra as the great Assam syntaxis formed at the edge of the northward-driving Indian continent, are inferred.) Continental-shelf and continental-slope strata were prograded from Sumatra and Java onto oceanic crust. These strata and their oceanic basement remained as part of the overriding continental plate when plate underthrusting became vigorous in late Oligocene time. Sedimentary material on the subducting plate was scraped off at the leading edge of the overriding plate, tilting it upward to form the outer-arc basin behind it. (I infer a similar origin for the analogous Cretaceous Great Valley synclinorial basin of California. There, strata prograded over oceanic crust, which is exposed now beneath the strata and is still attached to the continent, during a latest Jurassic and Early Cretaceous period of little or no subduction, before rapid subduction in middle Cretaceous time began to produce the Franciscan melange wedge.) The leading edge has since provided a deep buttress against which the melange wedge has accumlated, and over the top of which wedge material has flowed gravitationally toward the axis of the basin, to form thrusts and asymmetric folds.

The main volcanic belt of Java and Sumatra approximately follows a contour of 125 km on the Benioff zone, so proto-magmas must be generated by processes related to such a position of the subducting plate. Uniform oceanic crust enters the subduction zone, hence those deep proto-magmas likely are nearly constant in composition along the belt. The magmas that reach the surface however show bulk compositional changes reflecting crustal contamination: the bulk composition of volcanic rocks is approximately rhyodacite in Sumatra, where the basement consists of old sialic complexes; silicic andesite in west Java, where the basement is mostly old melange; and mafic andesite farther east, where there is no pre-Tertiary basement (fig. 2). At any one transect, and superimposed on the bulk-composition variations, the magmas that reach the surface display the usual correlation of K/Si ratios with height above the Benioff zone (Hatherton and Dickinson, 1969).

Banda Arc

The Java Trench and associated volcanic arc and Benioff zone continue eastward into the Banda Arc (figs. 2, 3). Trench and melange wedge bulge southward in front of the Sumba microcontinent, the outer-arc basin loses its identity in the Sumba sector, and major changes are now underway in the arc, but the general continuity is obvious. From Timor around the great bend to Seram, the trench is only 1.5-3.5 km deep, and it marks the dihedral angle of the sea floor between the subducting Australian-New Guinea continental shelf and the top of the accretionary wedge of imbricated rocks and melange. Figure 8 illustrates this relationship at Seram. Deep-penetration profiles show the same features at many other places around the arc (Beck and Lehner, 1974; Crostella and Powell, 1976; Montecchi, 1976). The identity of the strongly reflecting shelf strata beneath the trench on such profiles was confirmed by Hole 262 of the Deep Sea Drilling Project: south of Timor, shallow-water upper Pliocene carbonates lie beneath the trench fill (Veevers and Heirtzler, 1974, p. 193-278). Figures 9 and 10 show the relationship of the trench to other components of the Banda Arc system, but lack the penetration to show the bottom of the melange wedge.

The available descriptions of nearly all of the islands--Roti, Timor, Seram, the Kai and Tanimbar groups, etc.--of the outer-arc ridge of the Banda Arc show them to be composed of melange plus imbricated sedimentary and crystalline rocks, confirming interpretations based on reflection profiles, on continuity from the Mentawai Islands, and on the Vening Meinesz gravity pattern. Scaly clay forms a ubiquitous matrix, within which are chaotically enclosed clasts, lenses, and intercalated sheets, of all sizes up to tens of kilometres long, of shelf (Permian and younger), slope, and deep-water (Cretaceous and younger) sediments, and of igneous and metamorphic rocks, mostly of mafic and ultramafic compositions but locally including glaucophane schist and other high-pressure metamorphic rocks. Dips of all of these components are mostly steep to moderate. Virtually all mappers in these terrains have emphasized their chaotic character, while explaining their origin by a variety of imaginative tectonic and gravitational hypotheses. For example, Audley-Charles (as, 1968) and his associates (as, Carter,

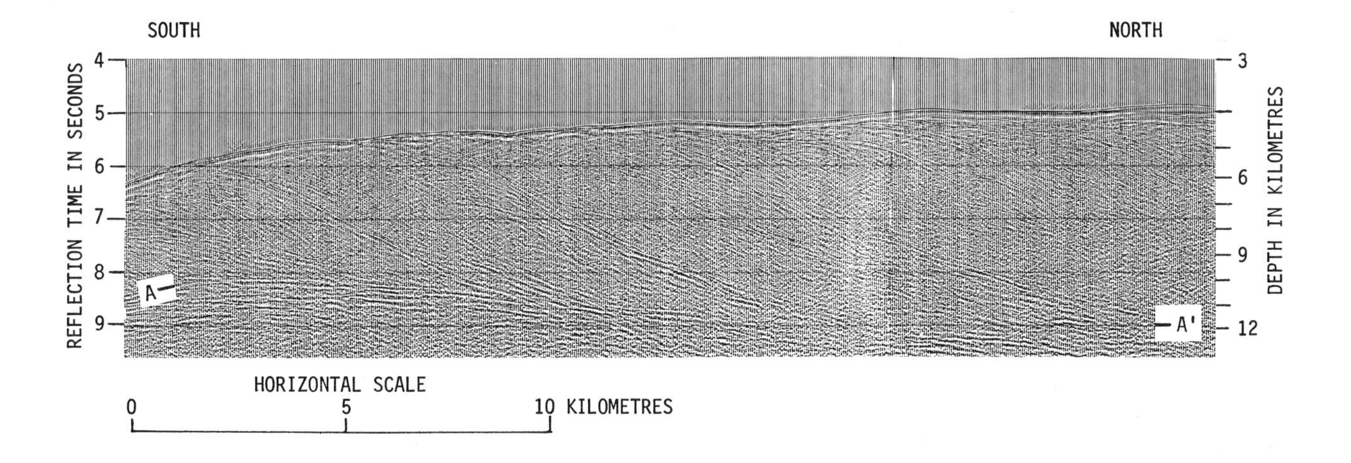

Fig. 6. Seismic-reflection profile across part of wedge south of Java. The Java Trench is to the left of this segment with a water depth of about 6 km (8 sec), and the crest of the outer-arc ridge is to the right, with a depth of less than 3 km. Imbricate structure dips moderately north, sharply disharmonic to the very gently dipping top (A-A') of the undersliding oceanic plate beneath. The depth scale, which applies to the right end of the profile only, is calculated from stacking velocities; seismic velocity increases downward within the wedge from 1.8 to about 5 km/sec. Profile provided by Shell Internationale Petroleum Maatschappij N. V.; see also Beck and Lehner (1974, fig. 22).

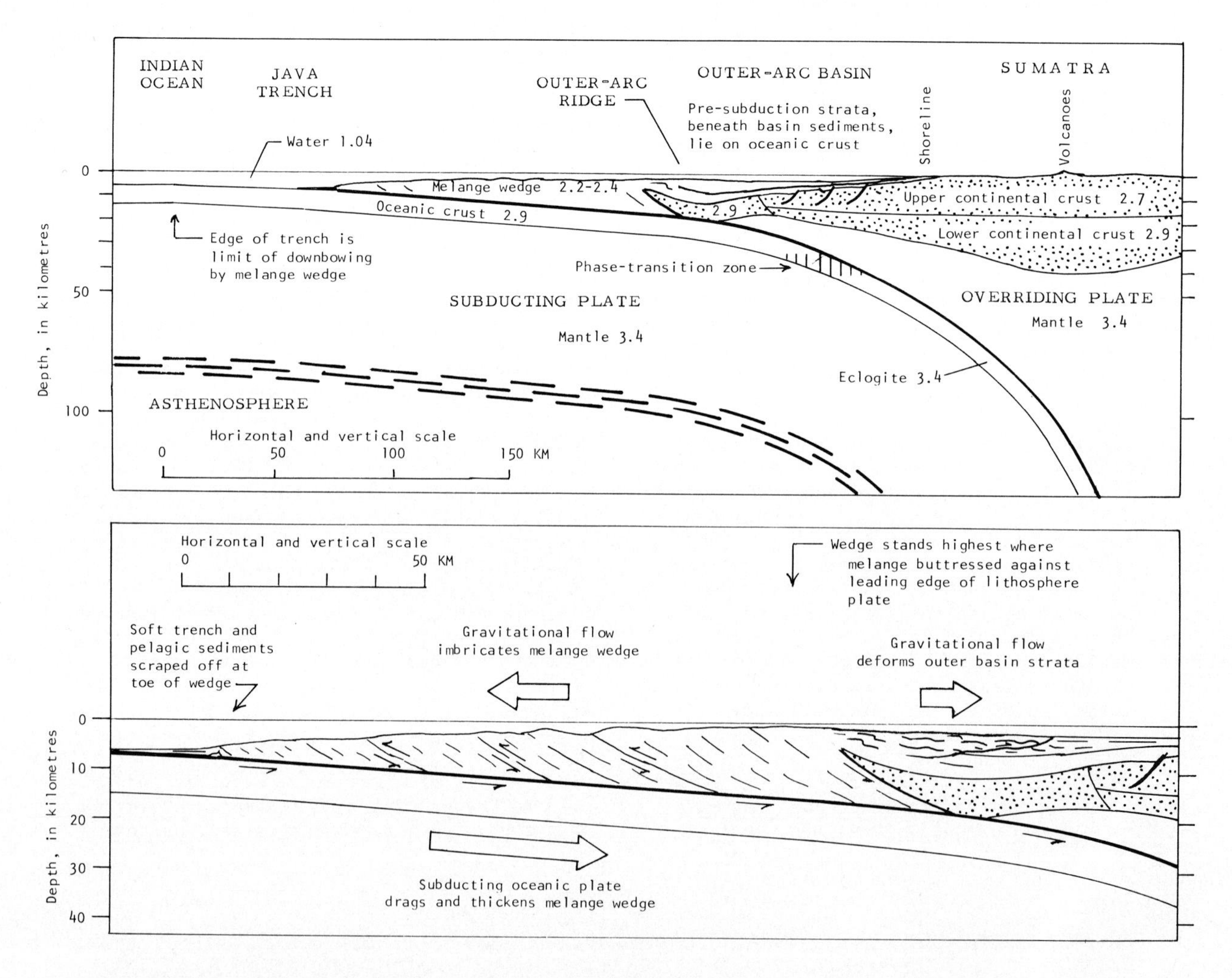

Fig. 7. Sections through the subduction system of the Java Trench, off southern Sumatra. A, top: major tectonic components, showing densities (in grams per cubic centimeter). B, bottom: mechanism of deformation of the melange wedge; components as in A.

Audley-Charles, and Barber, 1976) assume, in a long series of papers, that the melange is a surficial olistostrome sheet, deposited on top of a terrain which was supposed in their early papers to be autochthonous but is now recognized as itself chaotically allochthonous. I have studied the complexes in the field only in the medial part of western Timor; there, even strata as young as early Pleistocene are imbricated steeply with concordantly-foliated, scaly-clay polymict melange. Mapping by many geologists shows this to be the general relationship on the islands of the outer-arc ridge.

Permian to Paleogene shallow-water strata within the chaotic complexes are in Australian lithologic and faunal facies on Timor, and in west New Guinea ones on Seram and Kai. Shingling of shelf strata upward into the melange wedge as it ramped onto the continental shelf is inferred. Some of the Timor Permian is limestone containing fossils indicative of warmer water temperatures than any assemblages known on shore in Australia, and this has led to speculation (as, by Carter and others, 1976) that the limestone was swept in from some far-distant site. As the northwesternmost Permian drilled on the Australian shelf does however include abundant carbonate, and as the Timor Permian limestone faunas are closely similar both taxonomically and paleoclimatically to those of northern India and Pakistan (adjacent to the northwest Australian shelf before Jurassic rifting), I conclude that the Timor rocks in question formed on the warm, north margin of Gondwanaland, and are not now grossly out of place relative to Australia.

Gravity surveys suggest that continental crust extends into the Banda Sea, beneath the thickening arcuate wedge of melange and imbricated materials, almost as far seaward as the north coast of Timor (Chamalaun and others, 1976) and the south coast of Seram (Milsom, 1976), beyond which it may or may not be present beneath overriding oceanic crust. The leading edge of the overriding oceanic plate is inferred to produce a buttress behind the melange wedge. Quartz monzonite, cordierite gneiss, and other silicic rocks likely derived from upper continental crust occur in the tectonic chaos of Seram (Barber and Audley-Charles, 1976).

Between the outer-arc ridge of melange and the inner volcanic arc of the Banda System is an outer-arc basin which mostly is deeper than the trench itself. The Weber Deep, the east part of this basin, reaches depths greater than 7 km. Such depths are characteristic of trenches, and some geologists (as, Carter and others, 1976, and Crostella and Powell, 1976) base tectonic conjectures on the assumption that this basin is indeed a subducting trench. Reflection profiles across the basin, including those of figures 9 and 10, however demonstrate that sediments lap onto basement rocks on both sides, hence disprove this assumption. Further, the Benioff zone is deep within the mantle beneath the basin. Raising the leading edge of the overriding oceanic plate by melange stuffed beneath it, with corresponding elastic depression of the basin, probably have produced the basin.

The Banda Arc records the collision of an island arc with the continent of Australia and New Guinea, and the shoving up onto the continent of an imbricated pile of material derived partly from the continent itself, and partly by deep-water melange-wedge formation in front of the advancing arc. The shallow trench likely results from isostatic depression of the continental margin by the weight of melange pile. Deformation within the melange wedge is inferred to be due,

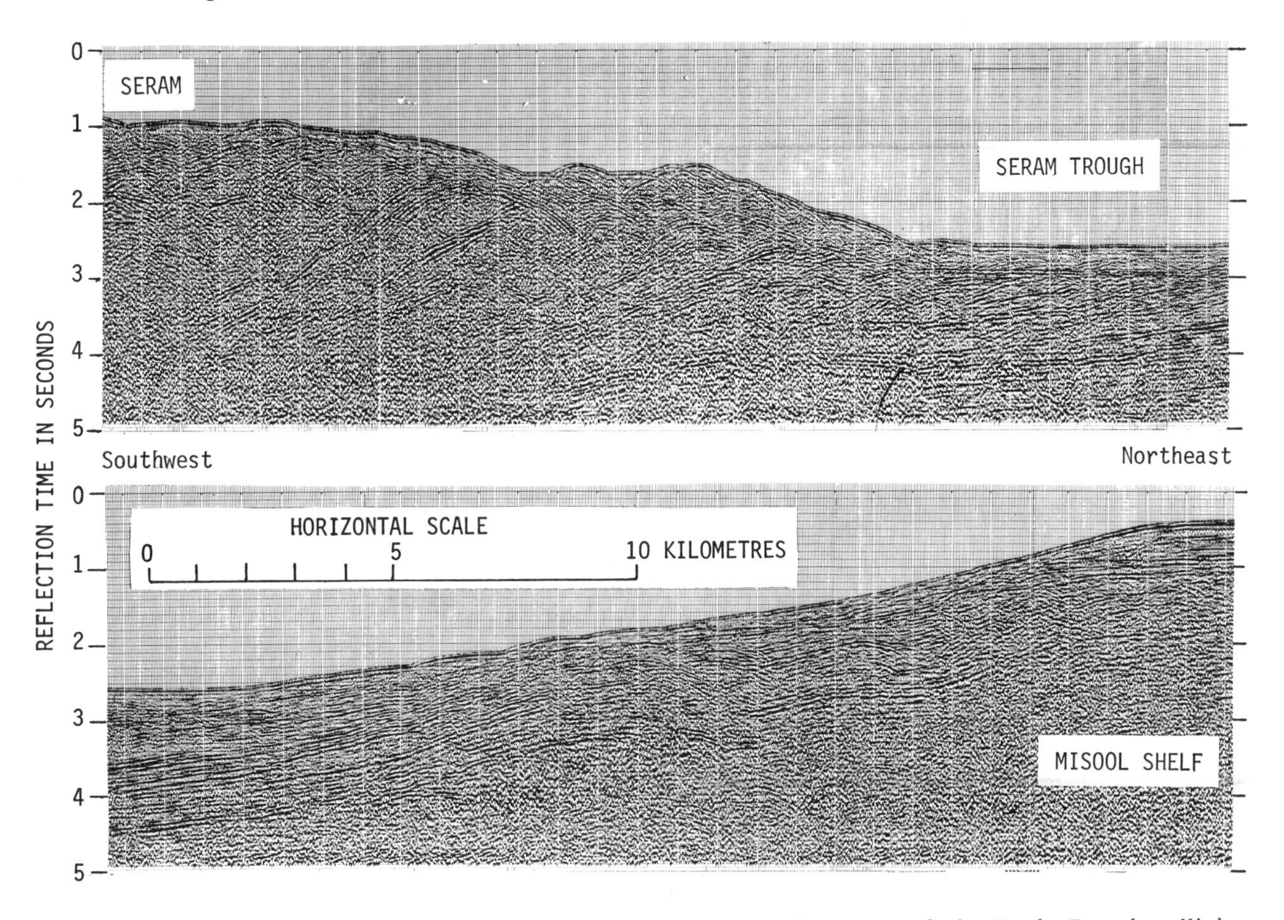

Fig. 8. Seismic-reflection profile across the Seram Trough sector of the Banda Trench. High-velocity (Vp = 3.4-5 km/sec) shallow water strata of the New Guinea continental shelf (near the island of Misool) show only slight deformation as they incline down the continental slope, beneath the very low-velocity turbidite fill of the trough, and pass under the toe of the wedge of low-velocity melange and imbricated sediments whose top rises toward Seram. The soft turbidites are scraped off against the toe, and are imbricated to moderate dips disharmonic to the shelf strata lying tectonically beneath them. Shelf strata, in New Guinea facies, are however imbricated into the wedge as exposed on Seram. Profile provided by Phillips Petroleum Company Indonesia.

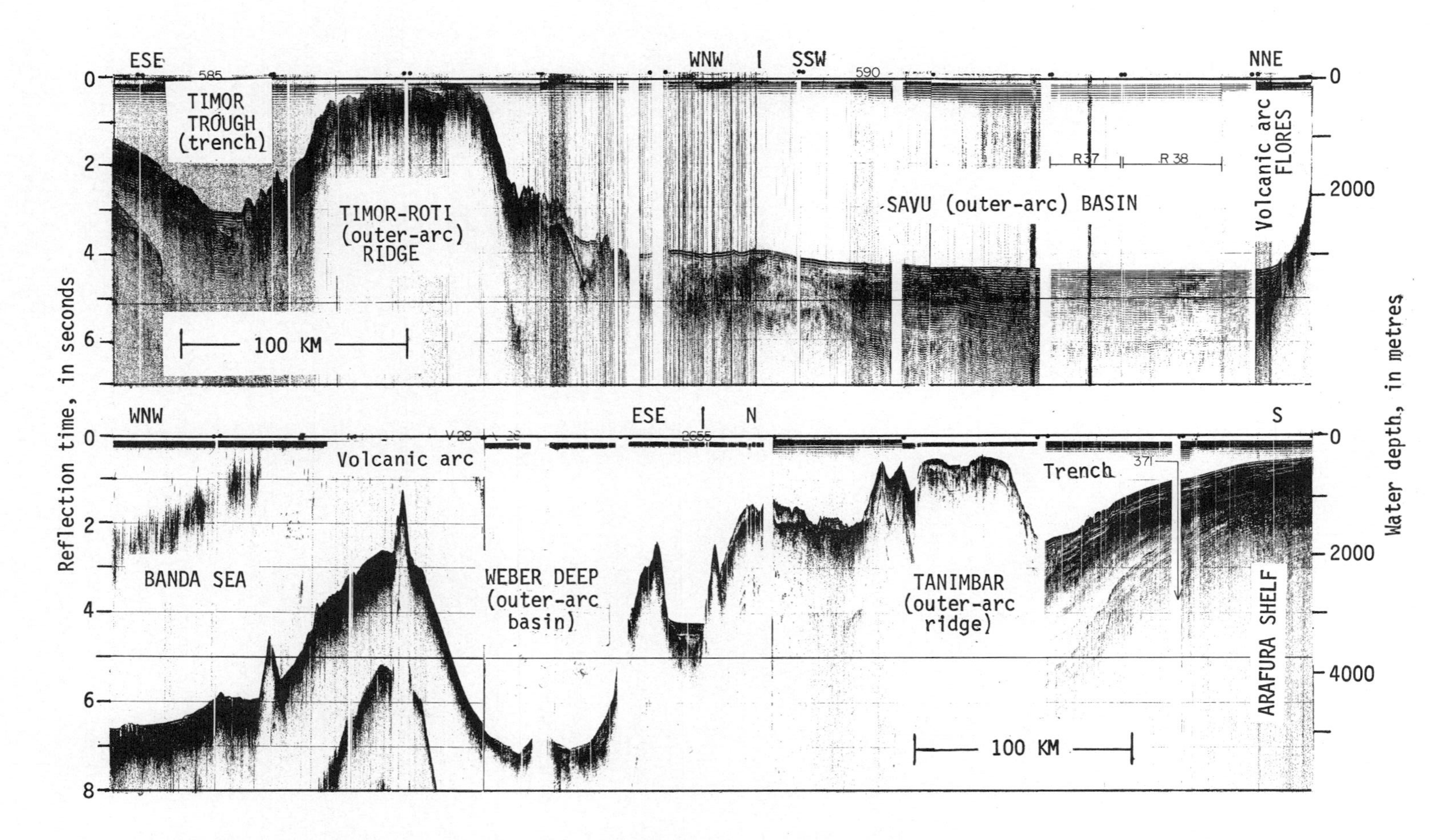

Fig. 9 (top). Seismic-reflection profile across the Banda arc system in the Timor sector. Australian shelf strata plane into the trench. The shallow structure of the Timor-Roti ridge is here anticlinorial. No trench is present within the Savu Basin. Profile from Lamont-Doherty Geological Observatory, R.V. Robert Conrad Cruise 14.

Fig. 10 (bottom). Seismic-reflection profile across the Banda Arc system near Tanimbar. Shelf strata, on the right, continue with gentle dip beneath the low-velocity acoustic basement of Tanimbar, although this can not be seen on this shallow-penetration line. There is no trench in the Weber Deep. Profile from Lamont-Doherty Geological Observatory, R.V. Vema cruise 28. Figures 9 and 10 view the arc from opposite directions.

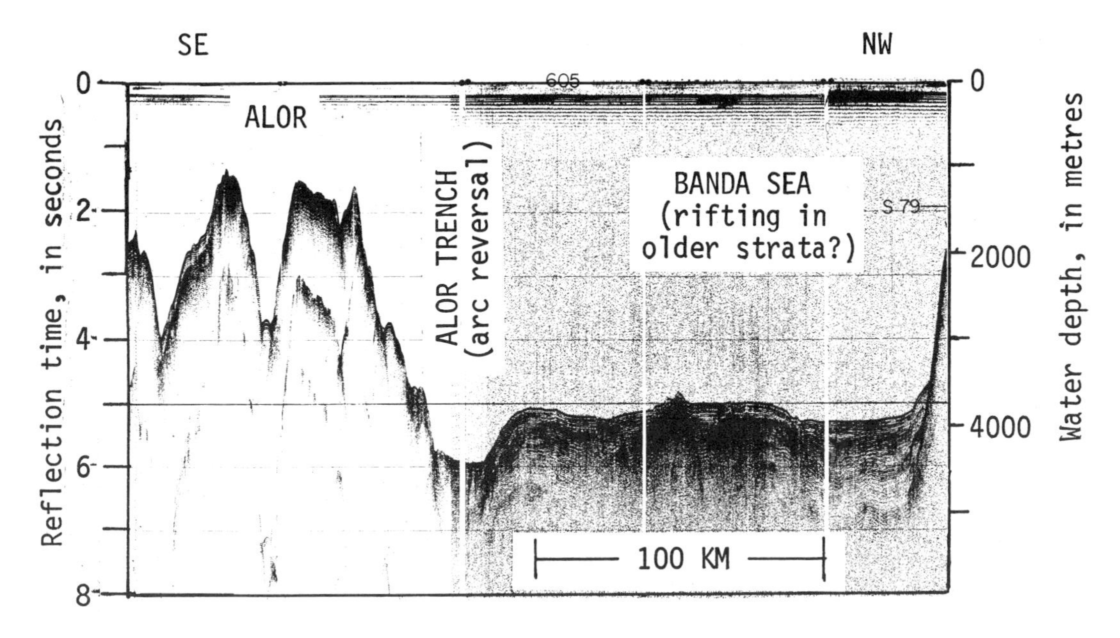

Fig. 11. Seismic-reflection profile across the inner part of the Banda Arc system north of Timor. Alor is a dormant part of the inner volcanic arc, formed above a north-dipping subduction system. Alor Trench at the north base of the volcanic ridge represents an arc reversal. The deeper strata visible in the Banda Sea show tilting, perhaps indicating rifting. Profile from Lamont-Doherty Geological Observatory, R. V. Robert Conrad Cruise 14.

as off Java and Sumatra, to interaction of subductive underflow and gravitational overflow with a buttress. My interpretation is that strata from the Australian and New Guinea shelves did not become incorporated into the melange now exposed on the outer-arc islands until middle or late Pliocene time, and hence that most of the growth of the wedge has occurred within the last 5 million years; but one can argue that on the contrary the exposed melange formed mostly more than 12 million years ago, before late Miocene time. A rapid Neogene migration of the Banda Arc front away from Sulawesi (which itself moved away from Borneo), opening the Banda Sea behind it, can be inferred from the stratigraphic, magmatic, and subduction histories of the islands around the Banda Sea. Reflection profiles across the Banda Sea show its structure to be complex, and deep-sea drilling plus many more profiles are needed to adequately constrain such speculation. A clockwise Neogene rotation of Vogelkop Peninsula, about an oroclinal pivot in the southeast, can be inferred from the abrupt inflection of structural and facies trends, including source directions for clastic sediments, at the base of the peninsula in western New Guinea. The south limb of the Banda Arc may have "unrolled" eastward against Australia as the arc has lengthened and migrated eastward, and the Vogelkop Peninsula may have retreated by rotating away from the advancing arc.

Rapid uplift characterizes the outer-arc ridge, and Pleistocene reefs typically are raised to altitudes of hundreds of metres on its islands. Correspondingly rapid growth of the entire melange wedge is inferred: the slopes of the top and bottom of the wedge represent a dynamic equilibrium and change little, and the rapidly increasing volume of accreting melange is matched by a sinking base and a rising surface as the wedge grows laterally and vertically.

Trench sediments scraped off at the toe of a wedge can be seen in many profiles (as, fig. 8 and, in another arc, fig. 12). Karig and Sharman (1975) are among those who view the growth of such a wedge as due mostly to this scraping, and who regard the rest of the wedge as a passive mass subject primarily to backward rotation as new packets are stuffed in at the front. The accreted-toe model of Karig and Sharman predicts that wedges should have flat, stable tops behind the frontal zone of scraping and rotation; they do not. Throughout the Banda arc, shelf strata slide smoothly beneath the toe of the wedge, yet reappear imbricated, along with other strata deposited on top of the wedge, into the melange exposed on the islands: much internal imbrication of the wedge is required. Abundant mud volcanoes along the ridge attest to the presence of unlithified sediments within even the thickest part of the wedge.

The inner volcanic arc is continuous around the Banda system, but there are no presently active volcanoes on that arc north of central and eastern Timor (fig. 3). Shallow subduction may have virtually stopped in this sector, even

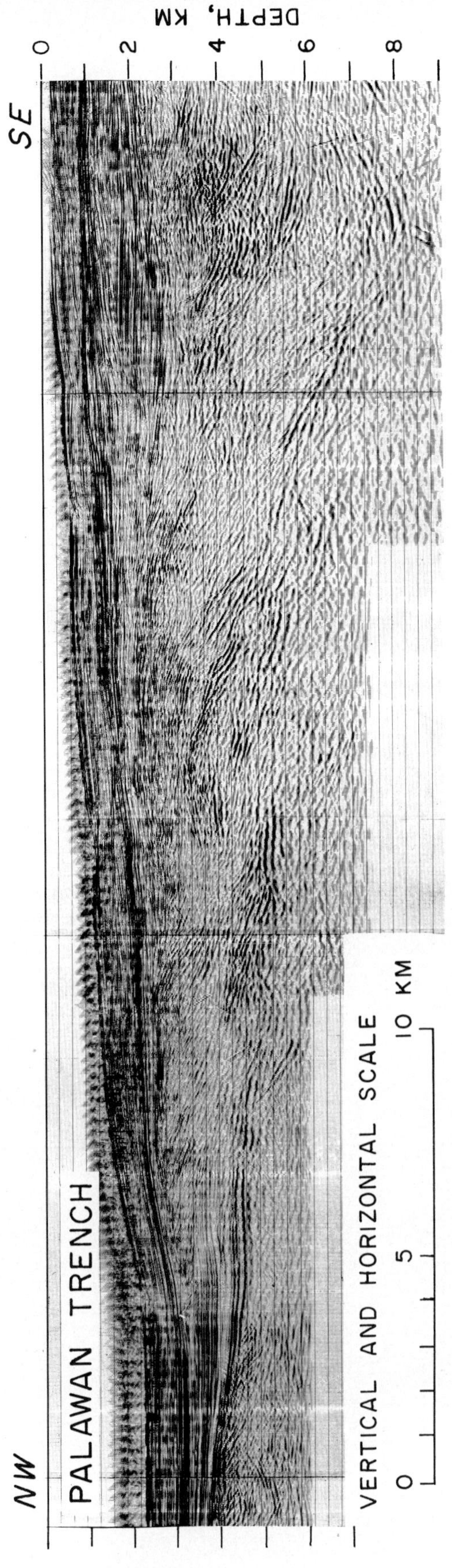

Fig. 12. Migrated and depth-coverted seismic-reflection profile across the buried melange wedge and trench northwest of southern Palawan, provided by Chevron Overseas Petroleum and Texaco Overseas Petroleum.

though there is still an active deep Benioff zone to the north (Hamilton, 1974b). A trench with southward-subduction geometry lies at the north base of the dormant volcanic ridge in this sector (fig. 11). A similar trench is present at the north base of the volcanic arc along western Flores, which is still active volcanically (fig. 3). Apparently the south limb of the Banda Arc is undergoing a reversal in subduction direction: future subduction will be southward beneath the Australian continent, as enlarged by the addition of the collided Banda Arc. The subduction direction presumably is now reversing because the light continental crust of Australia can not be subducted further. The new southward subduction system is not yet manifest in either mantle-earthquake patterns or a volcanic arc, and so it may have been inaugurated only recently.

Complex arcs of the northern Indonesian region

A number of subduction-arc systems display complexities of inauguration, extinction, collision, capture, and reversal in the region of northern Borneo, the Philippines, northern Sulawesi, and Halmahera. The following discussion treats only some of the complications of these arcs.

Northwest Borneo-Palawan subduction system

A subduction system deactivated at about the end of Miocene time is expressed by various onshore and offshore elements in the Palawan and northwestern Borneo area. A profile (fig. 12; ages of strata and presence of melange have been confirmed by drilling) across the offshore part of the system illustrates not only its late history but details of the subduction process as well. Undeformed Pliocene and Quaternary strata, about 1 km thick, are prograded across the top of the extinct melange wedge and trench. Pre-Pliocene trench sediments were scraped off against the underside of the toe of the melange wedge; the top sediments were scraped off first, and the basal sediments survived as coherent layers for 3 or 4 km from the landward edge of the trench, beyond which everything above the gently subducting oceanic plate was imbricated with moderate dips. A basin of basal to upper Miocene strata rests on the southeast half of the illustrated part of the wedge, and was deposited concurrently with subduction and internal imbri- of the wedge, for the basin strata were progressively more deformed downward.

A broad terrain of melange plus progressively older imbricated and broken formations extends landward from the fossil trench and the offshore part of the melange wedge. Onshore Palawan and northwest Borneo consist mainly of an Eocene subduction complex--the polymict melange of the "Danau," "Engkilili," and "Lupar" terrains, and the broken formations, imbricated sheets, and subordinate melange of the "Rajang," "Crocker,"

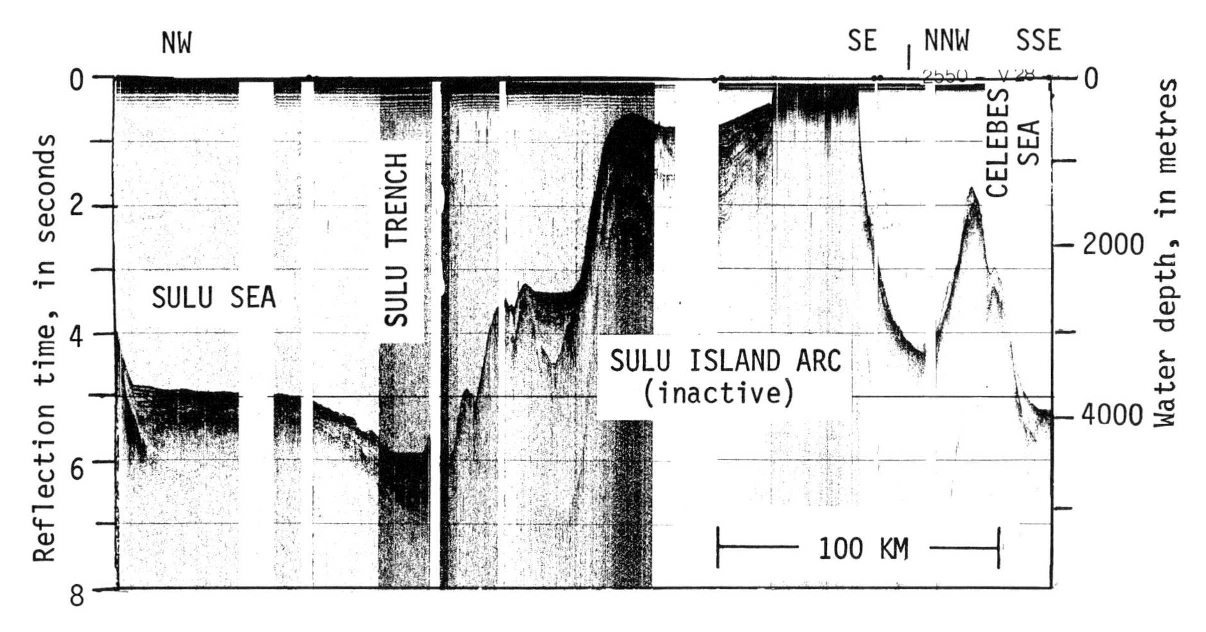

Fig. 13. Seismic-reflection profile across the inactive Sulu island arc and trench. There is no volcanism or Benioff-zone seismicity now active in this arc. Profile from Lamont-Doherty Geological Observatory, R. V. Vema Cruise 28.

"Embaluh," and other rock assemblages. The abyssal, quartzose clastic strata which dominate these complexes are irregularly younger, from early to late Eocene, northward toward the South China Sea across the crescentic outcrop belt (Keij, 1965), yet the coherent rocks mostly dip landward and are right-side up (Stauffer, 1968); thus growth of the great accretionary wedge by imbrication at its seaward side is indicated. The crescentic arc of the subduction complex curves through Sarawak (northwest Borneo) and projects toward the south China coast (fig. 3), which however it does not reach. The arc may have been produced by the counterclockwise rotation of Borneo and Palawan, hinged oroclinally in the west, over the subducting floor of the South China Sea.

The extinct Cagayan volcanic arc in the Sulu Sea (fig. 3) presumably is the magmatic counterpart of the Palawan subduction terrain. McManus and Tate (1976) infer a volcanic source for much of the sediment now in the northwest Borneo subduction complex, and perhaps many of the poorly delimited and vaguely dated igneous rocks of interior Borneo are paired to the northwest Borneo complex.

Sabah-Sulu-Zamboanga arc

The Sulu archipelago is another dormant island arc. It has a trench at its northwest base (fig. 13; Murphy, 1975, figs. 20 and 21), exposures of ultramafic and mafic rocks (presumably melange components) on its northwest and central crestal areas, and extinct, eroded volcanoes along the southeast part of its creast. The arc, with its paired melange and volcanic belts, comes ashore at both ends, as eastern Sabah in northeast Borneo and as Zamboanga Peninsula in western Mindanao. In Sabah, the volcanic-arc rocks range in age from late Oligocene to Pleistocene, and the polymict melange ("Wariu," "Trusmadi," and "Chert-Spilite" units of Borneo geologists) and imbricated and broken formations (much of the "Ayer," "Crocker," and other units) exposed onshore involve Upper Jurassic through lower Miocene materials and include blocks of glaucophane schist. Offshore to the north, however, strata as old as early Miocene are but slightly deformed; this, considered with the distribution of materials onshore, may indicate that the Sulu arc originally stood above a northwest-dipping subduction system, that the arc collided with Borneo, and that subsequently the direction of subduction reversed to the southeast-dipping system evident in the preserved geometry. In Zamboanga, the exposed melange terrain contains rocks as old as Cretaceous, and the paired volcanic belt is of late Oligocene or early Miocene and younger age.

Sangihe and Halmahera arcs

The Sangihe and Halmahera island arcs face each other and are now in the process of colliding in the south. In the north, the collision is complete. Conflicting reversals of subduction direction are underway.

The Sangihe arc connects northeastern Sulawesi and south-central Mindanao (fig. 3), has active volcanoes in its southern and central portions, and stands above a Benioff zone that dips moderately west to depths of more than 600 km beneath the Celebes Sea (Hamilton, 1974b). Across the Molucca Sea, the active Halmahera volcanic arc stands above a Benioff zone dipping more gently eastward to depths of about 200 km. Between the arcs is a broad submarine ridge, exposed as melange on Talaud (Sukamto, 1976), and shown to be underlain by very thick melange by gravity data (Vening Meinesz, 1954) and reflection profiling. The two Benioff zones intersect about 50 km beneath the center of the melange ridge (for example, Hatherton and Dickinson, 1969). There is a shallow trench at the east base of the Sangihe ridge (figs. 14 and 15), but the Sangihe Benioff zone projects to the surface far to the east of this trench. I have seen a deep-penetration seismic-reflection profile which shows unequivocally that this trench marks the trace of a thrust dipping gently eastward beneath the Molucca Sea melange. Suggestions of this thrust can be seen on figures 14 and 15. The trace of the Sangihe subduction zone has apparently been overridden, for a horizontal distance of about 100 km, by the melange wedge at the front of the Halmahera system, or else the Sangihe melange has been captured by the Halmahera arc. The inactive northern part of the Sangihe arc comes ashore in south-central Mindanao, and extensive middle Tertiary melange terrains are exposed farther east in that island. I infer that much of eastern Mindanao represents an aggregate of Sangihe and Halmahera arcs after collision.

There is now a trench, with east-dipping subduction, at the west base of the volcanically inactive northern part of the Sangihe ridge (figs. 3 and 15). As the volcanic ridge itself fits the Benioff zone dipping westward beneath it from the opposite side, it appears that the Sangihe arc has been captured and added to the Halmahera arc, and that a new subduction zone is breaking through on the west side of the aggregate. The Moro Gulf earthquake of 16 August, 1976, Richter magnitude 8.0, apparently occurred along this subduction zone, which continues westward along southern Mindanao.

North Sulawesi Trench

Another trench bounds the Celebes Sea against north Sulawesi (figs. 3 and 14). This trench ends in the west at the seismically defined projection of a very active left-lateral strike-slip fault system that curves through Sulawesi (fig. 3). Apparently northern Sulawesi is rotating clockwise, bounded against southwestern Sulawesi by the strike-slip fault, and against the Celebes Sea by a subduction zone along which plate convergence decreases eastward to zero at an orocline. The North Arm of Sulawesi represents the inactive continuation of the Sangihe volcanic arc, and the present eastward trend of all but the end of the arm is the product of this rotation.

Philippine Trench

The great Philippine (or Mindanao) Trench, which marks the trace of a subduction zone dipping westward, trends from the central Philippines to central Halmahera, along the east side of the Sangihe-Halmahera aggregate. Despite the impressive depth of this trench, and the severe deformation of even very young sediments shown in reflection profiles across it, only a shallow Benioff zone dipping westward from it is suggested by seismicity south of Mindanao (Hamilton, 1974b). Another arc-reversal is inferred to be in progress here: the colliding Sangihe and Halmahera arcs are nearly locked together, and a new subduction system is breaking through on the east side of the aggregate.

Philippines as a composite of arc fragments

Subduction systems that have produced melange and volcanic arcs which are now part of the southern Philippines include the Palawan, Sulu, Sangihe, Halmahera, and Philippine systems noted here. Still other large parts of the southern Philippines do not obviously belong to any of these arcs, and both disrupted and rearranged fragments of these arcs and components of yet more arcs likely are present. Farther north, the Philippine Trench, on the east side of the archipelago, dies out as an active structure, and the Manila Trench begins on the west side, representing a subduction-direction reversal of the entire northern Philippines (Murphy, 1973). The oppositely-facing subduction systems are not joined by a direct transform fault, and the broad S-curve of the central Philippines appears to be a double orocline serving a transform function. The left-lateral Philippine fault zone of the east-central Philippines (Allen, 1962) may be a product of flexural slip in the eastern orocline; it does not reach Mindanao as an active strike-slip feature, and is not now a major plate boundary.

Overview

The behavior of representative arc systems, mostly Neogene, has been discussed here. Many other complexities, in the Neogene as well as the older record, were not mentioned. Throughout Cenozoic time, Pacific plates have converged westward with Asia, and Indian-Australian ones northward, and the Indonesian-Philippine-Melanesian boundary region has responded with rapidly changing patterns of subduction, spreading, migration, strike-slip faulting, oroclinal folding, and rotation. After arcs have collided with each other or with continents, convergence continues at a new subduction zone on the outside

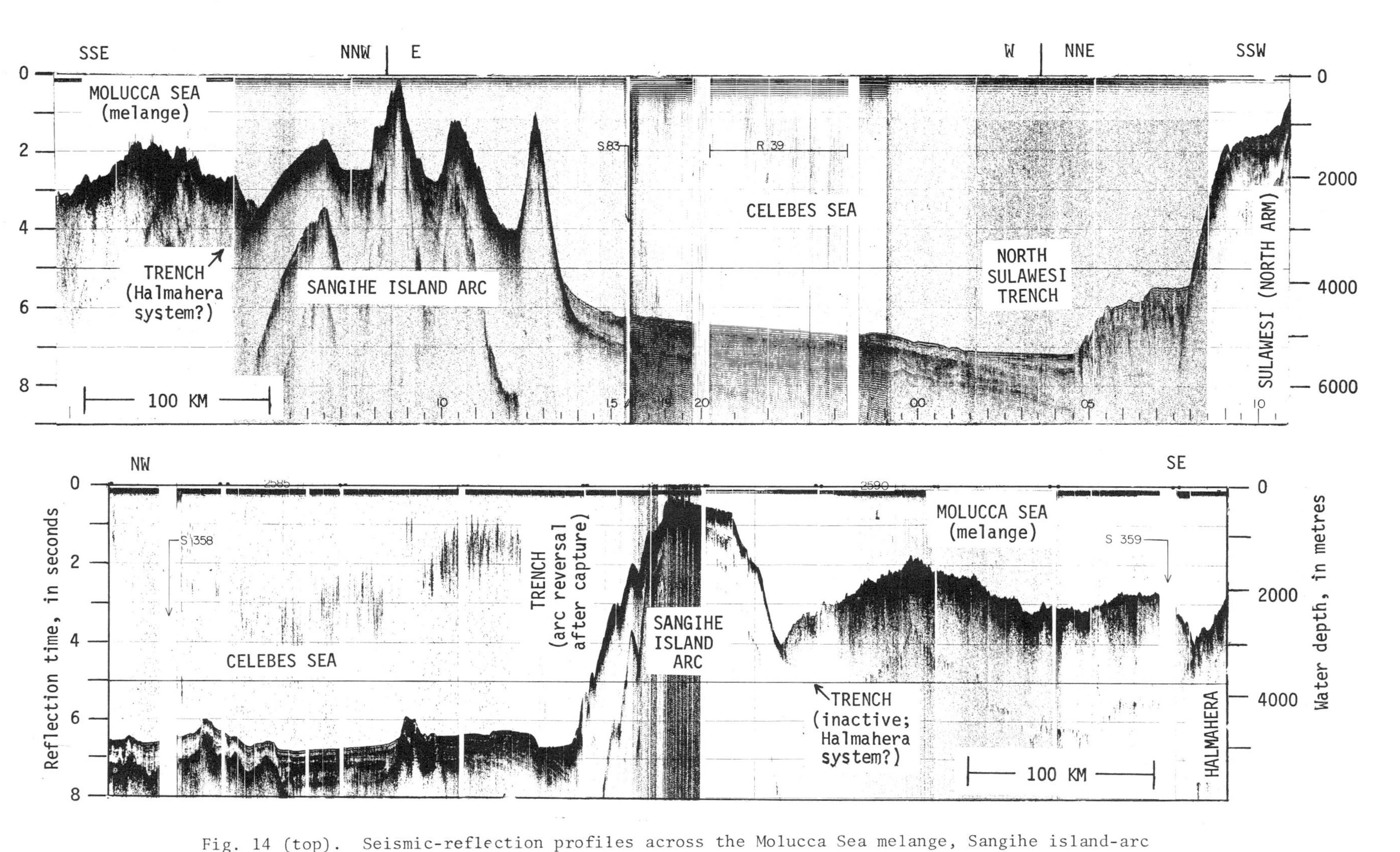

Fig. 14 (top). Seismic-reflection profiles across the Molucca Sea melange, Sangihe island-arc ridge, and North Sulawesi Trench. The volcanoes of the Sangihe arc tie to a Benioff zone dipping westward, buth the shallow trench at the east side of the arc marks the front of the opposed Halmahera arc system, which has overridden the trace of the Sangihe subduction zone. Profile from Lamont-Doherty Geological Observatory, R. V. Robert Conrad Cruise 14.

Fig. 15 (bottom). Seismic-reflection profile across the Sangihe arc, from the northern Celebes Sea to the soutnern Molucca Sea. (Note that figs. 14 and 15 view the arc from opposite directions) The Sangihe-Halmahera collision is here complete, and a new subduction zone has broken through on the west side of the aggregate. Profile from Lamont-Doherty Geological Observatory, R. V. Vema Cruise 28.

of the collision aggregate. Some arcs die without collision, some arcs reverse polarity, and other arcs are born. Subduction hinge lines migrate and change shape as spherical plates inflect through them. Such behavior, like that of mid-ocean ridges, can not be explained by any simplistic model of drive by thermal-convection cells or by gravity. Subduction and spreading appear to be products, not causes, of plate motions. Those motions might be produced by transfer of rotational inertia between lithospheric plates and the rest of the earth, in positive-feedback systems compensating for dynamic imbalances.

If present motions of the major lithosphere plates continue for another 30 million years or so, Australia and China will collide, and Indonesia and the Philippines will be compressed between them into a broad orogenic belt. Belts of complexes of Indonesian type lie between many of the small continents whose Phanerozoic aggregation produced Eurasia, and presumably these orogenic terrrains passed through stages comparable to that of modern Indonesia.

References

Allen, C. R., Circum-Pacific faulting in the Philippines-Taiwan region: J. Geophys. Res., 67, 4795-4812, 1962.

Audley-Charles, M. G., The geology of Portuguese Timor: Geol. Soc. London Mem. 4, 76 p., 1968.

Barber, A. J., and M. G. Audley-Charles, The significance of the metamorphic rocks of Timor in the development of the Banda Arc, eastern Indonesia: Tectonophysics, 30, 119-128, 1968.

Baumann, P., P. de Genevraye, L. Samuel, Mudjito, and S. Sajekti, Contribution to the geological knowledge of south west Java: Indonesian Petrol. Assoc. Proc., 2, 105-108, 1973.

Beck, R. H., and P. Lehner, Oceans, new frontier in exploration: Amer. Assoc. Petrol. Geol. Bull., 58, 376-395, 1974.

van Bemmelen, R. W., The geology of Indonesia: Govt. Printing Office, The Hague, 1A (732 p.), 1B (60 p. + portfolio), 2 (265 p.), 1949.

Bignell, J. D., and N. J. Snelling, Geochronology of Malayan granites: Overseas Geol. and Mineral Resources Bull. 47, U. K. Inst. Geol. Sciences, in press.

Carter, D. J., M. G. Audley-Charles, and A. J. Barber, Stratigraphical analysis of island arc-continental margin collision in eastern Indonesia: Geol. Soc. London Jour., 132, 179-198, 1976.

Chamalaun, F. H., K. Lockwood, and A. White, The Bouguer gravity field of eastern Timor: Tectonophysics, 30, 241-259, 1976.

Cooper, J. A., A. W. Webb, and G. Whitaker, Isotopic measurements in the Cape York Peninsula area, Queensland: Geol. Soc. Australia J., 22, 285-310, 1975.

Crostella, A. A., and D. E. Powell, Geology and hydrocarbon prospects of the Timor area: Indonesian Petrol. Assoc. Proc., 4-II, 149-171, 1976.

Elliott, D., The motion of thrust sheets: J. Geophys. Res., 81, 949-963, 1976.

Grow, J. A., Crustal and upper mantle structure of the central Aleutian arc: Geol. Soc. Amer. Bull., 84, 2169-2192, 1973.

Hamilton, W., Sedimentary basins of the Indonesian region: U.S. Geol. Survey Map I-875-B, 1974a.

______, Earthquake map of the Indonesian region: U.S. Geol. Survey Map I-875-C, 1974b.

Hatherton, T., and W. R. Dickinson, The relationship between andesitic volcanism and seismicity in Indonesia, the Lesser Antilles, and other island arcs: J. Geophys. Res., 74, 5301-5310, 1969.

Karig, D. E., and G. F. Sharman, III, Subduction and accretion at trenches: Geol. Soc. Amer. Bull., 86, 377-389, 1975.

Keij, A. J., Late Cretaceous and Palaeogene arenaceous Foraminifera from flysch deposits in northwest Borneo: Geol. Survey Borneo Region Malaysia Ann. Rept. 1964, 155-158, 1965.

McElhinny, M. W., N. S. Haile, and A. R. Crawford, Palaeomagnetic evidence shows Malay Peninsula was not a part of Gondwanaland: Nature, 252, 641-645, 1974.

McManus, J. and R. B. Tate, Volcanic control of structures in north and west Borneo: Southeast Asia Petrol. Explor. Soc., 1976 Offshore Conf., Paper 5, 13 p., 1976.

Milsom, J., Reconnaissance gravity of Seram, eastern Indonesia: Rept. Geophysics 76/1, Imperial College Science and Technology, London, 27 p., 1976.

Minster, J. B., T. H. Jordan, P. Molnar, and E. Haines, Numerical modelling of instantaneous plate tectonics: Geophys. J. Roy. Astron. Soc., 36, 541-576, 1974.

Montecchi, P. A., Some shallow tectonic consequences of 'subduction' and their meaning to the hydrocarbon explorationist: Amer. Assoc. Petrol. Geol. Mem. 25, 189-202, 1976.

Murphy, R. W., The Manila Trench-west Taiwan foldbelt--a flipped subduction zone: Geol. Soc. Malaysia Bull. 6, 27-42, 1973.

______, Tertiary basins of Southeast Asia: Southeast Asia Petrol. Explor. Soc. Proc., 2, 1-36, 1975.

Sclater, J. G., and R. L. Fisher, Evolution of the east central Indian Ocean, with emphasis on the tectonic setting of the Ninetyeast Ridge: Geol. Soc. Amer. Bull., 85, 683-702, 1974.

Stauffer, P. H., Studies in the Crocker formation, Sabah: Borneo Region Malaysia Geol. Survey Bull. 8, 1-13, 1968.

Sukamto, R., Singkapan 'melange' di Pulau Karakelang: Geol. Survey Indonesia Newsletter, 8, (19), 5-6, 1976.

United States Geological Survey, Bathymetric map of the Indonesian region: U.S. Geol. Survey Map I-875-A, 1974.

Veevers, J. J., J. R. Heirtzler, and others, Initial reports of the Deep Sea Drilling Project, 27: U.S. Govt. Printing Office, Washington, 1060 p., 1974.

Vening Meinesz, F. A., Indonesian archipelago--a geophysical study: Geol. Soc. Amer. Bull., 65, 143-164, 1954.

TECTONO-STRATIGRAPHIC EVOLUTION OF SUBDUCTION-CONTROLLED SEDIMENTARY ASSEMBLAGES

William R. Dickinson

Department of Geology, Stanford University, Stanford, California 94305

Abstract: Dispersal of sediment from uplifted belts along convergent plate junctures is partly into associated orogenic basins whose structural frameworks and depositional histories are tied closely to the evolution of nearby arc-trench systems. Salient types of basins include: (1) forearc basins in arc-trench gaps between subduction complexes and magmatic arcs, (2) remnant ocean basins on plates being consumed, (3) interarc basins between frontal arcs and remnant arcs, and (4) asymmetric foreland basins of two kinds: (a) peripheral basins formed between suture belts and cratons during crustal collisions, and (b) retroarc basins located between backarc fold-thrust belts and cratons. Except within rifted interarc basins, evolutionary configurations of most orogenic basins seemingly are controlled mainly by regional downward flexure of plate margins in response to local tectonic loading of plate edges beneath progressively accreted subduction complexes or successively stacked thrust sheets. Downward curvature of the upper surfaces of descending plates and downward flexure of plates under sedimentary loads also contribute to subsidence in orogenic basins. Basin evolution is thus a product of isostatic response to changing loads of rock masses that are continually redistributed by linked tectonic and sedimentary processes. The sedimentary assemblages of orogenic basins can be viewed, together with the volcano-plutonic complexes of magmatic arcs and the tectonic welts of subduction zones and suture belts, as part of a standing wave of diverse material thrown up above descending slabs of lithosphere by plate interaction along convergent plate junctures.

Introduction

Plate-tectonic theory places special emphasis on the horizontal translations of lithosphere. However, the relative elevations of plate surfaces are also direct results of plate interactions. For example, the contrast between the mainly subaerial continental blocks and the mainly subaqueous oceanic basins reflects the generation of new oceanic lithosphere and the conservation of old continental lithosphere. Moreover, major internal changes in the bathymetry of the oceans and the topography of the continents are also direct results of plate interactions. The time-dependent thermotectonic subsidence of oceanic lithosphere is now well established [Tréhu, 1975], and the relation of mountain belts to plate boundaries is also clear [Dewey and Bird, 1970]. Because uplifted areas tend to shed sediment and subsiding regions tend to attract sediment, plate tectonics exerts strong control over patterns of erosion and sedimentation at the surface [Dickinson, 1974]. In this paper I use a plate-tectonic framework to describe the key relations in space and time between stratigraphic and tectonic elements of orogenic belts tied to plate consumption at subduction zones.

Orogenic Basins

Orogenic belts produced by plate consumption at subduction zones [Figure 1) include arc orogens of both intra-oceanic and continental-margin types, and collision orogens where either intra-oceanic or continental-margin arcs may be involved [Coleman, 1975]. In arc orogens, oceanic lithosphere is continually consumed along a flanking trench, and an active magmatic belt forms the main spine of the orogen. An accretionary subduction complex develops in association with the subduction zone at the trench. By contrast, collision orogens develop only where all the intervening oceanic lithosphere once lying between two continental blocks has been consumed. Subsequent attempts to consume continental lithosphere are resisted by the buoyancy of continental crust [McKenzie, 1969], and the subduction zone is converted into a suture belt [Dewey and Burke, 1973]. Arc magmatism gradually subsides as the continental blocks involved are welded together.

Orogenic basins include two main types that form in front of the magmatic arc, and two main types that form behind it. Linked most closely to arc-trench systems are forearc basins that develop within the arc-trench gap between the trench slope break and the volcanic front of the magmatic arc. At least the outer parts of forearc basins are part of the accretionary prism

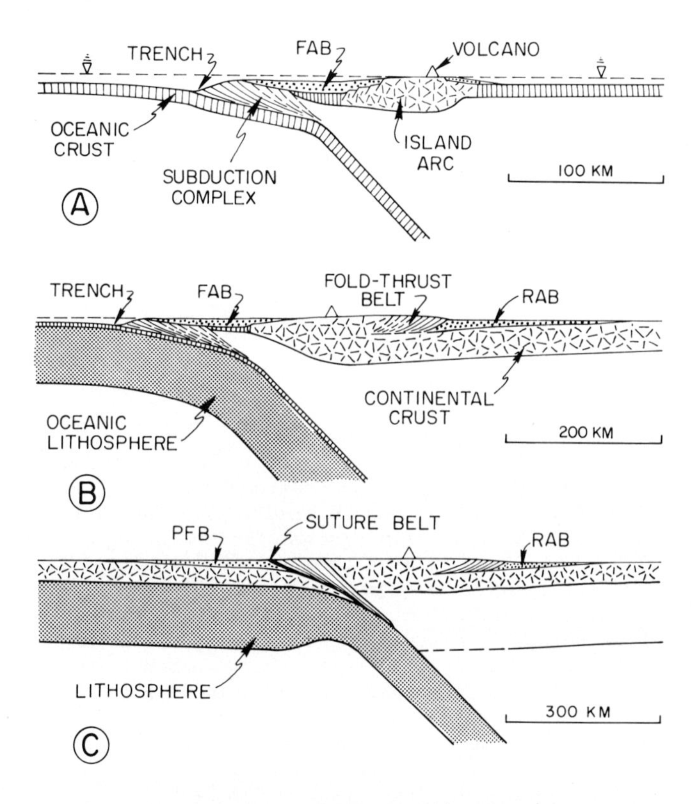

Figure 1. Schematic diagrams of orogen types: (A) intra-oceanic arc orogen, (B) continental-margin arc orogen, (C) intercontinental collision orogen. Horizontal and vertical scales are the same but differ for each diagram. (See scales on figures.) Types of orogenic basins shown by stipples include: FAB - forearc basin; RAB - retroarc foreland basin; PFB - peripheral foreland basin. (See text for explanations.)

[Karig and Sharman, 1975], but forearc basins are not part of the subduction complex as that term is used here.

Beyond the trench is the open oceanic region that becomes a remnant ocean basin subject to special sedimentary and tectonic influences as a collision orogen develops. When the lithosphere of a remnant ocean basin is finally consumed, the edge of the bounding continental block is drawn down against the subduction complex. A peripheral foreland basin is then formed on the adjacent surface of the partly subducted continental block. The deformed materials of the subduction complex form an ophiolitic suture belt that lies between this type of foreland basin and the volcano-plutonic complex of the magmatic arc.

Behind magmatic arcs, quite different tectonic styles may prevail. For either intra-oceanic or continental-margin arcs, deformation in the backarc area may be slight, as is the case for both the Aleutians and Java. Where intense, deformation may be either extensional to form interarc basins and remnant arcs, or contractional to form fold-thrust belts and retroarc foreland basins. The volcano-plutonic complex of the magmatic arc lies between this type of foreland basin and the ophiolitic complex of the subduction zone. Development of backarc fold-thrust belts and retroarc foreland basins apparently reflects overall kinematics across the arc structure opposite to that responsible for interarc basins. As discussed below, the foreland fold-thrust belt presumably develops where the edge of the rigid craton underthrusts the heated infrastructure along the rear flank of the magmatic arc. This partial subduction accompanies formation of the retroarc basin along the pericratonic belt between arc orogen and craton.

Tectonism and Subsidence

Subsidence within interarc basins stems from the cooling of newly formed lithosphere in a manner analogous to the behavior of open ocean basins [Karig, 1971]. For the other orogenic basins, however, active uplift and continuing subsidence occur in adjacent parallel belts. The flanks of the uplifted belts are marked by tectonic telescoping of structural elements in response to active contractional deformation. The sedimented belts are characterized by comparatively passive subsidence without strong tectonic deformation.

In such orogenic basins, subsidence cannot be ascribed primarily to any of the three main factors that control deposition in rifted basins and along rifted continental margins. These are crustal attenuation, thermal decay, and regional flexure from local sedimentary loading of plate interiors [Walcott, 1972]. Instead, two other influences are seemingly dominant. One is the sinking caused locally where plates bend to descend into the mantle. This kind of flexure is a direct effect of the subduction and may cause intense subsidence of finite amount, but may not be suitable to explain prolonged and incremental subsidence. The other is regional downward flexure of plate edges in response to local tectonic loads formed by telescoping of structural elements associated with subduction systems [Hamilton, 1973]. This kind of flexure is an induced effect of subduction, and seems capable of progressive development that could cause prolonged and incremental subsidence.

In the following three sections, the relations between tectonism and subsidence are discussed for forearc basins, foreland basins, and retroarc basins in that order.

Forearc Basins and Subduction Complexes

Figure 2 depicts the evolution of the accretionary prism of an arc-trench system.

In profile, trench slopes typically are underlain by many kilometers of deformed strata forming a tectonically thickened prism for which the term subduction complex is convenient. Reflectors that are best interpreted as imbricate thrusts are inclined gently landward within the subduction com-

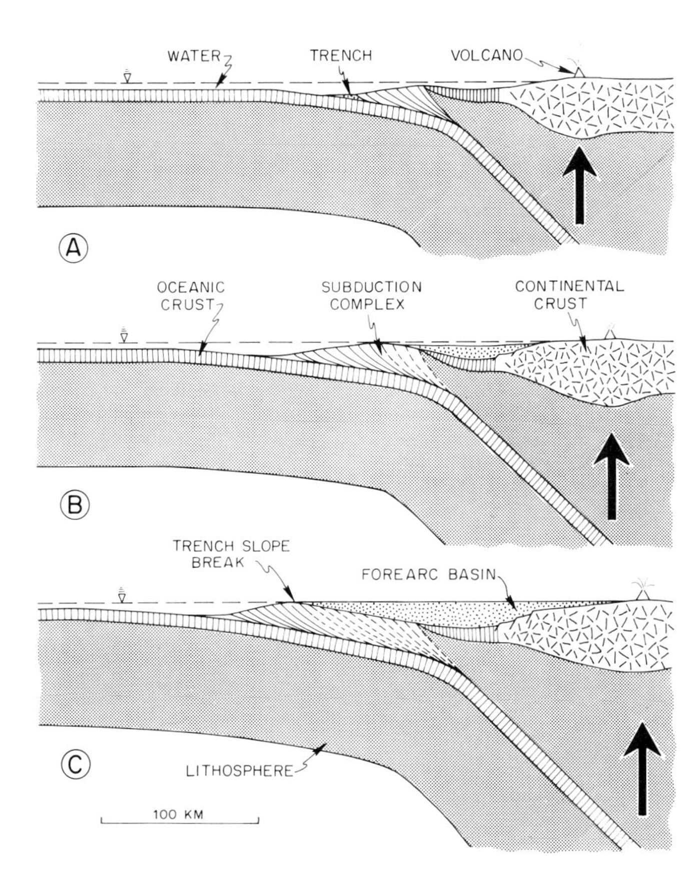

Figure 2. Idealized diagrams showing inferred evolution (A to B to C) of the accretionary prism [Karig and Sharman, 1975] of an arc-trench system: A, incipient stage; B, starved forearc basin; C, full forearc basin. Solid lines within subduction complex denote active imbricate zones, whereas dashed lines denote inactive structures. Large arrows show path of magma transit from seismic zone to magmatic arc. Vertical scale equals horizontal scale.

plex [Beck, 1972]. These reflectors are concave upward and merge successively downward with a more gently inclined surface of apparent décollement that marks the base of the subduction complex. To seaward, the décollement reflector is continuous updip laterally with the reflection from the igneous layer of the ocean floor. The underlapping wedges of deformed strata bounded by major imbricate thrusts within the subduction complex thus presumably take their initial form as successively underthrust packets of sediment stripped off the oceanic crust [Seely and others, 1974]. The materials in the subducted packets may be composed of trench-fill or open-ocean sediments, or of both in varying proportions.

The subduction complex is an accretionary tectonic feature that grows outward as wedges of material are underthrust beneath its flank at the trench axis, and upward as isostatic buoyancy progressively uplifts previously accreted wedges that are stacked vertically by tectonic imbrication. As uplift proceeds, the tendency for internal imbrication is enhanced by seaward gravitational spreading of the subduction complex [Hamilton, 1973]. The bulk strength of the subducted materials thus likely controls the inclination of the inner slope of the trench. As subduction tends to drag the toe of the slope landward, gravitational collapse of the growing subduction complex compensates by inducing décollement that detaches weak materials from the upper surface of the descending plate. This process allows the toe of the slope to slip relatively seaward above the surface of décollement as the descending plate slides beneath the subduction complex. Hamilton [1973] aptly compares the subduction complex, his "mélange wedge," to a standing wave that spreads gravitationally forward at the same speed with which it is dragged back by the subducting plate.

The trench slope break marks the approximate inner edge of the zone of uplift and active deformation, which declines gradually in intensity upward from the toe of the inner slope of the trench [Karig and Sharman, 1975]. Landward of the trench slope break, deposition of essentially undeformed sediment may occur within a forearc basin whose lateral extent commonly increases progressively with time as the arc-trench gap widens [Dickinson, 1973]. Sediments that are deposited seaward of the trench slope break are kneaded tectonically into the subduction complex, although the degree of their disruption and deformation may be quite variable. Deposits in this setting include those of small perched basins bounded by slope faults as well as true slope deposits.

The trench slope break forms the seaward threshold of the forearc basin and the arc structure forms its landward flank. Bathymetry within the forearc basin is a function of the relation between the rate of sediment delivery to the arc-trench gap and the rate of growth of structural relief between the floor of the basin and the uplifted crest of the subduction complex at the trench slope break. When the basin is filled, the total thickness of sediment present is a measure of the ultimate structural relief attained by combined subsidence of the basin floor and uplift of the subduction complex.

The nature of the substratum beneath the sediment fill of forearc basins in the arc-trench gap remains uncertain except along the landward flanks where strata lap across arc terranes. Available geophysical data is typically inadequate to distinguish between older parts of the subduction complex buried beneath the seaward flank of the forearc basin and pre-subduction crustal elements trapped between the subduction complex and the arc structure [Grow, 1973]. The substratum beneath old forearc basins can be observed locally where exposed by uplift in eroded arc-trench systems. The sedimentary sequences then occur in belts that lie between terranes of blueschists and mélanges representing subduction complexes on one side and terranes of metavolcanics and batholiths representing magmatic arcs on the other. Beneath the seaward flanks of thick forearc basins in California [Bailey and others, 1970], New Zealand [Blake and Landis, 1973], and Burma [Brunnschweiler, 1966], the lowest stratigraphic horizons of sedimentary sequences that occupy this general tectonic position rest concordantly on ophiolite sequences which are underthrust by subduction complexes. Where such thin oceanic crust is present beneath forearc basins, thick sedimentation can proceed by isostatic subsidence in a manner analogous to subsidence of the continental rise off rifted continental margins. The growing subduction complex serves merely as a dam to trap sediment within the forearc basin.

As the subduction complex widens by accretion, the trench slope break progrades seaward [Dickinson, 1973]. The seaward flank of the forearc basin then shifts seaward as well until the sediments overlie the subduction complex as part of the accretionary prism [Karig and Sharman, 1975]. The contact between the sedimentary sequence of the forearc basin and the underlying subduction complex is thus a time-gransgressive feature that records the space-time trajectory of the migrating trench slope break. Because the trench slope break is fundamentally a tectonic boundary marking the landward limit of major active deformation within the subduction complex, the time-transgressive upper contact of the subduction complex within the accretionary prism may evolve as a time-transgressive zone of tectonic dislocation rather than as an ordinary unconformity [Dickinson, 1975]. Presumably, this type of structural contact might develop most readily in cases where the forearc basin is kept full of sediment to the threshold. In other cases, an onlapping unconformity may develop.

Retrograde migration of the axis of arc magmatism commonly accompanies prograde migration of the trench slope break [Dickinson, 1973]. The landward flank of the forearc basin then transgresses across eroded roots of the magmatic arc [e.g., Ingersoll, 1975].

The roughly symmetric broadening of the forearc basin as it evolves probably can be ascribed to isostatic downflexure of the lithosphere under the sedimentary load of the forearc basin. On the landward side, the structural discontinuity that commonly develops between the forearc basin and the arc structure may reflect mechanical decoupling of the whole accretionary prism [Karig and Sharman, 1975]. The progressive subsidence of successive trench slope breaks as the upper surface of the subduction complex acquires a landward slope probably depends also upon continued sedimentation to augment the sedimentary load of the forearc basin.

The internal geometry and size of accretionary prisms along the trench flank of arc orogens are thus controlled by complex interplay between tectonic and sedimentary processes. The thickness of sediment in the trench at any given time depends upon the relation between rate of subduction and rate of sedimentation, and overfilled subduction zones may lack trench morphology. The rate of growth of the subduction complex depends upon the volume of combined trench and oceanic sediment drawn into the subduction zone. The potential thickness of sediment in the forearc basin depends upon the elevation of the trench slope break, as controlled by buoyant isostatic uplift and gravitational collapse of the growing subduction complex, and upon the amount of subsidence in the arc-trench gap, as controlled by the crustal thickness there. Overall basin configurations can be explained qualitatively by isostatic arguments without appeal to dynamic effects of plate descent. However, the apparently rapid buoyant uplift of subduction complexes where plate descent stops [e.g., Ernst, 1975] suggests that some unknown component of subsidence affecting the accretionary prism stems from plate dynamics. This post-subduction uplift turns the seaward flank of the forearc basin sharply upward to leave the remainder of the basin fill preserved as the keel of an asymmetric regional syncline.

Foreland Basins and Remnant Oceans

Figure 3 depicts the evolution of sedimentary basins associated with crustal collisions.

When ocean basins are destroyed by plate consumption, the remnant ocean basin that remains just prior to crustal collision becomes a special kind of depositional site [Graham and others, 1975]. In general, crustal collision cannot be simultaneous along the full length of a suture belt [Dewey and Burke, 1974], and collision orogens develop sequentially as gradually lengthening features (Figure 3). As their evolution proceeds, sediment shed longitudinally from orogenic highlands is deposited as turbidite fans that occupy the elongate remnant ocean basins. The immense Indus and Bengal Fans, built largely of sediment shed longitudinally from the Himalayan ranges, are modern examples [Curray and Moore, 1971]. As subduction continues along the flank of a remnant ocean basin, such fan turbidites are incorporated into subduction complexes of immense aggregate volume. Along the flank of the Bay of Bengal, the Andaman-Nicobar insular ridge west of the Andaman Sea and the Mentawai Islands along the trench slope break off Sumatra are composed largely of deformed Bengal Fan sediment, as are the Indoburman Ranges along strike on the Asian mainland beside the ancestral head of the Bay of Bengal [Curray and Moore, 1974]. Mapping on land [e.g., Brunnschweiler, 1966] and seismic reflection profiling of the insular ridges [Curray and Moore, 1974] indicate that the deformed terrane consists in general of isoclinally folded packets of sediment arranged in thrust plates or nappes that are separated in part by mélange bands and dip gently in imbricate fashion away from the remnant ocean basin.

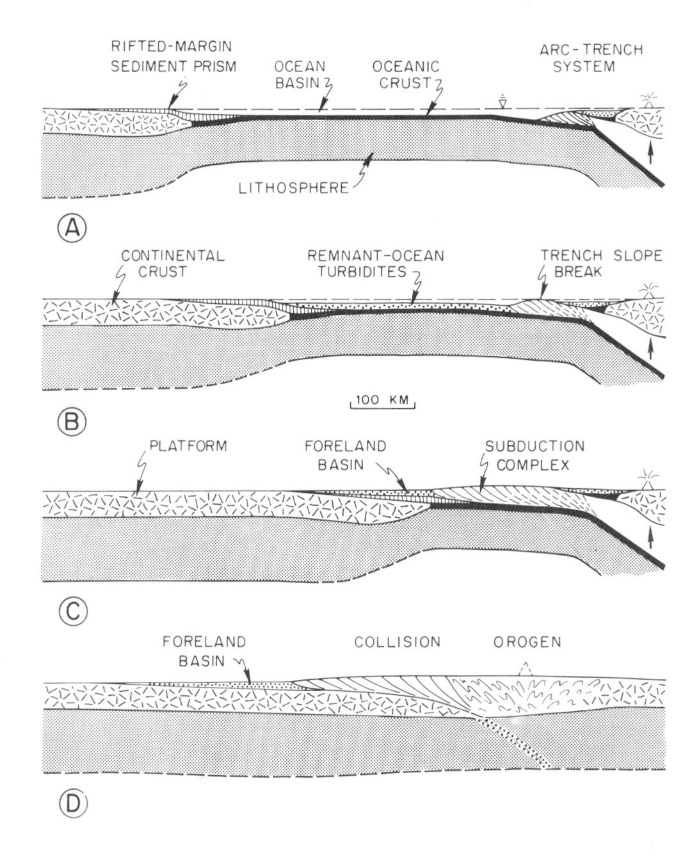

Figure 3. Idealized diagrams showing inferred evolution (A to B to C to D) of sedimentary basins associated with crustal collision to form cryptic intercontinental suture belt within collisional orogen (sketch D highly stylized). Diagrams represent a sequence of events in time at one place along a suture belt marked by diachronous closure. Hence, erosion in one segment (D) of the orogen where the sutured intercontinental join was complete could disperse sediment longitudinally past a migrating tectonic transition point (C) to feed subsea turbidite fans in a remnant ocean basin (B) along tectonic strike. Vertical scale equals horizontal scale.

When a remnant ocean basin is finally consumed entirely, the tectonically quiescent continental margin along the passive flank of the ocean presumably is drawn against and partly beneath the internally imbricated subduction complexes of remnant-ocean sediment. As the suture belt locks closed, strata of rifted-margin sediment prisms along the previously quiescent continental margin may also be involved in thrusting associated with the subduction zone. Varied behavior in detail can be expected as the relative and absolute volumes of remnant-ocean fans and rifted-margin sediments vary for particular cases. As partial subduction continues until the buoyancy of the subducted continental edge arrests plate descent, additional thrusting may also develop from gravitational spreading that induces wholesale imbrication of the flank of the uplifted orogen. Furthermore, Bird and others [1975] argue that the greater thickness of continental lithosphere in comparison to oceanic lithosphere will promote a shallower angle of plate descent at the subduction zone during collision events, and that this effect will cause additional thrusts to propagate through the flexed continental block ahead of the suture belt.

As inferred from isopachs, the configurations of foreland basins are remarkably simple. Typically, they are broad, asymmetric downwarps that begin as a feather edge on the cratonic flank, where they may merge with a thin platform cover. From that point, the substratum tilts smoothly downward beneath the fold-thrust belt at the flank of the orogen. Gentle folds or local faults and basement uplifts may rumple the floors of the basins, but these seem clearly to be secondary features. Only along the flank of the orogen, where both the fill and the substratum of the foreland basin may be involved in thrusting and uplift, is the broad sweep of a foreland basin commonly broken by intricate structural detail.

The consistent geographic association of the broad asymmetric downwarp with the suture belt suggests that isostatic downflexure of the lithosphere in response to tectonic loading by telescoped structural elements along the fold-thrust belt is the prime cause of subsidence. The dynamic effect of plate descent alone seems an inadequate explanation for subsidence whose cumulative effects persist without rebound long after plate descent has stopped. The sedimentary load of the forearc basin must be counted, of course, as contributory to the regional isostatic balance achieved by flexure.

Perhaps the simplest foreland environments in which to perceive the key processes that control subsidence in the foreland basin are those where the fold-thrust belt is composed mainly of oceanic rock assemblages as in the Ouachita orogenic belt [Wickham and others, 1976]. In such cases, imbricated thrust sheets and nappes of turbidites and other deep-water facies have been carried over coeval shelf and platform associations [Figure 3]. The foreland basin lies in front of and is partly overridden by the thrust front marked by the facies contrast. The overthrust oceanic masses apparently had a tectonic thickness of 5 to 10 km, and thus represent adequate loads to cause the subsidence in the adjacent foreland basins by downflexure of lithosphere to achieve regional isostatic balance (e.g., Price, 1973, p. 498).

Along other collision orogens, as in the Zagros Mountains of Iran [Haynes and McQuillan, 1974], the suture belt and mélanges of the subduction complex lie well within the orogenic highlands (Figure 4). The fold-thrust belt beside the late Cenozoic foreland basin along the Persian Gulf is

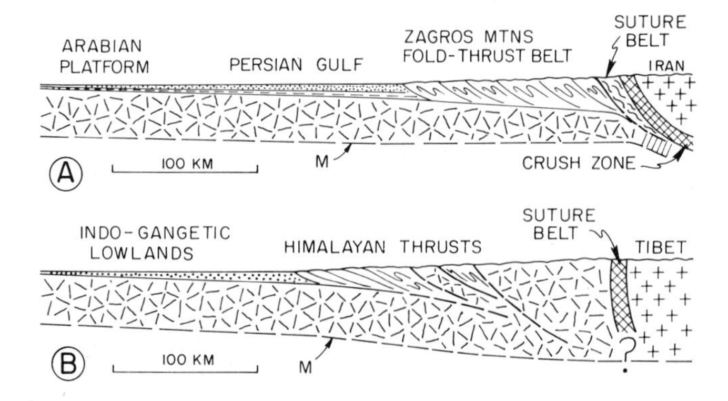

Figure 4. Schematic true-scale diagrams of peripheral foreland basins where fold-thrust belt is composed mainly of materials carried back from the continental edge. A, Zagros region beside Persian Gulf, where fold-thrust belt contains strata partly peeled by décollement off subducted continental margin (generalized after Haynes and McQuillan, 1974, Figure 2, and Bird and others, 1975, Figure 1); B, Himalayan region, where over-thrust masses include crystalline slices of continental basement (generalized after Powell and Conaghan, 1973, Figure 2, and Ramachandra Rao, 1973, Figure 7).

composed of rocks deposited and now deformed on the same side of the suture (Figure 4A). Thus, the tectonic load that causes subsidence within the foreland basin need not be an exotic rock mass, but can form by local décollement and structural telescoping along the subducted continental edge. In the Himalayas (Figure 4B), the whole thrust complex more than 100 km wide is composed of peninsular Indian crustal elements including basement rocks [Powell and Conaghan, 1973].

Retroarc Basins and Fold-Thrust Belts

Elongate retroarc basins that develop along the pericratonic belt between craton and orogen have an asymmetric profile similar to that of the foreland basins formed adjacent to the suture belts of collision orogens. Their geographic association with fold-thrust belts is also analogous. The tectonic load of stacked thrust sheets is thus probably responsible for subsidence, as Price [1973] has argued for the late Mesozoic and early Cenozoic retroarc basin in Alberta. This relationship is not as readily demonstrable, however, because no oceanic closure and crustal collision are involved locally in the evolution of the basins (Figure 5). The deformed strata of the fold-thrust belt are typically miogeoclinal strata thrust back toward the craton from which they were derived.

Burchfiel and Davis [1975] ascribe the development of the Cordilleran fold-thrust belt to underthrusting of the rigid lithosphere of the continental block beneath the inland flank of the arc structure. Armstrong and Dick [1974] earlier noted how hot lithosphere that is soft or thin might control the location of major regional thrusts behind arcs. The zone of crustal flowage behind the arc was probably located along the trend marked by syntectonic polymetamorphic rocks in the Cordilleran infrastructure [Armstrong and Hansen, 1966]. The infrastructural belt lies just west of the Cordilleran fold-thrust belt but still east of the main batholith belt. The position of the most active thrusting migrated away from the arc and infrastructure as successive thrust sheets were stacked one below the other [Armstrong and Oriel, 1965]. The keel of the retroarc basin ("fore-deep") similarly migrated eastward and deepened as the fold-thrust belt evolved [Bally and others, 1966].

These known and suggested relations of tectonic elements in space and time imply that the backarc fold-thrust belt is a type of subduction complex. Note, however, that the total amount of implied contractional motion is orders of magnitude less than for the subduction complexes associated with trenches, remnant oceans, and suture belts. The imbricated tectonic wedge of miogeoclinal strata peeled by décollement off underthrust basement beneath the Cordilleran fold-thrust belt records more than 100 km but less than 250 km of crustal shortening, rather than the aggregate thousands of kilometers inferred from plate consumption at typical trenches. The retroarc subduction can be viewed as a mechanism for structural stiffening of the orogenic belt, and is kept within limits by the buoyancy of continental crust. In this respect, however, the development of a retroarc foreland basin is similar to the development of a peripheral foreland basin once the remnant ocean basin is closed along the suture belt. Coney [1973] was thus able to discuss both kinds of foreland basins in terms of inferred absolute motions of plates with respect to underlying asthenosphere without making distinctions between the two types of forelands. Mechanisms of thrust movement in the two cases are probably similar [Elliott, 1976].

Structural relations along the eastern or inland flank of the well studied Canadian Cordillera appear to be in harmony with the views adopted here. Bally and others [1966] presented

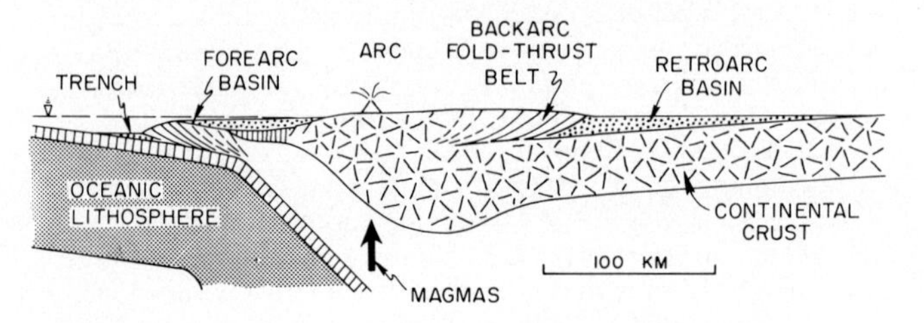

Figure 5. Idealized true-scale diagram showing tectonic relations of retroarc foreland basin associated with continental margin arc-trench system.

the case for major contractional translation of orogen with respect to craton. They concluded that the relative movement was concentrated mainly along the zone of décollement in the fold-thrust belt. They avoided the decision between overthrusting or underthrusting, but noted that the structural data are compatible with the concept of an underthrust craton proposed by Charlesworth [1959].

Price and Mountjoy [1970] added the concept of gravitational spreading of the orogen with its hot, mobile core. Imbricate structures of the fold-thrust belt thus reflect failure along the flank of an infrastructural welt. They left open the question of whether the horizontal supracrustal shortening across the fold-thrust belt is equivalent to the amount of fundamental crustal shortening or not. Campbell [1973] recently concluded that the visible geometry of geologic structures within the infrastructural belt precludes major extensional deformation at high structural levels. Consequently, the imbrication in the fold-thrust belt apparently cannot be due simply to upwelling and lateral flowage of a hot, mobile infrastructure punching through a stable, passive crust. Instead, the imbricated fold-thrust belt seemingly must be viewed, by analogy with other kinds of subduction complexes, as a standing wave spreading forward at the same speed that its leading edge tends to be dragged back by the underthrusting craton. The crustal shortening recorded by décollement and related tectonic processes along the flank of the orogen apparently is thus a rough measure of intracontinental contraction between the craton and the center of mass of the orogen.

Water depths within retroarc foreland basins are not known to be great, and fluvio-deltaic systems are common. Apparently, sedimentation roughly keeps pace with subsidence. Linkage between the growth of tectonic loads in the fold-thrust belt and the amount of subsidence in the basin may imply linkage as well between rates of subsidence in the basin and rates of uplift and erosion in highlands along the fold-thrust belt.

The treatment of backarc fold-thrust belts and retroarc foreland basins as tectonic elements of arc-trench systems changes some traditional views of intracontinental tectonics. The geodynamic effects of plate consumption along a continental margin may extend across roughly half the width of a typical continental block [Dickinson, in press]. Some supposedly epeirogenic transgressions and regressions on cratons may instead be distant effects of foreland deformation.

Conclusions

The orogenic systems that develop along convergent plate junctures are integrated engines in which magmatic, tectonic, and sedimentary mechanisms of mass transfer are linked by complex processes governed by plate tectonics. The net tendency is toward isostatic balance maintained by plate flexure to compensate for changing distributions of crustal masses. Transient gravitational effects stemming from overall plate dynamics also play an uncertain but likely subordinate isostatic role.

Plate tectonics is thus the key to understanding the stratigraphic evolution of orogenic basins. The sediment sources in orogenic uplifts and the sedimentary prisms of the associated basins together form a kind of standing wave of varied materials constructed above sites of plate descent by the geologic processes set in motion by plate convergence. An integrated quantitative evaluation of the relative influence of various magmatic, metamorphic, tectonic, and sedimentary processes on the evolutionary trends of the whole orogenic terrane stands as a challenge for the future.

References

Armstrong, F.C., and Oriel, S.S., Tectonic development of Idaho-Wyoming thrust belt, Am. Assoc. Petroleum Geologists Bull., 49, 1847-1866, 1965.

Armstrong, R.L., and Dick, H.J.B., A model for the development of thin overthrust sheets of crystalline rock, Geology, 2, 35-40, 1974.

Armstrong, R.L., and Hansen, E., Cordilleran infrastructure in the eastern Great Basin, Am. Jour. Sci., 264, 112-127, 1966.

Bailey, E.H., Blake, M.C., Jr., and Jones, D.L., On-land Mesozoic oceanic crust in California Coast Ranges, U.S. Geol. Survey Prof. Paper 700-C, C70-C81, 1970.

Bally, A.W., Gordy, P.L., and Stewart, G.A., Structure, seismic data, and orogenic evolution of southern Canadian Rocky Mountains, Bull. Can. Petroleum Geology, 14, 337-381, 1966.

Beck, R.H., The oceans, the new frontier in exploration, Austral. Petroleum Explor. Assoc. Jour., 12, 5-28, 1972.

Bird, Peter, Toksoz, M.N., and Sleep, N.H., Thermal and mechanical models of continent-continent convergence zones, Jour. Geophys. Res., 80, 4405-4416, 1975.

Blake, M.C., Jr., and Landis, C.A., The Dun Mountain ultramafic belt - Permian oceanic crust and upper mantle in New Zealand, U.S. Geol. Survey Jour. Res., 1, 529-534, 1975.

Brunnschweiler, R.O., On the geology of the Indoburman Ranges (Arakan Coast and Yoma, Chin Hills, Naga Hills), Geol. Soc. Australia Jour., 13, 137-194, 1966.

Burchfiel, B.C., and Davis, G.A., Nature and controls of Cordilleran orogenesis, western United States; extensions of an earlier synthesis, Am. Jour. Sci., 275-A (Rodgers Vol.), 363-396, 1975.

Campbell, R.B., Structural cross-section and tectonic model of the southeastern Canadian Cordillera, Can. Jour. Earth Sci., 10, 1607-1620, 1973.

Charlesworth, H.A.K., Some suggestions on the structural development of the Rocky Mountains in Canada, Alberta Soc. Petroleum Geologists Jour., 7, 249-256, 1959.

Coleman, P.J., On island arcs, Earth-Science Reviews, 11, 47-80, 1975.

Coney, P.J., Plate tectonics of marginal foreland thrust-fold belts, Geology, 1, 131-134, 1973.

Curray, J.R., and Moore, D.G., Growth of the Bengal deep-sea fan and denudation in the Himalayas, Geol. Soc. America Bull., 82, 563-572, 1971.

Curray, J.R., and Moore, D.G., Sedimentary and tectonic processes in the Bengal deep-sea fan and geosyncline, in The geology of continental margins, edited by C.A. Burk and C.L. Drake, p. 617-628, Springer-Verlag, N.Y., 1974.

Dewey, J.F., and Bird, J.M., Mountain belts and the new global tectonics, Jour. Geophys. Res., 75, 2625-2647, 1970.

Dewey, J.F., and Burke, K.C.A., Tibetan, Variscan, and Precambrian basement reactivation products of continental collision, Jour. Geology, 81, 683-692, 1973.

Dewey, J.F., and Burke, K.C.A., Hot spots and continental breakup: Implications for collisional orogeny, Geology, 2, 57-60, 1973.

Dickinson, W.R., Widths of modern arc-trench gaps proportional to past duration of igneous activity in associated magmatic arcs, Jour. Geophys. Res., 78, 3376-3389, 1973.

Dickinson, W.R., Plate tectonics and sedimentation, in Tectonics and sedimentation, edited by W.R. Dickinson, p. 1-27, Soc. Econ. Paleontologists and Mineralogists Special Pub. No. 22, 1974.

Dickinson, W.R., Time-gransgressive tectonic contacts bordering subduction complexes, Geol. Soc. America Abstracts with Programs, 7, 1052, 1975,

Dickinson, W.R., Sedimentary basins developed during evolution of Mesozoic-Cenozoic arc-trench system in western North America, Can. Jour. Earth Science, in press, 1976.

Elliott, David, The motion of thrust sheets, Jour. Geophys. Res., 81, 949-963.

Ernst, W.G., Systematics of large-scale tectonics and age progressions in Alpine and circum-Pacific blueschist belts, Tectonophysics, 26, 229-246, 1975.

Graham, S.A., Dickinson, W.R., and Ingersoll, R.V., Himalayan-Bengal model for flysch dispersal in Apallachian-Ouachita system, Geol. Soc. America Bull., 86, 273-286, 1975.

Grow, J.A., Crustal and upper mantle structure of the central Aleutian arc, Geol. Soc. America Bull., 84, 2169-2192, 1973.

Hamilton, Warren, Tectonics of the Indonesian region, Geol. Soc. Malaysia Bull., 6, 3-10, 1973.

Haynes, S.J., and McQuillan, Henry, Evolution of the Zagros suture zone, Iran, Geol. Soc. America Bull., 85, 739-744, 1974.

Ingersoll, R.V., Sedimentary and tectonic evolution of the Late Cretaceous forearc basin of northern and central California, Geol. Soc. America Abstracts with Programs, 7, 1127, 1974.

Karig, D.E., Origin and development of marginal basins in the western Pacific, Jour. Geophys. Res., 76, 2542-2561, 1971.

Karig, D.E., and Sharman, G.F., III, Subduction and accretion in trenches, Geol. Soc. America Bull., 86, 377-389, 1975.

McKenzie, D.P., Speculations on the consequences and causes of plate motions, Geophys. Jour. Roy. Astron. Soc., 18, 1-32, 1969.

Powell, C.McA., and Conaghan, P.J., Plate tectonics and the Himalayas, Earth and Planet. Sci. Lettrs., 20, 1-12, 1973.

Price, R.A., and Mountjoy, E.W., Geologic Structure of the Canadian Rocky Mountains between Bow and Athabasca Rivers, Geol. Assoc. Canada Special Paper No. 6, 7-25, 1970.

Price, R.A., Large-scale gravitational flow of supracrustal rocks, southern Canadian Rockies, in Gravity and tectonics, edited by K.A. Dejong, and Robert Scholten, p. 491-502, Wiley, N.Y., 1973.

Ramachandra Rao, M.B., The subsurface geology of the Indo-Gangetic plains, Geol. Soc. India Jour., 14, 217-242, 1973.

Seely, D.R., Vail, P.R., and Walton, G.G., Trench slope model, in The geology of continental margins, edited by C.A. Burk, and C.L. Drake, p. 249-260, Springer-Verlag, N.Y., 1974.

Tréhu, A.M., 1975, Depth versus $(age)^{1/2}$, a perspective on mid-ocean rises, Earth Planet. Sci. Lettrs., 27, 287-304, 1975.

Walcott, R.E., Gravity, flexure, and the growth of sedimentary basins at a continental edge, Geol. Soc. America Bull., 83, 1845-1848, 1972.

Wickham, John, Roeder, Dietrich, and Briggs, Garrett, Plate tectonics models for Ouachita foldbelt, Geology, 4, 173-176, 1976.

MULTIFOLD SEISMIC REFLECTION RECORDS FROM THE NORTHERN VENEZUELA BASIN AND THE NORTH SLOPE OF THE MUERTOS TRENCH

John W. Ladd, J. Lamar Worzel and Joel S. Watkins

Geophysics Laboratory, Marine Science Institute, University of Texas,
700 The Strand, Galveston, Texas 77550

Abstract. Multifold seismic reflection records over the Muertos Trench reveal details of the Carib beds and deeper reflectors in the Venezuela Basin as well as sediment units and folds within the north slope of the Muertos Trench that are similar to structures of inner slopes of Pacific trenches. Venezuela Basin reflectors can be traced landward of the trench axis for 40 km. Folds in the Carib beds near the trench and folds in the trench fill may suggest a compressional regime that may have created some of the structures of the north slope, in particular, the anticline that forms the trench slope break. The landward tilting of ponded sediments on north slope terraces particularly north of the trench slope break is also similar to occurrances in many Pacific trenches. Slumping of material from Hispaniola is an important process for the development of the sediment wedge of the north slope; whether or not slumping and similar processes of sediment dispersal together with isostatic adjustment can account for the features of the Muertos Trench and north slope is debated. The structures within the north slope may possibly be the result of late Tertiary subduction along the Muertos Trench even though late Tertiary volcanism is lacking in Hispaniola and present earthquake hypocenters suggest a southward dipping Benioff zone related to the Puerto Rico Trench. The Muertos Trench, Greater Antilles, and Puerto Rico Trench together form a unit which may be similar to other double trench structures such as the Philippines and Panama.

Introduction

The Venezuela Basin, which is bounded on the north by the Muertos Trench, is underlain by roughly one second of sediments which overlie Coniacian basalt (Edgar, Saunders, et al., 1973). The interface between the basalt and overlying sediments is a strong, smooth reflector that has been given the designation B". Overlying B" another reflector that is widespread in the Caribbean is designated A" (Ewing, J. et al., 1968; Edgar et al., 1971) and is correlatable with Middle Eocene siliceous limestone and chert. The entire section overlying B" has been called the Carib beds. Early single channel reflection profiling did not resolve reflectors beneath B", but recent single channel and multichannel reflection work have revealed reflectors down to at least one second beneath B" in parts of the basin (Garrison, 1972; Hopkins, 1973; Marine Science Institute work in progress), and DSDP drilling which encountered B" at two sites in the Venezuela Basin has indicated that in at least one place B" marks a dolerite sill.

The Muertos Trench is a topographic trough that runs eastward from the intersection of the Beata Ridge and Hispaniola to the longitude of St. Croix. The Enriquillo graben in Hispaniola may be a structural continuation to the west across the axis of the Beata Ridge. As a topographic feature the Muertos Trench is best developed south of the Dominican Republic and Mona Passage, and it is in this region that R/V IDA GREEN 15-3 made three crossings collecting multifold CDP seismic reflection data (Figure 1). Previously published data in this region show the reflectors A" and B" of the Venezuela Basin dipping northward beneath the southern slope of the trench (Edgar et al., 1971; Garrison, 1972). The sparker records published by Garrison indicate that one or both of these reflectors continues northward beneath the northern slope of the trench where they maintain their northerly dip. Other than this faint continuation of the Caribbean reflectors, published data have revealed very little of the structures

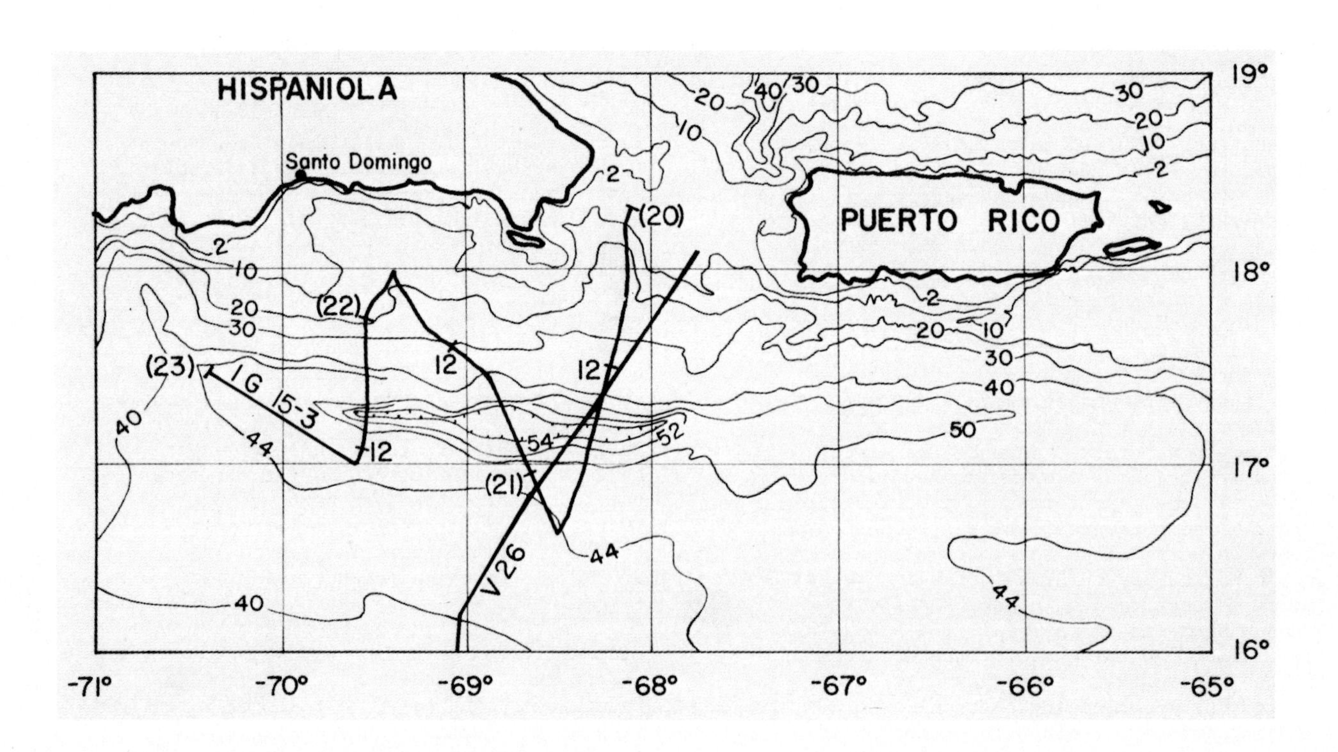

Fig. 1. Track of R/V Vema Cruise 26 and R/V IDA GREEN Cruise 15. Contours adapted from USGS Chart of the Caribbean in preparation. The four sections of the IDA GREEN track between major course changes are identified from east to west as VB1N, VB2N, VB4N, and VB5N.

beneath the north slope. In this paper the increased resolution of multichannel common depth point profiling reveals more of the structure of the north slope of the Muertos Trench as well as the nature of reflectors that underlie the Venezuela Basin. We also note similarities between Muertos Trench structures and Pacific Trench structures and propose alternative hypotheses for the tectonic development of the Muertos Trench.

Seismic Reflectors in the Vicinity of the Muertos Trench

Our stacked multichannel records give increased resolution of the Carib beds of the Venezuela Basin as well as their continuation beneath the north slope of the Muertos Trench. A record from Lamont VEMA cruise 26 crosses our line near the southern end of line VB2N (Figs. 1 and 2), and displays reflectors B" and A" as Edgar et al. (1971) saw them, allowing definite correlation with reflectors on our records that show other reflectors both above and below A" and B" (Figs. 3 and 4). The water bottom rises as an irregular surface southward from the axis of the Muertos Trench where it is 5.5 km deep to the central Venezuela Basin where it is about 4 km deep. Approximately 0.4 seconds subbottom A" appears as a strong though somewhat discontinuous reflector with weaker more discontinuous reflectors above and below. The discontinuous nature of A" is such that on line VB2N one cannot follow an individual amplitude peak in the vicinity of A" for more than a kilometer or so except immediately south of the Muertos Trench where it can be followed for about ten kilometers (Figs. 3 and 4). The terminations of finite segments of A" are often marked by small vertical offsets of 0.02 seconds possibly indicating minor faults. B" is similarly discontinuous in places, but no consistent relation between breaks in A" and breaks in B" is apparent. Farther west on line VB4N (Figs 1 and 5), A" is much more continuous, while B" is less continuous than on line VB2N. On line VB2N diffraction hyperbolas are seen immediately beneath A", particularly near the trench axis (Fig. 4). Generally some break in A" can be seen in association with the hyperbolas, though this is not always the case. On line VB4N diffraction hyperbolas are more commonly related to B". About midway between A" and the water bottom a weak reflector is seen on line VB2N. This reflector is more discontinuous than A" being composed of a series of reflecting segments each roughly 0.5 km long. This reflector is stronger and more continuous on VB4N and about ten kilometers from the southern end of VB4N individual segments of this reflector onlap each other toward the south possibly indicating that this reflector is older to the south. Between 0.7 and 0.8 seconds subbottom B" is seen as a fairly strong continuous reflector. As in the case of A", B" is really the strongest reflector within a group of reflectors. Weaker less continuous reflectors are seen above and below.

Garrison (1972) shows some clear sub B" reflectors just to the west of the Aves Swell, and Hopkins (1973) shows sub B" reflectors in the Aruba Gap area. Our data also resolve reflectors beneath B" in the Venezuela Basin. In the southern Venezuela Basin (not illustrated here) we see reflectors at least 1.5 seconds sub B". Velocities in this zone are greater than 4 km/sec. indicating at least 3 km of section beneath B". Between B" and about 1 second below B" discontinuous reflectors are seen on line VB2N (Fig. 4). The reflector segments, which in some cases are ten kilometers long, are generally parallel to B" and do not define any structures beneath B". The one exception is beneath the southern edge of the trench floor on line VB2N where sub B" reflectors define a wedge about 0.2 seconds thick and dipping landward. This wedge terminates to the south against some diffraction hyperbolas that may relate to a fold or a basement high. To the south of this diffraction zone sub B" reflectors are again seen with an apparent northward dip and an angular unconformity at B". Our line VB4N also shows a zone of discontinuous reflector segments that define a reflecting horizon about 0.2 secons beneath B" (Fig. 5). This zone fades away northward at the trench axis but continues southward to the end of VB4N. About ten kilometers from the south end of VB4N some reflectors with apparent northward dips are seen between B" and this sub B" reflecting zone. These dipping reflectors are truncated above by B" but their relation to the sub B" reflecting zone is obscure.

On line VB2N reflector B" can be followed with a fair amount of certainty northward beneath the turbidite fill of the trench axis. Just to the south of the trench axis the reflectors above B" including A" are distorted by a fold. A diffraction pattern associated with this fold obscures A" and it is not clear that A" can be followed beneath the trench fill. A diffraction pattern beneath this fold also distorts B"

As seen on line VB2N, the floor of the Muertos Trench is almost flat and is underlain by a lens of strong reflectors which are nearly horizontal except for a possible fold that does not affect the sea floor at the northern margin of the trench floor. This apparent fold could be a result of side echoes and may not be a fold at all. A reverse fault in the central part of the trench fill indicates a slight upward throw to the north. The thickest part of the lens of strongly reflecting turbidites extends to 0.4 seconds subbottom. Beneath the lens is a zone of weak reflectors that in a general sense is similar to the zone above A" further south in the Venezuela Basin, but the details of this zone are different - one cannot discern the reflector above A" separated from underlying strong reflectors by a

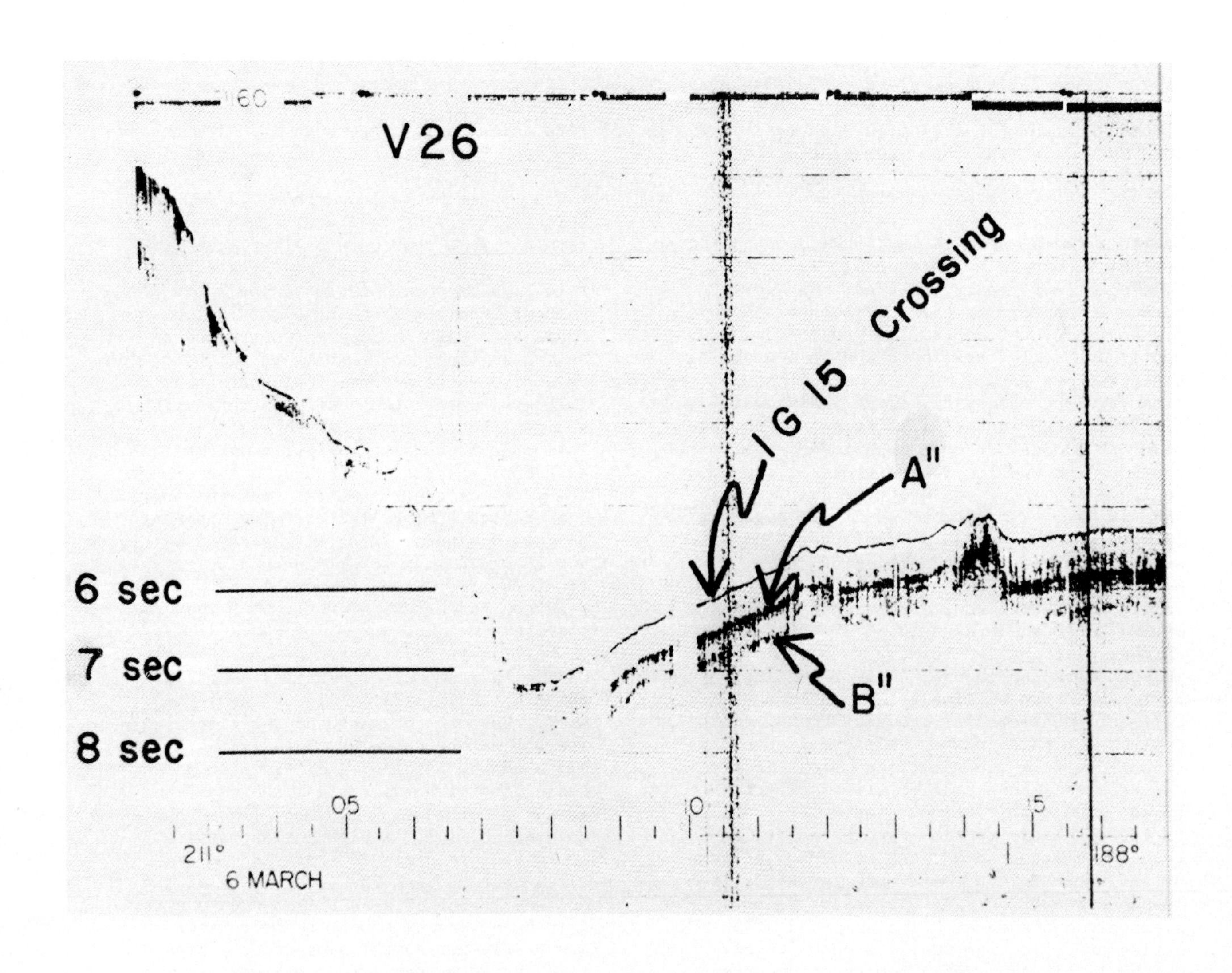

Fig. 2. Single channel reflection record of VEMA showing reflectors A" and B" at the point where the track is crossed by IDA GREEN. These serve to identify A" and B" in our new section.

transparent zone. Under the trench axis the zone of strong reflectors below the weak reflector zone is about 0.4 seconds thick down to the probable extension of B". The zone of weak reflectors seen between A" and B" to the south is absent. The significant differences in detail between the section below the strong turbidite reflectors and the section further to the south in the Venezuela Basin overlying B" is a matter of discussion.

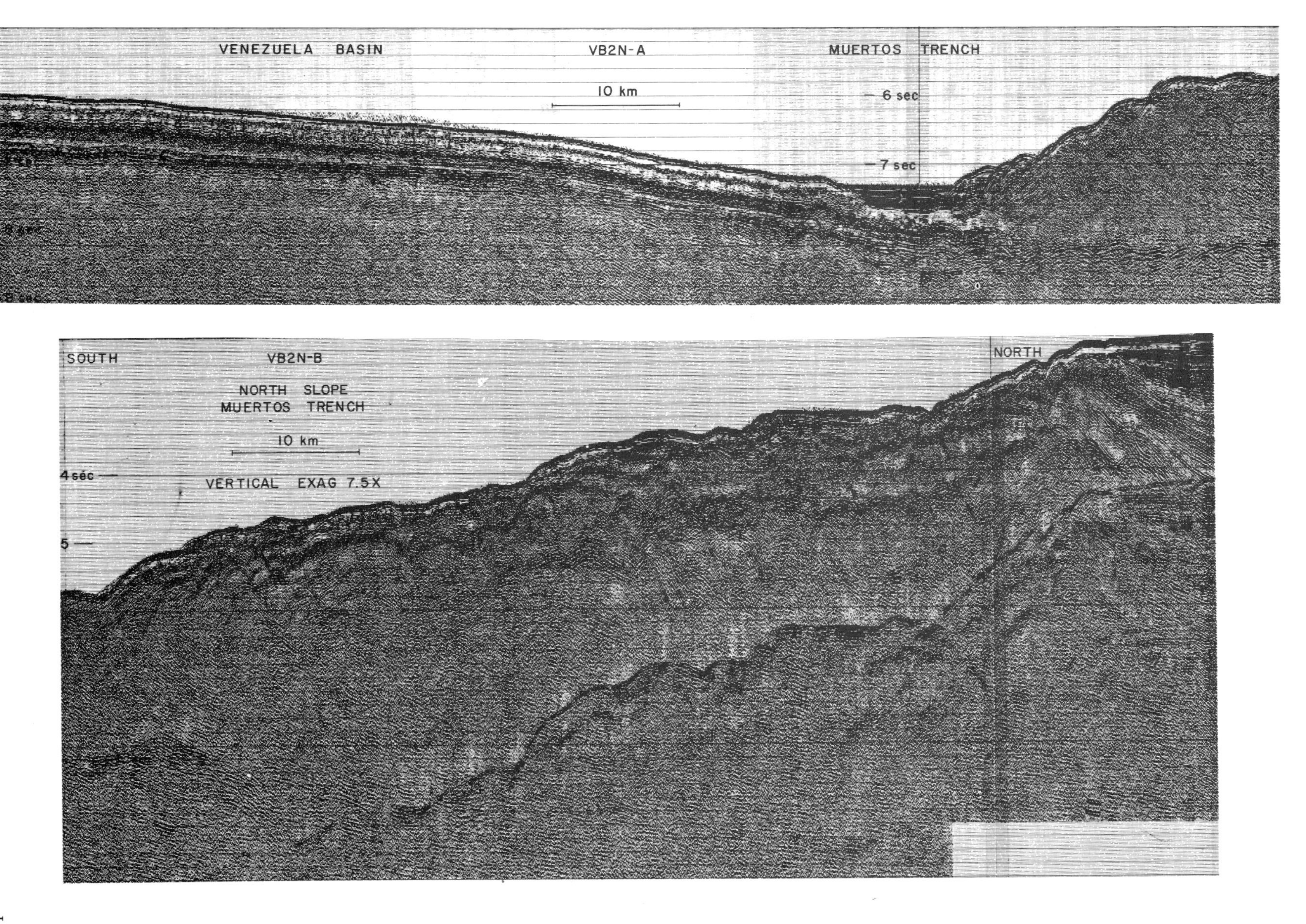

Fig. 3. 24-fold stacked seismic reflection records from line VB2N of IDA GREEN Cruise 15.

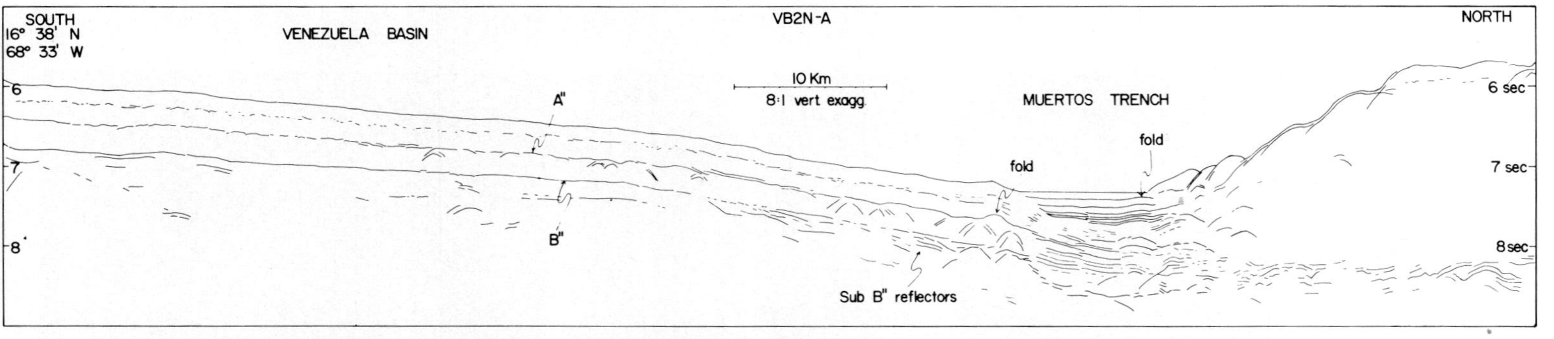

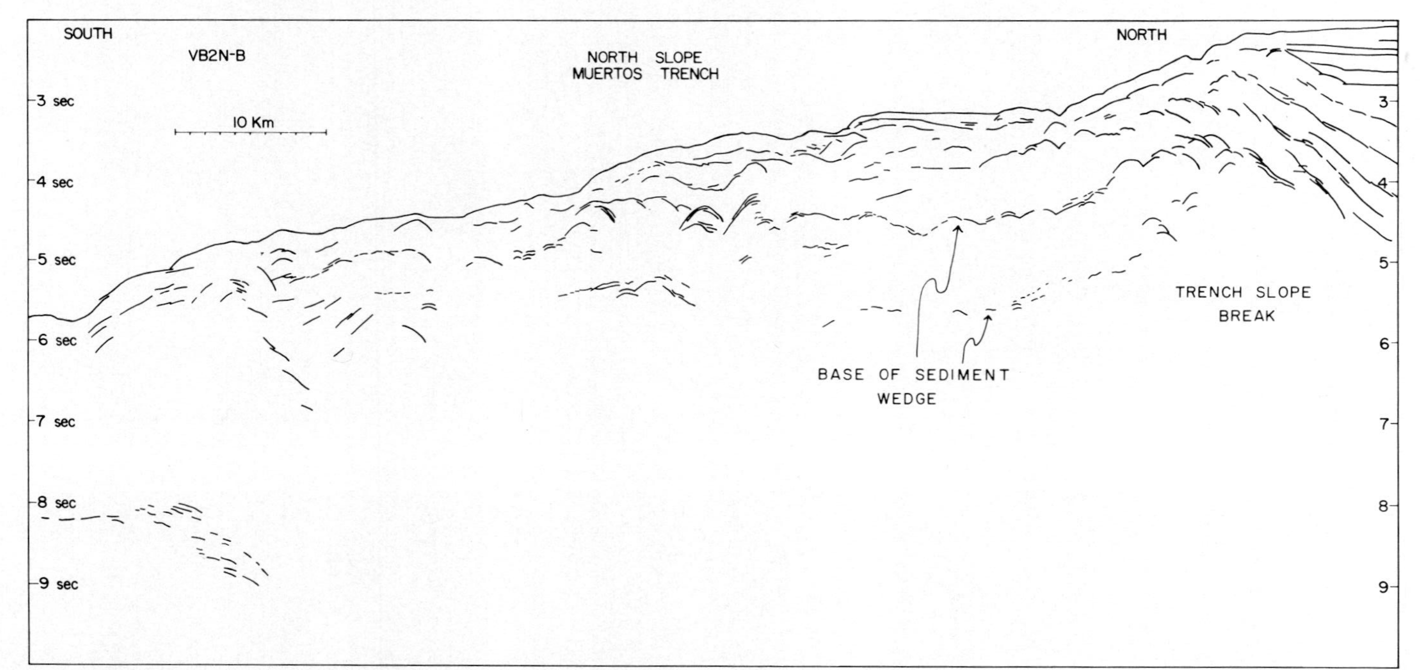

Fig. 4. Tracings of Fig. 3. Note particularly reflectors A'' and B'' and the continuation of reflectors beneath the north slope of Muertos Trench. Note also the reflectors within the north slope that define two units which appear folded at the trench slope break.

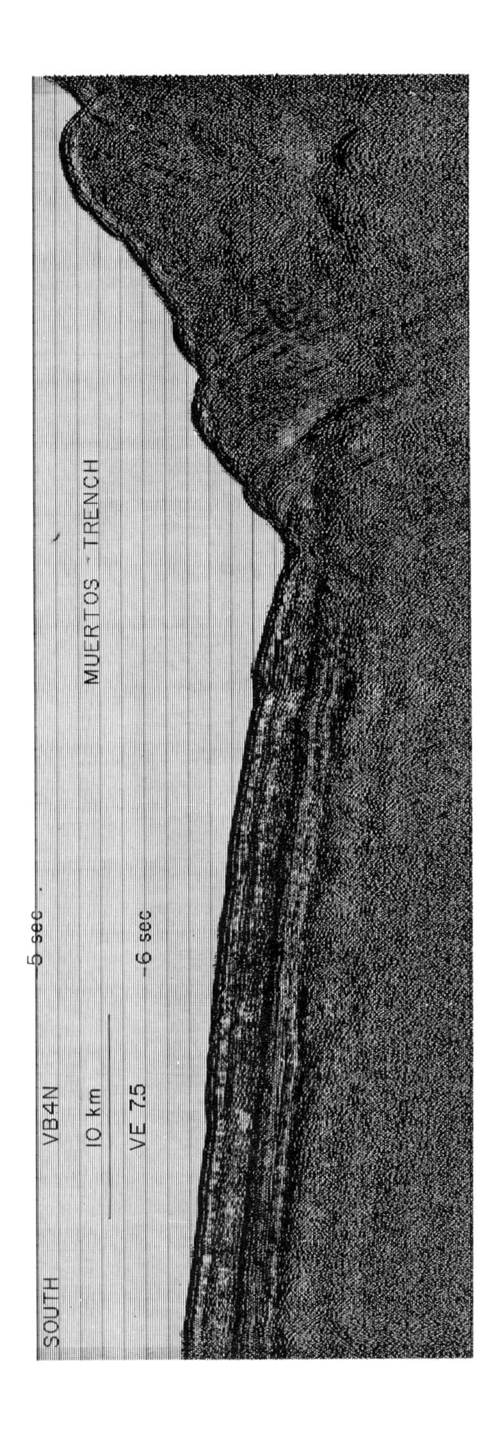

Fig. 5. 12-fold seismic reflection records from line VB4N of IDA GREEN Cruise 15. The prominent reflectors of A" and B" are easily seen to the south of the Muertos Trench and close inspection will reveal a reflection zone approximately 0.4 seconds beneath B".

These differences may suggest that the section overlying B" to the south is missing beneath the trench where instead trench axis fill extends 0.9 seconds beneath the floor of the trench; or these differences may be due to the tectonism and overburden in the trench region that has altered the normal section above B" after deposition; or the character of waves reflected from beneath the turbidite fill may have been altered by the presence of the fill.

The apparent fold at the northern edge of

the trench floor also affects reflectors beneath the strong turbidite reflectors. Within the anticline these reflectors become segmented and one can no longer follow individual reflectors northward. However, a zone of reflector segments continuous with the lower part of the deeper zone of strong reflectors can be traced 40 kilometers northward of the foot of the north slope on line VB2N. For about 30 km the top of this zone which probably is a continuation of B" parallels the 8.2 second time line and then dips to the north for another ten km where it finally becomes too weak to follow. With corrections for the velocity in the overburden (see below) and using data from Garrison (1972) as well as our data the dip of the reflector zone that follows the 8.2 second line is approximately 7°. The apparent dip of the reflecting zone on line VB2N where the zone dips from 8.2 to 9 seconds is about 14° after correcting for the velocity of the overburden.

The crossing of the trench on line VB4N indicates almost no trench fill, and the ocean floor there is 0.02 seconds shoaler than at VB2N. Further to the east on line VB1N unprocessed data indicate a northward dipping trench fill with the deepest water being about 0.02 seconds shoaler than the sea floor on line VB2N. On line VB4N a reflecting zone that is a northward continuation of the A" - B" interval can be followed about 25 km north of the trench axis. At the trench axis B" fades while A" continues strong for about 8 km northward where there is an apparent step in the reflecting zone. It is not clear whether this step marks the termination of A" and the continuation of B". Possibly B" is faint to invisible just north of the trench axis because the contorted nature of A" does not permit strong coherent returns from beneath it. When A" terminates to the north, we may be seeing B" again.

The north slope of the trench between the ponding of the trench floor and the ponding behind the trench slope break in 2 seconds of water has an irregular hummocky surface underlain by highly contorted sediment wedges. On line VB2N beneath a topographic terrace in 5.8 seconds of water preliminary velocity analysis using our CDP data and a cross correlation technique gives an average interval velocity between the water bottom and the continuation of the B" - A" reflectors of 3.1 ± 0.2 km/sec further up the slope a wedge of sediment with a lower contact that intersects the water bottom at 4.2 seconds of water (upper reflector indicated by an arrow on Fig. 4) has an interval velocity of roughly 2.8 ± 0.2 km/sec whereas pieces of reflector seen at about 5.6 seconds on the time section (lower reflector indicated by an arrow on Fig. 4) allow determination of velocities of 3.3 - 3.6 km/sec for the material forming a lower sediment unit. The base of the upper sediment wedge dips gently northward for at least 10 km before reversing dip and forming an anticline beneath the trench slope break. Immediately beneath the sea floor on most of the north slope is a sediment unit with an irregular lower contact that parallels the sea bottom. This uppermost unit yields velocities of 1.9 - 2.0 km/sec. Further downslope to the south other discontinuous series of reflectors can be seen that yield similar velocity profiles i.e. 2+ km/sec immediately beneath the water bottom underlain by 3+ km/sec material. North of the trench slope break on line VB2N all but the most shallow reflectors dip landward with increasing dip down in the section (see north end of Figs. 3 and 4). On line VB2N the water bottom north of the trench slope break dips slightly southward, but farther west on line VB4N even the water bottom dips northward indicating that here sedimentation has not kept pace with continued landward subsidence of this basin relative to the trench slope break, (Fig. 6).

The folds in the trench fill on VB2N as well as the thickness of the trench fill and the structures forming the irregular topography on the lower north slope may be important clues about the nature and timing of tectonic events on the north slope. The uppermost surface of the turbidite fill is not deformed by the apparent fold. However deeper reflectors are apparently deformed indicating that the fold formed prior to the deposition of the topmost part of the trench fill. Definition of the deepest turbidite reflector would allow an estimate of the age of the turbidite fill (Karig, Ingle, *et al.*, 1975). If A" and some of the section above A" could be traced beneath the trench fill on line VB2N this would indicate the trench fill in this part of the trench is younger than Eocene. One might even argue that the transparent zone between the lowermost strong turbidite reflector and the deeper strong reflectors in the trench axis is entirely correlative with the zone of weak reflectors above A" in the Venezuela Basin to the south. If this is so, the turbidite fill may be only a Pleistocene phenomena. The apparent folding at the landward margin of the turbidite fill in the Muertos Trench is similar to apparent folding seen in turbidite fill just north of Panama in the Columbia Basin (Edgar *et al.*, 1971, Fig. 17) where Case (1974) has suggested the existence of a subduction zone.

Discussion

A. Subduction and the origin of the Muertos Trench.

The topographic expression and internal

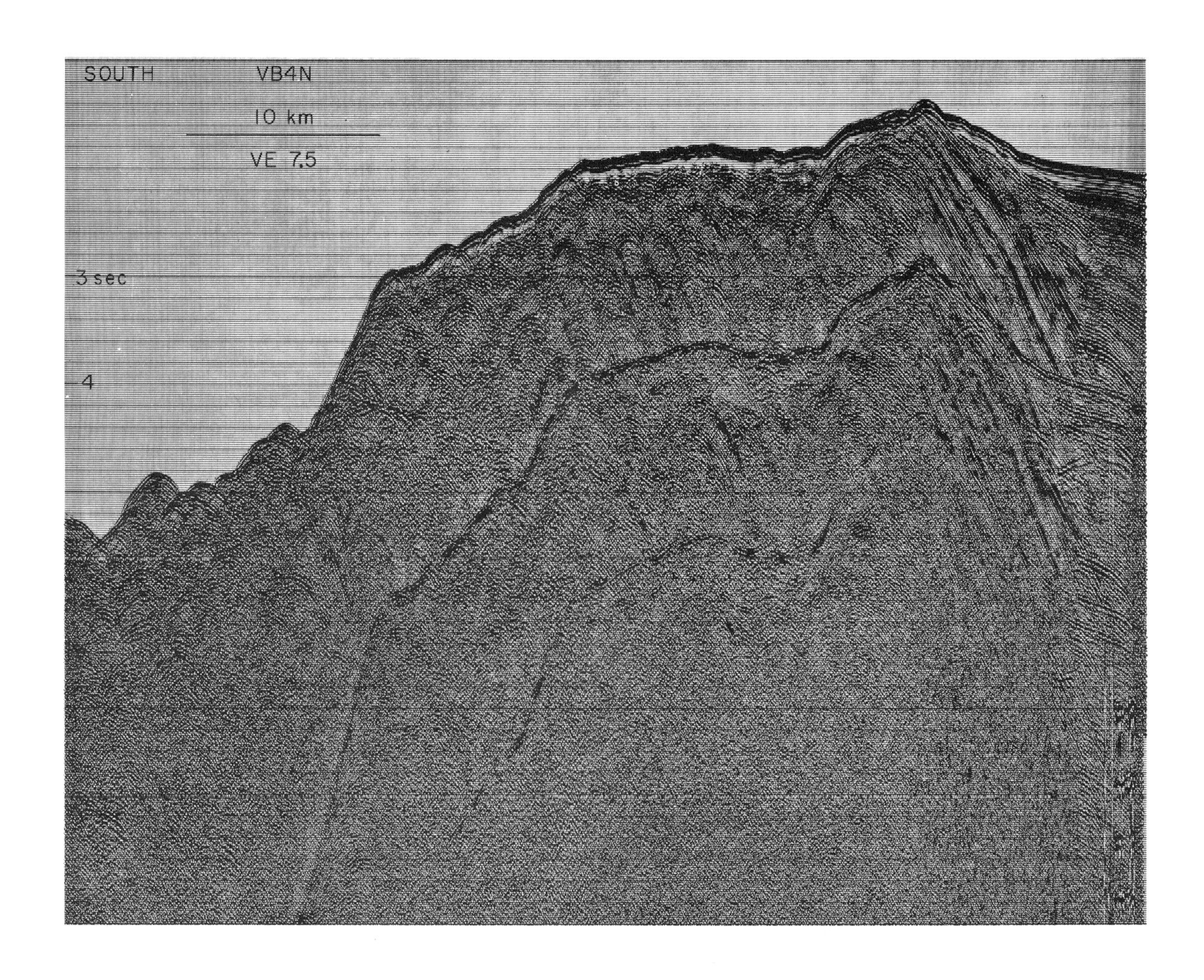

Fig. 6. Seismic reflection records from the north end of line VB4N. Note particularly the tilted ponded sediments north of the trench slope break.

structure of the Muertos Trench and its north slope are strongly similar to the form and structure of Pacific trenches and their related inner slopes as described by von Huene (1972), Beck and Lehner (1974), Seeley *et al.* (1974), and Karig and Sharman (1975). These similarities suggest that the Muertos Trench owes its existence to the same subduction processes that have been attributed to seismically active trenches around the Pacific (Isacks *et al.*, 1968; Seeley *et al.*, 1974). One element common to most of the Pacific trenches that is lacking in the Muertos Trench is a landward dipping Benioff zone. A plot of hypocenters using data from the National Earthquake Information Service for the period 1960 through 1975 suggests a southward dipping Benioff zone related to underthrusting along the Puerto Rico Trench as do one or two published focal mechanisms (Molnar and Sykes, 1969, 1971). Also the lack of late Tertiary volcanism in the Greater Antilles may indicate either that underthrusting is not active today along the Muertos Trench or that it is presently active at rates inappropriate for the production of active island arcs. Tectonic forces have, however, maintained the Muertos Trench since Late Cretaceous time as indicated by the paucity of turbidites in the central Venezuela Basin south of the trench axis (Edgar et al., 1971; Saunders et al., 1973). The duality of the Muertos Trench and the Puerto Rico Trench on either side of the Greater Antilles is also not a common feature of most Pacific trenches but is not unique. Case (1974) suggests a similar relationship across Panama, and Hayes and Ewing (1971) report an analogous situation in the Philippines with the Manila Trench to the west of Luzon and the Philippine Trench to the east.

Seismic reflection profiles over the Muertos Trench are similar to profiles over Pacific trenches. Multifold seismic reflection data have revealed a continuation of ocean basin reflectors dipping landward beneath the inner slopes of the Java Trench and the Middle America Trench in a manner similar to the Muertos Trench. The topographic high of the trench slope break which lies in 1 to 2 seconds of water on the Muertos Trench north slope is similar to the trench slope break and its shoreward ponded sediments of the Central Aleutians (Marlow et al., 1973). The landward tilting of sediments in the forearc basins behind the trench slope break and in sediment ponds on the lower slopes of trenches as seen on the Muertos Trench north slope indicates a differential rise of the trenchward portions of terraces on inner trench slopes which has been attributed to underthrusting and accretion of sediment at the base of the inner slope (Seeley et al., 1974; Karig and Sharman, 1975). DSDP leg 18 site 175 on the lower continental slope off Oregon cored sediments that were originally deposited on the adjacent abyssal plain but had been uplifted at least 200 meters and possibly 700 meters sometime between 0.3 and 0.45 m. y. BP. (Kulm, von Huene, et al., 1973). Site 176 on the outer shelf indicated similar uplift as did site 187 on leg 19 (Creager, Scholl, et al., 1973) which was drilled on the outer edge of the Aleutian terrace, a location similar to the trench slope break at the north end of line VB2N. Perhaps similar uplift is occurring on the north slope of the Muertos Trench.

Site 181 leg 19 of DSDP was drilled on the lower continental slope about 2000 meters above the Aleutian Trench. The lowest unit cored was a highly compacted mudstone with 10% water content and consolidation measurements suggesting burial of 1.5 km. Site 298 leg 31 (Karig, Ingle, et al., 1975) was drilled in an analogous position in the Nankai Trough and found a fold overturned seaward similar to the folds described by Moore (1973) in turbidites on Shumagin and Sanak Islands in the Aleutians. Moore suggested that the turbidites had originated in the trench axis and had subsequently been folded and uplifted within the inner trench slope. During this process of folding he suggested that the sediments were "strain hardened" and thus attained the compacted form found in site 181. The relatively high velocities of 3+ km/sec found within the north slope of the Muertos Trench may be a result of similar folding and hardening. Velocities of 4.4 km/sec were found at homologous depths under the inner slope of the Aleutian Trench (von Huene, 1972).

Recent compressional folding in the Muertos Trench is suggested by the apparent folds within the turbidite fill at the northern margin of the Muertos Trench and within the Carib beds to the south of the turbidite fill on line VB2N. The irregular topography on the lower slope may be indicative of more tightly folded sediment that has been thrust against the lower slope in a mild form of obduction. The sequence of curved topographic terraces at the base of the inner slope on line VB2N suggests possible thrust slices. The lack of strong continuous reflectors within portions of the north slope is probably due to extreme folding within the sediment pile. It is not due to a lack of energy penetrating the pile, for we do see the deep reflection at 8 seconds.

Slumping is undoubtedly an important process on the lower north slope of the Muertos Trench as it is in other trench environments (von Huene, 1972; Piper et al., 1973; Scholl, 1974; von Huene, 1974) and the load of sediments forming the inner slope will tend to isostatically depress the underlying lithosphere (Karig and Sharman, 1975); however, these processes alone cannot maintain the trench, else one would expect to find trenches on Atlantic margins where slumping and other sediment processes have built large sediment wedges. It is difficult to understand how material could slump across the forearc basin to form the anticlinal ridge of the trench slope break which is delineated by thick (approx. 1 km) sedimentary units that can probably be followed 40 km seaward from the trench slope break (Fig. 4).

The interpretation of the reflecting horizons immediately beneath the trench axis is critical in establishing rates of subduction. Although in detail the units underlying the upper strong turbidite reflectors are different from the Carib beds to the south, their overall general pattern allows one to argue that A" and much of the Carib bed above A" may exist beneath the turbidites of the trench axis. The fold in the Carib beds south of the trench fill disrupts the beds, but the top of the lower strongly reflecting zone beneath the trench fill aligns with the top of the A" reflecting zone to the south. The transparent layer below A" is found on both sides of the fold and ends abruptly under the southern edge of the trench fill where it is replaced by two strong reflectors. The nature of the transition is uncertain since the reflectors that bound this transparent zone above and below are not noticeably discontinuous across the boundary. The transparent zone above A" in the Venezuelan Basin may correspond in part to the transparent zone seen below the upper strong reflectors of the turbidite fill. If these correlations are right and A" and part of the overlying Carib bed does continue beneath the trench axis, then the turbidite fill is a late Tertiary addition; and possible Late Cretaceous and early Tertiary trench fill may have been incorporated into inner slope folds. This would suggest a much more rapid subduction process than would be possible if trench fill extended deeper at the trench axis, truncating not only late Tertiary Carib beds but early Tertiary and Late Cretaceous beds as well.

B. Sedimentation and Isostatic Adjustment.

An alternate hypothesis for the observed section is that an oceanic crust near to the island ridge was overlain by a great thickness of sediments from the nearby island ridge and the concomitant isostatic adjustment has produced the observed dips. In such a mass of overlying sediments, considerable slumping and gravity tectonics would have occurred.

The Beck and Lehner (1973) section of the Java Trench has been attributed to subduction with additions of the material above the oceanic crust beneath the shoreward wall as "tectonized sediments." Worzel (1976) showed that if these "tectonized sediments" were added from the Island ridge by any process whatever, but no doubt including gravity tectonics, to a previously horizontal ocean crust, that the resulting isostatic adjustment would have resulted in the same seismic section.

To see if a similar result might be obtained here, we may proceed as follows: Fig. 7 is a line drawing of the time section of VB2N. Velocities observed in 1955 by Atlantis and Caryn just seaward of the Muertos Trench (Officer et al., 1959) and on Vema 8 Wissama 1 (Ewing, Antoine and Ewing, 1960) are shown at their respective locations. Velocities determined from the multichannel data are in quite good agreement with these refraction results. Densities determined from the Nafe-Drake velocity density curve (Ludwig et al., 1970) are added within the layers near to the refraction results.

Figure 8 shows this same section converted to a depth section using these velocities. At the bottom, the depth section is shown without vertical exaggeration so that one can consider the true perspective.

By removing the layers of velocity 1.7 km/sec, 2.7 km/sec and 2.8 km/sec - those layers above B" and returning the section to isostatic equilibrium, the attitude of the sea floor is reconstructed as in Figure 9. The present sea floor surface is superimposed as a dotted line above the restored section for easy comparison.

The proper perspective is, of course, to view the sea floor of Turonian-Santonian time as the near monoclinal surface of the 3.9 km/sec layer in Figure 9. This surface is dipping gently (about 1°) towards the island ridge as it still does today in the northern half of the Venezuelan Basin. Subsequent deposition of sediments predominantly from the nearby island ridge of Dominica-Puerto Rico with gravity slumping and tectonism produced the present depth structure shown in Figure 8. The hummocky terrain of the north slope of the Muertos Trench is just that expected from a slump deformed slope as is the chopped up series of reflectors beneath this north slope. The trench basin filled with turbidites is caused by a regional adjustment to the sediment load of the north wall. The increasing landward dip of the deeper turbidites within this trench basin demonstrates the continuing isostatic adjustment. These layers were laid down as turbidites in horizontal layers which have since been tilted by the continuing regional isostatic adjustment to the sediments still accumulating on the flank of the island ridge. The slight folding along the north side of the trench fill is the encroachment of the continuing slumping of the tectonized sediments beneath the north wall. The small fault in the middle of the trench basin is another adjustment to this pressure. On the easternmost line, not yet processed, the landward dip of the present trench surface attests to very recent isostatic adjustment. The basin at the north ends of lines VB2N and

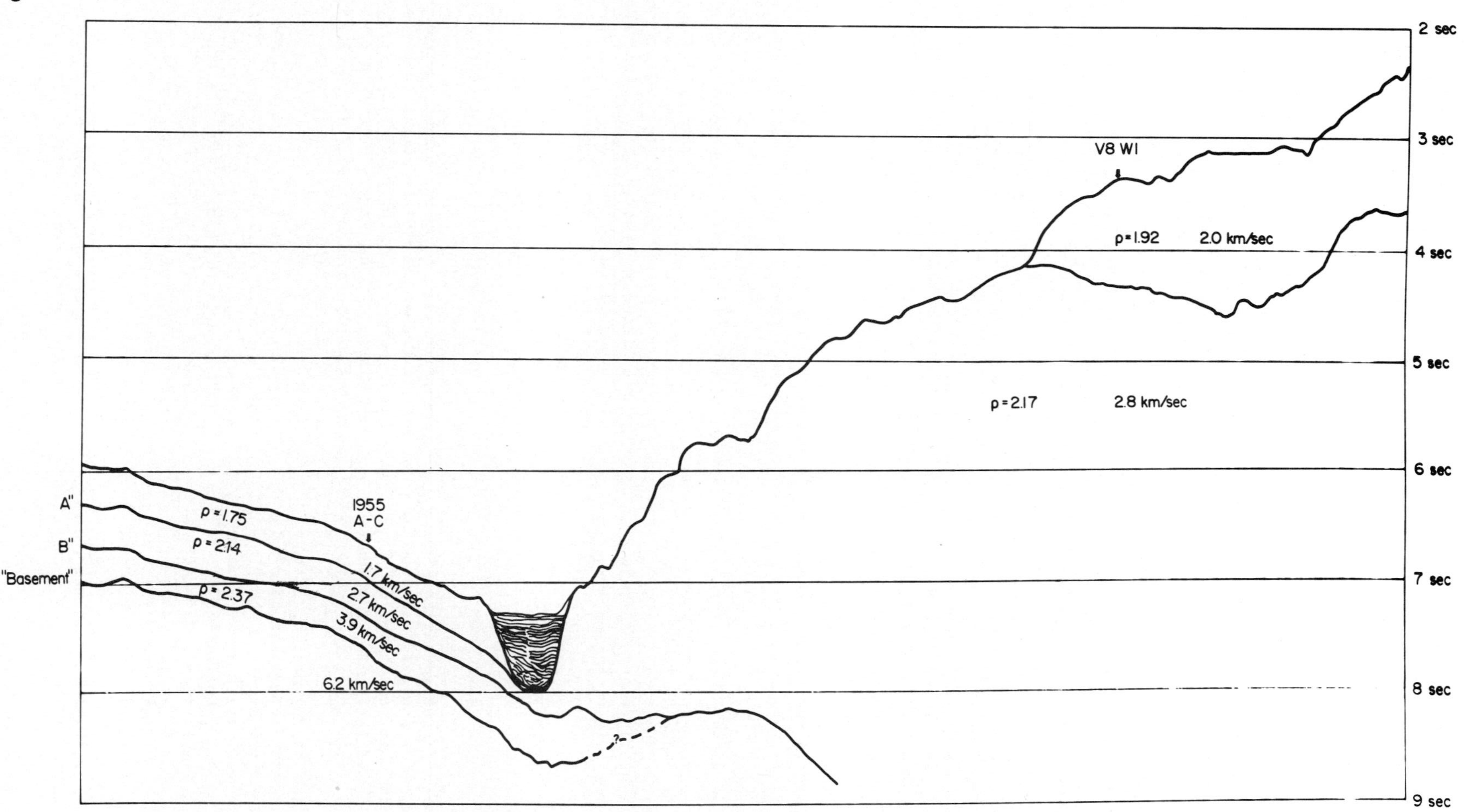

Fig. 7. Line drawing of time section of VB2N. Refraction measurements of Altantis-Carun, 1955 and Vema 8-Wissama 1 are indicated at their appropriate locations. Densities determined from the refraction velocities by the Nafe-Drake velocity-density curve are shown nearby. These densities are used in some of the calculations.

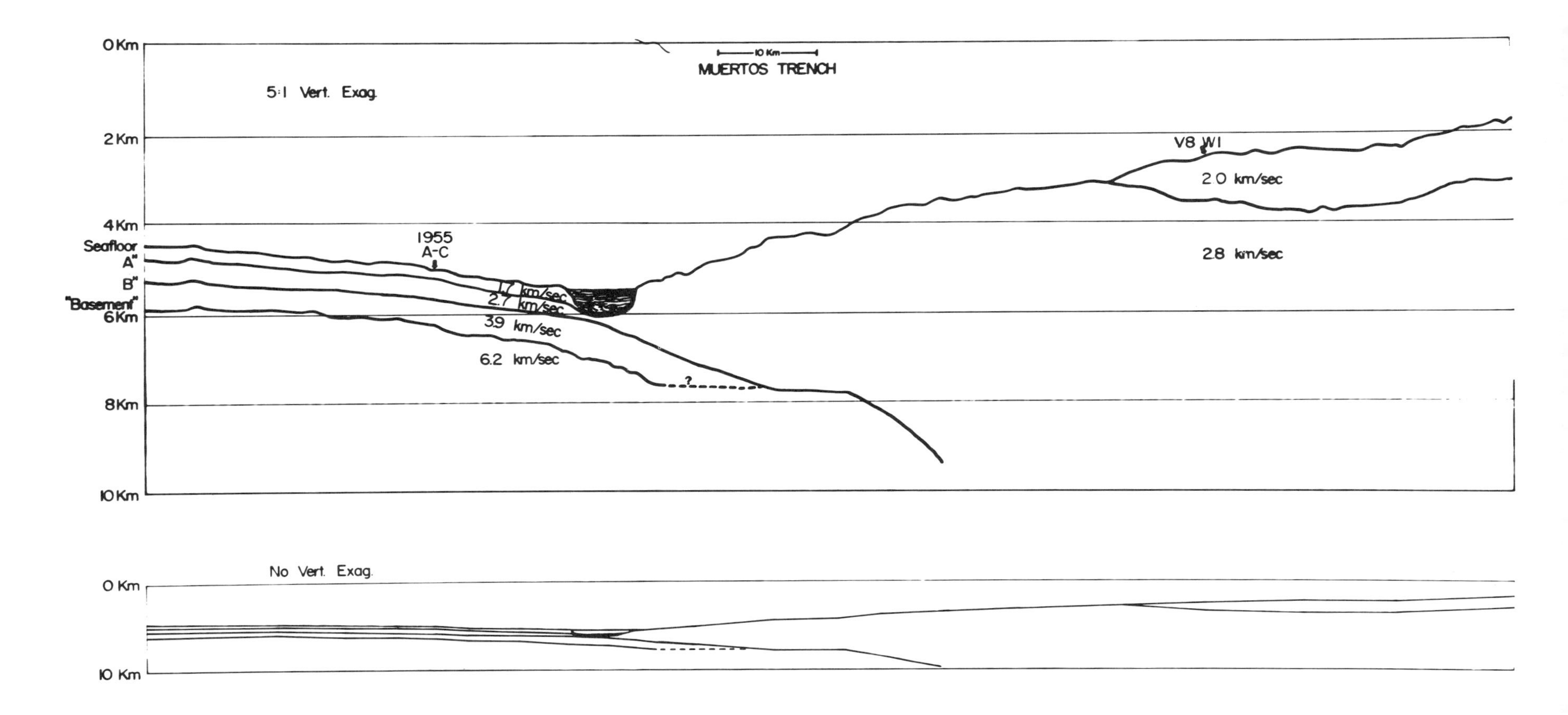

Fig. 8. Fig. 7 converted to a depth section. At the bottom is the same section without vertical exaggeration. Note that the 3.9 km/sec layer which is horizontal in the time section is dipping landward at about 7° beneath the north wall.

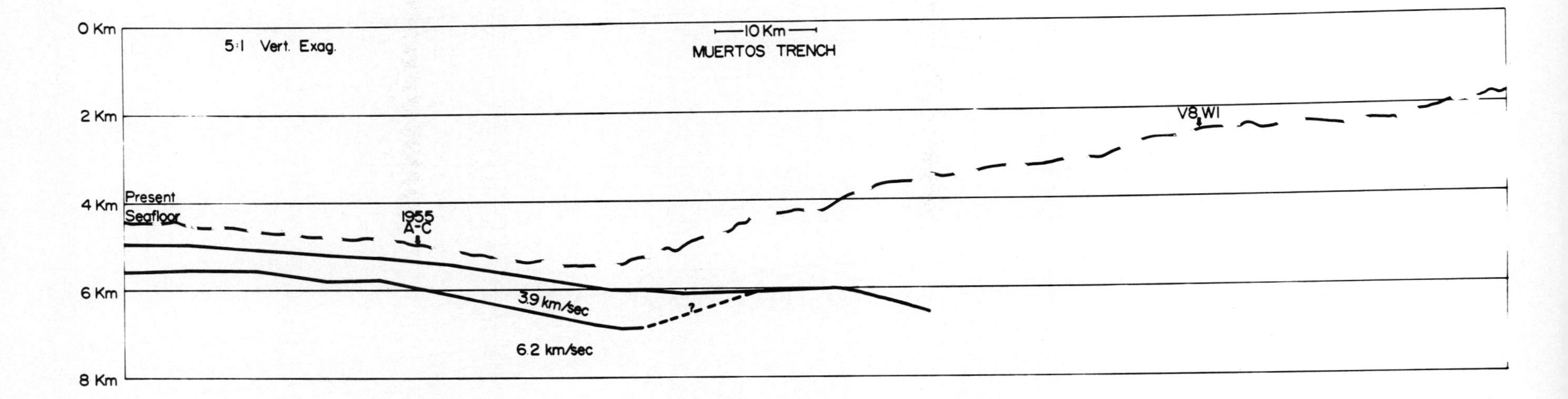

Fig. 9. Restored section to the epoch just after the 3.9 km/sec layer was laid down (probably Turonian-Santonian). Superimposed is the present sea floor, shown as a dashed line, for easy comparison. The surface of the restored sea floor is almost monoclinal dipping about 1° away from the center of the Venezuelan Basin towards the island ridge.

VB4N could well be a slump seen now receiving turbidites from nearer shore deposition on the east end of the Island of Dominica. Intermediate depth (70-140 km) earthquakes in this vicinity (Molnar and Sykes, 1969) may well have played a role in initiating or abetting the formation of such a slump scar and the embayment near Santo Domingo.

Conclusion

Farther to the east, at the longitude of San Juan, Worzel and Ewing (1954) showed that there is a negative gravity anomaly of 100 milligals centered on the Muertos Trench. Such an anomaly could result from either of the above hypotheses, but at least it does demonstrate that the crustal structure beneath the Muertos Trench is depressed beneath its isostatic equilibrium position, and that the process which produced the Muertos Trench is either still operational or is in its concluding phases.

Acknowledgments. Dr. Maurice Ewing founded the Geophysics Laboratory in Galveston in order to continue his work in marine geology and geophysics as well as his work in lunar geophysics. This paper on the Muertos Trench is a direct outgrowth of the interest of Dr. Ewing in applying multichannel seismic reflection techniques to the solution of fundamental tectonic problems in the Caribbean.

We wish to acknowledge the help given to us by the ship's crew of IDA GREEN, Otis Murray commanding, and the scientific staff. We are grateful to Cecil and Ida Green who made the ship and the basic equipment available for our work. Much geophysical equipment and valuable advice were provided by Chevron Oil Co., Continental Oil Co., Exxon Production Research, Mobil Oil Exploration, Shell Oil Co., and Texaco Inc.

This work was supported principally by NSF grant DES 75-06249. James Storm and our Industrial Associates provided additional support.

Marine Science Institute, Geophysical Laboratory Contribution No. 97.

References

Beck, R. H. and P. Lehner, Oceans, new frontier in exploration, Am. Assoc. Pet. Geol. Bul., 58, 376-395, 1974.

Case, J. E., Oceanic crust forms basement of eastern Panama, Geol. Soc. Am. Bul., 85, 645-652, 1974.

Creager, J. S., D. W. Scholl, et al., Initial Reports of the Deep Sea Drilling Project, 19, U. S. Govt. Printing Office 913 pp., 1973.

Edgar, N. T., J. I. Ewing, and J. Hennion, Seismic refraction and reflection in Caribbean Sea, Am. Assoc. Pet. Geol. Bul., 55, 833-870, 1971.

Edgar, N. T., J. B. Saunders, et al., Initial Reports of the Deep Sea Drilling Project, 15, U. S. Govt. Printing Office, 1137 pp., 1973.

Ewing, J., J. Antoine and M. Ewing, Geophysical measurements in the Western Caribbean sea and the Gulf of Mexico, Jour. Geophys. Res., 65, 4087-4126, 1960.

Ewing, J., M. Talwani and M. Ewing, Sediment distribution in the Caribbean Sea, Fourth Caribbean Geological Conf., Trinidad, 1965, 317-323, 1968.

Garrison, L. E., Acoustic Reflection Profiles Eastern Greater Antilles, National Technical Information Service, Dept. of Commerce, U. S. Geol Surv. document GD-72-004, 1972.

Hayes, D. E. and M. Ewing, Pacific boundary structure, in Maxwell (ed.), The Sea, 4(2), 29-72, 1971.

Hopkins, H. R., Geology of the Aruba Gap abyssal plain near DSDP site 153, in Edgar, N. T., J. B. Saunders, et al., Initial Reports of the Deep Sea Drilling Project, 15, U. S. Govt. Printing Office, 1039-1050, 1973.

Isacks, B., J. Oliver, and L. R. Sykes, Seismology and the new global tectonics, J. Geophys. Res., 73, 5855-5899, 1968.

Karig, D. E., J. C. Ingle Jr., et al., Initial Reports of the Deep Sea Drilling Project, 31, U. S. Govt. Printing Office, 927 pp., 1975.

Karig, D. E. and G. F. Sharman, III, Subduction and accretion in trenches, Geol. Soc. Am. Bul., 86, 377-389, 1975.

Kulm, L. D., R. von Huene, et al., Initial Reports of the Deep Sea Drilling Project, 18, U. S. Govt. Printing Office, 1075 pp., 1973.

Ludwig, William J., John E. Nafe, and Charles L. Drake, Seismic Refraction, in The Sea, 4(1), Arthur E. Maxwell, (ed.) Wiley-Interscience, 53-84, 1970.

Marlow, M. S., D. W. Scholl, E. C. Buffington, and T. R. Alpha, Tectonic history of the Central Aleutian Arc, Geol. Soc., Am. Bul., 84, 1555-1574, 1973.

Molnar, P. and L. R. Sykes, Tectonics of the Caribbean and Middle America regions from focal mechanisms and seismicity, Geol. Soc. Am. Bul., 80, 1639-1684, 1969.

Molnar, P. and L. R. Sykes, Plate Tectonics in the Hispaniola area: discussion, Geol. Soc. Am. Bul., 84, 2005-2020, 1973.

Moore, J. C., Complex deformation of Cretaceous trench deposits, southwestern Alaska, Geol. Soc. Am. Bul., 82, 2005-2020, 1973.

Officer, C. B., J. I. Ewing, J. F. Hennion, D. G. Harkrider and D. E. Miller, Geophysical investigations in the eastern Caribbean: Summary of 1955 and 1956 Cruises in Physics and Chemistry of the Earth, 3, Ahrens, Press, Rankama and Runcorn, (eds.) Pergamon Press, 17-109, 1959.

Piper, D. J. W., R. von Huene, J. R. Duncan, Late Quaternary sedimentation in the active eastern Aleutian Trench, Geology, 1, 19-26, 1973.

Saunders, J. B., N. T. Edgar, and T. W. Donnelly, Cruise Synthesis, in Edgar, N. T., J. B. Saunders, et al., Initial Reports of the Deep Sea Drilling Project, 15, U. S. Govt. Printing Office, 1077-1112, 1973.

Scholl, D. A., Sedimentary sequences in north Pacific trenches, in The Geology of Continental Margins, eds. C. A. Burk and C. L. Drake, Springer-Verlag, New York, 493-504, 1974.

Seely, D. R., P. R. Vail, and G. G. Walton, Trench slope model, in The Geology of Continental Margins, eds. C. A. Burk and C. L. Drake, Springer-Verlag, New York, 249-260, 1974.

von Huene, R., Structure of the continental margin and tectonism at the eastern Aleutian Trench, Geol. Soc. Am. Bul., 83, 3613-3626, 1972.

von Huene, R., Modern trench sediments, in The Geology of Continental Margins, eds. C. A. Burk and C. L. Drake, Springer-Verlag, New York, 207-212, 1974.

Worzel, J. Lamar and Maurice Ewing, Gravity anomalies and structure of the West Indies, Geol. Soc. Am. Bul., 65(2), 195-200, 1954.

Worzel, J. Lamar, Gravity investigations of the subduction zone in The Geophysics of the Pacific Ocean Basin and its Margin, G. H. Sutton, M. H. Manghnani, R. Moberly (eds.) AGU geophysical monograph 19, 1-16, 1976.

THE INITIATION OF TRENCHES: A FINITE AMPLITUDE INSTABILITY

D.P. McKenzie

Department of Geodesy and Geophysics, Cambridge University,
Madingley Rise, Madingley Road, Cambridge CB3 OEZ

Abstract. Kinematic plate evolution continuously reduces the number of plates, and therefore there must be mechansims which create new plate boundaries. Creation of new ridges is simple, but producing new trenches is not. Simple plate models are used to demonstrate that a compressive stress of at least 800 bars and a rate of approach of at least 1.3 cm/yr are required to form a new trench. Neither condition is easily satisfied. To illustrate these ideas three regions where new trenches may be starting are compared with the simple theoretical models.

Introduction

How does consumption of oceanic plate at trenches begin? This question is an important one and is not easily answered by direct observation. Ridges appear to start without much difficulty. The magnetic anomaly patterns which they produce show that more have been active in the last 80 m.y. than are now spreading, and there is still a considerably greater length of active ridge crest than there is of trench. Furthermore, the major trench systems of the Indian and Pacific Oceans have probably existed for more than 200 m.y., whereas most ridges are considerably younger. There are no major trenches in the Atlantic and Western Indian Oceans, whose margins were formed by rifting since the Jurassic. If the Pacific and North East Indian Ocean boundaries were also formed by rifting, it probably happened much earlier. Therefore geological observations suggest that ridges start easily, but trenches do not. This behaviour is surprising. Convection in liquids heated from below consists of rising and sinking sheets and jets of hot and cold fluid. The flow occurs because the boundary layers at the top and bottom of the fluid are unstable: their density differs from that of the main body of the fluid. Because the upper mantle of the Earth is heated from both within and below, the instability of the cold upper boundary layer should dominate the behaviour (McKenzie, Roberts and Weiss 1974). Though the geophysical observations show this argument is wrong, it is not at once obvious why it should be when the denser plates are floating on top of the lighter material below. This is the principal problem examined below. The observed behaviour can be understood if the elastic and friction forces are taken into account. These forces prevent trenches from arising spontaneously. Instead their formation becomes a finite amplitude instability. Considerable work must be done by some outside source to thrust one plate over another and produce a sinking slab about 180 km long before the instability becomes self sustaining. Such types of instability are well known in fluid mechanics.

Forces

Unfortunately, unlike the more familiar fluid dynamical instabilities, trench formation is not easily described by a high order partial differential equation. However, the existence of an elastic plate considerably simplifies the stability problem because the resultant force on it must act towards the trench if the instability is to be self maintaining.

It is important to distinguish between various definitions of plate thickness. The elastic plate thickness is that part of the plate which can maintain shear stresses elastically for geological times, and is estimated to be between 20 and 30km. Only this part of the plate stores elastic energy when the plate is deformed as it approaches a trench. Below this elastic layer is a region which creeps when a shear stress is applied. Though this region does not store elastic energy it does resist the relative motion between plates. Its thickness is not well known: the value used below, of 80 km for the combined thickness of both layers, T_e, is little more than a guess. Any thermal definition of the plate thickness includes both these layers and also the thermal boundary layer, and is much better determined. Since buoyancy forces are produced by temperature differences, it is the thermal plate thickness T which is involved in the driving forces.

The simple model in Fig. 1. shows how the gravitational driving forces arise. The force due to ridge pushing, D_R must be included because any instability must conserve the Earth's surface area, and hence any plate consumption must be balanced by plate generation. Both D_R and the force due to the sinking slab D_S may be expressed in terms of the

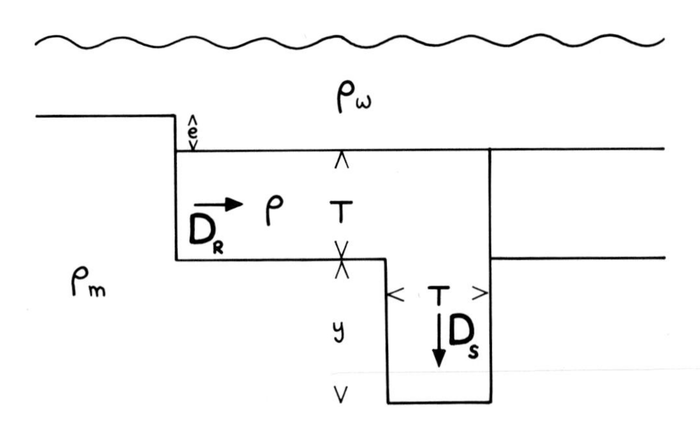

Fig. 1. Sketch to show how gravitational energy is released by trench formation.

elevation e of the ridge axis above old sea floor

$$D_R = eg(\rho_m - \rho_w)\frac{T}{2}$$
$$D_S = eg(\rho_m - \rho_w)y \qquad (1)$$

where

$$e = \frac{\rho - \rho_m}{\rho_m - \rho_w} T \qquad (2)$$

g is the acceleration due to gravity, ρ_m the density of the mantle, ρ ($>\rho_m$) that of the plate and ρ_w that of sea water. T is the thickness of the plate and y the depth of the tip of the slab below the base of the plate.

Our present knowledge of trenches suggests two resistive forces, shown in fig. 2, are important. The first of these F_E is well known and is caused by friction on the thrust plane.

$$F_E = \sigma T_e \cot \theta \qquad (3)$$

where σ is the shear stress on the fault plane and T_e is the thickness of the part of the plate which can maintain shear stresses. The best estimates of the shear stresses involved are derived from the seismic radiation from large earthquakes with thrust mechanisms beneath island arcs (see

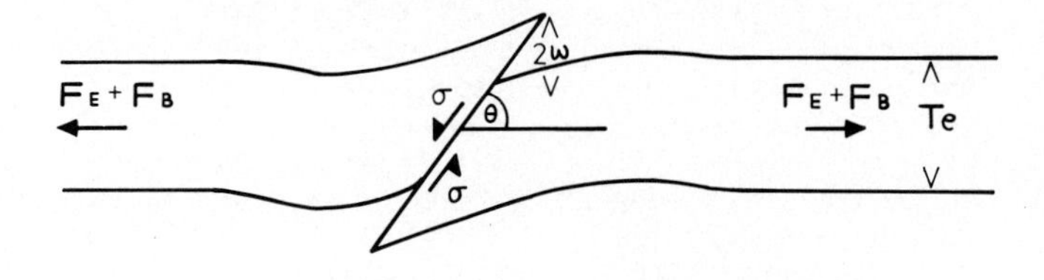

Fig. 2. The forces which resist trench formation arise through friction on the plate boundary and the need to bend both plates if subduction is to continue.

Richter and McKenzie 1976). Such studies give values of between 10 and 100 bars for the stress drop, and since the mean shear stress may be somewhat larger σ is taken to be 100 bars in the calculations below.

The other resistive force F_B is less well known and arises because work must be done against elastic and gravitational forces to form the topography of the trench and island arc. If the vertical force which maintains the shape of the trench is $F_N(x)$ and the vertical displacement of the surface of the plate from its original horizontal position is $w(x)$, the total work done W on one side of the trench is

$$W = -\frac{1}{2}\int_0^\infty F_N(x)\, w(x)\, dx \qquad (4)$$

where the x axis is horizontal with its origin at the trench. This expression is little use because $F_N(x)$ is unknown, but it can be converted to one which is useful if the bending is elastic. The relevant expression is given by Le Pichon et al. (1973)

$$F_N(x) = D d_x^4 w + S d_x^2 w + g(\rho_m - \rho_w)w \qquad (5)$$

where D is the flexural rigidity and S the external horizontal force/unit width. Whether w depends on S depends on whether

$$\varepsilon = \frac{S}{2\sqrt{Dg(\rho_m - \rho_w)}} \ll 1 \qquad (6)$$

The values in able 1 and a horizontal stress of 1 kilobar give $\varepsilon = 0.024$. Since the external stresses on the plates are unlikely to be much greater than this, $S d_x^2 w$ can be neglected. Substitution into (4) and integration by parts gives

$$W = -\frac{D}{2}\left[d_x w d_x^2 w - w d_x^3 w\right] - \frac{1}{2}\int_0^\infty \left(D(d_x^2 w)^2 + g(\rho_m - \rho_w)w^2\right) d_x \qquad (7)$$

Le Pichon et al. give the solution to (5) when

$$F_N(x) = 0$$
$$w = A\exp\left(-\frac{x}{\alpha}\right)\cos\left(\frac{x}{\alpha} + \phi\right) \qquad (8)$$

where A and ϕ are constants and

$$\alpha = \left(\frac{4D}{g(\rho_m - \rho_w)}\right)^{1/4} \qquad (9)$$

It is then simple to obtain a lower bound on W, W_L by considering only that part of the deformation which is seaward of the trench axis, where $F_N(x) = 0$. Substitution of (8) into (7) gives

$$W_L = A^2 g(\rho_m - \rho_w)\alpha \qquad (10)$$

when both sides of the trench are taken into account. The approximate relationship between the vertical motion $2w_o$ between the two sides of the trench is related to the horizontal motion X through the dip θ of the thrust fault

$$w_o = \frac{X}{2} \tan\theta = A(X) \cos\theta \qquad (11)$$

(8), (10) and (11) then give the horizontal force required to produce the bending

$$F_B = d_x W_L = \alpha g(\rho m - \rho w) \tan\theta \frac{A^2}{w_o} \qquad (12)$$

$$= \alpha g(\rho m - \rho w) \tan\theta \, w_o \left(2 + \frac{2\alpha(d_x w)_o}{w_o} + \frac{\alpha^2 (d_x w)_o^2}{w_o^2}\right)$$

where w_o and $(d_x w$ are the depth and the slope at the deepest part of the trench. If the shape of the trench is independent of X then

$$\frac{(d_x w)_o}{w_o} = \text{constant, and } F_B \propto X.$$

The most important assumption which has been made to obtain (12) is that the elastic deformation of the plate is complete when it reaches the axis of the trench. This assumption seems reasonable. Substitution of the values in Table 1 gives a radius of curvature there of about 200 km and a shear stress of 5 kilobars. There is no evidence for a smaller radius of curvature beneath the island arc, and greater stresses are likely to be relieved by fracture. Indeed, shear failure produced by bending often occurs on the ocean side of the trench (see McKenzie and Weiss 1975), suggesting that (12) may over estimate the force required. Therefore, these arguments suggest that (12) forms a reasonable estimate until more is known about the details of the deformation which occurs beneath island arcs.

Table 1.

e	=	3 km
g	=	10^3 cm sec^{-2}
$\rho_m - \rho_w$	=	2.3 gm cm^{-3}
T	=	120 km
T_e	=	80 km
w_o	=	3 km
$(d_x w)_o$	=	0.1
D	=	10^{30} dynes-cm
α	=	65 km
θ	=	10°
σ	=	100 bars

For an instability to grow we need

$$D_R + D_S > F_E + F_B \qquad (13)$$

Using the expressions above and the parameters in Table 1

$$4.1 \times 10^{15} + 6.9 \times 10^8 y > 4.8 \times 10^{15} + 2.3 \times 10^9 X \qquad (14)$$

where $y = x \sin\delta$
and δ is the dip of the slab.

This expression shows that, when $X = 0$, $D_R < F_E$. If this relationship did not hold then the compressive force produced by ridges would cause thrust faulting on pre-existing faults cutting oceanic plates. The reason why (14) can be satisfied when $X \neq 0$ is that F_B cannot exceed 8×10^5 dynes/cm when the trench depth reaches 3 km, whereas D_S can continue to increase. This behaviour is shown if Fig. 3. If $\delta = 90^o$ the inequality is satisfied if $X = X_o \simeq 130$ km, or 180 km if the slab dips at 45°. The instability will not grow when $X < X_o$ without an external energy supply, and is for this reason known as a finite amplitude instability. To cause the instability to grow also requires the compressive stress to exceed the greatest separation between σ_F and σ_D in Fig. 3, or about 800 bars (corresponding to a shear stress of 400 bars). One further condition must be satisfied. Because the density contrast between the sinking slab and the surrounding mantle is the consequence of their difference in temperature the consumption must occur sufficiently rapidly to maintain this contrast. The thermal time constant of the sinking slab is not known with any accuracy, but is probably about 10^7 years. If $\delta = 45^o$ and $X_o = 180$ km the consumption rate must exceed 1.8 cm/yr if the slab is to reach a depth of 130 km before warming up. Slower rates of convergence will not produce an instability. These three requirements are not easy to satisfy.

Observations

The most striking and important observation is a negative one: plates are mostly large and trenches are rare. Attempts to produce large aspect ratio convective rolls in fluids with constant or variable viscosity have not been successful. The upper boundary layer becomes unstable and the rolls break up into several smaller cells or rolls. This behaviour led Richter (1973)

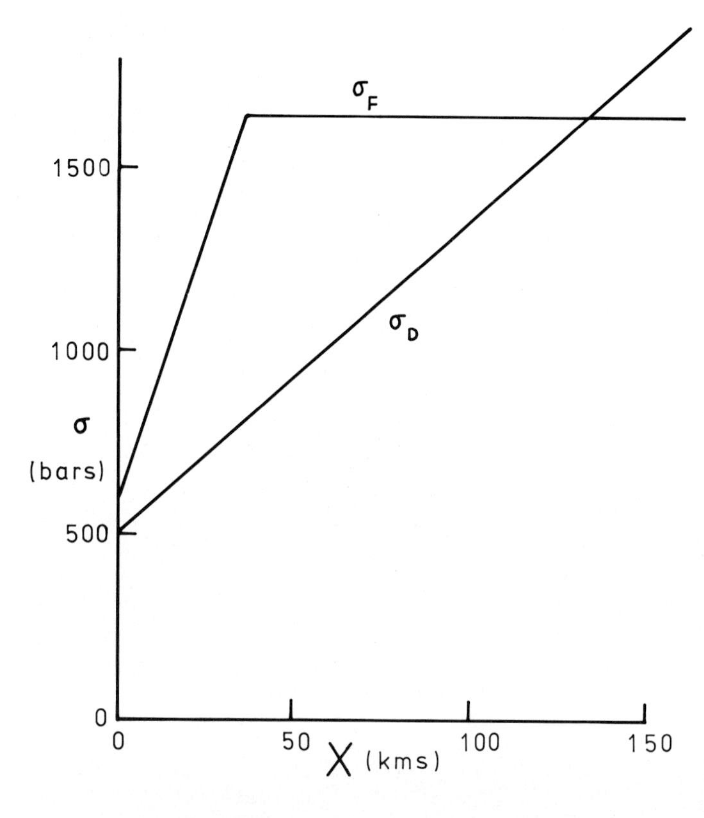

Fig. 3. The stress which arrises form the gravitational energy

$\sigma_D (= (D_s + D_R)/T_e)$

must exceed the resistive stress

$\sigma_F (= (F_E + F_B)/T_e)$

before the instability is self maintaining. Values obtained from (14) with $\delta = 90^o$.

and McKenzie and Weiss (1975) to argue that mantle convection must consist of two scales of motion, the plates forming part of the larger scale flow. Richter and McKenzie (1976) examined the energetics of the large scale circulation and demonstrated that the observed plate motions and other relevant geophysical observations were compatible with a simple convective model provided the lithosphere was decoupled from the mantle below by a thin low viscosity layer. However, they did not discuss the stability of the large scale flow. The expressions in the last section demonstrate that the large scale flow is only unstable to finite amplitude disturbances, a result which agrees with geophysical observations.

Despite the stabilizing effect of elastic and frictional forces new trenches are formed, and the force estimates above are of some help in understanding how this might occur. There are at least three places where trenches are believed to be in the process of forming, or to have recently done so. West of Gibraltar, the plate boundary between Africa and Eurasia consists of a thrust on which a number of large earthquakes have occurred (Fukao 1973) and is associated with large negative and positive gravity anomalies (Purdy 1975). Though consumption is in progress, the rate is only about 1.3 cm/yr and no sinking slab has been recognised. The plate boundary further west consists of a vertical strike slip fault, whereas to the east it is not clear whether the complicated tectonics (McKenzie 1972) resist or drive the motion of Aftica towards Europe. The other boundaries of the Aftican plate are ridges, as are those of the western part of the Eurasian plate. It is therefore likely that the thrust boundary west of Gibraltar is driven by a local stress concentration, and the work required to maintain the motion comes from the extensive ridges in the Atlantic and Indian Oceans.

Another example is the Aegean Arc in the Eastern Mediterranean, which probably started in the Miocene. This trench could have been started in the same way as that at present in progress west of Gibraltar, but another energy source is available. The difference in mean density between continental and oceanic crust produces a horizontal force in the same way as does the elevation of a ridge. If the difference in elevation between a continent and ocean is 7 km with 3 km above sea level and 4 km below, it can maintain a compressive stress of 850 bars. Whether or not it does so depends on whether the elevation difference is supported by this stress or by elastic forces within the continental lithosphere. But such stresses could initiate trenches, especially if regional metamorphism relaxed stresses elsewhere.

The third example is the diffuse line of epicentres which cross the Indian Ocean from Ceylon to Australia. Sykes (1970) has suggested they are produced by overthrusting, where a new island arc is developing. However, little is known about the mechanisms of the shocks or about the nature of the deformation, and further work is needed to establish if Sykes' suggestion is correct. If it is, the ideas discussed above will need to be modified, since there is no obvious source of energy which can develop a finite amplitude instability along a line 400 km long.

Once a sinking slab has been produced along part of a plate boundary, a source of work is available to drive instabilities elsewhere, and in this way extend the length of the island arc. A possible example of this process is the Macquarie Ridge, where the thrusting and bending can be maintained by slabs sinking beneath the New Zealand, Kermadec and Tonga arcs.

The arguments above have all been concerned

with the Earth, whose tectonics are now well known. It is, however, now clear that plate motions are not taking place on the surface of other planets in the solar system with the possible (but unlikely) exception of Venus. The argument above suggests why the Earth might be unique. The initiation of consumption depends on the shear stress on fault planes being as small as 100 bars, and on the existence of a low viscosity layer below the plates to permit their motions to be decoupled from the mantle. It is now believed that both are due to the existence of water, which is thought to reduce the friction on fault planes and to produce the low viscosity layer through partial melting of about 1% of the mantle. Similar quantities of water are not known to exist on other planets, and this difference could account for the differences in tectonic style. It does, however, seem inevitable that the interiors of all planets are convecting.

Conclusion

A simple model for the development of a new trench and sinking slab shows that it occurs through a finite amplitude instability. The compressive stress required is about 800 bars and the rate of approach must be greater than about 1.3 cm/yr. These conditions explain why new island arcs do not develop spontaneously in the middle of an ocean, yet under some conditions they allow new trenches to start. These calculations are consistent with a two scale model for mantle convection, and show that the large scale circulation of which the plate motions form part is stabilized by friction on plate boundaries and elastic forces within the plates.

Acknowledgements. This investigation arose from a conversation with Joseph Pedlosky and Stephen Davis during the summer program in Geophysical Fluid Dynamics at Woods Hole in 1975, and forms part of a general study of mantle convection at Cambridge supported by the Natural Environment Research Council.

References

Fukao, Y. 1973. Thrust faulting at a lithospheric plate boundary: the Portugal Earthquake of 1969. Earth Planet. Sci. Lett. 18 205.

Jeffreys, J. 1929. The Earth, 2nd edition, Cambridge U.P.

Le Pichon, X., Francheteau, J., and Bonnin, J. 1973 Plate Tectonics, Elsevier.

McKenzie, D.P. 1972. Active Tectonics of the Mediterranean region, Geophys. J. R. Astr. Soc. 30, 109 - 85.

McKenzie, D.P., Roberts, J.M. and Weiss, N.O. 1974. Convection in the Earth's mantle: towards a numerical simulation. J. Fluid Mech. 62 465.

McKenzie, D.P. and Weiss, N.O. 1975 Speculations on Thermal and Tectonic History of the Earth. Geophys. J.R. Astr. Soc. 42. 131.

Purdy, G.M. 1975. The eastern end of the Azores-Gibraltar Plate Boundary. Geophys J.R.Astr Soc. 43, 973.

Richter, F. 1973 Convection and Large Scale Circulation of the Mantle. J. Geophys. Res. 78, 8735.

Richter, F. and McKenzie, D. 1976 Simple Plate Models of Mantle Convection. Geophys J.R. Astr. Soc. (in press)

Sykes, L.R. 1970. Seismicity of the Indian Ocean and a Possible Nascent Island Arc between Ceylon and Australia. J. Geophys. Res. 75, 5041.

LITHOSPHERIC INSTABILITIES

D. L. Turcotte, W. F. Haxby, and J. R. Ockendon*

Department of Geological Sciences, Cornell University, Ithaca, New York 14853

Abstract. In this paper we define a mantle geoid. This is the height that hot solid mantle rock from the asthenosphere would attain if it were not confined by the lithosphere. The mantle geoid lies 3.25 km below the hydrogeoid (sea level). Hot mantle rock cannot entirely penetrate the continental lithosphere. One consequence of this partial penetration is rifting; as a result of rifting an accreting plate margin may be created. Hot mantle rock from the asthenosphere can penetrate through the oceanic lithosphere if the sea floor lies below the mantle geoid. Penetration of the oceanic lithosphere by this solid mantle rock is a necessary condition for the initiation of subduction. We argue that the same processes that are associated with rifting in continental lithosphere will be associated with behind arc spreading and the initiation of subduction in the oceanic lithosphere.

Introduction

There is conclusive observational evidence that the lithosphere bends and descends into the mantle at ocean trenches. Since the cold lithosphere is gravitationally unstable with respect to the hot, underlying mantle rock this subduction process is not surprising. This gravitational instability is not a sufficient condition, however, for the initiation of subduction. In terms of mantle convection the initiation of subduction can be treated as a fluid instability. A fluid layer heated from within or below becomes unstable when the critical Rayleigh number is exceeded. A fluid analysis, however, is not applicable to material with a finite yield strength. It would be expected that the finite strength of the lithosphere would result in a rigid outer shell as on the moon.

The mechanical stability of the lithosphere can also be considered. A lithospheric plate may buckle if the longitudinal compressional force is sufficiently large. The lithospheric plate may also fracture under compression. In either of these cases the hot mantle rock must override the lithosphere before subduction can be initiated.

In this paper we propose that subduction may be initiated by a tensional failure of the oceanic lithosphere. The tensional failure allows hot mantle rock to penetrate and override the lithosphere. This type of tensional failure is analogous to the failure of the continental lithosphere which can result in the formation of an accretional plate margin.

Fluid Instabilities

There are many similarities between plate tectonics and thermal convection in a fluid. It is expected that on geological time scales hot mantle rock will exhibit fluid behavior because of thermally activated creep processes. A fluid which is heated from within and cooled from above convects when the critical Rayleigh number is exceeded. An example of this type of convection is given in Figure 1a (Turcotte et al., 1973).

Because of the convection a thermal boundary layer develops on the upper boundary of the cell. This cold boundary layer separates from the upper boundary and forms a cold descending plume. The negative buoyancy force in the plume is primarily responsible for driving the flow.

There are important similarities between this flow and the flow in the mantle. The surface plates of plate tectonics are cold thermal boundary layers. The surface plates bend at ocean trenches and descend into the mantle similar to the cold descending plume in Figure 1a. The negative buoyancy of the descending plate is believed to play an important role in driving plate tectonics (Forsyth and Uyeda, 1975).

However, there are important differences between the flow illustrated in Figure 1a and the flow in the mantle. The most important of these concerns the rheology of the cold thermal boundary layer. In Figure 1a the fluid has a constant viscosity and the thermal boundary layer is not rigid; in particular, the fluid does not have bending rigidity. The surface plates, on the other hand, exhibit elastic behavior. At high temperatures the creep properties of mantle

---

* Permanent address: Mathematical Institute, University of Oxford.

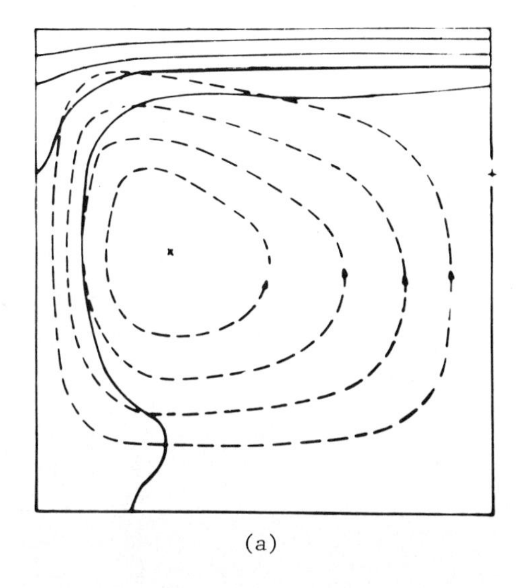

(a)

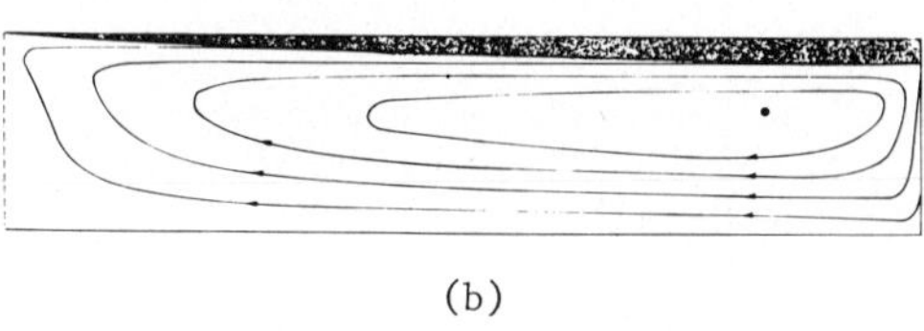

(b)

Figure 1. Convection in fluid layer heated from within and cooled from above. (a) Constant viscosity. Solid lines are streamlines; dashed lines are isotherms. (b) Temperature and depth-dependent viscosity with a rigid lithosphere. Solid lines are streamlines; shaded area the rigid lithosphere.

rocks are exponentially temperature-dependent. There is ample evidence that at temperatures below about 1000°C mantle rock exhibits plastic rather than fluid-like behavior and at a temperature below about 500°C mantle rock exhibits elastic-brittle behavior (Turcotte, 1974)

In order to better understand the role of a rigid thermal boundary layer, we have carried out a series of numerical calculations (Parmentier and Turcotte, 1976). In these calculations a fluid with a temperature- and pressure-dependent viscosity appropriate to the mantle (Torrance and Turcotte, 1971) is heated from within and cooled from above. In addition, fluid which is cooled below 1000°C is treated as being part of a rigid thermal boundary layer. One of these numerical calculations is shown in Figure 1b. An important difference between Figure 1a and Figure 1b is the aspect ratio of the flow. If the aspect ratio of the flow shown in Figure 1a is increased above about 1.5, the flow breaks up into two cells. The cold thermal boundary layer becomes gravitationally unstable and separates from the upper boundary of the cell. In Figure 1b this instability is suppressed. Aspect ratios as large as ten have been obtained without difficulty.

The flexural rigidity of the lithosphere is demonstrated by its behavior at ocean trenches. The topography of a typical ocean trench is shown in Figure 2 (Caldwell et al., 1976). It is compared with the elastic theory for a bending plate. Good agreement is obtained if the thickness of the elastic lithosphere is taken to be 30 km. This is the fraction of the thermal lithosphere that behaves elastically on geological time scales. Based on constant viscosity calculations, Richter (1973) and Richter and Parsons (1975) have suggested that transverse instabilities may occur beneath translating lithospheric plates. The elastic rigidity of the lithosphere would be expected to also suppress this type of instability. We suggest that in order to understand the initiation of subduction it is necessary to understand the elastic stability (instability) of the lithosphere.

Longitudinal Buckling

An elastic plate can be buckled by a horizontal load. A simple example is illustrated in Figure 3a. A plate of length ℓ and thickness h is rigidly constrained at one end. The vertical deflection of the plate w under an end load (per unit depth) P is governed by the equation

$$D \frac{d^4 w}{dx^4} + P \frac{d^2 w}{dx^2} = 0 \qquad (1)$$

where D is the flexural rigidity ($D = Eh^3/12[1 - \nu^2]$). The boundary conditions that must be satisfied are: $w = dw/dx = 0$ at $x = 0$, and $d^2w/dx^2 = d^3w/dx^3 = 0$ at $x = \ell$. This is an eigenvalue problem and the eigenvalue end load is

$$P = D\left(\frac{\pi}{2\ell}\right)^2 \qquad (2)$$

If the end load is less than this value no deflection occurs; at the critical load the plate buckles.

This simple analysis is not applicable to the lithosphere because the lithosphere experiences

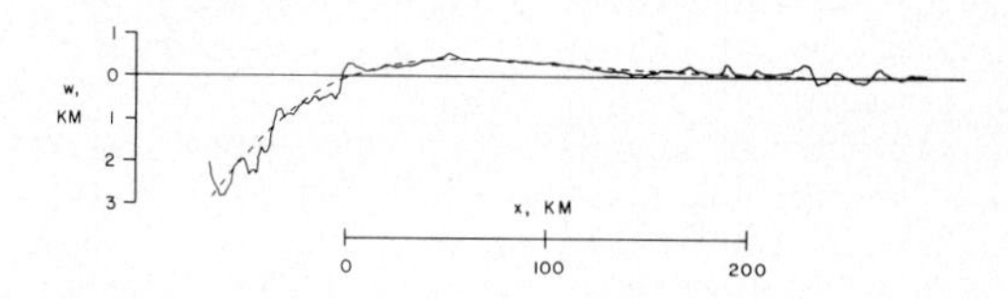

Figure 2. Comparison of the observed topography over the Bonin Trench (solid line) with the bending of an elastic plate under a vertical end load and bending moment (dashed line).

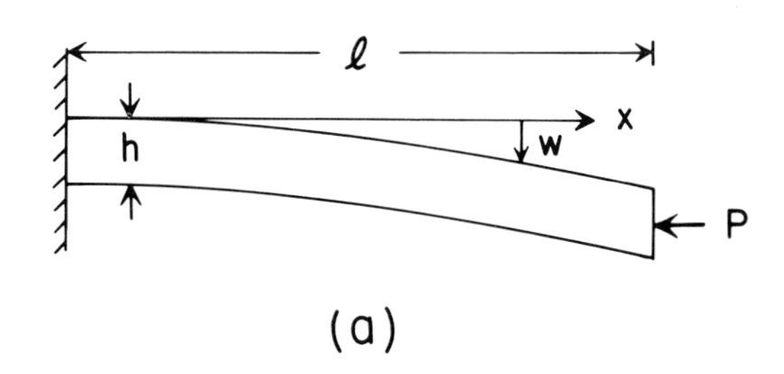

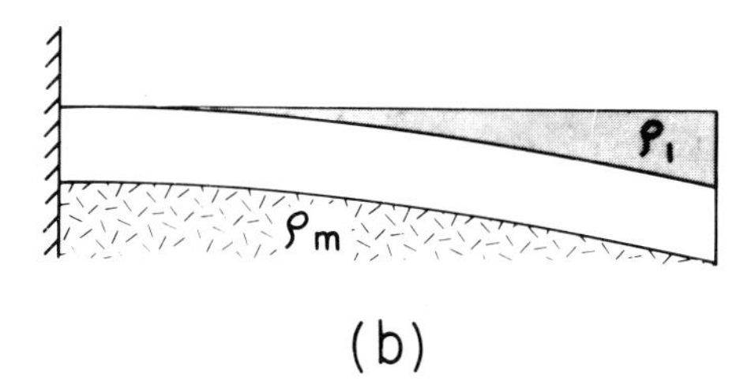

Figure 3. Instabilities of a cantilevered plate. (a) Longitudinal buckling due to an end load P. (b) Transverse buckling due to an overlying dense fluid, as the plate bends the dense fluid with density ρ_1 displaces the light fluid with density ρ_m.

a hydrostatic restoring force when it is deformed vertically. In this case the governing equation is

$$D \frac{d^4 w}{dx^4} + P \frac{d^2 w}{dx^2} + (\rho_m - \rho_w)\, gw = 0 \qquad (3)$$

It is assumed that the lithosphere is covered by water with a density ρ_w and that it displaces mantle rock with a density ρ_m. This problem was solved by Smoluchowski (1909) and for a plate of infinite length the horizontal eigenvalue buckling load is (Jeffreys, 1970)

$$P = (4D\,[\rho_m - \rho_w]\, g)^{1/2} = \left(\frac{Eh^3\,[\rho_m - \rho_w]\, g}{3\,[1 - \nu^2]} \right)^{1/2} \qquad (4)$$

The maximum stress that the lithosphere can transmit in compression is σ_m; if this stress is exceeded the lithosphere will fail plastically or will fracture. Taking $P = \sigma_m h$, the thickness of an elastic lithosphere which will buckle under a longitudinal load is

$$h = \frac{3\sigma_m^2\,(1 - \nu^2)}{E\,(\rho_m - \rho_w)\, g} \qquad (5)$$

Taking $\sigma_m = 10$ kb, $\nu = 0.25$, $E = 10^{12}$ dyne/cm^2, $\rho_m = 3.4$ gm/cm^3, and $\rho_w = 1$ gm/cm^3, we find that $h = 1.3$ km. Since it is known that the thickness of the elastic lithosphere is close to 30 km, we conclude that longitudinal buckling of the elastic lithosphere cannot occur.

Transverse Buckling

A plate can also buckle under a transverse load. A simple example is shown in Figure 3b. A horizontal plate is rigidly constrained at one end. When the plate is deformed downward it is assumed that the region between the initial and deformed positions of the plate is filled with a fluid of density ρ_1 and that the downward movement of the plate displaces a fluid of density ρ_m. We consider the case when $\rho_1 > \rho_m$ and determine whether the weight of the overlying dense fluid can cause the plate to deform downward without the application of other forces. The governing equation is

$$D \frac{d^4 w}{dx^4} - (\rho_1 - \rho_m)\, gw = 0 \qquad (6)$$

The required boundary conditions are the same as those used above: $w = dw/dx = 0$ at $x = 0$, and $d^2w/dx^2 = d^3w/dx^3 = 0$ at $x = \ell$.

No deformation occurs if the density of the overlying fluid is less than a critical value ρ_{1c}; if $\rho_1 > \rho_{1c}$ the plate buckles. This is an eigenvalue problem in which the density ρ_{1c} satisfies the relation

$$\cosh \left[\frac{(\rho_{1c} - \rho_m) g}{D} \right]^{1/4} \ell \, \cos \left[\frac{(\rho_{1c} - \rho_m) g}{D} \right]^{1/4} \ell = -1 \qquad (7)$$

And for the minimum eigenvalue the critical density ρ_{1c} satisfies the relation

$$\rho_{1c} - \rho_m = \frac{D}{g} \left(\frac{1.875}{\ell} \right)^4 \qquad (8)$$

If the density ρ_{1c} is less than this value no deflection occurs; at the critical density transverse buckling occurs. Taking $E = 10^{12}$ dyne/cm^2, $\nu = 0.25$ and $h = 30$ km, the critical density is given as a function of ℓ in Figure 4. When ℓ is greater than about 200 km, very small density excesses can induce instability.

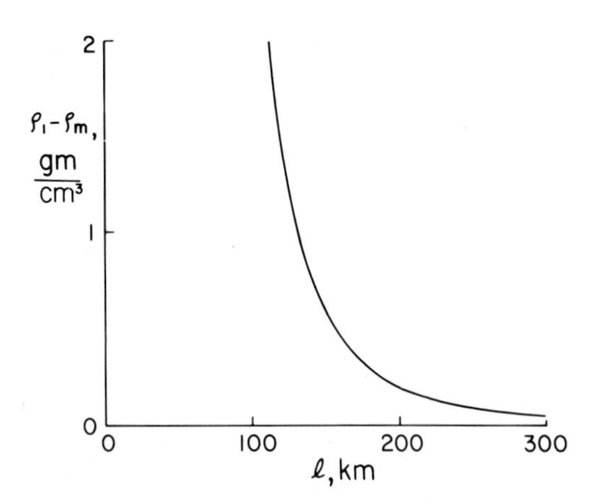

Figure 4. Density difference necessary for transverse buckling as a function of plate length for a plate thickness of 30 km.

The displacement of the plate is given by

$$w = w_{\ell}\ [0.500\ (\cosh 1.875\ \frac{x}{\ell}\ -\ \cos 1.875\ \frac{x}{\ell})$$
$$-\ 0.367\ (\sinh 1.875\ \frac{x}{\ell}\ -\ \sin 1.875\ \frac{x}{\ell})] \quad (9)$$

The amplitude of the displacement is not given by the linear stability analysis. This deformation is illustrated in Figure 5; it bears a considerable resemblance to the deformation of the lithosphere at a trench.

This analysis shows that a neutrally buoyant lithospheric plate will bend and descend into the mantle if it is overlain with rock which is only slightly more dense than the hot mantle rock. In fact, a large fraction of the oceanic lithosphere is gravitationally unstable; this instability will enhance the instability given above and will reduce ρ_{1c}. However, no matter how unstable this lithosphere is, it will not bend and descend into the mantle unless it is overlain with mantle rock. Sediments or basaltic volcanism can depress the lithosphere if they are piled onto it (i.e., at a continental margin or at a volcanic island); however, the amount of deformation is limited by the height of the pile. Sediments or basaltic volcanism cannot initiate subduction. The only way in which subduction can be initiated is for mantle rock to penetrate and override the lithosphere. This analysis also shows that several hundred kilometers of lithosphere must be overridden before subduction can be initiated because of the flexural strength of the lithosphere. The analogy is to a steel ship which will sink only if water is allowed to penetrate the hull.

Mantle Geoid and Manometer

In order to specify the stability of the lithosphere we derive a hot mantle geoid. The reference geoid for the earth is sea level. We will refer to this as the hydrogeoid. Clearly the hydrogeoid fluctuates as the amount of water stored in glaciers changes, as the volume of ocean ridges changes, and as the size of the continents increases. In continental areas sea level is defined as a continuation of the equipotential surface which defines the level of the oceans. Sea level in continental areas can also be defined by a hypothetical water manometer. If a hydromanometer is extended from an oceanic area to a continental area as illustrated in Figure 6, the height of the water in the continental area defines the hydrogeoid.

In this paper we wish to introduce the concept of a mantle geoid. This is the height to which hot mantle rock from the asthenosphere can rise defined in the same sense that the hydrogeoid is defined. Since the mantle rocks beneath the lithosphere exhibit fluid behavior on geological time scales, the concept of an equipotential mantle geoid should be meaningful. In order to determine the mantle geoid we consider a typical ocean ridge. This typical ocean ridge is at a depth of 2.5 km. It has an oceanic crust of hot basalt (ρ = 2.96 gm/cm^3) with a thickness of 4.5 km. Beneath the crust is hot mantle rock (ρ = 3.36 gm/cm^3). The mantle geoid is defined as the isostatic height of hot mantle overlain with sea water to the hydrogeoid. Therefore the depth of the mantle geoid below the hydrogeoid d_m is given by

$$d_m \times 1 + (7 - d_m)\ 3.36 = 2.5 \times 1 + 4.5 \times 2.96 \quad (10)$$

Solving for d_m gives d_m = 3.25 km. The mantle geoid lies 3.25 km below the hydrogeoid. Consider a hypothetical hot mantle rock manometer between a ridge crest and a typical ocean basin

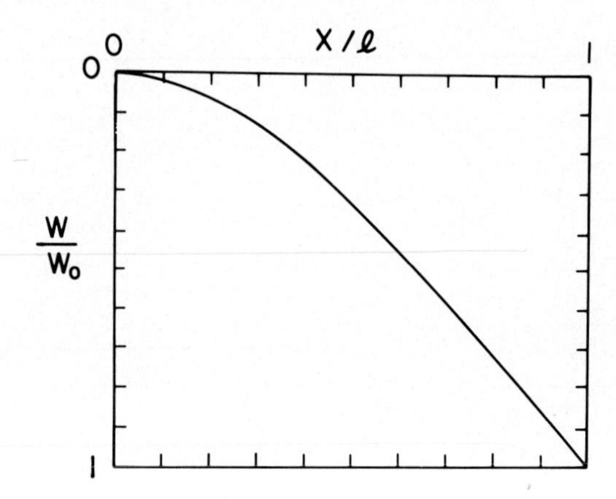

Figure 5. Plate deformation due to transverse buckling.

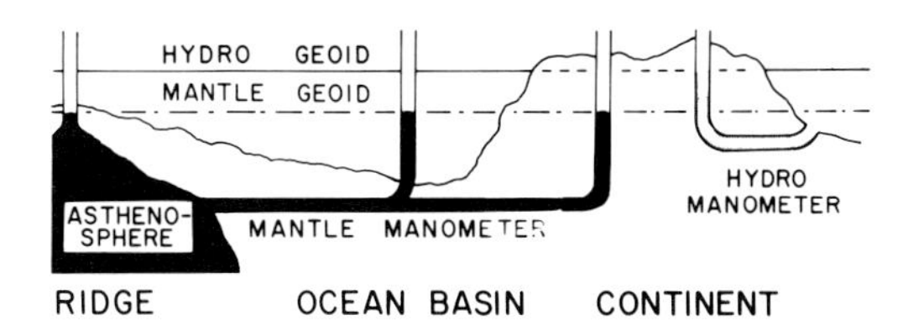

Figure 6. Illustration of mantle and hydromanometers relative to the mantle and hydrogeoids.

and continent as illustrated in Figure 6. This mantle manometer defines the mantle geoid in these areas. In basin areas the mantle geoid lies above sea floor. If the hot mantle rock from the asthenosphere can penetrate the oceanic lithosphere, it has sufficient hydraulic head to flow through it, cover the surface, and induce subduction. This is a direct consequence of the gravitational instability of the oceanic lithosphere. In continental areas the mantle geoid lies below the surface. Hot mantle rock from the asthenosphere can rise into the continental crust but cannot penetrate it. The continental lithosphere is stable.

It should be noted that in defining the mantle geoid it has been assumed that the hot mantle rock that has been depleted of basalt has the same density as undepleted mantle rock. If the depleted rock is taken to have a higher density the mantle geoid would approach the 2.5 km depth of the ocean ridges.

The depth of the sea floor in ocean basins is accurately represented by the cooling of a thermal boundary layer (Turcotte and Oxburgh, 1969). This depth d is given as a function of the age of the sea floor t by

$$d = d_r + \frac{2\alpha\rho_m}{(\rho_m - \rho_w)} \left(\frac{\kappa t}{\pi}\right)^{1/4} (T_m - T_o) \qquad (11)$$

where d_r is the depth of the ridge, α the volume coefficient of thermal expansion of mantle rock, κ the thermal diffusivity, T_m asthenosphere temperature, and T_o the sea-floor temperature. Taking $\alpha = 3 \times 10^{-5}\ °C^{-1}$, $\kappa = 10^{-2} cm^2/sec$, and $T_m - T_o = 1200°C$, the depth of the sea floor is given as a function of its age in Figure 7. It is seen that the mantle geoid lies above the sea floor for sea floor which is older than about 6 million years.

Initiation of Rifting and Subduction

The role of the mantle manometer is clearly illustrated in the initiation of rifting. This is illustrated in Figure 8. A tensional failure of the continental lithosphere initiates the formation of the typical graben valley and the diapiric rise of solid, hot mantle rock from the asthenosphere to near the continental moho (Figure 8a). As rifting continues the hot mantle rock penetrates the crust and rises to the mantle geoid. Either because of pressure release melting or frictional heating, some partial melting of the upwelling hot mantle rock occurs. As a result, the top of the rising (hot but solid) mantle rock is coated with a layer of basaltic magmatism. This is illustrated in Figure 8b and the Red Sea is a good example. If spreading continues, the axis of diapiric upwelling becomes a mid-oceanic ridge as illustrated in Figure 8c.

Now let us consider the same sequence of events in typical oceanic lithosphere. This is illustrated in Figure 9. Because the mantle geoid lies above the sea floor the rising hot, solid mantle rock from the asthenosphere can rise above the ocean crust and spread over the sea floor. The solid mantle rock would flow in a manner similar to that occurring in the Red Sea. Again some partial melting would occur and the mantle rock would be capped with a layer of basalt. The initial phase of this is illustrated in Figure 9a. Although the oceanic lithosphere is gravitationally unstable, a small amount of extruded dense mantle rock cannot induce subduction because of the flexural rigidity of the lithosphere. The stability analysis given in this paper shows that this rigidity will be overcome only when about 200 km of sea floor is overlain with dense mantle rock.

As the hydraulic head continues to drive mantle rock through the oceanic lithosphere, it continues to spread over the lithosphere as shown in Figure 9b. The weight of the dense, unstable lithosphere now causes it to bend and founder. The upwelling hot mantle rock replaces the foundering lithosphere. The result is the secondary, behind-arc spreading often observed behind subduction zones (Karig, 1970) as illustrated in Figure 9c.

The same sequence of events that led to rifting in a continental environment leads to the initiation of subduction in an oceanic environment. Upwelling of hot mantle rock in a continental environment cannot lead to subduction because the continental lithosphere is gravitationally stable; the mantle geoid lies below

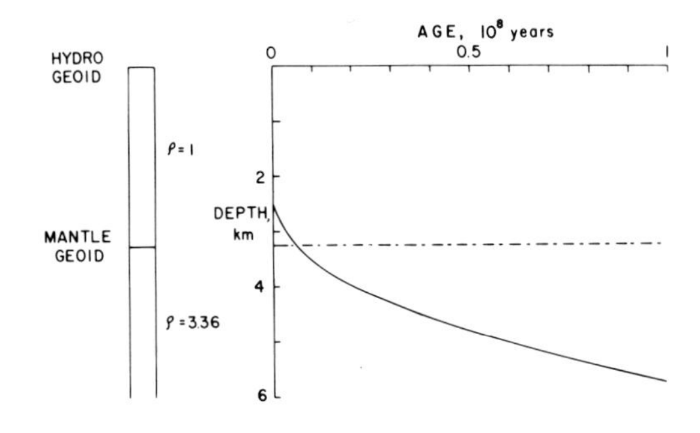

Figure 7. Comparison of the mantle geoid with the depth of the sea floor.

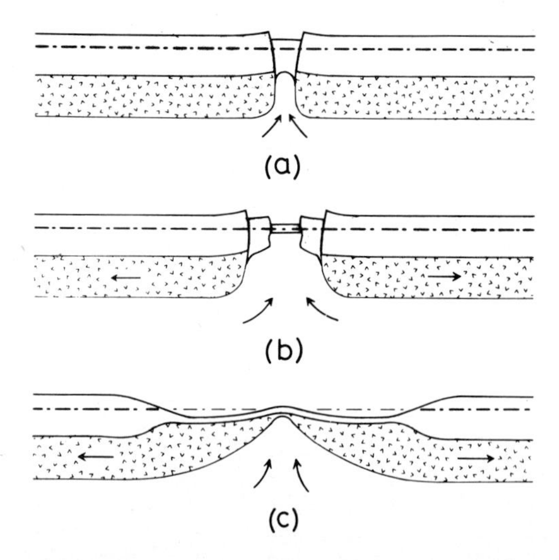

Figure 8. Illustration of rifting and the initiation of an accreting plate margin in a continental area.

the land surface. In ocean basins the mantle geoid lies above the sea floor; hot mantle rock can penetrate the oceanic lithosphere leading to the foundering of the gravitationally unstable oceanic lithosphere and the initiation of subduction.

Mantle Diapirism

There is ample observational evidence that diapirs of hot, solid mantle rock rise through the continental lithosphere. It is happening

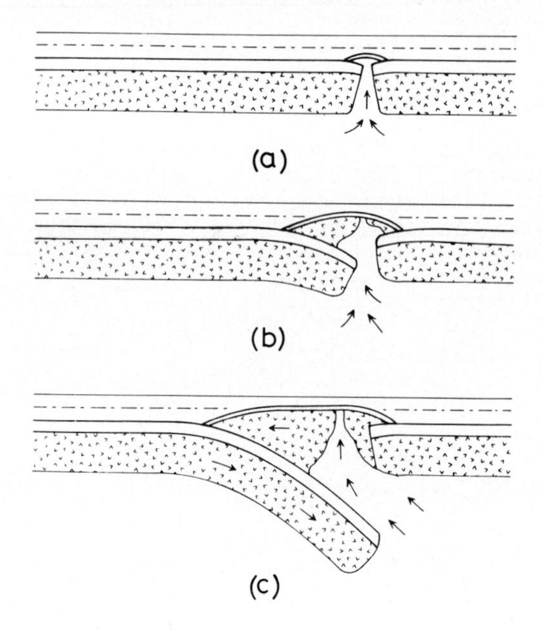

Figure 9. Illustration of the initiation of subduction in an oceanic area.

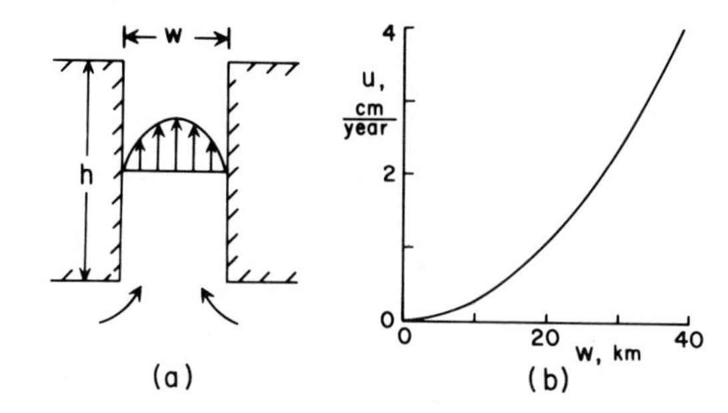

Figure 10. (a) Model for the diapiric penetration of the lithosphere by hot, solid mantle rock from the asthenosphere. (b) Velocity of penetration as a function of the width of the two-dimensional diapir.

today in the Red Sea. There is also indirect evidence that it is happening in the East African Rift and the Rhine Graben. We would use the South Sandwich Arc as a prime example of the diapiric rise of hot mantle rock behind a foundering lithosphere.

We consider a simple analysis of the flow of hot, solid mantle rock through the oceanic lithosphere. The mantle rock is assumed to behave as a Newtonian viscous fluid with viscosity η. The flow is assumed to take place through a crack in the lithosphere with a width w as illustrated in Figure 10a. The pressure (hydraulic head) available to drive the mantle rock through to oceanic lithosphere is

$$p = (\rho_m - \rho_w)\, g\, (d - d_m) \qquad (12)$$

where d is the depth of the sea floor. Taking d = 5 km we find that p = 400 b. The velocity of the hot rock through the gap is given by

$$u = \frac{w^2}{12\eta}\,\frac{p}{h} \qquad (13)$$

Taking the value of p given above, h = 100 km, and $\eta = 4 \times 10^{20}$ poise as a typical value for hot asthenospheric rock, the dependence of the velocity on the width of the crack is given in Figure 10b. It is seen that significant velocities (~ 1 cm/yr) require a crack width of about 20 km. Although this calculation is clearly very idealized in that it neglects heat transfer from the hot mantle rock, dissipative heating, pressure release melting, and other effects, the result that the rising diapir of hot mantle rock must have dimensions of 10's of kilometers is probably valid.

An important question is what initiates the diapiric rise of hot mantle rock. We have previously suggested that rifting is the result of the tensional failure of the lithosphere (Turcotte and Oxburgh, 1973). Possible sources of the required tensional stress include membrane stresses due to the movement of the surface plates over a non-spherical earth and thermal stresses. However, these mechanisms lead to relatively little horizontal strain. The probable cause of the diapiric upwelling beneath the Red Sea is the rotation of the Arabian plate away from the African plate. This type of rotation is often required by multiple plate interactions (Dewey, 1975). We suggest that such rotations in oceanic regions would initiate subduction. This mechanism could explain double trench systems such as that in the eastern Pacific. It could also explain isolated trench systems such as the Puerto Rican Trench and the South Sandwich Arc.

Acknowledgments. The authors would like to thank J. M. Bird for many helpful suggestions. This research has been supported by the Earth Sciences Section, National Science Foundation, NSF Grant DES 74-03259 and by the National Aeronautics and Space Administration, NASA Grant NSG 5060.

References

Caldwell, J. G., W. F. Haxby, D. E. Karig, and D. L. Turcotte, On the applicability of a universal elastic trench profile, Earth Planet. Sci. Lett., 31, 239-246, 1976.

Dewey, J. F., Finite plate implications: Some implications for the evolution of rock masses at plate margins, Amer. J. Sci., 275A, 260-284, 1975.

Forsyth, D., and S. Uyeda, On the relative importance of the driving forces of plate motion, Geophys. J. Roy. Astr. Soc., 43, 163-200, 1975.

Karig, D. E., Ridges and basins of the Tonga-Kermadec island arc system, J. Geophys. Res., 75, 239-254, 1970.

Jeffreys, H., The Earth, 5th ed., pp. 410-413, Cambridge, 1970.

Parmentier, E. M., and D. L. Turcotte, Studies of thermal convection beneath a rigid lithosphere, in preparation, 1976.

Richter, F. M., Convection and the large-scale circulation of the mantle, J. Geophys. Res., 78, 8735-8745, 1973.

Richter, F. M., and B. Parsons, On the interaction of two scales of convection in the mantle, J. Geophys. Res., 80, 2529-2941, 1975.

Smoluchowski, M., Uber ein gewisses stabilitatsproblem der elastizitatslehre und dessen beziehung zur entstehung von faltengebirgen, Bull. Int. Acad. Sci. de Cracovie, No. 2, 3-20, 1909.

Torrance, K. E., and D. L. Turcotte, Structure of convection cells in the mantle, J. Geophys. Res., 76, 1154-1161, 1971.

Turcotte, D. L., Are transform faults thermal contraction cracks?, J. Geophys. Res., 79, 2573-2577, 1974.

Turcotte, D. L., and E. R. Oxburgh, Convection in a mantle with variable physical properties, J. Geophys. Res., 74, 1458-1474, 1969.

Turcotte, D. L., and E. R. Oxburgh, Mid-plate tectonics, Nature, 244, 337-339, 1973.

Turcotte, D. L., K. E. Torrance, and A. T. Hsui, Convection in the earth's mantle, Meth. Comp. Phys., 13, 431-454, 1973.

MESOZOIC TECTONICS OF THE SOUTHERN ALASKA MARGIN

J. Casey Moore and William Connelly

Earth Sciences Board, University of California, Santa Cruz, CA 95064

Abstract. The southern margin of Alaska shows evidence of three major phases of magmatism and accretion of deep-sea deposits during the Mesozoic. Magmatism and subduction characterized both southwestern and southeastern Alaska from the Late Triassic into the Late Jurassic. During this time a mean subduction-slip vector of $N38 \pm 11^{\circ}W$ was recorded in the subduction complex of the Kodiak Islands. A two plate model which satisfies the extent of magmatism and the subduction-slip direction for the Late Triassic to Late Jurassic suggests that the Alaska orocline was significantly straighter than its present configuration. The concentration of Early to mid Cretaceous subduction and magmatism along the southeastern limb of the orocline coupled with permissive evidence for transform motion along the southwestern limb suggest that the Alaska orocline formed by the latest Jurassic. Late Cretaceous and Paleocene subduction and magmatism shifted to the southwestern limb of the orocline while strike-slip motion occurred along the southeastern limb, suggesting an orthogonal reorientation of relative plate motion. This Late Cretaceous pattern of plate motion agrees both with a subduction-slip vector of $N28 \pm 8^{\circ}W$ determined from accreted trench deposits on the Kodiak Islands, and the Kula-North America motion predicted by a purely plate-tectonic reconstruction.

Introduction

Unraveling the complex geological records of ancient convergent plate margins requires knowledge of polarity, duration, and extent of individual subductive events. An approximation of the direction of underthrusting is indicated by the relative orientation of paired metamorphic belts (Miyashiro, 1973) and by the position of a magmatic arc with respect to its associated subduction complex. Petrochemical studies of coeval igneous rocks of a magmatic arc show that magmas generated above progressively deeper levels of the Benioff zone show a systematic increase in K_2O/SiO_2 and related chemical ratios (Dickinson, 1975; Gill and Gorton, 1973), and therefore indicate subduction polarity.

These methods for determining polarity of ancient convergent margins provide only broad constraints on the specific slip vector between the down-going plate and the associated magmatic arc. For example, a comparison of patterns of modern volcanism (Katsui, 1971) with present convergence directions (Minster and others, 1974) indicates that volcanism ceases in all cases when the slip vectors approach 35° to the trend of the magmatic arc. Thus, knowledge of polarity may limit convergence direction only within 110° of arc. On the other hand, the structural fabric of a subduction complex preserves evidence of the slip vector that prevailed during its emplacement (Moore and Wheeler, in prep.) and may limit the ancient convergence direction within $20\text{-}30^{\circ}$ of arc. The careful interpretation of these polarity and slip direction indicators from ancient arc and subduction complex pairs provides critical information for placing complex orogenic belts into a plate-tectonic framework.

Repeated pulses of accretion and arc volcanism have occurred along the southern Alaska margin throughout the Phanerozoic. Magmatic activity has been concentrated along the Alaska Peninsula and inland portions of SE Alaska while deep-sea rocks were successively accreted seaward of these areas (Burk, 1965; Plafker, 1972a; Lanphere and Reed, 1973). Based on field work and a synthesis of regional literature (Connelly, 1976) we have attempted to unravel this complex superposition of magmatic arcs, forearc basin deposits, and subduction complexes, and to express them as a sequence of discrete events. We identify three magmatic arc and subduction complex pairs and infer their periods of activity to be Late Triassic to Late Jurassic, Early to mid Cretaceous, and Late Cretaceous to Paleocene. The extent, duration of activity, and polarity of these arc and subduction complex pairs is used in combination with subduction-slip vectors from structural analyses to outline a Mesozoic plate-tectonic framework for the southern Alaska margin. While we are confident about the existence of the arcs and subduction complexes, we recognize that our interpretations for their relative positions and timing may be modified in light of new data.

Each of the three arc and subduction complex

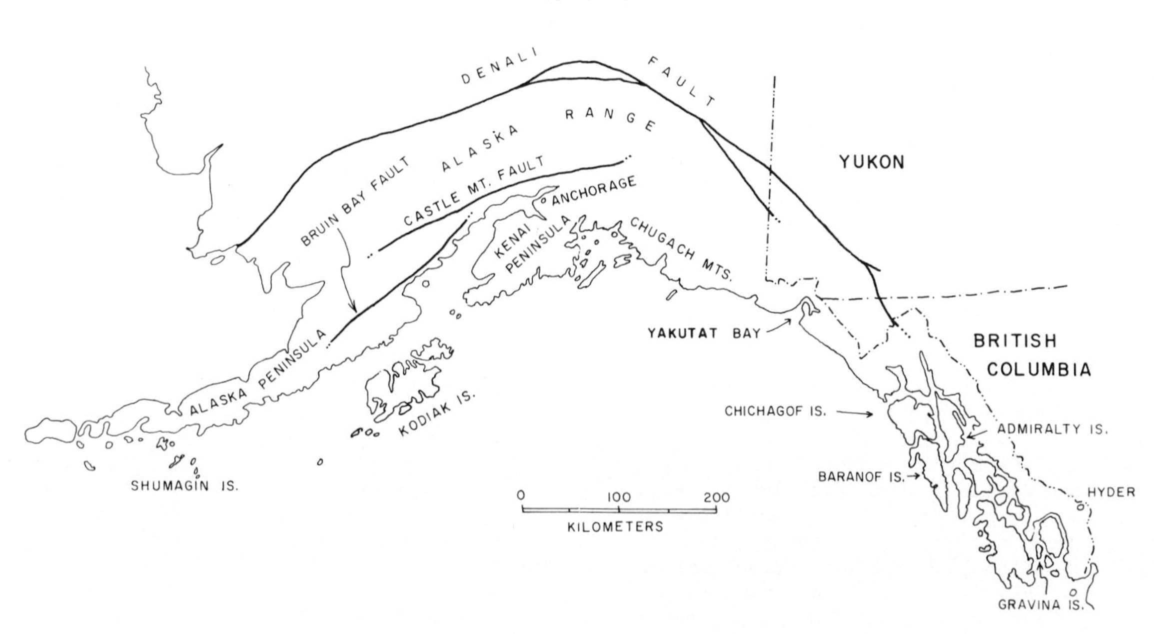

Figure 1. Index map showing distribution of major geographic localities.

pairs is discussed in relation to the Alaska orocline, the large-scale, right-angle flexure of the southern Alaska margin (Fig. 1). For our purposes, the southwestern limb of the orocline comprises the southern margin of southwestern Alaska including the Alaska-Aleutian Range, Alaska Peninsula, Shumagin-Kodiak Shelf, and the Kenai Peninsula-Anchorage area. The southeastern limb of the orocline comprises all of southeastern Alaska, parts of the Yukon, the eastern Alaska Range, and the Chugach Mountains between Valdez and Yakutat.

Talkeetna Arc and Uyak-McHugh Complex: Late Triassic to Late Jurassic

The direction of Late Triassic to Late Jurassic plate convergence can be interpreted from the orientation of slip lines determined from a structural analysis of the Uyak-McHugh complex. This subduction-slip vector combined with the present distribution of the associated Talkeetna magmatic arc constrains possible plate-tectonic interpretations and bears on the origin of the Alaska orocline (Fig. 2).

Geology

Uyak-McHugh Complex. The Uyak-McHugh complex of southwestern Alaska forms a well-defined northeast-trending belt of lithologically chaotic deep-sea rocks that extends over 600 km from the Kodiak Islands (Uyak Complex: Moore, 1969; Connelly, 1976; Connelly and others, this volume) through the Kenai Peninsula and Anchorage regions towards the vicinity of the Copper River (McHugh Complex: Clark, 1973; Beikman, 1974; Moore and Connelly, 1976; Magoon and others, 1976; Cowan and Boss, in prep.). Most of this belt is tectonic melange composed of blocks of ultramafic and gabbroic rocks, greenstone, radiolarian chert, and wacke, enclosed in a matrix of chert and tuffaceous argillite. Rocks are regionally metamorphosed to the prehnite-pumpellyite facies, but tectonic slabs of blueschist and greenschist facies rocks occur locally on the Kodiak Islands and near Seldovia. K-Ar ages of 180 to 190 myBP from the blueschists provide a measure for the time of emplacement of the melange (Forbes and Lanphere, 1973; Carden and others, 1976). Fossils collected in the Uyak Complex (as defined by Connelly, 1976) and McHugh Complex range in age from pre-Upper Permian to Upper Triassic. We interpret the Uyak-McHugh complex as a subduction complex and believe it was emplaced in early Mesozoic time.

Portions of the Kelp Bay Group and unnamed equivalent terranes of Chichagof and Baranof Islands provide the best possibility for the extension of the Uyak-McHugh complex in SE Alaska. The Kelp Bay Group and its equivalents include ultramafic and gabbroic rocks, greenstone, radiolarian chert, phyllite, wacke, and marble in chaotic tectonic juxtaposition (Loney and others, 1975). These rocks are metamorphosed to the prehnite-pumpellyite facies excepting local occurrences of blueschist (Reed and Coats, 1941). Recent collections of radiolaria from the southwestern (seaward) border of the Kelp Bay Group indicate that a portion of this terrane was

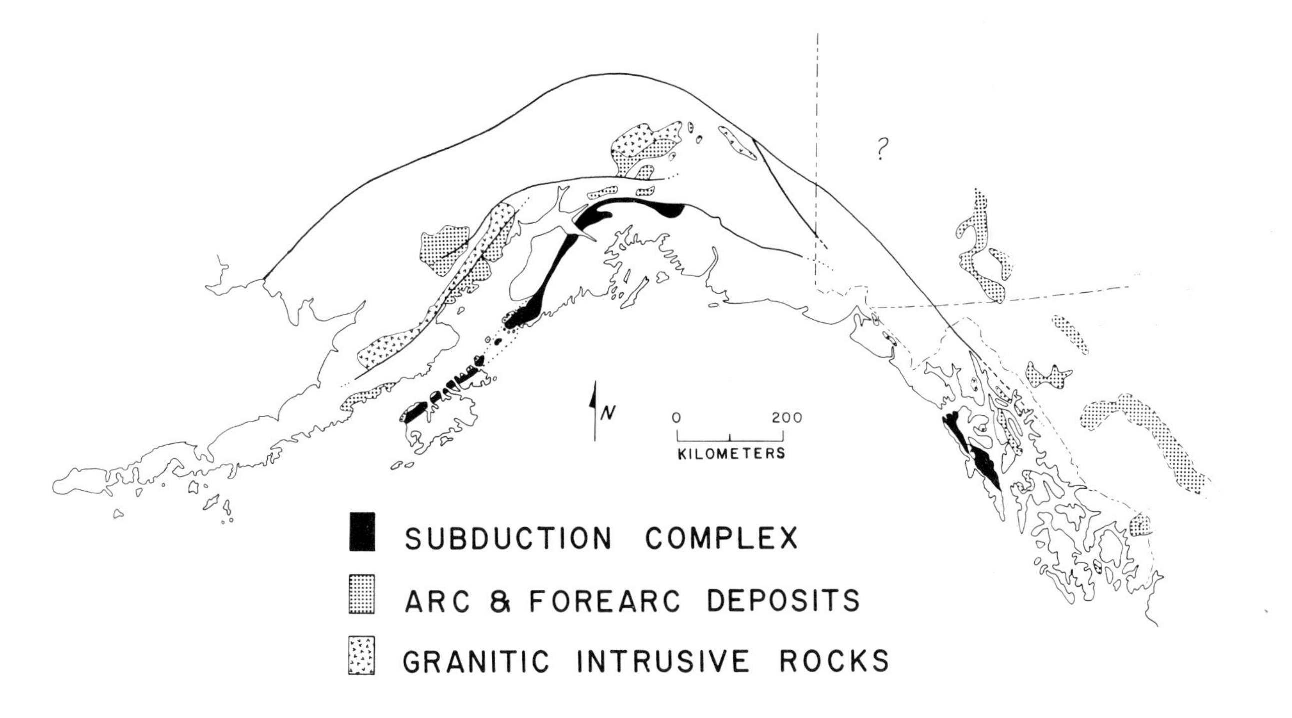

Figure 2. Distribution of Talkeetna magmatic arc, associated forearc deposits, and Uyak-McHugh subduction complex.

accreted in post-Valanginian time (Plafker and others, 1976). However a phyllitic unit located northeast (landward) of the Valanginian rocks is intruded by a mid-Jurassic pluton (Loney and others, 1975) and indicates that this part of the Kelp Bay Group was accreted by the early Mesozoic.

Similarities in lithology, metamorphism, and age of emplacement suggest a correlation between the melanges of southwestern, south-central, and southeastern Alaska described above. Similar terranes may exist in the gap between south-central and SE Alaska, but their delineation awaits further investigation.

Talkeetna Plutonic Arc. The roots of the Talkeetna magmatic arc in SW Alaska are represented by the oldest plutonic rocks of the Alaska-Aleutian Range batholith. This plutonic arc trends approximately parallel to its coeval subduction complex which lies some 140 km to the southeast. Concordant K-Ar ages from the granodioritic and quartz dioritic plutons forming this Jurassic arc range from 176 to 154 myBP (Reed and Lanphere, 1973).

The extension of the Talkeetna magmatic arc to the southeast limb of the orocline is evidenced by the occurrence of plutonic rocks of Late Triassic to Jurassic age in south-central and SE Alaska (Richter and others, 1975; Beikman, 1974, 1975; Loney and others, 1975). In the eastern Alaska Range, dioritic and quartic plutons yield K-Ar ages of 199 to 163 myBP. Plutonic rocks of quartz monzonite and quartz diorite composition crop out on Baranof and Chichagof Islands in SE Alaska and give radiometric ages ranging from 180 to 144 myBP. Taken together, age determinations from plutonic rocks of the composite Talkeetna magmatic arc span from the latest Triassic to the Late Jurassic but most are Early and Middle Jurassic in age.

Talkeetna Volcanic Arc and Associated Forearc Basin Rocks. Sequences of andesitic lava and associated volcaniclastic rocks best document the Talkeetna arc. Along the southwest limb of the orocline these rocks occur on the Alaska Peninsula at Puale Bay and in the Lake Iliamna-Kamishak Bay areas (Kamishak and Talkeetna Formations, Tuxedni Group, and probably the Shelikof and Chinitna Formations: Burk, 1965; Detterman and Hartsock, 1966), in the Talkeetna Mountains and Matanuska Valley region (Talkeetna Formation: Burk, 1965; Grantz and others, 1963), on the Kodiak Islands (Shuyak Formation: Moore and Connelly, 1976), the Barren Islands (Cowan and Boss, in prep.), and on the Kenai Peninsula (Martin, 1915). Biostratigraphically determined ages of these primary andesitic volcanic and volcaniclastic rocks indicate that the Talkeetna arc was active from Upper Triassic through the Middle Jurassic.

Southeastern Alaska and adjacent areas of British Columbia and the Yukon include numerous exposures of Upper Triassic rocks with prominent andesitic components. In SE Alaska these rocks occur in the Juneau and Admirality Island area (Gastineau Volcanic Group and Hyd Formation:

Lathram and others, 1965), on Kupreanof and Kuiu Islands (Hyd Group: Muffler, 1967), and in the Gravina Island area (Nehenta and Chapin Peak Formation: Berg, 1973; Buddington and Chapin, 1929). Similar andesitic volcanic and volcaniclastic rocks occur in northwestern British Columbia (Stuhini Group: Souther, 1971, in press) and in the Yukon (Lewes River Group: Wheeler, 1961). These Upper Triassic sections of SE Alaska and adjacent areas are terminated up-section by unconformities at most localities. However, evidence for volcanism continuing into the Jurassic is preserved in the Hyder area of SE Alaska where Buddington (1929) describes hornblende andesite, tuff, and breccia which have been correlated with widespread Jurassic volcanic sections in adjacent British Columbia (Souther, in press). Other areas that show evidence of continuing volcanism may include the Yukon (Laberge Group: Wheeler, 1961), northwestern British Columbia (Takwahoni Formation: Souther, 1971), and the Queen Charlotte Islands (Yakoun and Maude Formations: Southerland-Brown, 1968).

Summary. We interpret the Uyak-McHugh complex as a subduction complex that was emplaced coeval with volcanism and plutonism in the Talkeetna arc. Furthermore, we believe that this arc and subduction complex pair extends along the Alaska orocline from SW Alaska through south-central Alaska to SE Alaska and beyond. Stratigraphic evidence from SW Alaska shows that the volcanic arc was active from Late Triassic to Middle Jurassic, and radiometric ages from granitic intrusive rocks indicate that plutonism occurred from latest Triassic to Late Jurassic with most activity in the Early and Middle Jurassic. Available ages from blueschists of the Uyak-McHugh complex suggest that at least part of it was emplaced in the Early Jurassic.

Subduction-Slip Vector: Kodiak Islands

We have conducted detailed studies of the Kodiak Islands portion of the Uyak-McHugh complex and found that although lithologically chaotic, it shows a systematic orientation of small-scale structural features (Moore and Wheeler, 1975). Statistical analyses of the internal fabric of this tectonic melange provide important constraits on the early Mesozoic tectonics of the southern Alaska margin.

In the Kodiak Islands, the Uyak melange is composed of competent blocks of variable dimensions enclosed in a foliated cherty argillite matrix. The foliation planes are locally slickensided and oriented statistically parallel to the F_1 fold axial surfaces. The F_1 fold axes are distributed within a plane oriented parallel to fold axial surfaces and the foliation. A few examples of F_2 folds occur cross cutting and kinking the foliation.

Numerous (279) minor folds were measured in 5 large subareas along 200 km of structural strike in the Uyak melange in order to determine slip-planes and slip-lines and thereby outline the kinematics of the melange's emplacement. The best fitting great circle to the girdle of fold axes defines the slip-plane in each subarea (Hansen, 1971, chp. 3). Slip-lines were resolved in each subarea by the separation arc technique (Hansen, 1971, chp. 3) and the method of axial surface fabrics (Scott and Hansen, 1969). The mean slip-line is taken from eight individual slip-lines representing narrow separation arcs (less than 6°) or specific determinations by the method of axial surface fabrics. This mean slip-line plunges steeply to the northwest within the mean slip-plane which strikes northeast along the trend of the melange (Fig. 3). Fold axes with clockwise asymmetry (Hansen, 1971, p. 21) plunge predominantly to the northeast whereas fold axes with counterclockwise asymmetry plunge predominantly to the southwest, indicating that relative movement was northwest over southeast or that underthrusting occurred toward the Talkeetna magmatic arc.

The near-vertical orientation of slip-lines within this melange is inconsistent with the shallowly plunging slip vectors in modern subduction zones at depths less than 60 km (Stauder and Mualchin, 1976). It is probable that the internal structures of this melange have under-

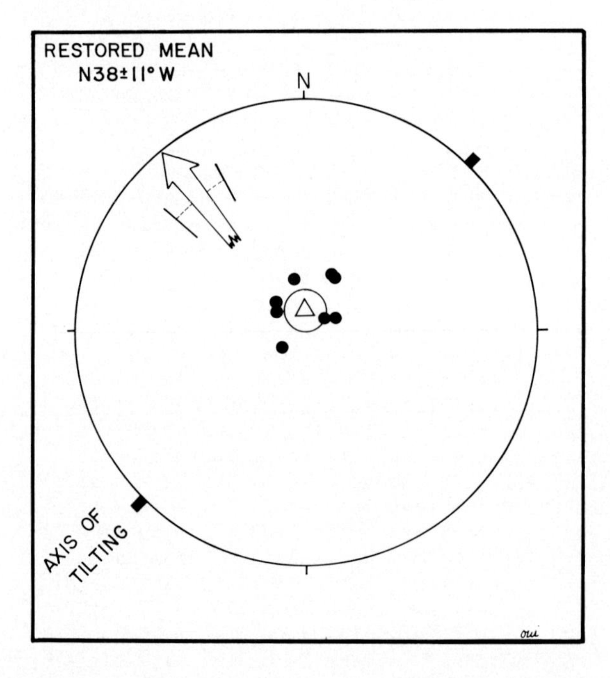

Figure 3. Slip vector determinations from Uyak-McHugh subduction complex (Uyak Complex) on Kodiak Islands. Solid circles represent individual slip vectors; triangle is vector mean; open circle is 95% confidence limit of mean calculated on basis of a Fisherian distribution. Horizontal arrow is azimuth of restored subduction-slip vector.

gone landward tilting as a result of subsequent accretion beneath its seaward margin (Seeley and others, 1974; Karig and Sharman, 1975). Accordingly, the mean slip-line has been rotated around a horizontal axis trending N44°E, or parallel to the structural fabric of the Uyak melange and the outcrop patterns of subsequently accreted deep-sea deposits. The nearly constant width of exposure of these accretionary belts suggests that landward tilting due to underplating was uniform; hence, the axis of tilting is presumed to have been nearly horizontal. The consistency of Uyak slip-lines along strike strongly suggests that they have not been tilted around significantly different axes. The restored mean slip-line is N38±11°W and represents the azimuth of movement between the upper and lower plates of the subduction zone.

It is possible that the mean slip-line (or subduction-slip vector) represents only one component of the total convergence vector and that the remaining component was accommodated by strike-slip movement within the magmatic arc (Fitch, 1972). In order to test this possibility we examined the major faults of the Alaska-Aleutian Range and found that the Bruin Bay fault shows a maximum left-lateral displacement of 19 km since the Late Triassic (Detterman and Hartsock, 1966) and that the Lake Clark-Castle Mountain fault displays a right-lateral offset of 13 km since the Triassic (Ivanhoe, 1962). Since these strike-slip displacements are small relative to the expected magnitude of underthrusting during the Late Triassic to Late Jurassic, the subduction-slip vector is probably a good approximation of the total convergence direction between the oceanic plate and the Talkeetna magmatic arc.

Plate-Tectonic Interpretation

If one accepts the subducton-slip vector as a valid indicator of relative plate movement, then the pole describing this motion would lie on a great-circle oriented normal to the slip vector. With this constraint it is possible to test which pole positions along the great circle yield convergence directions suitable for the production of magma along the composite Talkeetna magmatic arc.

Slip vectors for the convergence between two plates have been calculated for positions adjacent to the Talkeetna arc in SW and SE Alaska (Fig. 4). Since we have no rate information we have used the mean angular velocity (1.0°/my) of modern plate pairs with well-developed subduction zones (Minster and others, 1974) and our calculated velocities only have relative significance. For a pole at position 1 (Fig. 4) convergence would occur in SW Alaska but strike-slip motion would occur parallel to the SE Alaska margin; this strike-slip motion would not produce the observed magmatic arc along SE Alaska. The relative motion generated by pole 2 provides a sub-

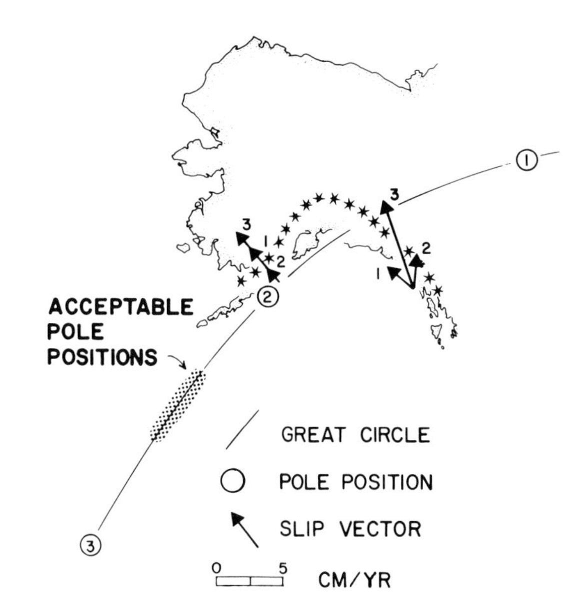

Figure 4. Calculated subduction-slip vectors for localities in SW and SE Alaska during Talkeetna event assuming present geometry of Alaska margin. Star pattern outlines extent of Talkeetna arc. Vectors correspond to respectively numbered pole positions; velocities only have relative significance.

stantial angle of convergence in SE Alaska but a negligible velocity of convergence in SW Alaska, and therefore does not account for the well-developed magmatic arc of the Alaska-Aleutian Range. Motion around pole 3 produces a substantial velocity in both SW and SE Alaska, but because the angle between the slip vector and the trend of the arc in SE Alaska (about 30°) is less than that required to generate magmatism in modern arcs, this pole is also unsatisfactory. With the present geometry of the Alaska margin, acceptable pole locations must be placed between positions 2 and 3 and are limited by the necessity of a sufficient velocity of convergence in SW Alaska and an adequate angle of convergence in SE Alaska. Since poles of rotation tend to change position during geologic time (for example, Pitman and Talwani, 1972), it is unlikely that a pole would remain in this restricted area during the tens of millions of years necessary for the emplacement of the Uyak melange.

In contrast, a wide range of pole positions are possible if the Alaska orocline is straightened (Fig. 5). As the great circle approaches parallelism with the magmatic arc the convergence vectors intersect the arc at even larger angles. With a straightened configuration, acceptable pole positions may be located anywhere beyond either termination of the magmatic arc.

Analogies to modern magmatic arcs strengthen the case for straightening of the orocline.

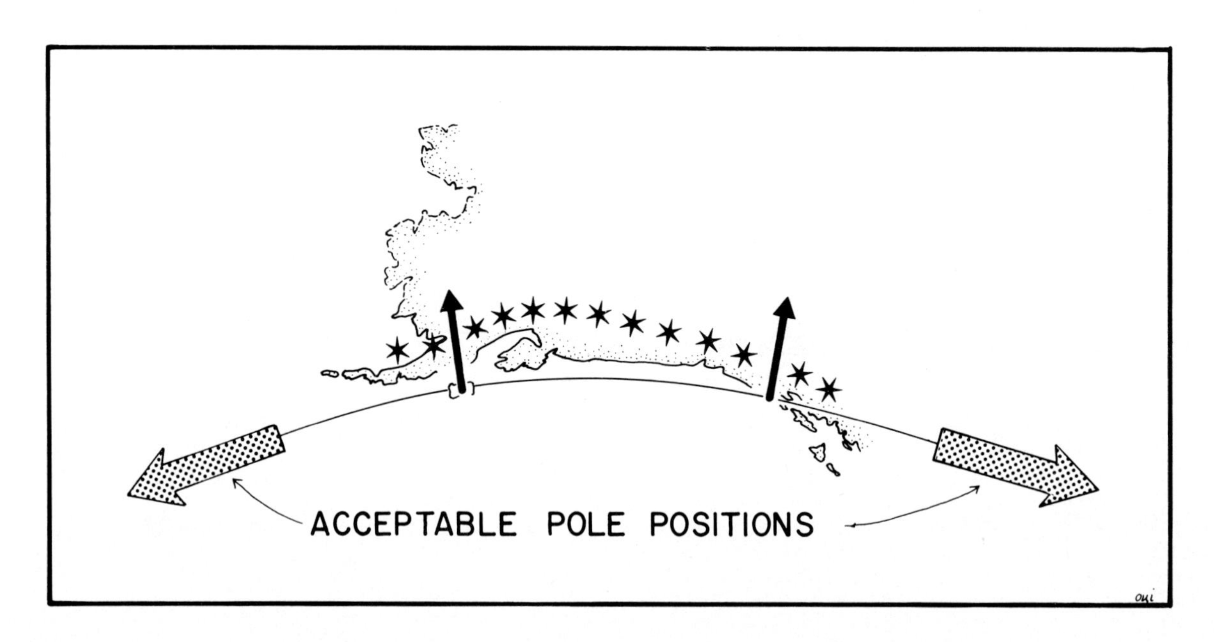

Figure 5. Generalized subduction-slip vectors for localities in SW and SE Alaska during Talkeetna event, assuming straightened Alaska margin. Stars outline extent of Talkeetna arc.

There is no re-entrant as sharp as the Alaska orocline in a convergence zone between two modern plates that has active volcanism along its entire extent. Therefore during magmatism in the Talkeetna arc, the Alaska margin was probably straightened relative to its present configuration.

The preceeding models utilize only two plates to account for the observed distribution of magmatism and the subduction-slip vector. Alternatively these facts might be explained by a three plate model with a spreading center extending into the apex of the orocline. Because we observe no arc-derived ash deposits in the radiolarian cherts (which accumulated over freshly extruded abyssal tholeiites at a spreading center), it is unlikely that a spreading center was close to the subduction zone and nearby arc as required by the three plate model.

Gravina-Nutzotin Arc and Baranof-Yakutat Complex: Early to Mid Cretaceous

Lower and mid Cretaceous rocks along the SE limb of the Alaska orocline evidence extensive magmatism and the accretion of deep-sea deposits (Fig. 6). In contrast, rocks of this age along the southwest limb of the orocline reflect neither significant magmatic nor accretionary activity. These facts provide fundamental constraints for the plate-tectonic interpretation of the southern Alaska margin during the Early to mid Cretaceous.

Geology

The Gravina-Nutzotin magmatic arc (Berg and others, 1972) extends southeastward from south-central Alaska (Nutzotin Mountain sequence and Chisana Formation: Berg and others, 1972; Richter and Jones, 1973) through the Yukon (Dezadeash Group: Kindle, 1953; Muller, 1967) to SE Alaska (Stevens Passage Group and Gravina Island Formation: Buddington and Chapin, 1929; Lathram and others, 1965; Berg, 1973), and consists principally of volcanic-rich sedimentary rocks with associated andesitic volcanic rocks. Although there may be some evidence of volcanism in the Late Jurassic, the main pulse of volcanic activity occurred during the Early Cretaceous. Radiometric ages of 117 to 89 myBP from mainly granodioritic plutons and zoned ultramafic intrusions along the Gravina-Nutzotin belt indicate that plutonism was approximately coeval with volcanism (Richter and others, 1975; Loney and others, 1975; Lanphere and Reed, 1973); many of these plutons are spacially associated with the volcanic arc and are likely cogenetic, while others are separate.

The Baranof-Yakutat complex is composed primarily of a turbidite sequence with minor greenstone and radiolarian chert. This composite tectonic unit extends along the SE Alaska margin from Baranof Island (Sitka Graywacke: Loney and others, 1975; and the seaward portion of the Kelp Bay Group, Plafker and others, 1976) to at least Yakutat Bay (Yakutat Group: Plafker, 1967; Plafker and others, 1976). Generally the rocks are pervasively sheared and complexly folded, and are metamorphosed to the prehnite-pumpellyite and low greenschist facies. Fossil evidence indicates that the Baranof-Yakutat complex was deposited in the Early Cretaceous. These rocks are interpreted as the subduction complex associated with the Gravina-Nutzotin magmatic arc (Berg and

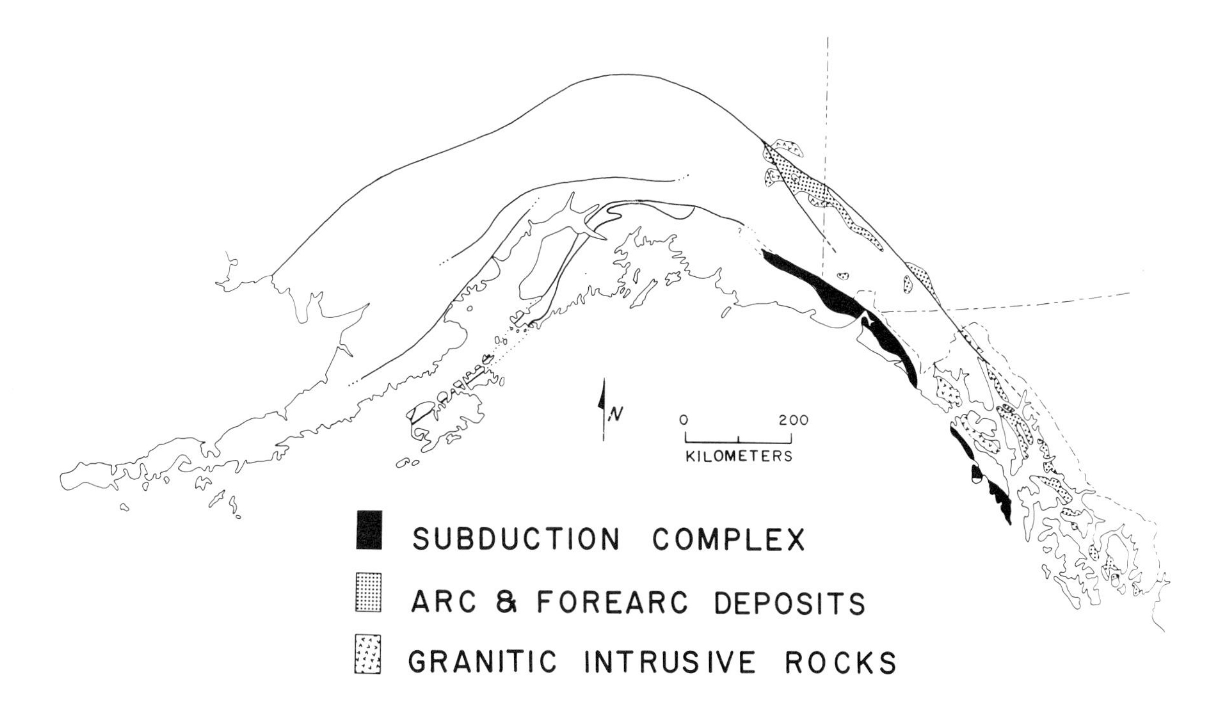

Figure 6. Distribution of Gravina-Nutzotin magmatic arc, associated forearc deposits, and Baranof-Yakutat subduction complex.

others, 1972; Plafker and others, 1976). One occurrence of Campanian rocks along the seaward border of the complex (Plafker and others, 1976) post-dates the cessation of Gravina-Nutzotin magmatism and may have accumulated in an inactivated trench or slope basin.

No volcanic or plutonic rocks of latest Jurassic to mid Cretaceous age are known to occur along the southwest limb of the Alaska orocline (Burk, 1965; Reed and Lanphere, 1973). Instead this was a period of quiescence when pre-existing plutonic rocks were uplifted and eroded to produce a thick sequence of arkosic sandstone and conglomerate along the Alaska Peninsula and in the Cook Inlet area (Naknek and Staniukovich Formations: Burk, 1965; Detterman and Hartsock, 1966). Only one small occurrence of possible Lower Cretaceous rocks are known from the accreted deep-sea terranes seaward of these sites of arkosic sedimentation (Cape Current terrane; Connelly, 1976). The absence of a magmatic arc and a significant volume of accreted deep-sea rocks of latest Jurassic to mid Cretaceous age along the southwest limb of the orocline closely constrains possible tectonic interpretations.

Plate-Tectonic Interpretation

The northeastward polarity of Early to mid Cretaceous underthrusting along the southeast limb of the orocline is indicated by the location of the magmatic arc to the northeast of the associated subduction complex. The absence of magmatism and significant accretion along the southwest limb of the orocline during this time interval suggests that appreciable convergence did not occur along that margin. If the Alaska orocline formed by the latest Jurassic, then convergence could have occurred along its southeast limb while simultaneous left-lateral transform faulting occurred along its southwest limb (Fig. 7). Moreover, a transform fault just seaward of the pre-existing Uyak-McHugh complex on the southwest limb would have conveyed sediments from that margin and accreted them along the southeast limb and thus explain the paucity of Upper Jurassic to mid Cretaceous clastic rocks in the accretionary deposits of the southwest limb. The consistency of structural data (Moore and Wheeler, in prep.) and lack of demonstrated offsets within the Uyak-McHugh complex suggests that internal rotations or translations did not disturb its existing structural fabric during this phase of transform faulting.

Iliamna-McKinley Arc and Shumagin-Valdez Complex: Late Cretaceous to Paleocene

During the Late Cretaceous, the sites of magmatic activity and the locus of accretion of deep-sea rocks switched from the southeast limb to the southwest limb of the orocline (Fig. 8). This shift in geologic symmetry apparently was in response to a nearly orthogonal reorientation in relative plate motion.

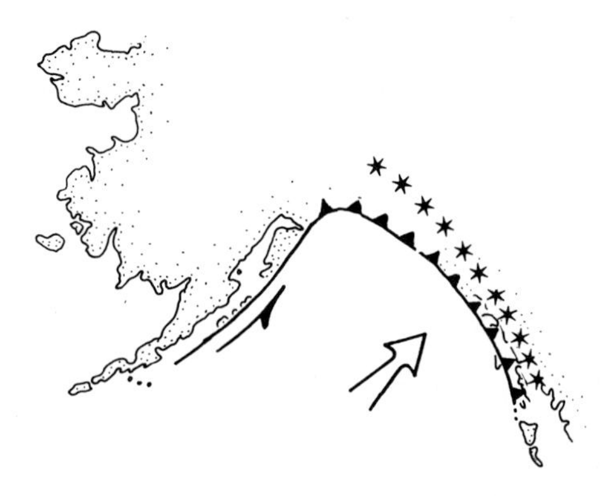

Figure 7. Plate-tectonic interpretation of Gravina-Nutzotin event. Stars and thrust symbol respectively indicate magmatic arc and subduction zone.

Geology

A Late Cretaceous to Paleocene arc and subduction complex extends from south-central Alaska to the southwest along the trend of the Alaska Peninsula. The Iliamna-McKinley magmatic arc is defined by the quartz dioritic and granodioritic plutons of the Alaska-Aleutian Range batholith which yield concordant K-Ar ages of 83 to 58 myBP (Reed and Lanphere, 1973). The Shumagin-Valdez complex is an Upper Cretaceous (Maastrichtian) turbidite sequence that forms a northeast-trending accretionary belt some 150 to 200 km seaward of this magmatic arc along the Shumagin-Kodiak Shelf (Shumagin Formation: Burk, 1965; Kodiak Formation: Moore, 1969) and the Kenai Peninsula and Anchorage areas (Valdez Group: Jones and Clark, 1973). It is characterized by seaward overturned folding and is locally deformed in the style of tectonic melange. Based on paleocurrent patterns, style of deformation, stratigraphy, and regional geology, this turbidite sequence has been interpreted as a trench and associated deep-sea sequence which was accreted to the SW Alaska margin during the Late Cretaceous and earliest Tertiary (Plafker, 1972a; Clark, 1973; Jones and Clark, 1973; Moore, 1973; Budnik, 1974). We interpret this accretionary belt as a subduction complex that was coeval with the Iliamna-McKinley magmatic arc to the northwest.

A complimentary magmatic arc of Late Cretaceous and Paleocene age apparently does not occur along the southeast limb of the orocline. No Late Cretaceous volcanic or plutonic rocks are known to post-date the 90 to 100 myBP cessation of Gravina-Nutzotin magmatism (Buddington and Chapin, 1929; Jones and MacKevett, 1969; Lanphere and Reed, 1973; Richter and others, 1975). Similarly, significant volumes of accreted deep-sea rocks do not generally occur along the southeast limb of the orocline.

Subduction-Slip Vector: Kodiak Islands

Numerous shear zones and associated broken formations pervade the Shumagin-Valdez complex.

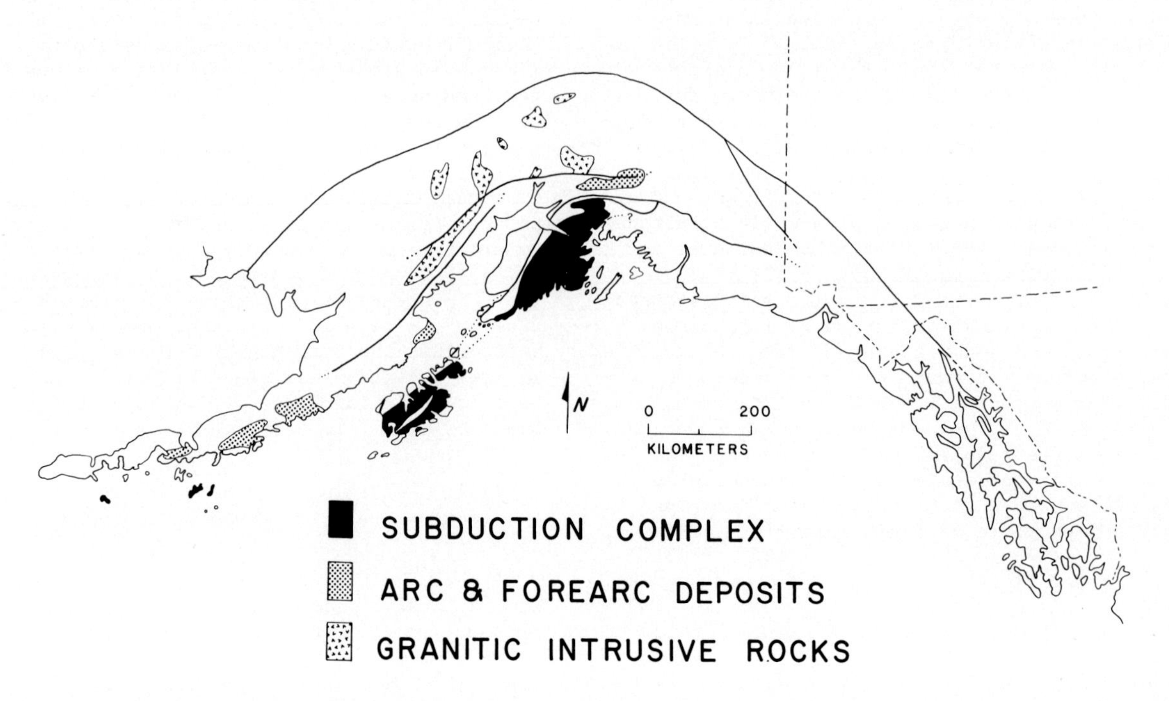

Figure 8. Distribution of Iliamna-McKinley magmatic arc, associated forearc deposits, and Shumagin-Valdez subduction complex.

In the Kodiak Islands, this complex underthrusts the older Uyak-McHugh complex along a well-defined fault. Adjacent to this fault the underthrust turbidite sequence is transformed into a kilometer-thick broken formation, whereas the more rigid overthrust plate is over-printed by a fracture cleavage inclined at about 45° to the thrust surface. This broken formation is identical in structural style to those occurring throughout the Shumagin-Valdez complex and apparently developed during the initial phase of underthrusting. The locally slaty nature of the fracture cleavage suggests that it formed perpendicular to the direction of maximum finite shortening, or parallel to the X-Y plate of the strain ellipsoid (Ramsay, 1967). As such, the cleavage would intersect the thrust surface along a line parallel to the Y-axis of the strain ellipsoid. The intersection of the plane normal to the Y-axis (that is, the X-Z plane) with the thrust surface (or the slip plane) defines the probable slip direction along the thrust surface (Moore, in prep.).

We have studied the thrust separating the Shumagin-Valdez complex from the older rocks along some 70 km of its extent in the Kodiak Islands. Using orientations of the thrust surface taken from map patterns along with individual cleavage measurements, we have utilized the technique outlined above to construct a series of slip-lines (Fig. 9). These slip-lines record the direction of movement along the thrust surface during emplacement of the complex. The mean slip-line has been restored to its presumed original azimuth by rotating around an axis of tilting which is parallel to the trend of the Shumagin-Valdez complex and subsequently accreted rocks in the Kodiak Islands. The restored mean azimuth of the subduction-slip vector is N28±8°W and represents the apparent convergence direction recorded by the hanging-wall of the subduction zone.

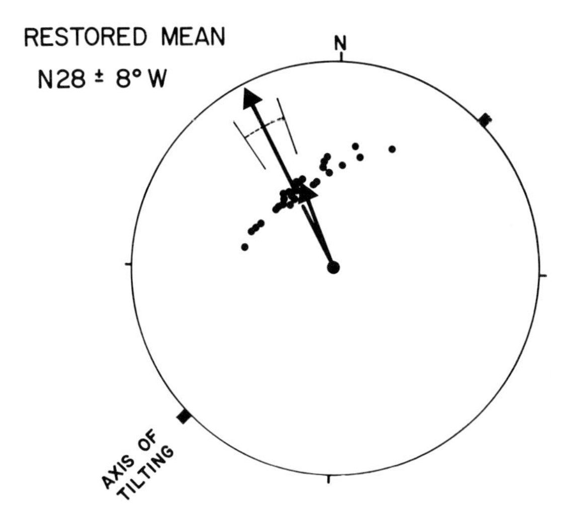

Figure 9. Slip vector determinations from Shumagin-Valdez subduction complex (Kodiak Formation) on Kodiak Islands. Solid circles represent individual slip vectors; inclined arrow is vector mean; horizontal arrow is restored azimuth of subduction-slip vector with arc indicating 95% confidence limit of mean (based on planar normal distribution).

It is possible that the derived subduction-slip vector represents but one component of the total convergence vector. However, the lack of large strike-slip offsets on major faults landward of the subduction complex suggests that the subduction-slip vector closely approximates the relative motion between the oceanic plate and the Iliamna-McKinley arc.

Plate-Tectonic Interpretation

The Iliamna-McKinley magmatic arc is flanked to the southeast by the coeval Shumagin-Valdez subduction complex, thus indicating that underthrusting was generally to the northwest. This determination of polarity is in agreement with the N28°W subduction-slip vector derived from structural studies. Both the absence of magmatism along the southeast limb of the orocline and the orientation of the Kodiak subduction-slip vector suggest that SE Alaska was primarily a zone of right-lateral strike-slip faulting during the Late Cretaceous (Fig. 10).

The plate-tectonic reconstruction of Cooper and others (1976) is based on marine magnetic anomaly and hotspot data (Fig. 10). Their calculated Kula-North American slip vector for the Bering Shelf edge of N23°W is within the error limits of our slip vector and is plate-tectonically consistent providing that the Kula-North American pole was located a substantial distance from both slip vectors. The absence of magmatism throughout SE Alaska in the Late Cretaceous suggests that the Kula-Farallon-North American triple junction and the Farallon-North American subduction zone was located south of this region and therefore did not complicate the two plate configuration. The agreement of Cooper and others' (1976) reconstruction with ours is particularly significant since each approach utilizes different data bases and makes substantial and different assumptions.

Summary and Discussion

Active subduction and arc magmatism occurred along most of the southern Alaska margin during the Late Triassic through early Late Jurassic. The extent of magmatism coupled with an estimated convergence vector and analogies to modern plate boundaries all suggest that the Alaska orocline was significantly straightened relative to its present configuration. The bending of Alaska must have occurred before the Early Cretaceous restriction of subduction and magmatism to the southeast limb of the orocline,

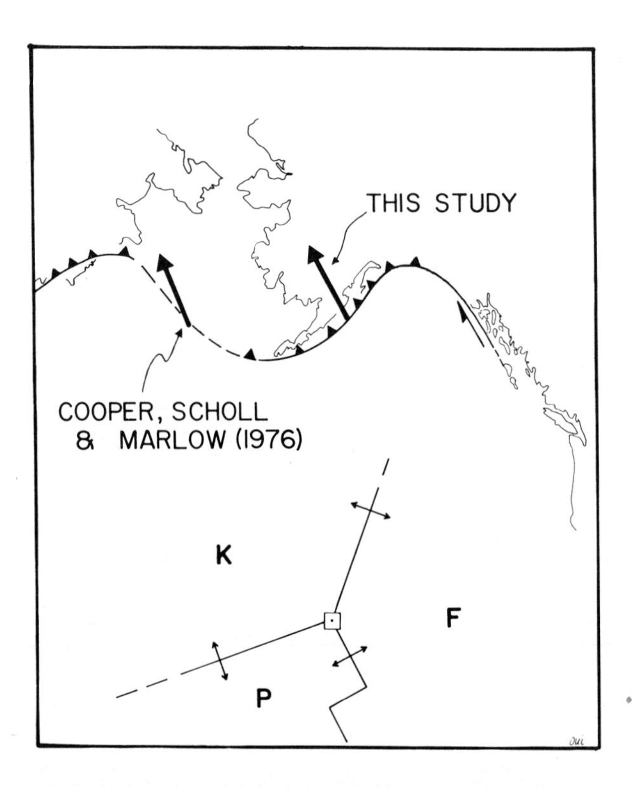

Figure 10. Plate-tectonic interpretation of southern Alaska margin during Iliamna-McKinley event. Plate geometry after Cooper and others (1976). Arrows indicate relative motion between Kula and North American plates.

probably in the latest Jurassic. During the Early and mid Cretaceous convergence along the southeast limb of the orocline was accompanied by left-lateral strike-slip motion along the seaward margin of the southwest limb. Following an orthogonal reorganization in relative plate motion, Late Cretaceous and earliest Tertiary subduction and magmatism was localized along the southwest limb of the orocline while right-lateral transform motion occurred along the southeast limb.

Several important strike-slip faults occur along the southeast limb of the orocline and have displaced rock units used in our analysis up to several hundred kilometers since the Middle Triassic (Plafker, 1972b; Ovenshine and Brew, 1972). However since these displacements are small relative to the length of the southeast limb they do not effect our interpretations.

Based on biostratigraphic and lithologic correlations we argue for the continuity of the Alaska margin since the Late Triassic, but cannot conclusively state whether or not it was physically attached to interior Alaska. Available paleomagnetic evidence suggest northward transport of the Alaska margin since Middle Triassic (Packer and Stone, 1974; Jones and others, 1976). However the lack of geographic and stratigraphic control in sampling and the lack of resolution at individual sampling sites prevent a rigorous paleomagnetic test of the continuity of the orocline, its time of bending, or its relationship to interior Alaska.

If the present-day margin of southern Alaska was attached to interior Alaska in the early Mesozoic then there are several possible explanations for the oroclinal bending. The bending could be related to the opening of the Yukon-Koyukuk basin (Carey, 1958; Patton, 1973) or some subsequently collapsed back-arc basin. It could also be related to compressional effects resulting from the collision between the North American and Eurasian plates along the Cherskiy fold belt and suture zone in Siberia (Churkin, 1973). Alternatively, if the southern Alaska margin was transported northward independent of the interior Alaska, then the oroclinal bending may reflect the molding (Csjetey, 1976) of the early Mesozoic arc and subduction complex against an older pre-existing continental margin.

Recently collected chert from the Uyak Complex of the Kodiak Islands has yielded Lower Cretaceous (upper Valanginian to Hauterivian) radiolaria; this necessitates two modifications of the above interpretations. 1) The age of emplacement of the Uyak-McHugh complex is now uncertain, but is definitely younger than the early Mesozoic age previously suspected. Because there was a hiatus in magmatism along the Alaska-Aleutian Range batholith between the early Mesozoic Talkeetna event and the Late Cretaceous Iliamna-McKinley event, we believe no subduction or accretion occurred at this time. This suggests that the Uyak-McHugh and Shumagin-Valdez complexes represent two phases (or facies?) of Late Cretaceous accretion. It is noteworthy that slip-vectors determined from each complex are indistinguishable at the 95% confidence level. 2) The Early Jurassic blueschist-bearing schist terranes previously included with the Uyak-McHugh must be genetically unrelated to the Uyak-McHugh. These schists must instead represent the only vestige of the subduction complex emplaced along southwestern Alaska during early Mesozoic Talkeetna magmatism. The confinement of the schists to a narrow zone along the northwest border of the Uyak-McHugh belt supports this interpretation.

Recent field work in the Kelp Bay Group of Baranof and Chichagof Islands by J. Carden, Wm. Connelly, and R. Forbes has substantiated the local occurrence of blueschist assemblages in the schists and phyllites exposed at Sister Lake (which appear to be an extension of the Pinnacle Peak Phyllite of Loney and others, 1975). These schists are being radiometrically dated at the University of Alaska.

Acknowledgments. We thank our co-workers, Betsy Hill, Malcolm Hill, and Jim Gill for their many contributions during our joint study of the lower Mesozoic rocks of Kodiak Islands and adjacent areas. Discussions with Henry Berg, George Plafker, David Jones, David Brew, and Robert Loney have been particularly helpful in developing our understanding of the geology of southeastern Alaska. We thank Alan Cooper for providing us with a pre-print of his paper. Ronald Bruhn, George Plafker, David Jones, and Mike Churkin provided many constructive comments on an early draft of this paper. This work was supported by the National Science Foundation (Grant GA-43266), the Alaskan Branch of the U.S. Geological Survey, and Exxon Production Research.

References

Beikman, H. M., Preliminary geologic map of the southeast quadrant of Alaska, U.S. Geol. Survey Misc. Field Studies Map MF612, Open-file, 1974.

Beikman, H. M., Preliminary geologic map of southeastern Alaska, U.S. Geol. Survey Misc. Field Studies Map MF673, Open-file, 1975.

Berg, H. C., Geology of Gravina Island, Alaska, U.S. Geol. Survey Bull. 1373, 41 p., 1973.

Berg, H. C., D. L. Jones, and D. H. Richter, Gravina-Nutzotin belt - Tectonic significance of an upper Mesozoic sedimentary and volcanic sequence in southern and southeastern Alaska, in Geological Survey Research 1972, U.S. Geol. Survey Prof. Paper, 800-D, D1-D24, 1972.

Buddington, A. F., Geology of Hyder and vicinity, southeastern Alaska, U.S. Geol. Survey Bull., 807, 124 p., 1929.

Buddington, A. F., and T. Chapin, Geology and mineral resources of southeastern Alaska, U.S. Geol. Survey Bull., 800, 398 p., 1929.

Budnik, R., The geologic history of the Valdez Group, Kenai Peninsual, Alaska: Deposition and deformation at a Late Cretaceous consumptive plate margin, Ph.D. thesis, Univ. of Calif., Los Angeles, 139 p., 1974.

Burk, C. A., Geology of the Alaska Peninsula - Island arc and continental margin, Geol. Soc. Amer. Memoir, 99, 250 p., 1965.

Carden, J. R., Wm. Connelly, R. B. Forbes, and D. L. Turner, Blueschists of the Kodiak Islands, Alaska: An extension of the Seldovia blueschist terrane, Geol. Soc. Amer. Absts. with Programs, 8 (in press), 1976.

Carey, S. W., The tectonic approach to continental drift, in Continental Drift - A sumposium, Carey, S. W. (ed.), Hobart, Tasmania Univ. Geology Dept., 177-355, 1958.

Churkin, M., Jr., Paleozoic and Precambrian rocks of Alaska and their role in its structural evolution, U.S. Geol. Survey Prof. Paper, 740, 64 p., 1973.

Clark, S. H. B., The McHugh Complex of south-central Alaska, U.S. Geol. Survey Bull., 1372-D, D1-D11, 1973.

Connelly, Wm., Mesozoic geology of the Kodiak Islands and its bearing on the tectonics of southern Alaska, Ph.D. thesis, Univ. of Calif., Santa Cruz, 1976.

Connelly, Wm., M. Hill, B. B. Hill, and J. C. Moore, The Uyak Complex, Kodiak Islands, Alaska: A subduction complex of early Mesozoic age, in Ewing Symposium Volume on Problems in the Evolution of Island Arcs, Deep Sea Trenches and Back-Arc Basins (this volume), 1976.

Cooper, A. K., D. W. Scholl, and M. S. Marlow, A plate-tectonic model for the evolution of the eastern Bering Sea Basin, Geol. Soc. Amer. Bull., 87, 1119-1126, 1976.

Csejtey, B., Jr., Tectonic implications of late Paleozoic volcanic arc in the Talkeetna Mountains, south-central Alaska, Geology, 4, 49-52, 1976.

Detterman, R. L., and J. K. Hartsock, Geology of the Iniskin-Tuxedni region, Alaska, U.S. Geol. Survey Prof. Paper, 512, 78 p., 1966.

Dickinson, W. R., Potash-depth (K-h) relations in continental margin and intra-oceanic magmatic arcs, Geology, 3, 53-56, 1975.

Fitch, T. J., Plate convergence, transcurrent faults, and internal deformation adjacent to southeast Asia and Western Pacific, J. Geophys. Res., 23, 4432-4460, 1972.

Forbes, R. B., and M. A. Lanphere, Tectonic significance of mineral ages of blueschists near Seldovia, Alaska, J. Geophys. Res., 78, 1383-1386, 1973.

Gill, J. B., and M. Gorton, A proposed geological and geochemical history of eastern Melanesia, in The Western Pacific: Island Arcs, Marginal Seas, Geochemistry, Coleman, P. (ed.), Perth, Univ. of Western Australia Press, 483-496, 1973.

Grantz, A., H. Thomas, T. W. Stern, and N. B. Sheffey, Potassium-argon and lead-alpha ages for stratigraphically bracketed plutonic rocks in the Talkeetna Mountains, Alaska, U.S. Geol. Survey Prof. Paper, 475B, 56-59, 1963.

Hansen, E., Strain Facies, Springer-Verlag, New York, 207 p., 1971.

Ivanhoe, L. F., Right-lateral strike-slip movement along the Lake Clark fault, Alaska, Geol. Soc. Amer. Bull., 73, 911-912, 1962.

Jones, D. L., and S. H. B. Clark, Upper Cretaceous (Maestrichtian) fossils from the Kenai-Chugach Mountains, Kodiak and Shumagin Islands, southern Alaska, U.S. Geol. Survey Jour. Res., 1, 125-136, 1973.

Jones, D. L., and E. M. MacKevett, Jr., Summary of Cretaceous stratigraphy in part of the McCarthy quadrangle, Alaska, U.S. Geol. Survey Bull., 1274-K, K1-K19, 1969.

Jones, D. L., E. A. Pessagno, and B. Csejtey, Jr., Significance of the Upper Chulitna ophiolite for the Late Mesozoic evolution of southern Alaska, Geol. Soc. Amer. Absts. with Programs (Cordilleran Sec.), 8, 385-386, 1976.

Karig, D. E., and C. F. Sharman, III, Subduction and accretion of trenches, Geol. Soc. Amer. Bull., 86, 377-389, 1975.

Katsui, Y., List of world active volcanoes, special issue, Bulletin of Volcanic Eruptions,

Volcanic Society of Japan, 160 p., 1971.
Kindle, E. D., Dezadeash map-area, Yukon Territory, Canada Geol. Survey Mem. 268, 68 p., 1953.
Lanphere, M. A., and B. L. Reed, Timing of Mesozoic and Cenozoic plutonic events in circum-Pacific North America, Geol. Soc. Amer. Bull., 84, 3773-3782, 1973.
Lathram, E. H., J. S. Pomeroy, H. C. Berg, and R. A. Loney, Reconnaissance geology of Admiralty Island, Alaska, U.S. Geol. Survey Bull., 1181-R, 48 p., 1965.
Loney, R. A., D. A. Brew, L. J. P. Muffler, and J. S. Pomeroy, Reconnaissance geology of Chichagof, Baranof, and Kruzof Islands, southeastern Alaska, U.S. Geol. Survey Prof. Paper, 792, 105 p., 1975.
Magoon, L. B., W. L. Adkison, J. A. Wolfe, J. S. Kelley, and D. L. Jones, Geologic framework of lower Cook Inlet, Alaska, with emphasis to onshore geology, Amer. Assoc. Pet. Geol. Abs. with Programs, Pacific Section, 12-13, 1976.
Martin, G. C., The western part of the Kenai Peninsula, U.S. Geol. Survey Bull., 537, 41-112, 1915.
Minster, J. B., T. H. Jordan, P. Molnar, and E. Haines, Numerical modelling of instantaneous plate tectonics, Geophys. J. R. Astr. Soc., 36, 541-576, 1974.
Miyashiro, A., Metamorphism and Metamorphic Belts, John Wiley and Sons, New York, 492 p., 1973.
Moore, G. W., New formations on Kodiak and adjacent islands, Alaska, U.S. Geol. Survey Bull., 1274-A, A28-A35, 1969.
Moore, J. C., Cretaceous continental margin sedimentation, southwestern Alaska, Geol. Soc. Amer. Bull., 84, 595-614, 1973.
Moore, J. C., and Wm. Connelly, Subduction, arc volcanism, and forearc sedimentation during the early Mesozoic, S.W. Alaska, Geol. Soc. Amer. Absts. with Programs (Cordilleran Sec.), 8, 397-398, 1976.
Moore, J. C., and R. L. Wheeler, Orientation of slip during early Mesozoic subduction, Kodiak Islands, Alaska, Geol. Soc. Amer. Absts. with Programs, 7, 306, 1975.
Muffler, L. J. P., Stratigraphy of the Keku Inlets and neighboring parts of the Kuiu and Kupreanof Islands, southeastern Alaska, U.S. Geol. Survey Bull., 1241-C, 52 p., 1967.
Muller, J. E., Kluane Lake map-area, Yukon Territory, Canada Geol. Survey Mem. 340, 137 p., 1967.
Ovenshine, A. T., and D. A. Brew, Separation and history of the Chatham Strait fault, southeast Alaska, North America (abs.), Internat. Geol. Cong., 24th, Montreal, Canada, Proc., 245-254, 1972.
Packer, D. R., and D. B. Stone, Paleomagnetism of Jurassic rocks from southern Alaska, and the tectonic implications, Canadian J. Earth Sci., 11, 976-997, 1974.
Patton, W. W., Jr., Reconnaissance geology of the Northern Yukon-Koyukuk Province, U.S. Geol. Survey Prof. Paper, 774-A, A1-A17, 1973.
Pitman, W. C., and M. Talwani, Sea-floor spreading in the North Atlantic, Geol. Soc. Amer. Bull., 83, 619-646, 1972.
Plafker, G., Geologic map of the Gulf of Alaska Tertiary Province, Alaska, U.S. Geol. Survey Misc. Geol. Inv. Map I-484, 1967.
_______, New data on Cenozoic displacements along the Fairweather fault system, Alaska, in Faults, fractures, lineaments, and related mineralization in the Canadian cordillera: Geol. Assoc. Canada, Cordilleran Section, Programme and abstracts, 30, 1972b.
_______, Alaskan earthquake of 1964 and Chilean earthquake of 1960: Implications for arc tectonics, J. Geophys. Res., 77, 901-925, 1972a.
Plafker, G., D. L. Jones, T. Hudson, and H. C. Berg, The Border Ranges fault system in the Saint Elias Mountains and Alexander Archipelago, U.S. Geol. Survey Circular, 733, 14-16, 1976.
Ramsay, J. G., Folding and Fracturing of Rocks, McGraw-Hill Book Company, New York, 568 p., 1967.
Reed, B. L., and M. A. Lanphere, Alaska-Aleutian Range Batholith: geochronology, chemistry, and relation to circum-Pacific plutonism, Geol. Soc. Amer. Bull., 84, 2583-2610, 1973.
Reed, J. C., and R. R. Coats, Geology and ore deposits of the Chichagof mining district, Alaska, U.S. Geol. Survey Bull., 929, 148 p., 1941.
Richter, D. H., and D. L. Jones, Structure and stratigraphy of eastern Alaska Range, Alaska, Amer. Assoc. Pet. Geol., 2nd Internat. Symposium Artic Geology, Mem., 19, 408-420, 1973.
Richter, D. H., M. A. Lanphere, and N. A. Matson, Jr., Granitic plutonism and metamorphism, eastern Alaska Range, Alaska, Geol. Soc. Amer. Bull., 86, 819-829, 1975
Scott, W. H., and E. C. Hansen, Movement directions and the axial plane fabrics of flexure folds, Annual Rept. of the Director, Geophys. Lab., 1967-1968, Carnegie Inst., Wash., 254-259, 1969.
Seeley, D. R., P. R. Vail, and G. G. Walton, Trench slope model, in The geology of continental margins, Burk, C. A., and Drake, C. A. (eds.), Springer-Verlag, New York, 249-260, 1974.
Souther, J. G., Geology and mineral deposits of Tulsequah map-area, British Columbia, Geol. Survey Canada Mem., 362, 84, 1971.
Souther, J. G., Volcanism and tectonic environments in the Canadian Cordillera - A second look, Geol. Assoc. Canada Special Volume (in press), 1976.
Stauder, W., and L. Mualchin, Fault motion in the larger earthquakes of the Kurile-Kamchatka Arc and of the Kurile-Hokkaido corner, J. Geophys. Res., 81, 297-308, 1976.
Sutherland-Brown, A., Geology of the Queen Charlotte Island, British Columbia, B. C. Dept. Mines, Bull., 54, 226 p., 1968.
Wheeler, J. O., Whitehorse map-area, Geol. Survey Canada Mem., 312, 1961.

MULTICHANNEL SEISMIC STUDY IN THE VENEZUELAN BASIN AND THE CURACAO RIDGE

Manik Talwani[1], Charles C. Windisch, Paul L. Stoffa, Peter Buhl and Robert E. Houtz

Lamont-Doherty Geological Observatory of Columbia University, Palisades, NY 10964

Preface. Maurice Ewing's work was marked by his success at developing new experimental techniques of looking at and below the ocean floor and using the results to gain new insights into the geology of the oceans. Multichannel seismic reflection at sea is a powerful experimental technique and it is only just beginning to be applied to problems of basic geology. We believed that the description of the results of our first multichannel seismic reflection experiments in the Caribbean would be a fitting tribute to Maurice Ewing.

Introduction

In spite of the fact that seismic reflection and refraction studies have been made in the Caribbean over a period of more than twenty years, fundamental questions about the distribution of sediments and the crustal structure of the Caribbean Sea still remain unanswered. A basic problem stems from the fact that earlier single channel seismic reflection studies using small sound sources were unable to map basement in the Colombian and Venezuelan Basins. Therefore the distribution and thickness of the sediments could not be established with confidence and the nature of the boundaries of these basins were also not fully understood. Our multichannel seismic results, by being able to map basement can make a valuable contribution to the solution of some of these problems. In addition the large sound source (used in conjunction with the multichannel work) proved also very useful in obtaining crustal and mantle refractions with sonobuoys. These results shed additional light on the complex question of crustal and mantle velocities in the Caribbean.

Cruise Plan

For our first multichannel seismic (MCS) experiment we selected sites off the east coast of the U.S., in the Puerto Rico Trench area, in the Caribbean Sea and in the area of the Lesser Antilles. The track of the CONRAD Cruise 19 during which multichannel data were obtained is shown in Figure 1. In this paper we discuss only the results obtained on lines 8, 9, and 10. Lines 8 and 10 were run from the center of the Venezuelan Basin to the Curacao Ridge. Line 9 was run roughly perpendicular to lines 8 and 10 and lies principally in the Los Roques Trench and on the Curacao Ridge.

Results

Reflector A" and the A"-B" Interval

In single channel seismic reflection measurements, two prominent reflectors, A" and B" stand out in the Caribbean Sea. When the margins of the Caribbean Basins are excluded, the thickness of the sediments above A" and within the A"-B" interval is remarkably constant. Edgar and others (1971) give these values at 0.55 sec and 0.4 sec double reflection time respectively. Our MCS profiles, however, extend to the southern and eastern margin of Venezuelan Basin and show large variations in thicknesses of sediment thickness above A" as well as in the A"-B" interval.

We must first identify layers A" and B" along our MCS lines 8 and 10. We discuss identification of B" in detail in the next section. A" can be identified at point A (see Figure 1 for location), which is close to the intersection of the single channel CONRAD-9 line and the MCS line 8 (Figure 2). We note the difference in appearance of A" on the single channel and in the MCS profile. We also call attention to the point that on the single channel record there appears to be an offset in A" shortly after 2200 hours. Sonobuoy 18 was run close to point A but it failed to pick up reflector A" which must lie within the part of the sediment column that has an average velocity of 1.97 km/sec obtained from wide angle reflections. In the compressed scale seismic sections (Figures 2 and 3) A" appears to lie below a sediment section that has a "scalloped" appearance. Horizon A" has been identified as a chert horizon of Middle Eocene age from samples obtained by piston coring (Talwani and others, 1966; Edgar and others, 1971). Saunders and others (1973) concluded on the basis of deep drilling results that "Horizon A" in the Caribbean results from

[1]Also Department of Geological Sciences, Columbia University.

Figures 2, 3, and 4 appear after page 86.

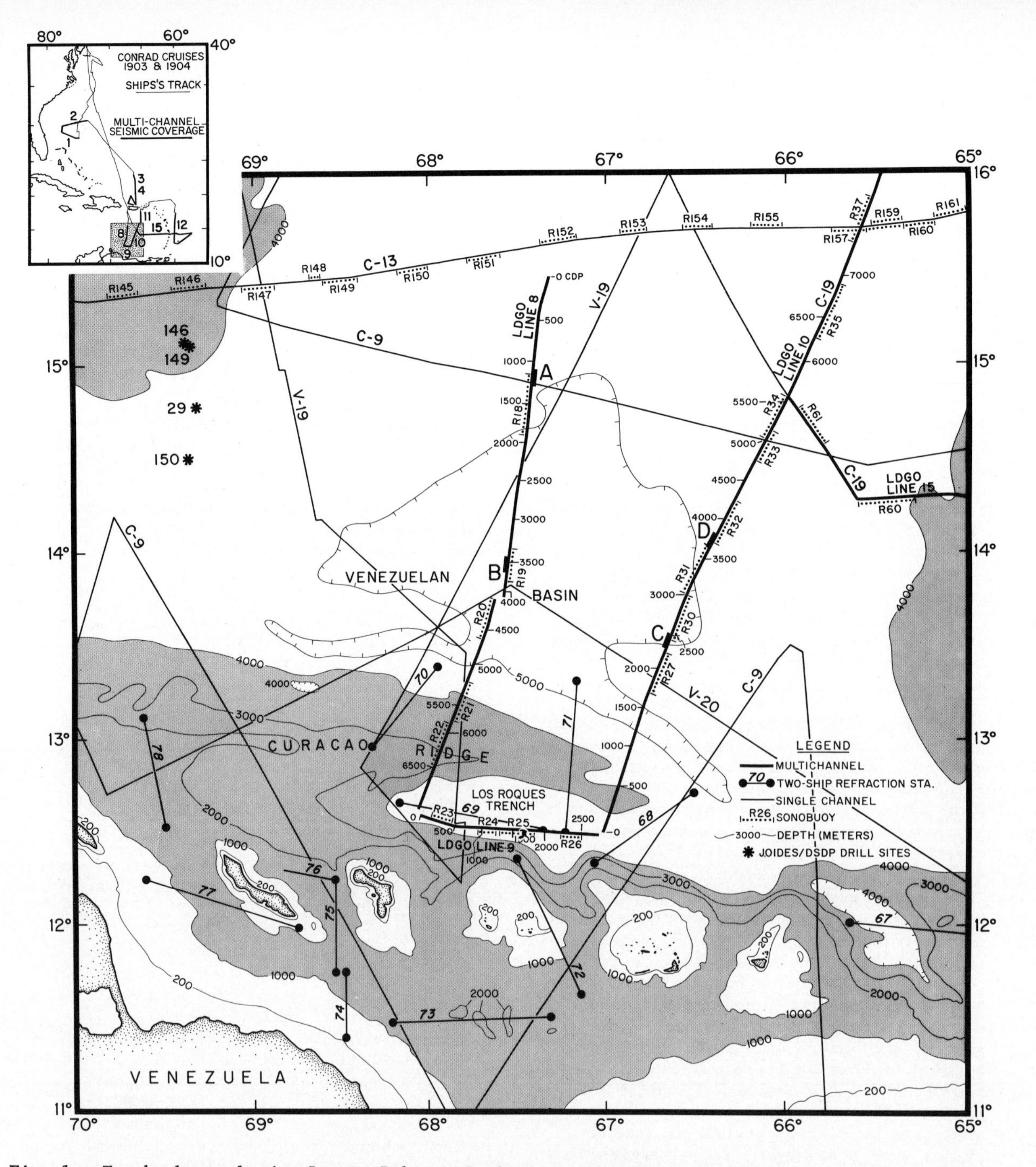

Fig. 1. Track chart showing Lamont-Doherty Geological Observatory multichannel seismic lines 8, 9, 10 in the Caribbean. This was a part of the program of the first two legs of CONRAD Cruise 19 in the North Atlantic and the Caribbean (see top left) during which multichannel seismic reflection profiling was carried out. The sonobuoys shot during the MCS reflection profiling and during the CONRAD 13 cruise are indicated by R18, etc. The positions of the earlier two ship seismic refraction stations are also indicated. Common Depth Point (CDP) numbers are indicated along the MCS track.

the impedance contrast between oozes and underlying lithified sediment, commonly associated with silicification. According to them, chert may be a minor or insignificant part of the overall lithologic "change." Saunders and others (1973) summarize the seismic velocities in the Venezuelan Basin as 1.63 km/sec down to Horizon A" and 2.55 km/sec for the A" to B" interval. In identifying A" in the various figures we were also guided by the observation of Ludwig and others (1975) that in the Venezuelan Basin velocities above A" were less than 2.50 km/sec and lay between 2.50 and 4.50 km/sec in the A"-B" interval. We had to keep in mind that there is an increase in velocity, as at sonobuoy station 27 (Figure 3), where there is a large thickness of overlying sediments and also that in the case of some sonobuoys the layering determined from the analysis of the sonobuoy does not exactly match the layering obtained from MCS reflections. We discuss the identification of B" in detail in the next section.

If our identification of layers A" and B" is correct, our results (Figures 2 and 3) clearly show that the major thickening of the sediment column southward as the Curacao Ridge is approached takes place in sediments younger than A". If the thickening is related to the bending down of basement due to subduction and preferential accumulation of terrigenous sediments, our results suggest that subduction has been going on in the post A" interval. However, from these results we cannot rule out that it has been going on prior to A" time also.

The A"-B" interval also thickens very slightly to the south in line 8, but in line 10 it actually thickens considerably to the north. Where lines 8 and 10 are compared it is quite clear that the A"-B" interval thickens to the east - that is towards the Aves Swell. Although Ludwig and others (1975) indicate some uncertainty about the identification of B" their results also show that the A"-B" interval thickens eastward towards the Aves Swell.

Reflector X

Lines 8 and 10, especially in the southern part of the Venezuelan Basin clearly show that the A"-B" interval is not uniform in character, but is divisible into two intervals by a reflector we have called X. There is a distinct velocity change at this reflector, which is indicated by the sonobuoy refraction experiments. The velocity below X and above B lies usually between 3.0 and 3.4 km/sec although the two sonobuoys closest to the Curacao Ridge on line 10 show velocities as high as 4.3 km/sec (Figure 3).

On line 8 the A"-X interval pinches out north of about CDP 3000 and X cannot be followed further north. Because this interval A"-X is absent at point A it might be argued that what we have identified as X on lines 8 and 10 is really A". This possibility exists but we have not adopted this identification for X because if we did so material with velocity 2.5 km/sec would lie above A" - a result that would be inconsistent with earlier measurements of velocities above A".

On line 8 the A"-X intervals thin southwards, and also thins to zero northwards. On line 10 it also thins southwards but increases northwards, suggesting that this thickening is associated with the Aves Swell.

Reflector B" and the Sub-B" Problem

Reflector B" is especially important in the understanding of the history of the Caribbean. It extends over a great part of the Venezuelan and Colombian basins and is the oldest horizon drilled in the Caribbean. When horizon B" was drilled on DSDP Leg 15 (Edgar and others, 1971), it was found to consist of tholeiitic basalt or sills of dolerite. Overlying sediments have a Late Cretaceous age (Coniacian to Companian).

Several factors have prevented the acceptance of layer B" as the true basement in the Colombian and Venezuelan Basin. Ewing and others (1967) remarked on the very smooth nature of the B" surface which is not typical of oceanic basement. Multichannel seismic results near JOIDES/DSDP drill hole 153 (Hopkins, 1973 from Exxon line) and near JOIDES/DSDP drill hole 150 (Saunders and others, 1973, from a Gulf line) show sub B" layering not generally found in oceanic basement. The Exxon line shown by Hopkins, which is near Aruba Gap, shows that the sub-B" layer is wedge shaped and a stacking velocity of 3.9 km/sec layer was obtained for it. Other earlier results also appeared to indicate a low sub-B" velocity. Edgar and others (1971) while indicating a large range for sub-B" velocities, placed the lower end of this range at 3.2 km/sec. The low velocities were considered indicative of sediment layers; thus it was generally felt that B" did not represent true basement but an igneous event which uniformly emplaced igneous material over a large area, and that layer B" is underlain and overlain by sediments. A natural corollary of these ideas is that true basement is older than B", that is, the Caribbean was formed at a time earlier than Late Cretaceous.

On the other hand, from a well determined line of sonobuoy stations across the Venezuelan Basin, Ludwig and others (1975) found that the velocities below layer B" were not lower than values expected from igneous basement. The location of the CONRAD-13 line of Ludwig and others (1975) is shown in Figure 1. Below B" they obtain velocities predominantly in the 4.5 to 5.3 km/sec range (from poorer data they obtained slightly lower sub-B" velocities in the Colombian Basin). The average compressional velocities of the igneous rocks taken from B" at sites 146 and 150 in the Venezuelan Basin (for location see Figure 1) as measured in the laboratory are 5.1 and 4.8 km/sec respectively, at roughly 0.5 lb pressure (Fox and Schreiber, 1973). Hence Ludwig and others (1975) reasoned that the layer immediately below B" must be dominantly igneous in composition and that intercalated sediments, if any, must have a comparably high velocity.

The results of Ludwig and others (1975) also show that the velocity of sediments between A" and B" is higher than 2.5 km/sec. We have made use of this result and the result cited earlier, that the velocity below B" is greater than 4.5 km/sec in identifying reflectors A" and B" in our multichannel profiles on the present cruise in the Venezuelan Basin.

The layers A" and B" were originally defined

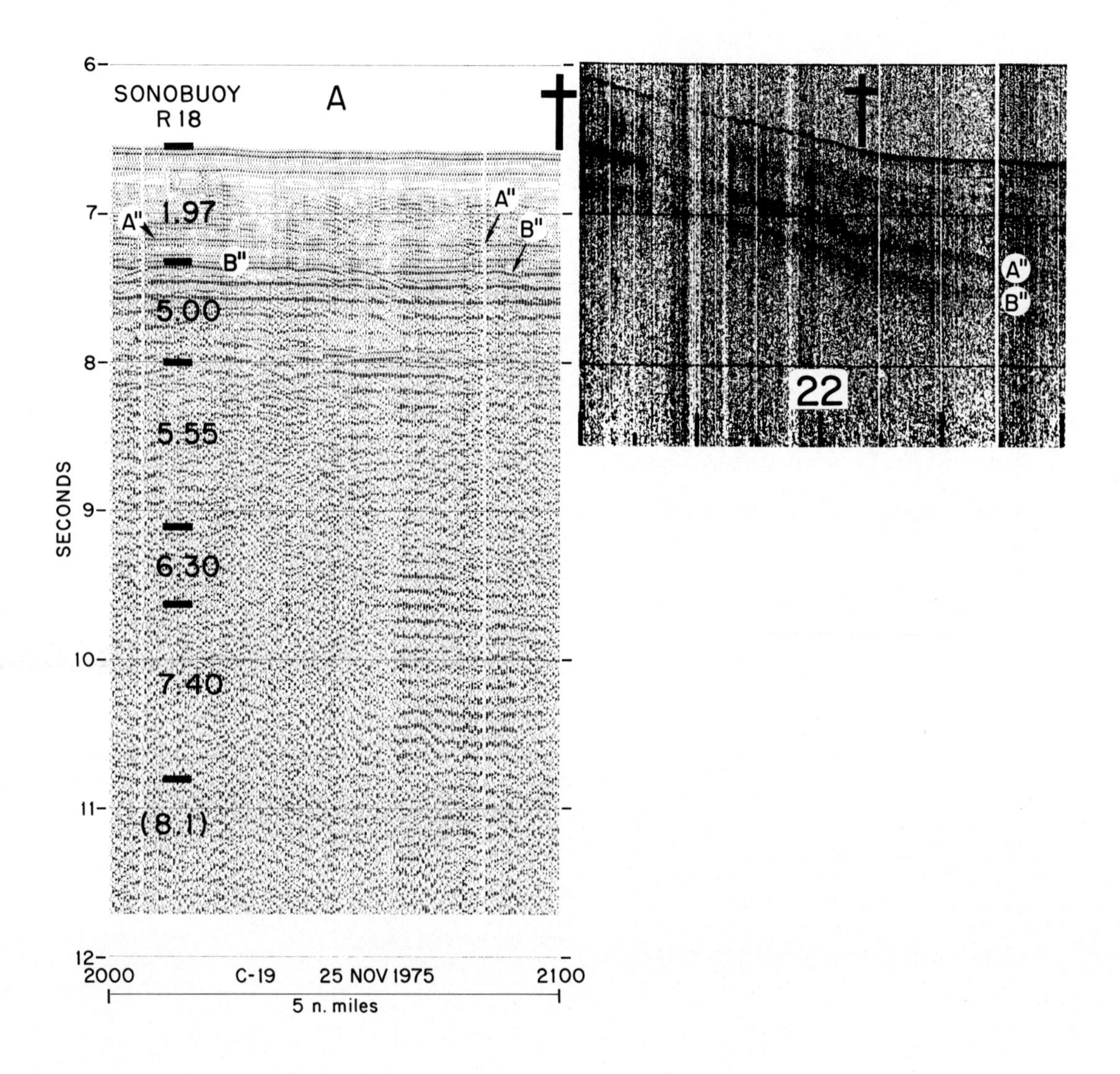

Fig. 5. Left: Detail of processed L-DGO MCS line 8 at Point A (Figure 1). Vertical exaggeration at sea bottom is approximately 3.5 to 1. See Table 1 for explanation of sonobuoy velocity determinations.

Right: Single channel seismic reflection data obtained on CONRAD Cruise 19 plotted at the same vertical scale. The crosses indicate the common point of the intersecting lines.

on single channel reflection records. Their appearance is somewhat different in the multichannel records primarily because of the lower frequencies used in the latter system. Older single channel lines cross the present multichannel lines at several places. However, layer B" is not clearly seen in the single channel lines as the Aves Swell is approached or near Curacao Ridge in the southern part of the Venezuelan Basin. The best intersection of a single channel track and the multichannel lines from the point of view of identifying B" in the MCS profiles is at point A (Figure 1). This enables us to identify B" on line 8. Note that sonobuoy 18 shows that the sub-B" velocity is 5 km/sec agreeing with the results of Ludwig and others (1975) elsewhere in the Venezuelan Basin. We also note that a sub-B" reflector, parallel to B" is also seen. The sonobuoy results indicate a velocity of 5.55 km/sec below the lower reflector. This "sub-B" reflector appears to be similar to the sub-B" reflectors obtained by other investigators using MCS equipment. In particular the Gulf line near JOIDES/DSDP site 150 (Saunders and others, 1973) shows a similar reflector.

We cannot be absolutely certain which reflector is B" as we go farther south and approach the Curacao Ridge. In making the correlation that we show in Figure 2, we have used the result that the layer below B" has a velocity that is greater than 4.5 km/sec. The nature of B" clearly changes along our lines. At point A it is very nearly flat. At point C, along line 10, the reflection hyperbolas are typical of basement. These results suggest that B" is basement or lies very close to basement. If our identification is incorrect, the only other reflector that could be B" is the prominent reflector lying above it (reflector X). But as noted earlier the velocities in the X-B" interval generally lie between 3.0 and 3.4 km/sec and do not exceed 4.5 km/sec as we would expect sub-B" velocities to do. Thus we find it difficult to identify any other reflector as B" than the one we had indicated.

However this identification raises several questions. Why does the surface of B" change from a rather smooth surface in the center of the Venezuelan Basin to a more rugged surface to the east as one approaches the Aves Swell (compare the surface of B" at B and at C in Figures 6 and 7), or to the south as it approaches the Curacao Ridge (compare A and B in Figures 5 and 6)? Is the ruggedness of the surface of B" related to the presence or absence of prominent underlying reflectors? Are these reflectors within basement?

If B" is basement or lies very close to it, the age of the Caribbean crust at the JOIDES/DSDP drill holes 146 and 150 must be close to Coniacian. If the Caribbean has been created by seafloor spreading, the age of the entire Caribbean crust must not differ too much from this age unless the rates of spreading have been extremely small. Christofferson (1973) has identified easterly trending magnetic anomalies in the Colombia Basin as being of Late Cretaceous age

We emphasize that our results are tentative. To confirm these it is essential to run MCS lines from the vicinity of the JOIDES/DSDP drill holes 146 and 150 eastward to intersect our lines 8 and 10.

Crustal and Mantle Velocities

Particularly on line 10 where the airguns were run at high pressures a prominent sub-basement reflector is often seen in the MCS reflection records (unfortunately it does not show well in reproductions; it is indicated by a dotted line in Figure 3).

Line 10 (Figure 3) during which a number of sonobuoys were shot shows a very curious result. The velocity below this reflector decreases progressively to the south from 8.2 km/sec in the center of the Venezuelan basin progressively to the south to 7.3 km/sec at sonobuoy 30. It is not absolutely clear whether this reflector continues uninterrupted further to the south; if it does, the velocity decreases further to 7.0 km/sec at sonobuoy 27. On consideration of the velocities determined below this reflector by the sonobuoy measurements, the possibility arises that this reflector represents the crust mantle interface. We also note that the velocities of the main crustal layer, as determined by the sonobuoys decrease southward from 7.15 km/sec to 6.24 km/sec.

The sonobuoy stations are unreversed; however the dips of the reflectors are small and unlikely to alter the results greatly. The stations on line 10 have been shot effectively updip, therefore the true velocities are slightly lower than the ones determined.

Earlier refraction studies in the Caribbean (Officer and others, 1959; Edgar and others, 1971) had revealed the presence of a lower crustal layer of velocity about 7.5 km/sec that lies over the M-discontinuity. The two-ship refraction station 25 of Officer and others (1959) lies in the vicinity of stations SB30 and 31 of this paper. Station 25 was re-interpreted by Edgar and others (1971) and the results are not very different from those at SB30 and 31 with one exception. The two-ship refraction station

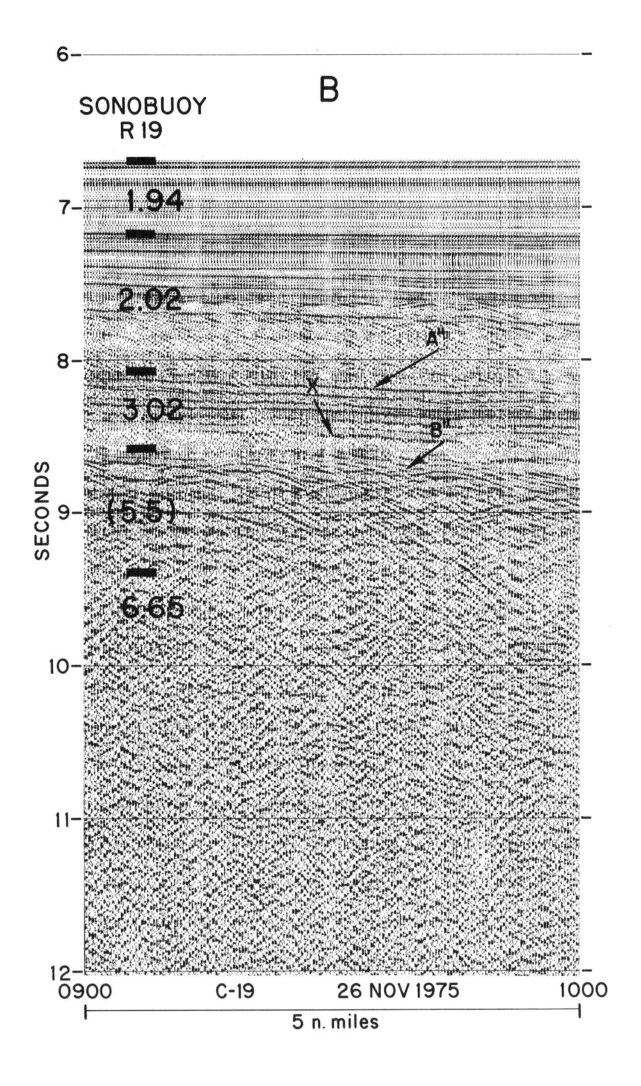

Fig. 6. Detail of processed L-DGO MCS line 8 at Point B. Vertical exaggeration at sea bottom is approximately 3.5 to 1. See Table 1 for explanation of sonobuoy velocity determinations.

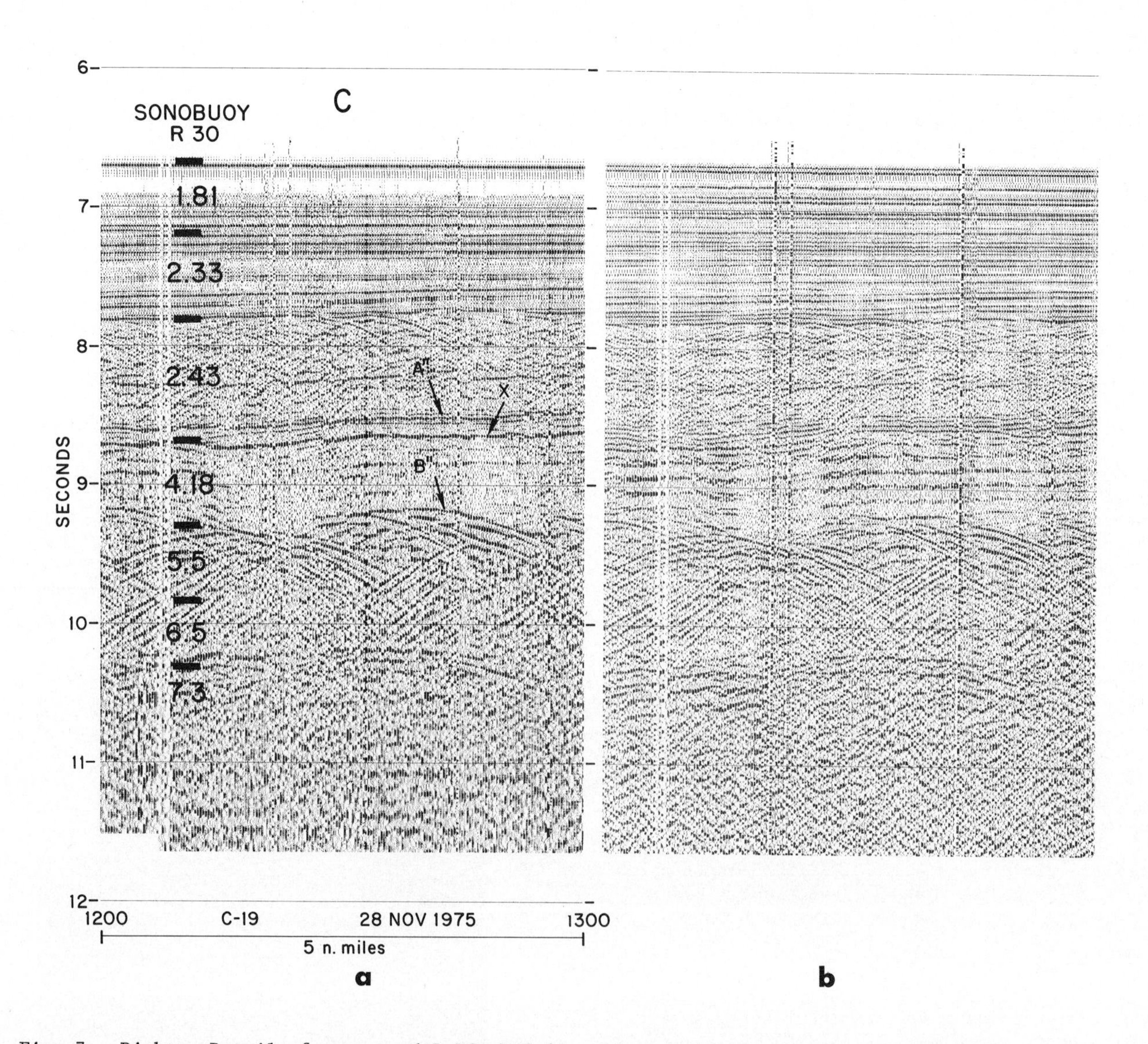

Fig. 7. Right: Detail of processed L-DGO MCS line 10 at Point C. Vertical exaggeration at sea bottom is approximately 3.5 to 1. No deconvolution applied.

Left: Same record as on right with deconvolution. Note how reflector X clearly emerges after application of deconvolution. See Table 1 for explanation of sonobuoy velocity determinations.

shows that material with velocity 8.2 km/sec underlies the 7.6 km/sec layer which lies below the reflector mentioned above. Thus the reflector is not the M-discontinuity. On the other hand, the results of sonobuoys 30 through 34 (Figure 3) appear to imply that the velocities below the same reflector vary from 8.2 km/sec to 7.3 km/sec and therefore it might be the M-discontinuity. In Table 1 and in Figure 3 we have considered this reflector to be the M-discontinuity. If this identification is correct, there is a large difference between the crustal thickness at sonobuoy station SB18 which lies roughly in the middle of the Venezuelan Basin and the crustal thickness at other stations lying to the east and the south. This poses the question whether there is a fundamental dividing line in the Venezuelan Basin which separates the area of flat basement with parallel sub-basement reflectors and a thick crust from an area to the east and the south where the basement is more rugged in character, the crust is anomalously thin, and the mantle velocities are lower.

Our new data has raised these questions. Further studies and experiments are necessary in order to answer them satisfactorily.

Curacao Ridge and the Los Roques Trench

Sediments are deformed on the Curacao Ridge and the Los Roques Trench. It is necessary to

TABLE 1. New Sonobuoy Results. For sonobuoys 20, 27, 30, 31, 32 and 34 we have assumed that the lowest layer is mantle for reasons explained in the text.

VENEZUELAN BASIN

Sonobuoy #	SB18		SB19		SB20		SB27		SB30		SB31		SB32		SB33		SB34		SB35	
	T km	V km/s	T km	V km/s	T km	V km/s	T km	V km/s	T km	V km/s	T km	V km/s	T km	V km/s	T km	V km/s	T km	V km/s	T km	V km/s
Water	4.91		5.01		5.00		4.98		4.99		5.00		4.98		4.87		4.83		4.59	
Sediments	0.77	1.97	0.45	1.94	0.37	1.70	0.46	1.60	0.46	1.81	0.40	1.76	0.36	1.76	0.65	1.87	0.56	1.66	0.59	1.83
			0.93	2.02	0.72	2.20	0.75	2.35	0.71	2.33	0.43	1.86	0.39	2.00	0.81	2.60	0.81	2.64	0.75	2.45
			0.74	3.02	0.97	2.49	0.45	2.48	1.08	2.43	0.93	2.84	0.83	2.72	0.83	3.37	0.89	3.30+		
					0.67	3.33	0.75	2.79	1.27	4.18+	0.89	3.08	0.79	2.99						
							0.76	2.98												
							1.19	4.31												
Layer 2	1.65	5.00	2.26	5.5*	0.76	4.63+	0.74	5.3*	1.49	5.50	1.14	5.3*	2.05	5.85			1.74	5.60		5.40
	3.11	5.55			0.78	5.20					0.78	5.80								
Layer 3	1.61	6.30		6.65	1.96	6.75	1.56	6.24	1.46	6.50	1.43	6.65	1.06	6.65			1.93	7.15		
	4.40	7.40						6.99												
Mantle		8.1*				7.40				7.30		7.60		7.90				8.20		
Total Crustal Thickness	16.45				11.23		11.64		11.46		11.00		10.46				10.76			

CURACAO RIDGE & LOS ROQUES TROUGH

Sonobuoy #	SB21		SB24		SB25		SB26	
	T km	V km/s	T km	V km/s	T km	V km/s	T km	V km/s
Water	4.51		4.53		4.63		4.64	
Sediments	4.82	2.42	0.60	1.84	0.72	1.73	0.38	1.67
					0.61	2.77	0.35	2.13
					0.74	3.3*	0.63	2.74
							1.04	3.35
							2.57	3.81
Layer 2						5.15		

COLUMNS ARE HEADED AS FOLLOWS:

km = layer thicknesses
km/s = layer velocities
* = assumed velocity (assumed mantle velocities tangent to observed mantle reflection curve)
+ = interval velocities with standard deviations > .25 km/s

Velocities smaller than 5 km/s are interval velocities obtained from reflection curves. Velocities of 5 km/sec and higher have been obtained from refraction lines.

migrate the MCS records in order to obtain details of the deformation. However, some preliminary results can be stated from the records which have only been subjected to normal moveout and stacking.

Both lines 8 and 10 (Figures 2 and 3) show that the Venezuelan Basin basement is apparently subducted under the northern end of the Curacao Ridge. Figure 9 clearly shows that the interval X-B" is also subducted under the Curacao Ridge. This leads to the observation that lithified sediments which presumably comprise this interval are capable of being subducted together with basement. On the other hand, overlying sediments presumably unlithified are deformed and folded to form the Curacao Ridge.

Sonobuoy 21 gives an average velocity of 2.4 km/sec for the deformed sediments very close to the base of the Curacao Ridge, which appears to be a reasonable value to expect for the post X sediments after they have been deformed and consolidated.

Figure 4 shows line 9 over the Los Roques Trench. In addition to the results from the sonobuoys shot during the current cruise we have added the results of earlier two-ship refraction stations (Edgar and others, 1971). The depth results at the southern ends of their stations 70 and 71 at the western end of station 68 and at both ends of their station 69 have been converted to two-way reflection time and plotted in Figure 4.

Within the Los Roques Trench, stratified sediments with velocities less than about 2.75 km/sec are seen. Clearly these relatively undeformed sediments are not related to the deformed Venezuelan Basin sediments. The latter must comprise the lower layer with velocity lying between 3 and 4 km/sec. Note that this velocity for the deformed sediments is higher here than the velocity of 2.4 km/sec (determined at sonobuoy 21) closer to the deformation front, which is located further north.

The earlier refraction results (Edgar and others, 1971) showed a great thickness of sediment on the Curacao Ridge and also great variability of thickness. Thus at the southern end of their profile 71, the thickness of the 4.0 km/sec layer is almost 9 km while at the eastern end of their profile 69 and at the western end of profile 68, the thickness of the lower velocity layer is much less. The new MCS reflection and sonobuoy refraction results support these earlier findings by demonstrating the large relief of the top surface of the 3+ km/sec layer and possibly giving some indication that the bottom of this layer may also have large relief.

It is interesting to note that the thickness of deformed sediments appears to be the least under the Los Roques Trench as if to suggest that the mode of deformation is such that when a thick pile of deformed sediments is accumulated it stands high and with a deep basement; if a thinner pile of deformed sediments is accumulated, it stands low with a shallower basement.

The flat-lying Venezuelan Basin sediments change at an almost vertical interface to the deformed Curacao Ridge sediments. It is clear from Figures 2 and 3 that at the edge of the Venezuelan Basin the turbidites are deformed and uplifted to form the Curacao Ridge sediments. Migration, which has not been applied in the figures shown, is necessary to trace the individual layers within the Curacao Ridge. Silver and others (1975) had also come to the conclusion from a number of single channel seismic reflection profiles across the Curacao Ridge - Venezuelan Basin boundary that the uppermost turbidite layers in the Venezuelan Basin are folded onto the south flank of the Curacao Ridge.

Two facts are remarkable about this subduction

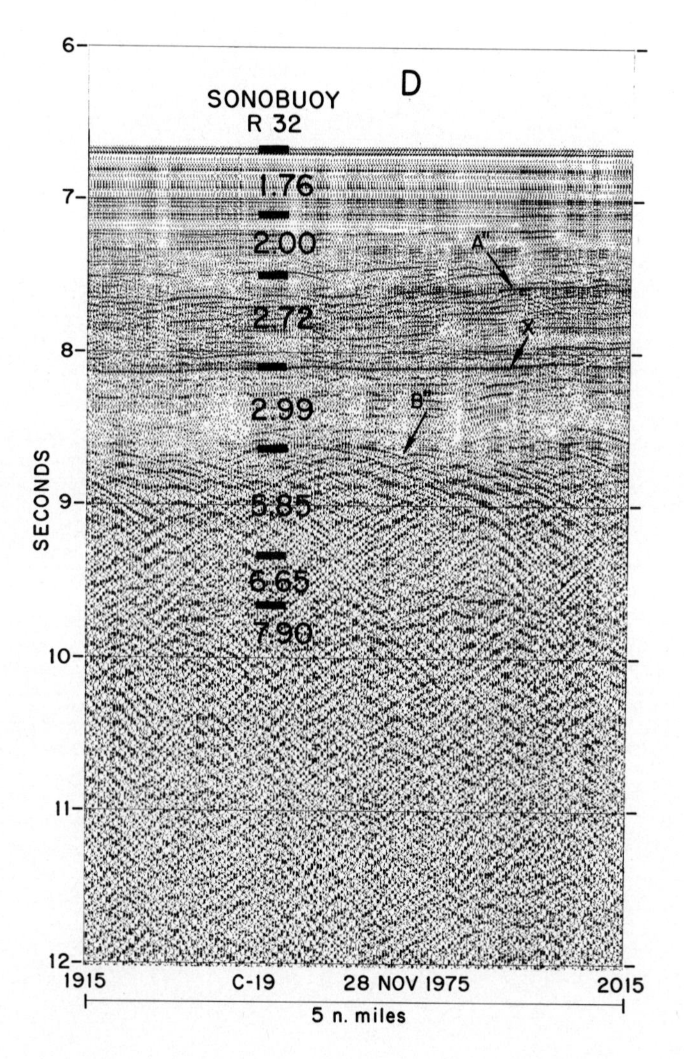

Fig. 8. Detail of processed L-DGO MCS line 10 at Point D. Vertical exaggeration at sea bottom is approximately 3.5 to 1. See Table 1 for explanation of sonobuoy velocity determinations.

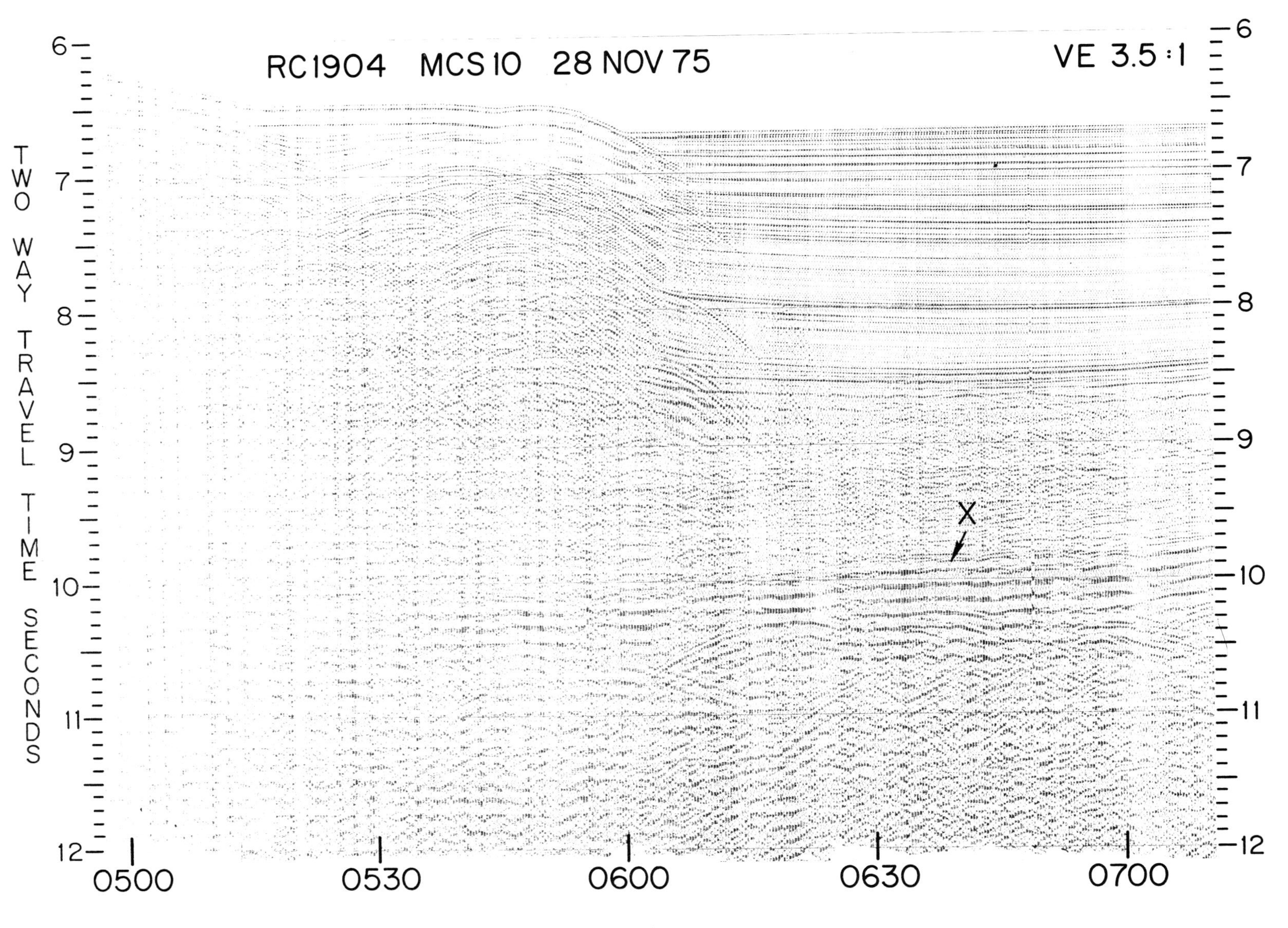

Fig. 9. Detail of processed L-DGO MCS line 10 at the boundary between the Curacao Ridge and the Venezuelan Basin. Note how the portion of the section underlying reflector X is subducted under the ridge. Vertical exaggeration at sea bottom is approximately 3.5 to 1.

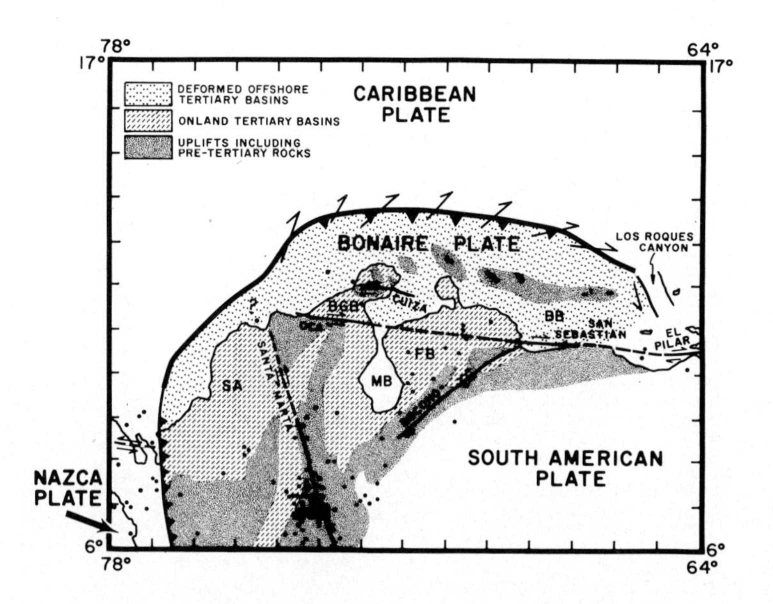

Fig. 10. Tectonic model for the southern margin of the Caribbean as presented by Silver and others (1975) with tectonic provinces as defined by Case (1974). Our results showing compression at the northern boundary of Curacao Ridge agree with this model as well as with the model of Jordan (1975) in Figure 11. More work along the southern boundary of Colombian and Venezuelan Basins is necessary to choose between these two and other models and establish the tectonic regime at the boundary.

boundary. The first is that the front is so sharp. This can be clearly seen in Figure 9 where the vertical exaggeration is small. There is almost no deformation in the Venezuelan Basin sediments seaward of the boundary with the Curacao Ridge. Secondly, although the deformation of the youngest turbidites onto the Curacao Ridge indicates present activity, as does the presence of a large isostatic negative gravity belt over the Curacao Ridge, there is no earthquake activity reported in connection with this boundary.

Case (1974) and Silver and others (1975) have emphasized that a belt of deformed sediments extends from the Colombian slope (Krause, 1971) all the way east along the northern continental margin of South America to the Curacao Ridge. This belt of deformed sediments coincides with a belt of negative isostatic anomalies (Talwani, 1965). If the deformation is due to compression in a north-south direction, it poses problems for postulated plate motions in the area since the motion between the Caribbean plate and the South American plate is predominately east-west (Molnar and Sykes, 1969). Silver and others (1975) find a way out of this difficulty by postulating a rotational movement for a small "Bonaire plate" which lies between the Caribbean plate and the South American plate (Figure 10). Jordan (1975) through an analysis of relative plate motions arrives at the result that the South American plate is moving along an azimuth N 75°W with respect to the Caribbean plate and that the motion between the two plates is taken up in the area lying between the front ranges of Venezuela and the southern margins of the Colombian and Venezuelan Basins. A component of compressive motion across the Curacao Ridge is compatible with this solution (Figure 11). Clearly a careful study of the nature of deformation of the marginal deformed sedimentary belt as well as the examination of the boundary between the Venezuelan and Colombian Basins and the continental slope of South America at a number of points is required to choose between these and other possible solutions and to understand in detail the nature of the plate motions at the Caribbean-South American plate boundary.

Principal Conclusions

(1) The basement of the Venezuelan Basin is subducted under the Curacao Ridge. Lithified sediments overlying basement extending up to

reflector X are subducted with the basement. Sediments younger than A" are deformed and uplifted and form the Curacao Ridge. The deformation front is very sharp and seaward of it the Venezuelan Basin sediments are almost completely undeformed. Landwards, the velocity of the deformed sediments increases away from the deformation front. No earthquakes appear to be associated with the presently ongoing subduction.

(2) The increased thickness of sediments in the Venezuelan Basin southwards towards Curacao Ridge is almost entirely due to the thickening of post A" sediments. However eastwards towards Aves Swell, sediments lying in the interval A"-B" thicken noticeably.

(3) The A"-B" interval is divided by a reflector X into two intervals with velocities about 2.6 km and 3.3 km/sec. The A"-X interval pinches out towards the center of the Venezuelan Basin but it increases appreciably as the Aves Swell is approached.

(4) Our tentative correlations suggest that B" is basement or lies close to its upper surface. The sub-B" reflectors found in the Venezuelan Basin are probably within basement. If these correlations are correct "true" basement in the Caribbean Basins cannot be much older than the age of sediments above B" that have been dated as Coniacian. Thus the concept of a much older "true" basement and of an "event" in the Coniacian which uniformly emplaced a thin layer of basalt over a large portion of the Venezuelan and Colombian Basins must be revised.

(5) Reflections from a prominent sub-basement reflector were obtained along line 10. Correlation with sonobuoy data suggests that the velocity below this reflector varies from 8.2 km/sec in the center of the Venezuelan Basin to 7.3 km/sec or even lower towards its southern border. The crustal velocities also decrease correspondingly from 7.2 km/sec to 6.2 km/sec. However if this reflector is identified as the M-discontinuity, we obtain a thickness of the Caribbean crust in the southeast part of the Venezuelan Basin equal to about 11 km, which is much smaller than the thickness at sonobuoy station 18 in the central Venezuelan Basin. It is also much less than the thicknesses determined by earlier refraction measurements in the Caribbean Sea. Our results pose two basic questions about the crust and mantle in the Caribbean. Is the 7+ km/sec layer a mantle layer rather than a crustal layer as thought earlier? Can the velocity below a reflector believed to be M-discontinuity change by such large amounts in a relatively short distance?

Appendix

Description of Data Acquisition and Processing System

In developing a multichannel seismic reflection system at Lamont-Doherty we have relied principally on commercially available equipment but the data processing method and programs have been developed here. Because we have made new innovations to develop a flexible facility that can be applied in a wide variety of situations and can serve both as an educational and as a research tool and because this is the first paper describing our multichannel seismic results, we describe our facility in some detail below.

Data Acquisition System

A 2.4 km long streamer from Seismic Engineering Co. consisting of 24 active sections, each

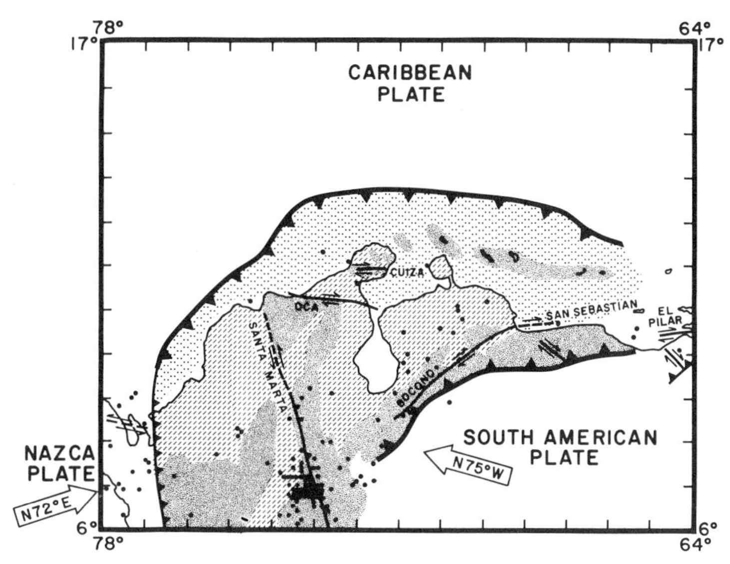

Fig. 11. Alternative tectonic model for the southern margin of the Caribbean by Jordan (1975).

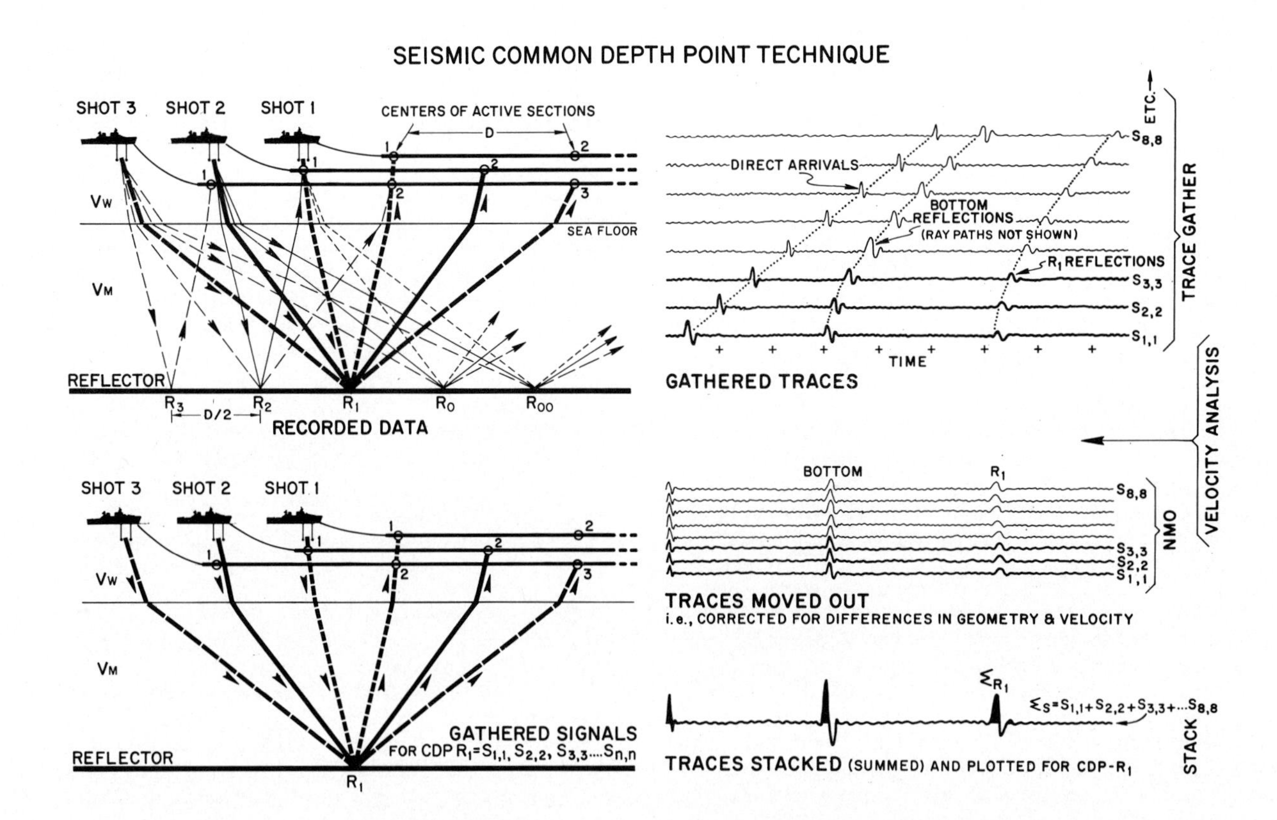

Fig. 12. Basic elements of the seismic Common Depth Point Technique.

100 m long, was employed. A Texas Instrument DFS IV was used as the digital recording system. The channel closest to the ship was recorded in analog fashion, wide-band, on a Raytheon recorder to obtain a monitor trace, and after suitable filtering on the drum recorders which are a part of standard Lamont single channel profiling system. Four Bolt airguns each with a capacity of 466 cu in, fired at 20 sec intervals, provided the source for the seismic signal. The total airgun volume was thus 1864 cu in which was reduced to 1748 cu in when wave shape kits were used. The airguns were to be used at an air pressure of 2000 psi provided by a two-stage Ingersoll/Bolt-Knese compressor system. Problems with the compressors prevented us from increasing the pressure beyond 1000 psi on line 8 (Figure 2). A maximum pressure of about 1700 psi was attained during portions of line 10 (Figure 3). Each shot was digitally recorded for a 12 sec interval; the sampling interval was 4 milliseconds.

Basic Elements of Seismic Common Depth Point Technique

The basic elements of the seismic Common Depth Point technique are illustrated in Figure 12 and are reviewed briefly below. A ship towing an airgun and a streamer is shown at left. Each of the 24 active sections of the streamer consists of a number of hydrophones which are wired together electrically, and the center of each section is represented by a circle in the illustration. (Only three sections are shown here.) If the distance between the centers of the active section is D, successive shots are fired by the ship at distance D/2 apart. Acoustic energy from each shot goes directly, and also after reflection, from various horizons to each active section and is recorded. Let us denote the signal received by active section 1 from shot point 1 by $S_{1,1}$. The various rays reflected at one particular reflector are shown in the top left drawing. Of these various rays some are reflected at the common reflection point R_1. If the streamer has 24 active sections, there will be 24 such rays. Three of these constitute parts of signals $S_{1,1}$, $S_{2,2}$, $S_{3,3}$ and are shown in the bottom left figure. The result of gathering all the signals reflected at a Common Depth Point is called a "gather." (In actual practice the signals are multiplexed before recording. Before gathering they must first be demultiplexed.) The reflections for the Common Depth Point R_1 are shown in the illustration on the right. The ray paths are different for each of these reflections and correspondingly the arrivals are at different times. If T_0 is the arrival time at zero offset (source receiver distance) for a reflector, the arrival time T for an offset, X is given by $T^2 = T_0^2 + X^2/V_{rms}^2$ where V_{rms} is the rms velocity down to the reflector. Thus the arrival time of the reflections if plotted against offset distance will lie along a hyperbola. In order to sum these traces the Normal Moveout Correction (NMO) which equals $T-T_0$ must be applied. In practice we determine V_{rms} as a function of T_0 from the gather, then use it in the above equation to determine $T-T_0$ for each point on the trace which is then moved upward by this amount. If we have calculated our NMO's correctly then reflections from a given horizon will appear at the same time as shown in the illustration. The final step in processing is simply to add all the corrected traces together to form a single trace. This process is known as stacking. The stacking velocity function, V_{rms}, can be used to obtain interval velocities. However, when the reflections are occurring at great depths $T_0 >> X^2/V_{rms}$, the NMO correction is small, and correspondingly, the derived interval velocity is not determined very accurately. In other words, where reflector depth is large the NMO correction is less important and velocities are poorly determined; conversely, when the reflector depth is small the NMO correction is important and the velocities

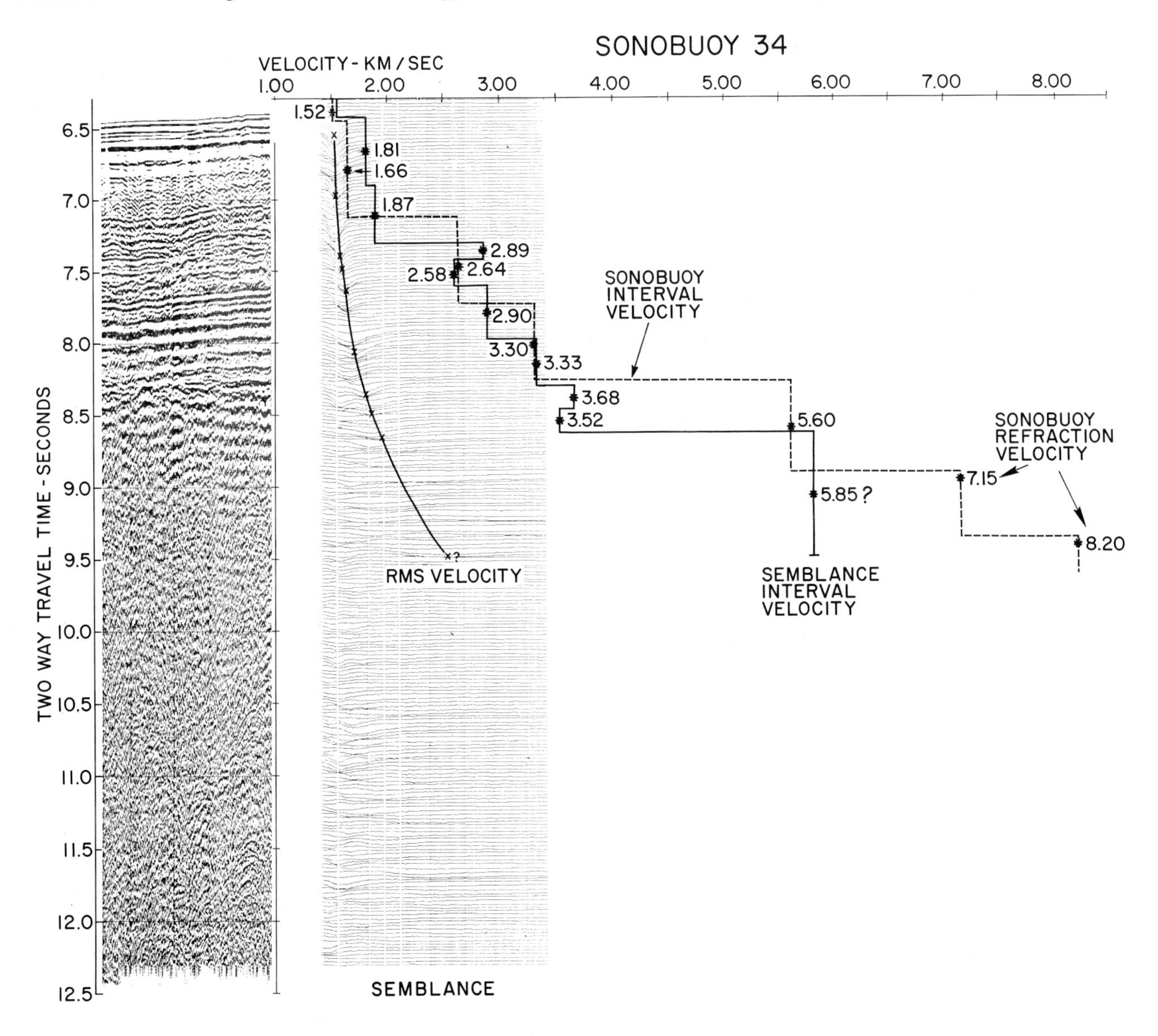

Fig. 13. Comparison of velocities obtained by sonobuoy 34 and by the semblance method on the MCS data. The semblance method gives more reliable results in the top part of the section. The sonobuoy determined velocities are more accurate for the lower part of the section.

are well determined. If large dips are present or if the layers are thin the interval velocity determination will again be inaccurate.

To determine the V_{rms} function we simply pick a zero offset reflection time, T_0, and a trial velocity, V, we then examine a small portion of each trace centered about a time T. T is determined for each trace from the equation given above by using the appropriate value of X. These trace portions are compared via a correlation technique known as semblance. If these portions resemble each other, then we get a high value for the semblance coefficient, and we most likely have guessed the correct velocity for this T_0. In practice we look at a range of velocities, stepping by some small increment. Having swept this T_0 we move down by a small amount to the next T_0 and repeat the velocity scan. We continue this practice for all T_0's. We now plot the semblance coefficients on a V_{rms} vs T_0 grid (Figure 13) and human judgement enters the process for the first time. Since several different velocities may give large semblance coefficient values for a given T_0, an interpreter using his geological knowledge must pick the most reasonable path through the V_{rms} vs T_0 grid. This path or V_{rms} function is then used to calculate the NMO correction for the gather. Two high semblance coefficients at a single T_0 are caused by two different reflections with the same zero offset arrival time. These events might, for example, be a primary reflection and a water column multiple. If the interpreter has chosen the V_{rms} associated with the primary event, then the primary will be reinforced by the NMO correction and stacking process, while the multiple will be suppressed. If the two velocities have a large difference, then the multiple attenuation will be large.

The techniques used in processing the MCS data are described in various publications. For a description of the CDP shooting pattern the interested reader should see Mayne (1962). Semblance velocity analysis is described in Tanner and Koehler (1969), and Tanner, Cook and Neidell (1970) discuss limitations of the CDP method. Stoffa and others (1974) and Buhl and others (1974) cover the application of homomorphic deconvolution to marine seismology.

Date Processing System

We describe our data processing in some detail because it was specially developed at Lamont-Doherty to give us flexibility and yet be economical without sacrificing speed.

The basic computer facility at Lamont-Doherty for the processing of the MCS data consists of two interconnected mini-computers (Data General) and a high speed floating point array transform processor (Floating Point Systems). By using this three computer configuration the computing is distributed between the processors and the total time for normal common depth point processing is minimized. One mini-computer controls the processing while the other reads and edits the large volume of tape data required. The array transform processor under the direction of the control processor carries out all repetitive algorithms in the minimum amount of time possible with today's technology.

The software required for Common Depth Point processing was developed at Lamont. All programs except one are written in Fortran with all computation bound algorithms being Fortran callable assembly language or array transform processor subroutines. Thus, the basic computations are performed quickly and efficiently while overall program logic can be altered at the Fortran level to perform various processing objectives. The one program which is not designed in this manner is our basic demultiplex-CDP gather program. Because of the stringent timing requirements to perform this operation, this program is written in assembly language. Consequently, we demultiplex and gather all field data collected immediately and archive a demultiplexed-gathered tape for future processing.

Once the data has been demultiplexed and gathered, it is necessary, as discussed above, to determine the stacking or RMS velocity which is used for the normal-moveout correction and subsequent stack. While we have developed software for velocity analysis based on trial constant velocity normal moveout, our interactive semblance velocity analysis program has proved more effective. The interpreter selects the Common Depth Points he wishes to analyze and selects the parameters which define the vertical two-way travel-time, T_0, velocity, V, grid. Since semblance is a normalized cross-correlation between all combinations of traces in a Common Depth Point along a specified travel-time curve, it would normally be a lengthy calculation. Using our array processor however, a typical semblance grid consisting of 8.68 million correlations takes less than 3 minutes. Once a semblance calculation has been performed, all the possible T_0, V combinations have been scanned and the correlation data is ready for presentation to the interpreter. To increase interpretation speed and accuracy, this data is graphically displayed on an interactive terminal in one second intervals of vertical two-way travel-time. As the interpreter selects possible semblance peaks, by using a cross-hair cursor, the vertical two-way travel-time, RMS velocity and corresponding interval velocity are displayed to aid him in his selection of the velocity function. As the T_0, V pairs are chosen, the velocity function is plotted on the screen to guide future selections. This interactive velocity analysis method is extremely easy to use and is accurate.

After velocity functions have been chosen by the interpreter, the normal moveout correction and stack are carried out along the line of multichannel data. Velocity functions are spatially interpolated for all Common Depth Points between velocity analysis points. This insures a smooth transition between velocity functions and a spatially continuous stacked section. Using our array processor for normal moveout and stack, we can process 24 channels with 12 seconds of 4 msec data at a rate of 1 km per minute. Our present limitation is the speed with which we can read the input Common Depth Point gathers. If this were not a factor, our total computation time would be on the order of 30 km per hour.

The output of our normal moveout and stack program is a stacked data tape. The data at this point is free from any artificial gain or amplitude distortion. The accuracy of the amplitude information provided by our field recording unit has been preserved. At the profiling stage, we have the option of applying any combination of trace scaling, time-varying gain and 1/R correction desired. A bandpass filter, based on the minimum side-lobe Hanning spectral window, can be selected. The profile can be variable area or one level, true amplitude or clipped, based on the geologist's choice. Traces can be mixed to bring out weak, usually sub-basement horizons. In addition to profiling stacked data, we can construct monitor profiles of any or all traces within a Common Depth Point gather. This allows us to construct a single channel variable area profile prior to normal moveout and stack to aid in identifying horizons for velocity analysis.

After normal moveout and stack, we can deconvolve the data using either time-varying predictive deconvolution or homomorphic deconvolution. Deconvolution provides improved temporal resolution and removes ambiguities due to the source signature.

Specific Processing Lines 8, 9 and 10

The multichannel seismic data for lines 8, 9 and 10 were demultiplexed and gathered to form 24-fold Common Depth Point data. All the velocities for normal moveout and stack were determined using an interactive semblance velocity analysis program. The data was then corrected for normal moveout and stacked. During normal moveout velocity functions at the Common Depth Points between the velocity analysis points were determined by interpolation. At profiling time the variable area stacked data (Figures 5 through 9) were bandpass filtered (10-30 hz) and a 500 msec time-varying gain function was applied starting at the sea floor. The data was then scaled by a factor of two and clipped where necessary for the final display. The one level non-variable area sections (Figures 2, 3 and 4) were displayed with the same bandpass filter and time-varying gain, but the scale factor was increased to five. Figure 7a shows a section of line 10 which was deconvolved after normal moveout and stack using the homomorphic deconvolution method described by Stoffa (1974). After deconvolution, the same display parameters for the variable area displays described above were applied.

Sonobuoys and Comparison of Sonobuoy Determined Velocities with Stacking Velocities

Sonobuoy refraction stations using SSQ 41A sonobuoys, supplied by the Office of Naval Research, were run in the usual manner (Le Pichon and others, 1968). Because of the large intensity sound source we were able to run the sonobuoy stations to large distances and in many instances were able to obtain subcrustal velocities.

During the sonobuoy interpretation, events were chosen which occurred at the vertical two-way travels corresponding to the major reflection events observed in the processed multichannel data. For several sonobuoys, we have qualitatively compared the interval velocities determined from the multichannel data from the semblance analysis, and the sonobuoy determined velocities. For the uppermost part of the section, one to two seconds subbottom, the semblance and sonobuoy results agree quite well (Figure 13), with the semblance technique giving slightly more time resolution. At depth, the sonobuoy technique is superior.

*Lamont-Doherty Geological Observatory Contribution No. 2411.

Acknowledgements. We thank the officers, crew and scientists of the R/V ROBERT D. CONRAD for assistance in this project which was principally supported by Grant DES 75-15865 from the National Science Foundation, and Contract N00014-75-C-0210 with the Office of Naval Research. The manuscript was critically reviewed by Eli Silver, Ed Driver and Bill Ludwig.

References

Buhl, P., P. L. Stoffa, and G. M. Bryan, The application of homomorphic deconvolution to shallow-water marine seismology - Part II: Real Data, Geophysics, 39, 417-426, 1974.

Christofferson, E., Linear magnetic anomalies in the Colombia Basin, central Caribbean Sea, Geol. Soc. Amer. Bull., 84, 3217-3230, 1973.

Edgar, N.T., J.I. Ewing and J. Hennion, Seismic refraction and reflection in Caribbean Sea, Amer. Assoc. Petrol. Geol., 55, 833-870, 1971.

Ewing, J. and M. Ewing, Sediment distribution on the mid-ocean ridges with respect to spreading of the sea floor, Science, 156, 1590-1592, 1967.

Fox, P. and E. Schreiber, Compressional wave velocities in basalt and dolerite samples recovered during Leg 15 in Initial Reports of the Deep Sea Drilling Project, 15, Washington, D.C., U.S. Government Printing Office, 1013-1016, 1973.

Hopkins, H.R., Geology of the Aruba Gap abyssal plain near DSDP site 153, in Initial Reports of the Deep Sea Drilling Project, 15, Washington, D.C., U.S. Government Printing Office, 1039-1050, 1973.

Jordon, T.H., The present-day motions of the Caribbean plate, J. Geophys. Res., 80, 4433-4440, 1975.

Krause, D.C., Bathymetry, geomagnetism and tectonics of the Caribbean Sea north of Colombia in Donnelly, T.W., ed., Caribbean geophysics, tectonics and petrologic studies, Geol. Soc. Amer. Mem. 130, 35-54, 1971.

Le Pichon, X., J. Ewing and R.E. Houtz, Deep-sea sediment velocity determination made while reflection profiling, J. Geophys. Res., 73, 2597-2614, 1968.

Ludwig, W.J., R.E. Houtz and J.I. Ewing, Profiler-sonobuoy measurements in Colombia and Venezuela Basins, Caribbean Sea, Amer. Assoc. Petrol. Geol., 59, 115, 1975.

Mayne, W.H., Common reflection point horizontal data stacking techniques, Geophysics, 27, Part II, 927-938, 1962.

Molnar, P. and L.R. Sykes, Tectonics of the Caribbean and Middle America regions from focal mechanisms and seismicity, Geol. Soc. Amer. Bull., 80, 1639-1684, 1969.

Officer, C., J. Ewing, J. Hennion, D. Harkinder, and D. Miller, Geophysical investigations in the eastern Caribbean - Summary of the 1955 and 1956 cruises, in Ahrens, L.M., F. Press, K. Rankama, and S.K. Runcorn, eds., Physics and Chemistry of the Earth, Pergamon Press, London, 3, 17-109, 1959.

Saunders, J.B., N.T. Edgar, T.W. Donnelly, and W.W. Hay, Cruise synthesis, in Initial Reports of the Deep Sea Drilling Project, 15, Washington, D.C., U.S. Government Printing Office, 1077-1111, 1973.

Silver, E.A., J.E. Case and H.J. MacGillvary, Geophysical study of the Venezuelan borderland, Geol. Soc. Amer. Bull., 86, 213-226, 1975.

Stoffa, P., The application of homomorphic deconvolution to shallow water marine seismology, Ph.D. Thesis, Columbia University, 1974.

Stoffa, P.L., P. Buhl and G.M. Bryan, The application of homomorphic deconvolution to shallow-water marine seismology - Part I: Models, Geophysics, 39, 401-416, 1974.

Talwani, M., Gravity anomaly belts in the Caribbean (Abs.), in Poehls, W.H., ed., Continental Margins and Island Arcs, Can. Geol. Survey Paper 66-15, 177, 1966.

Talwani, M., J. Ewing, M. Ewing, and T. Saito, Geological and geophysical studies of the Caribbean submarine escarpment (abs.), Geol. Soc. Amer. Special Paper 101, 217-218, 1966.

Tanner, M.T. and F. Koehler, Velocity spectra-digital computer derivation and applications of velocity functions, Geophysics, 34, 859-881, 1969.

Tanner, M.T., E.E. Cook and N.S. Neidell, Limitations of the reflection seismic method: Lessons from computer simulations, Geophysics, 35, 551-573, 1970.

GEOMETRY OF BENIOFF ZONES: LATERAL SEGMENTATION AND DOWNWARDS BENDING OF THE SUBDUCTED LITHOSPHERE

Bryan L. Isacks and Muawia Barazangi

Department of Geological Sciences, Cornell University, Ithaca, New York 14853

Abstract. The spatial distribution of earthquakes in the Benioff zones of four regions of lithosphere subduction are examined to determine the geometry of the descending lithosphere in relation to lateral segmentation, downwards bending, topographic features on the sub-oceanic plate and features of the upper plate such as volcanism and inter-arc spreading. Thicknesses of 20 to 30 km of the seismic zones in the mantle can often be related to fault-like features in the spatial distribution of hypocenters. In the Peru and the Kurile-Kamchatka regions the thickness of the seismic zone and focal mechanism solutions can be interpreted as an effect of bending. The upper 200-300 km of Benioff zones have a relatively uniform and simple structure within segments of arc length of about 500 to 1000 km. The junction between adjacent segments is in some cases characterized by a contorted but apparently continuous slab and in other cases by a break or tear in the descending slab. The line of active volcanoes most accurately marks the trends of the segments, and in many cases lies about 100 km above the descending slab. There is a remarkable correlation in western South America between the nearly flat Benioff zones of Peru and central Chile and the absence of Quaternary volcanoes, which suggests that generation of andesitic volcanism requires a wedge of asthenosphere between the subducted and upper plates. The Juan Fernandez ridge coincides with the inferred break between the gently dipping central Chile zone and the more steeply dipping one to the south, but the projection of the Nazca ridge is north of the well-defined tear in the subducted plate beneath southern Peru. At depths greater than about 200 km the trends of the inclined seismic zones often begin to diverge from the shallow trends and the geometry becomes more complex at depth. In two regions, Tonga and Izu-Bonin, the intermediate-depth zone appears to override the deeper zone. A close relationship in time and space is found between the geometry of the Tonga zone below 300 km, the opening of the Lau basin, and the subduction of the suboceanic Louisville ridge. Accumulated data offer increasing support for the interpretation that the deep earthquakes beneath Fiji and the northern Fiji plateau are part of a remnant slab that was subducted at a previously existing arc along the northern margin of the Fiji plateau.

Introduction

The expression of subduction zones on a map view of the oceanic trenches and the lineations of Quaternary volcanoes show generally well-defined segments in which the curvature of the surface features is relatively uniform and simple. Indeed, these are among the most uniform geometric forms on the earth's surface. The continuation of this simplicity of form in the distribution of mantle earthquakes is, of course, the basis for the recognition of Benioff zones and for the concept of the subduction of lithosphere. Large variations in the overall geometry of Benioff zones are now clearly recognized, but, as Sykes [1972] points out, the largest variations seem to occur deeper than about 300 km, while the upper 300 km appears more uniform in structure. Nevertheless questions are often raised about the coherence and integrity of the descending slab [e.g., Kanamori, 1971a and 1971 b; Carr et al., 1973; Abe, 1972a]. In this paper we consider examples of four major regions of lithosphere subduction and focus particularly upon the geometry of the upper 300 km of the Benioff zone. We find lateral segmentation of the descending lithosphere, but on a scale of 500-1000 km and, for the most part, this segmentation corresponds to the major segmentation apparent on map views. We examine the two-dimensional cross sectional geometry of the Benioff zones within the segments in relation to downwards bending of the lithosphere and volcanism, and then consider the three-dimensional structure in terms of the junction of segments and relationships to topographic and other features of the converging plates.

The data include compilations of the best available hypocentral locations for the Izu-Bonin-Mariana, the Tonga-Kermadec, and the Peru-Chile regions plus data of Vieth [1974] for the Kurile-Kamchatka region and Pascal et al. [1977] for the New Hebrides region. The data for the first three regions (Table 1) represent the best available locations based on standard techniques. We established criteria for selecting good quality data out of the hypocentral file. In this selection procedure all events reported by the ISS, ISC, and the large events of the USGS are examined and either rejected or selected depending on (1) the distances and the number of the local stations that recorded the events, (2) the consistency and the number of pP readings, and finally (3) the azimuthal distribution of the teleseismic stations. On the other hand, the data of Vieth and Pascal et al. are based on special methods of joint relocation in which the precision of relative location is substantially improved. The data are plotted on maps and sections by computer, and will be reported on in more detail in other papers [e.g., Barazangi and Isacks, 1976]. In this paper we review some of the major results of analysis of these data.

Table 1. Summary of the Data Used in This Study

Region	Period of Time	No. of Events	Source of Data
Tonga-Kermadec	1959-1961	364	Sykes (1966)[1]
	1962-1963	86	ISS
	1964-1973	918	ISC[3]
	1973-1975	40	EDR[5]
Izu-Bonin and Mariana	1961-1967	285	Katsumata and Sykes (1969)[2]
	1968-1972	378	ISC
Peru-Chile	1959-1963	129	ISS[4]
	1964-1973	1535	ISC
	1973-1975	36	EDR
New Hebrides	1962-1974	522	Pascal et al. (1977)
Kurile-Kamchatka	1964-1971	697	Vieth (1974)[6]

1 Sykes class "A" locations were selected.

2 Katsumata and Sykes classes "A" and "B" locations were selected.

3 Selection of locations from the Bulletins of the International Seismological Centre was based on three criteria: (1) depth control by pP-P times and (2) by readings at nearby stations and (3) epicenter control by azimuthal distribution of stations.

4 Selection of locations from the Bulletins of the International Seismological Summary after calculation of depth using listed pP arrivals; same procedure as for ISC.

5 Selection of locations from the Earthquake Data Reports of the U. S. Geological Survey, Department of Interior (previously NOAA, U. S. Department of Commerce); selection same as for ISC but applied only to events with magnitude larger than about 5.5.

6 Selection of events with computed errors in at least two of the three location parameters less than 10 km and the third not greater than 15 km.

Major Segmentation of Subduction Zones: Composite Cross-Sections

Thickness of Benioff Zones

With a limited set of data, it may not be clear whether the spatial distribution is best described by a thin contorted zone or by a thicker uncontorted zone. The thickness of the zone is thus crucial to the resolution of the geometry of Benioff zones. The apparent thickness will be a combination of a real thickness and the effect of errors of location.

There is also the problem of sampling. The activity is monitored during a very short period of time in terms of geological processes, and may not be representative of the long term average. The sampling problem is intensified by the tendency for mantle earthquakes to cluster in "nests". One might even think of these clusters as isolated pieces of lithosphere. However, if the overall collection of events forms a simple structure over large distances, it seems more reasonable to think of these clusters or nests as particularly active regions within a single slab of lithosphere. The results of our work support the latter interpretation.

Fault-like features or groups of fault-like features have been shown to cut across the overall slab-like geometry of the Benioff zone [Oike, 1971; Fukao, 1972; Billington and Isacks, 1975]. An example of recent work [Pascal et al., 1977] is shown in Figure 1 for the New Hebrides. In one case (earthquake 45), the inferred fault plane is vertical and strikes transverse to the arc, but is parallel to the trend of and along the down-dip projection of a major topographic feature being subducted there, the D'Entrecasteaux fracture zone (see also Figure 2). This feature in the descending plate causes an apparent reorientation of the prevalent down-dip extensional stress but does not appear to represent a major break in the plate. Another fault-like feature is identified near earthquake 46. The data shown here were relocated by the method of Joint Hypocenter Determination using a nearly constant network of stations, and the relative precision of location is quite good. The apparent thickness revealed by these data is thought to be too large to be explained by errors of location and indicates a thickness of the inclined zone of about 20 km.

Another example of a significant thickness of the seismic zone is reported by Vieth [1974]

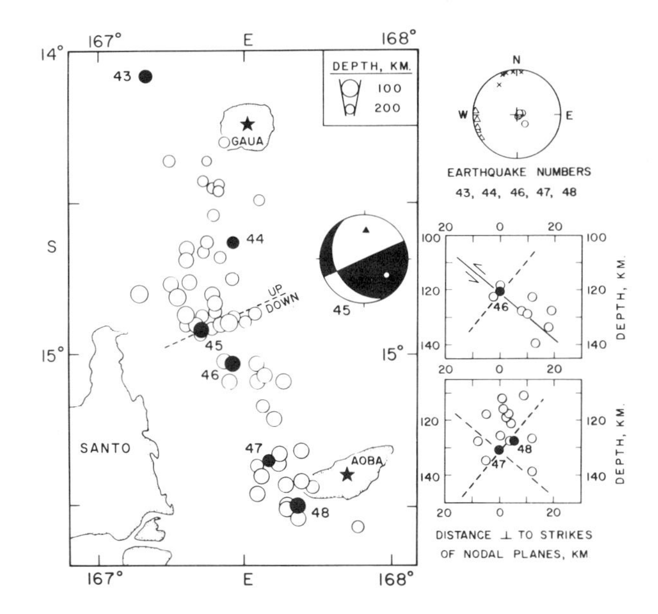

Figure 1. Intermediate-depth hypocenters and focal mechanisms in a small area of the New Hebrides, from Pascal et al. [1977]. Filled circles are earthquakes with focal mechanism solutions. Stars mark active volcanoes. Focal mechanism data shown on lower hemisphere plot, triangles show P axes, circles show T axes and X's show B axes. The two small cross sections on the right are vertical sections and are taken perpendicular to the B axes of the focal mechanism solutions of the specified earthquakes. The data plotted in the upper section are the cluster of events located to the southeast of event 46, and the data for the lower section are those events located to the southeast of event 47.

for the Kurile-Kamchatka region. We selected the best of Vieth's relocations as determined by the reported standard errors and plotted these data on narrow-width cross-sections and on maps for detailed examination. We performed this analysis specifically to see if any of the apparent thickness could be attributed, for example, to projection of a contorted three-dimensional structure onto a two-dimensional cross-section. The conclusion was that it could not. An example in which the apparent thickness is maintained in both halves of an already narrow section is shown in the south Kurile section of Figure 3. These Kurile-Kamchatka results offer compelling evidence that the apparent thickness is not due to mislocation errors. The Kurile-Kamchatka sections shown in Figure 3 illustrate the general feature of Vieth's results, that the shallow events, presumably occurring mainly in the thrust zone, are very tightly grouped in comparison with the intermediate depth events (see also Engdahl, this volume). Unless there is some very remarkable and unusual systematic error in Vieth's method of relocation, the shallow events should not be more precisely relocated than the intermediate-depth events; in fact, the opposite is more likely. Thus we feel a real thickness of about 30 km is supported by Vieth's data.

This thickness appears to continue for the plate subducted beneath northern Honshu. Uyeda [this volume] reports on work published in Japanese in which the mantle seismic zone beneath northern Honshu is resolved into two narrow zones separated by about 30 km. However, Vieth's data do not show a clear resolution into two bands.

In contrast, Engdahl [1971 and 1973] and Ansell and Smith [1975] find very thin zones in the western Aleutian and New Zealand, respectively, with thicknesses of about 10 km. The question of whether these results represent a problem of sampling, both in time and space, or real differences from place to place in the thickness of the seismically active part of the descending lithosphere, must await further data. For the present we take the Kurile-Kamchatka, Honshu, and New Hebrides results as well as the

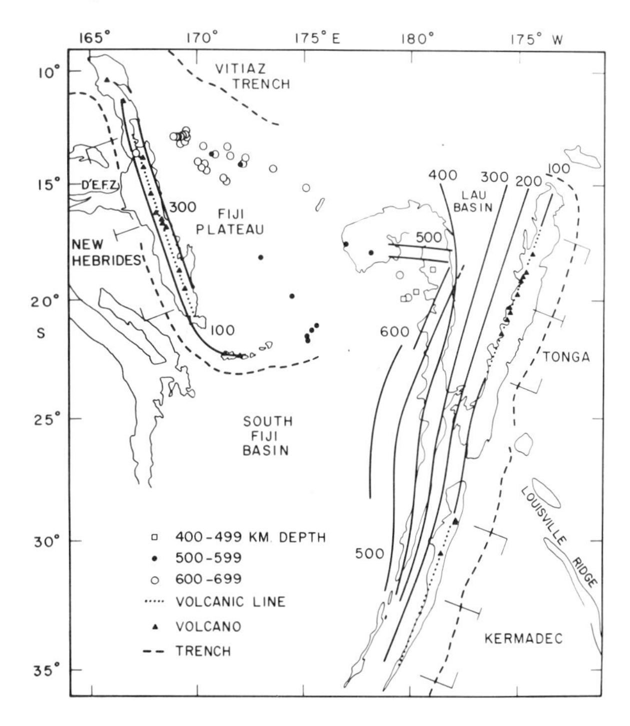

Figure 2. Map of Tonga, Kermadec and New Hebrides segments, showing contours of hypocentral depth to the top of the Benioff zone. Axis of trench follows deepest landward closed contour of the trench bathymetry as given by Mammerickx et al. [1971]. The direction and extent of the cross-sections shown in Figure 3 are indicated by the light lines. Volcanoes are taken from Richard [1962] and Fisher [1957].

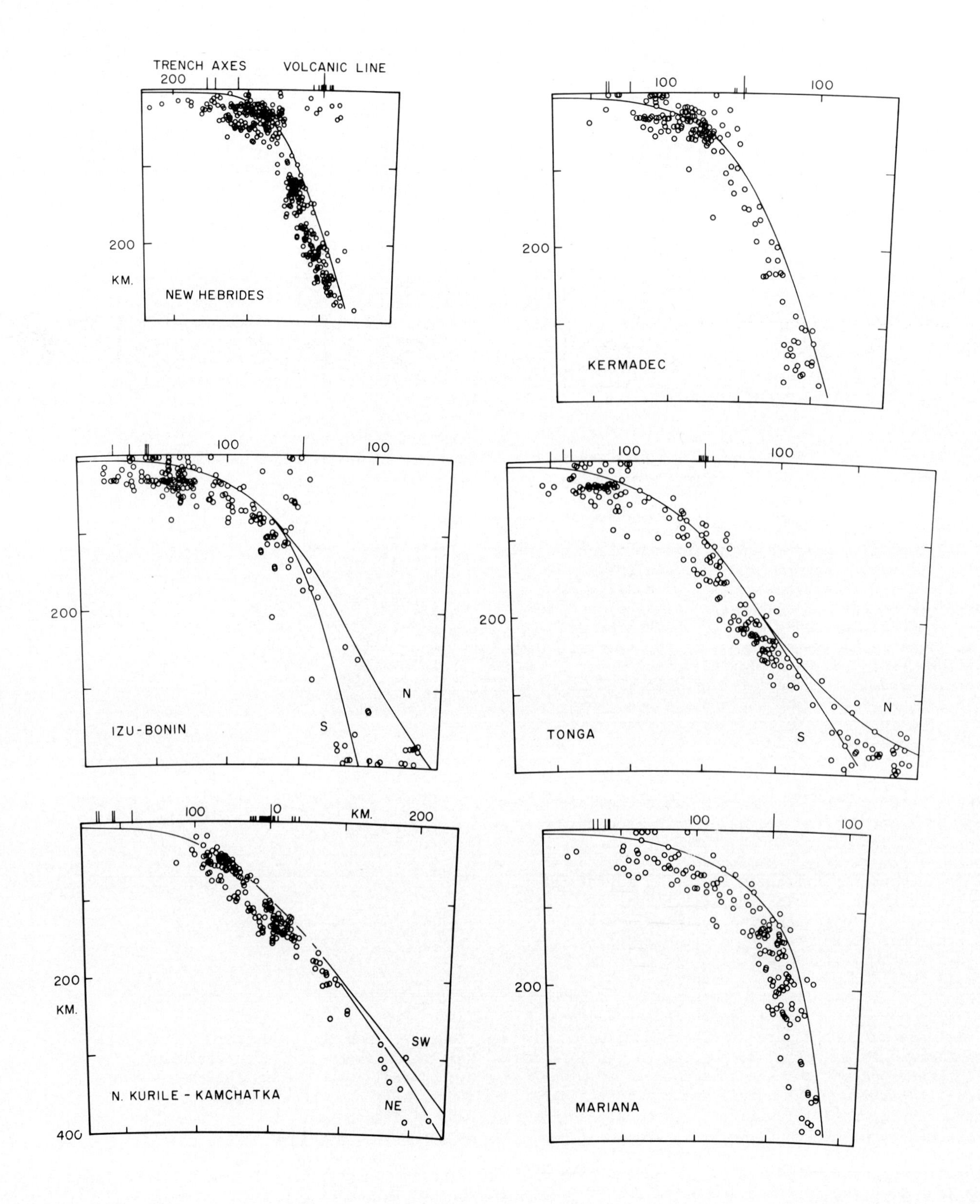

TRENCH AXES
VOLCANIC LINE
200
200
KM.
NEW HEBRIDES
100
100
200
KERMADEC
100
100
200
N
S
IZU-BONIN
100
100
200
N
S
TONGA
100
10
KM.
200
200
KM.
SW
NE
N. KURILE - KAMCHATKA
400
100
100
200
MARIANA

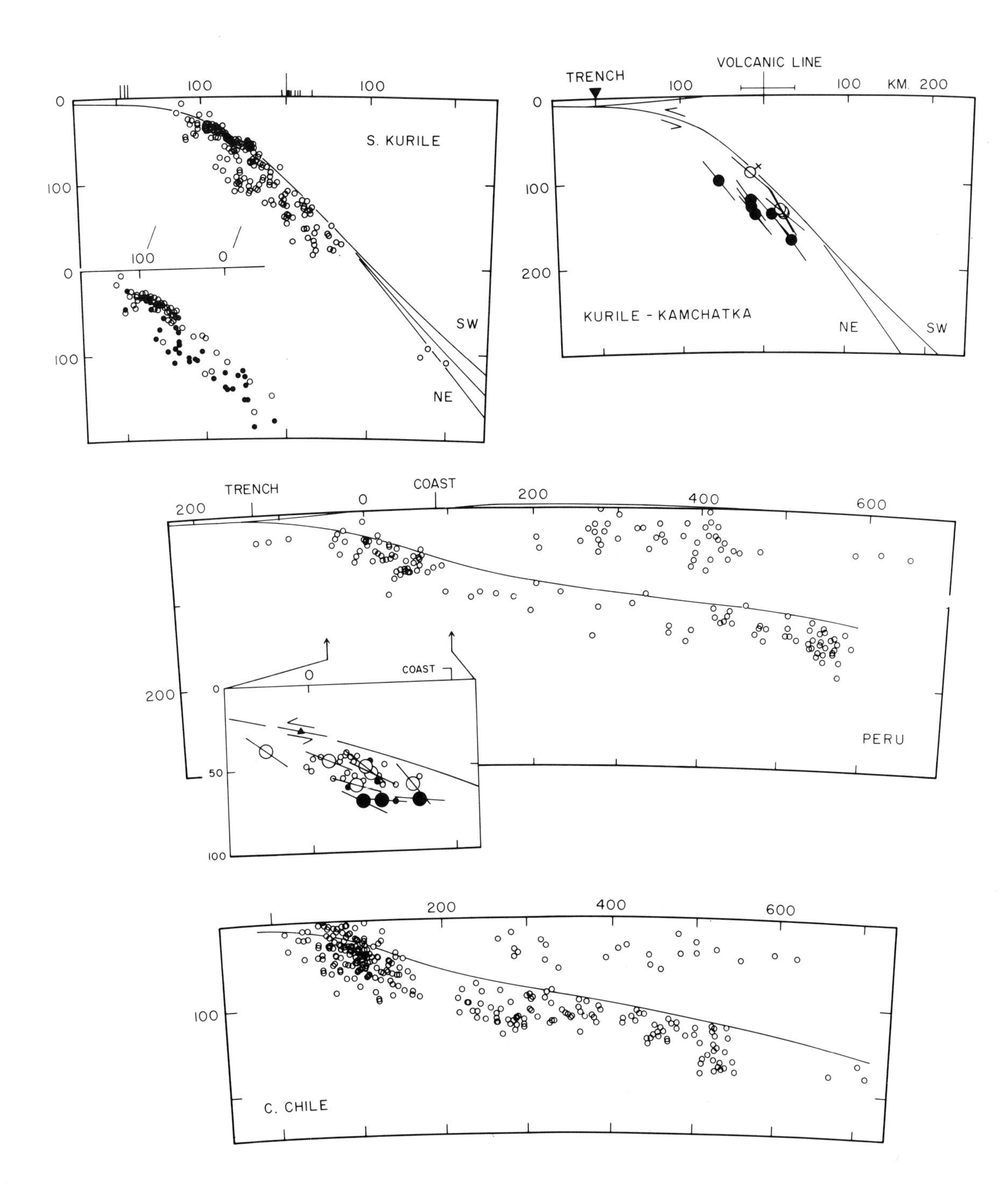
S. KURILE
SW
NE
TRENCH
VOLCANIC LINE
KM. 200
KURILE - KAMCHATKA
COAST
PERU
C. CHILE

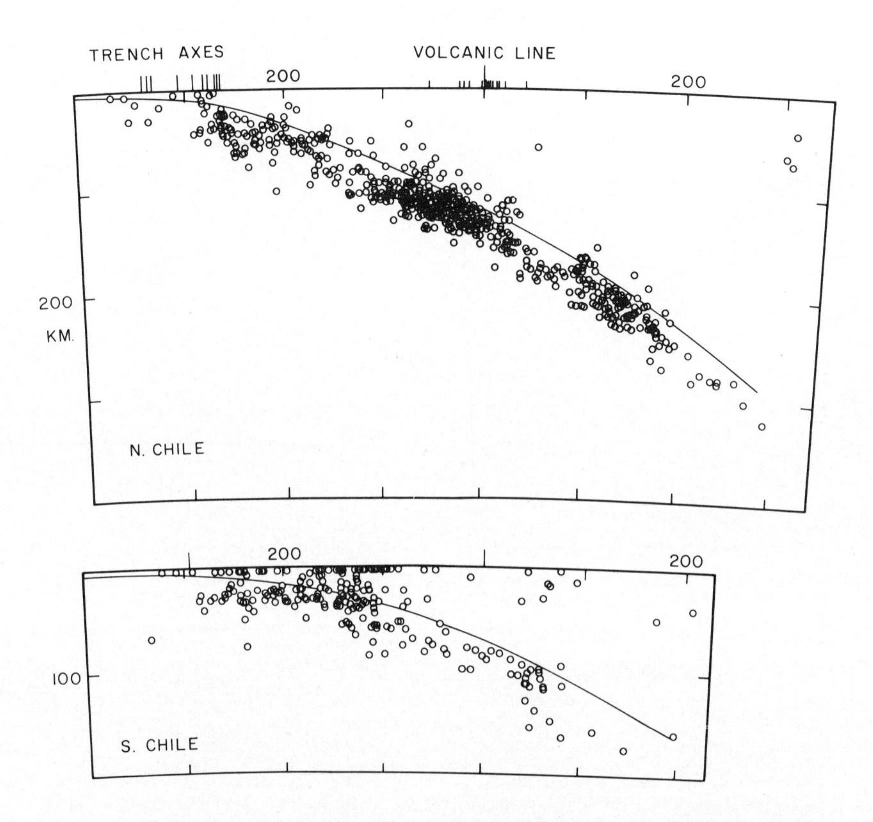

Figure 3. Composite cross-sections of earthquake hypocenters, for the segments indicated and named in Figures 2, 4, 5, and 6. The "0" km on the horizontal scale corresponds to the volcanic lines shown on the maps, and the projection of volcanoes and trench axes are shown by the vertical lines. The trench axes are for the narrow width sections used to make the composites. The focal mechanism data for the Peru and Kurile-Kamchatka sections show the projections of the P axis (open circles) or the T axis (filled circles) at the location of the particular earthquake in the cross-section. The small "x" in the Kurile-Kamchatka section indicates the location of one of the events, while the symbol is plotted within the inferred top of the lithosphere. The solid and open circles in the inset in the south Kurile section represent the most precisely located events of the two halves of the section. The small open and filled symbols in the Peru section (the inset) show inferred down-dip compression and extension, respectively, based on limited first motion data for aftershocks with accurate depths of the May 31, 1970 Peru earthquake. Most of the aftershocks of this event had a first motion pattern opposite to that of the initial event, and, apparently, stress relief on one side of the plate triggered earthquakes on the opposite side. The triangle indicates the thrust-type mechanism of the large event of October, 1966 [Abe, 1972b].

above cited studies of fault planes of deep earthquakes and data for Peru to be described in a later section of the paper (see Figure 3) to be convincing evidence that in many areas the seismically active part of the descending lithosphere is about 20 to 30 km thick. The apparent thicknesses revealed in the sections for the Izu-Bonin-Mariana, Tonga-Kermadec, and Peru-Chile regions are somewhat greater than this, and reflect errors in location which are larger than those for the relocations of Vieth and Pascal et al.

Given these real and apparent thicknesses of the seismic zone and the problems that come from a limited sample of a sporadic and clustered series of events, it becomes very difficult to infer small contortions or tears in the geometry of the descending plate unless a very large and precise data set is available. In particular we cannot find any substantial evidence for the pervasive offsets and changes in dip of the Benioff zone on the scale of 50-100 km that are claimed, for example, by Carr et al. [1973], Swift and Carr [1974], and Vieth [1974].

Composite Cross-Sections

Our procedure is first to examine the map plots and cross-sections for indications of substantial lateral changes, offsets or tears and then, together with the spatial distribution of volcanoes and trench bathymetry, define segments in which the structure is relatively uniform and simple. From this, narrow cross-sections are obtained perpendicular to the trend of the structure. The overall alignment of volcanoes best defines the trends of coherent segments, and on this basis a "volcanic line", either curved or straight, is defined on a map view and identified on each of the narrow sections. These sections are then composite together, shifting the sections if necessary to achieve a best fit of the overall distribution of hypocenters. If the structure is relatively uniform the shifts will be minimal and the composite sections will not have a thickness appreciably greater than that of any of the narrow sections. If substantial offsets or changes in dip exist, then the composite will be thick compared to the narrow sections, the shifting about will be substantial, or both. In all of the cases shown in Figure 3, the first situation prevailed, at least down to a certain depth. The shifts were within the errors of analysis or in some cases indicated a slight misprojection and led to a redetermination of the "volcanic line".

The depths below which the trend of the Benioff zone begins to diverge from the shallow part varies somewhat, as indicated in the sections, but is in most cases deeper than 200 to 300 km. As depth increases this divergence increases and the structure becomes considerably more complex, as shown by the map views of Figures 4, 5, and 6.

We conclude that at least down to depths of 200 km the sections shown in Figure 3 are good representations of a relatively uniform cross-sectional structure of the major segments considered, and that within these segments there are no clearly resolvable contortions or offsets involved in an intricate segmentation of the descending plate.

One possible exception to this which is under current investigation is the complex shallow structure of the New Hebrides in the region of the islands of Santo, Malekula and Efate, approximately adjacent to where the D'Entrecasteaux

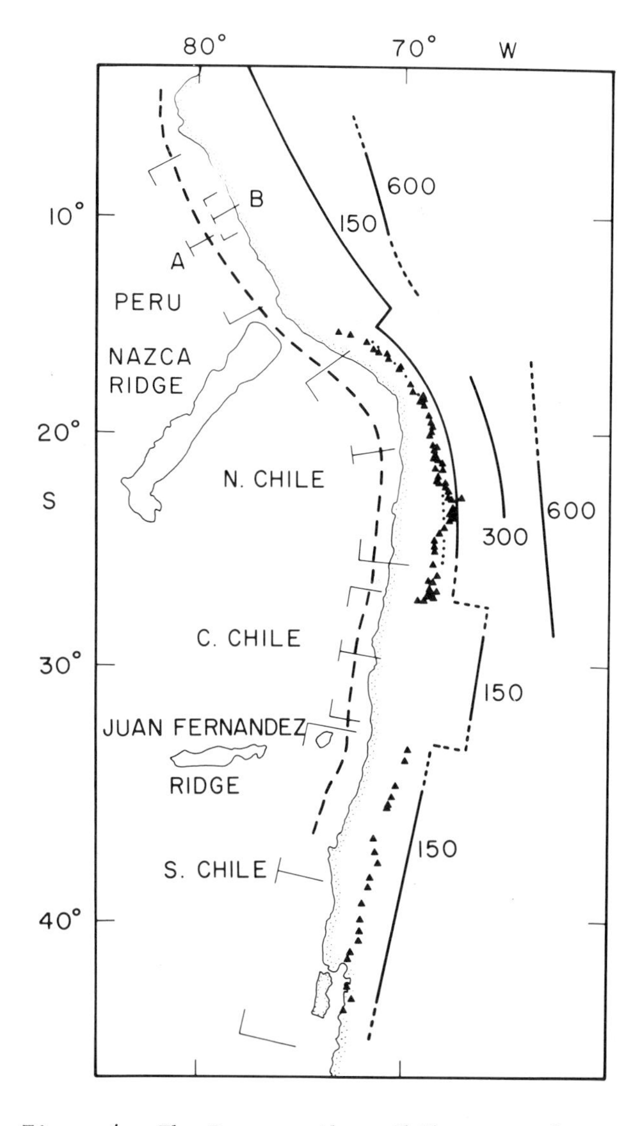

Figure 4. The Peru, northern Chile, central Chile, and southern Chile segments of western South America from data of Barazangi and Isacks [1976], and symbols as in Figure 2.

fracture zone intersects the arc [Pascal, 1974; Choudhury et al., 1975]. However, Pascal et al. [1977] and unpublished data of Cornell and the Office de la Recherche Scientifique et Technique Outre Mer (ORSTOM) from local seismic networks indicate that this complexity is probably more related to deformation of the upper plate than to segmentation of the descending plate.

In a map view most of the segments defined are nearly straight or have a slight curvature, convex seaward, as is implied by the straight lines drawn on the mercator projections of Figures 4, 5 and 6. In the Mariana and northern Chile, the curvature is significant and convex seaward for the Mariana and concave seaward for northern Chile. Note that the Kurile-Kamchatka arc, long taken as the classical example of simple curvature of island arcs, can be resolved into two more nearly straight segments.

New Results for South America

The compilation of data for South America deserves special emphasis since this region has been the subject of very diverse interpretation for some time. Barazangi and Isacks [1976] discuss the data in more detail. The data is now sufficient to define four major segments in Peru and Chile, as shown in Figures 3 and 4, two with very small dips and two with somewhat steeper dips. The flattish section beneath Peru was recognized by Isacks and Molnar [1971] and Sykes [1972] and is further developed in our new compilation. In addition, the flat zone in Chile, suggested by Stauder [1973], is now clearly delineated.

If the northern and central Chile segments are combined into one section, a diffuse wedge of seismicity results. This misleading effect, plus inclusion of more poorly located events may explain previous interpretations of the seismicity of western South America in terms of a wedge of seismicity in the upper mantle [James, 1971; Sacks and Okada, 1974], rather than in terms of a reasonably well-defined inclined seismic zone plus crustal seismicity

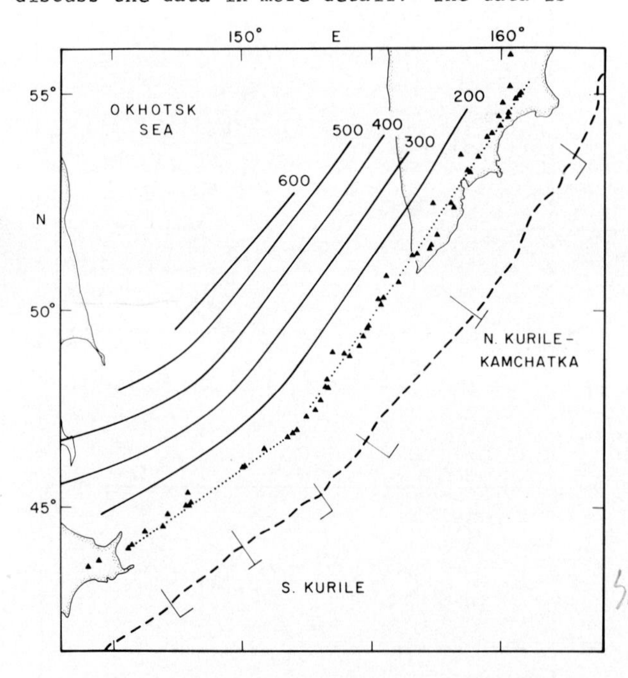

Figure 5. The Kurile-Kamchatka segments based on relocations of Vieth [1974] and volcanoes from Gorshkov [1958] and Vlodavetz and Piip [1959]. Trench axis from Chase et al. [1970]. Note that the northernmost volcano shown on the map is the southernmost one of a trend parallel to the volcanic line but offset to the northwest. The Benioff zone flattens considerably beneath the Hokkaido "corner" southwest of this map.

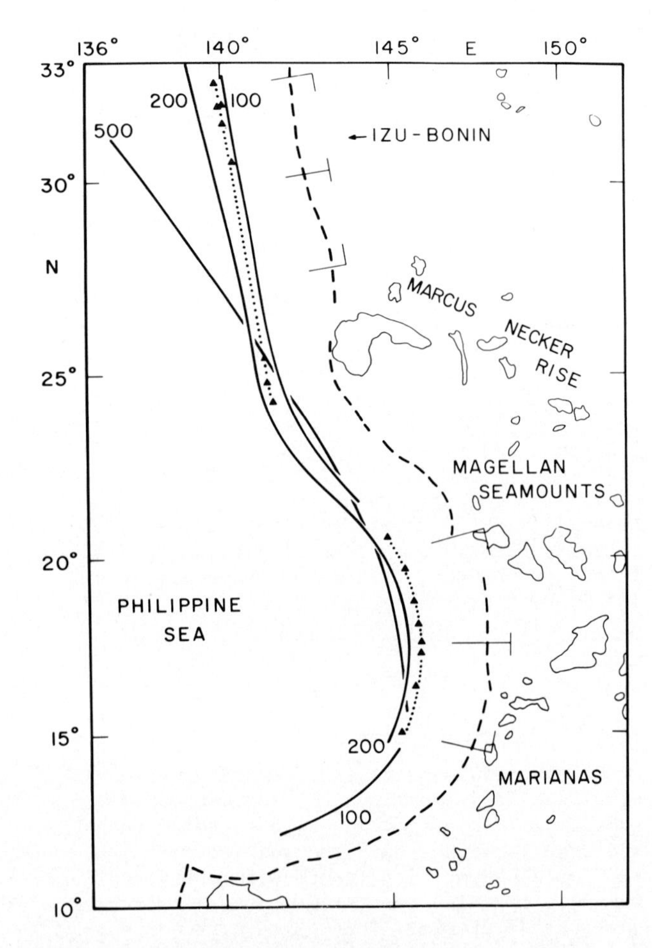

Figure 6. The Izu-Bonin and Mariana segments, as in Figure 2. Volcanoes from Kuno [1962], and trench axes from Chase et al. [1970]. Note that the contours drawn between the Izu-Bonin and Mariana segments are based on scanty data; although the overriding geometry seems clear, the structure of the zone could be considerably more complex than that shown here.

within the overriding South America plate. The flat zones effectively double the thickness of lithosphere in the region of Peru and central Chile, those regions would appear to have an anomalously thick lithospheric plate, as has also been suggested.

Comparison of Sections: Downwards Bending of the Descending Lithosphere

Upper Surface of the Descending Lithosphere

The curves shown in the cross-sections are estimates of the top of the descending lithosphere based upon the following constraints: (1) the seaward side of the oceanic trench is taken to be the upper surface of the plate just before thrusting beneath the landward plate; (2) the approximate location and dip of the major zone of thrust fault contact between the converging plates is indicated by the locations of shallow earthquakes and by the plunges of the slip vectors derived from focal mechanism solutions; and (3) the upper envelope of the spatial distribution of mantle earthquakes, with some interpretation exercised in the exclusion of points judged to represent scattering due to location errors.

Clearly such curves are somewhat subjective and only crude approximations, but they are nevertheless useful for obtaining a general idea of the shape of the downgoing plate and for comparisons between regions. One problem is that it is not known exactly where the seismically active part of the descending lithosphere is in relationship to the downgoing suboceanic lithosphere. Before subduction the lithosphere has an upper boundary at the top of the oceanic crust and a lower boundary perhaps 70 km deep at the top of the low velocity zone. The curves drawn in the cross-sections imply that the seismic zone is closer to the upper part of the lithosphere (see, for example, Engdahl et al., 1976). Another problem is the systematic error in locations caused by the high velocity slab. Vieth's method of relocation attempts to correct this error but the other data considered in this paper are subject to this error.

Figure 7 is a composite of the several curves given in the separate sections, with small differences ignored in order to emphasize the overall similarities of several of the regions and the large departures from these exhibited by certain other regions, notably New Hebrides and South America. Also included on this figure are curves from regions where particularly good data are available in the literature, including the Aleutian [Engdahl, 1971 and 1973], New Zealand [Hamilton and Gale, 1968; Ansell and Smith, 1975], Alaska [Lahr, 1975], and Central America [Dewey and Algermissen, 1974].

The results seem to group into several categories according to the steepness of dip and amount of curvature. The steepest and most curved are New Hebrides and Central America, which are also subducting the youngest oceanic lithosphere of all those considered. In the next group there are several which are quite similar except for the distance from the trench axis to the region of large downwards curvature. Karig and Sharman [1975] and Karig et al. [1976] show that this is probably attributable to a variable amount of material accreted onto the overriding plate. This group includes Kermadec, Izu-Bonin, Aleutian, Alaska, and New Zealand. The Mariana zone follows this curve for the upper part but steepens somewhat with depth, while Tonga and the Kurile-Kamchatka zones flatten somewhat with depth, Kurile-Kamchatka more so than Tonga. In striking contrast are the relatively flat zones of South America. Results of Katsumata [1967] and other sources indicate that northeast Honshu would be rather similar to the profile shown for northern Chile. The Peru and central Chile zones actually appear to have a recurved section located inland of the coast, with downwards convex curvature, and are nearly flat beneath the Andean cordillera.

Volcanoes

With the assumption than mantle earthquakes occur within the descending plate, the depths beneath the volcanoes to the tops of the Benioff zones shown in the sections of Figure 3 are upper bounds on the actual depths. Whatever thickness of plate lies above the earthquakes will reduce the depth by that amount. At face value, the results indicate that the "volcanic lines" often lie about 100 km above the descending plate, while individual volcanoes and groups of volcanoes located seaward of that line are at even smaller distances above the plate.

There appear to be some small but systematic differences among the arc segments. For example, for most of the volcanoes of the south Kurile segment the depths to the Benioff zone are deeper than those for the volcanoes of the north Kurile-Kamchatka segment. Similarly, the depths of the Benioff zone beneath the volcanoes of northern Chile are deeper than those for southern Chile.

The most notable deviation occurs in the New Hebrides where the volcanic line occurs about 180 km above the top of the descending slab. However, the very steep dip of the New Hebrides seismic zone would greatly amplify the effects of any systematic error of epicentral location, i. e., any systematic shift in hypocenters perpendicular to the arc.

Perhaps the most remarkable feature of the segmentation of the Nazca plate beneath western South America is the correlation of the flat

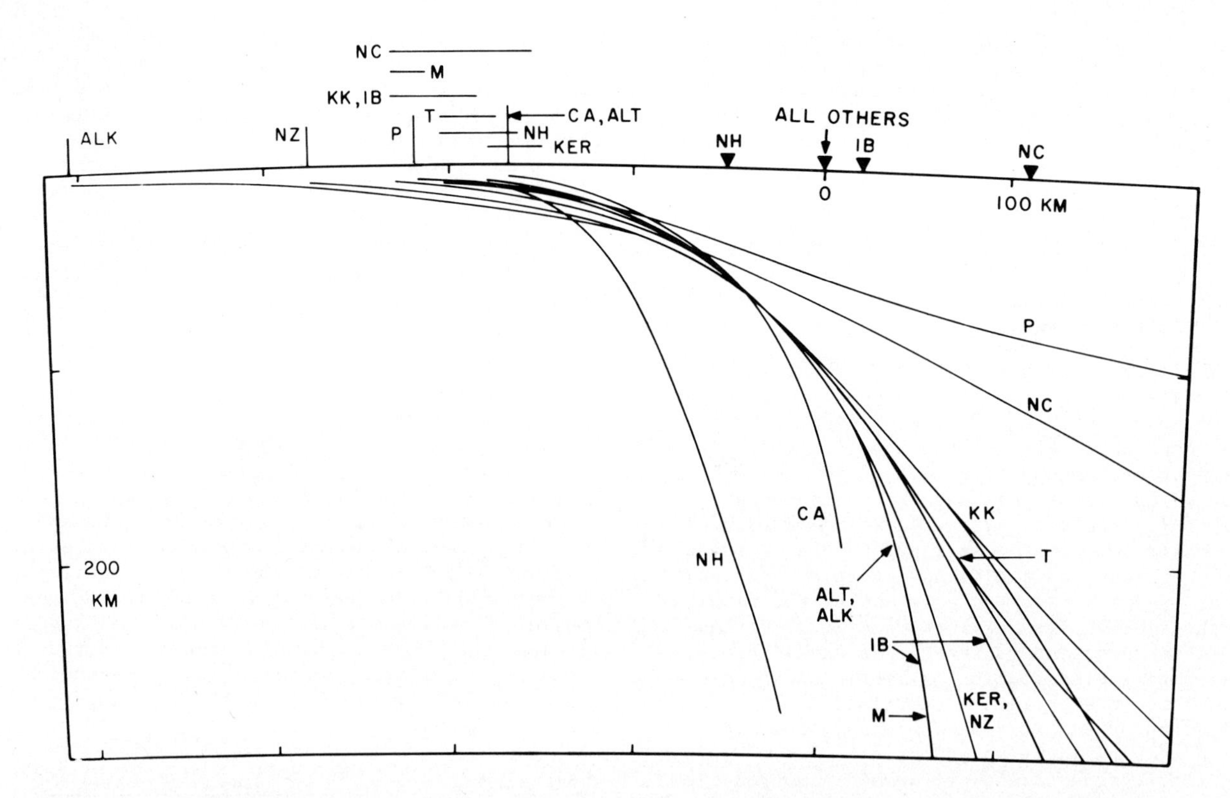

Figure 7. Comparison of the sections of Figure 3. The segment names are abbreviated for clarity (NH = New Hebrides, CA = Central America, ALT = Aleutian, ALK = Alaska, M = Mariana, IB = Izu-Bonin, KER = Kermadec, NZ = New Zealand, T = Tonga, KK = Kurile-Kamchatka, NC = North Chile, P = Peru). The sections are offset in certain cases with respect to the volcanic lines (solid triangles) to emphasize similarities and differences in the geometry. The central Chile and southern Chile segments are essentially identical to the Peru and northern Chile segments, respectively, and are not shown.

sections with the absence of Quaternary volcanism. This was recognized previously for Peru and it is now seen to be the case for the flat section in central Chile as well. The absence of volcanoes cannot be simply an effect of the depth reached by the descending slab because the earthquake zones reach depths greater than the 100 km or so above which volcanoes are found in other regions, including South America. The absence of volcanoes has also been explained by the effect of a high compressive stress in the overriding continental plate in the region of Peru [Sykes, 1972]. A strong coupling between the flat subducted slabs and the overriding continental plate could be an important factor in this process. The relatively high level of seismic activity in the crust beneath the eastern flanks of the Andes (the sub-Andean zone) may be correlated with the location of the flat slabs [Barazangi and Isacks, 1976] and with the region of high compressive stress.

Another explanation for the lack of volcanoes is the absence of a wedge of asthenosphere between the descending and upper plates. If the gently dipping slab were in contact with the overriding lithosphere, which may be the case if the continental South American plate has a lithospheric thickness of about 130 km, there would be no wedge of asthenosphere and thus no material hot enough to be partially melted by the infusion of water, for example, from the dehydration of material in the subducted plate [e.g., Anderson et al., 1976; see also Ringwood, this volume].

The more steeply dipping zone of northern Chile has above it the numerous Quaternary stratovolcanoes forming the curved line shown in Figure 4. The southern part of this line is bent first inland and then seaward. There is no evidence in our analysis of narrow cross-sections [Barazangi and Isacks, 1976] for a corresponding contortion of the descending plate. The Altiplano, the broad elevated region of northern Chile, western Bolivia and southern Peru, is also approximately coincident with and above the north Chile segment shown in Figure 4.

Bending of the Lithosphere

The shapes and similarities of the profiles illustrated in Figure 7, in addition to the uniformity of cross-sectional geometry within the large segments of arc, suggest to us that the mechanics of bending is important in the upper several hundred kilometers of subduction zones. The concept of elastic bending of a thin plate has proven to be quite successful in explaining the topography of the outer trench wall and outer rise. For example, Caldwell et al. [1976] derive thicknesses of an elastic plate of 20 to 30 km. Such a thickness is thought to correspond to the upper part of the lithosphere where temperatures are small enough for the bending stresses not to be relaxed by creep during the passage of the plate over the rise and into the trench. As the slab descends into the mantle, the upper part will remain cold to a substantial depth. It may be significant that the estimates of thickness of the seismic zone mentioned in a previous section are quite close to the elastic thickness calculated for the bending of the suboceanic lithosphere.

The curvature of the descending plate increases substantially from the region of small displacements and elastic deformation seaward of the trench. A radius of curvature of 200 km is typical, for example, and for a 20 km thickness this implies a fiber strain near the surface of the plate of 0.05. The corresponding elastic stress difference would be quite large, about 50 kilobars. It is thus probable that the bending of the plate landward of the trench involves plastic deformation. Specific evidence for this is described below.

Isacks and Molnar [1971] describe focal mechanism solutions of intermediate-depth earthquakes in the Kurile-Kamchatka region which indicate down-dip compression in some cases and down-dip extension in others. This pattern is not explained by the simple model of axial stresses induced by gravitational sinking of slabs. Stauder and Mualchin [1976] and Vieth [1974] provide new solutions which further develop that puzzling pattern. Vieth, however, shows that the two types of mechanisms occur in different parts of the cross-sectional profile of the seismic zone. The down-dip compressional types are nearer the top surface of the inclined seismic zone and the down-dip tensional types are on the bottom of the zone. In Figure 3 we show the mechanism solutions of Isacks and Molnar, and Stauder and Mualchin, but use the locations of Vieth where available and of the ISC otherwise, and plot the results on our sections. The top and bottom separation of the two types of mechanisms is quite clear and remarkable.

For Peru, focal mechanism solutions of Abe [1972b], Stauder [1975], and unpublished ones of our own, indicate four types of mechanisms, three of which are often found in other areas. The three common types are interpretable as: (1) underthrusting at shallow depths beneath the continental slope, directly representing relative motion between the converging plates; (2) down-dip extension within the subducted plate beneath the trench, the coast, and at intermediate depths inland; and (3) thrust and/or strike-slip mechanisms within the Andean crust, presumably indicative of horizontal east-west compression within the continental plate. The fourth type occurs beneath the coast at depths of about 40 to 60 km, is of the thrust type but with an orientation quite distinct from that of the typical underthrust type (1). The orientation of the compressional axes is approximately parallel to the dip of the inclined seismic zone, and thus parallel to the down-dip extensional stress axes of the solution for the other, somewhat deeper, coastal earthquakes. The Peru section in Figure 3 shows the relationship of these earthquake mechanisms to a cross-sectional view of the seismic zone, and shows the spatial relationships of the two types of earthquakes. The depths of all events depicted in the inserted section are based on redetermined pP-P times and are thus accurate. These events appear to occur within the subducted plate. We can thus interpret these data in terms of down-dip compression on top of the inclined seismic zone and down-dip tension on the bottom, in a manner similar to the interpretation of the Kurile-Kamchatka data, although the separation in space is not quite as clear in the case of Peru.

These two cases, schematically summarized in Figure 8, seem to be most easily explained in terms of bending effects. The orientation of stress is the opposite of that expected for simple elastic bending of a plate with the curvatures shown, so we must appeal to a kind of elastic "unbending" after a plastic deformation. In the case of Peru the curvature appears to reverse and become concave upwards, but this occurs somewhat inland of the earthquakes of interest. An oversimplified model for the unbending effect is shown in Figure 9. The fiber strains are estimated from the radii of curvature, and indicate that the deformation probably exceeds the range of elastic behavior. The elastic-plastic model shown in Figure 9 is given a yield stress of 5 kilobars only for the sake of illustration.

These results modify the conclusions of Isacks and Molnar [1971] regarding the unimportance of bending effects in interpreting the orientations of focal mechanism solutions for earthquakes in the descending lithosphere. Besides explaining the Kurile-Kamchatka and Peru results, bending or unbending may also account for the down-dip compressional mechanisms at intermediate depths in the Ryukyu and the South Sandwich arcs; the model of gravitational sinking of the slab did not account for those mechanisms.

The model shown in Figure 9 suggests that the orientation of stress along the top or the bottom of the plate may be a very sensitive function of the temporal change in curvature rather than of the curvature itself. It is also important to note that the interpretation as bending stresses depends upon having two opposite types of mechanisms on top and bottom of the seismic zone, respectively. With only one type, as is the case in most regions, one cannot distinguish bending stresses from the axial stresses which result from the slab being pulled or pushed downwards.

Three-Dimensional View: Relationship of Arc Segmentation to Structures in the Converging Plates

Junctions and Terminations of the Segments

The data considered in this paper are sufficient in only three cases to reveal whether, from one segment to another, the descending plate is laterally continuous or not. These cases include Tonga/Kermadec, south Kurile/north Kurile-Kamchatka, and Peru/north Chile. Figure 10 shows evidence that between the Tonga and Kermadec segments the intermediate-depth zone is continuous. Similarly, a map view of Vieth's data does not reveal any lack of continuity in the inclined zone between the northern and southern Kurile segments.

In contrast, the data for South America reveal a fairly definite and abrupt change in depth to the Benioff zone between the Peru and north Chile segments as is illustrated in Figure 11 from Barazangi and Isacks [1976]. This seems to be readily interpretable as a tear in the descending plate, and is close to the northern end of the line of Quaternary volcanoes in southern Peru. Although the data are not as clear, a similar tear is suggested between the central and southern Chile segments, as indicated in Figure 4. The transition between northern and central Chile segments is characterized by a gap

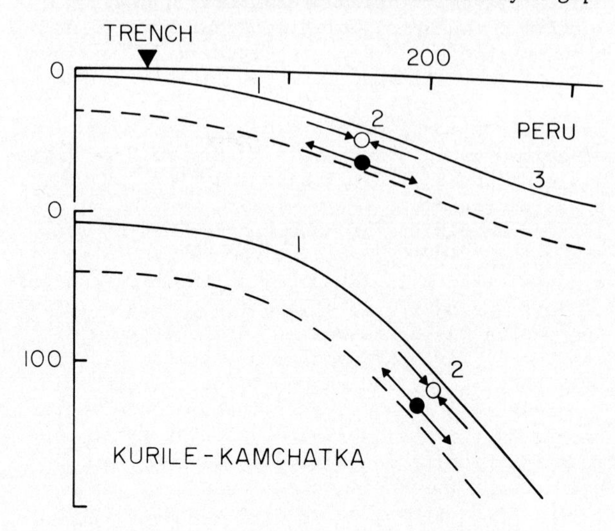

Figure 8. Schematic summary of focal mechanism data for Peru and Kurile-Kamchatka interpreted in terms of bending and illustrating results shown in Figure 3. The numbers correspond to points on the stress-strain curve of Figure 9.

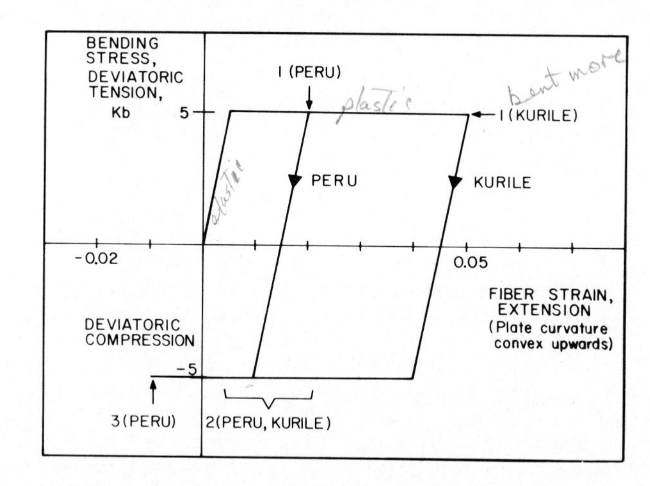

Figure 9. Idealized elastic-plastic stress-strain curve for material on top of the plate as it is transported down the subduction zone. The numbers correspond to different points in time as the material moves around the bend or to points in space for different parts of the plate as shown in Figure 8.

in intermediate-depth activity, and its character is not known. Limited evidence also suggests a steepening of the seismic zone at the beginning of the line of active volcanoes in Ecuador, north of the flat Peru segment, but the nature of the transition is not known.

Relationships to Features in the Converging Plates

The Mariana and Tonga segments coincide with presently active zones of inter-arc spreading, the Mariana trough in the one case and the Lau basin in the other. As one proceeds along strike from the Izu-Bonin to the Mariana, or from Kermadec to Tonga, the amount of Pliocene to Recent inter-arc spreading changes quite remarkably, but there is little expression of this in the cross-sectional shape of the Benioff zone at least down to about 300 km. The seaward offset of the trench and frontal arc of the Mariana (relative to the Izu-Bonin arc) is of course quite striking. At depths below 300 km, a marked flattening of the Tonga zone seems correlated with the fan shaped opening of the Lau basin (see Figure 2). In contrast, the Mariana zone steepens to a vertical dip beneath the seaward side of the Mariana trough (Figure 6). These results do not suggest a simple relationship between the cross-sectional shape of the descending plate and the opening of inter-arc basins.

The coincidence of the intersection of the Louisville ridge with the junction between the Tonga and Kermadec segments and with the apex of the triangular Lau basin is interesting, and suggests that the Pliocene-to-Recent opening of

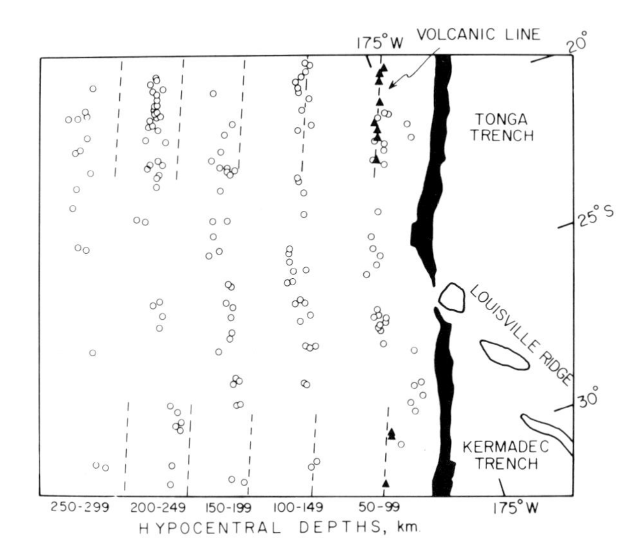

Figure 10. Plots of epicenters of earthquakes within specified intervals of depth. The map view is successively offset to the left for each interval of depth, for clarity, with the volcanic lines repeated as a reference. The data are interpreted to indicate a continuous zone between the Tonga and Kermadec segments.

the Lau basin [e.g., Karig, 1970; Gill, 1976] may be related to the subduction of the Louisville ridge. If this ridge is extended along the trend suggested by Larson and Chase [1972], it would have been subducted obliquely and its intersection with the arc would sweep rapidly southwards. Reconstruction of the motions of the plates in the region from the results of Molnar et al. [1975] and Weissel and Hayes [1972] for the Pacific versus Australian motions and with the closure of the Lau basin shows that the Louisville ridge would have swept down from the northern end of the arc to its present position essentially during the opening of the Lau basin. Projected downwards along the seismic zone (see also Vogt et al., 1976), the subducted part of the Louisville ridge would be located in the flat part of the seismic zone between the 300 and 400 km hypocentral depth contours shown in Figure 2. Thus the subduction of the Louisville ridge is closely related in time and space to the opening of the Lau basin and perhaps to the flattening of the Benioff zone below 300 km at the northern end of Tonga.

Unlike the Louisville ridge the Juan Fernandez ridge intersects Chile at a nearly right angle to the coast. This feature projects inland to the presumed break between the central and south Chile segments and close to the northern end of the Quaternary volcanoes of southern Chile. The Nazca ridge, however, intersects the coast and projects inland about 200 km north of the well-defined break in the slab beneath southern Peru. Whereas the structure in the plate responsible for the Juan Fernandez ridge may actually be a weakness along which the plate breaks, the Nazca ridge may be entirely within the flat Peru segment. These inferences, of course, depend upon the extrapolation of trends in the suboceanic plate down along the subducted plate. There is little evidence on how this should be done for the ridges of the Nazca plate.

The D'Entrecasteaux "fracture zone" also appears to intersect the New Hebrides arc at a nearly right angle, but as discussed before, appears to have little effect on the overall structure of the descending plate. The corre-

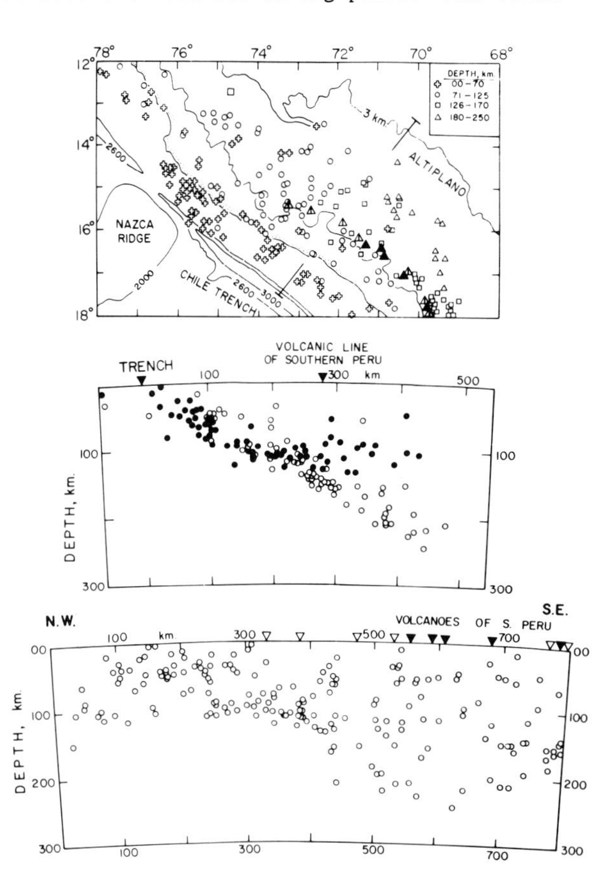

Figure 11. Map view, cross-section, and longitudinal section of the distribution of earthquakes in southern Peru and northern Chile, showing the abrupt increase in hypocentral depth in passing from the flat Peru segment to the more steeply dipping northern Chile segment. The filled and open circles in the middle cross-section are for events to the northwest and southeast, respectively, of the line indicated in the map view. The longitudinal section is perpendicular to that line. The solid triangles are the historically active volcanoes. The large open triangles are historically inactive stratovolcanoes thought to be of Quaternary age, (after Barazangi and Isacks, 1976).

lation of this topographic feature with the absence of a trench, the anomalous seaward locations of the islands of Santo and Malekula, the presence of the Aoba basin, and other anomalous features of the upper plate is quite striking.

The Magellan seamounts and Marcus Necker seamount chain are colliding with the Mariana and southern Izu-Bonin arcs, but unfortunately the seismicity between these two segments is still too sparse to obtain a clear view of the effects on the geometry of the descending plate. The continuous contours drawn in Figure 6 represent the available data but should not be taken to exclude a more complex or broken structure.

Deep Structure

The emphasis of this paper has been on the geometry of the plates above 300 km, and in this section we will only briefly review some of the features of deep structure which have emerged from our compilation of data, as illustrated by the contours of hypocentral depth shown in Figures 2, 4, 5, and 6.

In general the data show a warping and contortion of the deep zone relative to the shallow. In the Izu-Bonin region this warping seems to include a fan-shaped decrease in dip of the zone northwards and an overriding of the intermediate-depth zone over the deeper part in the region south of the Bonin islands.

As mentioned above, the northern part of the Tonga arc flattens at depths of 300 to 400 km. This zone may actually override the zone of deep earthquakes beneath the Fiji islands. The deep zone, represented by the nearly east-west contours in Figure 2, appears from the new data to be part of an overall west-northwest to east-southeast trending zone of deep earthquakes beneath the Fiji plateau. This trend can be extended to include recent well-located deep earthquakes beneath the eastern Solomon arc. The trend is approximately parallel to the supposed subduction system operative along the northern margin of what is now the Fiji plateau, as suggested by Chase [1970] and Gill and Gorton [1973], for example. That system would have included a New Hebrides arc rotated back to connect with Fiji and reversed in polarity. The Vitiaz trench is presumably a remnant of that system.

The recent deep earthquakes beneath the southwestern part of the Fiji plateau, forming a line roughly parallel to the New Hebrides arc, may be part of the plate subducted as the New Hebrides arc rotated around to its present position. It is also possible to place these events on the down-dip continuation of the flattened part of the Tonga zone, although the relationship of such a structure to the southern part of the Tonga zone is not clear.

Concluding Remarks

The segmentation described in this paper, plus the uniform geometry of the upper 300 km within the segments, suggest to us that the mechanics of bending, generally neglected in simple treatments of the deformation of spherical shells [e.g., Frank, 1968; Strobach, 1973; De Fazio, 1974], will be important and probably dominant in determining the geometry of the upper 300 km. The segmentation into separate flaps, such as in South America or in the region of Japan [Aoki, 1974] allows the plate to bend about a relatively straight axis.

The contrast between the South American sections and the others is quite remarkable, and suggests to us that the shape and thickness of the overriding plate may play a more significant role in that region than in other areas, where the overriding plates are thinner [e.g., see Jischke, 1975].

The evidence for elastic "unbending" stresses and for the thickness of the seismic zone, combined with the successful treatments of deformations of the lithosphere as a flexuring of an elastic plate all agree with Turcotte's [1974] view that the mechanics of the lithosphere is governed largely be an approximately 30 km thick zone in which stresses are not relaxed during times of several million years. The results of this paper suggest that this view should be applied to the descending plate at least down to 300 km.

Acknowledgments. We thank D. Turcotte, D. Karig, D. McAdoo, and J. Caldwell for useful discussions, and G. Cole, S. Billington, D. Chinn, J. Ni, N. Fitzpatrick for assistance on different aspects of the paper. This work is supported by National Science Foundation Grants DES75-14815 and DES74-03647.

References

Abe, K., Seismological evidence for a lithospheric tearing beneath the Aleutian arc, Phys. Earth Planet. Interiors, 5, 190,1972a.

Abe, K., Mechanisms and tectonic implications of the 1966 and 1970 Peru earthquakes, Phys. Earth Planet. Interiors, 5, 367,1972b.

Anderson, R. N., S. Uydea, and A. Miyashiro, Geophysical and geochemical constraints at converging plate boundaries-Part I: Dehydration in the downgoing slab, Geophys. J. R. Astr. Soc., 44, 333, 1976.

Ansell, J. H. and E. G. C. Smith, Detailed structure of a mantle seismic zone using the homogeneous station method, Nature, 253, 518, 1975.

Aoki, Plate tectonics of arc-junctions at central Japan, J. Physics of the Earth, 22, 141, 1974.

Barazangi, M. and B. L. Isacks, Spatial dis-

tribution of earthquakes and subduction of the Nazca Plate beneath South America, Geology, in press, 1976.
Billington, S. and B. L. Isacks, Identification of fault planes associated with deep earthquakes, Geophys. Res. Letters, 2, 63, 1975.
Caldwell, J. G., W. F. Haxby, D. E. Karig and D. L. Turcotte, On the applicability of a universal elastic trench profile, Earth Planet. Sci. Letters, 31, 239, 1976.
Carr, M. J., R. E. Stoiber, and C. L. Drake, Discontinuities in the deep seismic zones under the Japanese arcs, Bull. Geol. Soc. Amer., 84, 2917, 1973.
Chase, C. G., Tectonic history of the Fiji plateau, Bull. Geol. Soc. Amer., 82, 3087, 1970.
Chase, T. E., H. W. Menard and J. Mammerickx, Bathymetry of the North Pacific, Scripps Inst. Oceanography, La Jolla, California, 1970.
Choudhury, M. A., G. Poupinet and G. Perrier, Shear velocity from differential travel times of short-period ScS-P in New Hebrides Fiji-Tonga, and Banda Sea regions, Bull. Seismol. Soc. Amer., 65, 1787, 1975.
De Fazio, T. L., Island-arc and underthrust-plate geometry, Tectonophysics, 23, 149,1974.
Dewey, J. W. and S. T. Algermissen, Seismicity of the Middle America arc-trench system near Managua, Nicaragua, Bull. Seismol. Soc. Amer., 64, 1933, 1974.
Engdahl, E. R., Explosion effects and earthquakes in the Amchitka island region, Science, 173, 1232, 1971.
Engdahl, E. R., Relocation of intermediate depth earthquakes in the central Aleutians by seismic ray tracing, Nature Phys. Sci., 254, 23, 1973.
Engdahl, E. R., N. H. Sleep, and M. Lin, Plate effects in north Pacific subduction zones, Tectonophysics, in press, 1976.
Fisher, N. H., Catalogue of the Active Volcanoes of the World, Part V, Melanesia, Int. Volcanologist Association, Napoli, Italia, 1957.
Frank, F., Curvature of island arcs, Nature, 220, 363, 1968.
Fukao, Y., Source process of a large deep-focus earthquake and its tectonic implication - the western Brazil earthquake of 1963, Phys. Earth Planet. Interiors, 5, 61, 1972.
Gill, J. B., From island arc to oceanic islands: Fiji, southwestern Pacific, Geology, 4, 123, 1976.
Gill, J. and M. Gorton, A proposed geological and geochemical history of eastern Melanesia, in The Western Pacific: Island Arcs, Marginal Seas, Geochemistry, P. J. Coleman, ed., Univ. of W. Australia, 543, 1973.
Gorshkov, G. S., Catalogue of the Active Volcanoes of the World, Part VII, Kurile Islands Int. Volcanologist Association, Napoli, Italia, 99p., 1958.
Hamilton, R. M., and A. W. Gale, Seismicity and deep structure of North Island, New Zealand, J. Geophys. Res., 73, 3859, 1968.
Isacks, B. and P. Molnar, Distribution of stresses in the descending lithosphere from a global survey of focal-mechanism solutions of mantle earthquakes, Rev. Geophys. Space Phys., 9, 103, 1971.
James, D. E., A plate tectonic model for the evolution of the central Andes, Bull. Geol. Soc. Amer., 82, 3325, 1971.
Jischke, M. C., On the dynamics of descending lithospheric plates and slip zones, J. Geophys. Res., 80, 4809, 1975.
Kanamori, H., Great earthquakes at island arcs and the lithosphere, Tectonophysics, 12, 187, 1971a.
Kanamori, H., Focal mechanism of the Tokachi-Oki earthquake of May 16, 1968: contortion of the lithosphere at a junction of two trenches, Tectonophysics, 12, 1, 1971b.
Karig, D. E., Ridges and basins of the Tonga-Kermadec island arc system, J. Geophys. Res., 75, 239, 1970.
Karig, D. E., and G. Sharman, Subduction and accretion in trenches, Bull. Geol. Soc. Amer., 86, 377, 1975.
Karig, D. E., J. G. Caldwell and E. M. Parmentier, Effects of accretion on the geometry of the descending lithosphere, J. Geophys. Res., in press, 1976.
Katsumata, M., Seismic activities in and near the Japanese islands, J. Seismol. Soc. Japan (Zisin), 20 (in Japanese), 1, 1967.
Katsumata, M. and L. R. Sykes, Seismicity and tectonics of the western Pacific: Izu-Mariana-Caroline and Ryukyu-Taiwan regions, J. Geophys. Res., 74, 5923, 1969.
Kuno, H., Catalogue of the Active Volcanoes of the World, Part XI, Japan, Taiwan and Marianas, Int. Volcanologist Association, Napoli, Italia, 1962.
Lahr, J., Detailed seismic investigation of Pacific-North America plate interaction in southern Alaska, Ph.D. Thesis, Columbia University, New York, 141p, 1975.
Larson, R. L. and C. G. Chase, Late Mesozoic evolution of the western Pacific Ocean, Bull. Geol. Soc. Amer., 83, 3627, 1972.
Mammerickx, J., T. E. Chase, S. M. Smith and I. L. Taylor, Bathymetry of the South Pacific, Scripps Inst. Oceanography, La Jolla, California, 1971.
Molnar, P., T. Atwater, J. Mammerickx and S. M. Smith, Magnetic anomalies, bathymetry and the tectonic evolution of the South Pacific since the late Cretaceous, Geophys. J. R. Astr. Soc., 40, 383, 1975.
Oike, K., On the nature of the occurrence of intermediate and deep earthquakes, 3. Focal mechanism of multiplets, Bull. Disaster Prevention Res. Inst., 21, 153, 1971.
Pascal, G., Contribution a l'etude de la seis-

micite des Nouvelles-Hebrides, Thesis, University of Paris, O.R.S.T.O.M., Paris, 1974.

Pascal, G., B. L. Isacks, M. Barazangi, and J. Dubois, Seismotectonics of the New Hebrides island arc, to be submitted to Geophys. J. R. Astr. Soc., 1977.

Richard, J. J., Catalogue of the Active Volcanoes of the World, Part XIII, Kermadec, Tonga and Samoa, Int. Volcanologist Association, Napoli, Italia, 1962.

Sacks, I. S. and H. Okada, A comparison of the anelasticity structure beneath western South America and Japan, Phys. Earth Planet. Interiors, 9, 211, 1974.

Stauder, W., Mechanism and spatial distribution of Chilean earthquakes with relation to subduction of the oceanic plate, J. Geophys. Res., 78, 5033, 1973.

Stauder, W., Subduction of the Nazca Plate under Peru as evidenced by focal mechanisms and by seismicity, J. Geophys. Res., 80, 1053, 1975.

Stauder, W. and L. Mualchin, Fault motion in the larger earthquakes of the Kurile-Kamchatka arc and of the Kurile-Hokkaido corner, J. Geophys. Res., 81, 297, 1976.

Strobach, K., Curvature of island arcs and plate tectonics, Zeitschrift fur Geophysics, 39, 819, 1973.

Swift, S. and M. Carr, The segmented nature of the Chilean seismic zone, Phys. Earth Planet. Interiors, 9, 183, 1974.

Sykes, L. R., The seismicity and deep structure of island arcs, J. Geophys. Res., 71, 2981, 1966.

Sykes, L. R., Seismicity as a guide to global tectonics and earthquakes, Tectonophysics, 13, 393, 1972.

Turcotte, D. L., Are transform faults thermal contraction cracks?, J. Geophys. Res., 79, 2573, 1974.

Vieth, K. F., The relationship of island arc seismicity to plate tectonics, Thesis, Southern Methodist University, Dallas, Texas, 1974.

Vlodavetz, V. I. and B. I. Piip, Catalogue of the Active Volcanoes of the World, Part VII, Kamchatka and Continental Areas of Asia, Int. Volcanologist Association, Napoli, Italia, 110p. 1959.

Vogt, P. R., A. Lowrie, D. R. Bracey, and R. N. Hey, Subduction of aseismic oceanic ridges: Effects on shape, seismicity and other characteristics of consuming plate boundaries, Special Paper 172, Geol. Soc. Am., 59p, 1976.

Weissel, J. K. and D. E. Hayes, Magnetic anomalies in the southeast Indian Ocean, Antarctic Research Series, 19, Antarctic Oceanology II: The Australia-New Zealand Sector, ed. D. E. Hayes, Amer. Geophys. Union, 165, 1972.

Note added in proof by the authors: We apologize that the name of Dr. Karl F. Veith has been misspelled throughout the paper. The name is spelled as Vieth instead of Veith.

BATHYMETRIC HIGHS AND THE DEVELOPMENT OF CONVERGENT PLATE BOUNDARIES

John Kelleher[1] and William McCann[2]

Lamont-Doherty Geological Observatory of Columbia University, Palisades, New York 10964

Abstract. The distribution of great earthquakes along subduction boundaries shows a spatial correlation with the distribution of major bathymetric features of the underthrusting ocean floor. At locations where aseismic ridges, zones of seamounts or other bathymetric highs intersect active trenches, great earthquakes have occurred rarely if at all during recorded history. By contrast, nearly all segments of the margins opposite smooth, low-lying ocean floor have experienced at least one known large earthquake. In addition, locations where rises intersect trenches often show other evidence of modified subduction such as the absence or near absence of low-angle thrust-type focal mechanisms, gaps and irregularities in the line of active volcanoes and gaps in intermediate-depth hypocenters. One interpretation is that oceanic lithosphere varies significantly in average density and that aseismic ridges or other uplifted regions delineate relatively buoyant zones that resist subduction upon interaction with an active trench. Along some segments of convergent boundaries that are highly active as source regions of great earthquakes and tsunamis, there exists a profusion of ridges, scarps, deep-sea marine terraces or perched basins. Some evidence indicates that these morphologic features may be directly related to the occurrence of great shocks. Thus, the detailed bathymetry along the landward wall of trenches may constitute an invaluable guide to the long-term seismic and tsunamic regime.

Introduction

Along convergent boundaries wide variations are observed both in tectonic style and in the distribution of large shallow earthquakes. That is, along extensive segments of some island arcs great earthquakes were infrequent or absent during recorded history, and near many of these same segments Quaternary volcanism is largely absent.

Most of these areas of relative quiescence correspond spatially with locations where major bathymetric features of the underthrusting sea floor appear to be interacting with the convergent margin. Based, therefore, on a summary of evidence from most of the major subduction zones of the earth, we conclude that bathymetric highs within the sea floor are a major influence in the development of convergent plate boundaries.

Although the mechanism for this proposed interaction is not well understood, we prefer as a working hypothesis that aseismic ridges or other bathymetric highs may delineate zones of oceanic lithosphere that are buoyant relative to "typical" oceanic lithosphere. Such buoyant zones may resist subduction on interaction with an active trench. Under this hypothesis variations may exist in the average density or buoyancy of both slabs near the contact area. For example, an aseismic ridge might subduct partially to perhaps 100-150 km beneath an overthrusting continent whereas the same feature may subduct not at all on interaction with an oceanic island arc in the manner suggested by Vogt [1973] and Vogt et al. [1976]. While the evidence available to us does not fully establish the buoyancy hypothesis, this concept appears to explain at least some areas of severely modified subduction.

Modified Subduction

Western Pacific

The subduction zones along the western margin of the Pacific illustrate the variations in seismic-tectonic regimes that commonly occur where bathymetric features interact with active trenches. The distribution of great shocks, active volcanoes and major bathymetric features for the western Pacific are summarized in Figure 1.

From the figure wide contrasts are evident in both bathymetry and the rupture lengths of large earthquakes between the northern and southern parts of the plate boundary. Toward the north the ocean floor seaward of the trench is relatively low and smooth while the ocean floor east of the Mariana-Bonin zones is a region of nearly continuous seamounts, aseismic ridges and other bathymetric highs. Bathymetric charts with greater detail show an even more dramatic contrast than is indicated by the figure.

[1] Now At the Nuclear Regulatory Commission, Norfolk Avenue, Bethesda, Maryland 20015.

[2] Also Department of Geological Sciences of Columbia University.

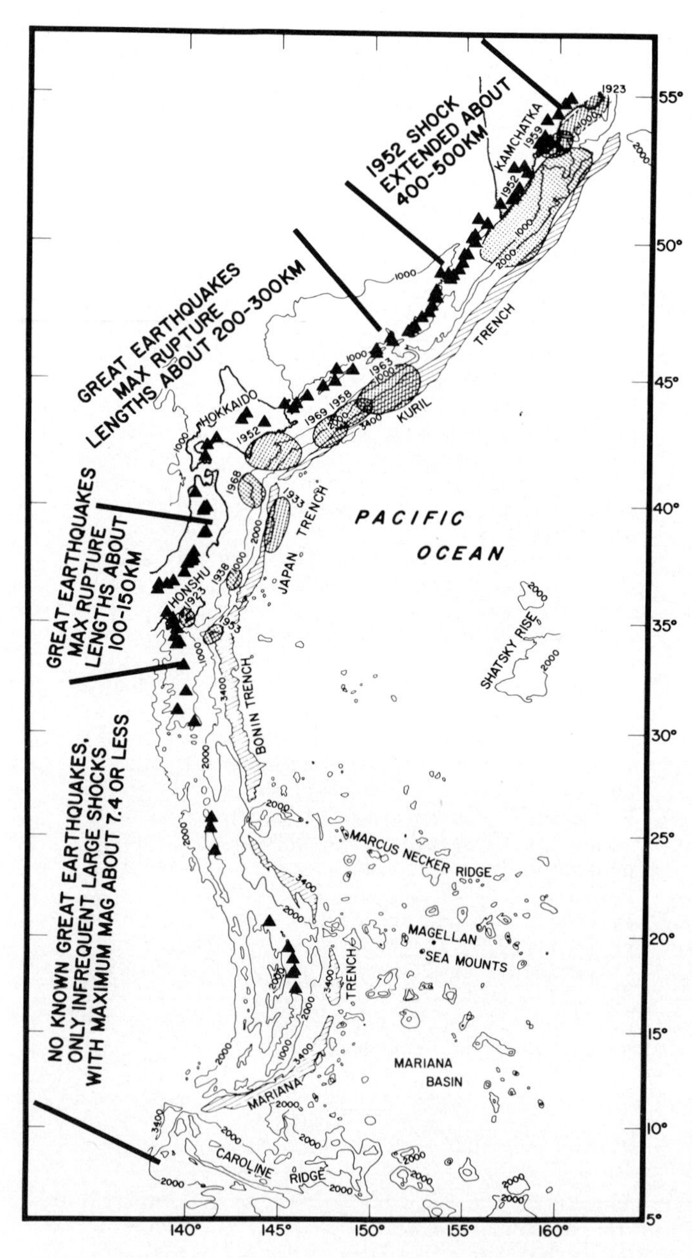

Figure 1. A comparison of bathymetry [Chase, 1970], rupture zones of large earthquakes (cross-hatching) and active volcanoes (solid triangles) along the western Pacific. Note the smooth ocean floor, large rupture zones and nearly continuous chain of volcanoes toward the north (Kamchatka-Kurile-Japan). Contrast these features with the irregular bathymetry, absence of great earthquakes and gaps and irregularities in the line of active volcanoes along the Bonin and Marianas arcs.

Near northern Japan-Kuriles-Kamchatka, opposite the low smooth sea floor, great shallow earthquakes are common, and many known shocks exceed 200 km in rupture length. Opposite the irregular sea floor toward the south earthquakes of such rupture lengths are unknown and even the shocks of magnitude about 7 are not frequent. (Richter [1958] lists one shock of magnitude 8.1 in 1902 but macroseismic evidence does not indicate that this event should be considered a great earthquake.)

Notice also from the figure the relatively continuous and linear chains of active volcanoes toward the north and contrast this with the gaps and offsets in volcanism in the south. This same pattern of gaps and offsets in active volcanism is observed near many other zones of interaction with ridges and seamounts.

Earthquakes, Tsunamis and Tectonics near the Japan Trench

Near the Japanese Islands, there exists a seismic-tsunamic record for about 1000 years or more which, while short geologically, is highly significant in that the record is equal to or greater than many recurrence intervals for great earthquakes. The record indicates distinctly different histories between adjacent regions of the Japan Trench i.e. regions which lie along the same plate margin. The area of northeastern Honshu, opposite the northern half of the trench, was subjected repeatedly to great earthquakes and destructive tsunamis. Toward the southern half of the trench, near a prominent zone of seamounts, the largest known shocks were only moderately large and destructive tsunamis originated rarely, if at all.

Figure 2 shows data for earthquakes, tsunamis and bathymetric features near the Japan Trench. The estimated rupture zones are characteristically larger north of about 38°N (Figure 2) and, while estimates are shown for only 50 years, the historic record clearly supports this interpretation [Musha, 1950; Utsu, written communication, 1975]. Tsunami source regions for about 70 years (Figure 2b) show a similar distribution which is also supported by extensive historic records [Iida et al., 1967; Hatori, written communication, 1975]. A prominent zone of seamounts appears to be interacting with the southern half of the trench while the sea floor tends to be smoother and low-lying opposite the northern half of the trench (Figure 2c). Thus, in this region where historic records extend back many recurrence intervals, there appears to be at least a spatial relationship between the morphology of the subducting sea floor and the distribution of great earthquakes and tsunamis.

Ryukyu Arc - Southwest Japan

For many hundreds of years, great shallow earthquakes have occurred regularly along the Nankai Trough near southwest Japan [Ando, 1975a, b]. Immediately southwest, however, along southern Kyushu and the central-northern

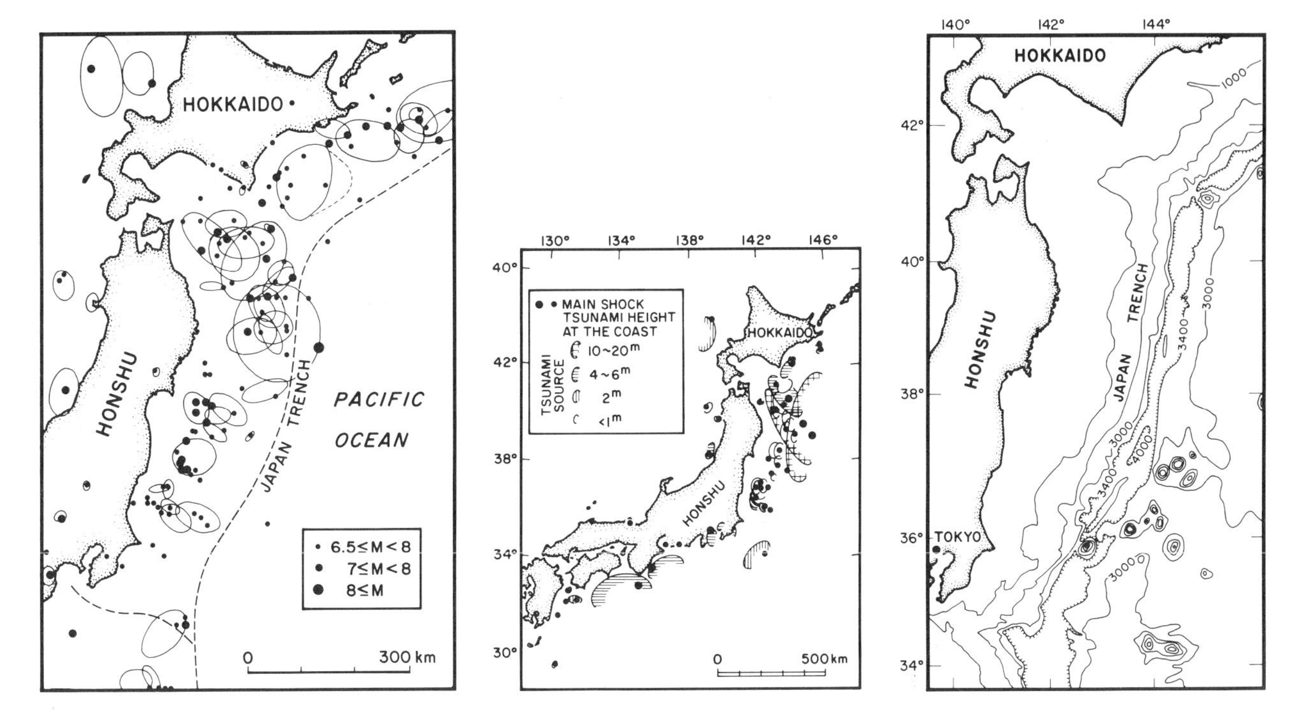

Figure 2. Seismicity and tectonics near the Japan Trench: 2a) Epicenters and focal regions of shallow earthquakes of magnitude 6.5 and larger from 1926 to 1973 after Utsu [1974]. Note the characteristically smaller focal regions near the southern half of the Japan Trench. 2b) Tsunami sources near Japan for 76 years (1893-1968) after Hatori [1969]. Distribution during past centuries was similar [Hatori, 1975]. Note large tsunami sources near northern half of Japan Trench. 2c) Bathymetry near Japan Trench after Chase et al. [1970]. Note zone of seamounts near southern half of trench and smooth ocean floor seaward of northern half of trench where great earthquakes and tsunamis have originated.

Ryukyu arc, great shallow earthquakes are unknown. Notice from Figure 3 that this striking change in the seismicity of great earthquakes occurs almost exactly where the Kyushu-Palau ridge intersects the plate margin near southeastern Kyushu. The change in seismicity is rather abrupt in space and consistent for 1000 years or more and suggests that the northeastern terminus of the Kyushu-Palau ridge is exerting a major influence on the tectonics of subduction near southern Kyushu.

The distribution of large earthquakes along the remainder of the Ryukyu arc also appears to be influenced by bathymetric features although the variations are less striking. Larger earthquakes, at least one (1771) of which was accompanied by a destructive sea wave and much loss of life, occurred near the southwestern end of the arc (Figure 3). By contrast the central portion of the plate boundary, from a point northeast of Okinawa to the intersection of the Kyushu-Palau ridge with the island of Kyushu, has experienced only occasional large earthquakes and these shocks provided no evidence of long rupture zones. Seaward of this central portion is a broad zone of irregular bathymetry with several aseismic ridges.

The active volcanoes along the island arc are, in this case, almost coterminous with the zone of irregular bathymetry on the sea floor seaward of the arc. This distribution is an exception to that found along most subduction zones and, perhaps, may be attributed to recent changes (1-2 my bp) in plate motion.

Other Subduction Zones

Most major subduction zones show the same distribution pattern as shown above: great earthquakes are infrequent or absent along those segments where significant bathymetric features appear to be interacting with the convergent margin [Kelleher and McCann, 1976]. By contrast most segments near regions of smooth, low-lying sea floor have experienced at least one large shallow earthquake.

The near absence of earthquake mechanisms indicating shallow-angle thrusting is another characteristic of convergent margins near major bathymetric highs. For example, Katsumata and Sykes [1969] found few such mechanisms along the Marianas-Bonin arcs whereas Stauder [1975], for a somewhat larger time interval, found numerous thrust-type mechanisms along the Hokkaido-Kuriles-Kamchatka zone (Figure 1).

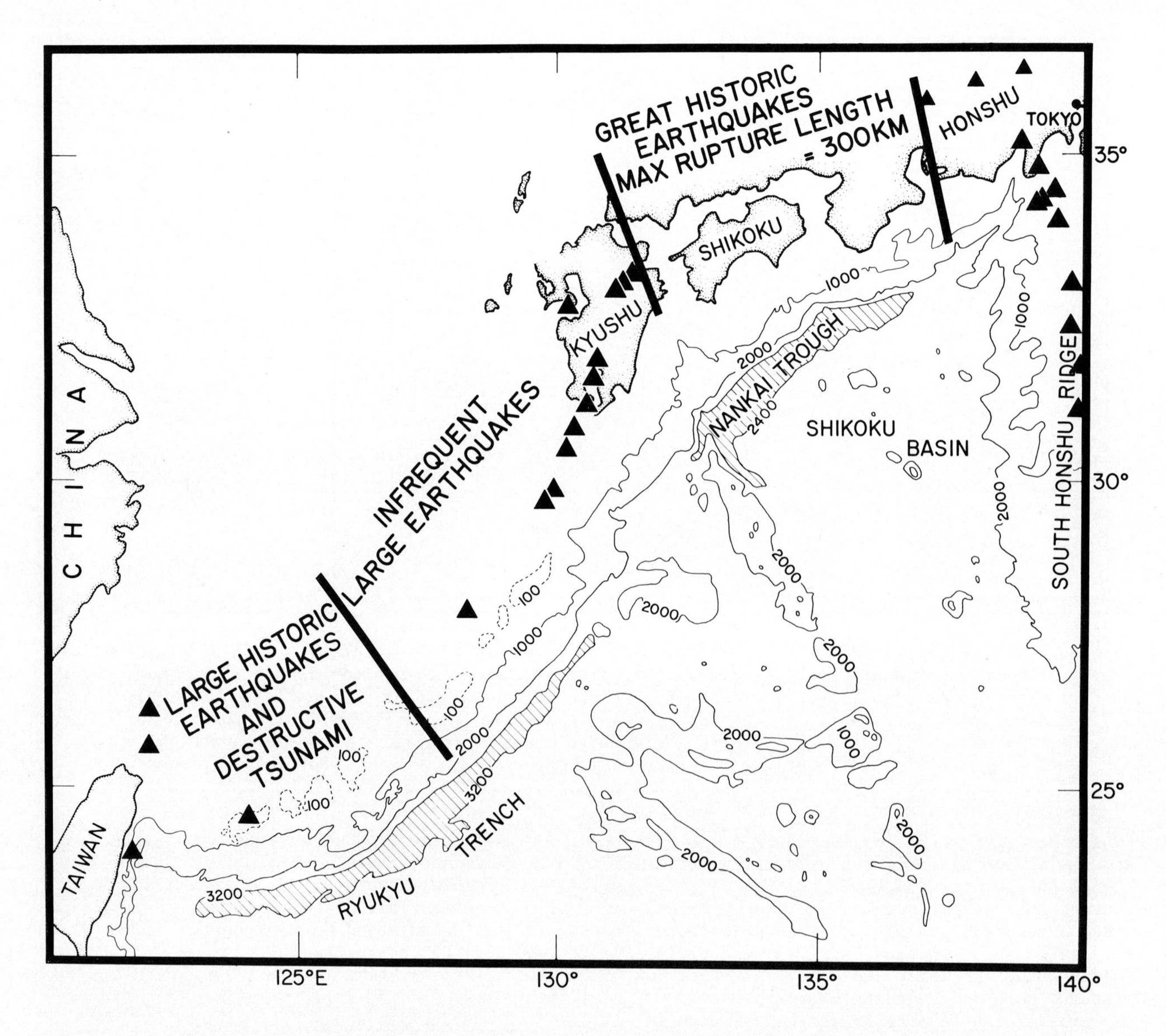

Figure 3. Bathymetry, large earthquake sources and active volcanoes along the Ryukyu arc-southwest Japan. Note the absence of great earthquakes near the zone of irregular bathymetry colliding at present with the central Ryukyus and part of Kyushu.

Crustal Roots and the Buoyancy Hypothesis

Nearly all aseismic ridges and other elevated features of the sea floor provide evidence of structural roots. Both the crustal velocities and depth to M discontinuity suggest a crustal structure intermediate between "oceanic" and "continental" regions of the crust. Because of the very small differences in density that are involved it is not possible to make a direct examination of the buoyancy hypothesis.

The data available at present, however, do not contradict the buoyancy hypothesis. In fact the typical dimensions observed for crustal roots appear to be consistent, based on current density estimates, with a tendency to resist subduction. Assuming an average difference of 0.7 gm/cc between oceanic crust and upper mantle, a 70 km thick lithosphere and no shearing of the ridge at the trench, then a difference in crustal thickness (i.e., increment caused by a crustal root) of 5 km or more would decrease average lithospheric density by about .05 gm/cc. This value, .05 gm/cc, is the approximate difference between subducting lithosphere and the surrounding asthenosphere by most estimates [Jacoby, 1970; Grow, 1973] and, thus, the buoyancy hypothesis is at least consistent with observations available at present.

Scarps, Ridges and Deep-Sea Terraces as Indicators of Seismic-Tsunamic Source Regions

Along some segments of subduction boundaries the morphology of the inner wall of the trench is distinguished by a highly developed series of ridges, scarps, deep-sea terraces or perched basins. Some of these same locations have an active seismic history including a record of one or more great shallow earthquakes accompanied by destructive sea waves. Thus, while the development of the inner wall may be dominated by the sediment supply, it is possible that large thrust earthquakes provide their own unique contribution to the evolving pattern of imbricate thrust faults described by Karig and Sharman [1975] and Seely et al. [1974].

The morphology of the inner wall, therefore, may yield important clues concerning the long-term seismic tsunamic history of a particular section of the plate margin. Specifically, an extremely rough topography with numerous fresh scarps and ridges, and with well-developed deep-sea terraces or perched basins may indicate that the region is a prime source area for large earthquakes and tsunamis.

Although tsunamis are generated by earthquakes, the implications of a tsunami record differ slightly from those of a seismic record. That is, the generation of a seismic sea wave requires substantial deformation at the sediment-water interface. Thus, tsunamic source regions may be intimately related to the changes in trench morphology wrought by an earthquake, i.e., such source regions may be the submarine equivalent of zones of surface faulting on land. The following examples provide insight into the potential relationships among great thrust earthquakes, the accompanying sea wave and deformations of the sea floor.

1964 Alaskan Earthquake

During the 1964 Alaskan earthquake, rupture propagated about 800 km parallel to the trench axis, a largely underwater ridge or scarp formed which probably extended about 400 km (about 10-12 m vertical deformation), and a Pacific-wide sea wave was generated. The examination of these features by Plafker [1969] is indicated in Figure 4; the profiles in the lower figure indicate the mechanism of the ridge formation. The ridge or scarp can be taken as the axis of a broad zone of uplift and the formation mechanism is interpreted here as displacement on an upper branch of the pattern of imbricate thrusts.

Of interest to the present discussion are two considerations: 1) the mechanism of the ridge formation is clearly suitable as a first step in the formation of a deep-sea terrace or perched basin and 2) the length of the ridge is a significant fractional part of the total rupture length. Certainly deep-sea terraces or perched basins may evolve through mechanisms other than that suggested here. Nonetheless, the observed deformation for the 1964 earthquake would act as a sediment trap leading to basin or terrace development and if the same branch of the thrust fault system is activated repeatedly during recurring great earthquakes, then an even larger ridge or scarp would form.

The second consideration is obvious but non-trivial: larger ruptures would tend to be associated with longer ridges. That is, despite complications, a greater extent of surface breakage tends to accompany larger rupture zones [Tocher, 1958; Thatcher and Hanks, 1973]. Thus, if deep-sea terraces are related to large thrust earthquakes as suggested above, then it is quite likely that the larger thrust earthquakes are associated with larger basins or terraces.

Deep-Sea Terraces along the Japan Trench

Numerous deep-sea terraces have developed along the inner wall of the Japan Trench near eastern Honshu. The size and distribution of these terraces according to Iwabuchi [1968] are shown in Figure 5. The most striking feature

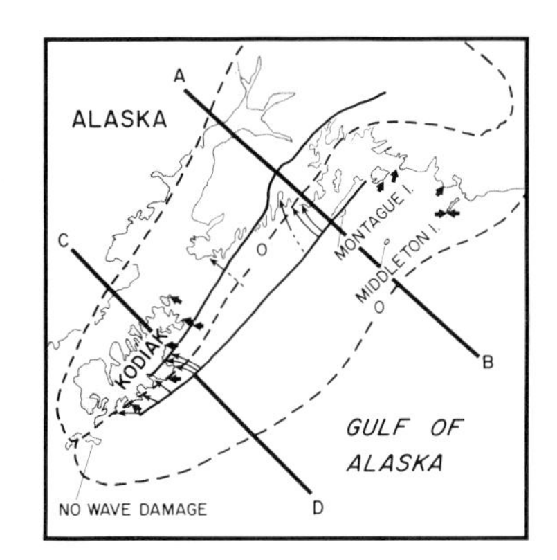

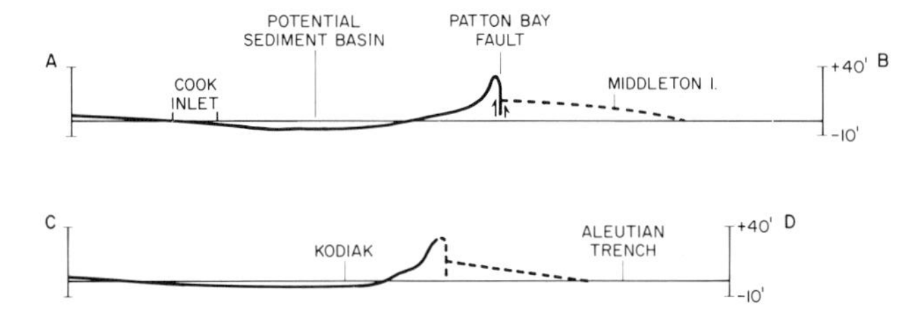

Figure 4. Land deformation and tsunami source region for the 1964 Alaskan earthquake after Plafker [1969]. In upper figure, zero isobase (dashed line) enclose axes of uplift and subsidence (solid lines). Direction (wide solid arrows) and estimated travel distance (narrow, larger arrows) of the sea wave are also indicated. Note that the ridge or scarp which formed during this shock (lower figure) is suitable as a first step in the formation of a deep-sea terrace.

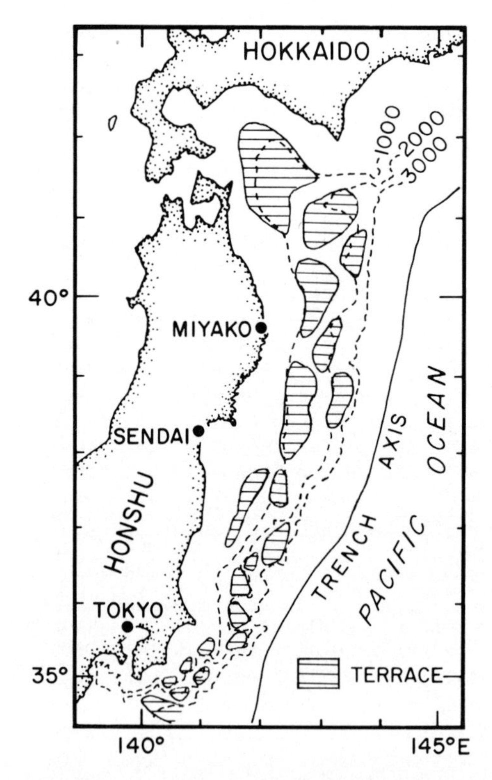

Figure 5. Deep-sea terraces near eastern Honshu after Iwabuchi [1968]. Note that characteristically larger terraces have formed near the northern part of the trench, the source for larger earthquakes and tsunamis (see Figure 2).

is the difference in size of terraces going from north to south. Most terraces north of about 38°N are significantly larger than those to the south.

Compare Figure 5 with the earthquake-tsunami distribution for the same region shown previously in Figure 2. The convergent margin near eastern Honshu presents a clear-cut example whereby larger seismic-tsunamic sources are at least spatially related to larger basins and terraces.

Other Examples of Possible Earthquake Morphology

The above discussion of the 1964 Alaskan earthquake and the margin near eastern Honshu are not isolated examples and numerous other instances can be cited where a profusion of ridges, scarps and terraces occur along margin segments that are active source regions for great earthquakes and tsunamis. The great Aleutian terraces [Marlow et al., 1973] occur near some of the largest known rupture zones [Sykes, 1971]. The source region of the tsunami generated by the 1946 Aleutian earthquake corresponds with the location of Unimak seamount. We suggest here that this "seamount" may be simply a striking indication of ridge formation. The great Mexican earthquake of 1932 (estimate of rupture zone in Kelleher et al. [1973]) appears to coincide with the large perched basin described by Ross and Shor [1965].

Along the Nankai Trough near southwest Japan a series of large deep-sea terraces (see Figure 19 of Yonekura, 1975) correspond almost exactly with the tectonic blocks which Ando [1975b] found to be associated with the recurring great earthquakes of southwest Japan. Recent detailed studies by A. Mogi and T. Sato of the Japanese Maritime Safety Agency (Mogi, 1976, unpublished data) show the inner wall of the Nankai Trough to be a succession of countless ridges and scarps which clearly diminish and possibly disappear near southern Kyushu. Notice from Figure 3 that great earthquakes and tsunamis are unknown near southern Kyushu where the Kyushu-Palau ridge interacts with the margin.

Summary and Conclusions

Major bathymetric features appear to exert a profound effect on the subduction process and may actually terminate the process either temporarily or permanently. Such features may delineate zones of lithosphere that are buoyant relative to typical oceanic lithosphere. Under this hypothesis of relative buoyancy the influence of bathymetric features on the subduction process is summarized in Figure 6.

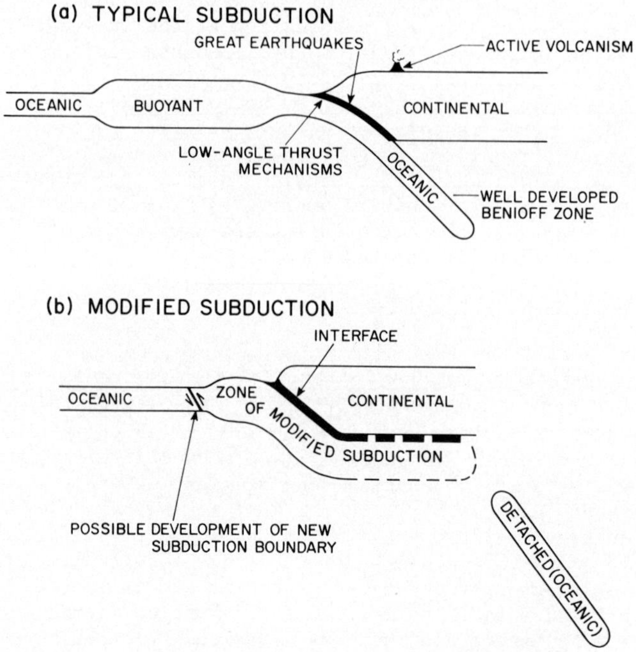

Figure 6. Speculative Summary of bathymetric features interacting with a subduction margin based on the buoyancy hypothesis. Note the inferred tendency toward quiescence upon interaction with buoyant zone i.e. large shallow shocks tend to be smaller (rupture length) and less frequent; volcanism may become inactive.

In Figure 6a the principal characteristics of typical subduction are indicated including active volcanism, great earthquakes, numerous low-angle, thrust-type mechanisms and a well-defined Benioff zone. Figure 6b suggests that the convergent margin is modified by interaction with a zone of buoyant lithosphere that resists subduction. A major effect is a tendency toward quiescence with large earthquakes becoming smaller and less frequent, and active volcanism diminishing or terminating. The leading, downdip slab of typical oceanic lithosphere may detach under these conditions and a new subduction boundary may eventually develop.

A rough topography along the inner wall including many scarps, ridges and terraces may indicate that the area is an active source region for great earthquakes and destructive tsunamis. Thus, the morphology of the inner wall may be a crucial guide to long-term seismic-tsunamic risk.

Lamont-Doherty Geological Observatory Contribution Number 2408.

Acknowledgements. We wish to thank K. Nakamura for many stimulating discussions and for his considerable help in obtaining and translating Japanese sources. The manuscript was critically reviewed by L. Sykes, T. Johnson and K. Nakamura. Thanks are also due to P. Lustig for preparing data and obtaining historic reports near the southwest Pacific, to L. Murphy for typing the manuscript and to K. Nagao for drafting the figures. This research was supported by National Science Foundation contracts DES-75-03640 and IDO-75-19794.

References

Ando, M., Possibility of a major earthquake in the Tokai District, Japan and its pre-estimated seismotectonic effects, Tectonophysics, 25, 69, 1975a.

Ando, M., Source mechanisms and tectonic significance of historical earthquakes along the Nankai Trough, Japan, Tectonophysics, 27, 119, 1975b.

Chase, T. E., H. W. Menard, and J. Mammerickx, Bathymetry of the North Pacific, Scripps Inst. of Oceanography, Inst. of Marine Resources, La Jolla, Calif., Charts Nos. 1-6, 1970.

Grow, J. A., Crustal and upper mantle structures of the Central Aleutian arc, Geol. Soc. Amer. Bull., 84, 2169, 1973.

Hatori, T., Dimensions and geographic distribution of tsunami sources near Japan, Bull. Earthquake Res. Inst., 47, 185, 1969.

Hatori, T., Sources of tsunamis generated off Boso Peninsula, Bull. Earthquake Res. Inst., 50, 83, 1975.

Iida, K., D. Cox, and G. Pararas-Carayannis, Preliminary catalog of tsunamis occurring in the Pacific Ocean, Hawaii Inst. of Geophysics, HIG-67-10, Honolulu, Hawaii, 1967.

Iwabuchi, Y., Topography of trenches east of the Japanese Islands, Jour. Geol. Soc. Japan, 74, 37, 1968.

Jacoby, W. R., Instability in the upper mantle and global plate movements, J. Geophys. Res., 75, 5671, 1970.

Karig, D., and G. Sharman, Subduction and accretion in trenches, Geol. Soc. Amer. Bull., 86, 377, 1975.

Katsumata, M., and L. R. Sykes, Seismicity and tectonics of the western Pacific: Izu-Mariana-Caroline and Ryukyu-Taiwan regions, J. Geophys. Res., 74, 5923, 1969.

Kelleher, J., L. Sykes, and J. Oliver, Possible criteria for predicting earthquake locations and their application to major plate boundaries of the Pacific and the Caribbean, J. Geophys. Res., 78, 2547, 1973.

Kelleher, J., and W. McCann, Buoyant zones, great earthquakes, and unstable boundaries of subduction, J. Geophys. Res. (in press), 1976.

Marlow, M., D. Scholl, E. Burrington, and T. Alpha, Tectonic history of the Central Aleutian arc, Geol. Soc. Amer. Bull., 84, 1555, 1973.

Plafker, G., Tectonics of the March 27, 1964, Alaska earthquake, U.S. Geol. Surv. Prof. Pap., 543-1, 1-74, 1969.

Richter, C., Elementary Seismology, 768 pp., W. H. Freeman, San Francisco, Calif., 1958.

Ross, D., and G. Shor, Reflection profiles across the Middle America trench, J. Geophys. Res., 70, 5551, 1965.

Seely, D., P. Vail, and G. Walton, Trench slope model in the geology of continental margins, C. Burk and C. Drake (ed.), 249-261, Springer-Verlag, New York, 1974.

Stauder, W., and L. Mualchin, Fault motion in the larger earthquakes of the Kurile-Kamchatka arc and of the Kurile-Hokkaido corner, J. Geophys. Res., 81, 297, 1976.

Sykes, L. R., Aftershock zones of great earthquakes, seismicity gaps, and earthquake prediction for Alaska and the Aleutians, J. Geophys. Res., 76, 8021, 1971.

Thatcher, W., and T. Hanks, Source parameters of Southern California earthquakes, J. Geophys. Res., 78, 8547, 1973.

Tocher, D., Earthquake energy and ground breakage, Bull. Seism. Soc. Amer., 48, 147, 1958.

Utsu, T., Space-time pattern of large earthquakes occurring off the Pacific coast of the Japanese Islands, J. Phys. Earth, 22, 325, 1974.

Vogt, P. R., Subduction and aseismic ridges, Nature, 241, 189, 1973.

Vogt, P. R., A. Lowrie, D. R. Bracey, and R. N. Hey, Subduction of aseismic oceanic ridges: Effects on shape, seismicity, and

other characteristics of consuming plate boundaries, Spec. Pap. 172, 60 pp., Geological Society of America, 1976.
Yonekura, N., Quaternary tectonic movements in the outer arc of southwest Japan with special reference to seismic crustal deformations, Bull. Dept. of Geography, Univ. of Tokyo, 7, 20, 1975.

IN SITU P-WAVE VELOCITIES IN DEEP EARTHQUAKE ZONES OF THE SW PACIFIC: EVIDENCE FOR A PHASE BOUNDARY BETWEEN THE UPPER AND LOWER MANTLE

Thomas J. Fitch

Research School of Earth Sciences, Australian National University, Canberra

Abstract. P-wave velocities below 600 km in the Tonga seismic zone and in the deep earthquake zone beneath the NW corner of the Fiji Plateau are about 1 km/sec (or 10%) higher than velocities in a "normal" mantle taken at these depths from a suite of earth models. This anomaly is significant at the 95% confidence level for these models. At shallower depths, between 500 and 580 km in the Tonga zone, *in situ* velocities either lie within the range of values predicted by these models or have 95% confidence limits that overlap with these values. The implied velocity gradient can only be explained by a transition to a denser mineral assemblage within the downgoing slab. The depth to this transition zone is not known elsewhere in the SW Pacific; however, in other regions comparably large velocity gradients exist in the depth range 600-700 km. This zone is often referred to as the "650 km discontinuity" and is taken as the boundary between the upper and lower mantle. In the slab this transition zone is centered at a depth of about 600 km and thus is likely to be elevated above the mean depth to this zone in the surrounding mantle. From thermal models the slab is expected to be cooler than its surroundings but possibly close to thermal equilibrium below 500 km. Therefore the slope, $\partial P/\partial T$, of the principal phase boundary is likely to be greater than zero. If the elevation of the tansition zone in the slab were as much as 50 km, then the attendant body force on the slab would be downward with nearly the same magnitude as that from the olivine-spinel transformation, which is expected in the mantle seismic zones in the depth range from 300 to 400 km. The possibility of a large thermal contrast between the slab and its surroundings consistent with an elevated phase boundary having a small $\partial P/\partial T$ encourages speculation that remnants of subducted lithosphere descend more than 100 km into the lower mantle, thereby accounting for some observed seismic anomalies below the depth of the deepest earthquakes at about 700 km.

Introduction

Fitch (1975) reported *in situ* compressional velocities in three deep earthquake zones that were higher by 5 to 10% than velocities in the same depths ranges taken from gross earth models. Here that work is extended to include estimates of *in situ* velocities in the depth range 500 to 650 km in the Tonga seismic zone.

Engdahl (1975) showed that travel time residuals from deep earthquakes in the Tonga seismic zone require a compressional velocity anomaly of plus 10% in or near the source region. Solomon and Paw U (1975) were unable to resolve the same anomaly from a different set of travel time residuals, although they claim to have resolved a velocity anomaly at a depth appropriate for the olivine-spinel transformation in this seismic zone. Engdahl's analysis scheme would appear to be more sensitive to near source anomalies and his results are consistent with the *in situ* velocities presented here.

It will be argued that a velocity transition is centered at a depth of about 600 km in the Tonga seismic zone and that this transition corresponds to the "650 km discontinuity" which must be primarily the result of one or more isochemical phase transformations. The configuration of the deep earthquake zone is not grossly different above and below the transition. Apparently, remnant lithosphere containing this seismic zone does not see this transition as a barrier to downward motion; consequently, this motion may continue below the depth of the deepest earthquakes.

A downward body force from an elevated transition zone supports thermodynamic arguments by Ringwood (1972) for convection across the "650 km discontinuity" in the form of sinking slabs. However, body forces arising from the thermal contrast between a cool slab and "normal" mantle may drive this form of convection even if the attendant phase transitions give rise to body forces that resist the downward motion (Schubert *et al.*, 1975). Phase boundaries in fluids of nearly constant viscosity are generally not barriers to convection because body forces arising from displacements of such boundaries will tend to be offset by forces arising from the latent heats of the transformations (Richter, 1973). However, the viscosity (or more appropriately the effective viscosity) of the mantle is likely to be strongly temperature and depth dependent (McKenzie, 1968; Smith and Toksöz, 1972; Toksöz et al., 1973; and McKenzie et al., 1974). Laboratory experiments by Turner (1973) have shown that convection in fluids in which viscosity is strongly temperature-dependent can exhibit features suggestive of large scale convection in the earth, which at the surface takes the form of plate motion.

Arguments have been put forward opposing the idea that convection occurs across the "650 km discontinuity"; however, these arguments are not considered conclusive. The earth's non-hydrostatic shape can be explained by a substantial increase in viscosity in the lower mantle (McKenzie, 1968). McKenzie and Weiss (1975) attribute this change to an increase in the activation energy for diffusional creep below the "650 km discontinuity". In contrast, Goldreich and Toomre (1969) show that any nearly spherical object in rotation can acquire a non-hydrostatic shape from mass anomalies that are evolving randomly. Cathles (1975) claims that the post-glacial uplift of the coast of eastern North America can only be explained by viscous flow in an upper and lower mantle of nearly uniform viscosity.

Convection across this transition zone would be resisted by body forces arising from a density change that results from a change in composition rather than the requirements of thermodynamics (Richter and Johnson, 1974). However, arguments for a compositional change between the upper and lower mantle, e.g., greater iron content in the lower mantle (Anderson, 1967), are often based on velocity-density systematics of Birch, which may be invalid across transition zones in which the coordination number of the primary cation changes from 4 to 6 fold (Liebermann and Ringwood, 1973). Phase transformations that are candidates to explain the "650 km discontinuity" involve such a change in Si^{+4} coordination (Ringwood, 1975).

Velocity transition between upper and lower mantle

The "650 km discontinuity" is observed as an increase in the apparent velocity, $(dT/d\Delta)^{-1}$, of P-waves measured at seismic arrays (Johnson,1967; Archambeau et al., 1969; Simpson et al., 1974; King and Calcagnile, 1976), underside reflections recorded as precursors to the core phase P'P' (Engdahl and Flinn, 1969; Whitcomb and Anderson, 1970), and wave forms of refracted P and S phases (Helmberger and Wiggins, 1971 and Helmberger and Engen, 1974 respectively). The amplitudes of short-period reflected signals (wave lengths of about 10 km) are large enough to require essentially a first-order seismic discontinuity, ΔVp of 10% in 4 km depth interval (Richards,1972). Whitcomb (1973) gives a maximum depth range of 630 to 670 km to the bottom edge of this reflector; the reflections he analyzed occurred beneath Antarctica and the Indian Ocean and not beneath active subduction zones.

Some $dT/d\Delta$ measurements are satisfied by large velocity gradients in a depth range appropriate for the "650 km discontinuity". Johnson (1967) and Archambeau et al.(1969) computed ΔVp's of 8 to 10% in a depth range of 600 to 680 km beneath the western United States. Some of the velocity models by Archambeau et al. have gradients that steepen from the top to the bottom of the transition zone. Simpson et al.(1974), in their model, SMAK I, pertaining to the mantle beneath NE Australia, show a ΔVp of 6% in the depth range from 680 to 695 km. Similarly, King and Calcagnile (1976) show that a first-order discontinuity of 4% in Vp at a depth of about 690 km beneath Fennoscandia and Western Russia satisfies data in the form of record sections compiled from explosions recorded at the NORSAR array in Norway.

In contrast, Helmberger's work with co-authors shows that refracted amplitudes of long-period P and S phases are more compatible with modest velocity gradients in the depth range from 625 to 700 km. They suggest an accumulated change in compressional and shear velocity of about 7% across this transition zone beneath the central and western United States.

Of particular interest to this study are Kanamori's (1967) $dT/d\Delta$ measurements that pertain to the mantle beneath the western margins of the north Pacific and the Philippine Sea, both of which are active plate margins with mantle seismic zones inclined from arc-like structures. These measurements are satisfied by a modest linear velocity gradient in the depth range 425 to 650 km. A large velocity gradient was resolved in a depth range appropriate for an olivine-spinel transition. It is difficult to reconcile Kanamori's results with the in situ velocities in Figure 1 unless an appeal is made to large compositional differences between the slab and surrounding mantle beneath either the NW or SW Pacific or both regions.

Transition Zone Chemistry

The olivine-pyroxene-garnet assemblage (pyrolite) of the uppermost mantle is transformed in the depth range from 200 to 1000 km to an assemblage that is about 25% more dense, when the comparison is made with densities reduced to ambient conditions (Ringwood, 1970, 1972, and 1975). Approximately 10% of this density increase is expected at P, T conditions appropriate for the velocity transition near 650 km: $200kbar < P < 230kbar$ and $1500^{\circ}C < T < 3000^{\circ}C$. The mineralogy of the low-pressure side of this transition zone, in simplified terms, can be divided into a olivine-spinel (A_2BO_4) fraction and a pyroxene-garnet (ABO_3) fraction, where A represents the least charged cations of which Fe^{+2} and Mg^{+2} predominate and B represents the principal cation, Si^{+4}.

A combination of the following transformations and disproportionations could account for a velocity transition at a depth appropriate for

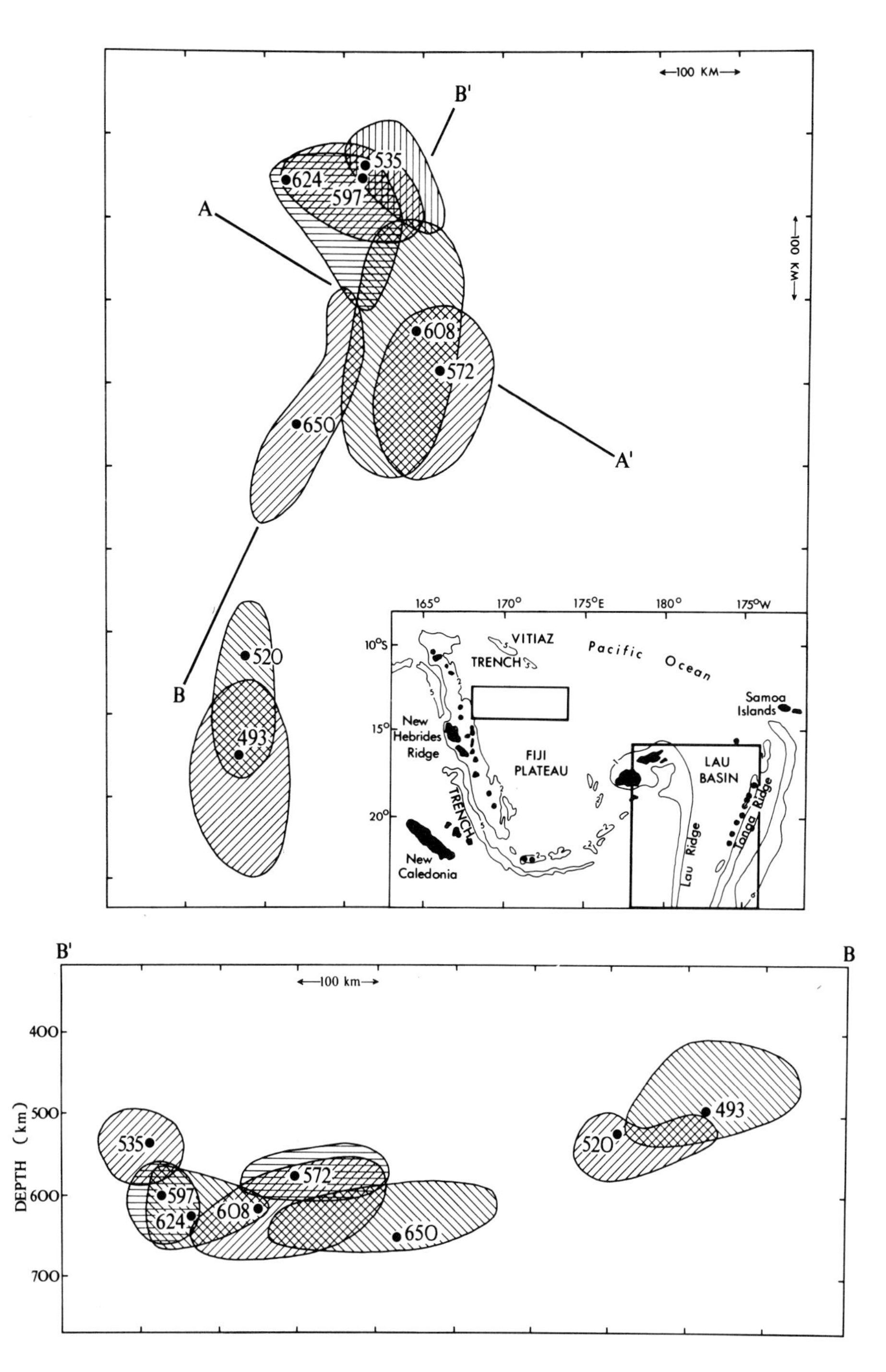

Fig. 1. Location and size of source regions line AA' refers to vertical section shown in Figure 6 of Mitronovas and Isacks (1971) and in the cartoon, Figure 8 part B.

the "650 km discontinuity" (Ringwood, 1975):

(1) A_2BO_4 (spinel) to 2AO (rocksalt)+BO_2(rutile)

(2) A_2BO_4 (spinel) to AO(rocksalt) + ABO_3 (ilmenite) followed at some greater depth by ABO_3(ilmenite) to ABO_3(perovskite)

(3) A_2BO_4(spinel) to AO(rocksalt) + ABO_3 (perovskite)

(4) ABO_3 ("garnet") to ABO_3(ilmenite)

(5) ABO_3 ("garnet") to ABO_3(perovskite)

Each of these transformations involves a change in coordination of Si^{+4} from 4 to 6 fold.

High pressure-temperature experiments in diamond anvil apparatus (Liu, 1974, 1975 and 1976) supports Ringwood's speculation that silicates with a perovskite structure are a major component of the lower mantle rather than a mixture of simple oxides (FeO, MgO and SiO_2, stishovite). In particular, Liu (1972) has shown that natural pyrope garnet (high Al garnet), enstatite ($MgSiO_3$) and forsterite (Mg_2SiO_4) are stable in the perovskite structure (or in the latter case perovskite plus rocksalt) at conditions expected in the upper part of the lower mantle. In the $MgSiO_3$ phase diagram an ilmenite stability field, that occurs on the low pressure side of the perovskite field, could coincide with P, T conditions at the "650 km discontinuity".

A large set of seismic data used in constructing the PEM models (Dziewonski et al., 1975) are compatible with material in the lower mantle that has higher density than the mixture of simple oxides with the stoichiometry of an olivine-pyroxene assemblage (Davies and Dziewonski, 1975). These conclusions are drawn from extrapolations from P, T conditions of the lower mantle to ambient conditions using fourth-order finite-straint equations. The argument is reinforced by a recent determination of the elastic properties of stishovite in which the bulk modulus was found to be unexpectedly low, 2.5 kbars at ambient conditions (Liebermann et al., 1976). They measured the P and S velocities in a hot-pressed specimen at 10 kbars and room temperature by a pulse superposition technique.

Significant compositional differences may exist between the Tonga seismic zone and its surroundings. If this is so the composition of the upper mantle beneath the SW Pacific must be different from that beneath the ridge crest where the slab was first differentiated to form oceanic lithosphere. Differentiation by itself cannot account for major differences in composition. This process extracts the more volatile basaltic fraction from the upper mantle to form oceanic crust and leaves behind a refractory residue that makes up the bulk of oceanic lithosphere; e.g., the olivine in pyrolite (average mantle rock) is $Fo_{89}\ Fa_{11}$ (Ringwood, 1966) and the refractory olivine of oceanic lithosphere derived from this "rock" would be depleted by only 1 to 2% in iron (Green and Liebermann, 1976). Only the refractory fraction (probably a harzburgite-lherzolite mixture) of the lithosphere can survive to great depths in the mantle. Compared with average mantle material at the same temperature the refractory material of remnant lithosphere requires greater pressures to transform to denser phases; however, this effect will be slight and possibly negligible. Furthermore, this effect in terms of a change in slope of the appropriate phase boundaries is not resolvable by state-of-the-art experiments at high pressures and temperatures.

Inversion Scheme for Computing In Situ Velocities

The accuracy with which one earthquake can be located with respect to another depends on an estimate of the average seismic velocity in the region between the events. Fitch (1975) showed that in situ compressional velocities and relative locations can be computed jointly from arrival times listed in standard seismological summaries. That analysis is used to compute the compressional velocities presented in Table 1. Each data set contained between 300 and 600 arrival times of P and pP phases from 10 to 15 closely spaced events. Figure 1 gives the location and size of each group, with the exception of the New Hebrides group which is shown in Figures 2 and 5 of Fitch (1975).

A master event is chosen so that the distance, L_j, between it and each of the secondary events in the group, is short in comparison with the shortest source-receiver distance. The formulation of the problem is illustrated in Figure 2. Each differential arrival time, ΔT_{ij}, between the master and the jth secondary event at the ith station obeys the equation:

$$\Delta T_{ij} = \Delta T_{Oj} - \frac{L_j \cos (S_{ij})}{V} \qquad (1)$$

ΔT_{Oj} is the differential origin time for these two events. S_{ij} is the angular separation between the ray path from the master to the ith and the vector separation between the master and the jth secondary event. V is the in situ velocity, which for P and pP arrivals is a compressional velocity.

The inversion is carried out by solving in a least squares sense a system of weighted linear equations. The weighting factors are estimates of the variance of the differential arrival times ΔT_{ij}, in accordance with Minster et al.(1973). Appendix 1 of Fitch (1975) gives the particular linearization of equation 1 that was found to be an effective estimator of in situ velocity.

In principle, the model includes all take-off angles and azimuths for ray paths from the master earthquake; however, errors in reading arrival times limit the number of angles that can be refined. The penalty for including too many angles, aside from a major increase in the computer time required for a solution, is an unstable solution resulting from the inversion of an ill-conditioned matrix. The selection of which angles to refine is based on the geophysics of the problem.

Ray paths to the island stations near the source region are likely to deviate significantly from paths predicted by any standard earth model. This results from lateral heterogeneity in the upper mantle associated with source regions of deep earthquakes. Take-off angles, IN_i, for ray paths to these stations (Figure 3) are refined in the solutions summarized in Table 1.
The distance to the farthest of these stations varies from 18° to 25° (arc distance) and is termed a cut-off distance (Figure 4). From Figure 3 it is apparent that station coverage is sparse in the distance range between cut-off and say 30°. The number of stations within the cut-off distance varies from 6 to 13. In the solutions reported here, azimuths for all ray paths

TABLE 1

IN SITU VELOCITIES

Solution	MASTER EARTHQUAKE DATE	LOCATION			# j's	# ΔT_{ij}'s	V_p km/sec Jeffreys (1939)	Refined*	χ^2	$\langle\chi^2\rangle$
					(as defined in eq. 1)					
1	3 Dec. 1966	24.80 S	179.97 E	493km	12	493	9.60	9.77 ± 0.28	455	419
2	22 Nov. 1968	23.65 S	179.96 W	520km	12	379	9.76	9.86 ± 0.56	238	321
3	10 Apr. 1965	17.84 S	178.40 W	537km	11	476	9.88	9.81 ± 0.43	470	424
4	9 Aug. 1969	19.83 S	177.99 W	572km	10	308	10.09	10.41 ± 0.36	194	260
5	24 Nov. 1969	18.01 S	178.40 W	597km	10	431	10.23	10.76 ± 0.26	431	383
6	10 Jun. 1967	19.40 S	178.24 W	608km	12	454	10.29	11.10 ± 0.23	454	398
7	28 May 1972	18.06 S	179.36 W	624km	13	622	10.37	11.22 ± 0.16	550	561
8	12 Oct. 1967	21.03 S	179.19 W	650km	11	587	10.47	11.06 ± 0.24	502	535
9A	10 Apr. 1965	13.45 S	170.30 E	641km	13	493	10.43	10.88 ± 0.19	487	434
9B	(see caption of Figure 6 for explanation)							11.14 ± 0.19	497	433

* Uncertainties are approximately one standard deviation

are constrained to lie in vertical planes and thus are fixed by the epicenter of the master earthquake and the station location.

Ray paths to stations more distant than the cut-off distance are assumed to be known *a priori* from a spherically symmetric earth model. The model used in this study is found in Table 2 of Jeffreys (1939) and there listed as a smoothed "solution for a loop". Each ray path in such a model is defined by a constant known as the ray parameter (Figure 4). Refinements to *in situ* velocity are computed so as to keep the ray parameters fixed; consequently, the coefficients of the ΔV terms become

$$\frac{L_j \cos(POL_j)}{V^2 \cos(IN_i)} \quad (2)$$

rather than the coefficient that is derived by simple differentiation of equation (1)

$$\frac{L_j \cos(S_{ij})}{V^2} \quad . \quad (3)$$

POL_j is the elevation angle for the vector separation between the master and the jth secondary event (Fitch 1975). The coefficients of the ΔV terms for data recorded within the cut-off distance are set to zero. Consequently, the uncertainty in *in situ* velocity is negatively correlated with the size of the data set. The larger sets are compiled from events that are better recorded at teleseismic distances.

The ray parameters are assumed to be known rather than the take-off angles because these parameters are in principle observable in that they can be measured directly at a seismic array or taken from the gradient of a plot of travel time versus distance. This assumption does not strongly tie the *in situ* velocities to a particular earth model (Fitch 1975). At first solutions were attempted in which no constraint was placed on the *in situ* velocities i.e. the coefficients of the ΔV terms were given by expression (3). These attempts were fruitless because they always resulted in the inversion of an ill-conditioned matrix.

In summary, consider the solution in Table 2 as typical. Ten secondary events are distributed within 130 km of a master event. 40 parameters are required for relative locations of these events and the differential origin times. In addition, there are 6 take-off angles, one for each station within a cut-off distance of 21°. With *in situ* velocity the model contains 47 variables. The data set, containing 431 differential arrival times, is more than sufficiently redundant for a solution.

Data

Arrival times were taken from the Bulletin of the International Seismological Center (BCIS). Impulsive signals have their arrival times given to the nearest 0.1 sec, whereas emergent signals and those without a quality indicator have times given to the nearest second. Chi-squared tests of residuals for the solutions in Table 1 show that standard deviations for the two classes of arrival times are approximately ±0.25 sec and

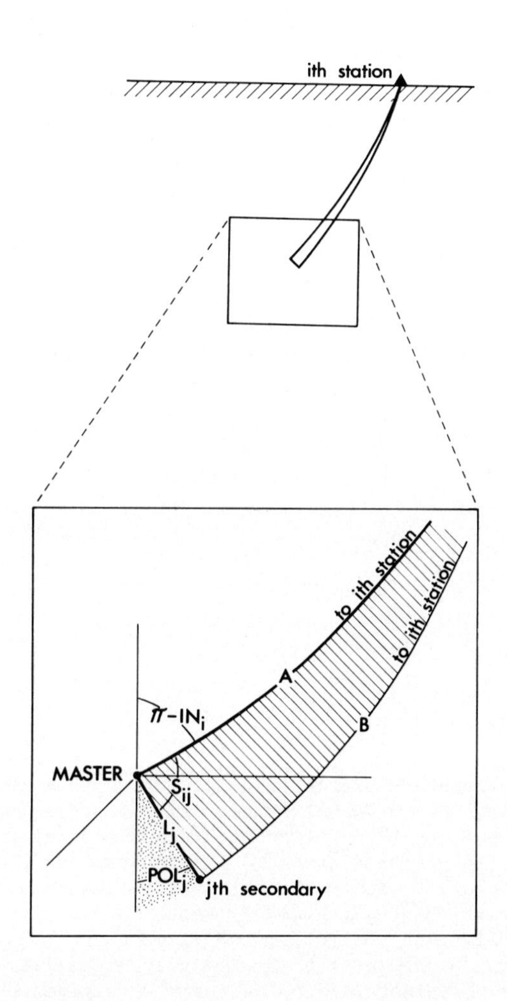

$S_{ij} = f\left(IN_i,\ theta_i,\ POL_j,\ AZ_j\right)$

$theta_i$: azimuth of a vertical plane containing master and ith station.

AZ_j : azimuth of vector separation between master and jth secondary event.

(From Appendix 1, Fitch 1975)

Maximum Approximation Error, ΔT:

If A and B were straight rays, the law of cosine yields

$$\frac{A}{B} = \sqrt{1 + \frac{L_j^2}{B^2} - 2\frac{L_j}{B}\cos\left(S_{ij}\right)}\ .$$

As $S_{ij} \rightarrow \frac{\pi}{2}$ the condition that L_j be sufficiently small for $\frac{L_j^2}{B^2}$ to be neglected in comparison to $2\frac{L_j}{B}\cos\left(S_{ij}\right)$ becomes untenable.

$$\frac{A}{B} \rightarrow \sqrt{1 + \frac{L_j^2}{B^2}} \sim 1 + \frac{1}{2}\frac{L_j^2}{B^2}$$

or in terms of travel times

$$\Delta T = \frac{A - B}{V} \sim \frac{1}{2}\frac{L_j^2}{VB}$$

Fig. 2. Geometry of the inversion problem and maximum approximation error.

±0.50 sec. These standard deviations are one half the values estimated for the differential arrival times, the ΔT_{ij}'s.

Each data set was made statistically equivalent by culling the raw data until the chi-squared test was satisfied, i.e., until the residuals yielded nearly the expected value of chi-squared. A comparison with the solutions in Fitch (1975) shows that the uncertainty assigned to data in that study was nearly two standard deviations. Consequently, the computed uncertainties in the model parameters in that study are approximately 95% confidence limits.

Bias

Repeated use of the arrival times from the master earthquake in generating the ΔT_{ij} imposes a strong but tolerable bias on the *in situ* velocities. In effect, these velocities pertain to the focal region of the master earthquake rather than the region enclosed by the group of secondary events. This bias is enhanced by the constraint on ray paths to stations more distant than the cut-off distance, because the depth of the master earthquake is required in order to compute the ray parameters for these paths. Solutions with master earthquakes from different depths within the same source region confirm this bias.

Focal depths for the master earthquake were taken from BCIS locations unless more accurate locations were available from the recent literature. Depths from the BCIS that are not controlled by the depth-sensitive phase pP are likely to be underestimates by as much as 20 km for deep earthquakes; e.g., Mitronovas and Isacks (1971) have accurately relocated the master event that occurred on 12 October 1967 (Table 1). Their depth of 650 km is 17 km deeper than the BCIS depth of 633 km. *In situ* velocities are underestimated by less than 1% as a result of errors in focal depth.

Ray parameters computed from a suite of earth models differ by so little that no more than a 2% bias in *in situ* velocity is expected by the exclusive use of a particular model (Fitch 1975). The ray parameters (equivalent to $dT/d\Delta$'s) from the Jeffreys model (1939) differ from more modern

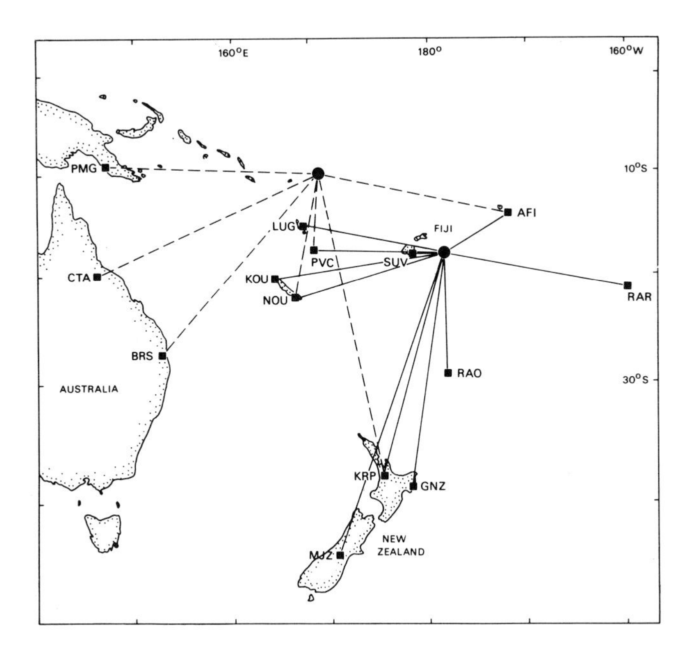

Fig. 3. Stations within and marginal to the south west Pacific.

earth models by 0.1 sec/deg in the distance ranges from 30° to 40° and from 60° to 85°. These differences are comparable to the standard errors in $dT/d\Delta$ measurements from the small UKAEA arrays (e.g. Simpson et al. 1974) and about twice the standard errors in apparent velocity from the extended Tonto Forest array in Arizona (Johnson 1969).

The validity of equation 1 depends on the L_j being sufficiently small. All stations included in the data sets for Tonga earthquakes, with the exception of the three closest stations SUV, AFI and RAO (Figure 3), are at least ten times more distant from the master event than are the secondary events. The inset in Figure 2 illustrates the least advantageous situation for satisfying the requirement that L_j be small. This situation arises when the angle S_{ij} approaches 90°. Excluding the three closest stations, the smallest source-receiver distance, B and A, is about 1500 km and the largest L_j is about 150 km. If the mean compressional velocity for ascending ray paths is about 8 km/sec the largest error, ΔT, is about 1/10 sec. This error is negligible by comparison with RMS residuals of about 1/2 sec.

However, from the 3 closest stations there could be data that do not meet the smallness condition on L_j. The smallest source-receiver distance for these stations is about 750 km so the maximum error, ΔT, is about 2 sec or twice the 95% confidence level of random errors in the ΔT_{ij}. Residuals this large are not permitted in the final solutions; however, for solutions numbered 1, 3, 4 and 8 in Table 1 the angle of incidence pertaining to SUV or RAO is more uncertain than the other refined angles of incidence, IN_i. The corresponding residuals are about twice the RMS value, i.e., about 1 sec. This suggests that approximation errors of about 1 sec may be present in no more than 1% to 2% of the data for these solutions. As more activity occurs in these source regions, a new selection of secondary events should eliminate this problem entirely.

The assumption of vertical ray paths is difficult to defend for particular source-receiver geometries such as those in which the path is likely to lie partly within an inclined seismic zone (Engdahl et al. 1976). To test for bias from this assumption, solutions were carried out with models that included the azimuths for ray paths from the master event to the closes stations, AFI, SUV and RAO. No significant bias in the in situ velocity was apparent from these solutions as might be expected because data from stations within the cut-off distance have their ΔV coefficients set equal to zero. Consequently, errors in the azimuths of those ray paths can only effect the in

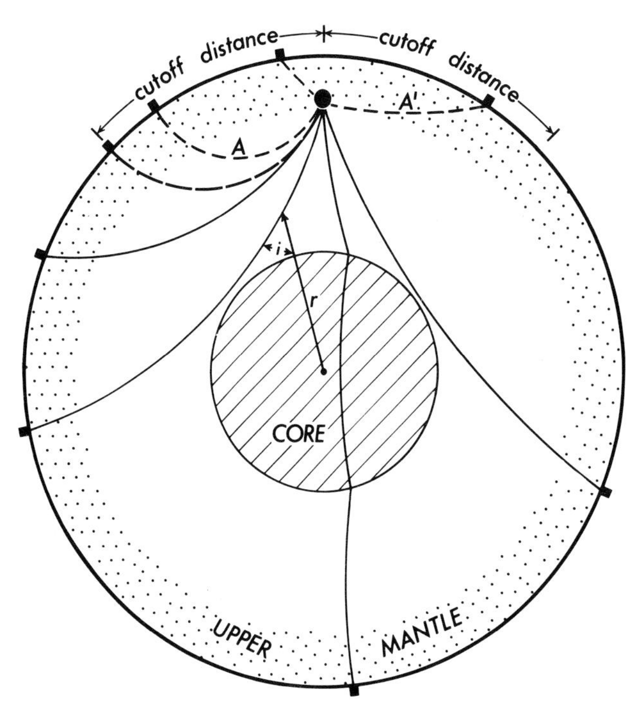

Fig. 4. Cut-off distance and constraints on ray paths.

Tonga 597

Event No.	Date	Azimuth (degrees)	Polar Angle (degrees)	Length (km)	Differential Origin Time (sec)	Standard Deviations (km) NS	EW	Z	No. of Data Points
1	15 Jan 65	214.0	43.1	56.1	15.7	2.1	7.1	1.2	28
2	9 Dec 65	101.4	24.4	55.6	16.0	3.6	3.2	2.0	52
3	25 Dec 65	265.0	68.2	83.8	18.1	2.4	3.1	1.7	54
4	30 Nov 67	78.2	19.9	45.4	15.6	2.6	3.9	2.0	42
5	22 Sep 68	246.3	35.7	49.3	15.1	1.9	6.6	1.8	32
6	1 Dec 69	132.5	100.1	76.4	15.8	2.9	0.3	3.5	39
7	12 Apr 71	60.6	75.5	24.8	12.5	1.4	2.8	2.4	57
8	30 Jun 72	302.9	132.0	35.7	12.3	1.2	1.7	5.0	44
9	5 Aug 72	309.5	82.8	14.6	12.3	3.4	0.7	3.0	41
10	4 Sep 72	116.8	120.0	20.0	11.6	2.2	0.7	3.7	42

Master Event: 24 Nov 1969, 18.01 S, 178.40 W, depth 597 km

Depth Range: 76 km

Station	Station Distance (degrees)	Station Azimuth (degrees)	Initial Take-Off Angle (degrees)	Final Take-Off Angle (degrees)
AFI	7.6	58.3	109.6	128.4 ± 7.2
RAO	11.2	177.8	92.7	104.3 ± 8.5
PVC	12.7	269.2	86.2	96.5 ± 5.1
LUG	14.1	278.0	81.1	94.8 ± 5.6
RAR	17.8	103.4	71.9	73.8 ± 4.5
KRP	20.5	193.8	67.7	56.6 ± 9.7
GNZ	20.8	187.9	67.4	65.5 ± 9.7

Jeffreys V_p at 597 km is 10.23 km/sec

Initial V_p = 10.5 km/sec

Final V_p = 10.76 ± 0.26 km/sec

Evaluation: χ^2 = 287

Number of data points = 431

Number of parameters = 48

Expected χ^2 = 383

RMS of Residuals = 0.536 sec

Average of Residuals = 0.005 sec

situ velocities through errors in the other parameters of the model.

A priori knowledge of the ray paths to distant stations may be suspect for particular stations in view of evidence for significant lateral heterogeneity in the lower mantle. All that can be hoped for is that these errors average to insignificance for data sets compiled from events that are well-recorded by a world-wide distribution of stations. Such averaging is expected to be more effective for variables that are common to a large part of the data set such as in situ velocity than for variables controlled by a small subset of the data such as the L_j's.

Ray Paths in the Source Regions

Differences between predicted and refined take-off angles in some cases can be explained by the presence or absence of anomalously high velocities along the ray paths. These differences pertaining to the Tonga source region are plotted in Figure 5. Source-receiver distances range from 3° for SUV (Fiji) to a cut-off distance between 18 and 24° for New Zealand stations (Figure 3). A bias toward negative differences shows that refinement generally rotates the take-off angles toward the upward vertical.

Paths to AFI (Samoa) are consistently steeper in the near source region than paths predicted from the Jeffreys model or for that matter any gross earth model. This can be explained by bodies of anomalously high velocity inclined more steeply in the source region than the predicted ray paths. The Tonga earthquake zone is known to contain material that has 6% to 7% higher P and S velocities, on the average, than "normal" mantle in the same depth range, and the inclination of the zone, although not constant, is greater than that of predicted ray paths to AFI (Mitronovas and Isacks, 1971).

The negative differences between take-off angles are particularly large for master earthquakes in the depth range from 500 to 575 km, which corresponds to a region of near "normal" in situ velocity (Figure 6). Large gradients encountered by rays descending and ascending from these depths may account for these differences, but 3-dimensional ray tracing (e.g. Jacob, 1970 and Toksöz et al. 1971) is required to test this speculation.

Ray paths to station KRP in the northwest corner of the north island of New Zealand and RAR (Raratonga) are not significantly different on the average from paths predicted by the Jeffreys model. Paths to KRP angle away from the Tonga-Kermadec seismic zone whereas paths to RAR are nearly normal to this zone and are likely to pass beneath it (Mitronovas and Isacks, 1971). Ray paths from the Tonga source regions to stations in the New Hebrides arc, PVC, ONE and LUG, and in New Caledonia, KOU and NOU, may or may not traverse the New Hebrides seismic zone, depending on the depth of the activity. Here again, 3-dimensional ray tracing could test the various possibilities.

Engdahl's Results

Arrival-time differences from earthquakes at depths greater than 300 km in the Tonga seismic

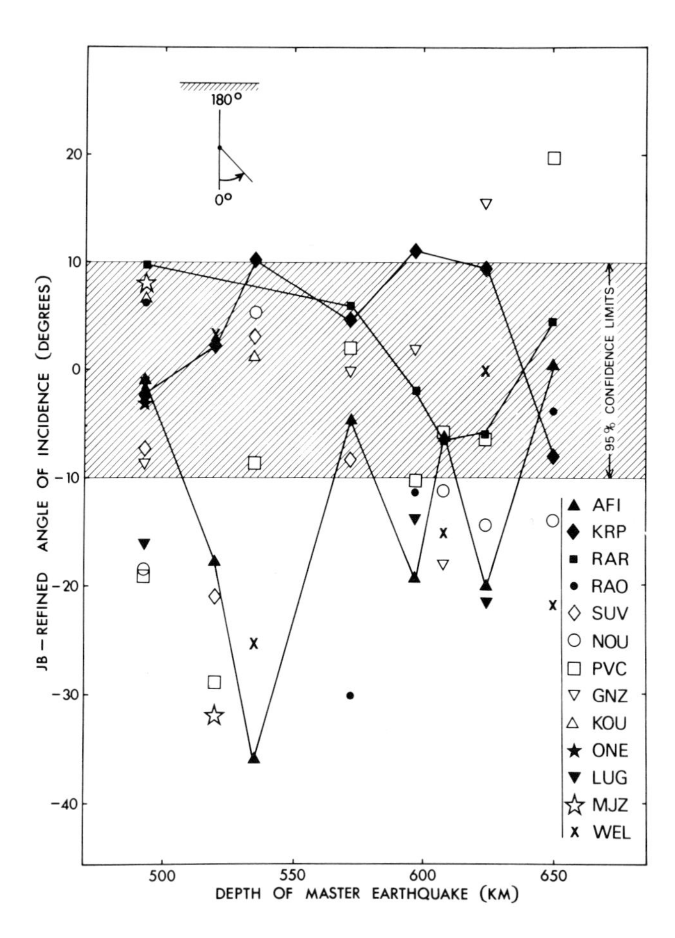

Fig. 5. Comparison of predicted and refined ray paths.

zone reveal a large travel-time anomaly in the source region (Engdahl, 1975). The arrival times were recorded at two stations in Alaska and one in the Aleutian Islands. Relative residuals were computed by subtracting from the observed arrival-time differences the values predicted from standard earthquake locations. Negative residuals, which imply anomalously high velocities, prevail for earthquakes toward the southern end of the zone. For this group the most negative residuals, less than -.5 sec, are either uniformly distributed with depth of focus or are distributed bimodally, at depths less than 400 km and greater than 500 km. The two distributions come from residuals from different station pairs.

For the interpretation of the results in Figure 6, Engdahl's residuals are important for three reasons. The large negative residuals do not trend toward more negative values at shallow focal depths. Such a trend would suggest a bias in the standard earthquake locations or a velocity anomaly that is uniform with depth and thus in conflict with the in situ velocities. He gives a rough estimate of 10% for the size of the anomaly, which agrees with the size of the anomaly shown in Figure 6. Lastly, the negative residuals persist for the deepest earthquakes, those in the depth range from 650 to 700 km, which suggests that anomalously high velocities are found at greater depths.

Fine Structure of Deep Seismic Zones

Abrupt steepening of mantle seismic zones may result from body forces imposed by phase transformations. However, possible competing effects from depth- and temperature-dependent viscosity can not be discounted, particularly below the

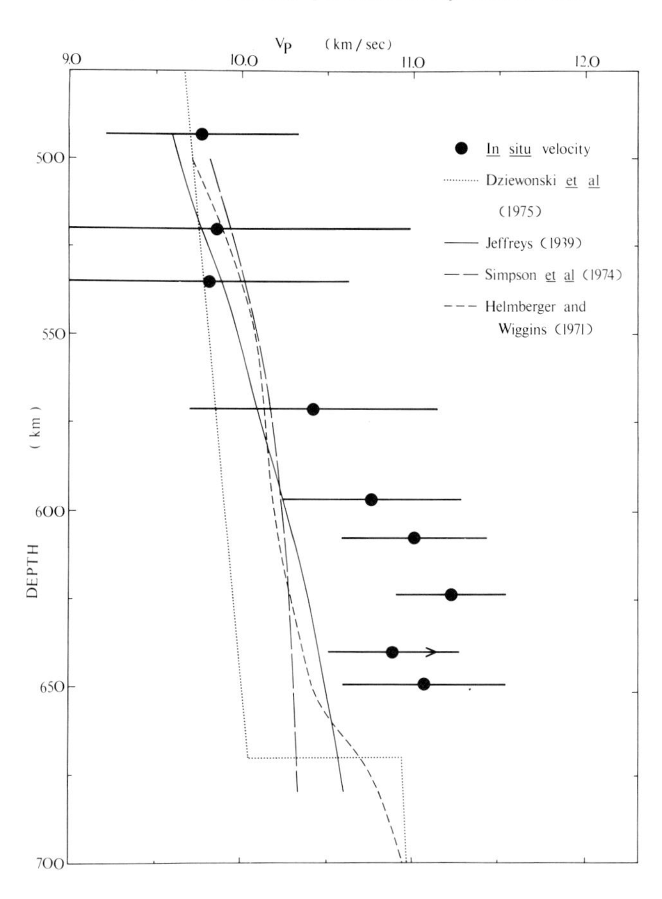

Fig. 6. In situ velocities and velocity-depth functions. The error bars are approximately 95% confidence limits. The arrow head pertains to solution 9B in Table 1. This solution oscillated about the velocity of 11.14 km/sec whereas solution 9A, in which station CTA is taken to be beyond the cut-off distance, converged to a lower velocity of 10.88 km/sec. Both solutions are obtained from arrival times from deep earthquakes beneath the NW corner of the Fiji plateau. These data sets could be improved judging from the large distances between master and several of the secondary events. The model by Dziewonski et al. differs from the others because the velocity jump at a depth of 670 km is a constraint on the solution rather than a consequence of it.

olivine-spinel phase boundary where experimental laboratory evidence has not constrained the rheological properties of mantle material.

A steepening of the Tonga seismic zone at a depth of about 400 km can be seen in some vertical sections transverse to those parts of the zone where activity extends below 600 km (Sykes 1966, Mitronovas and Isacks 1971). The opposite is seen in vertical sections across the northern end of the zone (Figure 6 of Mitronovas and Isacks 1971). This part of the zone is curved sharply toward the west, with the deepest part of the zone having the greatest curvature (Sykes 1966 and Oliver et al. 1973). As a result, the deeper part of the zone has a shallower dip. This curvature cannot be explained by forces imposed by phase transitions and consequently is of no further interest in this study.

Deep seismic zones in other parts of the world show a pronounced steepening below what appears to be a critical depth. That part of the seismic zone inclined from the Kurile arc to depths greater than 600 km is more steeply inclined below a depth of 400 to 500 km (Figures 17 and 18 of Veith 1974). Examples of seismic zones inclined more steeply from shallower depths are the Mariana zone in Figure 12 of Katsumata and Sykes (1969) and two deep earthquake zones beneath western South America (Isacks and Barazangi 1973). The south American deep earthquake may be detached, as are the deep earthquakes beneath the NW corner of the Fiji plateau (Pascal et al. 1973). These and other observed and hypothetical structures of mantle seismic zones are reviewed by Oliver et al. (1973).

The transition inferred from Figure 6 is consistent with a substantial increase in the downward body force. In fact, this increase may be as great as that resulting from the olivine-spinel transition (Ringwood 1972 and Schubert et al. 1975). This argument is based on the velocity-density systematics of Liebermann and Ringwood (1973). They suggest, from ultrasonic measurements on analogues of mantle silicates, that transformations with no change in principal cation coordination, such as olivine to spinel, have a $\Delta Vp/\Delta\rho$ of 3 to 4, whereas those resulting in an increase in coordination have a $\Delta Vp/\Delta\rho$ of about 2 (Figure 7). The latter transformations are likely to pertain to the "650 km discontinuity". Taking the velocity contrast from Figure 6 as 1 km/sec, a $\Delta Vp/\Delta\rho$ of 2 yields a $\Delta\rho$ of 0.5gmcm^{-3}. A maximum body force of 5×10^9 dyne cm^{-2} is estimated by integrating the $\Delta\rho$ over a slab thickness of 100 km. This force acts over a unit length parallel to the strike of the seismic zone and a unit depth along this zone. Schubert et al. (1975) computed a maximum body force of 2.6×10^9 dyne cm^{-2} from the transformation of olivine to spinel. The effects of latent heat and thermal expansion are included in their computations.

Figure 7 part B is a modification of Figure 12 of Schubert et al. (1975) showing the distribution of body forces as a function of depth in a hypothetical inclined slab. The transformation at 600 km has nearly twice the force/unit depth along the slab and one half the width of the olivine-spinel transformation. Integration of the body forces in Figure 7 yields the following pulling stresses on the lithosphere:

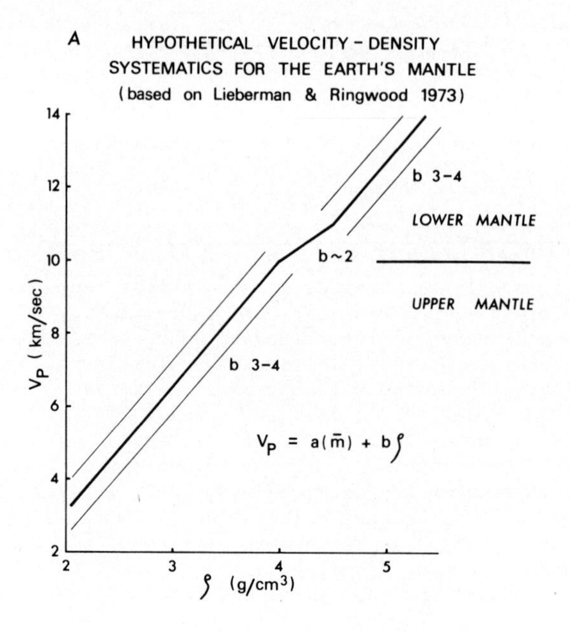

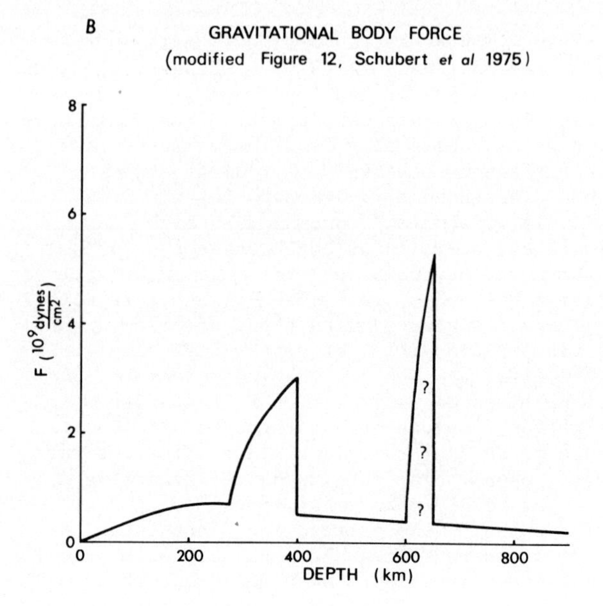

Fig. 7. A Velocity-density systematics. M refers to mean atomic weight.

B Body forces from thermal contrast and phase changes.

(1) 3.22 kbar X SIN (ϕ) from the thermal contrast
(2) 1.90 kbar X SIN (ϕ) from the olivine-spinel transformation
(3) $\sim$2 kbar X SIN (ϕ) from the elevated "650 km discontinuity"

ϕ is the inclination angle of the seismic zone from the free surface.

For comparison, McKenzie (1968) computed a pulling stress of 2.5 kbar X Sin(ϕ) from thermal effects alone. Schubert et al. assume a slip rate of 8 km/yr and a slab thickness of 100 km, whereas McKenzie assumes 10 cm/yr and 50 km for these values. The additional pulling stress from the phase-transformations is not large enough to change McKenzie's argument that the pulling forces cannot overcome the viscous drag of a large plate.

The inferred body forces from the phase transformations appear to be in conflict with focal mechanisms reported by Isacks and Molnar (1971). Mechanisms pertaining to the Tonga seismic zone show that axes of maximum compressive stress are aligned nearly parallel to the dip of the seismic zone at all depths. If the body forces in Figure 7 were the only forces acting on the seismic zone then down-dip extensions would prevail above each transition zone. Because these forces by themselves cannot drive the plate (McKenzie 1969) other forces must exist and these forces may load the seismic zone in compression. Resistance to downward motion could also prevent down-dip extension. However, the driving and resistance forces must be sufficiently small to preclude buckling of the seismic zone (Griggs, 1972) which is not apparent from earthquake locations (e.g., Sykes 1966).

Figure 8 illustrates a possible structural evolution for the Tonga seismic zone. Part B of this figure is taken schematically from profile AA' in Figure 6 of Mitronovas and Isacks (1971). The configuration of this zone, which includes an inflexion at a depth of about 550 km, is in part due to curvature of the northern end of the zone. The arrows parallel to the dip of the zone represent the summation of forces other than body forces from transition zones. Pushing and resistance forces and body forces from the thermal contrast are included in this summation. The downward arrows represent forces from transition zones. Only the major transition regions are illustrated. Changes in rheological properties such as effective viscosity may be abrupt across phase boundaries; e.g., increased resistance below the olivine-spinel transition may account for the inflexion in the seismic zone. If the downward velocity of the slab is stationary (Forsyth and Uyeda 1975), then at some depth the driving and resistance forces must be balanced.

Concluding Remarks

From Figure 6 and seismological evidence previously cited, a transition zone near the bottom of the Tonga seismic zone appears to be elevated but possibly by only a few km above a comparable zone in the surrounding mantle. Consequently, the slope, $\partial P/\partial T$, of the principal phase boundary is likely to be positive, assuming that the slab is cooler than its surroundings; e.g. thermal contrast of 800° from Schubert et al. (1975) and an elevation ranging from 20 to 50 km yields a $\partial P/\partial T$ in the range from 8 to 21 bars/^{o}C.

Small slopes are suggested by what little is known fo the transformations in silicates or their high pressure analogues in which the principal cation coordination changes from 4 to 6 fold (tetrahedral to octahedral). Slopes of such transformations are expected to be less than the slope of 30 bars/^{o}C for the olivine-spinel transformation in which Si^{+4} coordination remains tetrahedral (Akimoto and Fujisawa, 1966 and 1968 and Ringwood and Major, 1970).

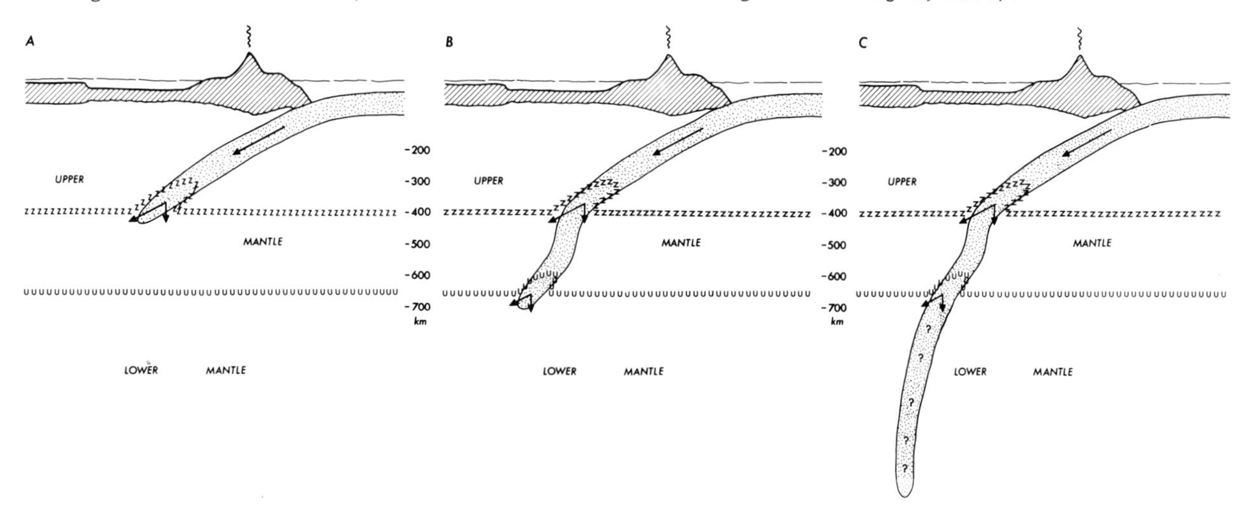

Fig. 8. Structural evolution of the Tonga Seismic Zone. The inclined slabs have a nearly uniform width and therefore the boundary of the slabs do not coincide with an isotherm.

Navrotsky and Kasper (1976) have discussed thermodynamic data pertaining to the slopes of transformations from spinel to ilmenite plus rock-salt structures and from ilmenite to perovskite structure. They argue that $\partial P/\partial T$, which is equivalent to the ratio of the change in entropy to the decrease in specific volume, $\Delta S/\Delta V$, will be positive for the spinel transformations and small and possibly negative for the ilmenite transformations. Small slopes are also predicted for the disproportionation of silicate spinel to a simple oxide mixture (Jackson et al. 1974; Navrotsky and Kasper, 1976); however, recent experimental evidence already cited suggests that the perovskite structure is predominate in the lower mantle.

In the depth range from 500 to 580 km in situ velocities fall within the range of values predicted by a suite of earth models, although the best of these data sets, which was compiled from the shallowest group of earthquakes (Table 1), yields a velocity that is 1% higher than that predicted by the J-B model (Figure 6). Data sets with master earthquakes at depths of 520, 537 and 572 km yielded in situ velocities with comparatively large uncertainties. This is explained by smaller contributions to these sets from stations further from the master earthquakes than the cut-off distances. These velocities may be biased and if so, the bias is likely to be toward low values because a slab that is up to 1000°C cooler than the surrounding mantle should show a positive V_p anomaly of up to 4%. This estimate is based on measurements of the elastic moduli of spinel, $MgAl_2O_4$, by Anderson et al. 1968, Chang and Barsch 1973 and Liu et al. 1974. This mineral is an iso-structural analogue of silicate spinels that are thought to account for more than 50% of the mantle in depth range of these earthquakes.

The anomaly pattern shown by the in situ velocities in Figure 6 suggests that the Tonga seismic zone extends at least 100 km below the top of the transition zone between the upper and lower mantle. The phase boundary is in the depth range 580 to 600 km whereas the seismic acitivity extends to a depth of about 700 km (Sykes 1966). If slab-like bodies exist below 700 km then the lack of earthquakes below this depth can be explained by the absence of major transition zones in the lower mantle (Ringwood, 1972). Alternatively, the temperature in the slab may rise above a critical value for earthquakes (McKenzie, 1969) or T/Tm may exceed a critical value, where T_m is melting temperature or more appropriately a solidus temperature (Griggs, 1972).

Studies of apparent velocities of P-waves measured at seismic arrays show conclusively that no transition with more than a few percent velocity change occurs in the lower mantle (Johnson, 1969; Chinnery, 1969). Although this evidence does not exclude the possibility of transformations in which the percentage density change is much greater than the velocity change, major transition zones comparable with those in the upper mantle seem unlikely.

Travel-time anomalies appear to be the key evidence needed to establish the existence of the slabs in the lower mantle. Recent studies by Jordan and Lynn (1974) and Engdahl (1975) show that anomalously high velocities exist in the lower mantle and could be extensions of deep seismic zones.

Acknowledgements. I am indebted to I.N.S. Jackson, R.C. Liebermann and L.G. Liu for numerous discussions pertaining to the mineralogy and the physical properties of the mantle. Lesley Hodgson prepared data for the computer and drafted the figures. J.R. Cleary and R.C. Liebermann have read the manuscript critically and offered suggestions for its improvement.

References

Akimoto, S.I. and H. Fujisawa, Olivine-spinel transition in system Mg_2SiO_4 at 800°C, Earth Planet.Sci.Lett., 1, 237-240, 1966.

Akimoto, S.I. and H. Fujisawa, Olivine-spinel solid solution equilibria in the system Mg_2SiO_4 - Fe_2SiO_4, J.Geophys.Res., 73, 1467-1479, 1968.

Anderson, D.L., Phase changes in the upper mantle, Science, 157, 1165-1173, 1967.

Anderson, D.L., C.G. Sammis and T.H. Jordan, Composition of the mantle and core, in: The Nature of the Solid earth, Eugene C. Robertson, ed., McGraw-Hill Inc. New York, 41-66, 1972.

Anderson, D.L., E. Schreiber, R.C. Liebermann and N. Soga, Some elastic constant data on minerals relevant to geophysics, Rev. of Geophys., 6, 491-524, 1968.

Archambeau, C.B., E.A. Flinn and D.G. Lambert, Fine structure of the upper mantle, J.Geophys. Res., 74, 5825-5865, 1969.

Cathles III, L.M., The Viscosity of the Earth's Mantle, Princeton University Press, Princeton, New Jersey, pp. 386, 1975.

Chang, Z.P. and G.R. Barsch, Pressure dependence of single crystal elastic constants and anharmonic properties of spinel, J.Geophys.Res., 78, 2418-2433, 1973.

Chinnery, M.A., Velocity anomalies in the lower mantle, Phys.Earth Planet. Inter., 2, 1-10, 1969.

Davies, G.F. and A.M. Dziewonski, Homogeneity and constitution of the Earth's lower mantle and outer core, Phys.Earth Planet.Inter., 10, 336-343, 1975.

Dziewonski, A.M., A.L. Hales and E.R. Lapwood, Parametrically simple earth models consistent with geophsyical data, Phys.Earth Planet. Inter., 10, 12-48, 1975.

Engdahl, E.R. and E.A. Flinn, Seismic waves reflected from discontinuities within Earth's upper mantle, Science, 163, 177-179, 1969.

Engdahl, E.R., Effects of plate structure and

dilatancy on relative teleseismic P-wave residuals, Geophys.Res.Lett., 2, 420-422, 1975

Engdahl, E.R., N.H. Sleep and M.T. Lin, Plate effects in North Pacific subduction zones, Tectonophys., 1976.

Fitch, T.J., Compressional velocity in source regions of deep earthquakes: an application of the master earthquake technique, Earth Planet. Sci.Lett., 26, 156-166, 1975.

Forsyth, D. and S. Uyeda, On the relative importance of the driving forces of plate motion, Geophys.J.R.Astron.soc., 43, 163-200, 1975.

Goldreich, P. and A. Toomre, Some remarks on polar wandering, Journ.Geophys.Res., 74, 2555-2567, 1969.

Green, D.H. and R.C. Liebermann, Phase equilibria and elastic properties of a pyrolite model for the oceanic upper mantle, Tectonophysics, 32, 61-92, 1976.

Griggs, D.T., The sinking lithosphere and the focal mechanism of deep earthquakes, in: The Nature of the Solid Earth, Ed. E.C. Robertson, McGraw-Hill, 361-384, 1972.

Helmberger, D. and R.A. Wiggins, Upper mantle structure of mid-western United States, J.Geophys.Res., 76, 3229-3245, 1971.

Helmberger, D.V. and G.R. Engen, Upper mantle shear structure, J.Geophys.Res., 79, 4017-4028, 1974.

Herrin, E., W. Tucker, S. Taggart, D.W. Gordon and S.L. Lobdell, Estimation of surface focus P travel times, Bull.Seism.Soc.Am., 58, 1273-1292, 1968.

Isacks, B. and P. Molnar, Distribution of stresses in the descending lithosphere from a global survey of focal-mechanism solutions of mantle earthquakes, Rev.Geophys.and Sp.Phys., 9, 103-174, 1971.

Isacks, B. and M. Barazangi, High frequency shear waves guided by a continuous lithosphere descending beneath western South America, Geophys.J.R.Astron.soc., 33, 129-139, 1973.

Jackson, I.N.S., R.C. Liebermann and A.E. Ringwood, Disproportionation of spinels to mixed oxides: significance of cation configuration and implications for the mantle, Earth Planet.Sci. Lett., 24, 203-208, 1974.

Jacob, K.H., Three-dimensional sersimic ray tracing in a laterally heterogeneous spherical earth, J.Geophys.Res., 75, 6675-6689, 1970.

Jeffreys, H., The times of P, S and SKS, and the velocities of P and S, Monthly Notices R. Astron.Soc.Geophys.Suppl., 4, 498-533, 1939.

Johnson, L.R., Array measurements of P velocities in the upper mantle, J.Geophys.Res., 72, 6309-6323, 1967.

Johnson, L.R., Array measurements of P velocities in the lower mantle, Bull.Seis.Soc.Am., 59, 973-1008, 1969.

Jordan, T.H. and W.S. Lynn, A velocity anomaly in the lower mantle, J.Geophys.Res., 79, 2679-2685, 1974.

Kanamori, H., Upper mantle structure from apparent velocities of P-waves recorded at Wakayama micro-earthquake observatory, Bull. Earthquake Res.Inst., 45, 657-678, 1967.

Katsumata, M. and L.R. Sykes, Seismicity and tectonics of the western Pacific: Izu-Mariana - Caroline and Ryukyu - Taiwan Regions, J.Geophys.Res., 74, 5923-5948, 1969.

King, D.W. and G. Calcagnile, P-wave velocities in the upper mantle beneath Fennoscandia and Western Russia,

Liebermann, R.C. and A.E. Ringwood, Birch's law and polymorphic phase transformations, J.Geophys.Res., 78, 6926-6932, 1973.

Liebermann, R.C., A.E. Ringwood and A. Major, Elasticity of Polycrystalline stishovite, Earth and Planet.Sci.Lett. sub.pub. 1976.

Liu, H.P., R.N. Schock and D.L. Anderson, Temperature dependence of single crystal spinel ($MgAl_2O_4$) elastic constants from 293K to 423K measured by light-sound scattering in the Raman-Nath region, Geophys.J.R.Astron. Soc., 42, 217-250, 1975.

Liu, L.G., Silicate perovskite from phase transformations of pyrope-garnet at high pressure and temperature, Geophys.Res.Lett., 1, 277-280, 1974.

Liu, L.G., Post-oxide phases of forsterite and enstatite, Geophys.Res.Lett., 2, 417-419, 1975

Liu, L.G., The high pressure phases of $MgSiO_3$, Earth Planet.Sci.Lett., 31, 200-208, 1976.

McKenzie, D.P., Speculations on the consequence and causes of plate motions, Geophys.J.R. Astron.Soc., 18, 1-32, 1969.

McKenzie, D. and N. Weiss, Speculations on the thermal and tectonic history of the earth, Geophys.J.R.Astron.Soc., 42, 131-174, 1975.

Minster, J.B., T.H. Jordan, P. Molnar and E. Haines, Numerical modelling of instantaneous plate tectonics, Geophys.J.R.Astron. Soc., 36, 541-576, 1973.

Mitronovas, W. and B.L. Isacks, Seismic Velocity Anomalies in the Upper mantle beneath the Tonga-Kermadec island arc, J.Geophys.Res., 76, 7154-7180, 1971.

Navrotsky, A. and R.B. Kasper, Spinel disproportionation at high pressure: calorimetric determination of enthalpy of formation of Mg_2SnO_4 and Co_2SnO_4 and some implications for silicates, Earth Planet.Sci.Lett., 31, 247-254, 1976.

Oliver, J., B.L. Isacks, M. Barazangi and W. Mitronovas, Dynamics of the down-going lithosphere, Tectonophys., 19, 133-147, 1973.

Richards, P.G., Seismic waves reflected from velocity gradient anomalies within the Earth's upper mantle, Zeitschrift für Geophysik, 38, 517-527, 1972.

Richter, F.M., Finite Amplitude Convection through a phase boundary, Geophys.J.R.Astron. Soc., 35, 265-276, 1973.

Richter, F.M. and C.E. Johnson, Stability of a chemically layered mantle, J.Geophys.Res., 79, 1635-1639, 1974.

Ringwood, A.E., The chemical composition and origin of the earth in Advances in Earth Sci. ed. P.M. Hurley, M.I.T. Press, Cambridge, Mass. 287-356, 1966.

Ringwood, A.E., Phase transformations and the constitution of the mantle, Phys.Earth Planet. Inter., 3, 109-155, 1970.

Ringwood, A.E. and A. Major, The system Mg_2SiO_4 - Fe_2SiO_4 at high pressures and temperatures, Phys.Earth Planet.Inter., 3, 89-108, 1970.

Ringwood, A.E., Phase transformations and mantle dynamics, Earth and Planet.Sci.Lett., 14, 233-241, 1972.

Ringwood, A.E., Composition and Petrology of the Earth's Mantle, McGraw-Hill Book Company, 618, 1975.

Pascal, G., J. Dubois, M. Barazangi, B.L. Isacks and J. Oliver, Seismic velocity anomalies beneath the New Hebrides arc: evidence for a detached slab in the upper mantle, J.Geophys. Res., 78, 6998-7004, 1973.

Schubert, G., D.A. Yuen, and D.L. Turcotte, Role of Phase Transitions in a Dynamic Mantle, Geophys.J.R.astr.Soc., 42, 705-735, 1975.

Smith, A.T. and M.N. Toksöz, Stress distribution beneath island arcs, Geophys.J.R.Astron.Soc., 29, 289-318, 1972.

Simpson, D.W., R.F. Mereu and D.W. King, An array study of P-wave velocities in the upper mantle transition zone beneath North-eastern Australia Bull.Seis.Soc.Am., 64, 1757-1788, 1974.

Snoke, J.A., I.S. Sacks, and H. Okada, A model not requiring continuous lithosphere for anomalous high-frequency arrivals from deep-focus South American earthquakes, Phys.Earth Planet.Inter., 9, 199-206, 1974.

Solomon, S.C. and K.T. Paw U, Elevation of the olivine-spinel transition in subducted lithosphere: Seismic Evidence, Phys.Earth Planet. Inter., 11, 97-108, 1975.

Sykes, L.R., The seismicity and deep structure of island arcs, J.Geophys.Res., 71, 2981-3006, 1966.

Toksöz, M.N., J.W. Minear and B.R. Julian, Temperature field and geophysical effects of a downgoing slab, J.Geophys.Res., 76, 1113-1138, 1971.

Toksöz, M.N., N.H.S. Sleep and A.T. Smith, Evolution of the downgoing lithosphere and mechanisms of deep focus earthquakes, Geophys.J.R.Astron.Soc., 35, 285-310, 1973.

Turner, J.S., Convection in the mantle: A laboratory model with temperature-dependent viscosity, Earth Planet.Sci.Lett., 17, 369-374, 1973.

Veith, K.F., The relationship of island arc seismicity to plate tectonics, Ph.D. Thesis, Southern Methodist University, Dallas, Texas 162pp., 1974.

Whitcomb, J.H. and D.L. Anderson, Reflection of P'P' seismic waves from discontinuities within the mantle, J.Geophys.Res., 75, 5713-5728, 1970.

Whitcomb, J.H., Part I A Study of the velocity structure of the earth by the use of core phases, Ph.D. Thesis, California Institute of Technology, Pasadena, California, pp. 1-173, 1973.

HEAT FLOW IN BACK-ARC BASINS OF THE WESTERN PACIFIC

T. Watanabe

Earthquake Research Institute, University of Tokyo
Tokyo, Japan 113

M. G. Langseth and R. N. Anderson

Lamont-Doherty Geological Observatory of Columbia University
Palisades, New York, 10964

Abstract. A review of heat-flow data in trench-island arc and back-arc basins of the Western Pacific shows that for basins associated with current subduction at a trench, profiles of heat flow normal to tectonic trends have common features: 1) the zone from the trench axis to the volcanic zone has below average heat flow; 2) the volcanic zone has high but variable values; and, 3) the back-arc basins have mean heat flows that depend on age. Basins younger than Miocene have high heat-flow values. Scattered very low values suggest that water circulation within the young oceanic crust may be a significant mode of heat transfer. Back-arc basins of Early Tertiary age with thick sediments all have mean heat flows of about 2.2 HFU and the variability is low. The Aleutian and Celebes basins have nearly normal mean heat-flow values. A diagram of Moho depth versus reciprocal mean heat flow illustrates that the properties and evolution of the lithosphere below marginal basins is significantly different from those of the North Pacific Plate. The present parameters of marginal basins may be best explained by a two-stage cooling history. Newly formed extensional basins cool toward a steady state with a thin lithosphere (40-50 km) and high heat flow (2.2 HFU). At some point in the evolution the additional source of heat is removed and the basin begins a second cooling stage toward values of flux typical of the old Pacific Plate. To explain the deeper Moho depths in back-arc basins a higher mean mantle density is required compared to the Pacific. Heat flux observed in basins at the first steady state point is nearly equal to the average heat flow of the Pacific Plate. This suggests that heat transfer processes associated with massive subduction in the Western Pacific taps the same mantle sources as the seafloor spreading process.

Introduction. High heat flow is often noted as one of the characteristic features of back-arc basins. Historically, however, the first heat-flow measurements in back-arc basins made in 1962 in the Bering and Caribbean seas gave normal or subnormal values (Gerard et al., 1962; Foster, 1962). The first above normal heat-flow observations in back-arc basins were made in the Sea of Japan (Yasui and Watanabe, 1965). Subsequent work further clarified the main features of the heat-flow distribution in the Sea of Japan (Yasui et al., 1968 a). At the same time, investigations in other marginal seas were carried out energetically by many investigators.

Of all marginal seas, the region surrounding Japan is the most intensely sampled with respect to heat flow; with data obtained (a) on land (Horai, 1964; Uyeda and Horai, 1964; (b) in the Japan Trench; and (c) in the deep Pacific Ocean (Vacquier et al., 1967). One of the major problems in island arc tectonics has been the plausible explanation of heat-flow variation across the Japan Arc (McKenzie and Sclater, 1968; Mc Kenzie, 1969; Hasebe et al., 1970; Minear and Toksöz, 1969; Sugimura and Uyeda, 1973). Observational constraints in most marginal basins require that additional heat supplied from below be included in both models to explain the present heat flow.

To explain the oceanic structure of the Sea of Japan various models have been proposed such as (a) the oceanization of the continental crust (Beloussov and Ruditch, 1961); (b) loss of the upper

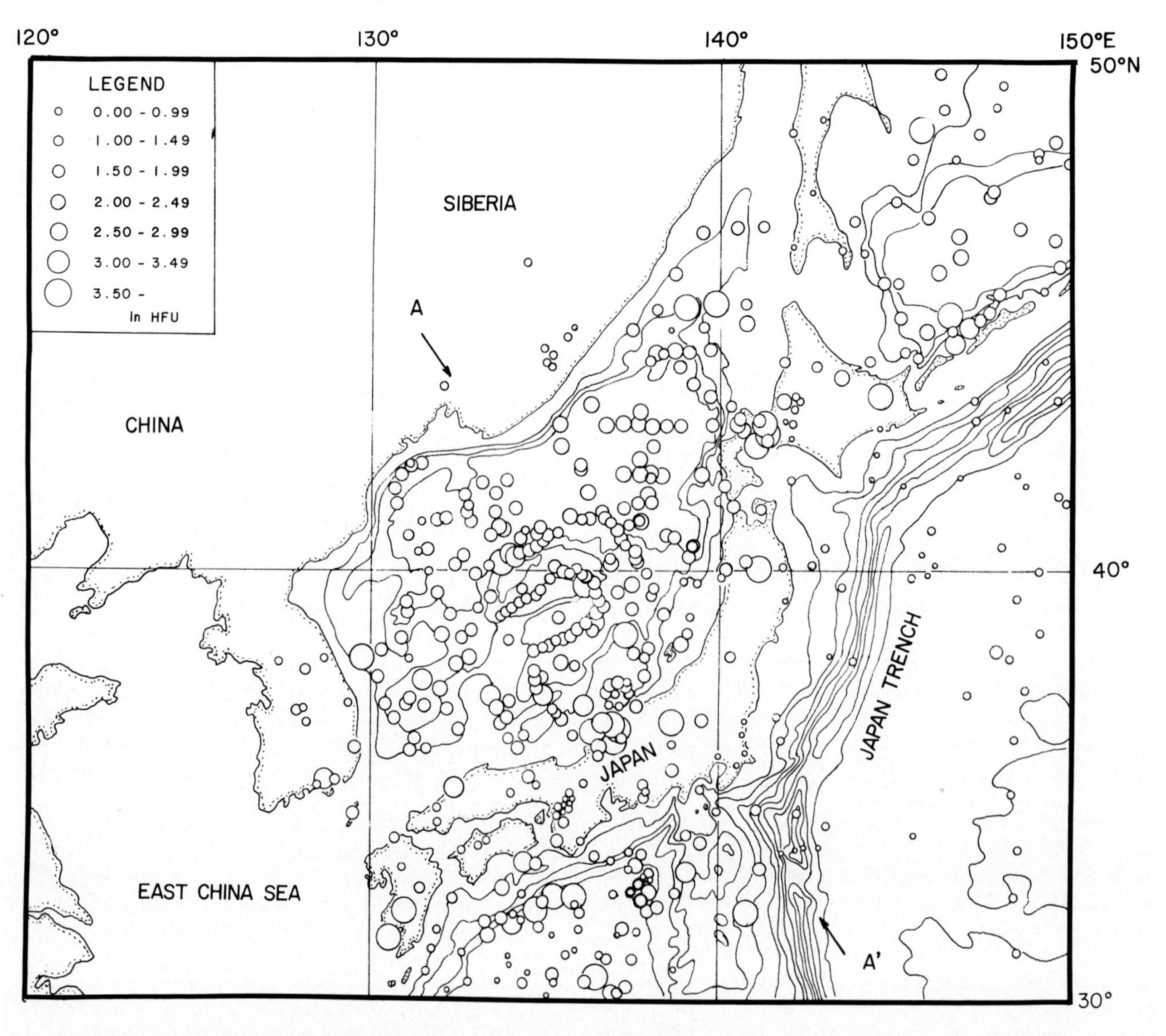

Figure 1. Heat-flow values in the Sea of Japan and surrounding areas: Circles indicate heat-flow stations and their size the intensity of values. The northeastern and southwestern part are divided by the A'A line. Data come from the references cited in the text, except for unpublished data which have been supplied by Ehara, Langseth, Nomura, Saki, Watanabe and Yasui.

low density layer of continental crust by erosion and uplift (Gorai, 1968; Minato and Hunahashi, 1970); (c) back-arc spreading and the creation of oceanic crust (Murauchi and Den, 1966; Karig, 1970, 1971 a, b); and, (d) trapped ocean (Shor, 1964; Vasilkovsky et al., 1971; Ben Avraham and Uyeda, 1973; Scholl et al., 1975). Besides the Sea of Japan, compilations and intercomparison of geological and geophysical features in many trench island arc-marginal basin systems have been made by various authors (Karig, 1971b; Sclater, 1972; Sugimura and Uyeda, 1973). The four proposed models for the Sea of Japan are continuously being tested in these other marginal basins, although the back-arc spreading and trapped ocean hypotheses have received more emphasis than the oceanization and erosion models.

In this paper, we present a survey of the distribution of heat-flow values within well sampled island arc-marginal basin systems. Much of the data presented here were accumulated many years ago; consequently, most stations have insufficient data for environmental disturbance evaluation. Therefore, no attempt has been made to correct the values for environmental effects. We have eliminated values at stations reported to be of low reliability with respect to instrumental or experimental problems. We examine the variation in heat flow in terms of its relation to major tectonic features. The mean heat flow and basement elevation in these basins are intercompared and related to a theoretical model of basin evolution.

Heat-flow distribution features in marginal seas having sufficient observations are examined using published values as well as unpublished measurements made by the Lamont-Doherty Geological Observatory (L-DGO) and by a number of Japanese groups (to be published as independent articles, now in preparation). Details of the geothermal data and local distribution will be discussed there. In this paper, we will describe only regional features and the results of intercomparisons of the major basins.

Distribution of Heat Flow in Marginal Seas

The Sea of Japan

The Sea of Japan is an often cited example of a typical marginal sea. Early results were compiled by Yasui et al. (1968 a). Udintsev et al. (1971) have presented additional data. One bottom hole heat-flow measurement was made in DSDP Hole 301 (Watanabe et al., 1975). A compilation of land heat-flow measurements has been made by Horai and Uyeda (Horai, 1964; Uyeda and Horai, 1964); subsequently, a considerable number of heat-flow stations have been measured in surrounding areas such as the Japanese Islands Ehara, 1971; Kono and Kobayashi, 1971; Uyeda et al., 1973; Nakagawa, 1975), Korea (Mizutani et al., 1970) and Siberia (Veselov et al., 1974). Some as yet unpublished values are used to make Figure 1, provided by Langseth, Ehara, Nomura, Watanabe, Saki and Yasui, which shows the currently available values in the Sea of Japan and surrounding areas.

Tectonically the Japanese Island arc-trench system can be divided into two distinct regions (Sugimura and Uyeda, 1973; Uyeda and Miyashiro, 1974). I. Northeastern Honshu, the Japan Trench and the northern basins of the Sea of Japan form what could be considered a classic trench-island arc system where present tectonic features extend southward along the Izu-Mariana Arc systems. II. Southwestern Honshu and the southern islands of Japan and adjacent seas have a less well developed pattern of subduction, although the Nankai Trough, seaward of Kyushu shows indications of being a juvenile trench

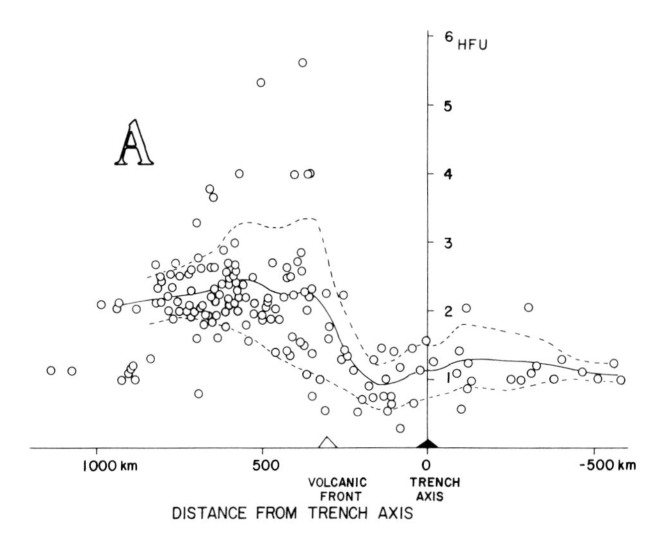

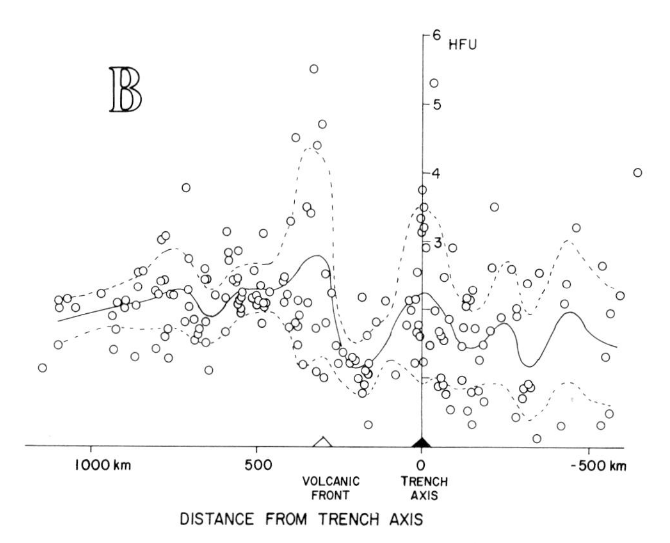

Figure 2. a) Heat-flow values (circles) versus distance from the axis of the trench: the Northeastern Sea of Japan. Solid and broken lines show the mean of values within ± 100 km window and its standard deviation respectively. These symbols are used throughout all of the following heat-flow profiles. The data sources are the same as for Figure 1.
b) Heat flow values versus distance from the axis of the trench: The Southwestern Sea of Japan. The data sources are the same as for Figure 1.

(Hilde et al., 1969; Kanamori, 1971; Karig, 1975). Clear features of subduction activity can be traced from the southwesternmost island southward along the Nansei Shoto (Figure 1). A deep seismic zone is associated with the Japan-Izu-Mariana Trench arc system (Matsuda and Uyeda, 1971).

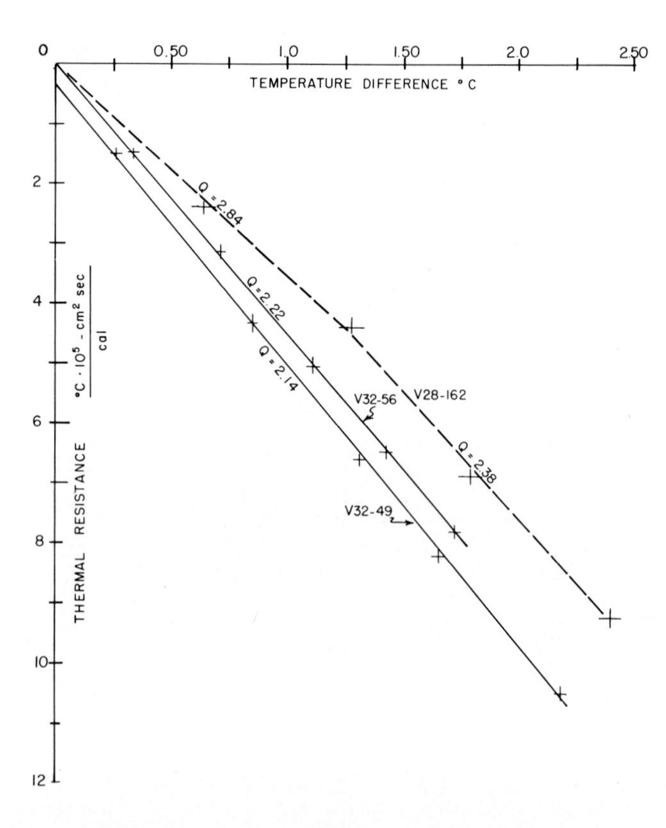

Figure 3. Sedimentary temperature increase, ΔT, above the bottom water, value versus thermal resistance, R_T, for three stations in the Japan Abyssal Plain. The estimated errors of each point are indicated by the crosses. The slope of the fitted lines, $\Delta T/R_T$, equals the heat flow. Two stations have a uniform heat flow within the measurement errors to deeper than 13 m. The station V28-162 indicates a 17% decrease in heat flow at a depth of about 8 meters. The mean thermal resistivity of the sediment is about 0.55°C/HFU.

Aoki (1974) has proposed that the subducting slab associated with the seismic activity may be split. The line A-A' in Figure 1 divides the two regions described above and is almost identical to the line of splitting of the subducted lithosphere proposed by Aoki. Heat-flow profiles roughly perpendicular to the tectonic trends are shown across the northeastern and southwestern regions in Figure 2 a and b. Heat flow in heat-flow units (HFU = 10^{-6} cal/cm^{2}sec = 41.87 mWm^{-2}) is plotted versus distance from the trench axis in km.

For the southwestern section, the Nankai Trough is assumed to be the trench axis which extends smoothly to the Japan Trench across the northernmost part of the Izu-Bonin Ridge. This line is almost identical to a proposed Cretaceous and Early Tertiary subduction system (Uyeda and Ben Avraham, 1972).

In the northeastern region (see Figure 2a) the heat flow is uniform, 1.1 HFU in the Northwestern Pacific Ocean floor. This value is typical of old oceanic crust (older than 100 m.y.). Low heat flow is observed between the landward wall of the trench and the volcanic front on Honshu. Scattered, very high values are observed near the volcanic zone. Uniformly high values generally greater than 2.0 HFU are distributed over the northeastern basins of the Sea of Japan. However, on the adjacent Siberian continent several measurements give low values of about 1.1 HFU.

In the southwestern part (Figure 2b) the distribution pattern is similar to that of the northwestern part, except that high values exhibiting much scatter are observed along the Nankai Trough and adjacent oceanic floor. Slightly depressed heat flow is observed over the Yamato Rise, located about 600 km landward of the trench axis, which seismic data show as a continental structure (Ludwig et al., 1973 a). As a whole, the present difference in tectonic styles of the two regions does not appear to affect the heat-flow distribution in the Sea of Japan basins or on the Japanese Islands.

The general heat-flow distribution features are correlatable with major topographic features. In the Japan Basin, oceanic floor deeper than 3000 m has an average heat flow of 2.22 $\pm$ 0.33 (s.d.) HFU and slightly higher values are observed deeper than 3500 m with an average of 2.32 $\pm$ 0.33 (s.d.) HFU. Values are higher in the eastern part of the basin. The generally high heat flow in the Sea of Japan basins continues smoothly onto the Japanese Islands, whereas heat flow abruptly decreases from high to low from the Sea of Japan onto surrounding lands such as the Korean Peninsula, Siberia and Sakhalin Island. The overall average of all 224 values in the Sea of Japan is 2.21 $\pm$ 0.51 (s.d.). The relatively low standard deviation compared to midoceanic ridge environments is probably due to the thick and uniform sedimentary cover in the Sea of Japan.

To account for part of the high heat flow observed in the Sea of Japan, Kobayashi and Nomura (1972) proposed possible additional heat generation from the oxidation of iron sulfides in the upper 7-8 meters of the seafloor sediment. They estimated that the added heat flow from this source could be as large as 1.0 HFU, and they further speculated that high heat flow in all marginal basins completely closed by shallow topography may be partly explained by this additional heat source.

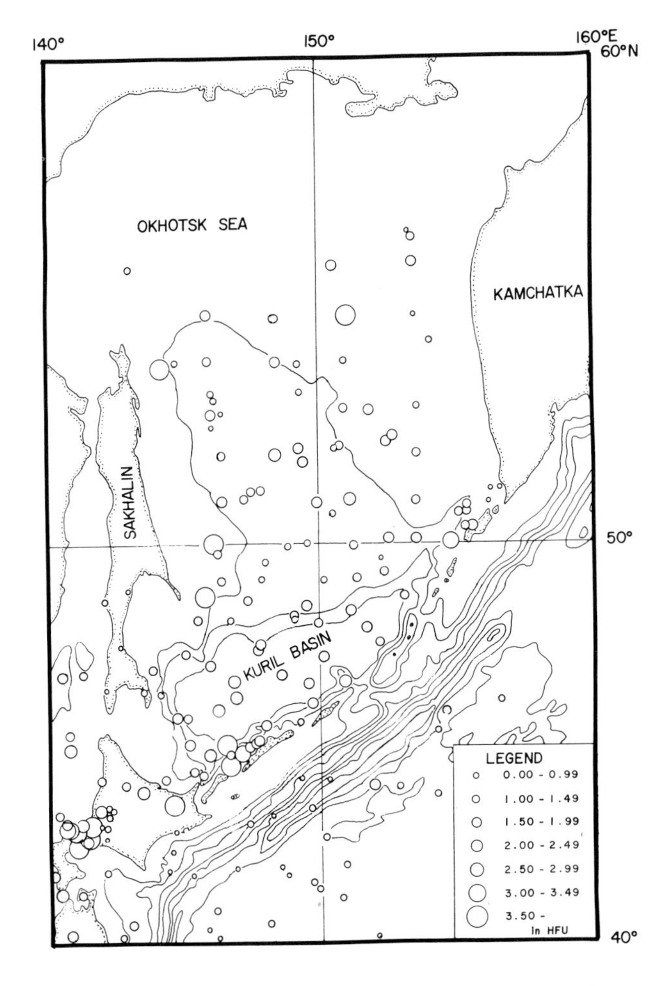

Figure 4. Heat-flow values in the Sea of Okhotsk and surrounding areas. The references are given in the text and additional unpublished data are provided by Langseth and Ehara.

During the 28th cruise of R. V. VEMA temperature measurements deeper than eight meters were attained at four stations, one of which is as shown in Figure 3 plotted versus thermal resistance of the sedimentary column. If heat flow Q is calculated as,

$$Q = \frac{T(z_1) - T(z_2)}{\int_{z_i}^{z_2} r(z)\, dz} \qquad (1)$$

where T(z) and r (z) are temperature and thermal resistivity respectively at depths z. The integral in Equation 1 is the thermal resistance of the sedimentary column which is calculated using the conductivity-depth relationship given by a smooth interpolation of measured values. The segments of the measurement shallower than eight meters yield a heat flow of 2.84 HFU, whereas those deeper than eight meters average 2.38 HFU. The heat-flow difference (0.46 HFU) may result from oxidation of iron sulfides. However, resolution of the measurement is not sufficient to determine if this difference is real or an experimental error. In 1975, during the 32nd cruise of R. V. VEMA, careful measurements were made to further test this hypothesis (Two examples are shown in Figure 3). The heat-flow values are 2.22 and 2.14 HFU and the linearity of temperature versus thermal resistance suggests no significant heat generation effect. During DSDP, Leg 31, heat flow was measured at Site 301. At the hole depth of 124.5 m, the heat-flow value is 2.02 HFU, which is in exact agreement with adjacent values from short-probe measurements (less than 2m) of 2.03 and 2.02 HFU. Although a possibility remains that heat-flow values higher than about 2.0 HFU may be influenced by heat generation by oxidation, the effect cannot be as large as previously suggested.

The Sea of Okhotsk

After initial studies by Yasui and others (Yasui et al., 1967, 1968b) intensive research has been carried out by Soviet scientists (Savostin and Vlasov, 1974), and further unpublished measurements were made during a USSR, US Japanese cooperative program by Langseth and Ehara. As shown in Figure 4, the station distribution is quite uniform. The heat-flow profile is shown in Figure 5. Deep Pacific Ocean values are similar to those off of Northeastern Japan. Low heat flow in the trench is well established. Heat flow

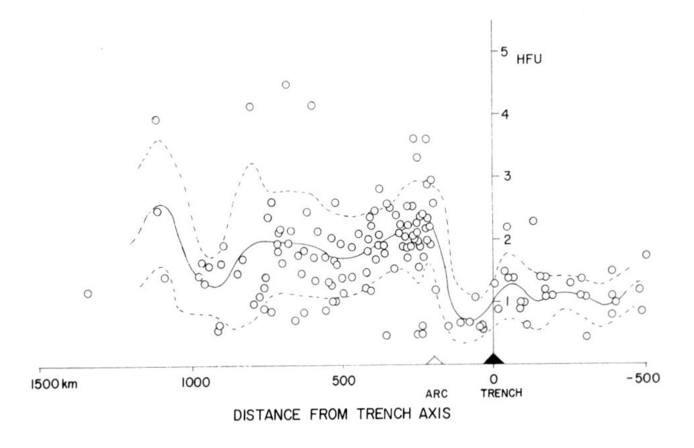

Figure 5. Heat flow versus distance from the axis of the Kuril Trench.

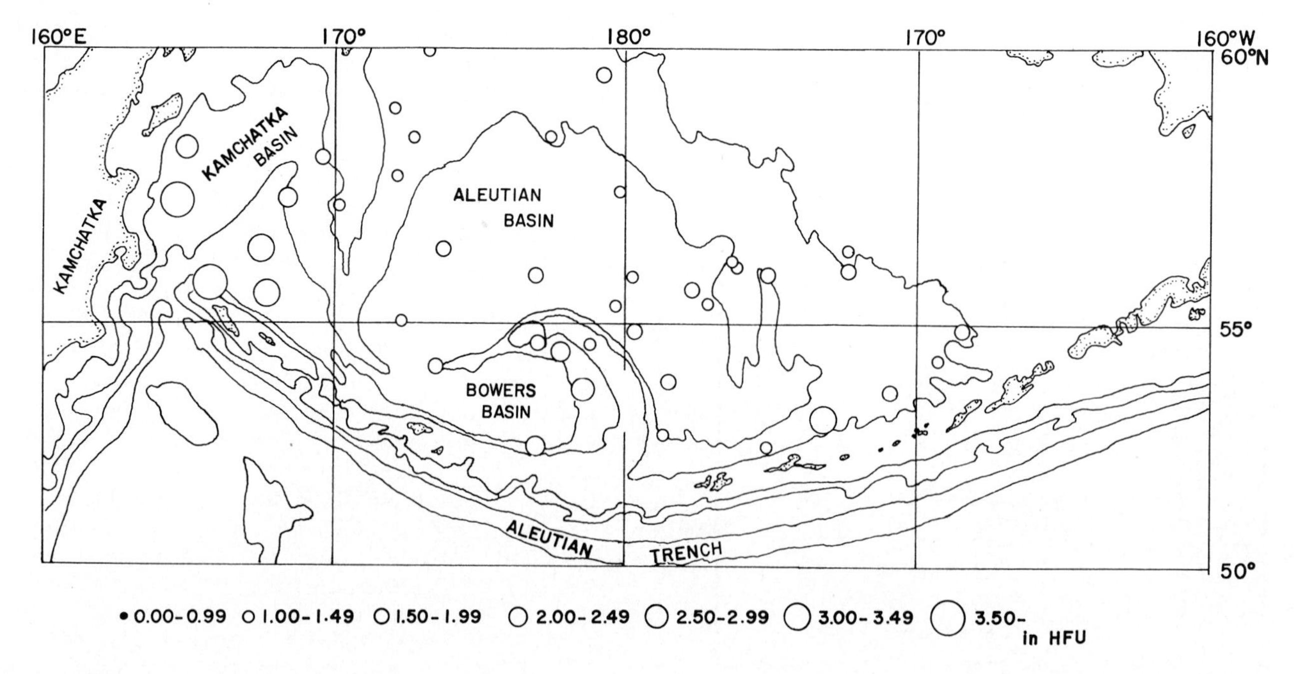

Figure 6. Heat-flow values in the Bering Sea and surrounding areas; Foster (1962), and unpublished values from Langseth and Watanabe.

increases sharply at the volcanic front, the Kuril Arc. Heat flow decreases towards the continent and is slightly depressed at the slope of the deeply submerged shelf. The depressed values are probably explained by the refractive effect of sub-bottom structure and the average heat flow is not significantly different from that of the Kuril Basin average (Savostin and Vlasov, 1974). Over the deeply submerged shelf, values show greater scatter but the mean is relatively high; 1.95 ± 0.87 (s.d.). This is the only example of a high heat-flow zone in the back-arc region which extends onto the adjacent continental structure, although it should be noted that the zone of deep earthquakes extends under the southern portion of the deeply submerged shelf. The overall average of 74 values in the Sea of Okhotsk is 1.93 ± 0.67 (s.d.) HFU. The average of the 14 Kuril Basin values taken in seafloor deeper than 3 km is 2.17 ± 0.31 (s.d.) HFU.

The Bering Sea

No data have been published since the initial work by Foster (1962). Most of the data used here were collected by Lamont-Doherty scientists on several cruises to the area. A detailed analysis of these data is in preparation.

The values are not as plentiful as in the Japan and Okhotsk basins, but they are uniformly distributed over the area (Figure 6). The heat-flow profile is shown in Figure 7. Geothermally, the Aleutian Basin is distinct from the Kamchatka Basin. The Aleutian Basin heat-flow values are near the earth's average and are uniform. The average of 11 values taken in seafloor deeper than 3660m is 1.44 ± 0.22 (s.d.) HFU. A few high values were observed near the Aleutian Arc and on Bowers Ridge.

The Kamchatka Basin has a high average, 2.88

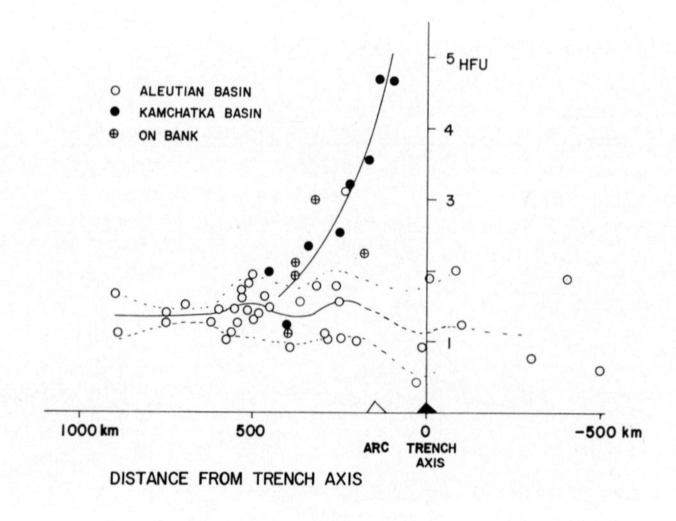

Figure 7. Heat-flow values versus distance from the axis of the Aleutian Trench. The data sources are the same as for Figure 6.

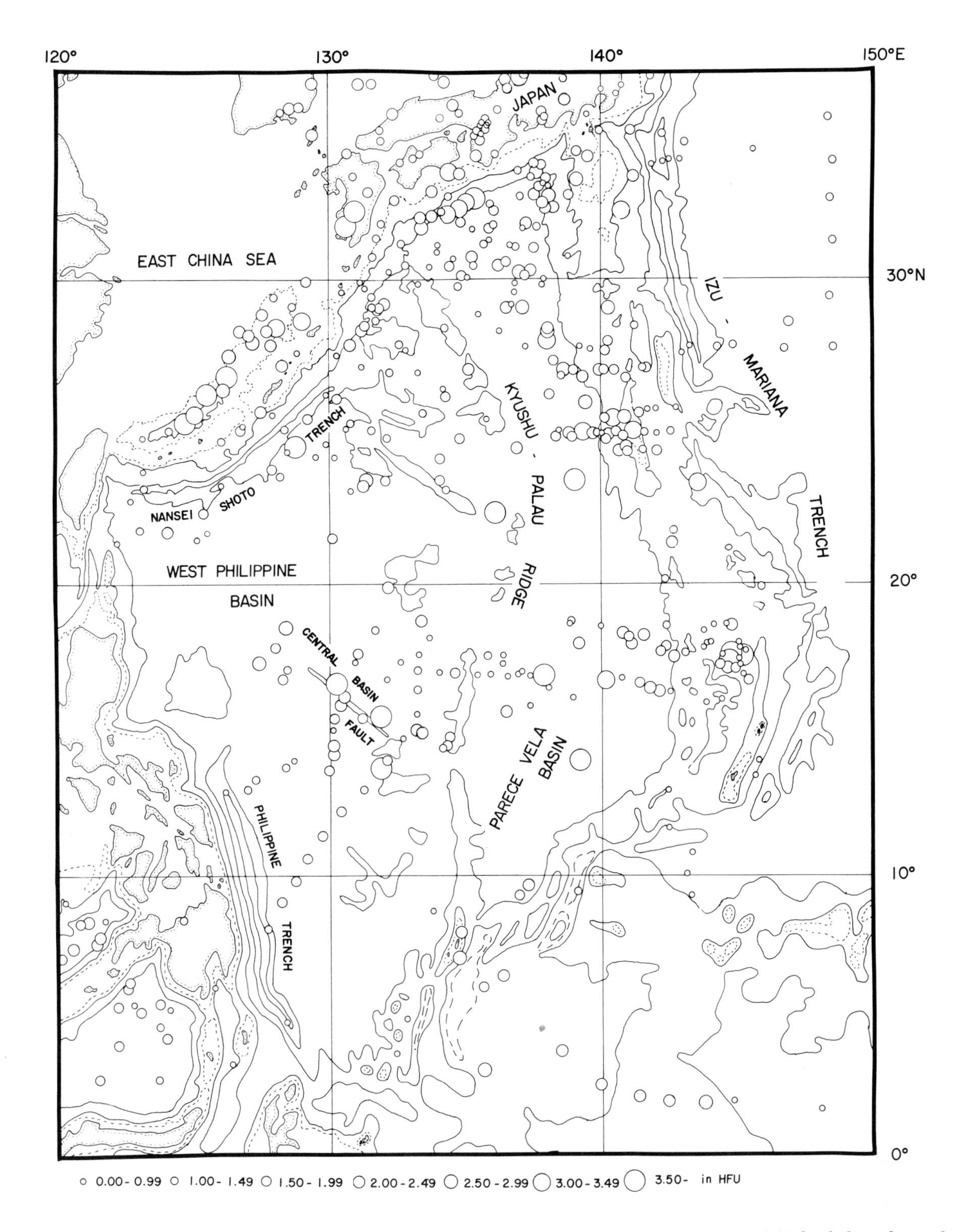

Figure 8. Heat-flow values in the Philippine Sea and surrounding areas. Unpublished data have been provided by Hilde, Isezaki, Kono, Langseth, Saki, Watanabe and Yasui.

± 1.07 (s.d.) HFU for four reliable stations in water deeper than 3660 m. A strong trend is evident. Heat flow decreases from the arc towards the northern part of the basin. A deep seismic zone exists only under the southeastern part of the Bering Sea and is missing beneath the basins (Kelleher, 1970).

Philippine Sea

Heat-flow results in the Philippine Sea have been published by various investigators (Yasui et al., 1963; Vacquier et al., 1967; Yasui et al., 1970; Veselov et al., 1974; Anderson, 1975; Sclater et al., 1976). An initial compilation was presented by Watanabe et al. (1970). Heat-flow values are shown in Figure 8, including recent unpublished values provided by Hilde, Langseth, Watanabe and Yasui. In general, the distribution of values is complicated. Many investigators have proposed that there is a major difference in age and possibly tectonic history between the basins east and west of the Palau-Kyushu Ridge (Ingle et al., 1975 a and b; Karig, 1975; Watts and Weissel 1975; Hilde et al., 1976). Therefore, these

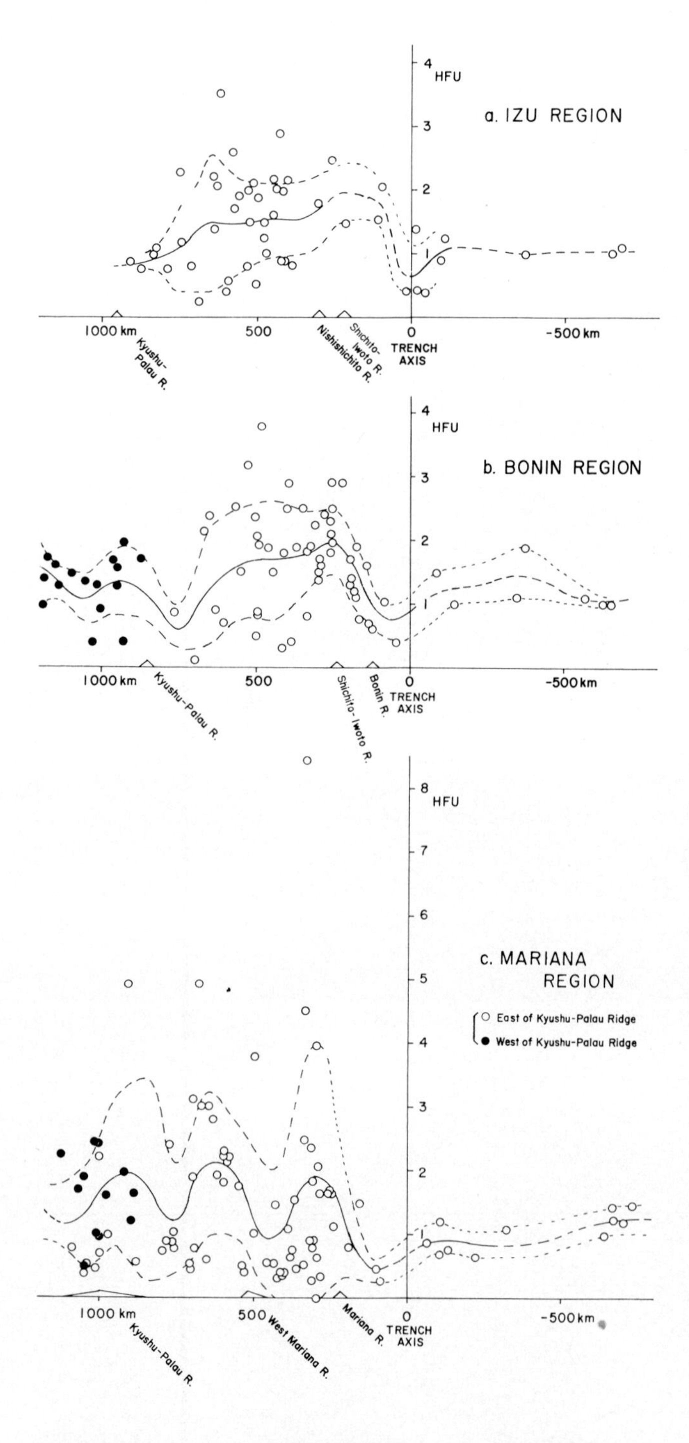

Figure 9. Three E-W profiles of heat-flow values versus distance from the trench axis for different zones along the Izu-Mariana Islands arc-trench system. Open circles are values east of the Kyushu-Palau Ridge whereas solid circles represent values west of the ridge.

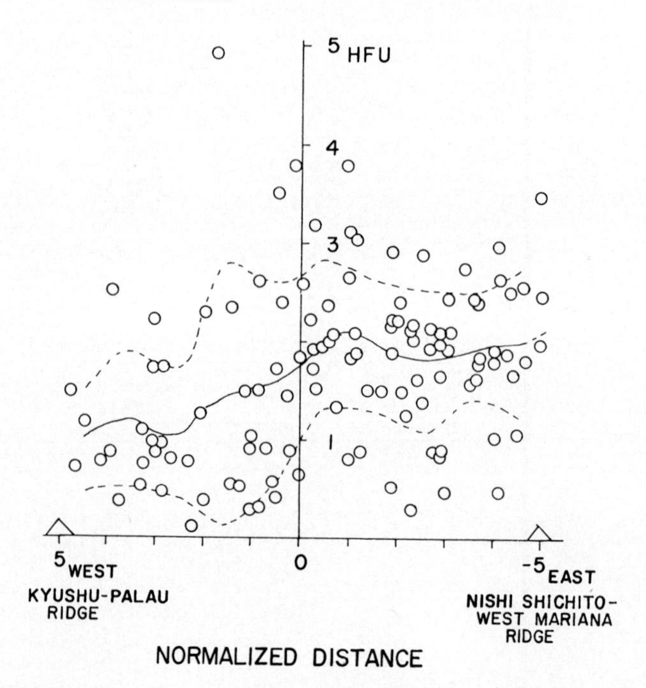

Figure 10. Heat-flow profile across the Shikoku and Parece Vela basins from east to west. The horizontal scale is normalized with respect to the distance between the Kyushu-Palau Ridge and the innermost ridge of the Izu-Mariana Arc.

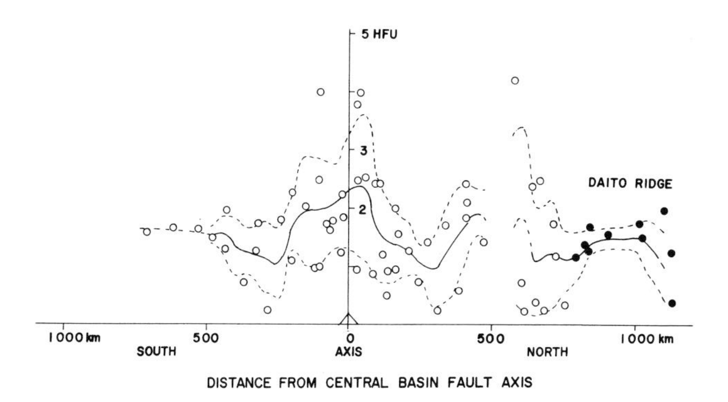

Figure 11. Heat-flow values versus distance from the axis of the Central Basin Fault. The unpublished data have been provided by Hilde, Isezaki, Kono, Langseth and Saki.

areas are examined separately. It must be noted however, that there is neither a clear discontinuity of the heat-flow distribution across the Palau-Kyushu Ridge nor an anomalous heat flow on the ridge.

Eastern Philippine Sea. It is widely thought that the eastern part of the Philippine Sea has evolved as the locus of subduction of the Pacific Plate (presently at the Izu-Mariana Trench) moves eastward relative to the Asian continent (Katsumata and Sykes, 1969; Santo, 1970). In the northernmost part of the eastern chain of basins along the Nankai Trough, high heat-flow values are measured. The average of 17 values in water deeper than 4500 m is 2.44 ± 1.12 (s.d.) HFU.

Analysis of the mantle's electromagnetic response over the northern part of the Shikoku Basin shows evidence of high temperatures at relatively shallow depths, less than 50 km; (Rikitake, 1969; Honkura, 1974) indicating that the surface heat flow reflects truly high mantle temperature gradients. Seismicity studies indicate that the floor of the Shikoku Basin basin is underthrusting the southwestern Japanese Arc. The average heat flow in this region has an intensity similar to that of the Sea of Japan and the Eastern Shikoku Basin to the south.

The alignment of ridges varies between the Izu, Bonin and Mariana arcs (Mogi, 1972). Profiles running east-west across the arcs relative to the trench axis are shown in Figure 9. Despite the scatter, the mean heat flow traces the same general relation between heat flow and tectonic province described across the Japanese Trench-arch and basin. One notable anomaly is the Mariana Trough (Figure 9c) where predominantly low values are seen amongst some isolated very high values. The heat-flow distribution in this trough has been discussed by Anderson (1975). This small basin is believed to have formed recently, within the last five million years, by the elevated central region. Thus, the likelihood of environmental disturbance is great and the mean heat flow of 1.45 ± 1.75 (s.d.) HFU is probably not representative.

The widths of the Shikoku and Parece Vela basins vary in the north-south direction so that the distribution is best represented on a profile with the distance normalized relative to a central axis running from the north to the south (Figure 10).

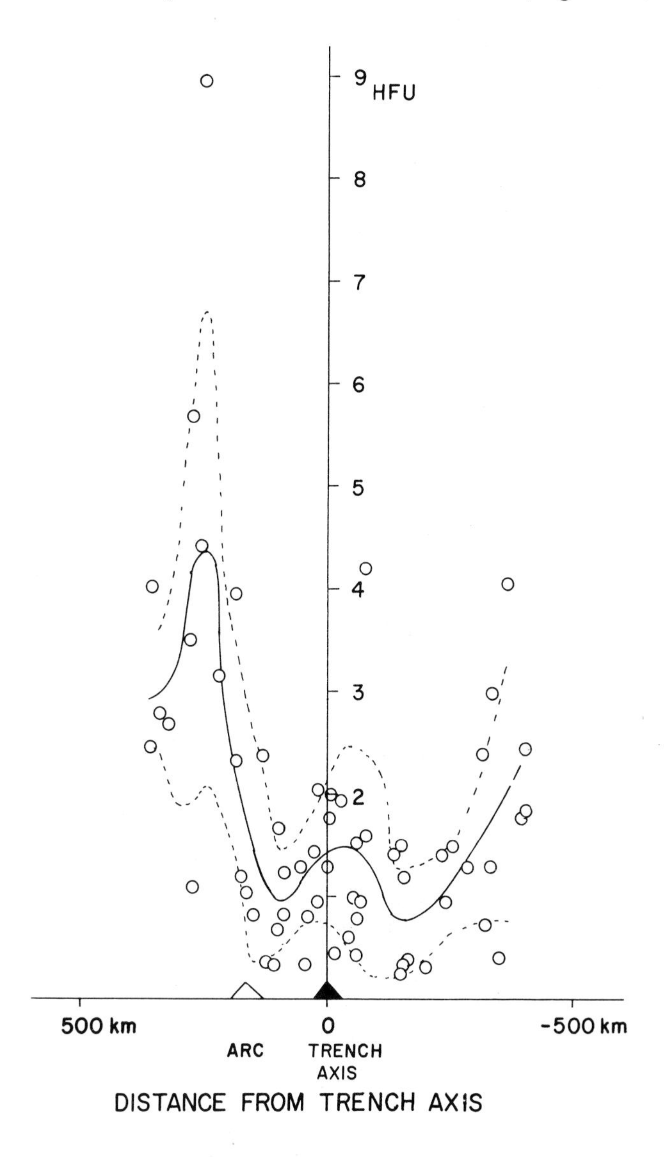

Figure 12. Heat-flow values versus distance from the Nansei-Shoto Trench axis.

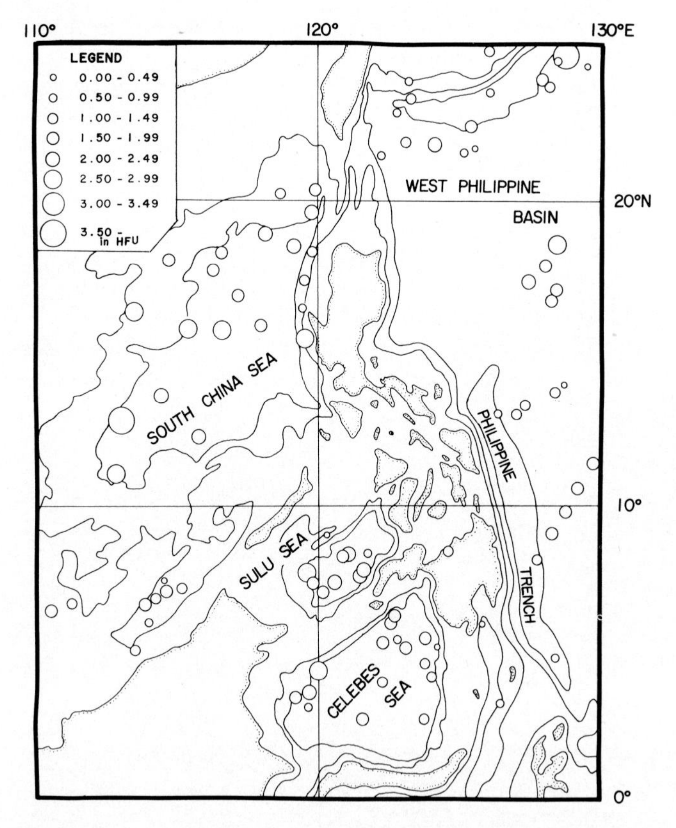

Figure 13. Heat-flow values in the South China Sulu and Celebes seas. The open circles show the heat-flow values from Nagasaka et al. (1970), Sclater et al. (1976), Sass and Munroe (1970), and Langseth (in preparation).

The eastern half of these basins, the youngest part have a heat flow averaging about 2 HFU, but large scatter is evident. Over the western half of this basin complex the mean heat flow decreases westward toward an average value of about 1HFU. There is large scatter in this region of the basins and many below normal values. This is probably related to the sparse sedimentary cover over much of this region, which may permit water circulation in the oceanic crust similar to that proposed to explain the zones of low heat flow in the midoceanic ridges (Lister, 1972, 1974). In general, environmental disturbances play a major role in producing the large scatter observed in the Philippine Sea. A careful analysis of these effects is required to better understand the true distribution of surface heat flow.

Western Philippine Sea. Station distribution in the Western Philippine Sea basins is not uniform and large areas are left unsampled. The average of all values in the Western Philippine Sea, excluding the trench areas, of 1.45 ± 0.81 (s.d.) HFU is slightly, but not significantly, lower than that of the eastern part. In the Philippine Basin the average of values measured in water deeper than 4000 m is 1.62 ± 0.92 (s.d.) HFU. This mean is significantly higher than that predicted for seafloor of the Philippine Basin age (greater than 50 m.y.; Karig, 1975; or 47 m.y. Watts et al., 1976b; or younger, Lauden, 1976); or depth (Sclater et al., 1976).

It has been suggested that the Central Basin Fault is a fossil spreading center which ceased activity more than 50 m.y. ago (Ben Avraham et al., 1972; Watts et al., 1976b). On the other hand, Karig (1975) has suggested that the Central Basin Fault is a fracture zone, a feature now quiet seismically (Shimamura et al., 1975). Deep water depths of the adjacent seafloor and the small crestal elevation, relative to the sur-

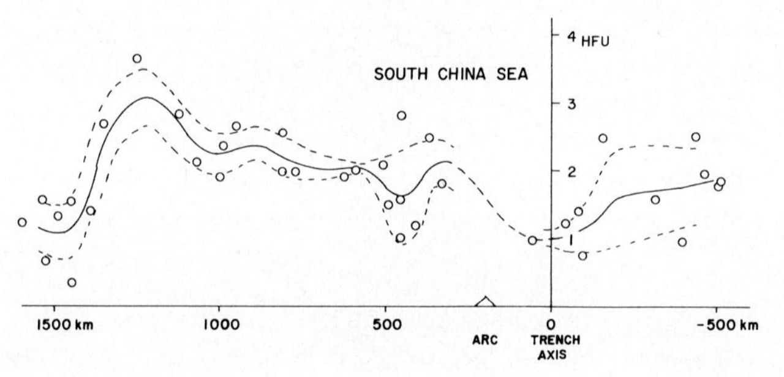

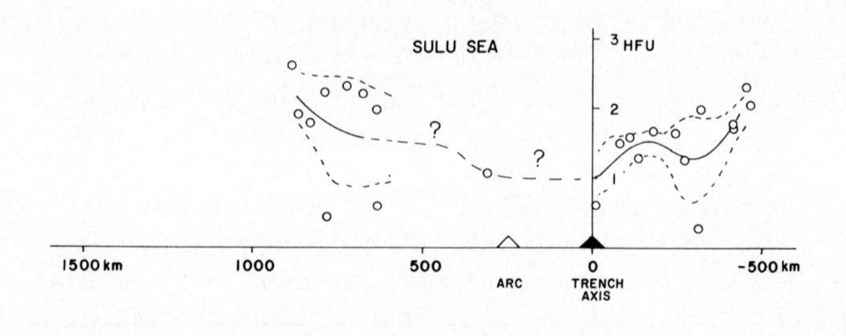

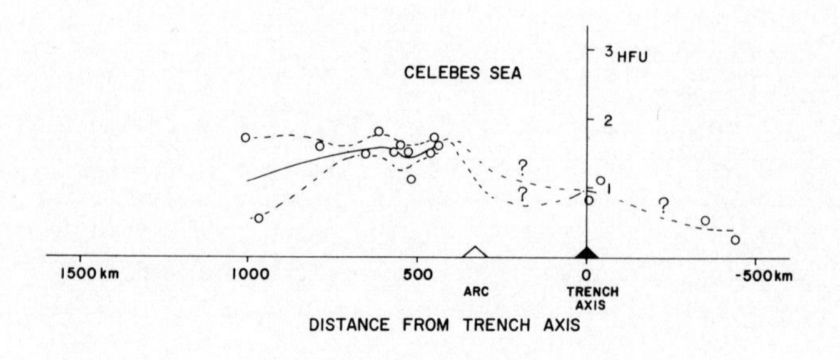

Figure 14. Heat-flow profiles, east to west across the South China, Sulu, and Celebes seas. The sources of data are the same as for Figure 13. IDOE Workshop (1974)

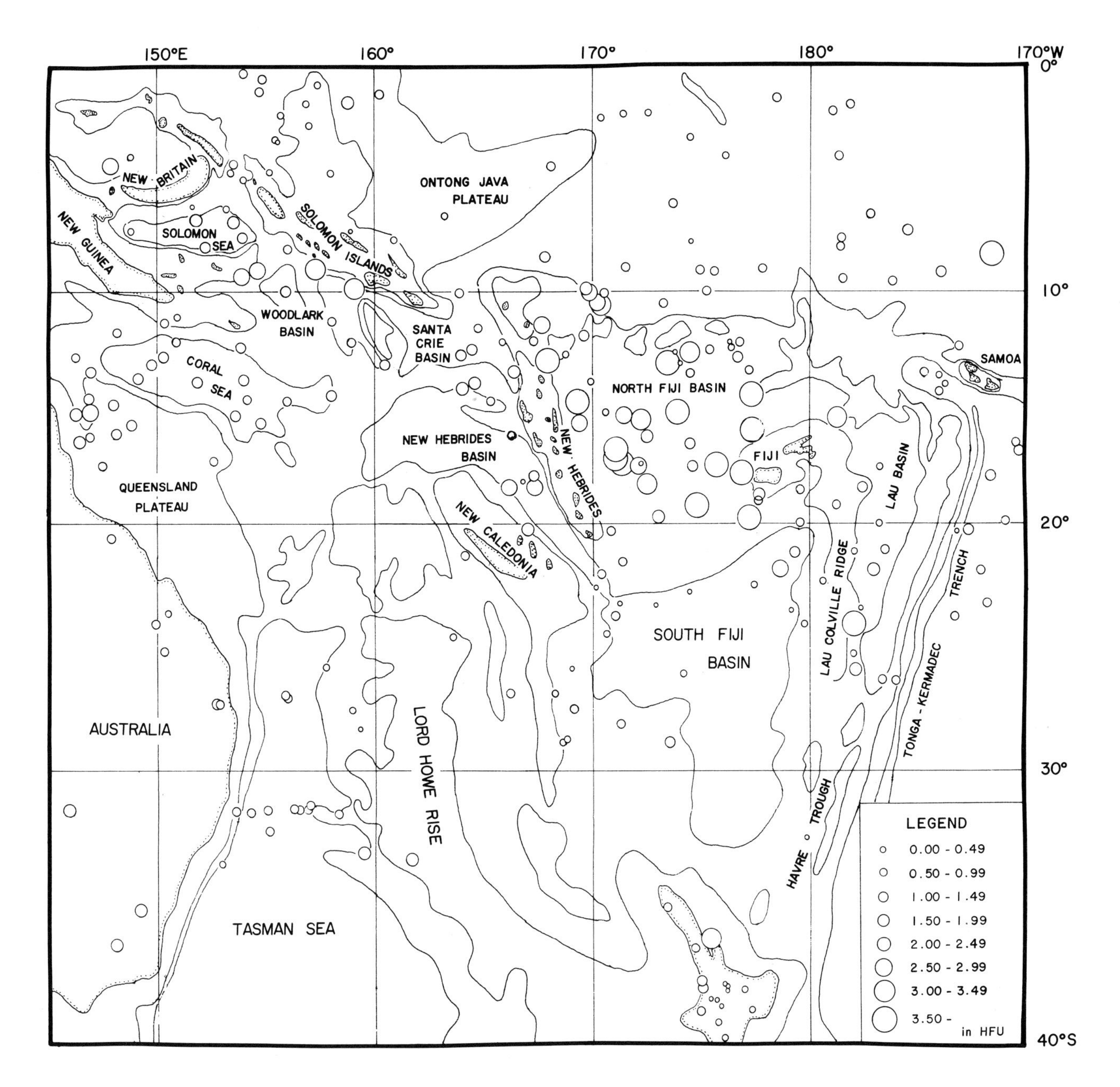

Figure 15. Heat-flow values in the Southwestern Pacific. The unpublished values have been provided by Watanabe, Popova and Udintsev.

rounding basin, suggest an old age for its formation.

On the other hand, the low seismic Q (Shimamura et al., 1975) thin sedimentary cover and several high heat-flow values near the fault are data supporting recent activity along the Central Basin Bault. Dredged basalt from the axis gives a radiometric age of 10 m. y. (Hilde et al., personal communication). By the use of unpublished data of Hilde, Kono, Saki and Isezaki, a heat-flow profile across the Central Fault can be made (Figure 11). Interestingly, the higher heat flow is nearly symmetrically distributed about the axis. Although environmental conditions may not be suitable for reliable heat-flow measurements, the true terrestrial heat flow is probably not lower since seafloor environmental effects typically reduce the observed heat flow. Therefore, the heat-flow data indicate that anomalously high temperatures remain below the Central Basin Fault.

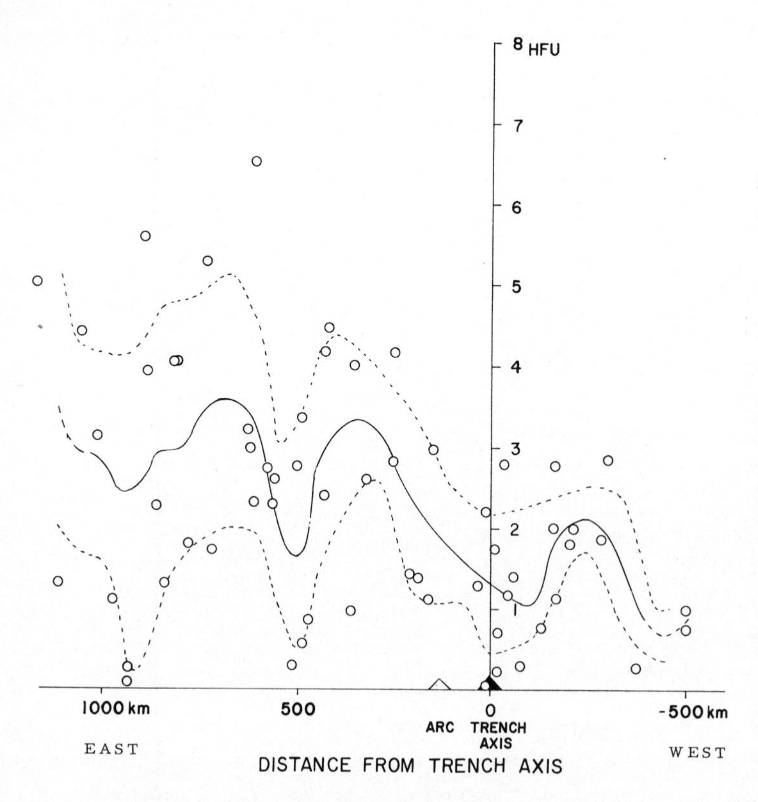

Figure 16. An east-west profile of heat-flow across the New Hebrides Arc and the North Fiji Basin.

East China Sea (Ryukyu Trough)

The heat-flow distribution in the East China Sea is shown in Figure 8. An initial description has been made by Yasui et al. (1970). Because the water depth is too shallow to attain reliable heat-flow measurements, stations in the back-arc region are limited to the Okinawa Trough. A heat-flow profile is shown in Figure 12.

Highly scattered values, distributed in seafloor seaward of the Ryukyu Trench, the Eastern Philippine Sea, decrease to a low heat flow between the trench and the inner arc, then dramatically increase at the inner arc, or volcanic front to a maximum value of 4.95 HFU on the trough's axis. Heat flow decreases toward the northwestern wall of the trough. More than 1.1 km of sediments are found in the trough floor (Murauchi et al., 1968; Ludwig et al., 1973b) providing a favorable geothermal environment. The deep earthquake zone extends inland as far as the continental margin of the trough. Geothermally, the Okinawa Trough seems to be an example of a young currently expanding zone behind an active arc (Karig, 1971a). The average of five values measured in seafloor deeper than 2000 m is 4.35 ± 2.57 (s.d.) HFU.

South China, Sulu and Celebes Seas

Not many values have been published in these areas (Nagasaka et al., 1970; Sass and Munroe, 1970; Sclater et al., 1976). The heat-flow distribution from the above sources, plus unpublished data provided by Langseth, Watanabe and Anderson are shown in Figure 13. The number of measurements is small but they are well distributed over the region and the scatter is low. Heat-flow profiles were constructed for each sea (Figure 14). The South China Sea is not presently a back-arc basin since the present direction of subduction seems to be from west to east at the Manila Trench (Murphy 1973). It has been proposed that the subduction direction was east to west from the eastern side of Luzon before ten m.y. ago (Karig, 1974).

South China Sea. In the northern portion of the South China Sea the seafloor is overlain by sediment 1 to 1.5 km thick. Heat flow in this region is uniform. The mean of seven values north of 16°N is 2.03 ± 0.19 HFU and in the southern portion of the basin a mean of 2.68 ± 0.39 HFU is observed and the scatter of values is greater.

Two rather narrow sediment filled troughs along the eastern margin of the South China Sea Basin, the Manila Trench in the north and the Borneo Trough in the south, are both characterized by numerous low values; the mean in the Borneo Trough is 1.22 ± 0.43 HFU.

Sulu Sea. Heat flow is uniform at about 2 HFU if two very low values, one of which was measured at a station on an elevated plateau, are ignored. The average of all the values and only basin values deeper than 4000 is 1.80 ± 0.69 (s.d.) HFU and 2.12 ± 0.25 HFU, respectively. Both the Sulu and South China seas are examples of enclosed basins in the Western Pacific having high heat flow (i.e. heat flow equal to the Sea of Japan or the Kuril Basin) and yet are not underlain by Benioff Zones.

Celebes Sea. In the Celebes Sea the measurements, though not numerous, are well distributed and are relatively uniform in value. In the western part of the basin which is not underlain by a Benioff Zone, three measurements have a mean value of 1.58 ± 0.08 (s.d.) HFU. A fourth low reliability station has been excluded. The eastern portion of the basin, which is underlain by a deep zone has a similar average. It is sampled by six stations with a mean of 1.59 ± 0.10. The basement depth in the Celebes Sea is significantly deeper than the South China Sea or

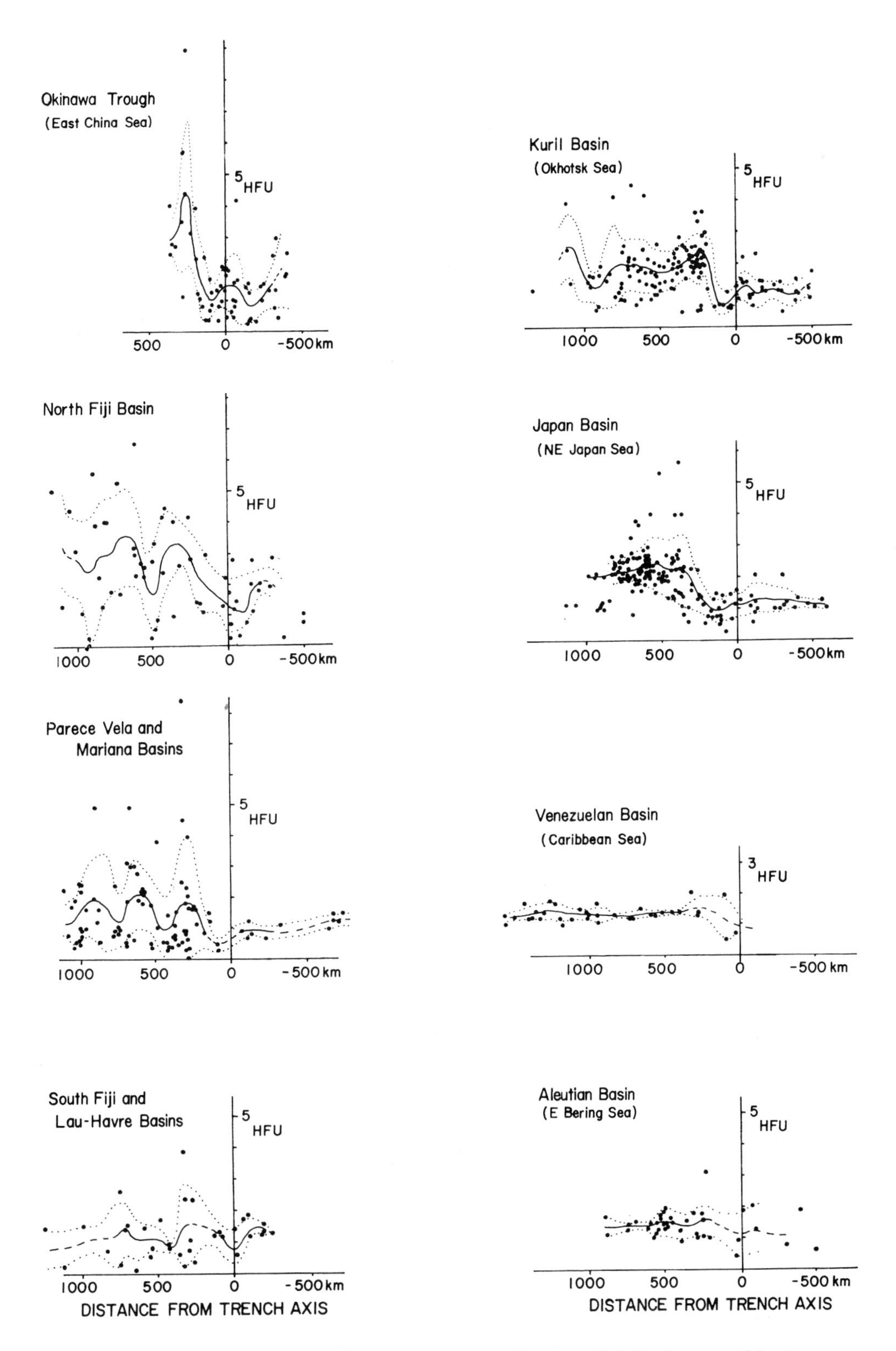

Figure 17. Heat flow versus distance across major trench island-arc and back-arc basins in the Western Pacific. The profiles are presented in order of estimated age beginning with the top of the left hand column.

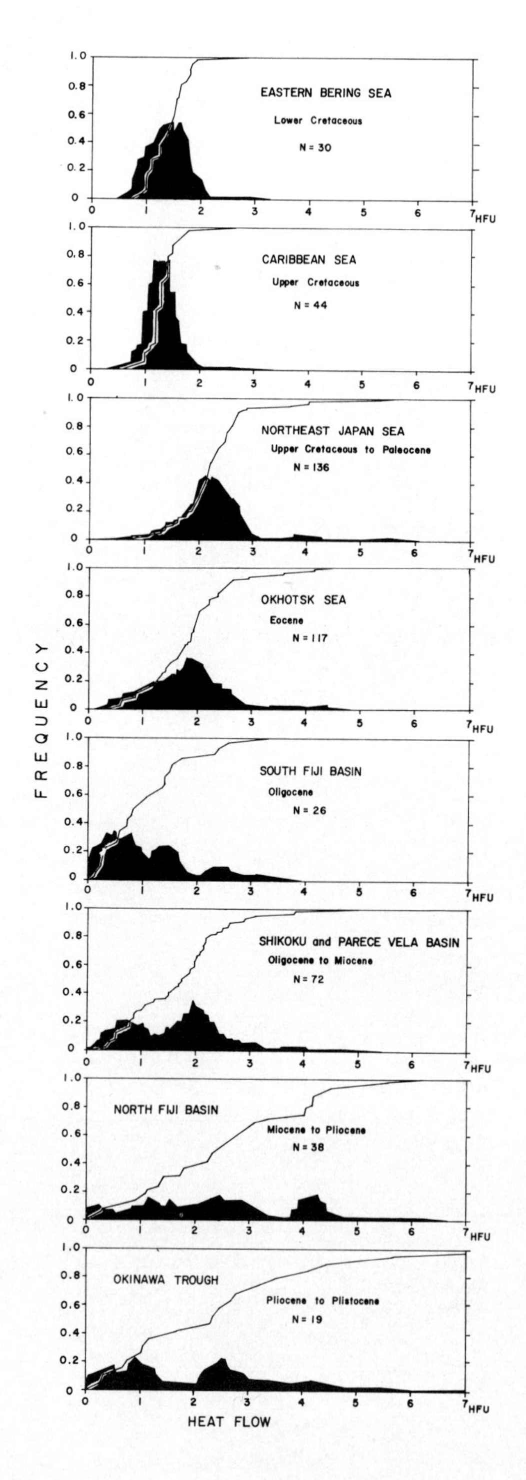

Figure 18. Histograms and cummulative curves for eight back-arc basins. The diagrams are arranged in chronological order starting from the bottom.

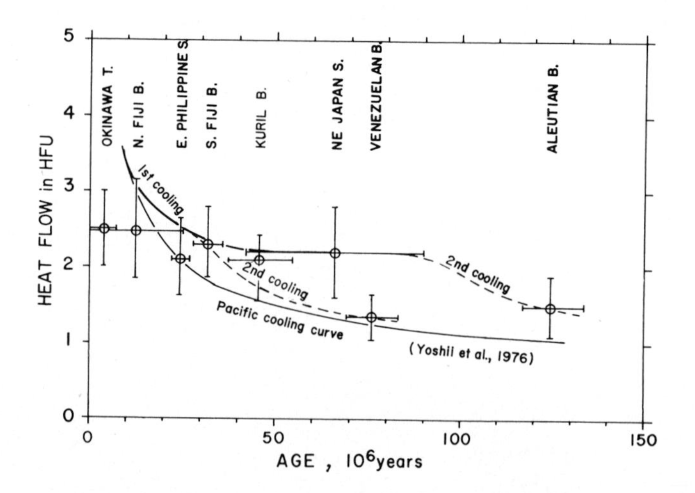

Figure 19. Mean heat-flow versus basin age for selected basins. The values are compared with; 1) the theoretical cooling curve of Yoshii et al. (1976) and a hypothetical two-stage cooling history. Firstly, the newly formed back-arc basin, shaped by extension, cools to a near-steady state at a heat flow of 2. 2 HFU. Secondly, the process bringing an additional one HFU to the surface stops and leads to a second cooling phase toward another steady-state value near 1. 2 HFU.

the Sulu basins, which supports the inference from the heat-flow results that a relatively low temperature mantle underlies the Celebes Sea.

Southwestern Pacific

There are many measurements in the complex basins south of the equator (Sclater et al., 1972; Halunen and Von Herzen, 1973; MacDonald et al., 1973; see Figure 15). Additional unpublished data are from Watanabe, Popova and Udintsev. However, considering the complexity of the tectonic setting of the region, the amount of data is not sufficient to define the heat-flow distribution in much of the area with any certainty. Although various interpretations have been made of the tectonic alignment in this area, all agree that classic trenc-arc-basin systems are developed in the New Hebrides Arc and Tonga-Kermadec Arc. In the New Hebrides-North Fiji Basin heat-flow stations are well distributed whereas observations are available only in the northern part of the South Fiji Basin and are sparse in the Lau Basin. In both areas the variability of heat-flow values is so high that it is unlikely that the present set of data sufficiently represents the actual geothermal features.

North Fiji Basin. A heat-flow profile across the New Hebrides Arc into the North Fiji Basin is shown in Figure 16. The most striking features are the high heat flow and variability in the

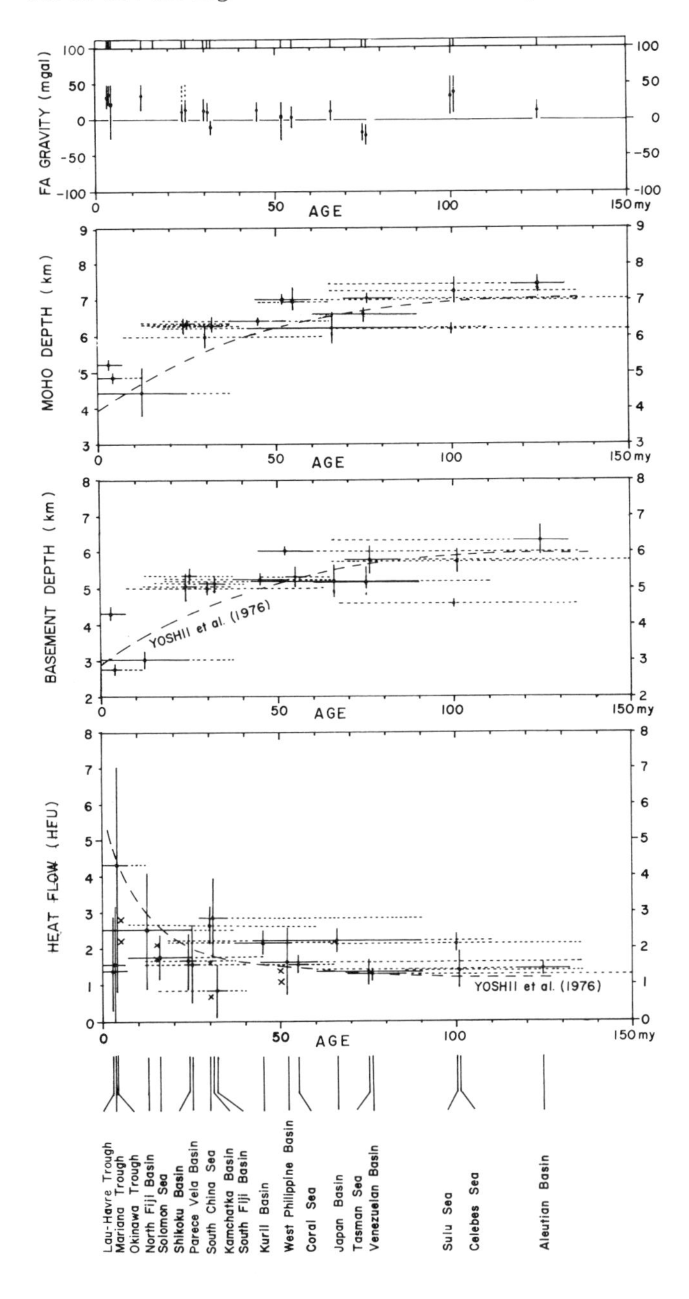

Figure 20. Free-air gravity anomaly, depth to the Mohorovičić discontinuity basement depth and heat flow versus age. The standard deviation in values are shown by the vertical bars and uncertainty in age estimate by the dotted horizontal lines.

basin. Because of the shallow water, less than 3500 m, temperature fluctuations in the bottom water may cause some of the scatter. In addition, a thin sedimentary cover is found over most of the basin and circulation of bottom water in the crust may play a significant role in creating the large scatter in this basin, which magnetic evidence indicates is relatively young (Chase, 1971). The data presented in Figure 16 show no clear trend relative to the proposed spreading axes (Chase, 1971; Sclater et al., 1972). The average of values at all stations in the North Fiji Basin deeper than 2000 m is 2.50 ± 1.61 (s.d.) HFU.

Discussion

Summary

The heat-flow profiles of back-arc basins that have been well sampled, and that are clearly associated with current subduction at an active trench, show common features in the relation of surface heat flow to tectonic province as summarized in Figure 17. The profiles have been arranged in chronological order based on current evidence as to the age of their oceanic crust. These features have been noted earlier on the basis of less data (McKenzie and Sclater, 1968; Yasui et al., 1970; Watanabe, 1972, 1974). The common features may be summarized as: 1) less than normal heat flow over the zone from the trench axis to the volcanic zone; 2) high, but variable, heat flow over the volcanic zone or island arc; and, 3) the mean heat flow in back-arc extensional basins appears to depend on age. In basins thought to be young the statistics of values are similar to those over young oceanic ridges, i.e. high variability and anomalously high and low values are observed; but no coherent spatial pattern is defined. In thickly sedimented basins probably of Early to Mid-Tertiary age, the mean heat flow is 2.0 to 2.2 HFU with relatively low variability (s.d. = 15%). In marginal basins thought to be formed during Mesozoic time, the mean heat flow is close to the world average based on observation (Lee and Uyeda, 1965) and the variability is low.

These features are also seen in the histograms and cummulative curves of selected back-arc basins shown in Figure 18. The mode of values less than one HFU which is strongly developed in the youngest basins, is probably the result of water circulation in the oceanic crust. Elder

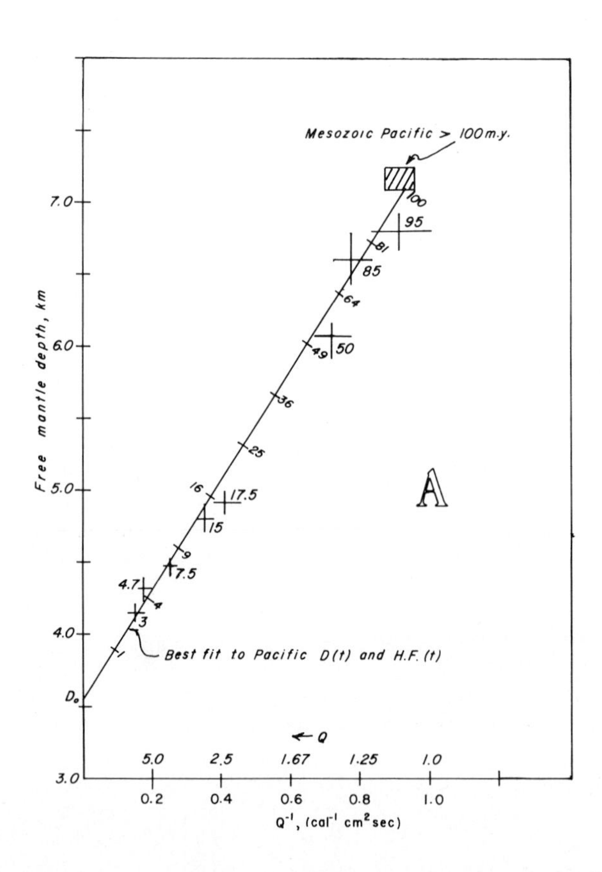

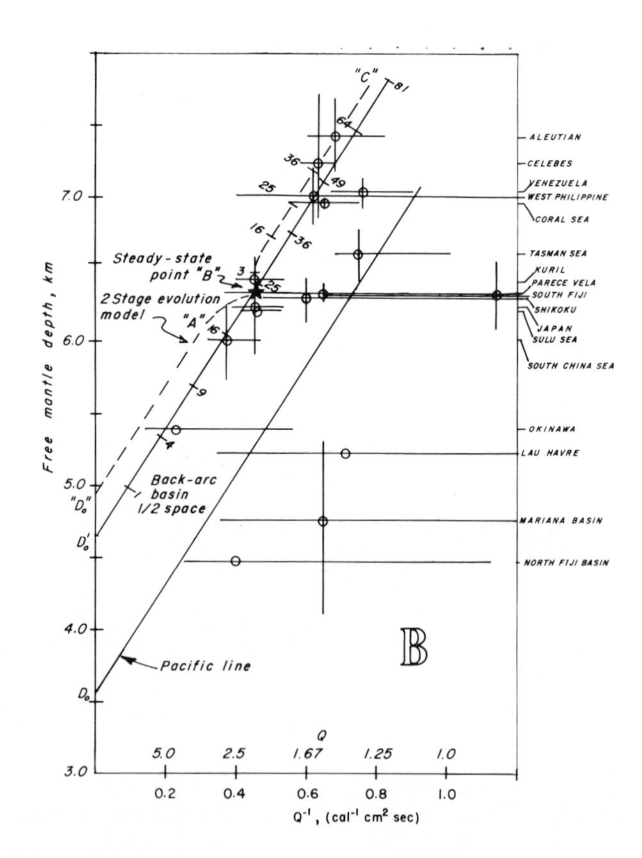

Figure 21. a) Free-mantle depth versus reciprocal of heat flow for reliable heat-flow values and elevations from the North Pacific (Parsons and Sclater, 1976). Ages along the evolution line are in millions of years. The ages of the observations are shown alongside each point.

b) Free-mantle depth versus the reciprocal of heat flow for back-arc basins. The crosses through the data points indicate the standard deviation. The North Pacific line, from Figure 21a, is shown for reference. The evolution best fitting older basin data has the same slope as the Pacific line.

(1965), Talwani et al. (1971), Lister, 1972, 1974 have proposed that the circulation of water in the fractured upper crust that is in good communication with the seawater due to a lack of sediment may be the main means of heat transfer, and the conductive component measured in sedimentary ponds may be only a fraction of the total heat loss.

Very high values, on the order of 4 HFU or greater are also frequently observed and form a distinct mode at the high end of the histogram for some basins. These high values are commonest in the most recent extensional basins, but are also measured in the thickly sedimented basins. These local regions of abnormally high heat flow are thought to be due to local magmatic activity in the oceanic basin. This possibility is supported by the occurrence of volcanic peaks in the marginal basins that are clearly younger than the oceanic crust (Ueno et al., 1974).

Between the modes associated with very low and very high values in young basins, there is a mode of values between 1.0 and 3.0 HFU. This 'middle' mode we think is composed of the most representative values of heat flow. However, in the youngest basins the loss of heat by water circulation may significantly mute the heat flow mean.

The mean value of the 'middle' mode varies systematically with basin age as shown in Figure 19. The heat-flow versus age relationship for these basins is significantly different from that of the midoceanic ridge. The thickly sedimented basins have heat flows well above those predicted by a simple cooling lithosphere model (e. q. Yoshii et al., 1976). This can be seen by examining the age relationship of a number of thermally related parameters. Figure 20 shows these parameters; heat flow, depth to basement, depth to Moho and the free-air gravity anomaly plotted versus estimated age. Each value has been derived as the mean of values observed inside the deepest depth contour line either in kilometers or kilofathoms, in the basin floor. The basement and Moho depths are adjusted for the loading effect of the layers lying above basement and Moho, using currently available seismic reflection and refraction data. The velocity to density relationship proposed by Nafe and Drake (1963) has been used to make these adjustments.

For comparison, the results of the cooling oceanic lithosphere model calculation by Yoshii et al. (1976) are shown in Figure 20. Essentially, this model is similar to those proposed by Sclater and Francheteau (1970), Parker and Oldenburg (1973) and Oldenburg (1975). In the younger thinly sedimented basins the mean heat flow lies well below the theoretical curve, probably due to the significant heat lost by water circulation, as discussed earlier.

The thickly sedimented basins where the heat flow is well established have mean values lying above the predicted cooling curve of thermal models. Yoshii (1973) and Sclater et al. (1976)noted that marginal basins of Oligocene or older are deeper than the oceanic model predicts. The Japan and Sulu basins have shallower depths for the estimaged age but are deeper than oceanic areas with similar heat flow. The younger basins have comparable basement depths but Moho depths are anomalous. The North Fiji Basin is anomalously shallow compared to all marginal basins. In the Okinawa Trough sedimentary cover is thick enough to assure a good environment for reliable heat-flow measurements but not in the Mariana and Lau-Havre Trough. The crustal structure is now well determined in these basins.

Heat Flow Versus Depth

The thermal evolution of the lithosphere is usually characterized in terms of heat flow, depth or lithospheric thickness versus age. However, the thermal evolution can also be characterized by the heat flow versus seafloor depth relation. This relation is particularly useful for back-arc basins since in some of these basins the heat flow is well determined, but the ages are poorly known. This is especially true of the thickly sedimented basins such as those of the Japan, Okhotsk and Bering seas.

The evolution of the lithosphere in terms of heat flow and depth is best illustrated on a plot of adjusted depth to the Mohoviritic discontinuity versus the reciprocal of the heat flow. The 'adjusted Moho depth' is the observed Moho depth adjusted upward for the effects of loading by sediments and the crust. The adjusted Moho depth is frequently referred to as the 'free mantle surface'. The depth to the 'free mantle surface' is determined for each basin based on seismic refraction measurements. By using this surface, the effects of variable crustal thickness are removed. The principal sources of refraction data used in this study are: Neprochnov et al., 1964; Murauchi et al., 1968; Ludwig, 1970; Ludwig et al., 1971; Hotta, 1972; Ludwig et al., 1973 a, Ludwig et al., 1973 b, Murauchi et al., 1973; Yoshii et al., 1973; and, Ludwig et al., 1975.

The usefulness of the depth versus reciprocal of heat flow relation stems from the observation that the thermal evolution of the oceanic lithosphere is described remarkably well by a simple cooling half-space model. In the oceanic areas with ages between 9 and 80 m. y., the depth versus age relation does not depart significantly from that predicted by a homogeneous half-space cooling from a uniform initial temperature. For a cooling half-space, the surface heat flow, as a function of time, is simply

$$Q(t) = \frac{K T_A}{\sqrt{\pi \kappa t}} \qquad (1)$$

where K is the thermal conductivity, T_A is the initial temperature, κ is the thermal diffusivity and t is the time.

The change in depth with time relative to some initial depth D_o is determined by multiplying the integrated temperature difference between the temperature profile at time t and the initial temperature T_A by the expansion coefficient α. Thus,

$$D(t) = D_o + \frac{2 \rho_o \alpha T_A \sqrt{\kappa t}}{(\rho_o - \rho_w)} \int_o^\infty \operatorname{erfc}(u)\, du$$

$$u = z/2\sqrt{\kappa t}$$

where z is the depth, ρ_o is the density at zero temperature and ρ_w is the density of water. The integral on the right hand side reduces to $\pi^{-1/2}$ so that

$$D(t) = D_o + \frac{2 \rho_o \alpha T_A}{(\rho_o - \rho_w)} \sqrt{\frac{\kappa t}{\pi}} \qquad (2)$$

Eliminating t, between equations 1 and 2, we see that

$$D(t) = D_o + \frac{2\rho_o \alpha T_A^2 K}{(\rho_o - \rho_w)\pi} Q(t)^{-1} \qquad (3)$$

Equation (3) shows that the depth is linearly proportional to the reciprocal of the heat flow. Thus, over much of its history the thermal evolution of the oceanic lithosphere will be described by a straight line on a depth versus reciprocal heat-flow plot.

In Figure 21A, on the left, we show 'reliable' measurements of heat flow from the North Pacific, compiled by Parsons and Sclater (in press) and the free-mantle surface. These results fit a straight line evolution very closely. The line is that defined by parameters that best fit the North Pacific elevation and heat-flow data. Notice that uniformly spaced time tic marks along the evolution line increase as the square of the age.

Thickly- Sedimented Basins. The data for the back-arc basins, in which we have reliable heat-flow determinations, fall well above these oceanic lines Figure 21B; i. e. in the direction of deeper basement depth for the same heat flow. This suggests a different D_o for these basins. This difference is not remarkable in itself, since the D_o of many segments of the midoceanic ridge varies by as much as a km around the value shown for the Pacific. These variations may reflect a small deep seated density variation in the mantle which will be discussed fully later.

More significant are the consequences of trying to fit a half-space evolution to the back-arc basins points. Many authors, cited earlier, have proposed that the oceanic lithospheres of back-arc basins were formed by extension, and consequently, we may expect them to follow a thermal evolution similar to the oceans. If we assume that the parameters of the mantle below these basins is not very different from the Pacific, then the slope of the half-space evolution line must be similar. A possible line is shown in Figure 21B on the right. The evolution predicts ages of 14-25 m. y. for the southern South China Sea, Kuril, Japan and Sulu basins, and ages of 40-50 m. y. for the Celebes and Aleutian basins. The ages of none of these basins are known with certainty. For the Japan Basin the average sedimentation rates (100 m/m. y.) measured at DSDP hole 301 (Karig, Ingle et al. , 1975) and total sedimentary thickness, about 2 km, support the possibility that all of the sediment was deposited in 20 m. y. which would agree with the predicted age.

Similarly, in the Aleutian Basin at DSDP hole 190, an average sedimentation rate of about 42m/m. y. was determined (Creager, Scholl et al. ,1973). The Aleutian Basin has an average sedimentary thickness of about 2. 5 km. Thus, the minimum age for the Aleutian Basin floor would be about 60 m. y. , which is in reasonable agreement with the age predicted for a half-space evolution.

Thus, we cannot rule out the possibly that these basins are following a simple half-space evolution subsequent to their formation by extension. However, greater ages for the Japan Basin have been proposed (see, for example, Kobayashi and Isezaki, 1976). The great lateral uniformity of heat-flow values over the Japan Basin also suggests an older age than 20 m. y. . A basin formed by accretion along a single axis should have significant lateral gradients in heat flow if the spreading stopped less than 20 m. y. ago. Mesozoic ages have been proposed for the Aleutian Basin by Scholl et al. , 1975 and Cooper et al. , 1976.

Older ages for these basins could conform to the observed heat flow and mantle depth only if some process intervenes to slow the evolution along the half-space line on a depth-heat flow diagram. The evolution could be slowed if heat is supplied from below in sufficient quantity to slow the drop in heat flow and free-mantle elevation. If additional heat is supplied from below for a sufficient period (about 20 m. y.) then the lithosphere would approach thermal steady state. Several investigators, McKenzie (1969) and Sclater (1972) have suggested that the lithosphere below some back-arc basins may be at steady state. For the basins, such as the Japan, Kuril and Sulu basins to be near steady state, heat flow of 2 HFU must be supplied to the base of the lithosphere.

Convection in the asthenosphere above the subducting slab has been proposed as one potential process that could transport the additional heat to the top of the asthenosphere, McKenzie (1969); Minear and Toksöz (1970); Oxburgh and Turcotte (1970); and, Toksöz and Bird (this volume). The lateral uniformity of heat flow suggests that an extensive lateral transfer of heat occurs at the base of the lithosphere, and, therefore, supports convection below the marginal basins. The ultimate source of additional heat is not known. Frictional heating along the top of the downgoing slab, as it shears along the mantle wedge below the marginal seas, has been proposed; but, it has been difficult to show how the high stress levels required can be maintained at asthenospheric temperatures.

Regardless of the source of additional heat, the evidence for convection and greater ages for these basins than a simple half-space cooling model predicts, suggests an alternative evolution as depicted in Figures 19 and 21. Initially, the basin floor is formed by extension and the newly formed lithosphere cools by conduction through the surface. During this phase, the evolution will closely follow that predicted by a cooling half-space model, $D_o'' \longrightarrow A$ on Figure 21 B. At about 10 m. y., additional heat from below begins to slow the increase of depth with time, while the surface heat flow continues to decrease. The evolution will follow a curved line $A \longrightarrow B$ toward a steady state condition with Q= 2.2 HFU and D = 6200 m, the point around which many of the basins are clustered. This condition could persist as long as additional heat is provided from depth. If the process contributing the additional heat is shut off, then the mantle will continue to evolve as shown by the line $B \rightarrow C$. Time tics along this part of the curve correspond to millions of years since the mantle was at thermal steady state.

When mantle is at steady state, it is easy to show that the depth would be related to the initial depth, D_o'', by

$$D = D_o'' + \frac{\rho_o \alpha K T_A^2}{2(\rho_o - \rho_w)} Q_{s.s}^{-1}$$

Thus, if we assume the basins that cluster 2.2 HFU and a free mantle depth of 6200 m are at steady state, we can calculate D_o'' and construct the curve shown in Figure 21B using the same thermal parameter as the Pacific lithosphere.

A two state evolution, during which the thermal evolution is arrested for a period of time would fit many of the observed gross characteristics of back-arc basins. For example, why many of the basins have similar heat flow and basement depth. The evolution plotted in Figure 21B could hold the lithosphere near the steady-state point for tens of millions of years. Notice that, even after the source of additional heat flux is removed, a period of ten million years is required before the heat flow continues to decrease. The two-stage evolution is also supported by the remarkable lateral uniformity of heat flow across these basins. An extended period with heat flow at the base of the lithosphere maintained at a nearly constant value would level the isotherms within the mantle.

Some form of convection is essential to bring the additional heat up near enough to the surface to affect lithosphere development. This convection has been directly related to subducting slabs either through viscous coupling of the slab motion with the slab (McKenzie, 1969; Toksöz and Bird, this volume) or as a result of heat generation along the upper surface of the slab due to viscous dissipation. However, it is clear that some of the basins, notably those of the South China and Sulu seas and the Kamchatka Basin are not underlain by deep Benioff Zones at present, yet they have mean heat flows and depths appropriate to basins that are over Benioff Zones. Figure 21B shows that if subducting slabs existed below these seas within the last 10-20 m. y. the thermal effects would still remain.

Alternatively, small scale convection in the asthenosphere below the Western Pacific Basin may not be so directly coupled to subducting plates, as has been hypothesized, Conceivably, small scale convection in the Western Pacific may result from thermally induced instabilities in the deep mantle which may play a role in the initiation of subduction and not vice versa.

The West Philippine Basin. Due to the high scatter of values, the mean heat flow in the Western Philippine Basin is not well determined. However, a relatively well defined set of magnetic anomalies has been found trending parallel to the Central Basin Fault. Various ages have been assigned to this sequence; initially Ben Avraham et al., 1972 proposed that they were of Mesozoic age. Later, Lauden (1976) and Watts et al. (1976b) proposed an Early Tertiary age in the range of 40 to 50 m. y. and JOIDES sites 290 to 295 indicated an Early Tertiary age for the basement. Figure 21B shows that, with an age of 40-50 m. y. and Moho depth of 7000 m, the lithosphere of the Western Philippine Basin could fit a cooling half-space model with D_o approximately one km deeper than the Pacific Plate, and similar in value to the other back-arc basins. The mean heat-flow value, even though it is poorly determined, is in accord with this evolution. If the ages, based on magnetic anomalies, are correct then the Western Philippine Basin does not appear to have passed through the hypothesized steady-state phase, but evolved in a manner similar to a typical midoceanic ridge. However, a higher mean mantle density than the Pacific is required to explain the deeper depths.

Marginal Basins that Have Been Dated. The floors of the Shikoku, South Fiji and Lau-Havre basins have been dated on the basis of magnetic anomalies (Watts et al., 1976a; Weissel et al.,

1976). Unfortunately, the heat flow in few of the basins has been well determined. For the Shikoku Basin an age of 19-24 m. y. has been deduced by Watts and Weissel (1975) and a Moho depth of 6100-6300 m has been determined from seismic profiler data. The point determined by these data could be placed on the half-space evolution line that best fits the other marginal basins. The mean heat flow which is probably underestimated would place it near the Pacific line, but at an age of 60 m. y., three times that determined from magnetic anomaly correlations. Similarly, the South Fiji Basin has been dated at between 27 and 38 m. y. and has a Moho depth of 6200 m. These data would place it on a half-space cooling line with a D_o intermediate between the North Pacific and the other back-arc basins. The mean heat flow in the South Fiji Basin is extremely low and most likely unrepresentative of the flux from the mantle beneath.

Very Young Marginal Basins. The North Fiji Basin, the Ryukyu Trough and the Lau-Havre Basin are thought to be very young inner-arc basins. It is not certain whether the obtained crustal structures are representative. The Moho depths are in reasonable agreement with the back-arc basin half-space cooling model, if we consider the uncertainty of the structure. More structural data in these basins are needed.

The Mariana Trough, on the other hand, has basement depths varying from 4300 m in the western part of the trough to 3200 m at what is thought to be an active spreading center (Anderson, 1975). If we assume typical oceanic crust, the adjusted Moho depth is 1.0 km deeper than the basement depth. Karig (1971a) argues that ages in the Mariana Trough are no more than 4-5 m. y. If the age of the trough ranges from 5 m. y. in the west to 0 at the active axis, this depth would be in good agreement with the marginal basin half-space evolution.

The Anomalous Depth of Marginal Basins

The back-arc basins in which we have accurate heat-flow values and the Mid to Early Tertiary basins in which we have reliable ages and Moho depths are all displaced approximately one km above the North Pacific evolution line. The free-air gravity anomalies in these basins are generally positive, but have values less than 50 mgal on average. Thus, they are near isostatic equilibrium. The most likely cause of the shift in D_o is a small but deep reaching density difference. For a layer about 300 km thick, a positive mean density difference of only 0.01 gm/cc is required. It is unlikely that the cool subducted slab beneath some back-arc basins contributes to this density difference since basins not underlain by subducting slabs show the greater depths as well. This density difference does indicate that the mantle below back-arc basins is not simply reheated Pacific Ocean mantle.

We can only speculate as to the source of this density difference. One mechanism could be a change in the viscosity versus temperature relation in the mantle behind island arcs, so as to allow efficient upward heat transfer (convection) at lower temperatures. In our model, this would be equivalent to decreasing T_A below the back-arc basins relative to the Pacific. To cause a 0.01 gm/cc change requires a ΔT_A of about -80°C.

The Source of High Heat Below Back-Arc Basins

Until we have more reliable determinations of the age of certain back-arc basins where the heat flow and crustal structure are well determined, we cannot exclude the possibility that the high heat flow results from a young age. If basins such as those of the Sea of Japan, Sea of Okhotsk and South China Sea were formed as recently as 20 to 25 m. y. ago, then their lithosphere may be following a simple thermal evolution through loss of heat through the surface (a cooling half-space evolution). However, there is strong evidence that these basins may be older than the ages given above. There are also strong physical arguments that some form of convection is taking place beneath back-arc basins. This convection is roughly at the scale of the basins, based on the uniformity of the heat flow. For the last two possibilities mentioned additional heat from below is required to arrest the evolution of the back-arc basins.

The source of this addional heat has been the subject of much debate. One observation relative to this discussion is worth making. Recent estimates of the average oceanic heat loss that include the heat lost by the seafloor spreading process, are about 2 HFU (Williams and Von Herzen, 1975). The average of all oceanic heat-flow measurements underestimates the heat flow because of the biasing of midoceanic ridge measurements to lower values. If these newer averages are valid then many of the back-arc basins have a similar average heat flow to the oceanic average.

The source of 2 HFU to drive the thermal convection of seafloor spreading in the principal oceanic plates is not known. The geochemistry of midocanic ridge basalts indicates that the mantle source regions do not have sufficient abundances of long-lived radioisotopes to produce the required quantity of heat. This leaves the possibility that either the deep mantle > 350 km as a possible region where sources are abundant enough, or the heat is stored in the deep mantle on the order of 10^8 years. If the heat is available in the deep mantle, then a simple explanation of the 'high' heat flow behind back-arc basins at steady state is to tap this heat by the convection taking place beneath these basins. By this argument, the heat flow behind island arcs is not high, but very close to the oceanic average. The principal difference then, between the open oceanic and the back arc basin thermal regime is the shorter time scale of the latter.

Lamont-Doherty Geological Observatory contribution # 2409.

Acknowledgements. This work was supported in part by the National Science Foundation through grant DES 74-24112. The work of T. Watanabe at Lamont-Doherty Geological Observatory was under a long-term overseas research grant from the Japanese Ministry of Education. The authors are grateful to many colleagues, especially M. Yasui, K. Horai, S. Ehara for the use of their unpublished heat-flow data. A. Watts and J. Weissel provided valuable information on the age of certain of the basins. Discussions with T. Yoshii and K. Nakamura are appreciated.

References

Aoki, H., Plate tectonics of arc-junction in Central Japan, J. Phys. Earth, 22, 141-161, 1974.

Anderson, R. N., Heat flow in the Mariana marginal basin, J. Geophys. Res., 80, 4043-4048, 1975.

Beloussov, V. V., and E. M. Ruditch, Island arcs in the development of the earth's structure, J. Geol., 69, 647-658, 1961.

Ben Avraham, Z., J. Segawa and C. Bowin, An extinct spreading center in the Philippine Sea, Nature, 240, 453-455, 1972.

Ben Avraham, Z. and S. Uyeda, The evolution of the China Basin and the Mesozoic paleogeography of Borneo, Earth Planet Sci. Lett., 18, 365-376, 1973.

Chase, C. G., Tectonic history of the Fiji Plateau, Geol. Soc. Amer. Bull., 82, 3087-3110, 1971.

Cooper, A. K., M. S. Marlow and D. W. Scholl, Mesozoic magnetic lineations in the Bering Sea marginal basin, J. Geophys. Res., 81, 1916-1934, 1976.

Creager, J. S., D. W. Scholl, et al., Initial Reports of the Deep Sea Drilling Project, Washington (U. S. Government Printing Office), 19, 371-411, 1973.

Ehara, S., Terrestrial heat flow in Hokkaido, Japan - Preliminary Report, Jour. Fac. Sci., Hokkaido Univ., Ser. VII, 3, 443-460, 1971.

Elder, J. W., Physical processes in geothermal areas, in, Terrestrial heat flow, Geophys. Monogr. 8, 211-239, Amer. Geophys. Union, Washington, D. C., 1965.

Foster, T. D., Heat-flow measurements in the Northeast Pacific and in the Bering Sea, J. Geophys. Res., 67, 2991-2993, 1962.

Gerard, R., M. G. Langseth, Jr., and M. Ewing, Thermal gradient measurements in the water and bottom sediment of the Western Atlantic, J. Geophys. Res., 67, 785-803, 1962.

Gorai, M., Some geological problems in the development of Japan and the neighboring island arcs, in, The Crust and Upper Mantle of the Pacific Area, Geophys. Monograph 12, edited by L. Knopoff et al., Amer. Geophys. Union, Washington, 481-485, 1968.

Halunen, A. J., Jr., and R. P. Von Herzen, Heat flow in the Western Equatorial Pacific Ocean, J. Geophys. Res., 78, 5195-5208, 1973.

Hasebe, K., N. Fujii and S. Uyeda, Thermal processes under island arcs, Tectonophys., 10, 335-355, 1970.

Hilde, T. W. C., S. Uyeda and L. Kroenke, Evolution of the Western Pacific and its margin, Tectonophys., in press, 1976.

Hilde, T. W. C., J. M. Wageman, and W. T. Hammond, The structure of the Tosa Terrace and the Nankai Trough off Southeastern Japan, Deep Sea Res., 16, 67-75, 1961.

Honkura, Y., Electrical conductivity anomalies beneath the Japan Arc, J. Geomag. Geoelectr. 26, 147-171, 1974.

Horai, K., Terrestrial heat flow in Japan (A summary of terrestrial heat flow measurements in Japan up to December, 1962), Bull. Earthquake Res. Inst., 42, 94-132, 1964.

Hotta, H., Crustal structure of the Pacific obtained by seismic refraction method, in, The Marine Physics, edited by T. Tomoda, Univ. Tokyo Press, 31-66, 1972

IDOE Workshop on Metallogenesis and Tectonic Patterns in East and Southeast Asia, Bangkok,

1973, The tectonic framework of Eastern Asia: A review of present understanding, Chapter III, in, Metallogenesis, Hydrocarbons and Tectonic Patterns in Eastern Asia, A Programme of Research, CCOP, IOC Report, 23-72, Thai Watana Panich PressCo., Ltd. Bangkok, 1974.

Ingle, J. C., Jr., D. E. Karig and S. M. White, Introduction and explanatory notes, in, Initial Reports of the Deep Sea Drilling Project, U. S. Government Printing Office, Wash., D. C., 31, 5-21, 1975a

Ingle, J. C., Jr., D. E. Karig, et al., Site Reports, in, Initial Report of the Deep Sea Drilling Project, U.S. Government Printing Office, Wash., D. C., 31, 49-468, 1975b.

Kanamori, H., Great earthquakes at island arcs and the lithosphere, Tectonophysics, 12, 187-189, 1971.

Karig, D. E., Ridges and basins of the Tonga-Kermadec Island Arc system, J. Geophys. Res., 75, 239-255, 1970.

Karig, D. E., Structural history of the Mariana Island Arc system, Bull. Geol. Soc. Amer., 82, 323-344, 1971a.

Karig, D. E., Origin and development of marginal basins in the Western Pacific, J. Geophys. Res., 76, 2542-2561, 1971b.

Karig, D. E. Evolution of arc systems in the Western Pacific, Ann. Rev. Earth Planet. Sci., 2, 51-75, 1974.

Karig, D. E. , Basin genesis in the Philippine Sea, in, Initial Reports of the Deep Sea Drilling Project, U.S. Government Printing Office, Wash., D. C., 31, 857-880, 1975.

Karig, D. E., J. C. Ingle, Jr., et al., Initial Reports of the Deep Sea Drilling Project, Washington (U.S. Government Printing Office) 31, 409-419, 1975.

Katsumata, M., and L. R. Sykes, Seismicity and tectonics of the Western Pacific: Izu-Mariana-Caroline and Ryukyu-Taiwan regions, J. Geophys. Res., 74, 5923-5948, 1969.

Kelleher, J. A., Space-time seismicity of the Alaska-Aleutian seismic zone, J. Geophys. Res., 75, 5745-5756, 1970.

Kobayashi, K. and N. Isezaki, Magnetic anomalies in the Sea of Japan and the Shikoku Basin, Possible tectonic implication, in, The Geophysics of the Pacific Ocean Basin and Its Margin, edited by G. H. Sutton et al., Geophys. Monogr. 19, Am. Geophys. Union, 235-251 1976.

Kobayashi, K. and M. Nomura, Iron sulfides in the sediment cores from the Sea of Japan and their geophysical implications, Earth Planet. Sci. Lett., 16, 200-208, 1972.

Kono, Y., and K. Kobayashi, Terrestrial heat flow in Hokuriku District, Central Japan, Sci. Rep. Kanazawa Univ., 16, 61-72, 1971.

Lauden, K. E., Magnetic anomalies in the West Philippine Basin, in, The Geophysics of the Pacific Basin and Its Margin, edited by G. H. Sutton et al., Am. Geophys. Union, Wash. D. C., 19, 253-267, 1976.

Lee, W. H. K., and S. Uyeda, Review of heat flow data, in, The Terrestrial Heat Flow, Geophys. Monogr. 8, Amer. Geophys. Union, Wash., D. C., 87-190, 1965.

Lister, C. R. B., On the thermal balance of a mid-ocean ridge, Geophys. J. R. astr. Soc., 26, 515-535, 1972.

Lister, C. R. B., Water percolation in the oceanic crust, Trans. Amer. Geophys. Union, EOS, 55, 740-742, 1974.

Ludwig, W. J., The Manila Trench and West Luzon Trough - III. Seismic refraction measurements. Deep Sea Res., 17, 553-571, 1970.

Ludwig, W. J., N. Den and S. Murauchi, Seismic reflection measurements of Southwest Japan Margin, J. Geophys. Res., 78, 2508-2516, 1973a.

Ludwig, W. J., S. Murauchi, N. Den, M. Ewing, H. Hotta, R. E. Houtz, T. Yoshii, T. Asanuma, K. Hagiwara, T. Sato and S. Ando, Structure of Bowers Ridge, Bering Sea, J. Geophys. Res., 76, 6350-6366, 1971

Ludwig, W. J., S. Murauchi, N. Den, P. Buhl, H. Hotta, M. Ewing, T. Asanuma, T. Yoshii and N. Sakajiri, Structure of East China Sea - West Philippine Sea Margin off Southern Kyushu, Japan, J. Geophys. Res., 78, 2526-2536, 1973b

Ludwig, W. J., S. Murauchi and R. E. Houtz, Sediments and structure of the Japan Sea, Geol. Soc. Amer. Bull., 86, 651-664, 1975.

MacDonald, K. C., B. P. Luyendyk and R. P. Von Herzen, Heat flow and plate boundaries in Melanesia, J. Geophys. Res., 78, 2537-2546, 1973.

Matsuda, T. and S. Uyeda, On the Pacific-type orogeny and its model. Extension of the paired belts concept and possible origin of marginal seas, Tectonophysics, 11, 5-217, 1971.

Mc Kenzie, D. P., Speculations on the consequence and cause of plate motions, Geophys. J. Roy. astr. Soc., 18, 1-32, 1969.

McKenzie, D. P., and J. G. Sclater, Heat flow inside the island arcs of the Northwestern

Pacific, J. Geophys. Res., 73, 3173-3179, 1968.
Minato, M., and M. Hunahashi, Origin of the earth's crust and its evolution, Hokkaido Univ. Fac. Sci. Jour., Ser. 4 (Geology and Mineralogy), 14, 515-561, 1970.
Minear, J. W., and M. N. Toksöz, Thermal regime of a downgoing slab and new global tectonics, J. Geophys. Res., 75, 1397-1419, 1970.
Mizutani, H., K. Baba, N. Kobayashi, C. C. Chang, C. H. Lee and Y. S. Kang, Heat flow in Korea, Tectonophysics, 10, 183-203, 1970.
Mogi, A., Bathymetry of the Kuroshio region, in, The Kuroshio, Univ. Tokyo Press, 53-80, 1972.
Murauchi, S., and N. Den, Origin of the Japan Sea, Paper presented at the Monthly Colloquium of the Earthquake Res. Inst., Univ. Tokyo, 1966.
Murauchi, S., N. Den, S. Asano, H. Hotta, T. Yoshii, T. Asanuma, K. Hagiwara, K. Ichikawa, T. Sato, W. J. Ludwig, J. I. Ewing, T. Edgar and R. E. Houtz, Crustal structure of the Philippine Sea, J. Geophys.Res., 73, 3143-3171, 1968.
Murauchi, S., W. J. Ludwig, N. Den, H. Hotta, T. Asanuma, T. Yoshii, A. Kubotera and K. Hagiwara, Structure of the Sulu Sea and the Celebes Sea, J. Geophys. Res., 78, 3437-3447, 1973.
Murphy, R. W., Diversity of Island arcs: Japan, Philippines, Northern Moluccas, Australian Petrol. Expl. Assoc., 13, 18-25, 1973
Nafe, J. E., and C. L. Drake, Physical properties of marine sediments, in, The Sea, 3, edited by M. N. Hill, Interscience Publ., New York, 794-815, 1963.
Nagasaka, K., J. Francheteau and T. Kishii, Terrestrial heat flow in the Celebes and Sulu seas, Marine Geophysical Res., 1, 99-103, 1970.
Nakagawa, I., Heat flow in Osaka Basin, Annual Meeting of Seismological Society of Japan, abstract, Fall, 1975.
Neprochnov, Yu. P., V. M. Kovylin, Ye. A. Selin, V. V. Zdorovenin and B. Ya. Karp, New data on the structure of the earth's crust in the Sea of Japan, Doklady Akad. Nauk SSSR, 155, 6, 1429-1431, 1964.
Oldenburg, D. W., A physical model for the creation of the lithosphere, Geophys. J. Roy. astr. Soc., 43, 425-451, 1975.
Oxburgh, E. R., and D. L. Turcotte, Thermal structure of island arcs, Geol. Soc. Amer. Bull., 81, 1665-1688, 1970.
Parker, R. L., and D. W. Oldenburg, Thermal model of ocean ridges, Nature, 242, 122, 137-139, 1973.
Parsons, B., and J. G. Sclater, An analysis of the variation of ocean floor heat flow and bathymetry with age, J. Geophys. Res., in press, 1976.
Rikitake, T., The undulation of an electrically conductive layer beneath the islands of Japan, Tectonophysics, 7, 257-264, 1969.
Santo, T., Regional study on the characteristic seismicity of the world. Part V. Bonin-Mariana Islands region. - Comparative study on the reliability of USCGS seismic data since 1963, Bull. Earthquake Res., Inst., 48, 363-379, 1970.
Sass, J. H., and R. J. Munroe, Heat flow from deep boreholes on two island arcs, J. Geophys. Res., 75, 4387-4395, 1970.
Savostin, L. A. and V. A. Vlasov, Results of geothermal research in the Sea of Okhotsk, in, The Geothermal Reports of Geothermal Research in the USSR, Moscow, 1974.
Sclater, J. G., Heat flow and elevation of the marginal basins of the Western Pacific, J. Geophys. Res., 77, 5705-5719, 1972.
Sclater, J. G. and J. Francheteau, The implication of terrestrial heat flow observations on current tectonic and geochemical models of the crust and upper mantle of the earth, Geophys. J. Roy. astr. Soc., 20, 509-542, 1970.
Sclater, J. G., D. E. Karig, L. A. Lawver, and K. Lauden, Heat flow, depth and crustal thickness of the marginal basins of the South Philippine Sea, J. Geophys. Res., 81, 309-318, 1976.
Sclater, J. G., V. G. Ritter and F. S. Dixon, Heat flow in the Southwestern Pacific, J. Geophys. Res., 77, 5697-5704, 1972.
Shimamura, H., Y. Tomoda and T. Asada, Seismographic observation at the bottom of the Central Basin Fault of the Philippine Sea, Nature, 253, 177, 179, 1975.
Scholl, D. W., E. C. Buffington and M. S. Marlow, Plate tectonics and the structural evolution of the Aleutian-Bering Sea region, in, The Contributions to the Geology of the Bering Sea Basin and Adjacent Regions, edited by R. B. Forbes, Geol. Soc. Amer. Spec. Paper 151, 1-32, 1975.
Shor, G. G., Jr., Structure of the Bering Sea and the Aleutian Ridge, Marine Geol., 1, 213-219, 1964.
Sugimura, A., and S. Uyeda, Island Arcs, Japan and its environs, in Developments in Geotec-

tonics, Vol. 3, Elsevier, Netherlands, 247, 1973.

Talwani, M., C. C. Windish, and M. G. Langseth Reykjanes Ridge crest: A detailed geophysical study, J. Geophys. Res., 76, 473-517, 1971.

Toksöz, M. N. and P. Bird, Formation and Evolution of marginal basins and continental plateaus, in, Ewing Symposium Volume on The Evolution of Island Arcs, Deep Sea Trenches and Back-Arc Basins, edited by M. Talwani and W. C. Pitman III, Amer. Geophys. Union Monogr. Ser., 1976

Udintsev, G. B., Yu. B. Smirnov, A. K. Popova, B. V. Shekhvatov and E. V. Suvilov, New data on heat flow through the floor of the Indian and Pacific Oceans, Doklady Akad. Nauk. SSSR, 200, 453-456, 1971.

Ueno, N., I. Kaneoka and M. Ozima, Isotopic ages and strontium isotopic ratios of submarine rocks in the Japan Sea, Geochemical J., 8, 157, 1974.

Uyeda, S., and Z. Ben Avraham, Origin and development of the Philippine Sea, Nature, 240, 176-178, 1972.

Uyeda, S., and K. Horai, Terrestrial heat flow in Japan, J. Geophys. Res., 69, 2121-2141, 1964.

Uyeda, S. and A. Miyashiro, Plate tectonics and the Japanese Islands: A synthesis, Geol. Soc. Amer. Bull., 85, 1159-1170, 1974.

Uyeda, S., T. Watanabe, N. Mizushima, M. Yasui and S. Horie, Terrestrial heat flow in Lake Biwa, Central Japan, Proc. Japan Acad. 49, 341-346, 1973.

Vacquier, V., S. Uyeda, M. Yasui, J. Sclater, C. Corry and T. Watanabe, Heat flow measurements in the Northwestern Pacific, Bull. Earthquake Res. Inst., 44, 1519-1535, 1967.

Vasilkovsky, N. P., G. B. Udintsev, B. G. Kulp and E. A. Monravova, The Japan Sea -Relict of Ocean, in, The Island Arcs and Marginal Sea, edited by S. Asano and G. B. Udintsev, Tokai Univ. Press, Tokyo, 57-64, 1971

Veselov, O. V., N. A. Volkova, G. D. Yeremin, N. A. Kozlov and V. V. Soinov, Heat flow measurements in the zone transitional from the Asiatic Continent to the Pacific Ocean, Doklady Akad. Nauk, SSSR, 217, 897-899, 1974.

Watanabe, T., Heat flow through the ocean floor, in, The Marine Physics, Tokai Univ. Press, Tokyo, 1-107, 1972.

Watanabe, T., Heat flow in the Western Pacific and marginal seas. Marine Science Monthly, 6, 7, 13-17, 1974

Watanabe, T., D. Epp, S. Uyeda, M. Langseth and M. Yasui, Heat flow in the Philippine Sea, Tectonophysics, 10, 205-224, 1970.

Watanabe, T., R. P. Von Herzen and A. Erickson, Geothermal studies Leg 31 Deep Sea Drilling Project, in, Initial Reports of the Deep Sea Drilling Project, U. S. Government Printing Office, 31, 573-576, 1975.

Watts, A. B. and J. K. Weissel, Tectonic history of the Shikoku marginal basin, Earth Planet. Sci. Lett., 25, 239-250, 1975

Watts, A. B., J. K. Weissel and F. J. Davey, Tectonic evolution of the South Fiji marginal basin, in, Ewing Symposium Volume on The Evolution of Island Arcs, Deep Sea Trenches and Back-Arc Basins, edited by M. Talwani and W. C. Pitman, III, Amer. Geophys. Union Monogr., Ser., 1976a

Watts, A. B., J. K. Weissel and R. L. Larson, Seafloor spreading in marginal basins of the Western Pacific, Tectonophysics, in press, 1976b

Weissel, J. K., Evolution of the Lau Basin by the growth of small plates, in, Ewing Symposium Volume on The Evolution of Island Arcs, Deep Sea Trenches and Back-Arc Basins, edited by M. Talwani and W. C. Pitman, III, Amer. Geophys. Union, Monogr. Ser., 1976.

Williams, D. L., and R. P. Von Herzen, Heat loss from the earth: New estimate, Geology, 2, 327-328, 1975.

Yasui, M., D. Epp, K. Nagasaka and T. Kishii, Terrestrial heat flow in the seas around the Nansei Shoto (Ryukyu Islands), Tectonophysics, 10, 225-234, 1970.

Yasui, M., K. Horai, S. Uyeda and H. Akamatsu, Heat flow measurements in the Western Pacific during the JEDS-5 and other cruises in 1962 aboard M. S. RYOFU MARU, Oceanogr. Mag., 14, 147-156, 1963.

Yasui, M. T. Kishii and K. Sudo, Terrestrial heat flow in the Sea of Okhotsk, 1. Oceanogr. Mag., 19, 87-94, 1967.

Yasui, M., T. Kishii, T. Watanabe and S. Uyeda, Heat flow in the Japan Sea in, The Crust and Upper Mantle of the Pacific Area, edited by L. Knopoff et al., Amer. Geophys. Union Monogr. 12, 3-16, 1968a

Yasui, M., K. Nagasaka, T. Kishii and A. J. Halunen, Terrestrial heat flow in the Sea of Okhotsk, 2. Oceanogr. Mag., 20, 73-86, 1968 b.

Yasui, M., T. Watanabe, Terrestrial heat flow in the Japan Sea, 1, Bull. Earthquake Res. Inst., 43, 549-563, 1965.

Yoshii, T., Upper mantle structure beneath the North Pacific and the marginal seas, J. Phys. Earth, 21, 313-328, 1973.

Yoshii, T., Y. Kono and K. Ito, Thickening of the oceanic lithosphere, in, The Geophysics of the Pacific Ocean Basin and its Margin, Amer. Geophys. Union, Monogr. Ser., 19, 423-430, 1976.

Yoshii, T., W. J. Ludwig, N. Den, S. Murauchi, M. Ewing, H. Hotta, P. Buhl, T. Asanuma, and N. Sakajiri, Structure of Southwest Japan Margin off Shikoku, J. Geophys. Res., 78, 2517-2525, 1973.

SEISMIC AND ASEISMIC SLIP ALONG SUBDUCTION ZONES AND THEIR TECTONIC IMPLICATIONS

Hiroo Kanamori

Seismological Laboratory, California Institute of Technology, Pasadena, California 91125

Abstract. Results of detailed mechanism studies of great earthquakes are used together with their repeat times to determine the amount of seismic slip along various subduction zones. Comparison of the seismic slip with the rate of plate motion suggests that, in Chile, and possibly Alaska, the seismic slip rate is comparable to the rate of plate motion while, in the Kuriles and Northern Japan, the seismic slip constitutes only a very small portion, approximately 1/4, of the total slip. In the Sanriku region, and to the south of it, the relative amount of seismic slip is even smaller. These results suggest that in Chile and Alaska the coupling and interaction between the oceanic and continental lithosphere are very strong, resulting in great earthquakes with a very large rupture zone, and in break-off of the undergoing lithosphere at shallow depths. In the Kuriles and Northern Japan, the oceanic and continental lithosphere are largely decoupled, so that the slip becomes largely aseismic, and the rupture length of earthquakes reduced. The reduced interaction at the inter-plate boundary may allow the oceanic lithosphere to subduct more easily and to form a continuous Benioff zone extending to depths. It may also facilitate ridge subduction beneath island arcs, which may play an important role in the formation of marginal seas such as the Japan Sea. The decoupling is also evidenced by silent or tsunami earthquakes [e.g., the 1896 Sanriku earthquake], great intra-plate normal-fault earthquakes [e.g., the 1933 Sanriku earthquake], and crustal deformation. A natural extension of this concept of inter-plate decoupling is the spontaneous sinking of the oceanic lithosphere with a consequent retreating subduction. Retreating subduction may be an important mechanism in the formation of marginal seas such as the Philippine Sea, and explains the complete lack of major shallow earthquake activity along some subduction zones such as the Izu-Bonin-Mariana arc.

Introduction

In the theory of plate tectonics, lithospheric subduction is one of the major tectonic processes related to the formation and evolution of island arcs. The deep and intermediate earthquake zones along island arcs delineate the geometry of the subducting lithosphere, and major shallow earthquake activity along island arcs is a manifestation of the mechanical interaction between the subducting and the overriding lithospheres. The geometrical agreement of focal mechanisms of earthquakes along the Circum-Pacific belt led McKenzie and Parker [1967] to the concept of rigid plates dividing the earth's surface. Isacks and Molnar [1969] used the geometry of the compression and tension axes of deep and intermediate earthquakes to infer the stress distribution in the descending lithosphere. Stauder [1968] found tensional mechanisms for earthquakes along the trench axis in the Aleutians and interpreted them as fractures due to bending of the oceanic lithosphere. Besides these geometrical arguments, studies of physical processes associated with large earthquakes provided important information concerning the dynamics of plate subduction [Plafker, 1972; Kanamori, 1971b; Abe, 1972]. Kanamori [1971b] proposed a model of gradual thinning and weakening of the ocean-continent lithospheric boundary to account for the differences in the maximum dimension of rupture zones among different island arcs. It was suggested that such differences in the coupling between the oceanic and continental lithospheres may control various tectonic processes at island arcs. The argument was based on detailed studies of great earthquakes through analysis of long-period surface waves (periods of 200 to 300 sec). Such long-period waves represent the overall crustal deformation at plate boundaries more directly than conventional short-period seismic waves. Kelleher et al. [1974] found a good correlation between the maximum dimension of rupture zones of great earthquakes and the width of the area of lithospheric contact at various subduction zones. This paper extends the previous paper [Kanamori, 1971b] by including more recent data from great earthquakes. Inclusion of these recent data strengthens the conclusions of the previous paper. It is almost certain that the degree of mechanical coupling (or decoupling) between the oceanic and continental lithospheres varies among different island arcs; along some subduction zones the coupling is very strong, so that the plate motion is almost entirely taken up by seismic slip. It is relatively weak elsewhere and the seismic slip represents only a very minor part of the plate motion. Along some subduction zones, the boundary is entirely decoupled and the plate motion seems to take place aseismically. It is proposed that such coupling and decoupling of plates may play an essential role in the evolution of island arcs and marginal seas.

Summary of Characteristics of Great Earthquakes along the Circum Pacific Belt

Figure 1 shows the rupture zones and the mechanisms of major earthquakes for which detailed studies were made by using long-period surface waves. Rupture zones of other major earthquakes have been mapped by Fedotov [1965], Mogi [1968a, b], Sykes [1971], and Kelleher et al. [1973]. Except for the 1933 Sanriku earthquake and the 1970 Peruvian earthquake, the mechanisms of these earthquakes suggest low-angle thrust faulting, which is consistent with the plate motion along the Circum Pacific belt. The 1933 Sanriku earthquake is an exceptionally large normal-fault earthquake which occurred beneath the axis of the Japan Trench [Kanamori, 1971a]. It is interpreted as a lithospheric normal-fault which cuts through the entire thickness of the lithosphere. The 1970 Peruvian earthquake is another normal-fault event [Abe, 1972] within the oceanic lithosphere. These two events are therefore intra-plate earthquakes and do not represent slip between the oceanic and continental lithospheres.

Two other important features in Figure 1 are (1) a remarkable regional variation of rupture lengths of great earthquakes despite their nearly identical earthquake magnitude, and (2) a nearly complete lack of large shallow activity along the Izu-Bonin-Mariana arc, despite its typical island arc features such as a deep trench, volcanic activity and a Benioff zone. Regarding (1), it is now widely known [e.g., Kanamori and Anderson, 1975] that earthquake magnitude M_S is not a meaningful parameter for very large earthquakes. The seismic moment M_o which represents the

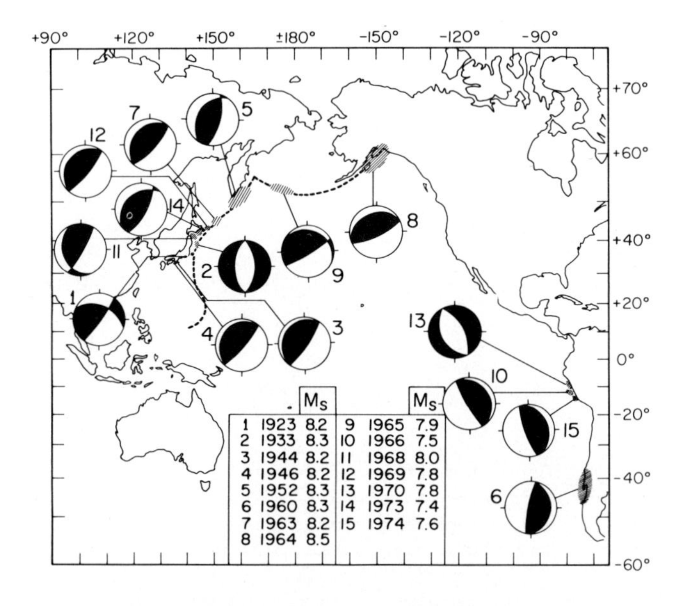

Figure 1. Mechanisms and rupture zones of large earthquakes along the Circum Pacific belt for which detailed studies of long-period waves have been made. Mechanism diagrams show the stereographic projection of the lower hemisphere. Dark and white quadrants indicate compression and dilatation respectively. References are: 1923 Kanto [Kanamori, 1971c]; 1933 Sanriku [Kanamori, 1971a]; 1944 Tonankai [Kanamori, 1972a]; 1946 Nankaido [Kanamori, 1972a]; 1952 Kamchatka [Kanamori, 1976]; 1960 Chile [Kanamori and Cipar, 1974]; 1963 Kurile Is. [Kanamori, 1970a]; 1964 Alaska [Kanamori, 1970b]; 1965 Rat Is. [Wu and Kanamori, 1973]; 1966 Peru [Abe, 1972]; 1968 Tokachi-Oki [Kanamori, 1971d]; 1969 Kurile Is. [Abe, 1973]; 1970 Peru [Abe, 1972]; 1973 Nemuro-Oki [Shimazaki, 1975a]; 1974 Peru [unpublished].

overall amount of displacement at the source is a more adequate parameter for the present discussion. Great earthquakes in Chile, Alaska, the Aleutians, and Kamchatka have very large rupture lengths and very large M_o ranging from 10^{29} to 10^{30} dyne-cm, while earthquakes in Peru and Japan have small rupture lengths and M_o, about 10^{28} dyne-cm. Kanamori [1971b] and Kelleher et al. [1974] argued that this remarkable difference in the characteristics of great earthquakes reflects regional differences in the contact zone between the oceanic and continental lithospheres. As regards (2), two mechanisms are possible. First, as a result of weakening and decoupling at the interface, the subduction has become nearly completely aseismic. In this case, the subduction is still taking place but without any major seismic activity. The second mechanism involves a buoyant oceanic lithosphere [Vogt, 1973; Kelleher and McCann, 1976]; part of the oceanic lithosphere is less dense than elsewhere and is not capable of subducting under the opposing lithosphere, and little or no subduction is now taking place. This mechanism is attractive in that it explains the arcuate feature of island arcs, some of the characteristic distributions of large earthquakes and volcanoes along island arcs and the regional variations in the shape of the Benioff zone [Kelleher and McCann, 1976]. However, these geometrical arguments alone are not enough to fully evaluate these possibilities. It is hoped that recent progress in long-period seismology will provide more direct clues to the understanding of these problems.

Seismic and Aseismic Slip

Among various subduction zones, historical earthquake data is most complete for southwest Japan. It is well known that along the Nankai trough in southwest Japan (Figure 2), large earthquakes have occurred very regularly in time (about once every 125 years) and space [Imamura, 1928; Ando, 1975a]. This regularity in time and space may justify the use of relatively recent data in estimating the seismic slip rate along various other subduction zones.

For Chile, along the rupture zone of the 1960 great Chilean earthquake, historic records suggest a repeat time of the order of a century [Lomnitz, 1970, Kelleher, et al., 1973]; large earthquakes occurred in 1970, 1837, 1737, and 1575. If we assume that these earthquakes involved a slip which is comparable, on the average, to that of the 1960 event (about 20 to 25 m), we can estimate the seismic slip rate. The amount of slip associated with the 1960 event is estimated on the basis of geodetic data [Plafker, 1972] and long-period surface-wave data [Kanamori and Cipar, 1974]. Dividing the seismic slip in each event by the repeat time gives the seismic slip rate. Figure 3a compares the seismic slip rate thus determined and the rate of relative plate motion between the South American plate and the Nazca plate, which is estimated to be about 11 cm/year [Morgan et al., 1969; Minster et al., 1974]. An uncertainty of ± 30% is attached to the estimate of the seismic slip. Although the plate slip rate seems to be slightly smaller than the seismic slip, this discrepancy is not significant in view of the large uncertainty in the magnitude of historical events. To the first order of approximation, we may conclude that the seismic slip rate is about the same order of magnitude as the rate of the plate motion.

For Alaska, there are no historical earthquake data from which we can estimate the repeat time. Plafker [1972] suggests on the basis of geomorphological data a repeat time

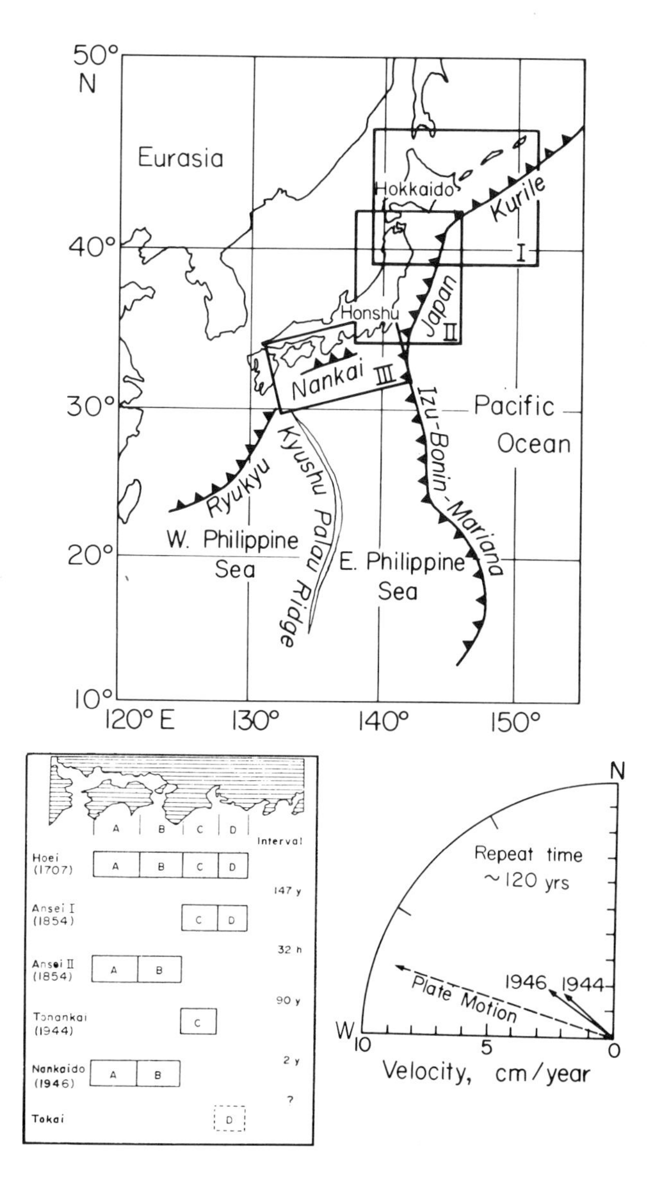

Figure 2. Index map (left) and historical seismicity along the Nankai trough, southwest Japan (Box III in the index map). Major earthquakes occurred in 1498, 1605, 1707, 1854, and 1944 to 1946 along this plate boundary. The last three sequences are shown in the upper right figure [after Ando, 1975a]. Arrows indicate the direction and the rate of seismic slip and the plate motion.

of 900 to 1350 years. There was a sequence of large earthquakes around the turn of the century near the rupture zone of the 1964 Alaskan earthquake [Sykes, 1971], but it is not very clear whether they occurred on the same fault as the 1964 event or not. If they indeed occurred on the same fault, the repeat time may be as short as 60 years, but Plafker's argument does not support it.

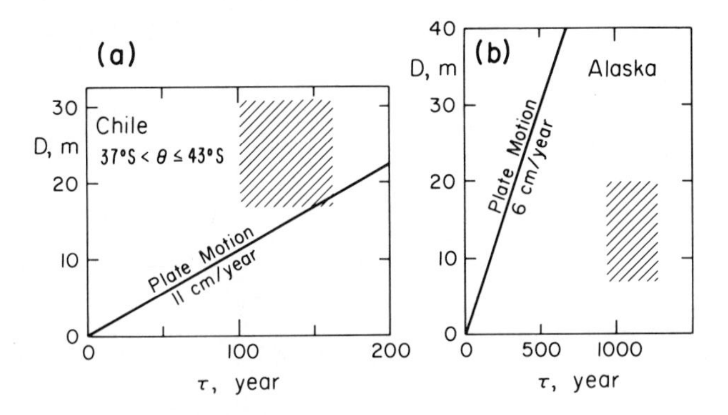

Figure 3. Seismic slip, repeat time of major earthquakes and the rate of plate motion for Chile and Alaska. Hatching indicates the range of seismic slip and the repeat time.

Along the Aleutian islands, the seismic activity is very high, and there seems to be two major seismic sequences during the past 60 years or so [Sykes, 1971; Anderson, 1974], one around the turn of the century and the other between 1957 and 1965. In view of this very high activity, a repeat time of about 1000 years seems somewhat too long. Figure 3b compares the seismic slip and the plate motion for Alaska. Plafker's [1972] repeat time of 930 to 1350 years is used. With this repeat time, the slip rate of the plate motion, 6 cm/year [LePichon, 1968; Minster et al., 1974] is several times larger than the seismic slip. However, if the repeat time is 60 years, the seismic slip becomes too large. If the repeat time is about 200 years or so, the seismic slip and the plate motion become comparable. In view of the fairly high seismic activity along the Aleutians, a repeat time of 200 years or so does not seem unreasonable, but the lack of reliable historical data does not permit more detailed discussion.

For the region from the Kurile Islands to Hokkaido, Japan, the historical data are more complete (Box I in Figure 2). As shown in Figure 4, during the 21 years from 1952 to 1973, six large earthquakes (M_S > 7.5) occurred along the arc, filling the entire seismic belt. Except for the 1958 event, detailed mechanism studies have been made for these events (for reference, see caption for Figure 1). These studies indicate that each of these earthquakes represents a slip of 2 to 3 m on a low angle thrust fault dipping 10 to 20 degrees NW (Figure 4). The slip direction agrees approximately with that of the Pacific plate with respect to the Eurasian plate. Thus, it is suggested that the Pacific plate subducted during the last 21 year period by 2 to 3 m along this plate boundary. Historical seismicity in this area shows that there was another sequence of major earthquakes along this zone from 1843 to 1918, as shown in Figure 4, indicating a repeat time of approximately 100 years [Utsu, 1972; Usami, 1966]. The seismic slip rate is therefore estimated to be 2.5 m/100 years = 2.5 cm/year, which is only 1/4 of the rate of the plate motion in this region, 9 cm/year [Minster et al., 1974]. Thus we may conclude that approximately 3/4 of the total slip must be taken up by aseismic slip, if the plate motion is uniform on this time scale. Independent evidence for such aseismic slip is found in the crustal deformation in Hokkaido. Figure 5 shows the subsidence in a coastal area of Hokkaido during the period from 1900 to 1955. Shimazaki [1974a] interpreted this subsidence to be the result of the drag caused by the underthrusting oceanic lithosphere. Shimazaki found that this subsidence can be explained if the continental lithosphere beneath Hokkaido is dragged at a rate of 2.7 cm/year. Since the rate of plate motion is 9 cm/year, this result suggests that about 3/4 of the slip takes place without causing crustal deformation, the interface being largely decoupled mechanically. Although the estimate of the seismic slip and the amount of the drag may be subject to some error, it is almost certain that a large part of the plate motion is taking place aseismically in the Kurile-Hokkaido region.

Farther to the south, in the Sanriku region (Box II in Figure 2), the evidence is more striking. In this region, two major earthquakes occurred in recent years, in 1896 (many large aftershocks occurred for several years following this event) and 1933 (Figure 6). Before 1896, only three major earthquakes are known to have occurred, in 1677, 1611, and 869 [Usami, 1966]. Although the historical data may not be very complete, it is clear that there is not a continuous zone of frequent thrust earthquake activity between the trench and the coast in this region (see Figure 6). As mentioned earlier, the 1933 event is a normal fault event with-

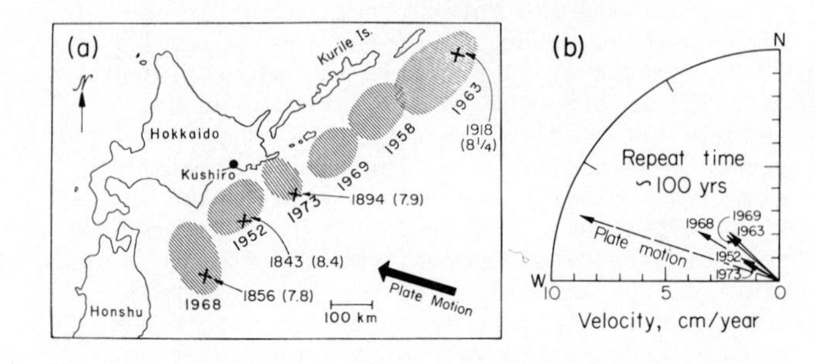

Figure 4. Seismicity, seismic slip and the plate motion in the Kurile Is.-Hokkaido region (Box I in Figure 2). Arrows indicate the direction and the rate of seismic slip and the plate motion.

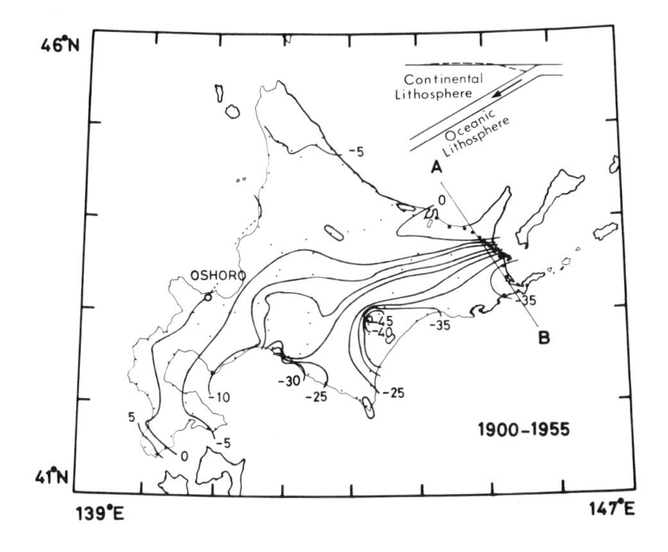

Figure 5. Vertical crustal deformation in Hokkaido, Japan, during the period 1900 to 1955 (unit:cm). Shimazaki's [1974a] model is schematically shown in the inset. Dotted curve shows the deformation due to the drag of the undergoing oceanic lithosphere.

in the oceanic lithosphere, which does not represent slip on the inter-plate boundary. The location of the 1896 Sanriku earthquake is not known accurately, but macro-seismic data and some instrumental data strongly suggest that this is a tsunami earthquake, or, in more general terms, a silent earthquake [Kanamori, 1972b]. A tsunami earthquake refers to an earthquake which generates very extensive tsunamis but relatively weak seismic waves, indicating that a very slow source process is involved. The 1896 event generated one of the most disastrous tsunamis in Japanese history while its earthquake magnitude is only 7 to 7 1/2. It is suspected that this earthquake represents a large aseismic event. It is interesting to note that tsunami earthquakes occasionally occur in the Hokkaido-Kurile region too, although they are not as striking as the 1896 Sanriku earthquake. Shimazaki [1975b] found a very low stress drop earthquake (therefore of aseismic character) which occurred in 1968 within the fault zone of the major 1969 Kurile Island earthquake (Figure 4), thereby contributing significantly to the overall plate motion along the Kurile arc. Another striking example is shown in Figure 7. On June 10, 1975, a M_s = 7.0 (m_b = 5.8) earthquake occurred within the rupture zone of the 1969 Kurile Island earthquake. This earthquake was followed by an aftershock on June 13, 1975, which occurred essentially at the same location as the main shock. Figure 7 compares the seismograms of the main shock and the aftershock recorded by short-period, intermediate period, and ultra-long period seismographs at Pasadena. On the short-period records, the aftershock is slightly larger than the main shock, but on the ultra-long period records, the main shock shows a very clear long-period phase while the aftershock has no energy in this frequency range. Similar results were found at Japanese stations [Suzuki and Sugimoto, 1975; Nagamune and Chiurei, 1976; Tsujiura, 1975] and WWSSN stations [Shimazaki, personal communication, 1976]. The long-period excitation of the main shock was anomalously large for an earthquake of this magnitude ($m_b \sim 5.8$). This earthquake generated a tsunami as high as 90 cm along the Japanese coast which is very anomalous for a $m_b \sim 5.8$ event. This event may be considered as a tsunami earthquake which reflects a weakened coupling between the oceanic and continental lithosphere there.

To the south of the Sanriku region, a series of moderate-size ($M_s \sim 7.1$ to 7.7) earthquakes (Shioya-Oki earthquakes) occurred in 1938 (see Figure 6). A recent detailed study by Abe [1976] showed that this sequence consists of three thrust and two normal-fault earthquakes having a total seismic moment of about 2 X 10^{28} dyne-cm. However, historical data suggest that there was no major earthquake in this region at least for the past

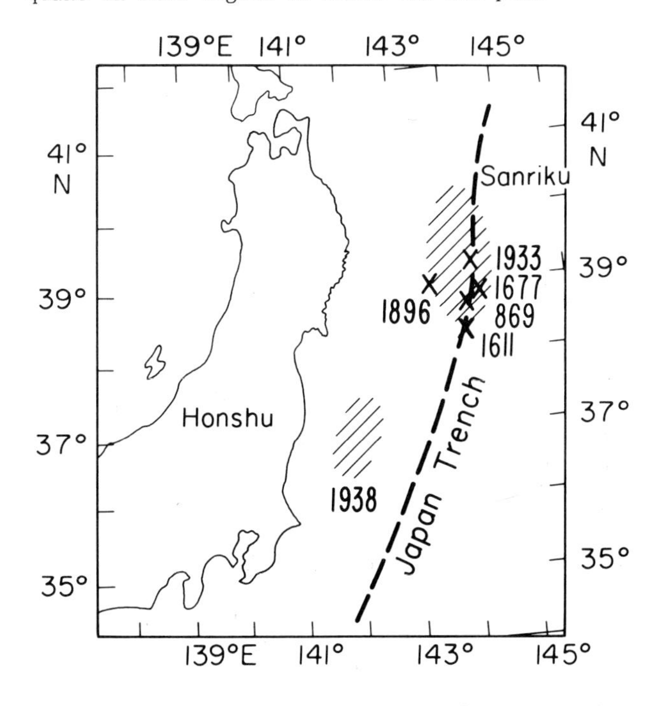

Figure 6. Sanriku earthquakes of 1896 and 1933 and Shioya-Oki earthquakes of 1938 [Abe, 1976] (Box II in Figure 2). Historical earthquakes larger than $M_s > 8.0$ between 36°N and 40°N are also included.

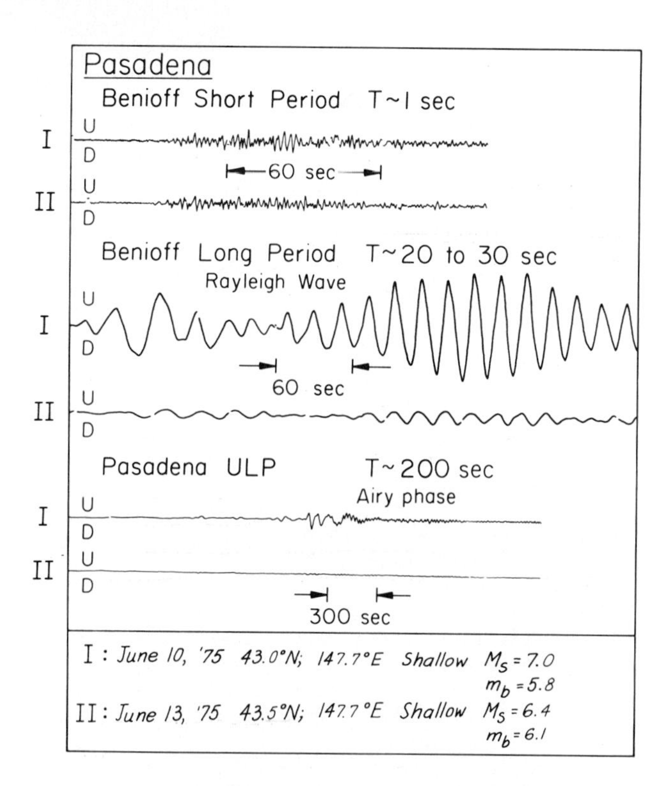

Figure 7. Seismograms of a tsunami earthquake recorded at Pasadena. The main shock on June 10, 1975 (tsunami earthquake) and one of the aftershocks on June 13, 1975 (regular earthquake) are compared at three different periods. Note that the amplitude ratio of the main shock to the aftershock becomes progressively larger as the period increases.

800 years. Abe [1976] concluded that the seismic slip rate there is about 0.4 cm/year suggesting that the plate motion is largely aseismic. This finding is very consistent with the above model of plate decoupling.

Along the Izu-Bonin-Mariana arc, there is no evidence for great earthquakes at shallow depths during this century [Gutenberg and Richter, 1954]. This lack of shallow activity, together with the foregoing arguments may lead one to the conclusion that the plate motion there is entirely aseismic. However, one difficulty arises. The Izu-Bonin-Mariana arc constitutes a boundary between the Pacific plate and the Philippine Sea plate. The Philippine Sea plate is subducting underneath southwest Japan along the Nankai Trough and the Ryukyu arc (Figure 2). The rate of the plate motion between the Pacific and the Eurasia plates has been determined to be about 9.2 cm/year [LePichon, 1968; Minster et al., 1974]. However, neither the slip rate between the Pacific and the Philippine Sea plates nor that between the Philippine Sea and the Eurasia plates is known independently. It is therefore necessary to consider the Nankai trough and the Marianas simultaneously. As shown in Figure 2, historical seismicity in this region shows a remarkably regular repeat time of major earthquake sequences along the Nankai trough [Imamura, 1928; Ando, 1975a]. The last three sequences are shown in the figure. The average repeat time is about 120 years. The amount of slip for each seismic event is about 2 to 5 m [Kanamori, 1972a, Ando, 1975b] indicating a seismic slip rate of about 3.5 cm/year. Since the seismic slip rate along the Marianas is practically zero, this value represents the total seismic slip rate between the Pacific and the Eurasia plates. Since the rate of plate motion is about 9 cm/year, the difference, 5.5 cm/year, must be absorbed either at the Marianas or along the Nankai trough as aseismic slip. If we include the possible inter-arc spreading [Karig, 1971] along the Marianas, the amount of aseismic slip would be even larger. Two extreme cases are possible: (1) The slip along the Nankai trough is largely seismic and that along the Marianas is aseismic. (2) There is little slip along the Marianas and the slip along the Nankai trough consists of 3.5 cm/year seismic and 5.5 cm/year or more of aseismic slip. There will also be cases which lie between these two extreme cases. In any event, it is important to note that a substantial part of the slip must be aseismic, if the plate motion is uniform on this time scale. If, as proposed by Kelleher and McCann [1976], the Pacific basin to the east of the Marianas is buoyant and the subduction along the Marianas has recently either decelerated or ceased, (2) would be the case. However, in view of the evidence for gradual decoupling of the plates in the Hokkaido and Sanriku regions, it seems natural to postulate complete decoupling as an important mechanism for the lack of major shallow activity along the Marianas. Although we prefer case (1) on these grounds, it is still possible that the buoyancy of the subducting lithosphere is playing an important role in modifying the mode of subduction along the Marianas. This point will be discussed further in relation to the origin of the Philippine Sea.

Model of Plate Coupling and Decoupling

On the basis of the seismological results presented in the previous section, a model of plate coupling and decoupling is proposed in an attempt to understand the fundamental physical mechanisms operating in various subduction zones. This model is basically the same as that proposed by Kanamori [1971b], but some refinements and modifications have been made on the basis of more recent data. The idea is schematically shown in Figure 8. Figure 8a shows the oceanic lithosphere which is underthrusting beneath the continental lithosphere and opposed by the latter. Because of its strength, the oceanic lithosphere is unlikely to bend very sharply and low-angle (10° to 20°) thrusting occurs. The stress in the oceanic lithosphere is compressive. We consider that this stage corresponds to the Chilean and possibly the Alaska type structure (if the repeat time is much shorter than the geomorphological evidence suggests). At this stage, the width of the contact zone is very large and the coupling is very strong, so that when slippage occurs it results in a major earthquake such as the 1960 Chilean earthquake, involving an extensive crustal deformation and

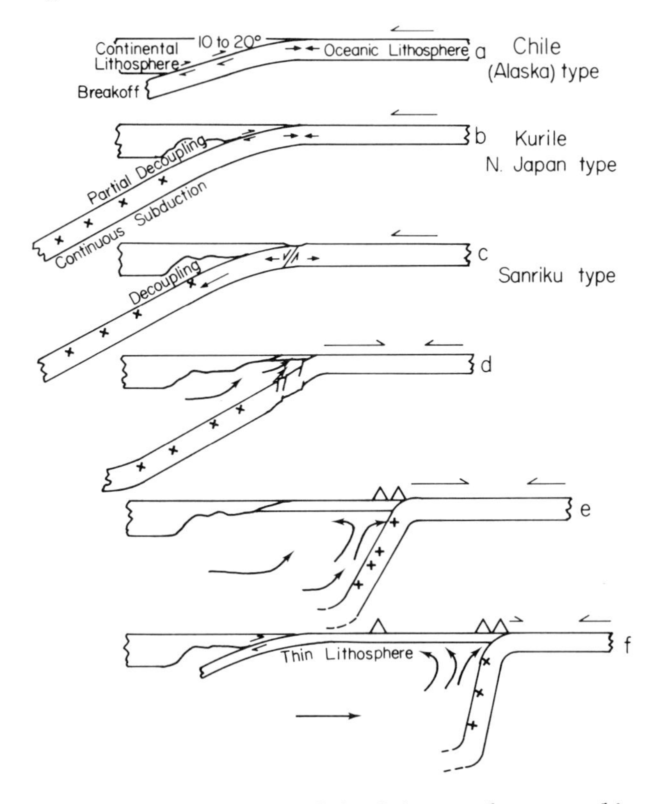

Figure 8. Schematic model of inter-plate coupling and decoupling, sinking and retreating subduction. a. Strong coupling between the oceanic and continental lithospheres results in great earthquakes and break off of the subducting lithosphere at shallow depths. b. Partial decoupling results in smaller earthquakes and continuous subduction. c. Further decoupling results in aseismic events and intra-plate tensional events. d. Sinking plate results in retreating subduction and formation of a new thin lithosphere. e. Episodic retreat and formation of ridges. f. Decelerated retreat and commencement of new subduction.

rupture zone. The lack of deep earthquakes below 200 km in these regions may indicate an absence of the downgoing plate at these depths. It is possible that where the interplate coupling and interaction are very strong, the downgoing oceanic lithosphere itself is substantially fractured and tends to break off, at relatively shallow depths, into pieces which sink into the mantle.

As the underthrusting proceeds further, both the strength and width of the interface decrease because of various effects of the plate interaction such as extensive fracturing, formation of gouge, water injection, and possible partial melting (Figure 8b). This stage corresponds to the Kurile to Hokkaido region where the reduced coupling results in smaller rupture zones of earthquakes, large aseismic slip, and occasional occurrence of silent or tsunami earthquakes. The stress in the oceanic lithosphere is still compressive. Because of the reduced interaction at the inter-plate boundary, the oceanic plate can subduct smoothly, without breaking off into pieces, to form a well defined continuous deep seismic zone. Thus a continuous Benioff zone may be an indication of partially or completely decoupled lithospheric plates.

Subduction of a ridge, which is often considered to be a major tectonic event [e.g., Wilson, 1973; Uyeda and Miyashiro, 1974], may take place more easily under these conditions. When the interface is further weakened, the continental and the oceanic lithospheres are nearly completely decoupled so that no major thrust earthquake can occur along the interface (Figure 8c). Because of the reduced coupling, the tensional force caused by the gravitational pull of the denser downgoing lithosphere may be transmitted to the oceanic lithosphere and may cause a large intra-plate normal-fault earthquake. We consider that this stage corresponds to the Sanriku region. The 1896 tsunami earthquake and the 1933 lithospheric normal-fault earthquake reflect, respectively, the decoupling of the interface and the intra-plate tensional fracture [Kanamori, 1972b]. At this stage the stress within the oceanic lithosphere is largely tensional. The above process is in general consistent with the model proposed by Kelleher [1974], who showed that the width of the contact zone controls the maximum length of the rupture zones of great earthquakes in various island arcs.

After the plates are decoupled, several modes of deformation of the underthrusting lithosphere are possible. First, a complete detachment may take place in the form as suggested by Kanamori [1971b]. Second, the oceanic plate, having lost the mechanical support of the opposing continental lithosphere, may start sinking from the leading edge. Third, the above process may occur by discontinuous shear faulting, as suggested by Lliboutry [1969]. The overall pattern of the deformation may be similar between the second and third cases. In any event, once the plates are decoupled, there is no reason for them to be in contact. Although there is no seismological evidence that favors any particular one of these possibilities, the second or third one seems to explain more naturally the transition from the Sanriku type structure to the Izu-Bonin-Mariana type structure. In this case, the subduction zone retreats in the direction opposite to the direction of the plate motion as the leading edge sinks and falls off (Figure 8d). A counter flow may fill the opening between the continental lithosphere and the retreating subduction zone. This counter flow may take place in the form of episodic interarc spreading, with upwelling material eventually forming a thin lithosphere between the continental lithosphere and the retreating subduction zone (Figure 8e). Such a thin lithosphere has been found for the Philippine Sea by surface-wave studies [Kanamori and Abe, 1968; Seekins and Teng, 1976]. As this newly created lithosphere cools and becomes rigid, a new episode of subduction may commence at the boundary between this new lithosphere and the continental lithosphere (Figure 8f). The subduction along the Nankai trough, which is believed to have commenced very recently (1 to 4 m years ago) [Fitch and Scholz, 1971; Kanamori, 1972a], may correspond to this type. Once the relatively rigid lithosphere is formed, the coupling between this lithosphere and the oceanic lithosphere may be restored. Retreating subduction may be an important element in the formation of the Philippine Sea, and will be discussed more fully in the next section.

The above model is very consistent with the topographic and gravity highs seaward of trenches [Walcott, 1970; Hanks, 1971; Watts and Talwani, 1974]. A recent extensive study of Watts and Talwani [1974] is particularly intriguing in this context. Figure 9 shows some of the representative profiles perpendicular to the Aleutians, Kuriles, Bonin and Marianas. The topographic high and the gravity high are most conspicuous for the Aleutians and the Kuriles, but are not very obvious for the southern Bonin and Marianas. Watts and Talwani [1974] explained these features in terms of a flexural bending of the oceanic lithosphere due to compression and vertical loading at the trench. They found that a large compressive stress is required to explain the topographic and gravity highs for the Aleutians and Kuriles, but for the southern Bonin and Marianas, they can be explained without horizontal compressive stress; only vertical loading is necessary. Although the details may vary according to the assumptions, their results are qualitatively very consistent with the present model in which compressional coupling gradually weakens from the Aleutians to the Kuriles, and farther south to the Bonin-Marianas, where the plates are decoupled.

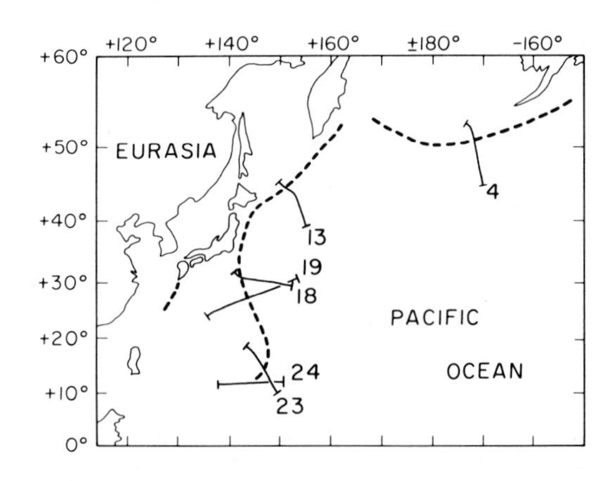

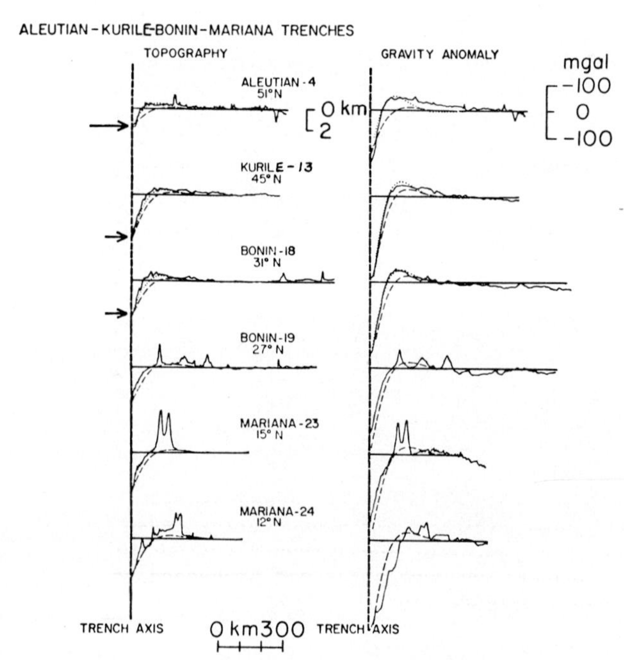

Figure 9. Topography and gravity anomaly along representative cross sections in the Pacific shown in the top figure. Solid curves show the observed profiles. Dashed curves are the profiles computed for a model with only vertical loading at the trench. Dotted curves are the profiles computed for a model with vertical loading and compression. The length of the arrows indicates the magnitude of the compressive stress [after Watts and Talwani, 1975].

Retreating Subduction and the Philippine Sea

The Philippine Sea is a marginal sea between the Bonin-Mariana arc and the Ryukyu arc both of which are dipping westward (Figure 2). Several models have been proposed for the origin of the Philippine Sea [e.g. Karig, 1971; Uyeda and Ben-Avraham, 1972; Uyeda and Miyashiro, 1974]. Here we show how the concept of retreating subduction can be employed as one of the fundamental physical mechanisms of formation of island arcs and marginal seas. We take Uyeda and Ben-Avraham's [1972] model shown in Figure 10. Uyeda and Ben-Avraham [1972] postulated that the Kula-Pacific ridge subducted beneath Japan sometime in the late Cretaceous about 100 m years ago. As the Pacific plate changed its direction of motion from a NNW to a WNW direction, about 40 m years ago, subduction started along a transform fault which connected the Kula-Pacific ridge with the Philippine ridge (Figure 10). Subsequent interarc spreading as suggested by Karig [1971] along this subduction zone formed the present day Philippine Sea. It is assumed in this model that the direction of motion of the Kula plate did not change when the Pacific plate changed its direction of motion to allow subduction of the Pacific plate beneath the Kula plate. This assumption seems somewhat ad hoc. Introduction of a retreating subduction zone may offer an alternative model which is shown in Figure 11. Before 40 m years ago, both the Pacific and the Kula plates were subducting beneath the Kurile-Japan-Ryukyu arc. We assume that the plate decoupling began from the southwest end. Then a retreat of the subduction zone began and proceeded as shown in Figure 11b to form the Philippine

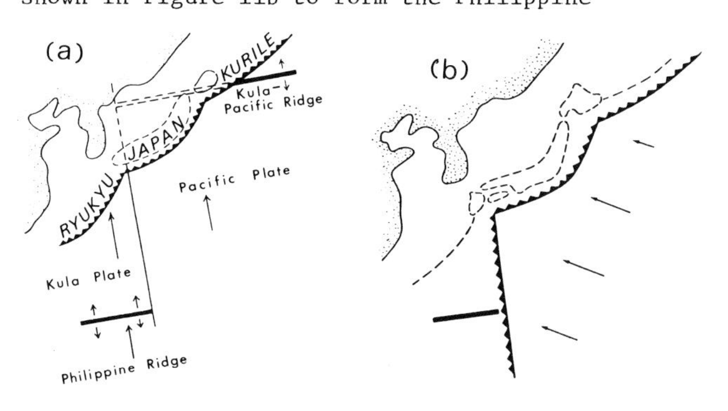

Figure 10. Schematic sketch of possible history of the Philippine Sea [after Uyeda and Ben-Avraham, 1972]. a. Kula-Pacific ridge descended beneath Japan 80 to 90 m.y. ago. b. About 40 m.y. ago, the direction of motion of the Pacific plate changed. The Kula plate did not change its direction of motion and subduction started along the transform fault.

Sea. The upwelling counter flow took place in the form of inter-arc spreading, filled the opening, and cooled to form the Philippine Sea plate (Figure 11c). When a reasonably rigid lithosphere was formed, a new episode of underthrusting began along the Nankai trough and the Ryukyu trench. The retreat and the upwelling may have been episodic resulting in various NW-SE trending features in the western Philippine Sea and N-S trending features in the eastern Philippine Sea.

Although the change in the direction of the Pacific plate is not essential in this model, it may be related to the formation of the Kyushu-Palau ridge (Figure 2) which marks the boundary between the Western and Eastern Philippine Sea. When the direction of motion of the Pacific plate changed from NNW to WNW, the component of the velocity perpendicular to the subduction zone must have increased. Since the rate of retreat is determined by a balance between the rate of fall-off of the plate at its leading edge and the plate motion perpendicular to the subduction zone, a sudden increase in the plate velocity may have resulted in a substantial difference in the nature of the marginal sea formed by this process. An increase in the velocity of plate motion would also result in an increase of the dip of the Benioff zone. The nearly vertical Benioff zone in the southern end of the Marianas [Katsumata and Sykes, 1969] may have been caused by either an increase in the effective plate velocity, a decrease in the fall-off rate or both.

Discussion and Conclusions

Although an attempt is made in the previous section to explain some of the details of the geological features of the Philippine Sea, the emphasis of the present paper is on the fundamental physical mechanisms operating in the formation of geological features. In constructing a model for a specific area, it would be necessary to arrange these fundamental mechanisms in appropriate temporal and spatial order. We emphasize that inferences from detailed studies of great earthquakes strongly suggest that inter-plate coupling and decoupling play a fundamental role in the formation and evolution of island arcs. Strong inter-plate coupling results in great earthquakes and disrupted Benioff zones. Partial and complete decoupling results in aseismic slip, continuous Benioff zones, gravitational sinking of the subducting lithosphere and finally, retreating subduction zones. Retreating subduction would imply a more disruptive process for formation of marginal seas than for ordinary ocean basins which are formed by spreading from a linear oceanic ridge. This difference may explain

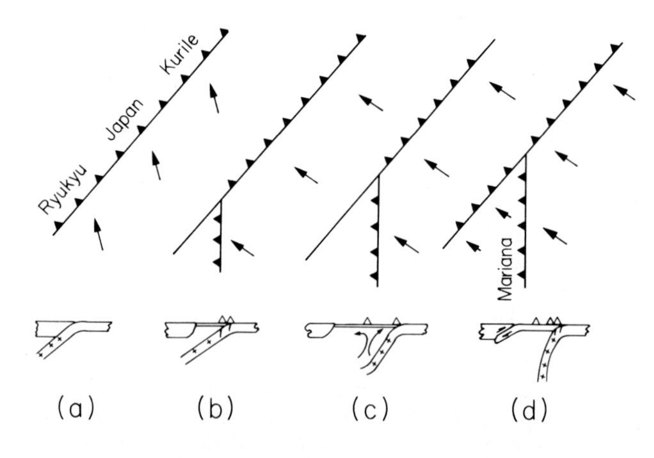

Figure 11. Schematic figure of possible history of the Philippine Sea. a. Before 80 to 90 m.y. ago. b. Before 40 m.y. ago. Retreat started and the Pacific plate changed its direction. c. Further retreat and formation of the eastern Philippine Sea. d. New subduction along the Nankai trough and Ryukyu arc and the present day Marianas.

the less distinct linear features in magnetic stripes and heat flow distribution in most marginal seas and the disparity in the water depth versus age relation [Sclater et al., 1976].

Although we believe that the inter-plate coupling and decoupling are the major factors that affect the mode of subduction and evolution of island arcs, aseismic ridges and buoyant lithospheres suggested by Vogt [1973] and Kelleher and McCann [1976] may still play an important role in modifying the mode of subduction. For example, if the lithosphere to the east of the Marianas is indeed buoyant, as suggested by Kelleher and McCann [1976], it may have prevented further fall-off of the lithosphere along the Marianas thereby terminating the retreat of the subduction.

Acknowledgements. I benefited greatly from discussions held with J. Tuzo Wilson while he was visiting Caltech as a Fairchild Distinguished Scholar.

Discussions held with John Kelleher, Kazuaki Nakamura and Seiya Uyeda at the Ewing Symposium were very useful in writing the final manuscript.

I thank Katsuyuki Abe, Kunihiko Shimazaki, and Gordon Stewart for giving me permission to use a part of their results before publication.

Research supported by the Earth Sciences Section National Science Foundation Grant No. (EAR72-22489) and No. (EAR76-14262).

Contribution No. 2757, Geological and Planetary Sciences, Seismological Laboratory, California Institute of Technology, Pasadena, California 91125.

References

Abe, K., Mechanisms and tectonic implications of the 1966 and 1970 Peru earthquakes, Phys. Earth Planet. Interiors, 5, 367-379, 1972.

Abe, K., Tsunami and mechanism of great earthquakes, Phys. Earth Planet. Interiors, 7, 143-153, 1973.

Abe, K., Mechanisms of the 1938 Shioya-Oki earthquakes and their tectonic implications, to be submitted to Tectonophysics, 1976.

Anderson, D. L., Accelerated plate tectonics, Science, 187, 1077-1079, 1975.

Ando, M., Source mechanisms and tectonic significance of historical earthquakes along the Nankai trough, Japan, Tectonophysics, 27, 119-140, 1975a.

Ando, M., Long-duration faulting in the 1946 Nankaido earthquake (abstract), EOS, 56, 1067, 1975b.

Fedotov, S. A., Regularities of the distribution of strong earthquakes in Kamchatka, the Kurile Islands and northeastern Japan, Trans. Acad. Sci. U.S.S.R., Inst. Phys. Earth, 36, (203), 66-93, 1965 (in Russian).

Fitch, T. J., and C. H. Scholz, A mechanism for underthrusting in southwest Japan: A model of convergent plate interactions, J. Geophys. Res., 76, 7260-7292, 1971.

Gutenberg, B., and C. F. Richter, Seismicity of the Earth, 2nd Edition, Princeton University Press, Princeton, N.J., 310 pp., 1954.

Hanks, T. C., The Kuril trench-Hokkaido rise system: Large shallow earthquakes and simple models of deformation, Geophys. J., 23, 173-189, 1971.

Imamura, A., On the seismic activity of central Japan, Jap. J. Astron. Geophys., 6, 119-137, 1928.

Isacks, B., and P. Molnar, Mantle earthquake mechanisms and the sinking of the lithosphere, Nature, 223, 1121-1124, 1969.

Kanamori, H., Synthesis of long-period surface waves and its application to earthquake source studies-Kurile Islands earthquake of October 13, 1963, J. Geophys. Res., 75, 5011-5027, 1970a.

Kanamori, H., The Alaska earthquake of 1964: Radiation of long-period surface waves and source mechanism, J. Geophys. Res., 75, 5029-5040, 1970b.

Kanamori, H., Seismological evidence for a lithospheric normal faulting-The Sanriku earthquake of 1933, Phys. Earth Planet. Interiors, 4, 289-300, 1971a.

Kanamori, H., Great earthquakes at island arcs and the lithosphere, Tectonophysics, 12, 187-198, 1971b.

Kanamori, H., Faulting of the Great Kanto earthquake of 1923 as revealed by seismological data, Bull. Earthquake Res. Inst. Tokyo Univ., 49, 13-18, 1971c.

Kanamori, H., Focal mechanism of the Tokachi-

Oki earthquake of May 16, 1968: Contortion of the lithosphere at a junction of two trenches, Tectonophysics, 12, 1-13, 1971d.

Kanamori, H., Tectonic implications of the 1944 Tonankai and the 1946 Nankaido earthquakes, Phys. Earth Planet. Interiors, 5, 129-139, 1972a.

Kanamori, H., Mechanism of Tsunami earthquakes, Phys. Earth Planet. Interiors, 6, 346-359, 1972b.

Kanamori, H., Re-examination of the earth's free oscillations excited by the Kamchatka earthquake of November 4, 1952, Phys. Earth Planet. Interiors, 11, 216-226, 1976.

Kanamori, H., and K. Abe, Deep structure of island arcs as revealed by surface waves, Bull. Earthquake Res. Inst. Tokyo Univ., 46, 1001-1025, 1968.

Kanamori, H., and D. L. Anderson, Theoretical basis of some empirical relations in seismology, Bull. Seismol. Soc. Am., 65, 1073-1095, 1975.

Kanamori, H., and J. J. Cipar, Focal process of the great Chilean earthquake, May 22, 1960, Phys. Earth Planet. Interiors, 9, 128-136, 1974.

Karig, D. E., Origin and development of marginal basins in the western Pacific, J. Geophys. Res., 76, 2542-2560, 1971.

Katsumata, M., and L. R. Sykes, Seismicity and tectonics of the western Pacific: Izu-Mariana-Caroline and Ryukyu-Taiwan regions, J. Geophys. Res., 74, 5923-5948, 1969.

Kelleher, J., and W. McCann, Buoyant zones, great earthquakes and dynamic boundaries of subduction, J. Geophys. Res., 81, 1976 (in press).

Kelleher, J., J. Savino, H. Rowlett, and W. McCann, Why and where great thrust earthquakes occur along island arcs, J. Geophys. Res., 79, 4889-4899, 1974.

Kelleher, J., L. Sykes, and J. Oliver, Possible criteria for predicting earthquake locations and their application to major plate boundaries of the Pacific and the Caribbean, J. Geophys. Res., 78, 2547-2585, 1973.

LePichon, X., Sea-floor spreading and continental drift, J. Geophys. Res., 73, 3661-3697, 1968.

Lliboutry, L., Sea-floor spreading, continental drift and lithosphere sinking with an asthenosphere at melting point, J. Geophys. Res., 74, 6525-6540, 1969.

Lomnitz, C., Major earthquakes and tsunamis in Chile during the period 1535 to 1955, Geol. Rundschau, 59, 938-960, 1970.

Mckenzie, Dan P. and R. L. Parker, The north Pacific: An example of tectonics on a sphere, Nature, 216, 1276-1280, 1967.

Minster, J. B., T. H. Jordan, P. Molnar, and E. Haines, Numerical modelling of instantaneous plate tectonics, Geophys. J., 36, 541-576, 1974.

Mogi, K., Development of aftershock areas of great earthquakes, Bull. Earthquake Res. Inst. Tokyo Univ., 46, 175-203, 1968a.

Mogi, K., Some features of recent seismic activity in and near Japan, 1, Bull. Earthquake Res. Inst. Tokyo Univ., 46, 1225-1236, 1968b.

Morgan, W. J., P. R. Vogt, and D. F. Falls, Magnetic anomalies and sea floor spreading on the Chile rise, Nature, 222, 137-142, 1969.

Nagamune, T., and M. Chiurei, On the magnitude of an earthquake off the east coast of Hokkaido which occurred on June 10, 1975, Quart. J. Seism., 40, 105-107, 1976 (in Japanese).

Plafker, G., Alaskan earthquake of 1964 and Chilean earthquake of 1960: Implications for arc tectonics, J. Geophys. Res., 77, 901-925, 1972.

Sclater, J. G., K. Karig, L. A. Lawver, and K. Louden, Heat flow, depth, and crustal thickness of the marginal basins of the south Philippine Sea, J. Geophys. Res., 81, 309-318, 1976.

Seekins, L. C., and Ta-Liang Teng, Lateral variations in the structure of the Philippine Sea plate, J. Geophys. Res., in press, 1976.

Shimazaki, K., Pre-seismic crustal deformation caused by an underthrusting oceanic plate, in eastern Hokkaido, Japan, Phys. Earth Planet. Interiors, 8, 148-157, 1974a.

Shimazaki, K., Nemuro-Oki earthquake of June 17, 1973: A lithospheric rebound at the upper half of the interface, Phys. Earth Planet. Interiors, 9, 314-327, 1975a.

Shimazaki, K., Low-stress-drop precursor to the Kurile Islands earthquakes of August 11, 1969, EOS, 56, 1028 (abstract), 1975b.

Stauder, W., Tensional character of earthquake foci beneath the Aleutian trench with relation to sea-floor spreading, J. Geophys. Res., 73, 7693-7701, 1968.

Suzuki, S., and H. Sugimoto, 1975 Nemuro-Oki earthquake, abstract, Annaul Meeting of Seism. Soc. of Japan, 112, 1975.

Sykes, L. R., Aftershock zones of great earthquakes, seismicity gaps, and earthquake prediction for Alaska and the Aleutians, J. Geophys. Res., 76, 8021-8041, 1971.

Tsujiura, M., Spectral ratio of two earthquakes off the east coast of Hokkaido, on June 10 and 14, 1975, abstract, Annual Meeting of Seism. Soc. of Japan, 114, 1975.

Usami, T., Descriptive table of major earthquakes in and near Japan which were accompanied by damages, Bull. Earthquake Res. Inst. Tokyo Univ., 44, 1571-1622, 1966 (in Japanese).

Utsu, T., Large earthquakes near Hokkaido and the expectancy of the occurrence of a large earthquake off Nemuro, Rep. Coordinating Committee for Earthquake Prediction, 7, 7-13, 1972 (in Japanese).

Uyeda, S., and Z. Ben-Avraham, Origin and development of the Philippine Sea, Nature, 240, 176-178, 1972.

Uyeda, S., and A. Miyashiro, Plate tectonics and the Japanese Islands: A synthesis, Geol. Soc. Am. Bull., 85, 1159-1170, 1974.

Vogt, P. R., Subduction and aseismic ridges, Nature, 241, 189-191, 1973.

Walcott, R. I., Flexural rigidity, thickness, and viscosity of the lithosphere, J. Geophys. Res., 75, 3941-3954, 1970.

Watts, A. B., and M. Talwani, Gravity anomalies seaward of deep-sea trenches and their tectonic implications, Geophys. J., 36, 57-90, 1974.

Wilson, J. T., Mantle plumes and plate motions, Tectonophysics, 19, 149-164, 1973.

Wu, F. T., and H. Kanamori, Source mechanics of February 4, 1965, Rat Island earthquake, J. Geophys. Res., 78, 6082-6092, 1973.

GROWTH PATTERNS ON THE UPPER TRENCH SLOPE

Daniel E. Karig

Department of Geological Sciences, Cornell University, Ithaca, New York 14853

Abstract. Comparison of morphology and shallow structure of arc systems in different stages of development indicates that new trenches form close to the base of continental or insular slopes. As accretion proceeds, the prism of accreted material grows outward and upward in front of this older slope to produce the step-like profiles commonly seen in arcs accreting primarly sedimentary material. Because the rear flank of the active accretionary prism appears to remain fixed during a subduction episode, widening of the upper slope or forearc basin implies that it lies primarily on a basement of accreted material. Accretion also causes depression of the descending lithospheric plate, but the coupling of this depression with subsidence in the upper slope area is not straightforward, and implies that other important processes control vertical displacements across this part of the arc system. Some reasonable possibilities include densification of accreted material by dewatering and metamorphism, additional loading by upper slope sediments, and compression across the rear flank of the upper slope area. The mode of trenchward migration of the upper slope basin, and the relationship of these basin sediments to the underlying accretionary material is observable along the Sumatra sector of the Sunda Arc. In that region, a steep flexure on the rear side of the ridge along the trench slope break drops older parts of the accreted prism downward beneath the upper slope basin. Sediments in that basin in part lap onto this flexure and in part are involved in it. Similar geometries can be interpreted from data in other arcs with similar step-like morphology. However, the style of transition from upper to lower slope regimes may differ where accretion of oceanic crustal material predominates, and in some areas there may be no sharp contrast between upper and lower sectors of the slope.

Introduction

The distribution of shallow structural deformation across the inner trench slope indicates that this region can usually be divided into two mechanical or morphotectonic regimes [Karig, 1970; Dickinson, 1971; Seely et al., 1974]. The boundary is the trench slope break which can assume a wide range of morphologies. Recently, most attention has been focused on the lower slope, where major underthrusting and uplift associated with accretion of trench and ocean floor materials has been observed [Kulm and Fowler, 1974; Seely et al., 1974]. Continued deformation, but at a decreasing rate, appears to persist from the trench to the trench slope break as observed in the deformation of sediments deposited on the lower trench slope.

The upper slope region lies between the trench slope break and the frontal arc and is characterized by sedimentation without significant deformation. The term upper slope is used here to cover the range of morphologies, from a subtle slope decrease to a deep basin, that this feature can assume. More specialized or local terms such as terrace, forearc basin [Seely et al., 1974], interdeep [van Bemmelen, 1949], and outer arc basin [Hamilton, 1973] have also been used for the upper slope.

The variability in morphology of the trench slope can reasonably be attributed to variations in the rates of sediment fed to the trench and to the upper slope and to the previous subduction history [Karig and Mammerickx, 1972; Dickinson, 1973; Grow, 1973a]. In particular, there is a striking difference in the morphologic pattern and apparent mode of growth between arc systems accreting primarily sediments and those accreting a large percentage of oceanic crustal rocks [Karig, 1974]. Other postulated processes, such as tectonic erosion, or strike-slip faulting resulting from oblique subduction or from end-on arc collision would be expected to degrade these simpler morphologic patterns.

Outward Growth of the Accretionary Prism

Although accretion of trench fill, ocean floor sediments, or oceanic crustal rocks to the inner slope as the cause of growth of the accretionary prism is reasonably substantiated in many arcs, there is little agreement concerning the method by which the accretionary prism behind the trench slope break evolves. Equally uncertain is the relationship of accreted material to oceanic and continental crust beneath and behind the prism. Oceanic crust is commonly assumed to underlie most upper trench slopes [Scholl and Marlow, 1974; Dewey, 1976]. This postulate is supported by the ophiolite sheets that are exposed beneath some uplifted slope sediments [Dickinson, this volume] and by the shallow magnetic anomalies beneath the upper slope of the Aleutian arc [Grow, 1973a]. In several arcs, continental crust has been suggested to underlie the upper slope and to extend nearly to the trench [Scholl et al., 1970; Hussong et al., 1976].

In this paper, the growth patterns of accretionary prisms involving primarily sediments are examined in an attempt to learn how the upper slope area and its flanking structures evolve. The assumption used is that variations among arcs involving similar materials can be interpreted, with caution, as equivalent to an evolutionary trend. Although only systems with well developed upper slope basins are discussed, it should be noted that this type is very common. In addition to dominating the arc systems used in this paper and shown in Figure 5 of Karig and Sharman [1975], this step-like morphology, with a sharply bounded trench slope break and upper slope discontinuity, predominates in such areas as southern Chile, Middle America, west Luzon, and Java. The model developed here is strongly biased by observations along the Sumatra sector of the Sunda arc, where the trench slope break is above sea level and where a large quantity of proprietary reflection and drilling data have been collected during petroleum exploration. Extrapolation of this model to other arcs where only morphology and shallow penetration reflection data are available is more speculative.

This analysis begins with the identification of an incipient or nascent subduction zone because in that situation the relation of the new plate boundary to the pre-existing morphologic and tectonic units can be most clearly seen. The nascent condition of an arc system refers to the initiation of a discrete episode of plate convergence along a given boundary. Good examples of young systems, with accretionary prisms built of sediments, can be identified in the Shikoku arc, along the northeast coast of Luzon, and off central Mexico.

Along the eastern Luzon zone, a very small accretionary prism formed when the apron of sediments flanking Luzon was collapsed against the continental margin [Karig and Sharman, 1975, p. 380]. A very small upper slope basin is now forming on and behind this deformed mass. In this situation it is quite apparent that the new, or rejuvenated, trench was formed very close to the base of the pre-existing continental slope and that most of this slope still remains behind the presently evolving accretionary prism [Figure 1].

The contemporary episode of subduction in the Shikoku (southwest Japan) arc began in the Miocene, following the disruption of an older convergent boundary by the opening of the Shikoku Basin [Karig, 1975]. The subduction rate appears to have increased since the Miocene [Karig, Ingle et al., 1975] but is still low. The size of the accretionary prism increases southwestward away from the rotation pole near northeast Japan. Seismic reflection profiles [Ludwig et al., 1973; Karig, 1975], especially near the northeast end, suggest that the new prism has been built against the outer edge of the late-Mesozoic-early Cenozoic Shimato complex formed by earlier episodes of subduction in the same sense.

Along the central Mexican sector of the Middle America trench, oblique subduction in the Miocene (Schilt and Truchan, 1976; in preparation] and probably earlier is apparently responsible for the truncation of the continental margin [Karig et al., 1975; in preparation]. Since the late Miocene, perpendicular subduction

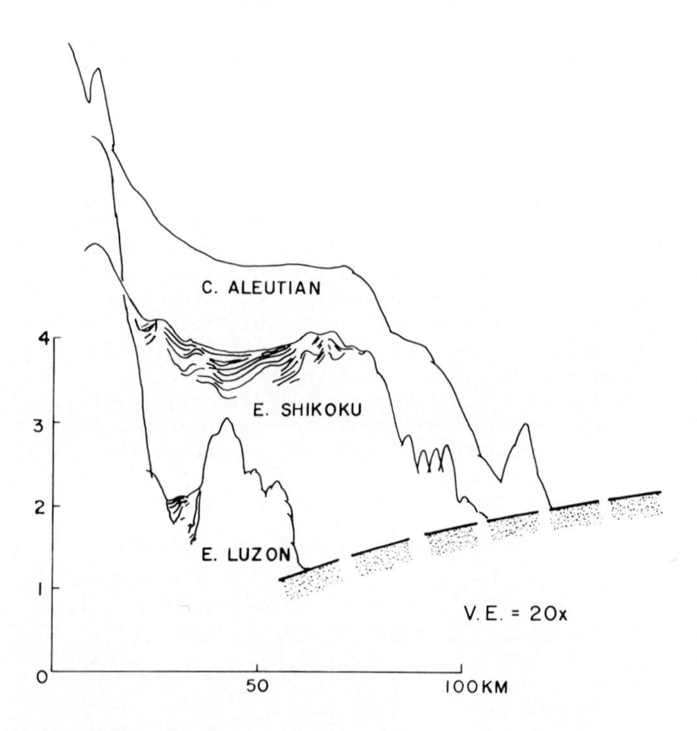

Figure 1. Comparative profiles across the accretionary prisms of arc systems with step-like trench slope morphology illustrating the upward and outward growth away from an older continental slope. These three profiles are fitted to a single downgoing plate configuration, but it is recognized that even small prisms will somewhat depress this plate and modify its shape. The rearward tilts in sediments of the upper slope basin are shown in the eastern Luzon (Karig and Sharman, 1975) and eastern Shikoku (Karig, 1975) profiles. No reflection data exist for the central Aleutian profile but other profiles in that area (e.g. Grow, 1973) show similar tilts.

has been building a new accretionary prism and upper slope sediment accumulation. In this example, the seaward edge of continental crust is easily mapped because of its distinctive magnetic field pattern, and lies at the projected base of the pre-existing continental slope [Figure 2].

Progressive accretion would be the cause of trenchward growth of the accretionary prism and widening of the upper slope area, as suggested by stacked morphologic profiles [Figure 1], if it can be assumed that the rear flank of the upper slope does not migrate during a subduction pulse [Karig and Sharman, 1975, p. 381]. If this assumption holds, then the basement beneath most of the upper slope region should consist of accreted material which becomes progressively younger toward the trench. In some wider accretionary prisms this assumption might be questioned. For example, in Sumatra, pre-Tertiary metamorphic basement crops out on promontories along the west coast, and point to the existence of continental crust beneath at least the landward section of the continental shelf. Published drilling results on this sheld [Bowman, 1974; Caldwell, 1975; Hariadi and Soeparjadi, 1975; and unpublished company reports] show up to 2 km of Miocene and younger strata unconformably overlying basement. However, reflection profiles such as Figure 6 suggests, and other profiles confirm, that the structural edge of the basin lies near the outer edge of the shelf, where there is on the order of 4 km of relief on the basement. Littoral to non-marine early Tertiary sediments cropping out on offshore islands as protrusions above the general level of the unconformity indicate that the vertical displacements are at least in part an oscillation rather than simple subsidence. Although the shoreline has migrated back and forth off Sumatra during much of the Tertiary, the rear structural flank of the upper slope basin has remained approximately fixed.

In some cases, ophiolite slabs may underlie parts of upper slope basins [e.g., Dickinson, this volume], but these observations do not invalidate the model proposed above. A discontinuous zone of oceanic crust, between the continental crust and accretionary basement, could result from initiation of subduction along an irregular continental margin. It is likely that the downgoing lithosphere, with its strongly elastic character, could not follow the irregularities of that margin and that pieces of oceanic crust would be trapped in the re-entrants.

Early development of the accretionary prism seems primarily one of upgrowth and outgrowth from a pre-existing continental or insular slope [Figure 1]. Strata within the upper slope basin during this early stage tend to have a rearward tilt which might best be attributed to continued uplift of the seaward flank of the basin by accretionary processes.

Vertical Displacements on the Upper Trench Slope

As growth and widening of the prism continue, absolute subsidence of the surficial sediments of the upper slope occurs. Although this can be documented at present in only a few cases [e.g., Seely et al., 1974] proprietary well data demonstrate that this behavior is common. The

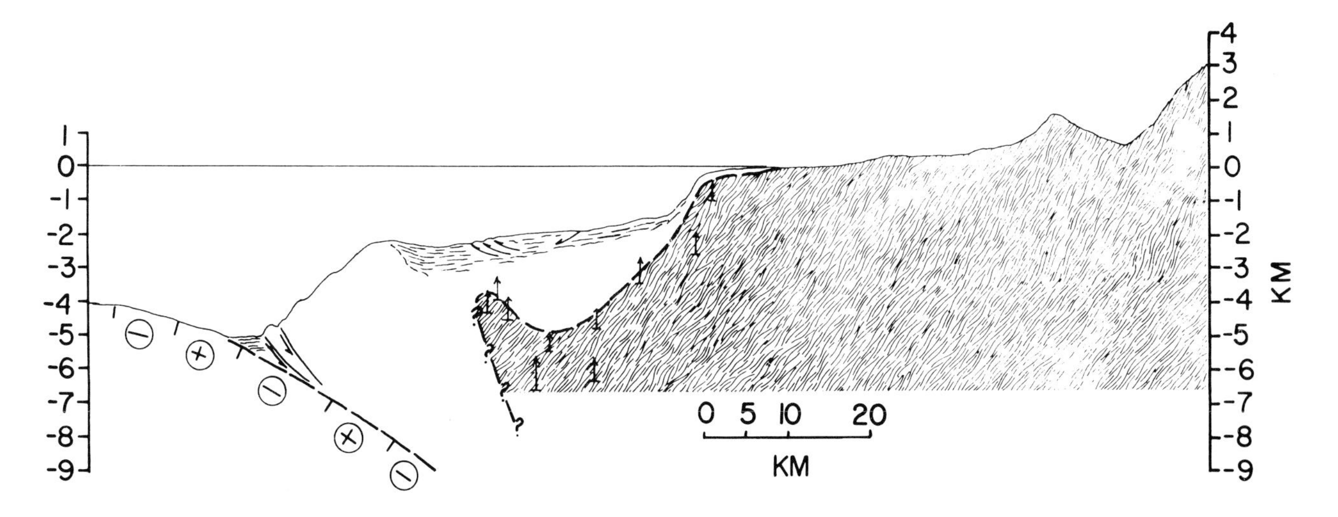

Figure 2. Cross-section through the Middle America arc system just northwest of Acapulco (from Karig et al., 1975; in prep.) based on a detailed seismic reflection and magnetic survey. Maximum depth to the magnetic anomalies associated with the Paleozoic(?) metamorphic complex are shown by the barred arrows. The Cocos plate magnetic anomalies and polarities are indicated where that pattern was observed. Between is a body of non-magnetic presumably accreted material. Note slumping on the upper slope, shown schematically by arrows.

tendency for rearward tilting of sediments in the upper slope basin continues as the basin widens so that the basin axes remain close to the structural rear edge of the upper slope [e.g., Ross and Shor, 1965; von Huene et al., 1971; Seely et al., 1974; Coulbourne and Moberly, in press; Figure 6 of this paper].

In some more developed arcs there also appears to be subsidence of the upper slope with respect to the frontal arc, but the nature of these displacements is not clear. In the southern Middle America trench, upper slope sediments dip landward toward outcrops of Paleozoic metamorphics [Ross and Shor, 1965] requiring either faulting or flexuring or both. In the eastern Aleutians, a discrete fault system separates uplifted early Tertiary and Cretaceous trench sediments from the upper slope basin [von Huene et al., 1971]. The seismic reflection profiles suggest that the dips of the fault planes are quite high [ibid]. Movement during the 1964 Alaskan earthquake on the Patton Bay fault, which is a part of this zone, was high-angle reverse [Plafker, 1969] and seismologic data from the same tectonic setting in the central Aleutians suggest a similar sense of motion [Murdock, 1969; Grow, 1973a].

One logical cause of subsidence of the upper slope area is the loading of the downgoing lithosphere by the mass of material incrementally added close to the trench and the more broadly distributed depression due to that load. That the applied load is capable of causing incremental flattening of the upper seismic zone can be quantitatively demonstrated [Karig et al., in press], but the distribution of the displacements across the upper slope is not simply attributable to this model.

If the incremental load is added primarily between the trench and trench slope break, subsidence of the previously deposited upper slope sediments should decrease toward the continent and tilts should be trenchward. Because these sediments tilt landward, it is clear that other processes must be active. Moreover, the rapid changes from subsidence to uplift across the rearward basin flank require a different explanation.

Several processes can be suggested, all of which remain speculative until there are sufficient data to test them. One is progressive densification of material within the accretionary prism. The few sets of refraction data across obviously accreting arc systems [Shor and von Huene, 1972; Yoshii et al., 1973] show a steady landward increase in seismic velocities at any given structural level. The refraction units in these studies have usually been drawn parallel to the velocity structure, but at shallow levels these units sharply cross-cut the structural units. The landward increase in density can be attributed in part to progressive dewatering in the accreted material and, at greater depths, to metamorphic processes. Although the process undoubtedly occurs and total reduction in volume of a given piece of rock could be large, it is doubtful whether the distribution of densifying processes is such to produce landward increasing subsidence.

Yet another process which must contribute to the rearward tilt is the additional load of upper slope sediments [see also Dickinson, this volume] which reaches a maximum thickness near the rear flank of the basin. This load distribution would tend to propagate further subsidence in the area of maximum sedimentation and so preserve the locus of maximum subsidence.

If the subsidence and rearward tilt of the upper slope is associated with uplift of the adjacent frontal arc, and the two units are separated by steep reverse faults, some sort of compressional mechanism might be suspected. The downdragging of the basement beneath the upper slope by shear along the plate interface [Seely et al., 1974, Figure 12; personal communication, 1976; Hussong et al., 1976] is attractive, but this mechanism does not explain the uplift of the frontal arc and such downdragging of crustal material would be opposed by buoyant forces. Alternatively, the compression within the upper plate that would result from interplate shear may cause high-angle reverse faulting beneath the rear edge of the upper slope. This faulting would be accompanied by depression of the footwall block and uplift of the hanging wall, as often observed. Moreover, the region could remain in isostatic equilibrium, with a positive-negative gravity anomaly pair developed over the fault region.

Forward Edge of the Upper Trench Slope

The evolution of the forward edge of the upper slope basin is a more tractable problem because of recent scientific investigations and petroleum exploration in that tectonic setting. Several different relationships between upper slope sediments and the accreted "melange" have been proposed. Hamilton [1973] and Scholl and Marlow [1974] have suggested that upper slope sediments become progressively deformed and are incorporated into the "melange" near the trench slope break by one process or another. Dickinson [1975] has postulated that the contact between the two units is a transgressive thrust. Yet another alternative is that upper slope sediments transgress trenchward over accreted material as the prism widens [e.g., Figure 4 of Karig, 1974].

The strongly uplifted and partly emergent trench slope break of the Sunda arc west of Sumatra provides an excellent example with which to explore these alternatives. Close control of the critical relationships is afforded by geologic mapping on the islands [van Bemmelen, 1949; Verstappen, 1973; work in progress by Karig and

G.F. Moore] and by a large amount of proprietary data, both on and offshore, collected during exploration for petroleum resources. Unfortunately, these latter data cannot be shown. Most of the conclusions presented here come from the area of Nias, the largest and most extensively mapped island of the chain [Figure 3].

Over most of Nias, a strip-like pattern of melange (Oyo complex) and lower trench slope sediments is exposed [Figure 4] which represents the uplift and erosion of slope basins to different structural levels. Along nearly the entire eastern coast of the island, a sharp flexure drops these Miocene and older units several

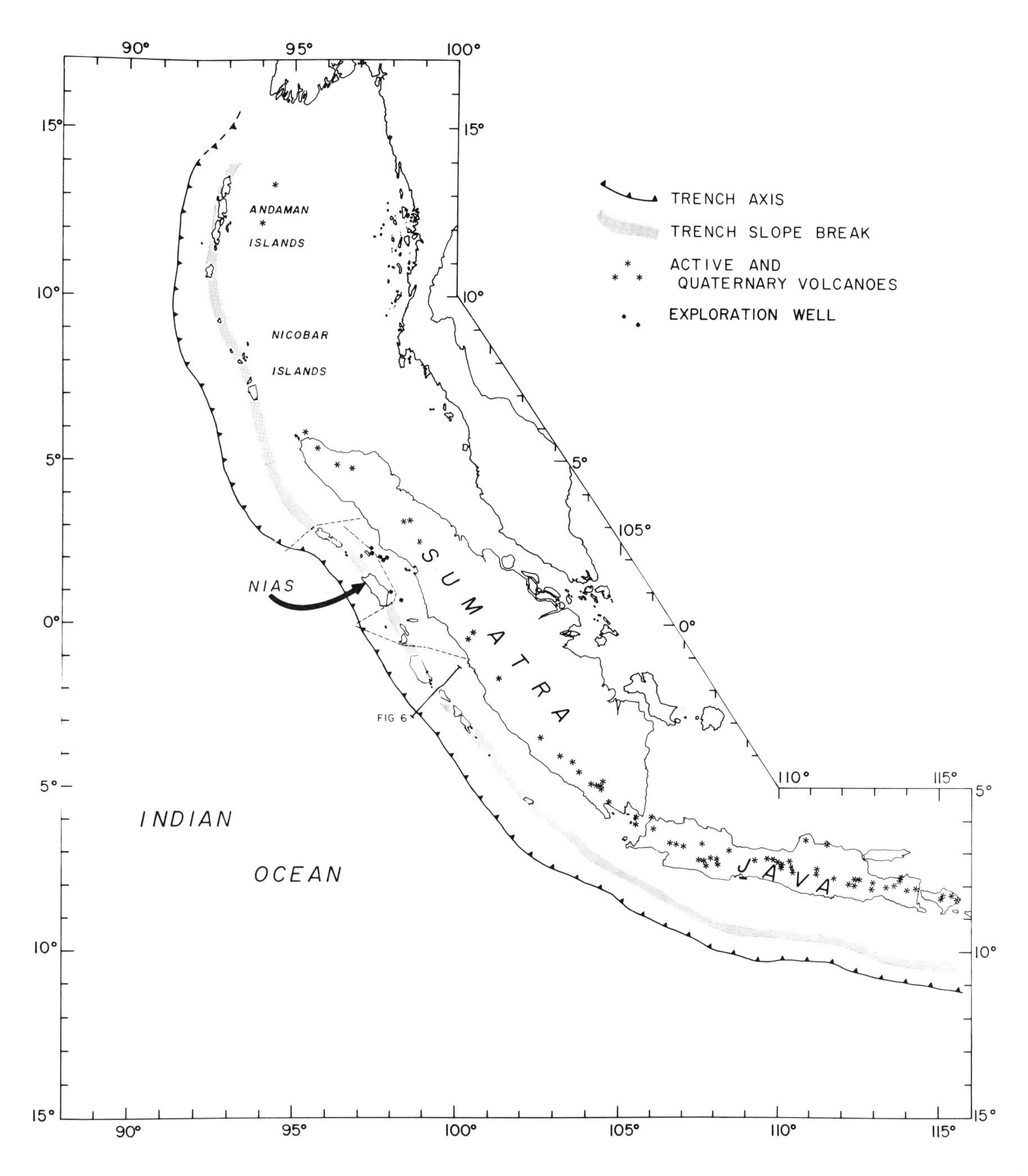

Figure 3. Index map of the Sunda Arc showing some of the control available in the Nias area. The reflection profiling coverage of Scripps Inst. Oceanography is shown by dashed lines and in Fig. 6. Only those exploratory wells near Nias are indicated.

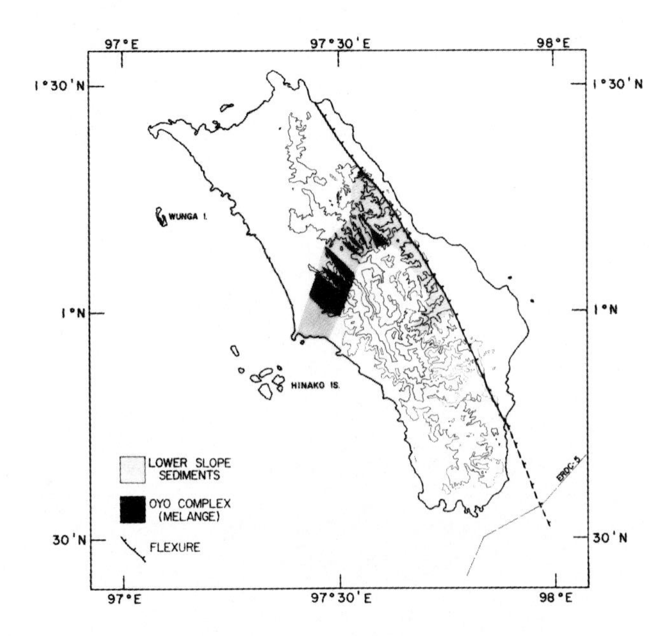

Figure 4. Outline map of Nias showing the general distribution of units in the area mapped by Karig and G.F. Moore. Contours at 200m intervals are added to show the relation of the flexure with the relief. The flexure is schematically shown but its position is well controlled by aerial photography and by previous mapping. The cross section in Fig. 5 is a composite across the northern end of the mapped area. The track of S.I.O. Eurydice Cruise is also shown off the south end of the island.

kilometers down to the east [Figures 4, 5]. This flexure, in which several units of very resistant bioclastic limestones have dips of from 45° to near vertical, is very clearly expressed in aerial photos and the topography, because it forms the mountain front.

Along the several transects mapped across the flexure, Pliocene and younger sediments are in part flexed with and in part lie with angular unconformity on the older, lower slope sediments. The younger sediments in the flexure flatten rapidly eastward and form the trenchward flank of the upper slope basin. Petroleum exploration in the Nias area has been concentrated in this area between the flexure and the deep water part of the basin.

This flexure on Nias can be traced to sea at both ends, where it appears on reflection profiles [e.g., SIO Eurydice, leg 5]. Comparison of seismic reflection profiles with onshore geology demonstrates that acoustic basement includes moderately deformed lower slope sediments as well as melange. On profiles over the flexure, all that is generally seen is an offset in acoustic basement, because no coherent energy is returned from the steep and somewhat sheared strata within the flexure.

Although the shallow structure across the rear of the trench slope break at Nias is quite clear, the deep structure and the relationship of the flexure to the thrusts which dominate the lower slope are not yet known. In the area most closely studied, the flexure lies approximately the same distance (7-10 km) from the first major thrust ridge as these ridges seem to be separated from each other on the upper part of the lower slope. This suggests that the flexure may represent normal displacement along one of the zones which previously were high-angle reverse faults forming the rear boundaries of the lower slope basins.

The Plio-Pleistocene reef cap of the east coast of Nias regressed across the flexure without suffering any obvious deformation in several localities, indicating that there has been little Quaternary vertical offset along this zone. Apparently, during the Quaternary, the entire trench slope break area rose relative to sea level. The present eastern coastline is transgressing across a 6 m terrace, implying stability or slight submergence [Verstappen, 1973; personal observation]. In contrast, the west coast is more strongly emergent. The offshore Hinako and Wunga islands [Figure 4] are recently uplifted, undissected reef tracts, and lie on a ridge which seems to mark the present site of most rapid uplift. Preliminary data from Nias and from offshore wells indicate that there has been a progressive stepwise westward migration of the trench slope break since at least the mid-Miocene, but that this progression has been overprinted with more general, low amplitude vertical oscillations of the region.

Similar or compatible morphologic and structural relationships are observed along other parts of the trench slope break off Sumatra. Verstappen [1973] notes that the east coasts of many of the islands are strongly submergent, citing the historical disappearance of several small islands. Along most of these coastlines there are sharp drops in acoustic basement as shown on seismic reflection profiles. Figure 6 shows one of the few reproducible profiles, across the trench slope break near Siberut, that demonstrates most of the generally observed features in this setting. This section of the arc differs from that near Nias in that the water depth in the upper slope basin is more than 1 km greater. The configuration of acoustic basement along the trenchward flank of the upper slope basin is readily observed, whereas east of Nias it is buried under several kilometers of basin fill. In both areas the acoustic basement drops basinward in steps. Where these steps are more deeply buried beneath the basin fill, they do not cause any offset in the shallow part of the basin section. Because of their similar structural setting and similar appearance on reflection profiles, these acoustic steps and the exposed flexure on Nias are equated. This correlation is further extrapolated to suggest that the

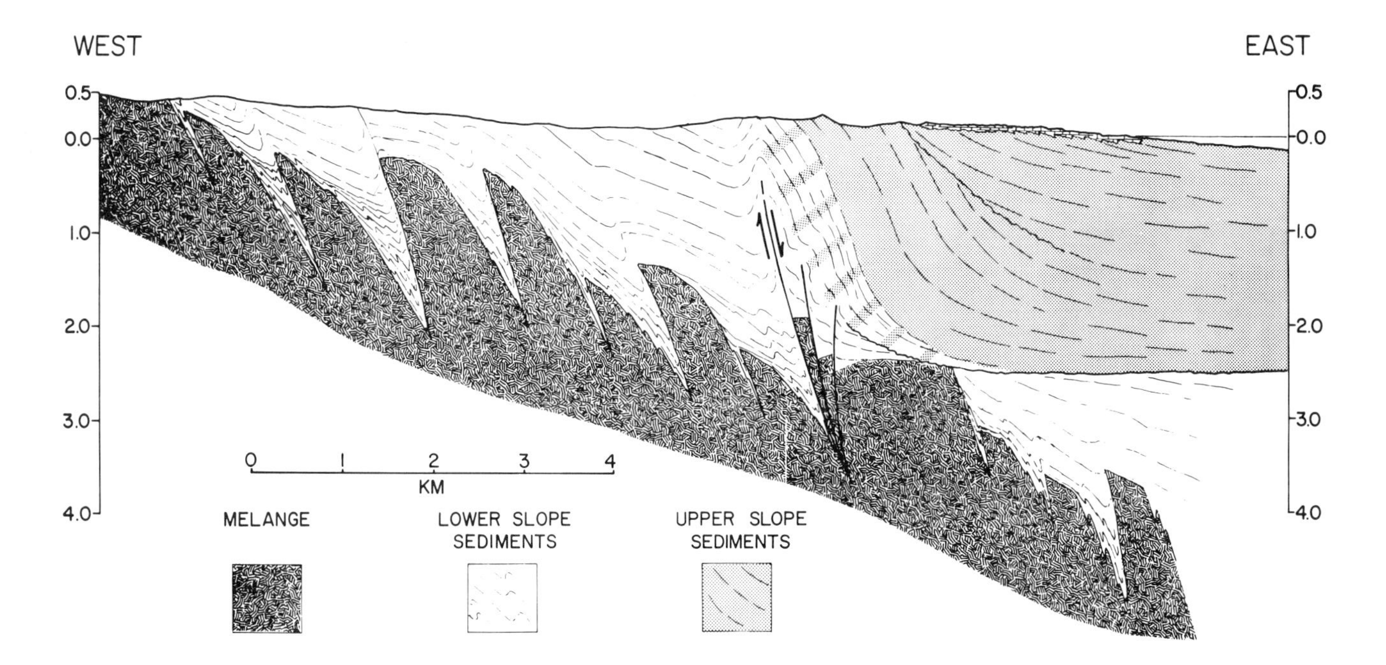

Figure 5. Composite section across the eastern flank of Nias showing the structural relationship of geologic units at the rear of the trench slope break. The surface geology is controlled by our field mapping except that the structures in the lower slope basin on the west are schematic. The deeper structure of this slope basin is based on mapping further west where erosion and greater uplift has exposed it. Proprietary reflection profiles and an exploratory well guide the structure east of the flexure above the unconformity. The deep structure of the flexure and the geology below the deep unconformity to the east are hypothetical.

flexures become successively younger westward, and that after they de-activate they become overlapped by younger basin sediments.

There is no evidence in the geological or geophysical data off Sumatra for any thrusting at the trenchward flank of the upper slope basin or of conversion of upper slope sediments to anything approaching even broken formation in intensity of deformation. The loss of acoustic coherence, both downward and trenchward of sediments at the forward edge of the upper slope basin [Hamilton, 1973], does not require that these sediments become highly deformed. The same acoustic pattern can be produced by the proposed combination of transgression and flexuring.

The trenchward growth of the upper slope basin west of Sumatra thus appears to be a transgression of upper slope basin sediments over a zone of differential vertical movements (flexures) at the trench slope break. The displacements on these flexures are possibly much larger than usual because of the large uplift of the trench slope break associated with the accretion of Bengal fan sediments since the late Miocene [Curray and Moore, 1974]. Nevertheless, a similar pattern of transgression plus deformation of upper slope sediments at the trench slope break can be inferred in other arc systems.

In the central Aleutian arc system, several reflection profiles [Grow, 1973b] and the results of DSDP holes 186 and 187 [Creager, Scholl et al., 1973] show that acoustically resolvable upper slope sediments thin markedly and are folded into low-amplitude ridges as they approach the trench slope break [Figure 7]. These sediments also appear to transgress an acoustic basement which can be interpreted from the drilling results to be only moderately folded. Grow [1973b] suggested that this basement was comprised of older upper slope sediments, but here they are thought to be lower slope sediments. It is further suggested that these folds are not compressional features, but more the effect of differential vertical movement over flexures. The major difference between this example and that off Sumatra is the amplitude of the flexures.

Random reflection profiles across arcs with well defined upper slope basins tend to show one of two types of sediment-acoustic basement relationships at the rear of the trench slope break [Figure 8]. In one type, upper slope sediments transgress across a relatively steeply dipping basement high and show little folding. In the second type, the older upper slope sediments are being tilted up at the basin flank and the youngest strata are ponded behind them. This latter mode appears to reduce the width of the basin and could not operate continuously, or else the basin flank would migrate arcward. The

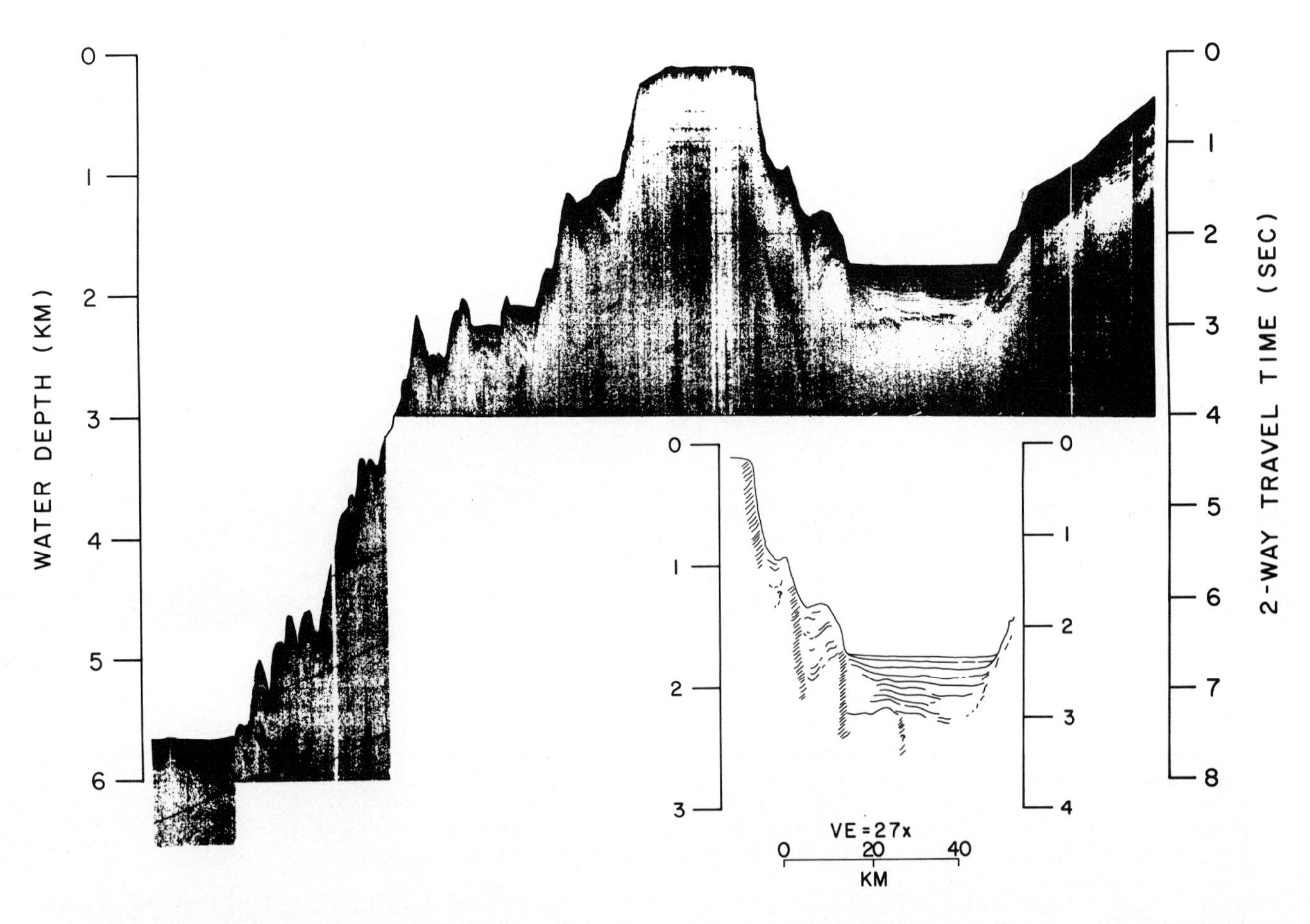

Figure 6. Seismic reflection profile across the Sunda Arc near Siberut (see Fig. 3) from SIO Antipode Cruise, with an interpretive line drawing of the upper slope area. The apparent edge of the continental crust lies beneath the prograded upper slope on the right.

observations off Sumatra and the central Aleutians would lead to the interpretation that the two modes act primarily in sequence at a given locality. Tilting of the sediments at the basin flank would accomodate the differential vertical motion and would be followed by transgression of younger basin sediments over the flexure. Later a new tilted zone, or flexure, would develop trenchward of the earlier one and the process would be repeated. It is clear, however, that the two modes are not entirely separated in space and time and also that more than a single flexure can be active at one time.

Conclusions

Development of an accretionary prism appears to involve a number of geologic processes and the boundaries between morphotectonic units seem to mark balance points between regimes where certain processes are dominant. Although many of these processes are not yet understood, their effects on the shallow structural levels are strikingly systematic. Material accreted at the trench passes sequentially across the lower slope, trench slope break, and then becomes basement for the upper slope sediments as the morphotectonic units move outboard. The data presented here indicate that this migration proceeds continuously during a subduction pulse, but that on a finer scale, the transitions occur along discrete structures with separations of several to over 10 km. Moreover, the rate of migration and the style of slope morphology that is produced may vary with time, depending on changing subduction and material feed rates.

One important implication of such systematic behavior is that it should be preserved in the age relationships between accreted material and lower slope sediments and between these two and upper slope sediments.

In situations such as that presently occurring off Sumatra, where the trench slope break emerges, the upper slope sediments should be separated from lower slope sediments and accreted material by angular unconformities. However, much more commonly the trench slope break lies several kilometers deep, and even off Sumatra, most of its length is below sea level. In these cases, there may not be extensive erosion over the

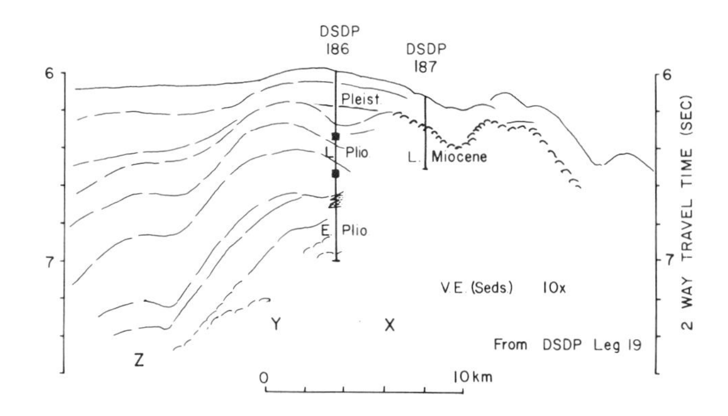

Figure 7. Composite section across the trench slope break of the central Aleutian arc using seismic reflection profiles and drilling results as shown in Creager, Scholl et al. (1973). The upper slope sediments on the left thin and terminate against acoustic basement which consists of moderately deformed Late Miocene strata. The anticlines are interpreted as the shallow reflection of flexures at the rear of the trench slope break.

trench slope break and the transition from lower to upper slope deposition may be more subtle. The differentiation might be obvious on the scale of reflection profiles, but obscure in the outcrop.

Another point to stress again is that the observations discussed and model proposed here do not necessarily apply to all arc systems. In arcs accreting oceanic crustal rocks, as the Tonga and Mariana, the upper slope area does appear to widen by growth of its forward edge [Karig, 1971], but tilting and subsidence do not seem to occur. Flexures at the rear of the trench slope break are not obvious but may account for the sharp blockage of several surveyed canyons which cross the inner trench slope.

A more fundamental variant would be required if there were no sharp transition from lower to upper slope tectonic regimes. Such a conclusion is reached by Coulbourn and Moberly [in press] and by Kulm [this volume] in studies of the Andean arc, where the morphotectonic division of the inner trench slope is not clearly observed. However, Coulbourn and Moberly [in press] do not recognize that both lower and upper slope basins show strata which tilt progressively landward

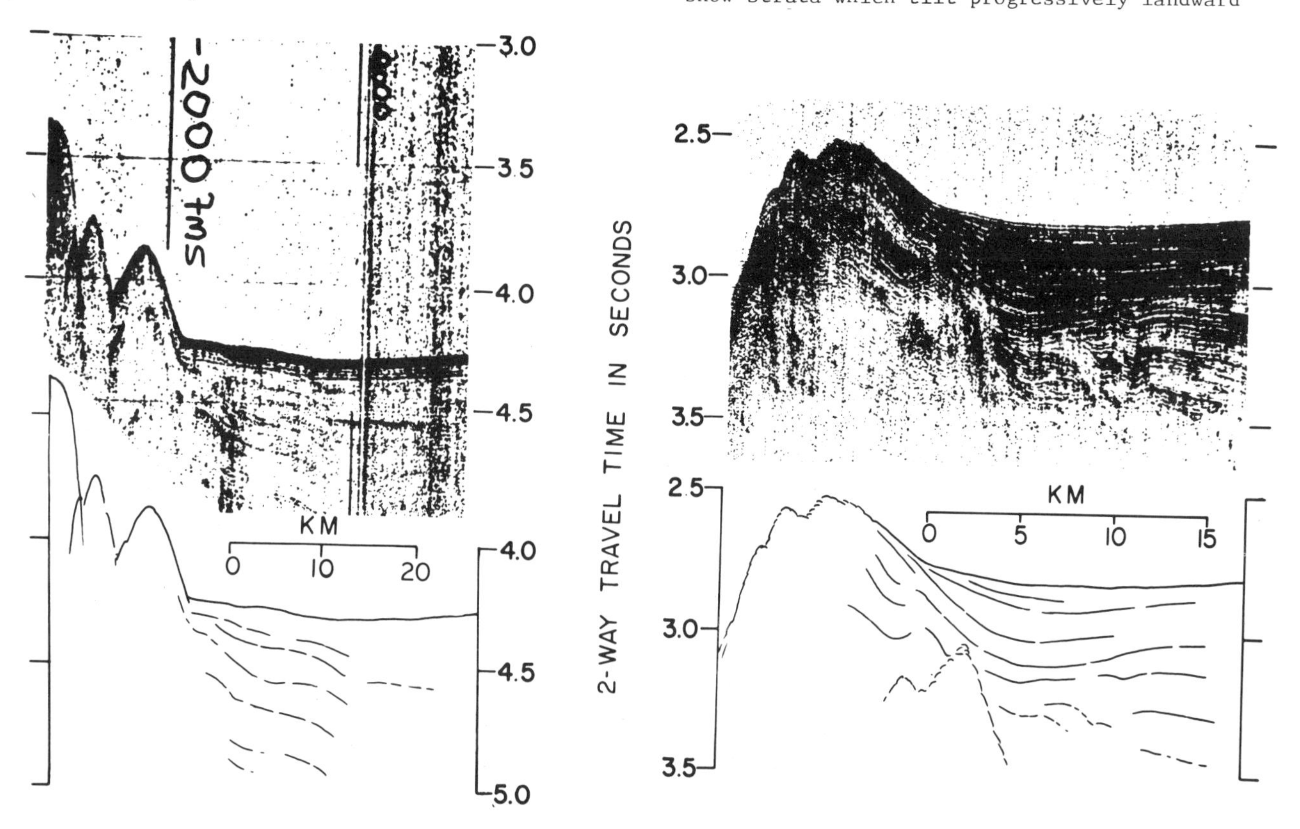

Figure 8. Reflection profile across trench slope breaks illustrating two modes of deformation at the forward edge of the upper slope basin. The profile on the left, from SIO Antipode cruise across the West Luzon arc, shows primarily a ponding and transgression behind an acoustic basement high. The profile on the right is across the trench slope break in the Shikoku arc and shows primarily upwarping of upper slope sediments at the outer edge of the upper slope basin.

with depth [e.g., Figure 6] and use that characteristic to define forearc basins. Because there are superficial similarities between lower and upper slope basins on single-channel reflection profiles, only more sophisticated and integrated studies will be able to determine the nature of deformation and changes in this deformation across the inner trench slope.

Acknowledgments. The data and conclusions presented here were derived from several projects supported by the National Science Foundation under grants DES 75-04015, DES 75-04018 and IDO75-19348. I am especially grateful to the geologists of Union Oil Corporation who shared their information in western Sumatra with us and who helped us in countless other ways during our field work in the area.

References

Bowman, J.D., Petroleum developments in Far East in 1973, Am. Assoc. Petrol. Geol. Bull., 60, 2124-2156, 1974.

Caldwell, R.D., Petroleum developments in Far East in 1974, Am. Assoc. Petrol. Geol. Bull., 59, 1977-2010, 1975.

Coulbourn, W.T., and R. Moberly, Structural evidence of the evolution of fore-arc basins off South America, Canadian J. Earth Sci., in press.

Creager, J.S., D.W. Scholl, et al., Initial Reports of the Deep Sea Drilling Project, v. 10, U.S. Government Printing Office, Washington, D.C., 913 pp., 1973.

Curray, J.R., and D.G. Moore, Sedimentary and tectonic processes in the Bengal deep-sea fan and geosyncline, in The Geology of Continental Margins, edited by C.A. Burke and C.L. Drake, Springer-Verlag, New York, 1009 pp., 1974.

Dewey, J.F., Ophiolite obduction, Tectonophysics, 31, 93-120, 1976.

Dickinson, W.R., Clastic sedimentary sequences deposited in shelf, slope, and trough settings between magmatic arcs and associated trenches, Pacific Geology, 3, 15-30, 1971.

Dickinson, W.R., Widths of modern arc-trench gaps proportional to past duration of igneous activity in associated magmatic arcs, J. Geophys. Res., 78, 3376-3389, 1973.

Dickinson, W.R., Time-transgressive tectonic contacts bordering subduction complexes, Geol. Soc. Am. Abst. with Programs, 7, 1052, 1975.

Grow, J.A., Crustal and upper mantle structure of the central Aleutian arc, Geol. Soc. Am. Bull., 84, 2169-2192, 1973a.

Grow, J.A., Implications of deep sea drilling, sites 186 and 187, on island arc structure, in Initial Reports of the Deep Sea Drilling Project, v. 19, edited by J.S. Creager and D.W. Scholl, et al., U.S. Government Printing Office, Washington, D.C., 1973b.

Hamilton, W., Tectonics of the Indonesian region, Geol. Soc. Malaysia Bull., 6, 3-10, 1973.

Hariadi, N., and R.A. Soeparjadi, Exploration of the Mentawai Block - West Sumatra, Proc. 4th Ann. Mtg. Indonesian Petrol. Assoc., 19 pp., preprint.

Hussong, D.M., P.B. Edwards, S.H. Johnson, J.F. Campbell, and G.H. Sutton, Crustal structure of the Peru-Chile Trench: 8°-12°S latitude, AGU Geophy. Mon. 19 (Woollard Symposium), 71-86, 1976.

Karig, D.E., Ridges and basins of the Tonga-Kermadec island arc system, J. Geophys. Res., 75, 239-255, 1970.

Karig, D.E., Evolution of arc systems in the western Pacific, Ann. Rev. Earth Planet. Sci., 2, 51-75, 1974.

Karig, D.E., Basin genesis in the Philippine Sea, in Initial Reports of the Deep Sea Drilling Project, v. 31, edited by J.C. Ingle, D.E. Karig, et al., U.S. Government Printing Office, Washington, D.C., 927 pp., 1975.

Karig, D.E., R. Cardwell, G. Moore, and S. Schilt, Late Cenozoic subduction and continental truncation in the Middle America Trench, Geol. Soc. Am. Abst. with Programs, 7, 1139, 1975.

Karig, D.E., J.C. Ingle, Jr., et al., Initial Reports of the Deep Sea Drilling Project, v. 31, U.S. Government Printing Office, Washington, D.C., 927 pp., 1975.

Karig, D.E., and J. Mammerickx, Tectonic framework of the New Hebrides island arc system, Marine Geology, 12, 187-205, 1972.

Karig, D.E., and G.F. Sharman, Subduction and accretion in trenches, Geol. Soc. Am. Bull., 86, 377-389, 1975.

Kulm, L.D., and G.A. Fowler, Oregon continental margin structure and stratigraphy: a test of the imbricate thrust model, in The Geology of Continental Margins, edited by C.A. Burk and C.L. Drake, pp. 261-283, Springer-Verlag, New York, 1974.

Ludwig, W.T., N. Den, and S. Murauchi, Seismic reflection measurements of southwest Japan margin, J. Geophys. Res., 78, 2508-2516, 1973.

Murdock, J.N., Crust-mantle system in the central Aleutian region, Bull. Seism. Soc. Am., 59, 1543-1558, 1969.

Plafker, G., Tectonics of the March 1964 Alaska earthquake, U.S. Geol. Surv. Prof. Paper 543-I, 74 pp., 1969.

Ross, D.A., and G.G. Shor, Jr., Reflection profiles across the Middle America Trench, J. Geophys. Res., 70, 5551-5572, 1965.

Schilt, F.S., and Marek Truchan, Plate motions of the northern part of the Cocos plate (abst.), Trans. AGU, 59, 333, 1976.

Seely, D.R., P.R. Vail, and G.G. Walton, Trench slope model, in The Geology of Continental Margins, edited by C.A. Burk and C.L. Drake, pp. 249-260, Springer-Verlag, New York, 1974.

Scholl, D.W., M.N. Christensen, R. von Huene, and M.S. Marlow, Peru-Chile Trench sediments and sea floor spreading, Geol. Soc. Am. Bull., 81, 1339-1360, 1970.

Scholl, D.W., and M.S. Marlow, The sedimentary sequence in modern Pacific trenches, and the apparent variety of similar sequences in deformed circumpacific eugeosynclines, in Modern and Ancient Geosynclinal Sedimentation, Soc. Econ. Paleontologists and Mineralogists Sp. Pub. 19, pp. 193-211, 1974.

Shor, G.G., Jr., and R. von Huene, Marine seismic refraction studies near Kodiak, Alaska, Geophys., 37, 697 700, 1972.

van Bemmelen, R.W., The Geology of Indonesia, v. 1A, Government Printing Office, The Hague, 732 pp., 1969.

Verstappen, H.Th,, A geomorphological reconnaissance of Sumatra and adjacent islands (Indonesia), Verhandelingen of the Royal Dutch Geographical Society n. 1, Walters-Noordhoff Publ., Groningen, 182 pp., 1973.

von Huene, R., E.H. Lathram, and E. Reimnitz, Possible petroleum resources of offshore Pacific-margin Tertiary basin, Alaska, Am. Assoc. Petrol. Geol. Mem. 15, 136-151, 1971.

Yoshii, T., W.T. Ludwig, N. Den, S. Murauchi, M. Ewing, A. Hotta, P. Buhl, T. Asanuma, and N. Sakajiri, Structure of southwest Japan margin off Shikoku, J. Geophys. Res., 78, 2517-2525, 1973.

THE SIGNIFICANCE OF LANDWARD VERGENCE AND OBLIQUE STRUCTURAL TRENDS ON TRENCH INNER SLOPES

D. R. Seely

Exxon Production Research Company, Houston, Texas 77001

Abstract. Laboratory models and theory, as well as the modern trench inner slopes off Washington and in the Gulf of Alaska, Gulf Coast occurrences of overpressured shale, and deep ocean mud diapirs, suggest that landward vergence results primarily from the presence of low-strength, probably overpressured, zones in trench and slope sediments. These zones are also an important control of slope steepness and the sizes of thrust plates. Seaward-dipping anisotropy near the base of the slope may trigger landward vergence, and similar anisotropy observed in underlying buried basalt "hills" may lead to obduction of oceanic crust. Landward vergence on the upper parts of trench slopes and the obliquity of structural trends across these slopes indicate that plate convergence is the primary cause of slope structures rather than gravitational potential created by the presence of the bathymetric slope.

Landward Vergence

Studies of active modern trench slopes provide a new opportunity to develop an understanding of the geologic conditions controlling the direction of vergence and other structural attributes found in overthrust belts. The presence of landward-dipping thrust planes in association with trenches was postulated on the basis of earthquake seismology before the advent of plate tectonics (see, for example, Wilson, 1954). The model was expanded by plate tectonicists (Isacks et al., 1968) and documented with reflection seismology in modern trenches by Beck (1972) and by Seely et al. (1974). The occurrence of such thrust planes or fold overturning toward the sea (Figure 1) will be referred to as seaward vergence in this paper. Evidence is convincing, however, that overthrusting toward the land, or landward vergence, has emplaced oceanic crust onto some adjacent land areas in the past, a process referred to as "obduction" by Coleman (1971) and recently examined by Dewey (1976). In the following section, models and theory are used to explain the landward vergence at the foot of a modern slope, and speculations are made on some conditions favoring overthrusting from an ocean basin onto an adjacent land area.

Laboratory Models and Theory

Conditions analogous to seaward vergence were modeled in a sandbox by Hubbert (1951). In his analogy, the sediments in the model of Figure 2a would be abyssal plain or trench sediments moving toward the left and colliding with the inner edge of the trench, represented by the left end of the box. Horizontal stresses are at a maximum at the point of collision at the left end of the box and die out toward the right.

The shear stresses at the base of the sand, or sediments, result from friction at the contact or from the shear strength of the basal layer. The effect of these stresses is to bend the stress trajectories and introduce an asymmetry into the deformation, as shown by Hafner (Figure 2b). The leftward- (or landward-) dipping thrust faults are thus more gently dipping than the rightward- (or seaward-) dipping faults. As motion continues, displacement on the landward-dipping faults increases because motion on them is most nearly parallel to the motion between the sea floor and the inner edge of the trench.

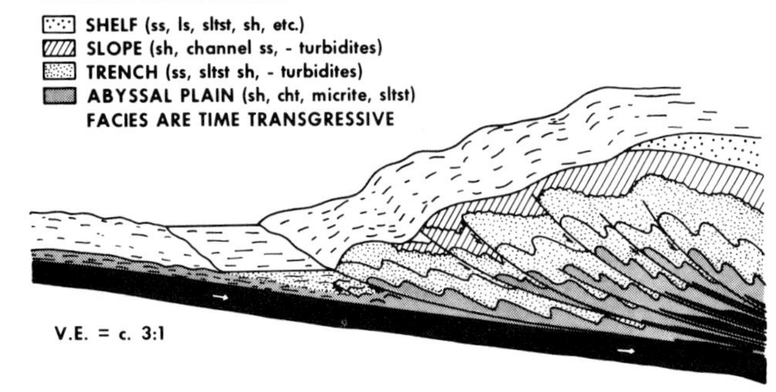

Figure 1. Generalized model of trench inner slope (Seely et al., 1974).

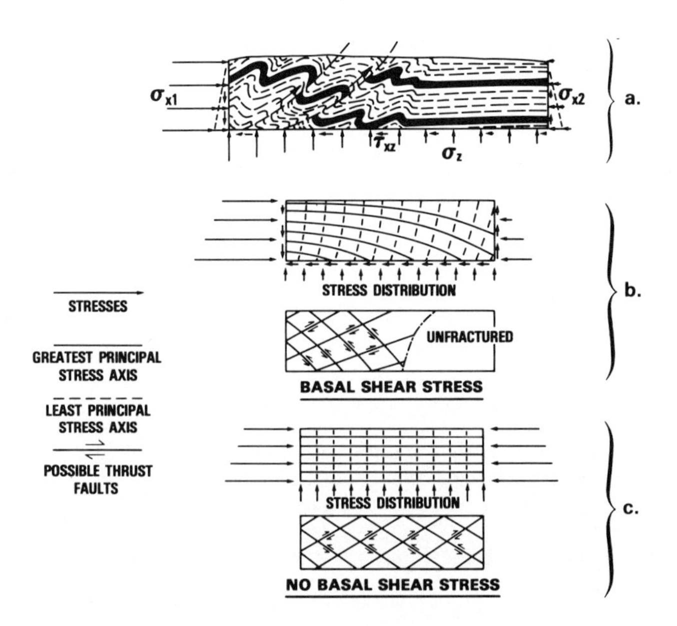

Figure 2. The effect of basal shear stress on stresses and faults under conditions of horizontal compression.

a. Two-dimensional stresses acting on boundaries of compressed block (Hubbert, 1951).

b. Stress trajectories and faults in a compressed block with a basal boundary shear stress (Hafner, 1951).

c. Stress trajectories and faults in a compressed block without a basal boundary shear stress (Billings, 1972, after Hafner, 1951).

Similar results are obtained with clay, as seen in a typical model (Figure 3). The homogeneous clay rests on a conveyor belt moving from right to left against the board barrier at the left edge. The scale is in centimeters, and the grooves in the side of the clay can be used for determining fault displacements. As in Hubbert's sandbox, the thrust faults dip toward the barrier like typical seaward vergence at the foot of trench inner slopes.

In Figure 2c, Billings (1972) shows the stress trajectories and resulting possible thrust orientations under idealized conditions of no basal friction, such as when there is a thin basal layer having no shear strength. Deformation is symmetrical--the conditions are those of pure shear--and the development of either fault set is equally likely.

These conditions can be approached experimentally by placing a watery layer at the base of the clay, as was done in the experiment pictured in Figure 4. The clay is moving from right to left against the wood barrier. Vergence is both toward and away from the barrier, and some structures are nearly symmetrical. Note also that the thrust plates have much

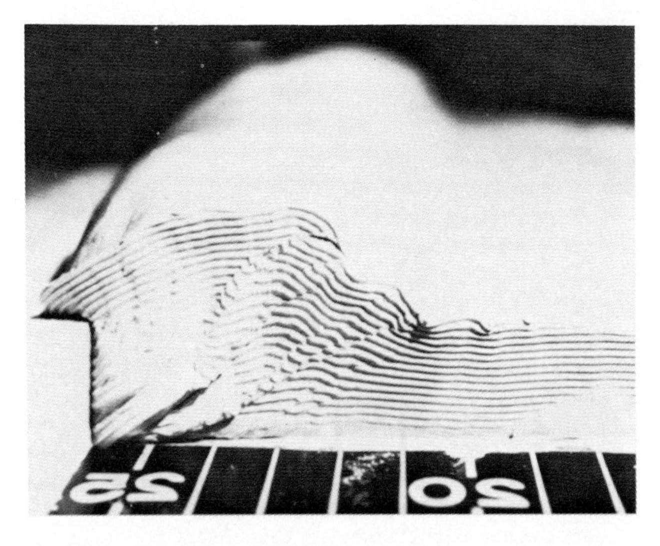

Figure 3. Clay model of thrusting with basal boundary shear stress.

Figure 4. Clay model of thrusting with no effective basal boundary shear stress.

greater breadth than in the previous model, as would be expected, because shear stresses within the clay near the barrier are reduced by the near absence of shear stress at its base. The greater breadth was predicted by Hubbert and Rubey (1959) and discussed in a series of subsequent papers reviewed by Roberts (1972).

Cascadia Basin Example

In nature, conditions approximating the second model appear to have existed during Pleistocene-Holocene time along the eastern edge of the Cascadia basin off Washington and northwestern Oregon (Figure 5). The basin floor has been moving toward the lower slope probably in a generally northeastwardly direction during this time, as determined by Atwater (1970), although at times the motion was complicated by the presence of small plates according to Silver (1971). Ocean floor magnetic anomalies extend from the basin to points landward from the base

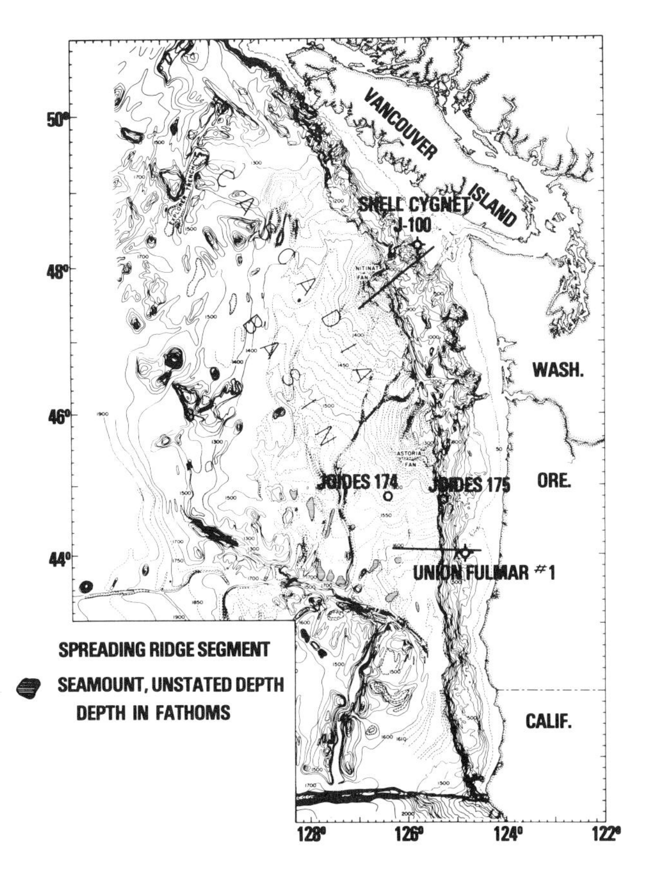

Figure 5. The Cascadia basin region. Seismic line southwest of Vancouver Island discussed in this paper. Seismic line at 44°N discussed by Seely et al., (1974). Bathymetric map compiled by McManus (1964) and modified by R. J. Page (pers. comm., 1975).

of the slope, indicating continuity of oceanic basement to points beneath the slope (Silver, 1972). A prominent lower-slope terrace has formed.

Sediments beneath the basin floor may have had (and may still possess) a basal sequence analogous to the watery layer at the base of the clay model. Their equivalents were penetrated at JOIDES hole 174 located near the lower edge of Astoria fan, where the sequence is believed to be like that near the eastern edge of Cascadia basin, although thinner and with fewer coarse components. The section consists of 284 meters of late Pleistocene, rapidly deposited, medium-to-fine sand turbidites resting on 595 meters of more slowly deposited, thin-bedded, mud-silt abyssal plain turbidites of Pliocene to late Pleistocene age. Water-retentive seventeen-angstrom mixed-layer clays constitute between 17 and 27 percent of the clay fraction in Pleistocene cores and between 26 and 35 percent in Pliocene cores, as reported by Hayes (1973).

Conditions of rapid deposition of water-retentive clays and overlying sands favor the development of overpressuring of the basal clay sequence. This is because water cannot escape from the clays rapidly enough to maintain hydrostatic pressures as overburden accumulates. As pore pressures approach geostatic pressures, the shear strength of the clay due to the geostatic pressure approaches zero, and once cohesion is broken, the clay behaves as a low-viscosity fluid.

Rates of Pleistocene deposition were several times greater in the eastern part of Cascadia basin toward the apices of Pleistocene fans than at JOIDES site 174. Where the Washington CDP seismic line is located (Figure 5), calculated depositional rates for the Pleistocene vary from a minimum of 410 meters (1350 feet) per million years to a maximum of about 690 meters (2250 feet) per million years, depending on the placement of the base of the Pleistocene by correlation with site 174. These rates compare with rates of 200 to 300 meters per million years and 500 meters per million years, respectively, for the Frio of South Texas (R. S. Bishop, personal communication) and the Neogene of Louisiana (Bredehoeft and Hanshaw, 1968). Both of the latter sections overlie overpressured shales.

Lancelot and Embley (1975) recently described deep ocean mud diapirs and attributed them to three factors: basement peaks, a thin pelagic cover, and a massive accumulation of young, terrigenous deposits on the pelagics. The existence of such diapirs argues for the presence of overpressuring in the deep ocean basins created in a manner similar to that postulated for offshore Washington.

The sediments thickening toward the right on the seaward end of the Washington seismic line (Figure 6a) are interpreted to comprise the downlapping Pleistocene fan complex. The Nitinat

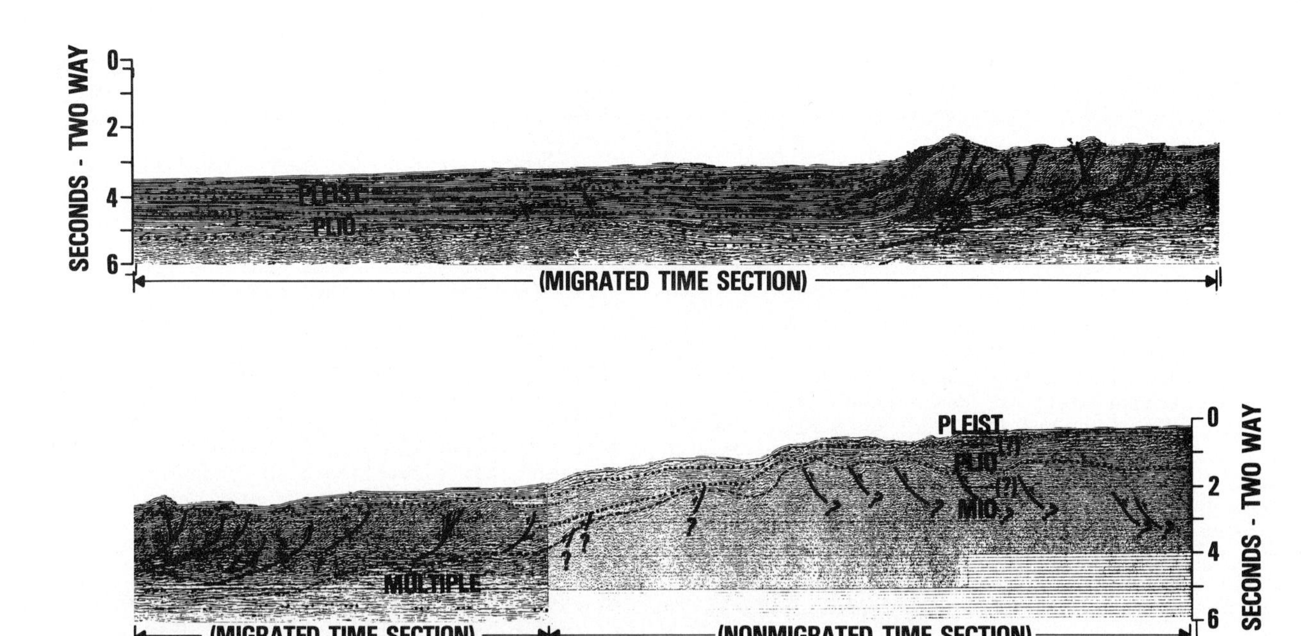

Figure 6a. Northwest Washington seismic line. Regional line across shelf, slope, and abyssal plain. See Figure 5 for location. Dots show interpreted bedding configuration.

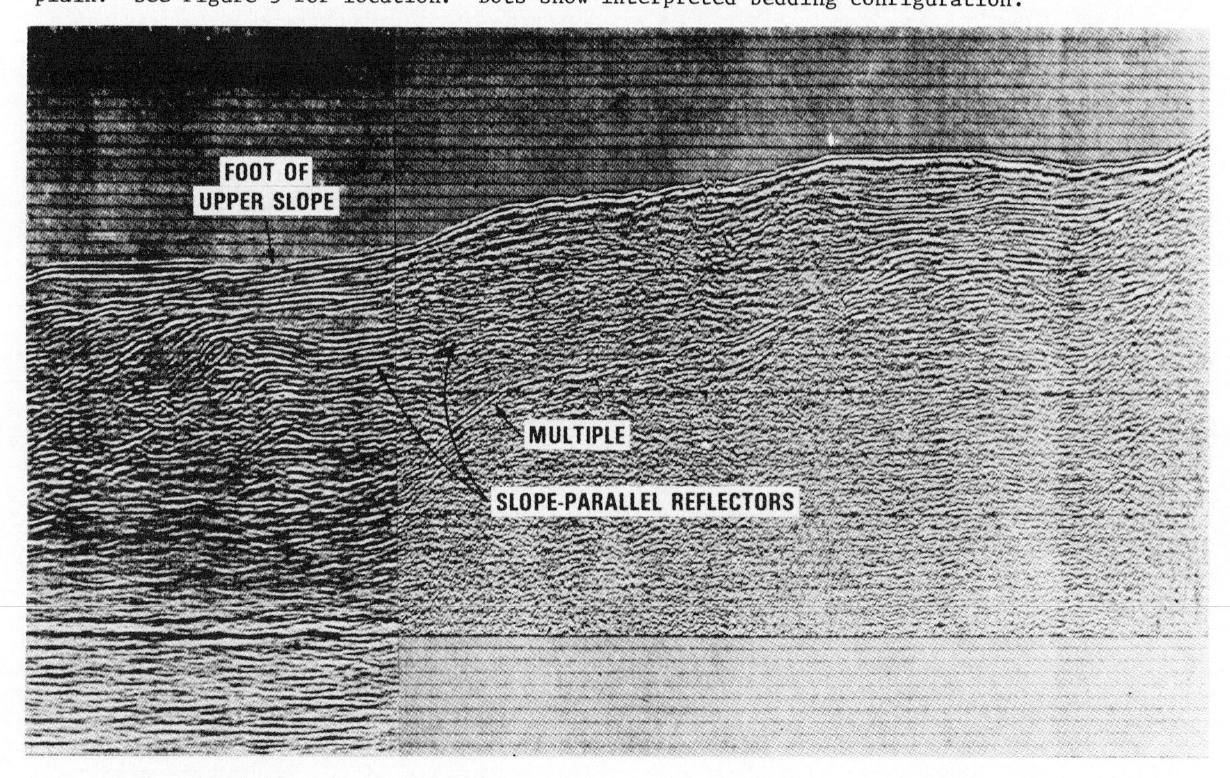

Figure 6b. Northwest Washington seismic line. Pleistocene(?) slope-parallel reflectors at the foot of the upper slope.

fan (Figure 5) at the top of the section is its final, most seaward product. The marked landward thickening of the Pleistocene contrasts with the relatively uniform thickness of the Pliocene, which lies between it and the strong reflector interpreted to be the top of the basalt. The Pliocene sediments were described as seismically transparent pelagics by Hayes and Ewing (1970), on the basis of seismic profiler records. On Figure 6a, however, these sediments can be seen to have parallel, onlapping reflectors, which led Kulm and Fowler (1974) to the conclusion that they are turbidites.

The landward vergence clearly seen at the foot of the slope (Figure 7) has been previously illustrated with arcer data by Silver (1972), Barnard (1973), Kulm et al. (1973), and Carson and Myers (1974). Its occurrence coincides with the extent of the lower-slope terrace between Vancouver Island and northwestern Oregon. Its existence as far south as 47°N is easily traced on Silver's profiles, and a good example of landward vergence at 45°N is shown by Kulm et al. At 44°N, where the terrace has terminated, seaward vergence can be seen in the CDP seismic section located there (Figure 5 and Seely et al., 1974) and also on the arcer sections published by Kulm et al. To the north of the terrace, seaward vergence is evident on Silver's arcer sections at 48°15'N.

On the basis of the continuity of Pleistocene reflectors from the abyssal plain across the lower terrace and from examination of dredge samples and isopachs of the terrace sediment cover (Figure 7), both Silver and Barnard considered the lower terrace to have accreted during the Pleistocene. To quote Barnard (1973): "Near the beginning of the Pleistocene, for some unknown reason, folding, thrusting, uplift and westward-progressing accretion of Cascadia basin sediment began along the base of the continental slope. This ..., has been relatively continuous throughout the Pleistocene."

The hypothesis presented here is that building of the lower terrace and development of landward vergence resulted from rapid deposition of the fan system on the landward-moving Cascadia basin floor. These phenomena may have been caused primarily by Pleistocene climatic conditions and do not require a change in convergence rates or directions for their explanation. Contemporaneous seaward vergence was developing on the outer part of the continental shelf (Figure 6a) near the beginning of the Pleistocene. Projected tops from the Shell Cygnet well are shown on the seismic section for approximate age dating.

A clay model was constructed to approximate the conditions controlling vergence in the area of our Washington seismic section in accordance with the above hypothesis. A strip parallel to the barrier and constituting the central third of the clay cake is underlain by watery clay. After approximately 7 centimeters of shortening and development of vergence facing away from the barrier, the inner edge of the water-based clay is located at 20.5 centimeters on the ruler

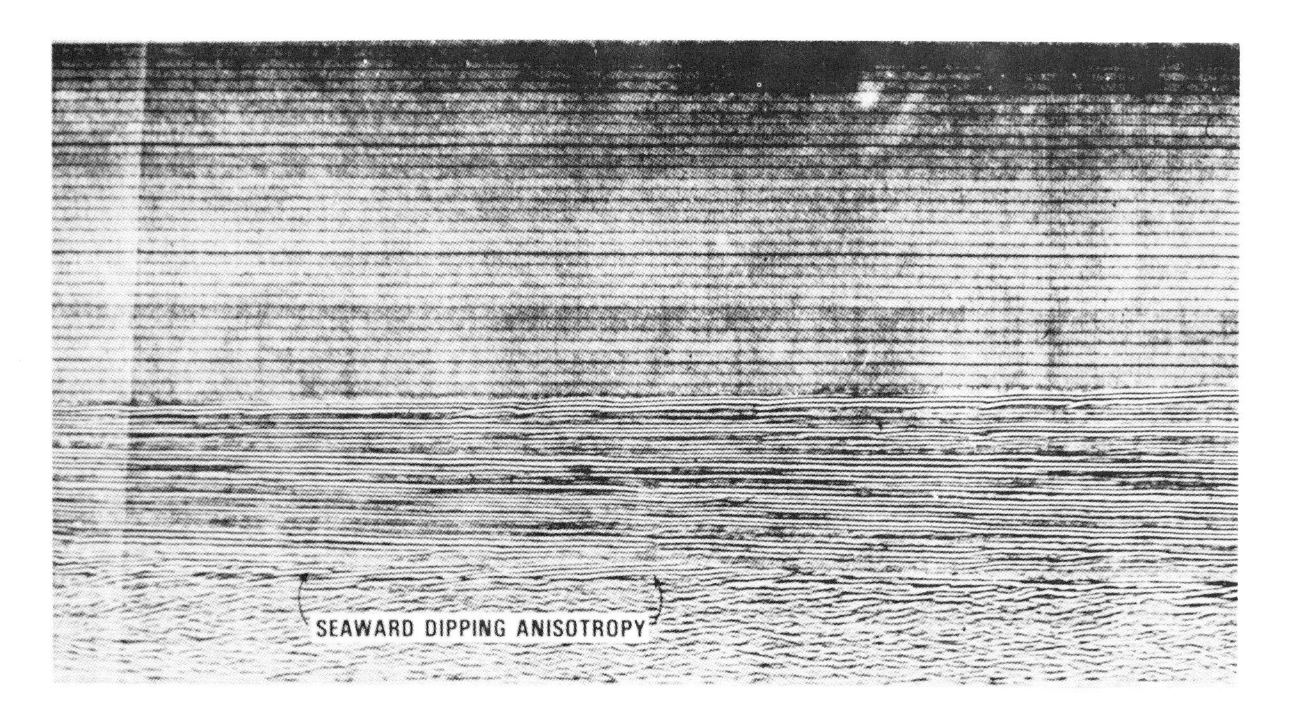

Figure 6c. Northwest Washington seismic line. Seaward-dipping anisotropy in basalt(?) at the west end of the line-possibly generated by normal faulting at the spreading ridge.

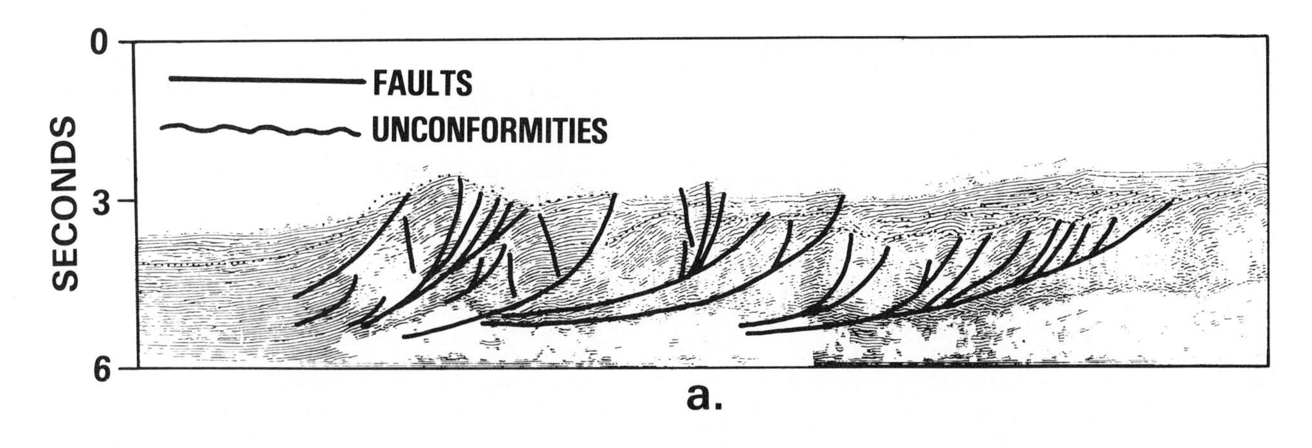

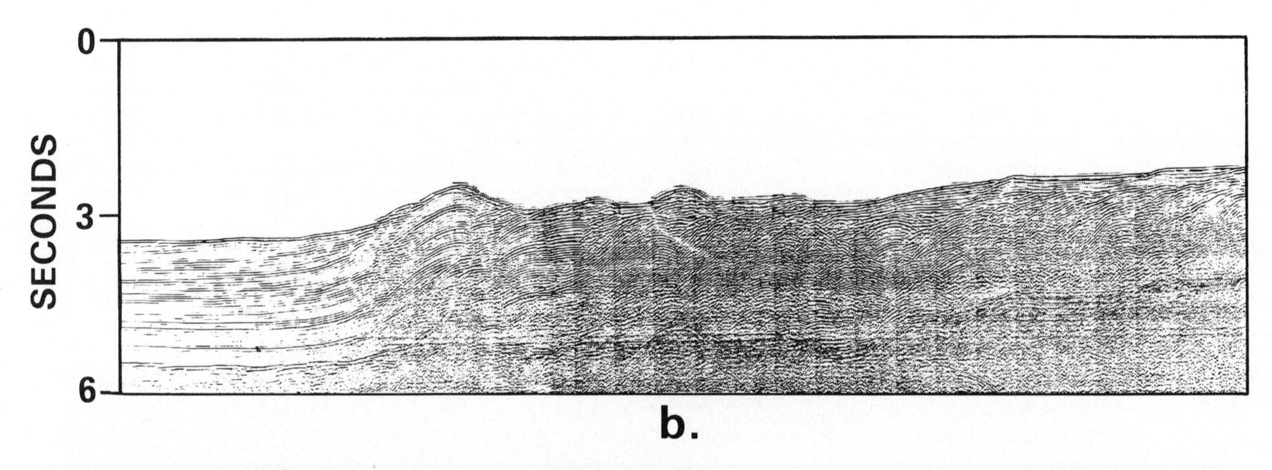

Figure 7. Detail of lower-slope part of Figure 6.
a. Migrated time display. Fault and unconformity interpretation by P. R. Vail (pers. comm., 1976).
b. Nonmigrated time display of Figure 7a.

(Figure 8). Vergence toward the barrier is beginning to develop, the conditions being perhaps similar to those off Washington at the beginning of the Pleistocene. After continued motion, an assemblage of landward-verging structures forms, producing a terrace--again perhaps similar to that now off Washington--at the foot of the earlier formed slope.

It should be noted that the terrace and steeper slope developed in clay of relatively uniform thickness, which suggests that the presence of basal zones having low shear strengths can have a marked effect on the steepness of trench slopes. This control of steepness is in addition to that related to changes in the thicknesses of incoming sediments, which is illustrated by a model with no basal watery layer (Figure 9). The structures are smaller and the slope steeper in association with the thinner clay on the near side of the model than in association with the thick clay on the far side.

Although the properties of the clay models are here considered appropriate for modeling vergence, as well as the relative sizes of structures and steepness of slopes, the models are not scaled reductions of nature. The clay viscosities, for example, are too high to satisfy the requirements of similitude for generation of Washington slope structures.

Possible Effects of Anisotropy

The presence of initial seaward-dipping anisotropy could facilitate the development of landward vergence where abyssal plain/trench sediments overlie a basal zone having low shear strength. Two examples of such anisotropy are diagrammed in Figure 10. In the upper diagram, rapid Pleistocene deposition is assumed to have onlapped the Washington slope and produced a seaward-dipping unconformity with a marked contrast in rock physical properties above and below. In the lower diagram, Pleistocene deposition is shown to have accumulated on the lower slope with slope-parallel bedding surfaces

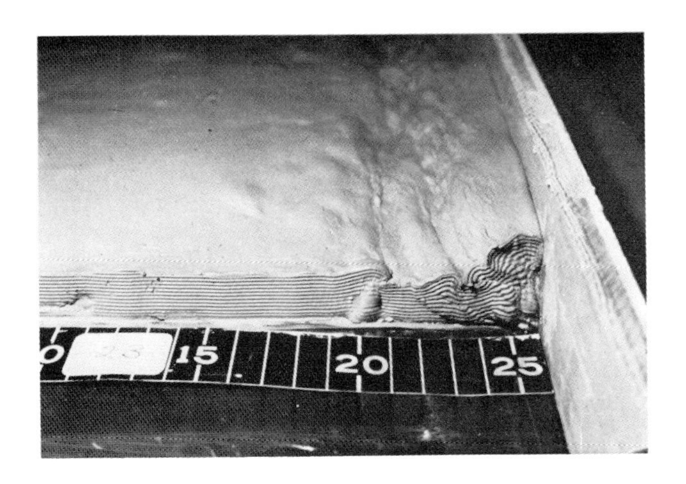

Figure 8. Clay model generating vergences and a terrace such as seen in Figure 6 (not to scale).

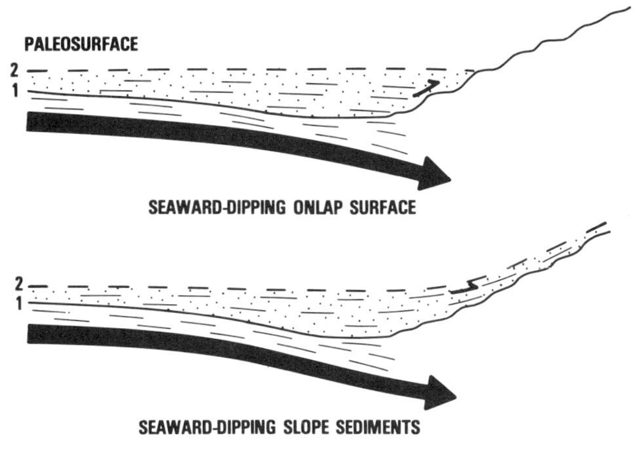

Figure 10. Two types of sedimentary anisotropy possibly influencing vergence.

forming anisotropy having a seaward dip. The Washington seismic line (Figure 6b) indicates that this latter condition may have been the case when landward thrusting began near the beginning of the Pleistocene, as can be seen by reflectors near the inner edge of the terrace.

A combination of factors that might lead to eventual obduction of oceanic crust on the Washington coast is presented in Figure 11: (1) Landward vergence in lower-slope sediments; (2) propagation of the basal, or sole, thrust into the buried basalt hill as convergence proceeds; (3) seaward-dipping anisotropy in the basalt to deflect the basal thrust down into the crust; (4) a gentle slope having low total bathymetric relief (which implies a young oceanic crust), so less work and lower thrust plate strengths are required to lift the oceanic crust onto the continental margin.

The first factor is documented in this paper (Figures 6 and 7) and the earlier papers cited. The second factor requires that the postulated overpressured zone continue to exist so that the present sole fault will grow westward. It also requires that the buried basalt hill project above the zone of overpressuring so that detachment, instead of following the basalt-sediment contact over the hill, will tend to

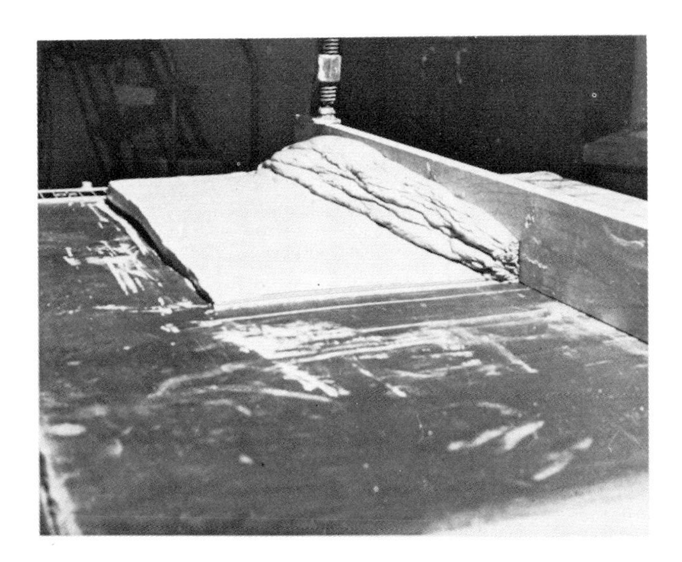

Figure 9. Clay model showing effect of thickness on slope steepness.

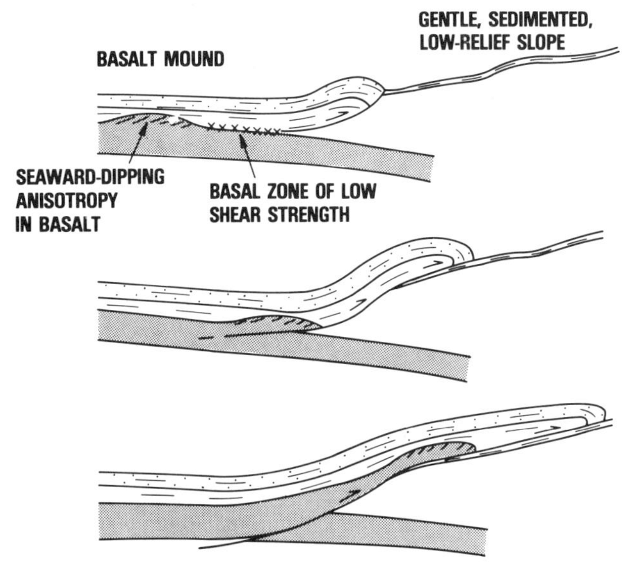

Figure 11. Some conditions possibly favoring development of obducted ophiolites.

exert a shear force on the hill. The third factor is necessary so the basal thrust will propagate downward into the basalt and underlying ultramafics rather that clip the hill off to form a large "flake." Seaward-dipping reflectors near the interpreted basalt-sediment contact are prevalent from the west end of Figure 6a (Figure 6c), westward into the Cascadia basin. Their physical significance is unknown, as is the degree to which they represent the required physical anisotropy. The fourth requirement is met by the unusually low total bathymetric relief (2600 meters) and gentle average slope (about two degrees) off Washington.

The Nongravitational Origin of Trench-Slope Structure

Both seaward and landward vergence are presently developing below the inner slope of the eastern Aleutian trench. To document the vergence, three deep-penetration seismic sections from that area are presented, beginning with the westernmost and ending with the easternmost.

In the westernmost section (Figure 12), which has depth as its vertical scale, the lowermost part of the slope displays evidence for landward vergence in the slight seaward tilting of sediments at the edge of the trench. The tilting is seen in the strong shallow reflector shown by the upper dotted line on the figure. The remainder of the slope, however, indicates seaward vergence that is manifested in the prevailing landward dips. The sawtooth lower edge of the seismic section is caused by relatively strong landward-dipping reflectors used by seismic processors to calculate the velocities necessary to convert time to depth.

The centrally located section (Figure 13) is also a depth section. Seaward vergence predominates beneath all except the top of the slope, where landward vergence is shown by small-displacement thrusts cutting upper-slope sediments. A basalt "hill" similar to one beneath the abyssal plain appears to have been underthrust below the lower slope, and continuation of the basalt farther landward is indicated by a distinct, low-frequency, discontinuous reflection.

The easternmost section (Figure 14) has time as its vertical scale and is of particular interest because it clearly shows landward vergence in the upper slope, reaching more than

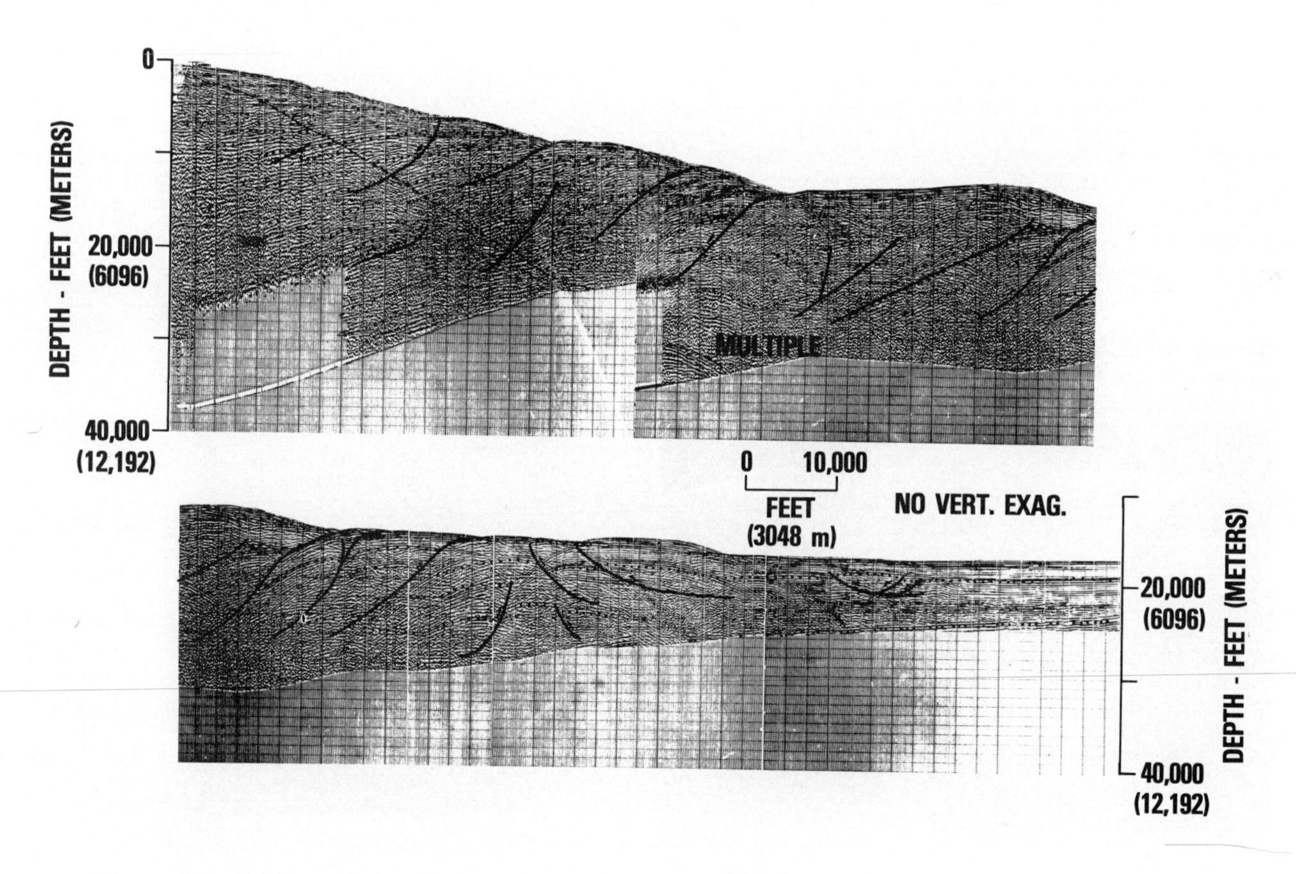

Figure 12. Westernmost seismic section of eastern Aleutian trench and slope. Nonmigrated depth section. Dots show interpreted bedding configuration.

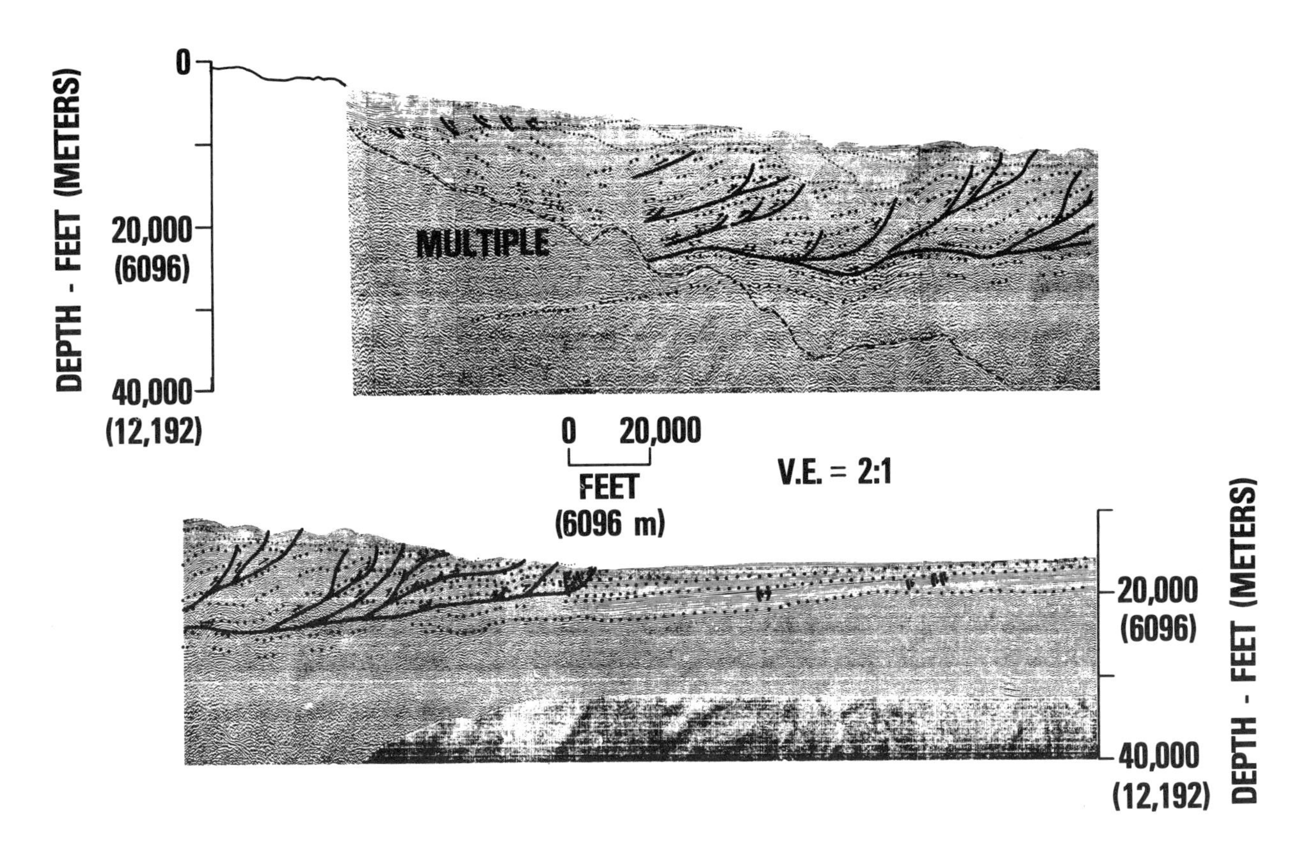

Figure 13. Central seismic section of eastern Aleutian trench and slope. Nonmigrated depth section. Dots show interpreted bedding configuration.

60 km from the trench, and contemporaneous seaward vergence in the lower slope. The upper-slope landward vergence has bathymetric expression (Figure 14). It directly refutes the conclusion recently stated by Elliott (1976) that "thrusts always move in the direction of surface slope." It further suggests that the driving force for these thrusts is "compression and drag from the downgoing slab," to use Elliott's words, and is not a result of an "increased surface slope produced by processes associated with the magmatic arc," as Elliott has suggested as a gravitationally based alternative cause for them. Their landward vergence may be developing as a result of a low-shear-strength layer beneath the slope or may be caused by rejuvenation of older, deeper structures having similar vergence.

The mechanics of the formation of structures beneath trench inner slopes appear to be best described by Chapple (1975), who applied them to overthrust orogenic belts. The importance of contrasts in the physical properties of the compressed sediments emphasized by Hubbert and Rubey (1959) is one of the parameters used by Chapple. It is a factor which the present study also suggests is of prime importance. Chapple's approach has the additional advantage that it does not require a correlation between slope direction and vergence or slope position and the occurrence of extensional or compressional structures. (Except for bedding-plane gliding in surficial upper-slope sediments, the upper parts of trench inner slopes generally are underlain at relatively shallow depths by compressional-not extensional-structures.)

Additional evidence that underthrusting and not gravity is the primary cause of trench-slope structuring is the nonparallelism between slope structures and the trend of the slope. An en echelon pattern is referred to by Barnard (1973), and such a pattern emerges from an interpretation of the lower-slope faults mapped by Silver (1971) off Washington (Figure 15). The pattern is also suggested by the relief created by the young, lower-slope structures evident in the bathymetry illustrated by McManus (1964).

A more obvious lack of parallelism between slope structures and slope trend is evident off northwestern Colombia (Figure 16). Here the active fold belt on land to the west of the Magdalena River continues across the shelf and diagonally down the slope to the edge of the abyssal plain. Gravitational body forces acting because of the slopes on which they occur can hardly have generated these folds. Perhaps a clay model of oblique convergence (Figure 17) comes closer to explaining the en echelon lower-slope patterns observed off Washington and Colombia than does the gravitational model. The

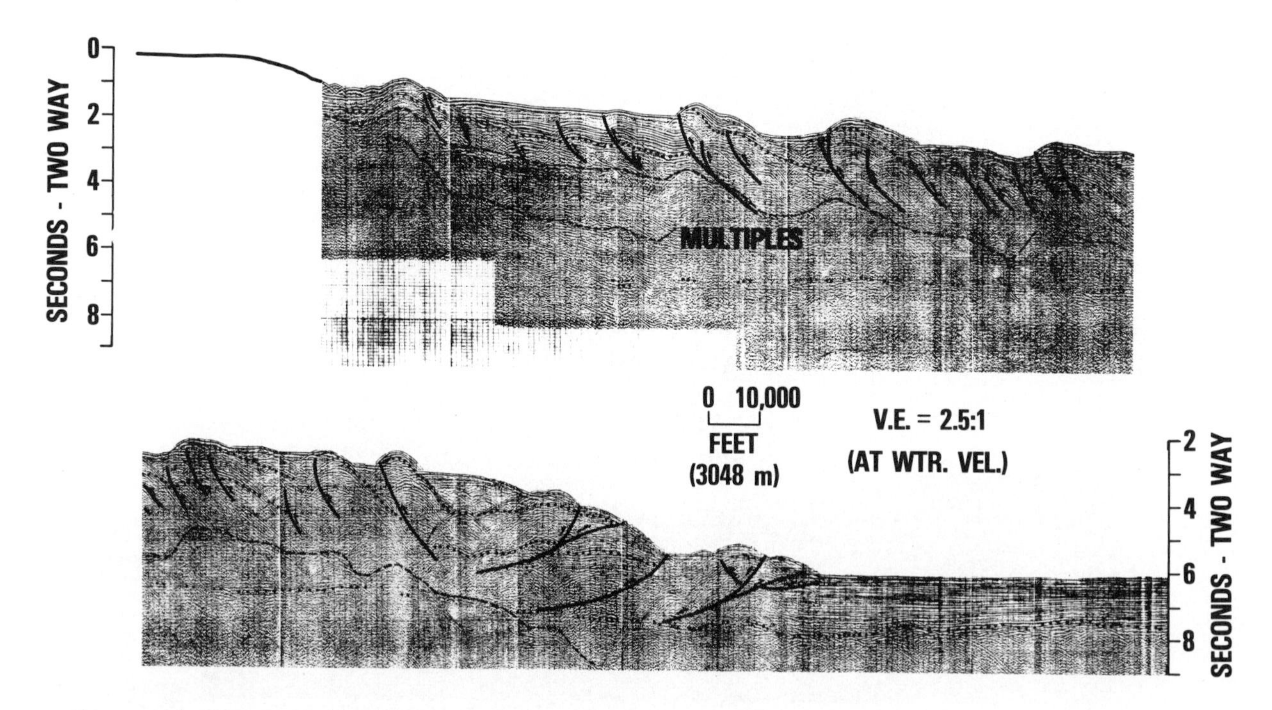

Figure 14. Easternmost seismic section of eastern Aleutian trench and slope. Time section. Dots show interpreted bedding configuration.

clay moved from left to right, converging on the barrier at an angle of 50 degrees. Viewed from above, the clay surface shows a high forming against the barrier with en echelon structures branching from it. These structures appear to terminate in the unstructured clay in a manner similar to the terminations of lower-slope structures against the abyssal plain described above.

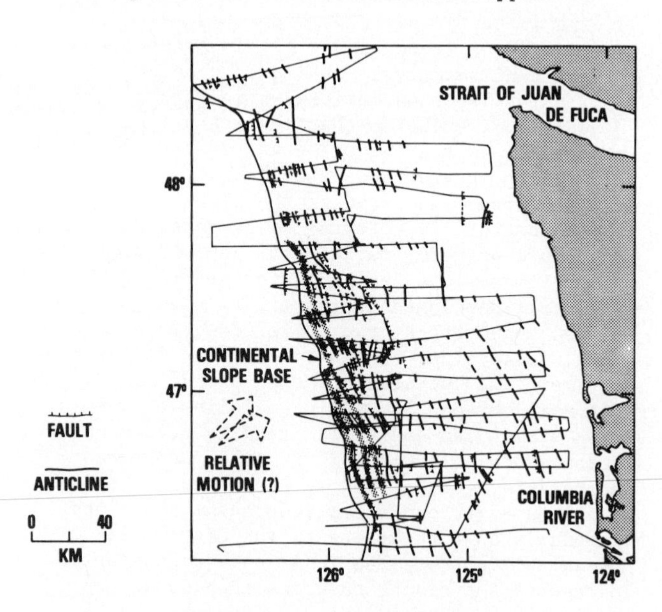

Figure 15. Interpreted en echelon pattern of thrust-anticlines near base of Washington continental slope. Map of faults and anticlines by Silver (1973).

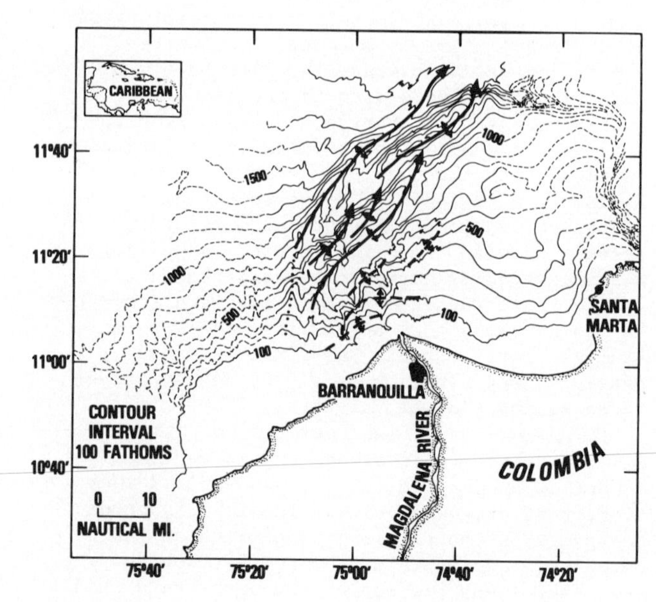

Figure 16. Growing folds having bathymetric expression on the slope off the mouth of the Magdalena River. R. A. Hoover (pers. comm., 1974), based on seismic data.

Figure 17. Clay model of en echelon patterns produced by oblique convergence (not to scale).

Summary

Zones of low shear strength in trench and slope sediments play a major role in the potential development of landward vergence. These zones may be caused by overpressuring due to rapid deposition of trench sediments. They also play a potentially important role in the determination of slope steepness. Under certain conditions, landward vergence may lead to obduction of oceanic crust. Other conditions possibly leading to obduction, such as compression within the oceanic plate, were not discussed. Landward vergence on the upper parts of trench slopes and the obliquity of structural trends across these slopes can be explained by plate-tectonic underthrusting and not as purely gravitational effects.

Acknowledgments. The writer would like to express his appreciation to R. S. Bishop and R. J. Page for assistance in preparing this paper, to D. E. Karig and D. E. Hayes for critically reading it, and to Exxon Company, U.S.A. and Exxon Production Research Company for encouraging its release.

References

Atwater, T., Implication of plate tectonics for the Cenozoic tectonics of western North America, Geol. Soc. Amer. Bull., 81, 3513-3536, 1970.

Barnard, W., Late Cenozoic sedimentation on the Washington continental slope, Univ. of Washington dissert., p. 82, 84-88, 132, 1973.

Beck, R. H., The oceans, the new frontier in exploration, Australian Petroleum Exploration Assoc., 12, pt. 2, p. 5-28, 1972.

Billings, M. P., Structural geology third ed., p. 232-233, Prentice-Hall, New Jersey, 1972.

Bredehoeft, J. D., and B. B. Hanshaw, On the maintenance of anomalous fluid pressures: I. Thick sedimentary sequences, Geol. Soc. Amer. Bull., 79, 1097-1106, 1968.

Carson, B., J. Yuan, and P. B. Myers, Jr., Initial deep-sea sediment deformation at the base of the Washington continental slope: a response to subduction, Geology, 2, 561-564, 1974.

Chapple, W. M., Mechanics of thin-skinned fold and thrust belts (abstract), EOS Trans. AGU, 56, 457, 1975.

Coleman, R. G., Plate tectonic emplacement of upper mantle periodites along continental edges, J. Geophys. Res., 76, 1212-1222, 1971.

Dewey, J. F., Ophiolite obduction, Tectonophysics, 31, 93-120, 1976.

Elliott, D., Motion of thrust sheets, J. Geophys. Res., 81, 949-963, 1976.

Hafner, W., Stress distributions and faulting, Geol. Soc. Amer. Bull., 62, 373-398, 1951.

Hayes, D. E., and M. Ewing, Pacific boundary structure, in The Sea, edited by A. E. Maxwell, Wiley, New York, 29-72, 1970.

Hayes, J. B., Clay petrology of mudstones, leg 18, Deep Sea Drilling Project, in Kuhn, L. D., R. von Huene, et al., Initial reports of the Deep Sea Drilling Project, 18, Washington (U.S. Gov't. Printing Office), 903-914, 1973.

Hubbert, M. K., Mechanical basis for certain familiar geologic structures, Geol. Soc. Amer. Bull., 62, 355-372, 1951.

Hubbert, M. K., and W. W. Rubey, Role of fluid pressure in mechanics of overthrusting faulting, Geol. Soc. Amer. Bull., 70, 115-166, 1959.

Isacks, B. L., J. Oliver, and L. R. Sykes, Seismology and the new global tectonics, J. Geophys. Res., 73, 5855-5899, 1968.

Kulm, L. D., and G. A. Fowler, Cenozoic sedimentary framework of the Gorda-Juan de Fuca plate and adjacent continental margin - a review, in Modern and ancient geosynclinal sedimentation; edited by Dott, R. H., Jr., and R. H. Shaver, SEPM Spec. Pub. 19, 212-229, 1974.

Kulm, L. D., R. A. Prince, and P. D. Snavely, Jr., Site survey of the northern Oregon continental margin and Astoria fan, in Kulm, L. D., R. von Huene, et al., Initial reports of the Deep Sea Drilling Project, 18, Washington (U.S. Gov't. Printing Office), 979-986, 1973.

Lancelot, Y., and R. W. Embley, Mud diapirism in deep oceanic basins (abstract), Geol. Soc. Amer. Ann. Mtg., 1159, 1975.

McManus, D. A., Major bathymetric features near the coast of Oregon, Washington and Vancouver Island, Northwest Sci., 38, 65-82, 1964.

Roberts, J. L., The mechanics of overthrusting faulting: a critical review, in International Geological Congress, 3, 593-598, 1972.

Seely, D. R., P. R. Vail, and G. G. Walton,

Trench slope model, in Geology of Continental Margins, edited by C. A. Burk and C. L. Drake, Springer, New York, 249-260, 1974.

Silver, E. A., Transitional tectonics and late Cenozoic structure of the continental margin off northernmost California, Geol. Soc. Amer. Bull., 82, 1-22, 1971.

Silver, E. A., Pleistocene tectonic accretion of the continental slope off Washington, Marine Geology, 13, 239-249, 1972.

von Huene, R., and L. D. Kulm, Tectonic summary of leg 18, in Kulm, L. D., R. von Huene, et al., Initial reports of the Deep Sea Drilling Project, 18, Washington (U.S. Gov't. Printing Office), 961-976, 1973.

Wilson, J. Tuzo, The development and structure of the crust, in The earth as a planet, edited by G. P. Kuiper, Chicago, Univ. Chicago Press, 138-214, 1954.

SEDIMENT SUBDUCTION AND OFFSCRAPING AT PACIFIC MARGINS

David W. Scholl, Michael S. Marlow and Alan K. Cooper

U.S. Geological Survey, Office of Marine Geology, 345 Middlefield Road, Menlo Park, CA 94025

Abstract. The concept of subduction of oceanic lithosphere requires that pelagic beds deposited in oceanic areas, and terrigenous sequences laid down in trenches and adjacent abyssal plains, are either consumed at trenches or accreted to inner trench slopes via offscraping. Offscraping is generally considered to be the dominant process, which has resulted in the concept that many Pacific margins have been uplifted to form foldbelts of offscraped deep sea deposits (i.e., eugeosynclinal sequences). In some trench areas offscraping appears to play a major role in forming the adjacent landward margins. However, other evidence implies that subduction rather than offscraping of sedimentary masses may occur at inner slopes.. The bulk of many margins and flanking foldbelts may, therefore, consist of continental margin sequences deformed in situ. For example, (1) the turbidite sequences of many Pacific trenches are unusual ice-age deposits formed at times of low sea level. Similar terrigenous sequences must have formed only infrequently in the past. It is unlikely, therefore, that many Pacific margins and adjacent foldbelts are substantially formed of offscraped trench deposits. (2) Thick sequences of deformed rocks underlie the inner slopes of many trenches, but these sequences can exist whether or not a thick (1-2 km) section of "offscrapable" deposits overlies the adjacent sea floor. (3) Most Pacific trenches contain pelagic rather than terrigenous deposits. If offscraping was the dominant margin-forming process in the Mesozoic and Tertiary, then, considering that more than 7000 km of lithosphere have underthrust most Pacific margins during this time, large volumes of pelagic beds should be exposed in coastal mountains. In fact, they contain only minor amounts of these deposits, which must have been lost via subduction. (4) Pelagic deposits that do occur in coastal mountains typically are the same age as associated terrigenous beds. However, if pelagic beds are oceanic offscrapings, then some of these outcrops should occur as tectonic blocks that are as much as 100 m.y. older than enclosing terrigenous masses. (5) The dominantly terrigenous deposits of many Pacific abyssal plains are interbedded with significant amounts (10-50%) of pelagic units. It is questionable, therefore, that offscraped abyssal plain sequences form any significant part of the exposed rocks of coastal foldbelts. (6) Some coastal mountains that have been underthrust during most of Mesozoic and Tertiary time (e.g., Chile, southern Peru, and Mexico) are underlain by Paleozoic and older continental-type basement complexes. The presence of these older rocks implies that sedimentary masses reaching adjacent trenches were subducted rather than offscraped, and that tectonic erosion, not accretion, has occurred at the base of the margin.

Introduction

Investigations of the tectonically active continental and island-arc margins that rim much of the Pacific have provided relatively few unambiguous clues clarifying their formative histories (Burk and Drake, 1974). In recent years the conceptually formulated (or paradigmatic) axioms of sea-floor spreading and plate tectonics have provided more insight into the explanation of geological and geophysical observations at active Pacific margins than have the observations themselves. Sediment offscraping and subduction, which have opposite geologic ramifications, are the two most commonly called upon axioms.

Offscraping involves the tectonic skimming of sedimentary and igneous rocks from the upper part of oceanic lithosphere as it is subducted beneath the leading edge of an active margin (fig. 1). The skimmed slices or piles of rock and sediment are folded against and partially thrust beneath older underthrust slices. This accretionary process forms part of a subduction complex and causes the margin to thicken and grow seaward with time. Offscraping provides a mechanism for the continuous upward displacement of oceanic sediment and igneous rock to form new continental or island-arc crust. The geologic significance of the offscraping process, and evidence favoring it, have been cogently present by Hamilton (1969), Dickinson (1971 a,b,c), Hsu (1971), Dickinson (1973), Seely and others (1974), Kulm and Fowler (1974), Karig and Sharman (1975), and Moore (1975), as well as others.

Subduction is the tectonic process whereby a segment of lithosphere is inserted partially or

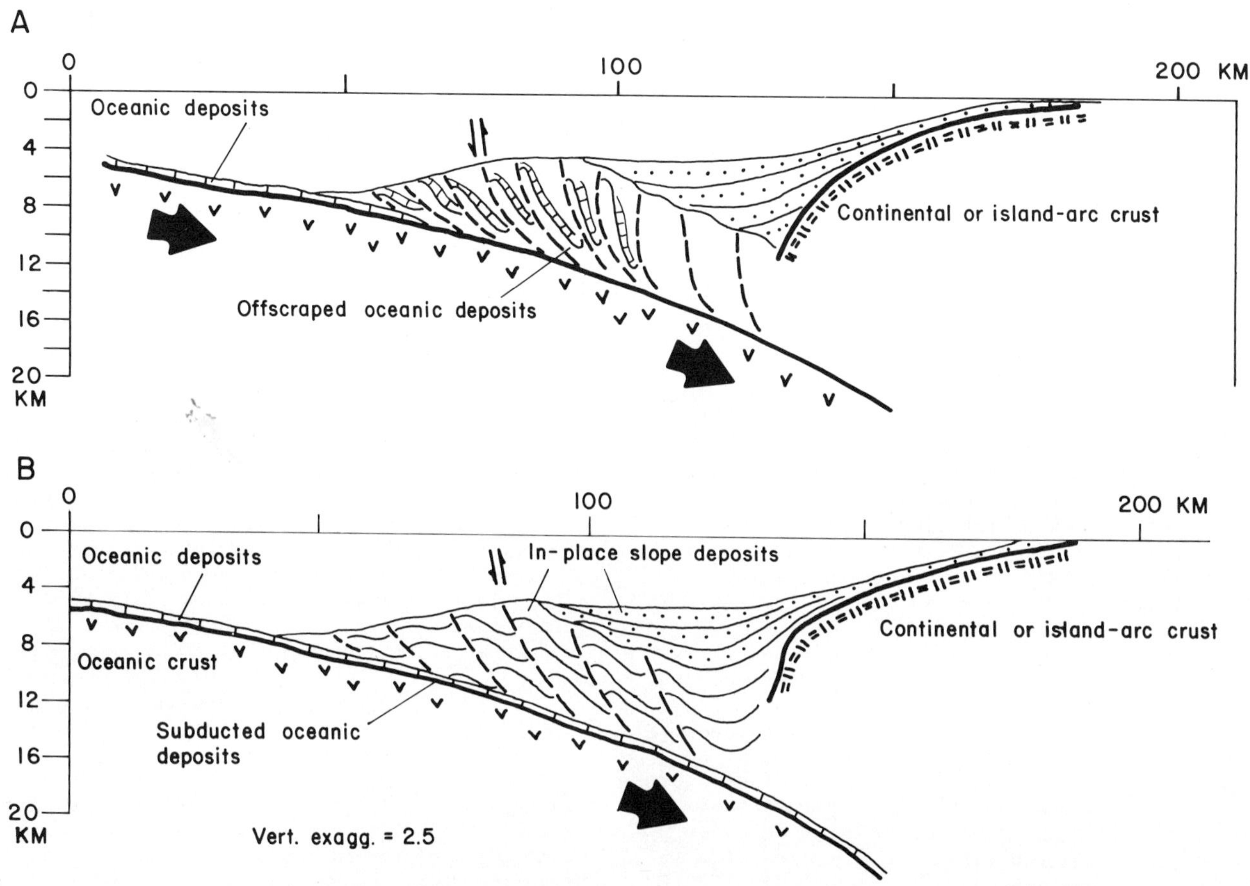

Fig. 1. Conceptual drawings showing how the structural-stratigraphic framework of margins underthrust by oceanic lithosphere (arrows) can be formed chiefly as the result of (A) the offscraping of oceanic beds deposited on the underthrusting crust, or (B) the subduction of these beds and the in-place deformation of continental or island-arc slope deposits. Offscraped oceanic deposits include pelagic beds, and the largely terrigenous sequences of abyssal plains and trench wedges.

totally below that of an adjacent one.* At active Pacific margins subduction involves the sinking or underthrusting of oceanic lithosphere beneath that of an adjacent continent or island-arc (fig. 1). Under certain circumstances, the sedimentary deposits overlying subducting oceanic crust may, therefore, be subducted along with the underlying lithospheric rocks.

Sediment subduction provides a mechanism for the downward displacement of oceanic or deep-sea deposits of depths of ten's or more kilometers beneath active margins. Because subducted sediments can reach mantle depths, they need not significantly contribute to the thickening or seaward growth of an active margin. Presumably, subducted sediment can return to the earth's surface only as highly disrupted and metamorphosed rocks. Scholl and Marlow (1974 a,b,c) point out many of the geologic implications of the subduction of oceanic sedimentary deposits, and direct attention to the likelihood that it occurs at many Pacific margins. Coats (1962) was among the first to conceptualize the process at Pacific margins.

*The tectonic termed "subduction" is generally construed to mean all of the direct as well as the indirect processes and consequences of a descending lithosphere (i.e., it encompasses both sediment offscraping and subduction, White and others, 1970, p. 3431; Davis and others, 1974, p. 20). However, in this paper sediment subduction refers specifically to only those processes that involve the downward dragging or deep subcrustal insertion of sedimentary rocks below insular or continental margins. Subduction, as a complex tectonic process, should not be confused with the concept of a subduction zone White and others, (1970). Insofar as this paper is concerned, the subduction zone comprises that part of an active margin (see Dickinson, 1973, 1974) composed of offscraped deep-sea deposits or deformed continental slope deposits that have been tectonically affected by the subduction of oceanic lithosphere.

The relative dominance of offscraping and subduction of deep-sea sedimentary deposits at modern Pacific margins bears greatly on concepts about the depositional and tectonic settings of the so-called Pacific eugeosynclinal complexes, many of which are underlain by oceanic crust (i.e., the ensimatic or thalassogeosynclinal rocks of Bogdanov, 1969). The exposed rocks of these geosynclinal masses are principally deformed graywacke and related finer-grained terrigenous beds of late Paleozoic, Mesozoic and Paleogene age. It is commonly assumed that these rock assemblages, typified by the Franciscan complex of coastal California, represent uplifted subduction complexes; that is, the elevated rock sequences of ancient, active Pacific margins. If offscraping dominated at these margins, then the deformed sedimentary sequences are chiefly the tectonically stacked accumulations of trenches and adjacent abyssal plains (Dickinson, 1973). If subduction dominated, then the folded beds are the in-place deformed depositional sequences of either former continental and island-arc margins, or those of rear- or intra-arc basins (Scholl and Marlow, 1974a,b).

Subduction of deep-sea deposits at Pacific margins

Six arguments are presented below that imply oceanic or deep-sea deposits (i.e., those that accumulate in oceanic areas above igneous oceanic crust) do not contribute significantly to either (1) the geologic framework of many active margins, or (2) the exposed rocks of many Pacific-rimming eugeosynclinal belts. These two conclusions are interdependent, although not exclusively so, and favor the proposition that subduction of deep-sea deposits occurs commonly at active margins. These arguments are not advanced to disprove the offscraping hypothesis, which is favored by geological and geophysical data in areas of thick oceanic deposits (Silver, 1972; Seely and others, 1974; Kulm and Fowler, 1974; Curray and Moore, 1974). We believe, however, that the offscraping model, the more popular of the two, has been applied too widely and liberally. Our intention is to sharpen the focus of debate about the relative importance of the two subduction related processes by emphasizing that the "tectonic coin" has two sides.

Modern trench wedges--their special significance

Many sectors of the trenches bordering the Pacific contain a seaward-thinning wedge of late Cenozoic turbidite deposits (figs. 2, 5; Scholl and Marlow, 1974a; Scholl, 1974). These terrigenous beds are bonafide or in-place trench deposits, i.e., they are not allochthonous units tectonically conveyed there by sea-floor spreading. Turbidite wedges are thickest, typically 500-1500 m, adjacent to the trench's inner wall.

Investigations of trench wedges indicate their exceptional thickness and lateral extent in western and northern Pacific trenches is a reflection of the special conditions brought about by episodes of Quaternary alpine and continental glaciation (Scholl and others, 1968, 1970; Scholl and Marlow, 1974a,b; von Huene, 1974). Thick, laterally continuous wedges, are also virtually restricted to trench sectors lying seaward of extensively glaciated coastal terranes (Scholl and Marlow, 1974a, fig. 1). The special conditions imposed by global glaciation are greatly increased subaerial erosion rates and repeated world-wide lowerings of sea levels (fig. 2). These two factors, especially eustatic sea level changes, significantly increase the amount of terrigenous detritus that can bypass shelf basins and the continental margin generally and accumulate in deep sea areas. In late Cenozoic time, the geologic and geomorphic effects of the accelerated transfer of coarse terrigenous debris from continental to deep-sea areas were profound and spectacular (Moore and Curray, 1974).

The argument is therefore made that because the special conditions imposed by global glaciation have occurred only infrequently in the geologic past, trench wedges of geologically significant thickness and length also form only infrequently. As a consequence, there is little reason to suppose that thick (>1000 m) turbidite wedges could have contributed significantly to the formation of either modern or ancient active margins.

Trench and abyssal plain sequences--Incidental relation to deformed inner slope sedimentary masses

Geophysical data, especially gravity and seismic reflection and refraction profiles, reveal that the inner slopes of many (most?) trenches are underlain by complexly deformed sedimentary rocks (Beck, 1972; von Huene, 1972; Silver, 1972; Marlow and others, 1973; Grow, 1973; Buffington, 1973; Beck and Lehner, 1974; Seely and others, 1974; Curray and Moore, 1974). This fact has been confirmed at several DSDP (Deep-Sea Drilling Project) sites (Kulm, von Huene and others, 1973; Creager, Scholl and others, 1973; Ingle, Karig and others, 1975). Many of these inner slopes lie adjacent to well-formed turbidite wedges (or older abyssal plain deposits), which, it is fairly argued, could have been offscraped to form a subduction complex of deformed deposits beneath the inner slopes.

Significantly, subduction complexes of deformed sedimentary deposits also lie adjacent to trenches that contain sedimentary sections only a few hundred meters in thickness (fig. 3). Good examples of this relation have been documented off Java (Beck and Lehner, 1974, figs. 20, 21) and the Guatamalan segment of the Middle America Trench (Seely and others, 1974, figs. 5, 6a, 7b). Both examples are especially instructive because neither trench area has a turbidite wedge of terrigenous deposits nor is it flanked by abyssal plain deposits of turbidite beds (fig. 3).

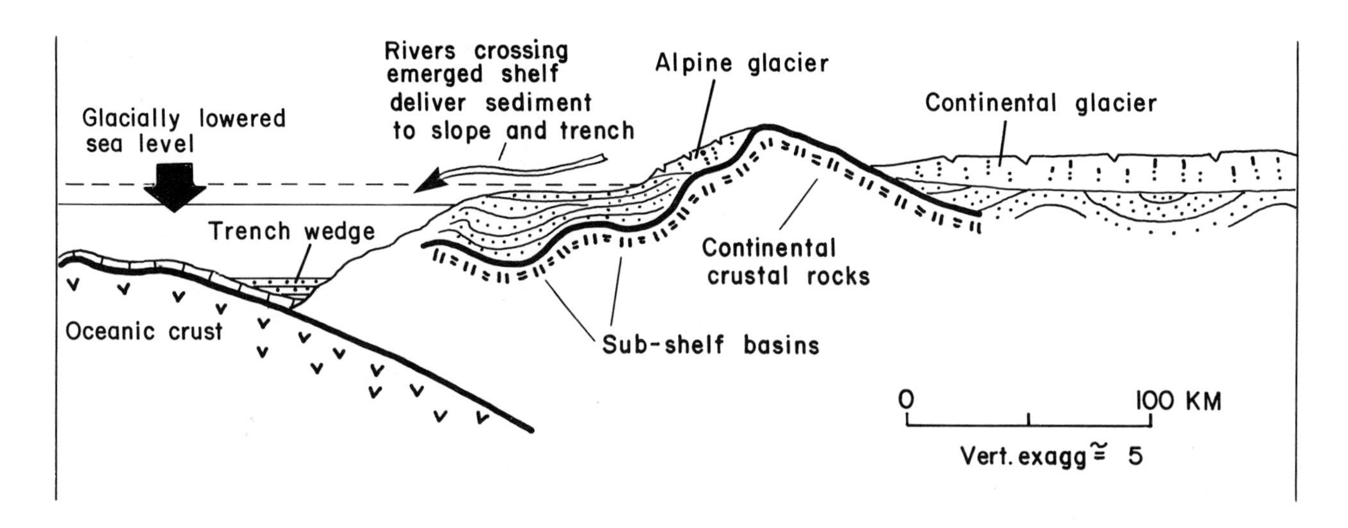

Fig. 2. Special conditions imposed by episodes of global glaciation in late Cenozoic time formed the characteristic turbidite wedges of many Pacific trenches. Continental and coastal alpine glaciation increased onshore erosion rates and eustatically lowered sea level, which allowed sediment-burdened rivers to bypass emerged shelves and contribute terrigenous detritus directly to continental or island-arc slopes. Downslope sliding of the rapidly deposited slope masses formed the trench wedge of mostly turbidite beds. Presumably, thick (>500 m), laterally extensive trench wedges were not typical of Pacific trenches throughout most of Mesozoic and Tertiary time.

These observations direct us to question the notion that the continuous offscraping of a few hundred meters of deep-sea deposits can form a tectonized pile of sediment thick enough (10-20 km) to form the bulk of an active margin--a subduction complex that ultimately may be elevated above sea level as a coastal fold belt of eugeosynclinal rocks. We argue that it is more reasonable to suppose that the thin mantle of oceanic deposits is swept tectonically downward beneath the active margin via subduction processes (fig. 1; Scholl and Marlow, 1974c). Accordingly, where they lie adjacent to thinly mantled oceanic crust, a common circumstance (J. Ewing and others, 1968; M. Ewing and others, 1969; Hayes and Ewing, 1970; Scholl and Marlow, 1974a), the accretionary wedges of Karig and Sharman (1975) and the accretionary fan-structures of Seely and others (1974) may be formed by the in-place deformation of typical continental or insular slope deposits.

Oceanic pelagic deposits--abundant on oceanic crust but uncommon in Pacific fold belts

Pelagic deposits are composed chiefly of particulate or chemically precipitated matter derived from the water column (including circulating hydrothermal fluids) by either organic or inorganic processes or by the settling of finely comminuted detritus derived from distant land areas. As emphasized by Jenkyns and Hsu (1974), pelagic beds can form in shallow-water as well as deep-sea areas, but oceanic pelagic deposits are those that accumulate in deep-water areas above oceanic crust. Most of the Pacific sea floor is overlain by oceanic pelagic beds, none, apparently, older than Triassic(?) (Larson and Chase, 1972).

The oceanic crust underlying Pacific trenches is typically overlain by several hundred meters of pelagic deposits (Scholl and Marlow, 1974a). However, along the eastern margins of the Pacific, adjacent to North and South America, the pelagic sequence is commonly overlain by a 500-1500-m-thick wedge of Pleistocene turbidites (Scholl and Marlow, 1974a, Table 1).

During the past 100-200 m.y. most active margins of the Pacific have been underthrust by as much as 7000-10,000 km of Pacific lithosphere (Larson and Chase, 1972; Larson and Pitman, 1972). Potentially, several million cubic kilometers of pelagic beds could have been offscraped along each 1000-km long segment of active margin (Scholl and Marlow, 1974a). These are geosynclinal volumes. However, the typical Pacific eugeosynclinal sequence, the raised subduction complex of many, comprises only a few percent (of exposed rocks) of pelagic-type deposits (Bailey and others, 1964; Burk, 1965; Bogdanov, 1969; Kimura, 1973, 1974; Kanmera, 1974; Blake and others, 1974).

If offscraping has been a dominant tectonic process at active Pacific margins, then tens of millions of cubic kilometers of oceanic pelagic deposits should be present in Pacific-rimming fold belts (Scholl and Marlow, 1974a). Their low abundance in these fold belts is critical evidence attesting to subduction, rather than offscraping, at many active Pacific margins.

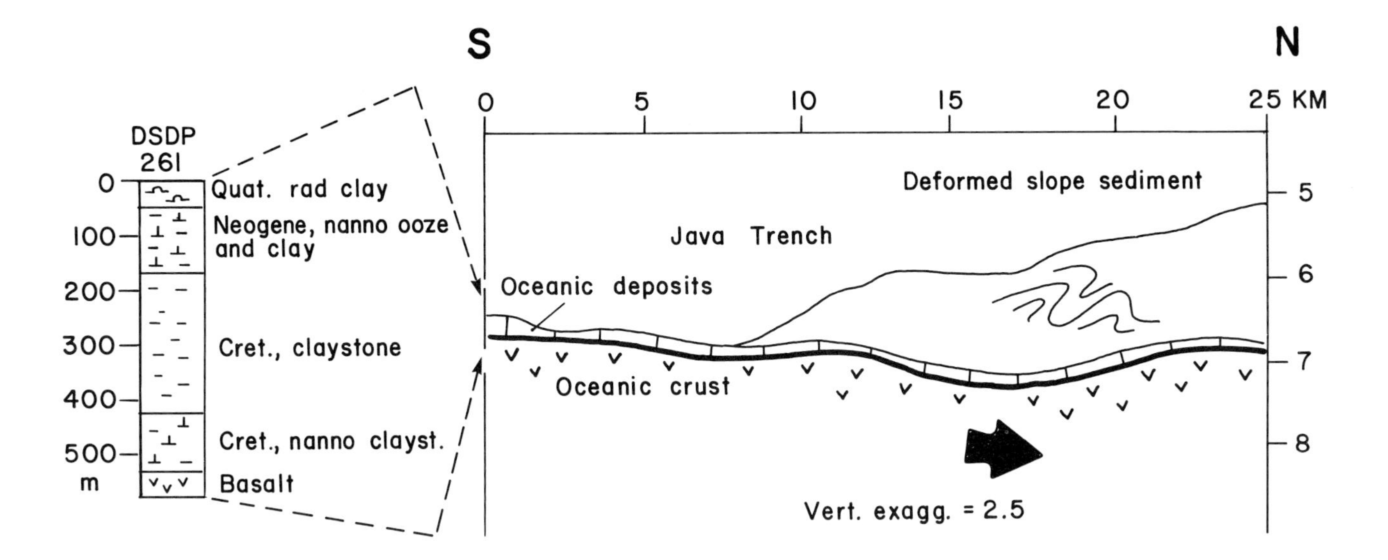

Fig. 3. Schematic structural cross-section of Java Trench based on a seismic reflection profile published by Beck and Lehner (1974, fig. 21). If the deformed beds underlying the inner or north slope of the trench are offscraped oceanic deposits, then considering the absence of a trench wedge of turbidite deposits and the lithologic composition of the stratigraphic sequence penetrated at relatively nearby DSDP site 261 (Veevers, J. J., and Heirtzler, J. R. and others, 1974), much of the mass of the inner slope may be constructed of tectonically stacked pelagic deposits. Alternatively, as this illustration portrays, the oceanic deposits may be subducted along with the underthrusting (arrow) oceanic crust. This model implied that the inner slope is underlain by continental slope deposits deformed in place.

Pelagic deposits in Pacific fold belts--age equivalent with associated terrigenous units

Most of the outcrops of pelagic rocks exposed in Pacific eugeosynclinal belts are chert and limestone beds associated with age-equivalent terrigenous sedimentary deposits. This relation is typical of the Franciscan complex of western United States (Raymond, 1974; Blake and Jones, 1974; Blake and others, 1974), perhaps the type example of an elevated subduction complex (Hsu, 1971), and the generally similar rocks of the Kamchatka-Koryak area (Gladenkov, 1964; Bogdanov, 1970; Avdeiko, 1971), and New Zealand (Landis and Bishop, 1972). Some of these deposits accumulated at water depths thousands of meters shallower than those characteristic of oceanic areas (Matthews and Wachs, 1973; Wachs and Hein, 1975).

Although the pelagic deposits of Pacific eugeosynclinal assemblages occur interbedded with terrigenous and volcanic sequences, in the Franciscan complex chert occurs more abundantly as isolated blocks within incoherently or chaotically bedded mélange sequences (fig. 4; Bailey and others, 1964; Blake and Jones, 1974; Blake and others, 1974; Raymond, 1974). Because these Franciscan chert masses are allochthonous bodies, they can be viewed as scraps of oceanic beds stripped from subducted lithosphere, and, therefore, as evidence favoring the concept that the Franciscan rocks are a subduction complex of offscraped trench and abyssal plain deposits. The merits of this important idea have been emphasized repeatedly by Hamilton (1969), Hsu (1971), Chipping (1971), Raymond (1974), as well as many others.

The Franciscan complex, like many other Pacific-rimming eugeosynclinal accumulations, was underthrust by at least 7000 km of oceanic crust during a formative period that lasted approximately 100 m.y. During most of this time the associated spreading center stood more than a thousand kilometers seaward of the Franciscan margin (Larson and Pitman, 1972). If offscraping had dominated at the Franciscan trench, then scraps of oceanic pelagic deposits ten to more than 100 m.y. older than associated terrigenous deposits should have been incorporated from time to time into the growing tectonic pile of the Franciscan subduction complex. However, Pessagno (1973) has made the startling discovery that the age of the Franciscan chert masses are virtually all the same, Late Jurassic (Tithonian) to possibly Early

Cretaceous (Neocomian); and that this narrow age range is equivalent to that of chert beds associated with fragments of oceanic crust (ophiolite successions) preserved in the California Coast Ranges (fig. 4). Blake and Jones (1974) conclude from this evidence that the bulk of the chert masses do not represent oceanic offscrapings but rather basal eugeosynclinal rocks that have been repeatedly disrupted and injected into younger structural and stratigraphic levels.

Expanding on the contention of Blake and Jones (1974), we argue more generally that the structural, stratigraphic and age relations of the pelagic deposits of most Pacific fold belts imply that their pelagic beds are typical geosynclinal deposits. These deposits accumulated in both deep and shallow-water areas in the immediate vicinity of an active Pacific margin or rear-arc basin. Because pelagic limestone and chert beds can accumulate depositionally within a Pacific-rimming eugeosynclinal environment (Kanmera, 1974; Wachs and Hein, 1975; Garrison, 1974; Garrison and others, 1975), it is not necessary to presume that they formed in oceanic areas hundreds and thousands of kilometers seaward of Pacific margins. The virtual total absence in Pacific fold belts of tectonic scraps of oceanic pelagic deposits representing ages considerably older than associated terrigenous beds is, therefore, powerful evidence that subduction of sedimentary beds occurs beneath Pacific margins (Scholl and Marlow, 1974a; Moore, 1975).

Pacific abyssal plains--form only in special areas and commonly include abundant pelagic beds

The sedimentary sequences of abyssal plains include many terrigenous turbidite beds, hence, it is generally supposed that offscraped abyssal plain sequences can contribute in a major way to the formation of a subduction complex and, ultimately, a Pacific-rimming eugeosynclinal complex. However, it is important to emphasize that abyssal plains overlie only a small fraction of the Pacific sea floor (Menard, 1964). Those that occur are restricted to the northeast Pacific and the Bellingshausen Abyssal Plain (southwestern Pacific) off the tectonically inactive margin of Antarctica (Ewing and others, 1968; Hollister, Craddock and others, 1974). The restricted occurrence of abyssal plains reflects the trapping of coarse-grained terrigenous detritus shed toward the Pacific by marginal seas, in shelf and slope basins lying upslope of Pacific trenches, and, especially during the Pleistocene, in these trenches (Scholl and Marlow, 1974a, b). Abyssal plains can form seaward of active margins if either (1) the trench is overwhelmed by a surfeit of turbidite deposition, or (2) the plain is fed by deep-sea channels and fans that do not cross an active trench (see Scholl and Marlow, 1974a, fig. 6). Overwhelming occurs when coastal drainages, especially of youthful mountain belts are extensively glaciated during periods of low sea level (e.g., southcentral Chile Trench) or by the

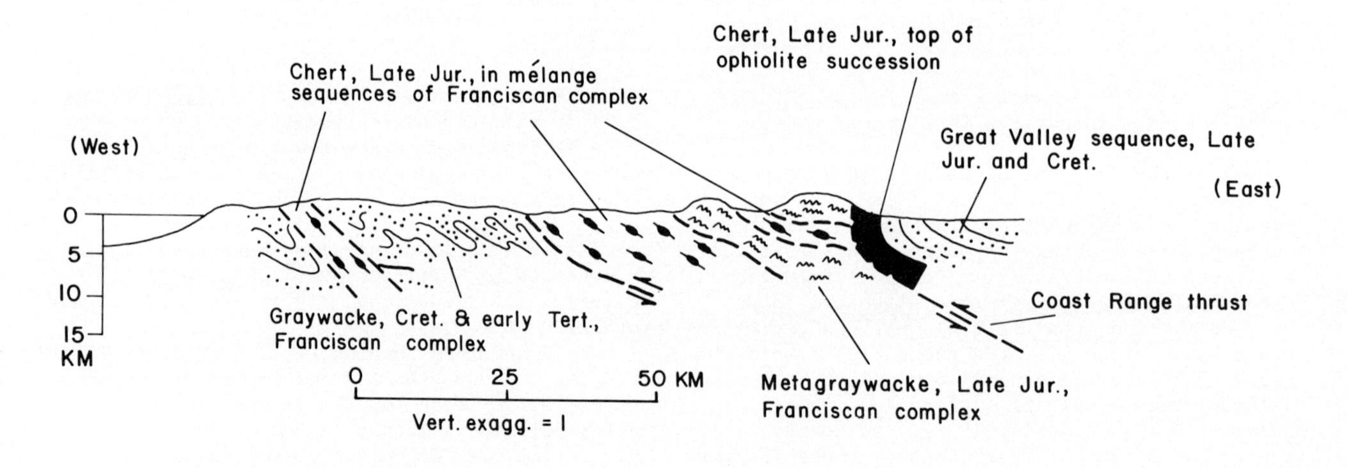

Fig. 4. Diagrammatic structural cross-section of the northern California Coast Ranges; adapted from Figure 3 of Blake and others (1974) and including data from Gucwa (1975). Allochthonous chert masses in Franciscan mélange sequences are virtually all of Late Jurassic to possibly Early Cretaceous age, the same age as associated terrigenous deposits and chert beds capping ophiolite successions (Passagno, 1973). The common age of the cert implies that the mélange chert bodies are disrupted and tectonically recycled geosynclinal deposits derived from a basal ophiolite succession (Blake and others, 1974). There are, therefore, no compelling reasons to believe that the chert masses represent scraps of oceanic pelagic beds skimmed from thousands of kilometers of oceanic crust that underthrust the Franciscan margin from Late Jurassic to early Tertiary time.

contributions of a major river (e.g., the Columbia River and the pre-Pleistocene turbidites of the Cascadia Basin).

Although dominated by terrigenous deposition during the past 5 m.y., the early Miocene through Quaternary sedimentary sequence of the Alaskan Abyssal Plain includes many interbedded pelagic deposits (we estimate 10-15 percent; Kulm, von Huene and others, 1973). The terrigenous sedimentary sequence of Cascadia Basin is also approximately 5 m.y. old, but because of the nearness of the Columbia River and the small size of the basin, the bulk of the section is more richly terrigenous (Kulm, von Huene and others, 1973). In contrast, approximately 50 percent of the section of the areally much larger Aleutian Abyssal Plain consists of oceanic pelagic deposits of Oligocene and younger age. They overlie turbidite beds that formed the plain in Eocene and early Oligocene time (Hamilton, 1967; Creager, Scholl and others, 1973; Hamilton, 1973; Stewart, 1976). About one-quarter of the mostly Cenozoic beds of the Bellingshausen section are pelagic deposits interbedded with terrigenous sand and mud (Hollister, Craddock and others, 1974).

We argue, therefore, that because abyssal plains form infrequently off active Pacific margins, and because prior to subduction those that do form will include abundant oceanic pelagic beds, that off-scraped abyssal plain sequences have not contributed significantly to the formation of many Pacific-rimming eugeosynclinal complex or active margins. We argue further that except where abyssal plain deposits reach unusual thicknesses (>500-1000 m), most Pacific abyssal plain sequences are subducted along with adjacent masses of pelagic and hemipelagic oceanic deposits.

Pacific Fold belts--many not underlain by subduction complexes

Coastal ranges lacking subduction complexes of Mesozoic and Tertiary rocks flank lengthy sectors of the Pacific rim. Some of these sectors have been active margins during much of the past 200 m.y. An excellent example is the coastal area of the eastern Pasific margin bordering southern Peru and virtually all of Chile. The bedrock core of the coastal ranges consists of an early Paleozoic (or older) complex of metamorphic rock and younger plutonic and volcanic masses of Mesozic age (fig. 5). This coastal belt of basement rocks, which includes typical continental crustal rocks and possibly ancient subduction complexes (Ernst, 1975), crops out within 100 km of the Peru-Chile Trench, and, undoubtedly, underlies much of the continental shelf and slope (Scholl and others, 1970). The nearness of this basement complex to the trench and the implications of paleogeologic reconstructions have led many geologists to propose that a strip of continental crust several hundred kilometers wide has disappeared from the western margin of South America (Miller, 1970 a,b Rutland,1971; Katz, 1971; Cobbing and Pitcher, 1972). A similar argument for the foundering and destruction of a coastal strip of active margins can be made for southern Mexico; here a broad terrane of Paleozoic and older basement rock (see King and

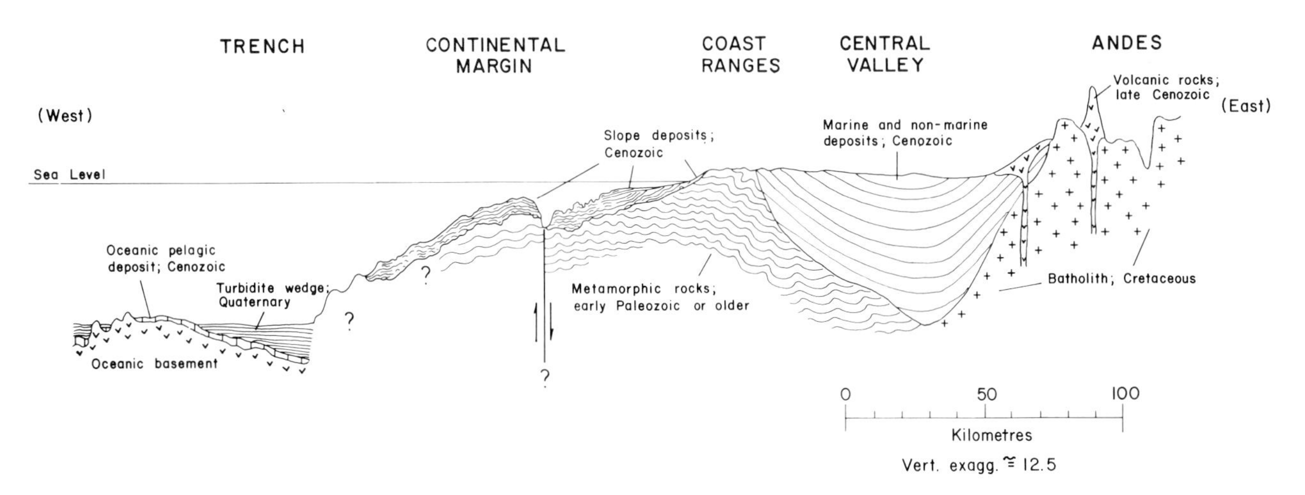

Fig. 5. Diagrammatic structural cross-section from the Peru-Chile Trench eastward to the Chilean Andes (Lat. 41°S). This section (adapted from Scholl and others, 1970, fig. 3) stresses the likelihood that metamorphic basement rock of pre-Mesozoic age underlies much of the Chilean margin, which has been underthrust throughout much of the past 200 m.y. The apparent absence of subduction complexes of Mesozoic and early Tertiary age implies that oceanic deposits have been subducted and that underthrusting has caused tectonic erosion of the leading edge of the continental crust.

others, 1969) is cut off by the trend of the continental slope and adjacent Middle America Trench (C. Burk, personal comm., 1975).

We argue, therefore, that exposed coastal basement complexes in lieu of deep-water subduction complexes are evidence that subduction of oceanic lithosphere has caused tectonic erosion of the leading edge of the continental crust (Scholl and Marlow, 1974b). Subduction of deep-sea deposits evidently accompanied the tectonic reduction of continental crustal rocks.

Overview and concluding remarks

We offer the following generalizations as contributions to the general debate on the tectonic development of active Pacific margins:

(1) Offscraping of terrigenous deposits laid down on oceanic crust is an important factor in forming active margins and Pacific-rimming foldbelts only where a fairly thick and volumetrically large sedimentary sequence (>500-1000 m) is swept tectonically against an island arc or continental margin.

(2) Subduction of deep-sea deposits takes place at those active margins flanked by igneous oceanic crust thinly (<300 m) mantled with sedimentary beds. This relation is typical of most western Pacific margins. Prior to the Pleistocene, when most Pacific trenches lacked thick, laterally extensive turbidite wedges, thin sedimentary sequences may have been typical of most peripheral sectors of Pacific crust.

(3) As argued by Moore (1975), a mechanical or strength-related mechanism may cause the selective subduction of pelagic beds and the corresponding selective offscraping of the terrigenous deposits of trench and abyssal plain sequences. However, oceanic pelagic beds even within thick sections, are commonly as weakly lithified as many deep-sea terrigenous beds (Moore, 1972; Schlanger and Douglas, 1974). We cannot, therefore, envision a tectonic mechanism that depends upon the physical properties of oceanic terrigenous and pelagic deposits that would consistently separate them at active margins throughout most of Mesozoic and Cenozoic time. Both pelagic and terrigenous deposits must have been commonly subducted at Mesozoic and Tertiary margins.

(4) Karig and Sharman (1975) show how offscraped trench and abyssal plain deposits can be wedged upward to form the characteristic trench-

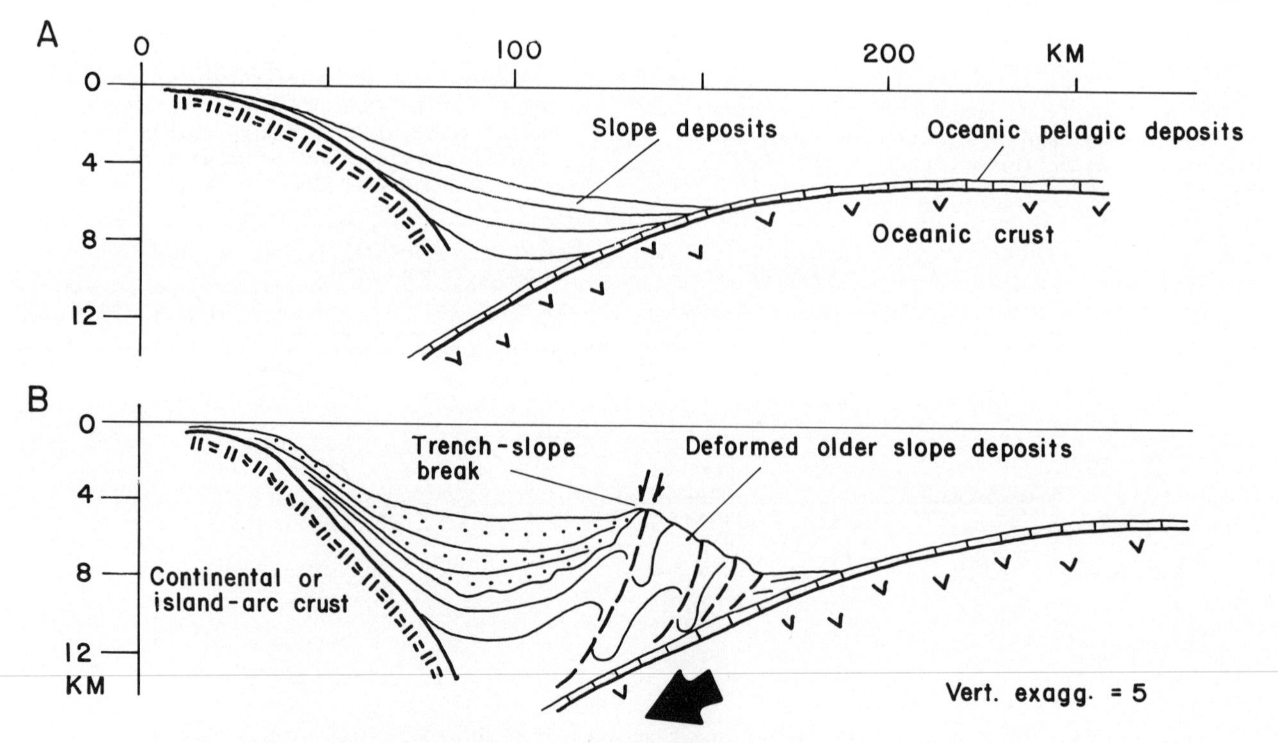

Fig. 6. Prominent trench-slope breaks and slope basins can be formed as a consequence of sediment offscraping (Karig and Sharman, 1975). However, as indicated in this drawing, the trench-slope break can also form as the result of in-place deformation of an apron of sediment deposited at the base of a continental or island-arc slope. As shown here, the sediment apron forms during a period of little or no underthrusting and is deformed and wedged upward during a subsequent phase of oceanic underthrusting.

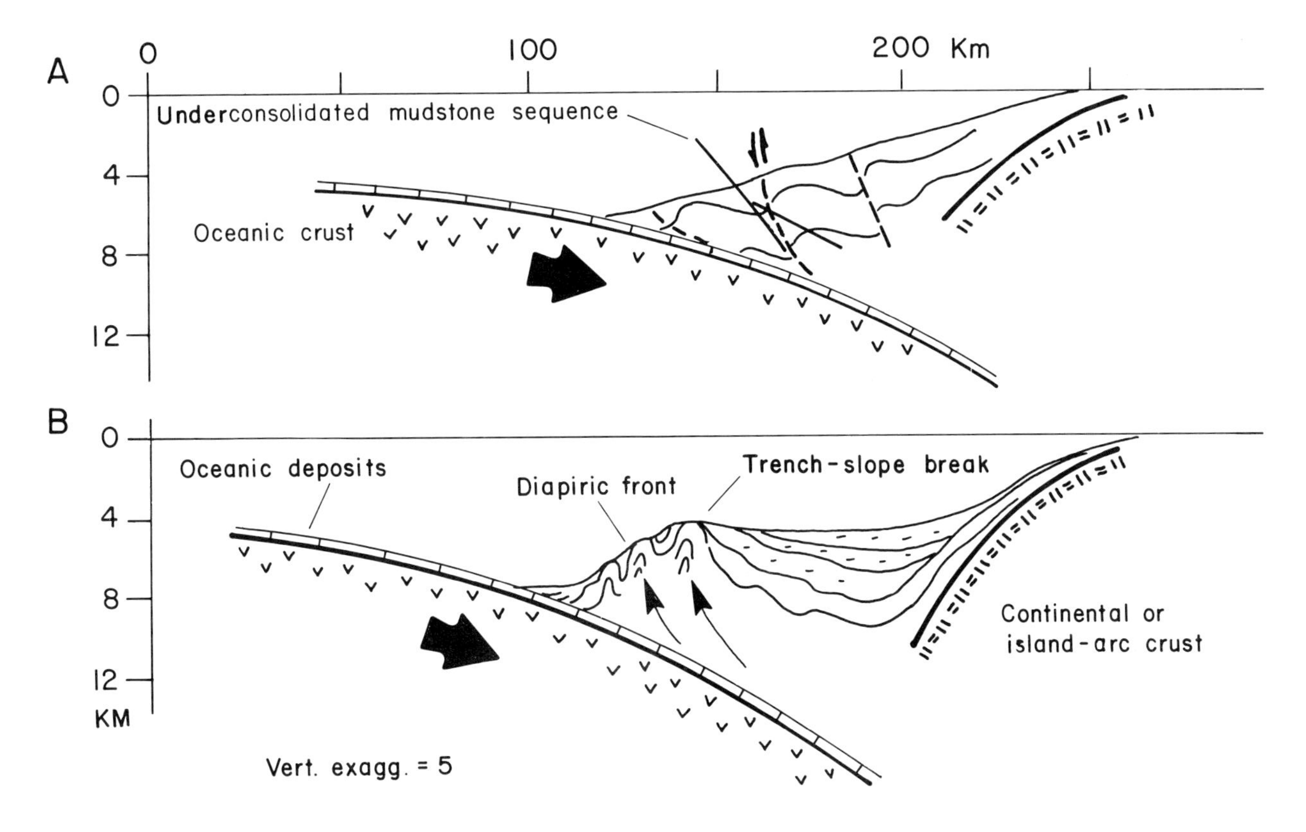

Fig. 7. In this model the trench-slope break forms when a mass of rapidly deposited and underconsolidated mudstone migrates outward and upward to form a diapiric front. This model, like that shown in Figure 6, calls for the in-place deformation of continental or island-arc slope deposits and the subduction of oceanic beds swept into the base of the margin by underthrusting oceanic lithosphere.

slope break (Dickinson, 1973) of many active margins. However, the in-place deformation of typical continental or island arc slope deposits can also explain the trench-slope break. Models of slope evolution illustrating this idea are shown on Figures 5 and 6 (see also Herron, et al, this volume, for an additional model).

REFERENCES

Avdeiko, G. P., Evolution of geosynclines on Kamchatka, Pacific Geology, v. 3, p. 1-4, 1971.

Bailey, E. H., Irwin, W. P., and Jones, D. L., Franciscan and related rocks, and their significance in the geology of western California, Calif. Div. Mines and Geol. Bull. 183, 177 p., 1964.

Beck, R. H., The oceans, the new frontier in exploration, Australian Petrol. Explor. Assoc. Jour., v. 12, p. 1-21, 1972.

Beck, R. H., and Lehner, P., Oceans, new frontier in exploration, Amer. Assoc. Petroleum Geologists Bull., v. 58, p. 376-395, 1974.

Blake, M. C. Jr., and Jones, D. L., Origin of Franciscan mélanges in northern California, in Dott, R. H. Jr., and Shaver, R. H. (eds.), Modern and ancient geosynclinal sedimentation Soc. Econ. Paleontologists and Mineralogists, Spec. Pub. 19, p. 345-357, 1974.

Blake, M. C. Jr., Jones, D. L., and Landis, C. A., Active continental margins, contrasts between California and New Zealand, in Burk, C. A. and Drake, C. L. (eds.), The geology of continental margins, Springer-Verlag, New York, p. 853-872, 1974.

Bogdanov, N. A., Thalassogeosynclines of the Circum-Pacific ring, Geotectonics, no. 3, p. 141-147, 1969.

Bogdanov, N. A., Some unusual features in the tectonics of the east of Koryak Uplands, Akad. Nauk, SSSR, Doklady, v. 192, p. 607-610, 1970.

Buffington, E. C., The Aleutian-Kamchatka Trench convergence and investigations of lithospheric plate interaction in the light of modern geotectonic theories (thesis), Los Angeles, Univ. Southern California, Dept. Geol. Sci., 1973.

Burk, C. A., Geology of the Alaska Peninsula--Island arc and continental margin, Geol. Soc. America Mem. 99, 250 p., 1965.

Burk, C. A., and Drake, C. L., Continental margins in perspective, in Burk, C. A., and Drake, C. L. (eds.), The geology of continental margins, Springer-Verlag, New York, p. 1003-1009, 1974.

Chipping, D. H., Paleoenvironmental significance of chert in the Franciscan Formation of western California, Geol. Soc. America Bull., v. 82, p. 1707-1712, 1971.

Coats, R. R., Magma type and crustal structure in the Aleutian arc, in McDonald, G. A., and Kuno, H. (eds.), The crust of the Pacific basin, Amer. Geophys. Union Geophys. Mon. 6, Natl. Acad. Sci-Natl. Research Council Pub. 1035, p. 92-109, 1962.

Cobbing, E. J., and Pitcher, W. S., Plate tectonics and the Peruvian Andes, Nature Physical Sci., v. 240, p. 51-53, 1972.

Creager, J. S., Scholl, D. W., and others, Initial reports of the Deep Sea Drilling Project, v. 19, Wash. D.C., U.S. Govt. Printing Office, 913 p., 1973.

Curray, J. R., and Moore, D. G., Sedimentary and tectonic processes in the Bengal deep-sea fan and geosyncline, in Burk, C. A., and Drake, C. L. (eds.), The geology of continental margins, Springer-Verlag, New York, p. 617-627, 1974.

Davis, G. A., Burchfiel, B. C., Case, J. E., and Viele, G. W., A defense of an "old global tectonics", in Kahle, C. F. (ed.), plate tectonics, assessments and reassessments, Amer. Assoc. Petroleum Geologists Mem. 23, p. 16-23, 1974.

Dickinson, W. R., Plate tectonics in geologic history, Science, v. 174, p. 107-113, 1971a.

Dickinson, W. R., Clastic sedimentary sequences deposited in shelf, slope and trough settings between magmatic arcs and associated trenches, Pacific Geology, v. 3, p. 15-30, 1971b.

Dickinson, W. R., Plate tectonics models of geosynclines, Earth and Planetary Sci. Letters, v. 10, p. 165-174, 1971c.

Dickinson, W. R., Widths of modern arc-trench gaps proportional to past duration of igneous activity in associated magmatic arcs, Jour. Geophys. Research, v. 78, p. 3376-3389, 1973.

Dickinson, W. R., Sedimentation within and beside ancient and modern magmatic arcs, in Dott, R. H. Jr., and Shaver, R. H. (eds.), Modern and ancient geosynclinal sedimentation, Soc. Econ. Paleontologists and Mineralogists, Spec. Pub. 19, p. 230-239, 1974.

Ernst, G. W., Systematics of large-scale tectonics and age progressions in alpine and Circum-Pacific blueschist belts, Tectonophysics, v. 26, p. 229-246, 1975.

Ewing, J., Ewing, M., Aitken, T., and Ludwig, W. J., North Pacific sediment layers measured by seismic profiling, in Knopoff, L., and others (eds.), The crust and upper mantle of the Pacific area, Amer. Geophys. Union Mon. 13, p. 147-173, 1968.

Ewing, M., Ewing, J., Houtz, R. E., and Leyden, R., 1969, Sediment distribution in the Bellingshausen Basin, in Robin, G. (ed.), Symposium on Antarctic Oceanography, Santiago, Chile, Internatl. Union Geodesy and Geophysics, p. 89-100, 1969.

Ewing, M., Houtz, R., and Ewing, J., South Pacific sediment distribution, Jour. Geophys. Research, v. 74, p. 2477-2511, 1969.

Garrison, R. E., Radiolarian cherts, pelagic limestones, and igneous rocks in eugeosynclinal assemblages, in Hsu, K. J., and Jenkyns, H. C. (eds), Pelagic sediments--On land and under the sea, Interntl. Assoc. Sedimentologists Spec. Pub. 1, p. 367-399, 1974.

Garrison, R. E., Schlanger, S. O., and Wachs, D., Petrology and paleogeographic significance of Tertiary nannoplankton-foraminiferal limestones, Guam, Paleogeography, Paleoclimatology and Palaeoecology, v. 17, p. 49-64, 1975.

Gladenkov, Yu. B., On the tectonics of the eastern part of the Koryak Uplands, Acad. Nauk, USSR, Geol. Inst., Trudy, v. 113, p. 7-23 (in Russian), 1964.

Grow, J. A., Crustal and upper mantle structure of the central Aleutian arc, Geol. Soc. America Bull., v. 84, p. 2169-2192, 1973.

Gucwa, P. R., Middle to Late Cretaceous sedimentary mélanges, Franciscan complex, northern California, Geology, v. 3, p. 105-108, 1975.

Hamilton, E. L., Marine geology of abyssal plains in the Gulf of Alaska, Jour. Geophys. Research, v. 72, p. 4189-4213, 1967.

Hamilton, E. L., Marine geology of the Aleutian Abyssal Plain, Marine Geology, v. 14, p. 295-325, 1973.

Hamilton, W., Mesozoic California and the underflow of Pacific mantle, Geol. Soc. America Bull., v. 80, p. 2409-2430, 1969.

Hayes, D. E., and Ewing, M., Pacific boundary structure, in Maxwell, A. E. (ed.), New concepts of sea floor evolution, pt. 2, regional observations concepts, The Seas, v. 4, New York, Wiley-Interscience Publishers, p. 29-72, 1970.

Hollister, C. D., Craddock, C., and others, Deep drilling in the southeast Pacific basin, Geotimes, August, p. 16-19, 1974.

Hsu, K., Franciscan mélange as a model for eugeosynclinal sedimentation and underthrusting tectonics, Jour. Geophys. Research, v. 76, p. 1162-1170, 1971.

Ingle, J. C. Jr., Karig, D. E., and others, Initial reports of the Deep Sea Drilling Project, v. 31, Wash. D.C., U.S. Govt. Printing Office, 1975.

Jenkyns, H. C., and Hsu, K. J., Pelagic sediments--On land and under the sea; an introduction, in Hsu, K. J., and Jenkyns, H. C., (eds.), Pelagic sediments--On land and under the sea, Interntl. Assoc. Sedimentologists Spec. Pub. 1, p. 1-10, 1974.

Kanmera, K., Paleozoic and Mesozoic geosynclinal

volcanism in the Japanese Islands and associated chert sedimentation, in Dott, R., H. Jr., and Shaver, R. H. (eds.), Moderns and ancient geosynclinal sedimentation, Soc. Econ. Paleontologists and Mineralogists, Spec. Pub. 19, p. 161-173, 1974.

Karig, D. E., and Sharman, G. E. III, Subduction and accretion in trenches, Geol. Soc. America Bull., v. 86, p. 377-389, 1975.

Katz, H. R., Continental margin in Chile--is tectonic style compression or extensional?, Amer. Assoc. Petroleum Geologists Bull., v. 55, p. 1753-1758, 1971.

Kimura, T., The old "inner" arc and its deformation in Japan, in Coleman, P. J. (ed.), The western Pacific, island arcs, marginal seas, geochemistry, New York, Crane, Russak and Co., and Univ. Western Australia Press, p. 255-273, 1973.

Kimura, T., The ancient continental margin of Japan, in Burk, C. A. and Drake, C. L. (eds.), The geology of continental margins, Springer-Verlag, New York, p. 817-829, 1974.

King, P. B., and others, Tectonic map of North America, U.S. Geol. Survey, Washington, D.C., 1969.

Kulm, L. D., von Huene, R., and others, Initial reports of the Deep Sea Drilling Project, Wash. D.C., U.S. Govt. Printing Office, 1077 p., 1973.

Kulm, L. D., and Fowler, G. A., Oregon continental margin structure and stratigraphy--A test of the imbricate thrust model, in Burk, C. A. and Drake, C. L. (eds.), The geology of continental margins, Springer-Verlag, New York, p. 261-283, 1974.

Landis, C. A., and Bishop, D. G., Plate tectonics and regional stratigraphic-metamorphic relations in the southern part of the New Zealand geosyncline, Geol. Soc. America Bull., v. 83, p. 2267-2284, 1972.

Larson, R. L., and Chase, C. G., Late Mesozoic evolution of the western Pacific Ocean, Geol. Soc. America Bull., v. 83, p. 3627-3644, 1972.

Larson, R. L., and Pitman, W. C., III, World-wide correlation of Mesozoic magnetic anomalies, and its implications, Geol. Soc. America Bull., v. 83, p. 3645-3662, 1972.

Marlow, M. S., Scholl, D. W., Buffington, E. C., and Alpha, T. R., Tectonic history of the western Aleutian arc, Geol. Soc. America Bull., v. 84, p. 1555-1574, 1973.

Matthews, V., and Wachs, D., Mixed depositional environments in the Franciscan geosynclinal assemblage, Jour. Sed. Petrology, v. 43, p. 516-517, 1973.

Menard, W. H., Marine geology of the Pacific, New York, McGraw-Hill Book Co., 271 p., 1964.

Miller, H., Das problem des hypothetischen "pazifischen Kontinentes" gesehen von der chilenischen Pazifikküste, Geol. Rundschau, v. 59, p. 927-938, 1970a.

Miller, H., Vergleischende Studien an prämesozoichen Gesteninen Chiles unter besonderer Berücksichtigung ihrer Kleintektonik, Geotektonische Forschungen, no. 36, 64 p., 1970b.

Moore, C. J., Selective subduction, Geology, v. 3, p. 530-532, 1975.

Moore, D. G., and Curray, J. R., Midplate continental margin geosynclines--Growth processes and Quaternary modifications, in Dott, R. H. Jr., and Shaver, R-. H. (eds.), Modern and ancient geosynclinal sedimentation, Soc. Econ. Paleontologists and Mineralogists, Spec. Pub. 19, p. 26-35, 1974.

Moore, T. C., Jr., DSDP--Successes, failures, proposals, Geotimes, July, p. 27-31, 1972.

Pessagno, E. A. Jr., Age and geologic significance of radiolarian cherts in the California Coast Ranges, Geology, v. 1, p. 153-156, 1973.

Raymond, L. A., Possible modern analogs for rocks of the Franciscan complex, Mount Oso area, California, Geology, v. 2, p. 143-146, 1974.

Rutland, R. W. R., Andean orogeny and ocean floor spreading, Nature, v. 233, p. 252-255, 1971.

Schlanger, S. O., and Douglas, R. G., The pelagic ooze-chalk-limestone transition and its implications for marine stratigraphy, in Hsu, K. J., and Jenkyns, H. C. (eds.), Pelagic sediments--On land and under the sea, Interntl. Assoc. Sedimentologists Spec. Pub. 1, p. 117-148, 1974.

Scholl, D. W., Sedimentary sequences in the North Pacific trenches, in Burk, C. A., and Drake, C. L. (eds.), The geology of continental margins, Springer-Verlag, New York, p. 493-504, 1974.

Scholl, D. W., Christinsen, M. N., von Huene, R., and Marlow, M. S., Peru-Chile Trench, sediments and sea-floor spreading, Geol. Soc. America Bull., v. 81, p. 1339-1360, 1970.

Scholl, D. W., and Marlow, M. S., Sedimentary sequences in modern Pacific trenches and the deformed circum-Pacific eugeosyncline, in Dott, R. H. Jr., and Shaver, R. H. (eds.), Modern and ancient geosynclinal sedimentation, Soc. Econ. Paleontologists and Mineralogists, Spec. Pub. 19, p. 193-211, 1974.

Scholl, D. W., and Marlow, M. S., Global tectonics and the sediments of modern and ancient trenches--Some different inerpretations, in Kahle, C. F. (ed.), Plate tectonics, assessments and reassessments, Amer. Assoc. Petroleum Geologists Mem. 23, p. 255-272, 1974b.

Scholl, D. W., and Marlow, M. S., Discussion, mixed depositional environments in the Franciscan geosynclinal assemblage, by Vincent Matthews III and Daniel Wachs, Jour. Sed. Petrology, v. 43, p. 516-517, Jour. Sed. Petrology, v. 44, p. 591-593, 1974c.

Scholl, D. W., von Huene, R., and Ridlon, J. B., Spreading of the ocean floor--undeformed sediments in the Peru-Chile Trench, Science, v. 159, p. 869-871, 1968.

Seely, D. R., Vail, P. R., and Walton, G. G., Trench slope model, in Burk, C. A., and Drake, C. L. (eds.), The geology of continental margins, Springer-Verlag, New York, p. 249-260, 1974.

Silver, E. A., Pleistocene tectonic accretion of the continental slope off Washington, Marine Geology, v. 13, p. 239-249, 1972.

Stewart, R. J., Turbidites of the Aleutian Abyssal Plain--Mineralogy, provenance, and constraints for Cenozoic motion of the Pacific plate, Geol. Soc. America Bull., v. 87, p. 793-808, 1976.

Veevers, J. J., Heitzler, J. R., and others, Initial reports of the Deep Sea Drilling Project, v. 27, Wash. D. C., Govt. Printing Office, 1060 p., 1974.

von Huene, R., Structure of the continental margin and tectonism at the eastern Aleutian Trench, Geol. Soc. America Bull., v. 83, p. 3613-3626, 1972.

von Huene, R., Modern trench sediments, in Burk, C. A. and Drake, C. L. (eds.), The geology of continental margins, Springer-Verlag, New York, p. 207-211, 1974.

Wachs, D., and Hein, J. R., Franciscan limestones and their environments of deposition, Geology, v. 1, p. 29-33, 1975.

White, D. A., Roeder, D. H., Nelson, T. H., and Crowell, J. C., Subduction, Geol. Soc. America Bull., v. 81, p. 3431-3432, 1970.

ST. GEORGE BASIN, BERING SEA SHELF: A COLLAPSED MESOZOIC MARGIN[1]

Michael S. Marlow, David W. Scholl and Alan K. Cooper

U. S. Geological Survey, 345 Middlefield Road, Menlo Park, California 94025

Abstract. St. George basin is a long (>300 km), narrow (30-50 km) graben whose long axis strikes northwestward parallel to the continental margin of the southern Bering Sea. Located near the Pribilof Islands, and beneath the virtually featureless Bering Sea shelf, the basin is filled with more than 10 km of upper Mesozoic(?) and Cenozoic sedimentary deposits. These sedimentary rocks are ruptured by normal faults associated with the sides of the graben; these ruptures commonly correlate with offsets in the basement surface. Offset along these faults increases with depth, implying that they are growth-type structures. The basement rocks flooring and flanking St. George basin are part of an assemblage of Mesozoic geosynclinal (forearc) rocks that probably extends from southern Alaska to eastern Siberia beneath the Bering Sea margin and outer shelf. A companion parallel belt of igneous rocks of late Mesozoic and earliest Tertiary age may also extend from western Alaska to northeastern Siberia beneath the inner Bering Sea shelf. Both the forearc deposits and the igneous rocks of the arc are thought to be the direct and indirect products of oblique convergence and subduction between the Kula(?) and North American plates along the Bering Sea margin during most of Mesozoic time. Underthrusting apparently ceased by the end of Mesozoic or in earliest Tertiary time. Tectonic deactivation evidently resulted in isostatic uplift of the margin and adjacent shelf, and, as a consequence, deep subaerial erosion. Since tectonic deactivation, the outer Bering Sea shelf has undergone extensional rifting and regional subsidence. Differential subsidence has resulted in the formation of a series of basement ridges and basins whose axes parallel the Bering Sea margin. Some of these basins are very large to gigantic in size, e.g., St. George Basin, and involve crustal subsidence exceeding 10 km. Such large-scale crustal collapse suggests deep crustal or upper mantle processes, such as thermal metamorphism or stress-induced crustal migration. Whichever process(es) caused the extensional collapse of the Mesozoic framework of the outer Bering Sea shelf and its adjacent margin, it has operated over a span of time of perhaps 50 to 70 m.y.

[1]Publication authorized by the Director, U. S. Geological Survey. The authors express their appreciation to K. Bailey, L. Bailey, M. Castain, C. Carpenter and P. Swenson for their help in the manuscript preparation. Reviewed by T. Vallier and J. Gardner.

Introduction

Recent interest in the resource potential of the Bering Sea shelf, especially hydrocarbon prospects, has stimulated reconnaissance geologic and geophysical mapping of the shelf and margin. Since 1965 the U. S. Geological Survey has conducted five expeditions to the Bering Sea, and, in conjunction with the University of Washington, has collected more than 24,000 kilometers of single-channel seismic reflection data in the shelf area. In 1975, 600 kilometers of 24-channel and single-channel seismic reflection, gravity, magnetic and bathymetric data was collected in the southern shelf area over St. George basin, between the Pribilof Islands and the western end of the Alaska Peninsula (110 km of trackline is shown in Fig. 1; Figs. 1 and 2, Marlow and others, 1976), using a tuned, five-airgun array (1326 cu. in.; 21,723 cc) in conjunction with an 80 kj sparker sound source.

Our data establish St. George basin as one of great size and depth. The origin of St. George basin is discussed here relative to a speculative plate tectonic model that suggests tectonic deactivation and subsequent collapse of the southern Bering Sea shelf and margin. It is argued that this basin, and both larger and smaller ones to the northwest, probably formed as a consequence of the formation of a new subduction zone and the resulting tectonic stagnation of the Bering Sea margin.

Geologic Setting

The oldest rocks exposed in the vicinity of St. George basin are Upper Jurassic silt-

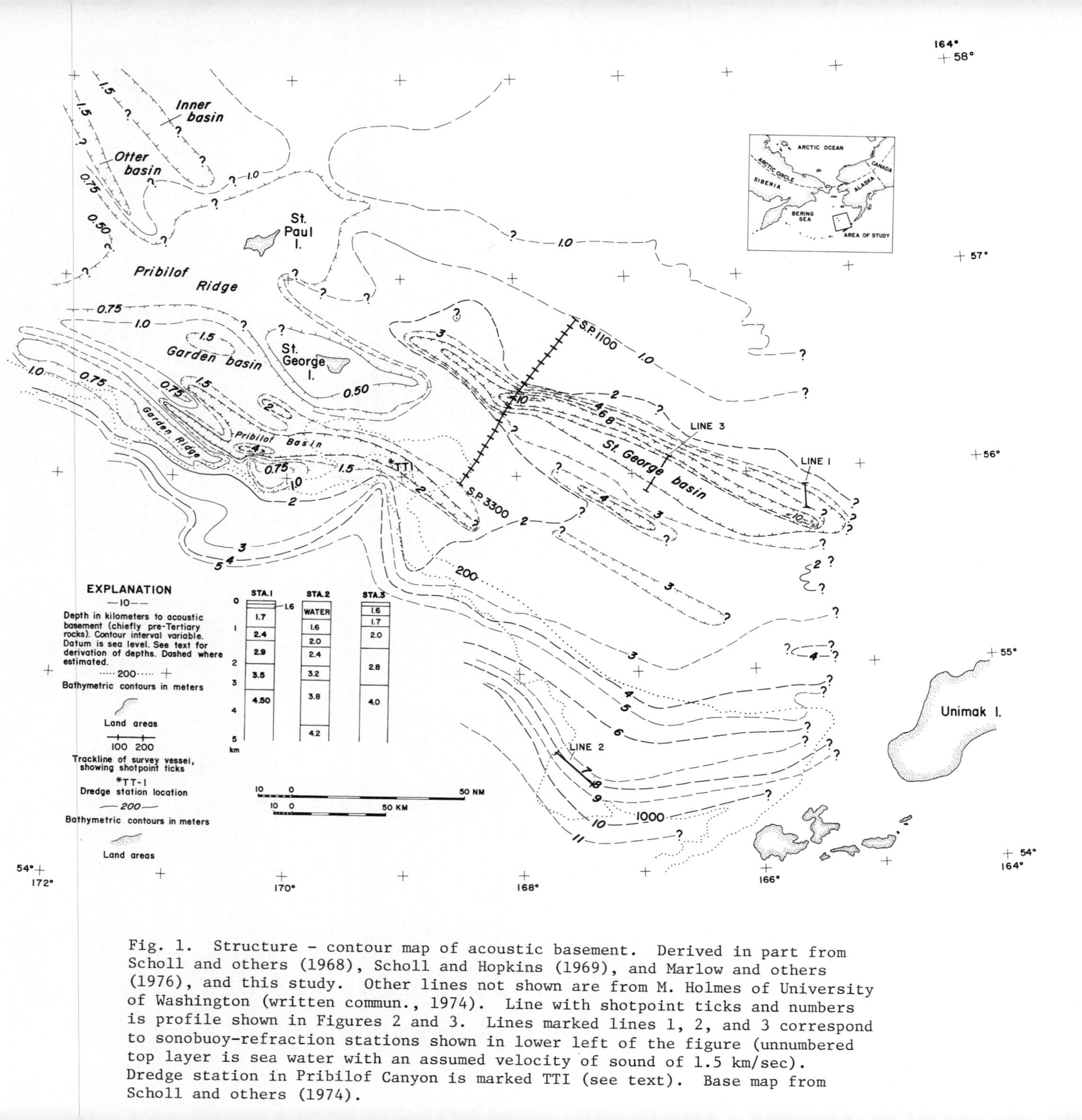

Fig. 1. Structure - contour map of acoustic basement. Derived in part from Scholl and others (1968), Scholl and Hopkins (1969), and Marlow and others (1976), and this study. Other lines not shown are from M. Holmes of University of Washington (written commun., 1974). Line with shotpoint ticks and numbers is profile shown in Figures 2 and 3. Lines marked lines 1, 2, and 3 correspond to sonobuoy-refraction stations shown in lower left of the figure (unnumbered top layer is sea water with an assumed velocity of sound of 1.5 km/sec). Dredge station in Pribilof Canyon is marked TTI (see text). Base map from Scholl and others (1974).

Figure 4 appears after page 214.

stone and sandstone of the Naknek Formation that crop out on the western end of the Alaska Peninsula in the Black Hills (Burk, 1965). Geophysical evidence suggests that these deformed rocks extend offshore northwestwardly and may connect with the north side of Pribilof ridge (Fig. 4, Marlow and others, 1976). The oldest sampled offshore rocks are deformed and lithified mudstone, siltstone, and sandstone of Late Cretaceous (Campanian) age that were dredged from acoustic basement first recognized on low-power, single-channel reflection profiles (Fig. 1; Scholl and others, 1966; Hopkins and others, 1969; Marlow and others, 1976). A more complete discussion of the geology of the eastern Bering Sea can be found in Marlow and others (1976).

Burk (1965) first noted the lithologic similarity of the deformed Mesozoic eugeosynclinal rocks cropping out in the Kodiak-Shumagin shelf area of southern Alaska to rocks exposed in the Koryak area of eastern Siberia. He suggested that these two fold belts are connected structurally beneath the Bering Sea shelf, a supposition later supported by Scholl and others (1966, 1968, 1975), Hopkins and others (1969), Hopkins and Scholl (1970), Patton and others (1974, 1976), and Marlow and others (1976). We assume that the seismically-resolved acoustic basement of the outer southern Bering Sea shelf and adjacent margin consists mainly of deformed eugeosynclinal rocks of Mesozoic age (see below).

In the interior region of southwestern Alaska, Mesozoic geosynclinal rocks strike southwestward toward the Bering Sea shelf (Payne, 1955; Hoare, 1961; Gates and Gryc, 1963; Patton, 1973; Fig. 2, Marlow and others, 1976). These geosynclinal trends either terminate beneath the inner shelf or turn northwest and merge with trends of the deformed eugeosynclinal assemblages that underlie the outer shelf and margin (Scholl and others, 1975; Marlow and others, 1976).

Seismic Reflection Data

Earlier work by Marlow and others (Figs. 2 and 7, 1976) delineated an elongate, sediment-filled basin, St. George basin, underlying the flat, shallow Bering shelf and trending northwest from the vicinity of the southern Alaska Peninsula toward the Pribilof Islands. In that study, structure contours on acoustic basement derived from 3,000 km of single-channel seismic reflection data show this basin to be 300 km long, 50 km wide, and filled with more than 6.5 km of Cenozoic sedimentary rocks (Marlow and others, 1976). Multichannel seismic reflection data, recently acquired and in part described below, show that this graben is much deeper and that it represents a mammoth extensional rift in the crustal rocks of the southern Bering Sea shelf.

Profile 8B.

A 24-channel seismic reflection profile across St. George basin, Figure 2, shows subshelf structures representative of the southern Bering Sea shelf and St. George basin axis; the profile is diagrammatically interpreted on Figure 3. Conversion of seismic travel-time to depths is based on Figure 6 of Marlow and others (1976).

The flat acoustic basement underlying the southwest end of profile 8B (Fig. 3; between shotpoints 3200 and 2700) is Pribilof ridge, which is overlain by a relatively undisturbed, layered sequence 1.3-1.4 km thick (1.3 - 1.4 sec). Within the acoustic basement, a number of gently dipping reflectors suggest the basement includes folded sedimentary beds. Farther north, between shotpoints 2700 and 2200, the basement descends in a series of down-to-basin steps, plunging to a maximum subbottom depth of about 5.4 seconds (over 10 km) beneath the axis of St. George basin. The overlying reflectors are broken by at least three major normal faults that dip toward the basin axis and appear to be related to offsets in the acoustic basement. Within the diverging basin fill, the offset along the faults increases with depth, implying that these are growth structures. Reflectors or strata are synclinally deformed about the basin's structural axis. The free-air gravity anomaly reaches a minimum of -1 mgal over the basin axis (a decrease of 55 mgal from the 54 mgal high over the basement high near shotpoint 1860).

From shotpoints 2200 to about 1860, the acoustic basement rises rapidly to a minimum depth of 0.55 second (0.5 km; Figs. 2 and 3). Again, the overlying reflectors are broken by normal faults that dip down to the basin axis.

A gentle swale in the basement extends from shotpoints 1860 to 1080; the maximum thickness of overlying strata is about 1.7 seconds (1.8 km) near shotpoint 1330. A corresponding relative gravity low of 41 mgal is centered over the swale. Intra-basement reflectors were resolved dipping at a gentle angle to the surface of the acoustic basement (shotpoints 1200 to 1400). Reflectors in the lower part of the overlying sedimentary section wedge out against the acoustic basement (shotpoints 1500 and 1720).

Structure Contours

Our new multichannel data show that St. George basin is filled near each end with more than 10 km of sedimentary section. We do not know whether the section thickens

LINE DIRECTIØN

VELØCITY FUNCTIØN
DIRECTIØN
LINE INTERSECTIØN
WATER DEPTHS

STATIØNS

U.S.G.S.

LINE 8B-SW
SP 1043 TØ 3294
BERING SHELF
24 FØLD STACK

INPUT REEL HEADER INFØRMATIØN

REEL NUMBER
DATE CREATED
NUMBER SAMPLE/TRACE 3000
SAMPLE RATE IN MILLS 2
PRØCESSØR DRESSER-ØLYMPIC
LINE NUMBER
JØB NUMBER
SECTIØN NUMBER
PRØCESSING STEP

FIELD INFØRMATIØN

ACQUIRED BY	U.S.G.S.	VESSEL	S.P. LEE
INSTRUMENTS	GUS	TYPE GAIN	AGR
SAMP RATE	2 MS	REC LENGTH	6.0 SEC
FILTER	5-100 HZ	NØ CHANNELS	48-24 BLANK
SØURCE	5 AIR GUNS	AVG DEPTH	35 FEET
CU IN	1326	PRESSURE	1900
PØP INTER	50 METERS		
GRØUP INTER	100 METERS	CABLE DEPTH	30-50 FEET
NØ. GRØUPS	24	CABLE LENGTH	2400 METERS
DIST TR 1	8474 FEET	DIST TR 24	930 FEET
BØAT DIR	NE-SW	DATE	SEPTEMBER 17,1975

PRØCESSING SEQUENCE

HØUSTØN DATA PRØCESSING CENTER

1) EDIT + DEMULTIPLEX

2) V2T GAIN RECØVERY

3) VELØCITY ANALYSIS AT 4 KM INTERVALS

4) APPLY NMØ

5) MUTE

6) STACK 24 FØLD

7) DECØNVØLUTIØN

AUTØ 2ND ZERØ CRØSSING PRED LENGTH
220MS ØPERATØR LENGTH
GATE START 100MS END 3500MS

8) SPACE AND TIME VARYING FREQUENCY FILTER

12-36 HZ
10-30 HZ
6-24 HZ

9) DIGITAL AGC 1000MS LENGTH

****** FILMING PARAMETERS ******

PERCENT GAIN 20
HØRIZØNTAL SCALE 50. TR/IN
VERTICAL SCALE 1.250IN/SEC
FILMING DIRECTIØN R/L
PERCENT BAR 0
PØLARITY BLACK+VE

DATA PRØCESSED BY
DRESSER ØLYMPIC

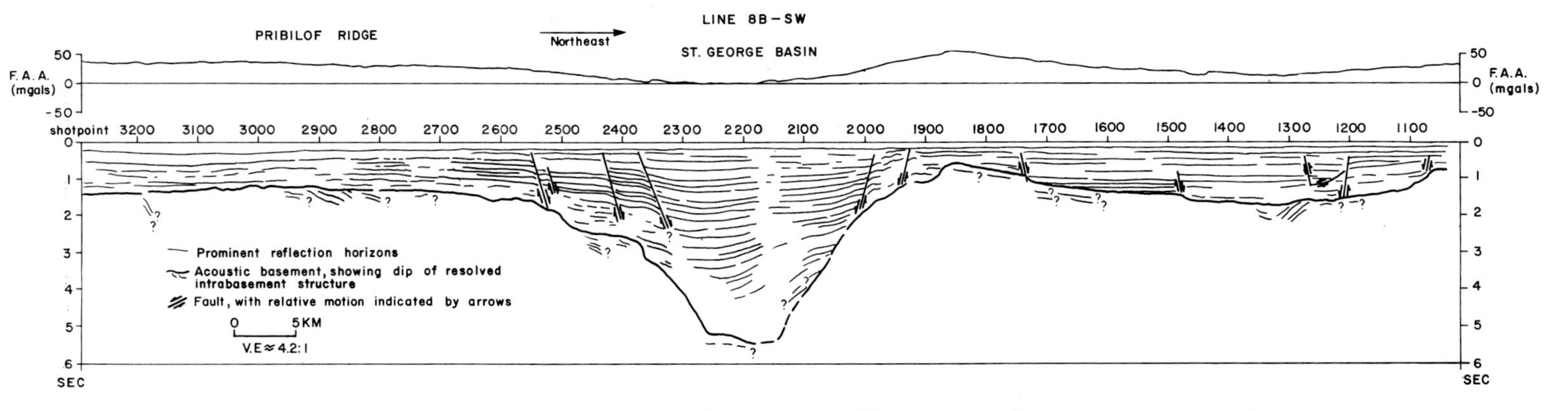

Fig. 3. Interpretative drawing of seismic reflection profile shown in Fig. 2. Note that the vertical exaggeration applies only to the water layer (assumed velocity of sound in sea water of 1.5 km/sec) and will decrease with depth in the section. For location of profile, see Fig. 1.

toward the center of the basin. Two smaller basins south of and parallel to St. George basin contain 3 to 4 km of layered fill (Fig. 1).

Northwest of St. George basin, the shelf is underlain by a complex of smaller basins and ridges, also parallel to the margin (Fig. 1; Marlow and others, 1976). Most of these basins, like St. George basin, are structural grabens or half-grabens bordered by normal faults. The largest structural high or ridge in the western basement complex, Pribilof ridge, is subaerially exposed in the Pribilof Islands (Fig. 1). To the southeast, the ridge flanks the southern side of St. George basin and extends toward Unimak Island (Fig. 1; Figs. 2 and 7, Marlow and others, 1976).

South of St. George basin, the acoustic basement deepens monoclinally toward the Aleutian Island arc, reaching depths greater than 11 km within 20 km of the Aleutian Ridge (Fig. 1; Figs. 2 and 7, Marlow and others, 1976). Unfortunately, we do not have seismic reflection data sufficient to determine the structural relation between the basement complex of the Bering Sea margin and the presumably younger one of the Aleutian Island arc (Scholl and others, 1975).

Sonobuoy Data

Three sonobuoy refraction lines in the area of St. George basin are shown in Figure 1. Two of the lines, stations 1 and 3, are over the basin, but not over its thickest sedimentary section. The depth to acoustic basement on both lines is about 3 km, and the corresponding seismic velocity of the basement rock is 4.0 to 4.5 km/sec (Fig. 1). Station 1 shows a layer about 0.7 km thick with a velocity of 3.5 km/sec, presumably part of the basin fill, that is apparently absent beneath the central area of the basin at station 3. Along the southern line (station 2) in deeper water over the upper continental slope, acoustic basement has a velocity of 4.2 km/sec. Here, two layers above basement have velocities of 3.2 and 3.8 km/sec and are 0.53 km and 1.54 km thick, respectively (Fig. 1).

Geopotential Data

Gravity Data

A free air gravity anomaly map of the St. George basin area is outlined in Figure 4. A regional map of the Bering Sea region that includes this area has been published by Watts (1975).

The irregular gravity low extending from the tip of the Alaska Peninsula northwestward toward the Pribilof Islands outlines St. George basin (Fig. 4). The southeastern end of the low, adjacent to the peninsula, may actually be the signature of a separate basin, Amak basin (not shown; Marlow and others, 1976). However, we have no seismic reflection data near the peninsula and, thus, Amak basin could actually be part of St. George basin. Immediately north of the Amak basin gravity low, a linear, 50-70 mgal gravity high extends about 70 km west of the Black Hills region of the Alaska Peninsula (Fig. 4). This high anomaly appears to turn and trend northwest along the northern flank of the gravity low associated with St. George basin, suggesting that the Jurassic rocks of the Black Hills area extend northwest and north of St. George basin. Furthermore, the structural and gravity data (Figs. 1 and 4) imply that Pribilof ridge may extend from the Pribilof Islands along the south flanks of St. George basin, and that the ridge may connect to a belt of younger Mesozoic rocks that trend into the Bering Sea from the Kodiak-Shumagin shelf of southern Alaska (Burk, 1965).

Magnetic Data

Figure 5 is a reproduction of a portion of a total-field magnetic anomaly map of the Bering shelf published by Bailey and others (1976). The St. George, Amak, and Bristol Bay basin complex is characterized by low-frequency, low-amplitude anomalies. This band of anomalies extends northwestward from the Alaska Peninsula to the Pribilof Islands area. A pronounced high of 400 gammas south of the low-frequency, low-amplitude zone (near 55.3°N, 167°W, Fig. 5) may be the expression of the southern extension of the Pribilof ridge, an extension probably underlain by folded eugeosynclinal rocks.

North of the low-amplitude, low-frequency anomaly belt, an arcuate zone of high-amplitude, high-frequency anomalies is traceable from east to west (Fig. 5). This belt of high anomalies is part of a larger, arcuate zone of similar anomalies that swings across the central and inner Bering Sea shelf (Marlow and others, 1976).

Discussion and Conclusions

Mesozoic Structural Trends

The gravity, magnetic, and seismic reflection data suggest that the Upper Jurassic shallow-marine rocks exposed in the Black Hills on the Alaska Peninsula may extend west to northwest along the north flank of St. George basin and may connect with the northern part of Pribilof

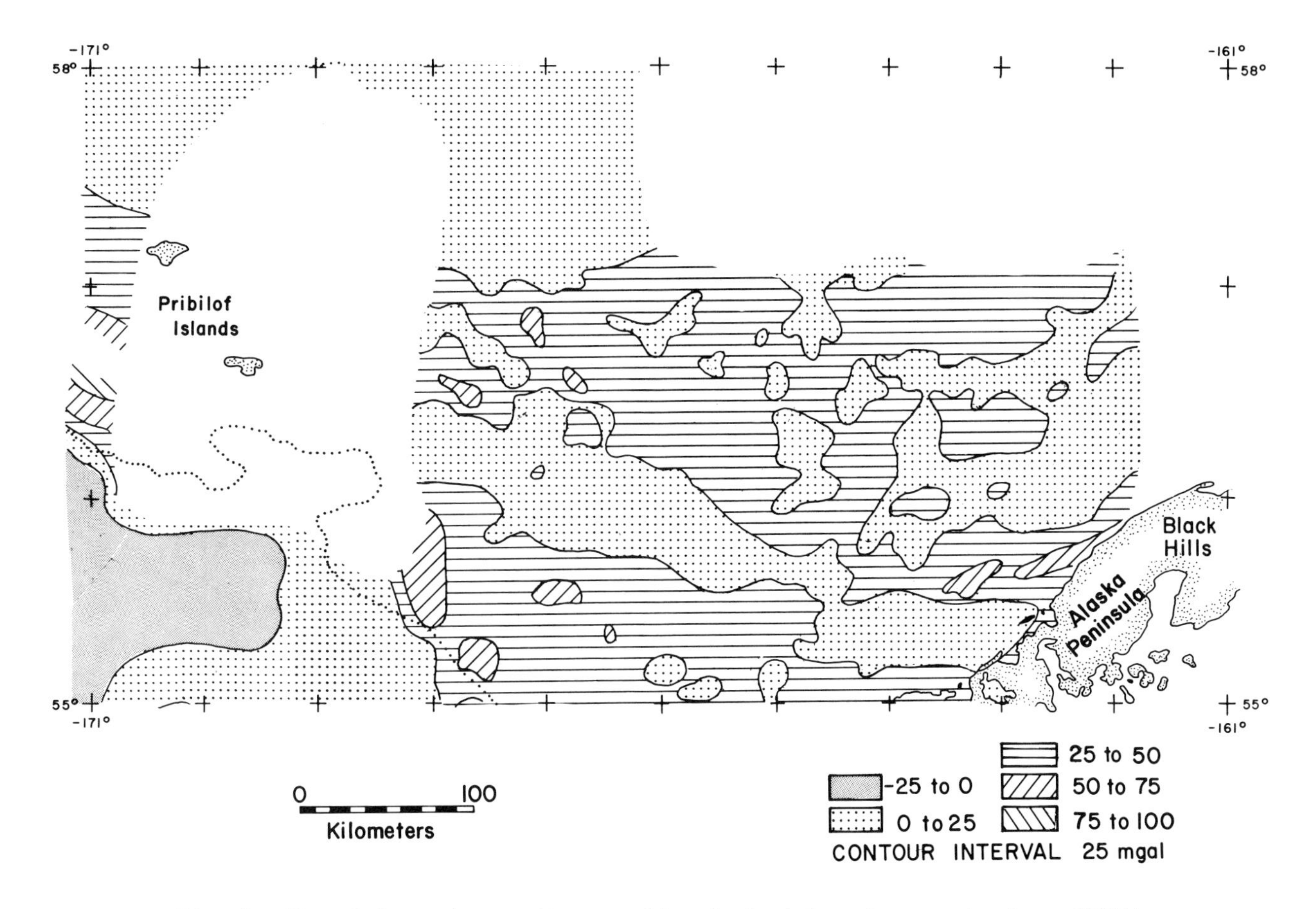

Fig. 4. Map of free air gravity anomalies derived from Pratt and others (1972) and Watts (1975). Land areas shown by stipple pattern. Two hundred meter bathymetric contour shown by dotted line.

ridge (Figs. 1, 2, 4, and 5). In addition, Upper Cretaceous (Campanian) rocks were dredged from the southern flank of this ridge in nearby Pribilof Canyon (Fig. 1; Hopkins and others, 1969). In agreement with the speculations of Burk (1965), we believe these deformed and intruded Mesozoic rocks are part of a second, younger continental margin eugeosynclinal (deep-water forearc area) assemblage that extends parallel to the inner belt of shallow-marine Jurassic rocks from southern Alaska via the outer Bering shelf to eastern Siberia (Scholl and others, 1975; Fig. 10, Marlow and others, 1976). We further speculate that a Jurassic, Cretaceous and earliest Tertiary magmatic arc extended parallel to and inside (landward) both the deep-water and shallow-water forearc depositional troughs. This igneous belt is characterized by high-amplitude, high-frequency magnetic anomaly signatures (Fig. 5). The magmatic arc is exposed as calc-alkalic volcanic and intrusive rocks of late Mesozoic and earliest Tertiary age on St. Matthew and St. Lawrence Islands on the Bering shelf and as similar rocks in southern and western Alaska and eastern Siberia (Fig. 6; Patton and others, 1974, 1976; Reed and Lanphere, 1973; Scholl and others, 1975; Marlow and others, 1976).

Deactivation and Crustal Collapse

The arcuate geosynclinal forearc and magmatic trends are probably the result of convergence and subduction of oceanic lithosphere (Kula(?) plate) beneath the Bering Sea continental margin during most of Mesozoic time (Cooper and others, 1976a, b; Scholl and others, 1975; Marlow and others, 1976). Convergence and subduction apparently ceased with the formation of the Aleutian Island arc in late Mesozoic or earliest Tertiary time, when the subduction zone shifted from the Bering Sea margin to a site near the present Aleutian Trench, thereby trapping a large section of oceanic plate (Kula?) within the abyssal Bering Sea.

Cessation of subduction tectonically deactivated the Bering Sea margin, presumably resulting in the isostatic rebound of the relatively light oceanic crust that had been thrust into mantle rocks beneath the margin. Rebound or uplift caused extensive erosion of the

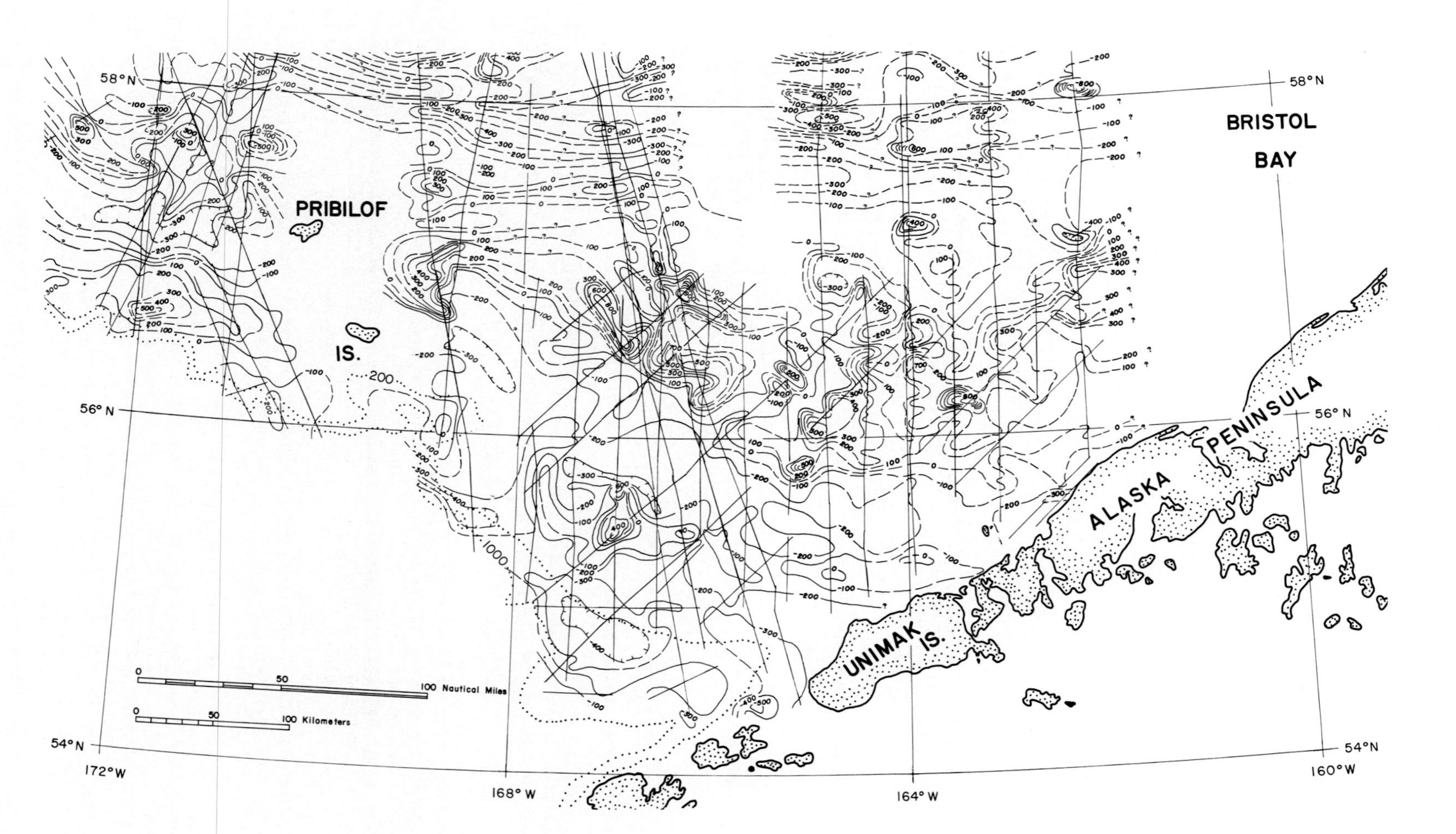

Fig. 5. Map of total-field magnetic anomalies derived from Bailey and others (1976). Contour interval is 100 gammas. Individual profiles have been upward continued to an elevation of one kilometer. Bathymetry in meters shown by dotted lines.

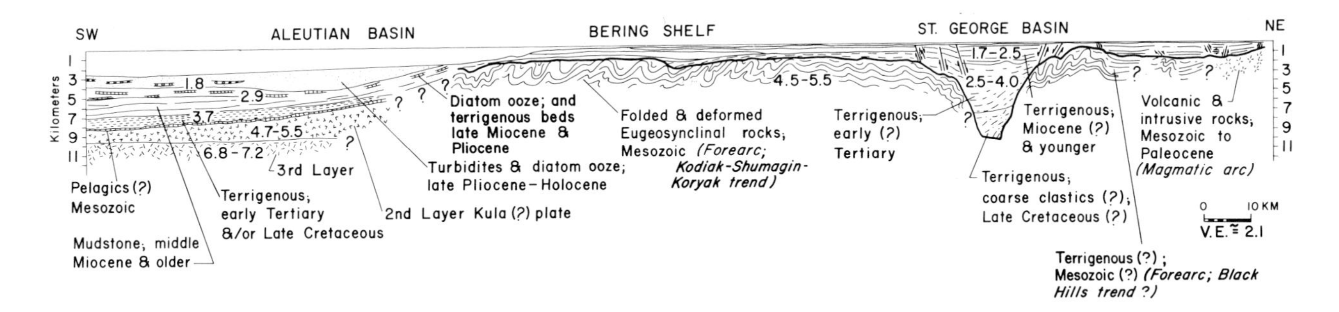

Fig. 6. Generalized cross section across the Bering Sea margin. Derived from Fig. 2 and 3 and from Scholl and others (1975) and Marlow and others (1976).

deformed Mesozoic rocks of the former active margin and adjacent geosynclinal areas. Seismic reflection data have revealed many truncated folds underlying the surface of the acoustic basement of the Bering Sea shelf (Figs. 2, 3, and 6; Figs. 3, 4, and 5, Marlow and others, 1976).

After uplift and erosion in the early Tertiary, and in part in conjunction with these processes, the margin underwent extensional collapse and differential subsidence during most of Cenozoic time. Elongate basins of great size and depth, exemplified by St. George basin, formed in the vicinity of the modern outer Bering Sea shelf (Figs. 1, 2 and 6; Marlow and others, 1976). Extensional deformation of the folded rocks of the Mesozoic basement has continued to the present, as evidenced by normal faults that flank the outer shelf basins and are growth-type structures which commonly rupture the entire Cenozoic basin fill. Collapse of the outer shelf and adjacent margin may have been aided by Cenozoic sediment-loading of the adjacent oceanic crust (Kula? plate) flooring the abyssal Bering Sea (Fig. 6).

Margin Models

Bott (1971) presents a generalized model that accounts for regional collapse of continental margins. He envisions lower crustal material flowing away from the margin owing to the differential loading of the continental crust and the adjacent, deep oceanic crust. Falvey (1974) offers an alternate model, in which normal heat flow (1.0 to 1.6 HFU) over a period of about 50 m.y. induces thermal metamorphism in the lower crust. Metamorphism results in volumetric shrinkage of the lower crust and this, in turn, causes regional subsidence of the margin and adjacent shelf. Although Falvey's model applies to a divergent (spreading) plate boundary, and although we have no heat flow data from the Bering Sea margin, the time span (50 - 70 m.y.) implied in his model is appealing for the eastern Bering Sea region. Uplift and subsequent collapse of the margin and nearby outer Bering Sea shelf apparently began at the end of Mesozoic or in earliest Tertiary time. Collapse and rifting have continued to the present over a time span of 50 to 70 million years (Fig. 6).

References

Bailey, K. A., A. K. Cooper, M. S. Marlow and D. W. Scholl, Preliminary residual magnetic map of the eastern Bering shelf and parts of western Alaska: U. S. Geol. Survey Misc. Inv. Map I-716, 1976.

Bott, M. H. P., Evolution of young continental margins and formation of shelf basins: Tectonophysics, v. 11, p. 319-327, 1971.

Burk, C. A., Geology of the Alaska Peninsula Island arc and continental margin: Geol. Soc. America Mem. 99,250 p. 1965.

Cooper, A. K., M. S. Marlow, and D. W. Scholl, Mesozoic magnetic lineations in the Bering Sea marginal basin: Jour. Geophys. Res., v. 81, p. 1916-1934, 1976a.

Cooper, A. K., D. W. Scholl, and M. S. Marlow, Plate tectonic model for the evolution of the eastern Bering Sea basin, in press, 1976b.

Falvey, D. A., The development of continental margins in plate tectonic theory: The Australian Petroleum Exploration Association Journal, p. 95-196, 1974.

Gates, G. O., and G. Gryc, Structure and tectonic history of Alaska, in Backbone of the Americas: AAPG Mem. 2, p. 264-277, 1963.

Hatten, C. W., Petroleum potential of Bristol Bay basin, Alaska, in Future petroleum provinces of the United States; their geology and potential, v. I: AAPG Mem. 15, p. 105-108. 1971.

Hoare, J. M., Geology and tectonic setting of lower Kuskokwim-Bristol Bay region, Alaska: AAPG Bull., v. 45, p. 594-611, 1961.

Hopkins, D. M., and D. W. Scholl, Tectonic development of Beringia, late Mesozoic to Holocene (abs.): AAPG Bull., v. 54, p. 2487-2488, 1970.

_____, et al., Cretaceous, Tertiary and early Pleistocene rocks from the continental margin in the Bering Sea: Geol. Soc. America Bull., v. 80, p. 1471-1480, 1969.

Marlow, M. S., D. W. Scholl, A. K. Cooper, and E. C. Buffington, Structure and evolution of Bering Sea shelf south of St. Lawrence Island, AAPG Bull., v. 60, p. 161-183, 1976.

Patton, W. W., Jr., Reconnaissance geology of the northern Yukon-Koyukuk province, Alaska: U. S. Geol. Survey Prof. Paper 774-A, 17 p. 1973.

____, M. A. Lanphere, T. P. Miller, and R. A. Scott, Age and tectonic significance of volcanic rocks on St. Matthew Island, Bering Sea, Alaska (abs.): Geol. Soc. America Abs. with Programs, v. 6, p. 9905-9906, 1974.

_____ _____ _____ _____, Age and tectonic significance of volcanic rocks on St. Matthew Island, Bering Sea, Alaska: U. S. Geol. Survey Jour. Research, v. 4, p. 67-74, 1976.

Payne, T. G., compiler, Mesozoic and Cenozoic tectonic elements of Alaska: U. S. Geol. Survey Misc. Geol. Inv. Map I-84, Scale 1:5,000,000, 1955.

Pratt, R. M., M. S. Rutstein, F. W. Walton, and J. A. Buschur, Extension of Alaska structural trends beneath Bristol Bay, Bering shelf, Alaska: Jour. Geophys. Research, v. 77, p. 4994-4999, 1972.

Reed, B. L., and M. A. Lanphere, Alaska-Aleutian range batholith: Geochronology, chemistry, and relation to circum-Pacific plutonism: Geol. Soc. America Bull., v. 84, p. 2583-2610, 1973.

Scholl, D. W., E. C. Buffington, and D. M. Hopkins, Exposure of basement rock on the continental slope of the Bering Sea: Science, v. 153, p. 992-994, 1966.

_____ _____ _____, Geologic history of the continental margin of North America in the Bering Sea: Marine Geology, v. 6, p. 297-330, 1968.

____ T. R. Alpha, M. S. Marlow, and E. C. Buffington, Base map of the Aleutian-Bering Sea region: U. S. Geol. Survey Misc. Geol. Inv. Map I-979, scale 1:2,500,000, 1974.

_____ _____ _____ _____, Plate tectonics and the structural evolution of the Aleutian-Bering Sea region: In: R. B. Forbes, ed., Contributions to the geology of the Bering Sea Basin and adjacent regions, Geological Soc. America Special Paper 151, p. 1-31, 1975.

Watts, A. B., Gravity field of the northwest Pacific ocean basin and its margin; Aleutian Island arc - trench system, Geol. Soc. of America Map - Chart Series MC-10, 1975.

GEOLOGICAL CONSEQUENCES OF RIDGE SUBDUCTION

Stephen E. DeLong and Paul J. Fox

Department of Geological Sciences, State University of New York at Albany, Albany NY 12222

Abstract. A simple model of the topographic and thermal consequences of subducting an oceanic spreading center at an island arc predicts three geological effects: (1) Shoaling and sub-aerial emergence of the crest of the arc as it moves "uphill" along the cooling-decay curve of the approaching ridge flank, and submergence of the crest during "downhill" motion of the arc along the opposite ridge flank as the trailing plate is subducted. Significant changes in arc-crest sedimentation should accompany this emergence/submergence. (2) Decrease or cessation of subduction-related magmatism as progressively younger and hotter oceanic lithosphere is subducted. This may be due to reduced frictional heating from a loss in temperature-dependent viscosity contrast, to shallow escape through relatively permeable oceanic crust of water released by dehydration, to buoyancy effects, or to a combination of these factors. Analogous arguments have been advanced previously to explain the absence of seismicity in areas where young lithosphere is being subducted. (3) Regional, low-grade thermal metamorphism of the arc rocks due to modest heating ($\Delta T \simeq 100$-$300°C$) during rapid passage (~5-10 m.y.) of the former ridge crest beneath the arc. A detailed test of this model has been made for the insular Aleutian arc, where all three of these phenomena are recorded geologically, with the following history and interpretation indicated: (1) reduction of magmatism on approach of the Kula Ridge in the mid-Eocene (~45 m.y.); (2) shoaling and emergence of the crest of the Aleutian arc in the Late Eocene to Oligocene; (3) subduction of the Kula Ridge and greenschist metamorphism of the arc at about 30-35 m.y.; (4) submergence of the arc crest in the mid-Oligocene to Miocene; and (5) resumption of arc magmatism at about 15 m.y. North-south discontinuities in seismicity, volcanism, and topography have also been recognized in southern Chile in the vicinity of the South America-Nazca-Antarctic triple junction (46°S) that may be related to subduction of the Chile Rise.

Introduction

Only a few examples of ridge subduction are well documented by marine magnetic anomalies that become younger toward land. The first instance to be recognized [Pitman and Hayes, 1968] was the western limb of the Great Magnetic Bight south of the Aleutian arc. This pattern of anomalies is interpreted [Grow and Atwater, 1970] to indicate successive subduction at the Aleutian Trench of the now-vanished Kula Plate and Kula Ridge, and some part of the Pacific Plate. Other examples, most of which are indicated on a map of world-wide anomalies [Pitman et al., 1974], include southwestern North America, southern Chile, Indonesia, Japan, and the Antarctic Peninsula. There are even fewer localities known where ridge subduction may occur in the near future - the Pacific Northwest-British Columbia region, southern Mexico, northern Chile, and the southern Scotia arc.

Despite the dearth of modern examples, the process of ridge subduction merits examination, because understanding of the associated geological phenomena would permit recognition of past occurrences of the process in older orogenic belts. Sutures and ophiolites along many of these belts record closing of oceans, a major aspect of the Wilson cycle [Dewey, 1975]. For most conceivable plate arrangements, a likely corollary to the Wilson cycle is the following: every time an ocean closes at least one ridge must be subducted. We suggest below that the sequence of geological events associated with ridge subduction is quite distinctive and that it may be relatively easy to recognize. The process may thus be of widespread interest and applicability in the geological record.

Previous discussions of ridge-trench interaction have (1) concluded that production of new sea floor ceases when a ridge arrives at a trench [Pitman and Hayes, 1968; Hayes and Heirtzler, 1968; Hayes and Pitman, 1970; Atwater, 1970; Marlow et al., 1973]; (2) assumed that enhanced magmatism and uplift would occur in the overlying arc as the ridge was subducted [Grow and Atwater, 1970; Bird and Dewey, 1970; Uyeda and Miyashiro, 1974]; and (3) examined possible geometric interactions between a ridge and a subduction zone pertinent to ophiolite obduction [Dewey, 1976] and to triple junction-transform evolution [Dickinson and Snyder, 1975]. We present here the results of a more detailed study of the geological consequences of ridge subduction, with emphasis on prediction of topographic and thermal effects and on development of a testable model. We first discuss two simple geometric considerations of plate motion that provide some constraints on models and interpretations of ridge subduction, then describe the model that we have developed, and finally summarize tests of the model as it may apply in the Aleutians and southern Chile.

Plate-Motion Constraints

Consider a three-plate system such as that in Figure 1a. If plate C is arbitrarily fixed, and the velocities of A and B relative to C are given by the arrows, then the A/B and B/C boundaries are immediately specified as a ridge and trench, respectively. We use the terms "leading plate" (B) and "trailing plate" (A) to refer to the plates that precede and follow the ridge, respectively, toward the trench.

We assume that the forces moving A and B

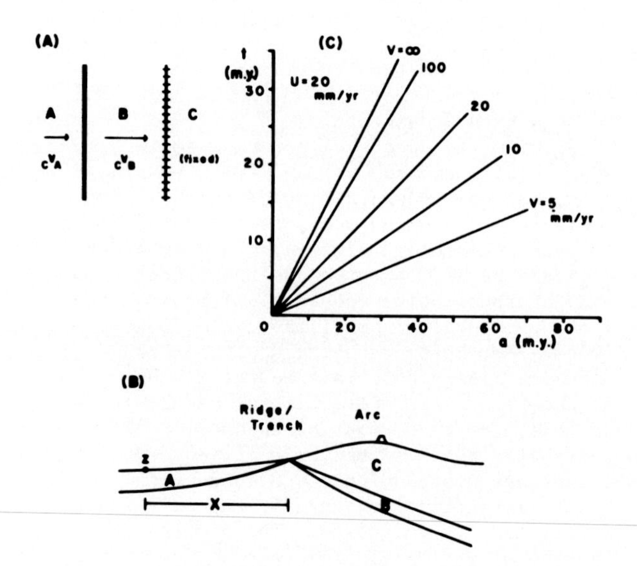

Figure 1. (A) Three-plate system, with velocities of A and B with respect to C given by $_Cv_A$ and $_Cv_B$, respectively. (B) Ridge arrives at trench. (C) Age of ridge subduction, t, as a function of age of anomaly nearest trench, a, for various half-rates, V, and subduction rate U = 20 mm/yr.

toward C at the specified rates are not located solely along the ridge or the trench. It follows from this assumption that subduction of the A/B ridge should have little effect on the convergence of A toward C and, hence, that subduction should continue at the trench, although at a slower rate. It is possible that mechanical difficulties may arise from attempting to subduct young oceanic lithosphere [Uyeda and Miyashiro, 1974], but eventually subduction should be re-established. This is confirmed by observation of the Aleutians, Japan, southern Chile, and Indonesia where ridge subduction has previously occurred. In all of these arcs, the trailing plate is underthrusting today.

Subduction of a ridge in a three-plate system such as Figure 1a could therefore provide a very brief hiatus in subduction at the trench or continuous subduction. In either of these situations, the age (a) of the youngest magnetic anomaly adjacent to the trench will always be older than the time (t) of ridge subduction. Determination of the value of t depends on a; the current subduction rate, U; and the spreading half-rate prior to ridge subduction, V, as recorded by magnetic anomalies on the remainder of the trailing plate. Figure 1b illustrates for time t the location (at distance X from the trench) of the anomaly (z) which is today just at the trench. Assuming that the trailing plate continued to subduct immediately after the ridge reached the trench and at the same convergence rate (i.e., $U = {_C}v_A$), then

$$X = Ut \tag{1}$$

and

$$X = (a-t)V, \tag{2}$$

from which

$$t = \frac{aV}{U+V}. \tag{3}$$

This equation shows that the only situation in which a = t will occur for U = 0, when the motion of A relative to C is parallel to the B/C trench. Then the A/C boundary after ridge subduction becomes a transform fault, as described by Atwater [1970] for southwestern North America.

Figure 1c compares t and a for various spreading half-rates, V, and U = 20 mm/yr, the present convergence rate between the Antarctic and South American Plates [Forsyth, 1975] along the coastline south of 46°S where part of the Chile Rise has been subducted. This plot shows that t is likely to be much younger than a for slow to moderately fast-spreading ridges, and that t will approach a only for extremely large half-rates. (Rearrangement of eq. (3) gives V = Ut/(a-t), confirming a > t for positive values of V.)

The second geometric consideration involves a potentially important hole in the lithosphere that may develop under an arc as a ridge is subducted. Dickinson and Snyder [1975] have described similar relationships and calculated numerical examples for the more complicated plate geometry and resultant hole that occurred during the subduction of the East Pacific Rise off California. One means of developing such a feature is shown in Figure 2, where plates A and B are moving toward C at the rates indicated by the arrows. As the portion of plate B between the central ridge segment and the trench is subducted, part of the ridge reaches the trench. By our previous assumption that driving forces are not localized only at ridges or trenches, the surficial remnants of plate B, labeled B' in Figure 2b, will continue to underthrust and subduction of plate A will be initiated.

The B' remnants are presumably still attached to the underthrust remainder of plate B, and there is no reason to believe that the central section of B should detach from the rest of the plate along the down-dip fracture zone extensions. Hence, as subduction continues, plate B should continue to converge on plate C faster than does plate A. This means that points I and II, representing the last oceanic crust formed on plates A and B, respectively, just as the ridge reached the trench, should continue to diverge down-dip. The resulting gap (stipled in Figures 2c and 2d) cannot remain empty, but should be filled by hot asthenosphere from beneath the slab.

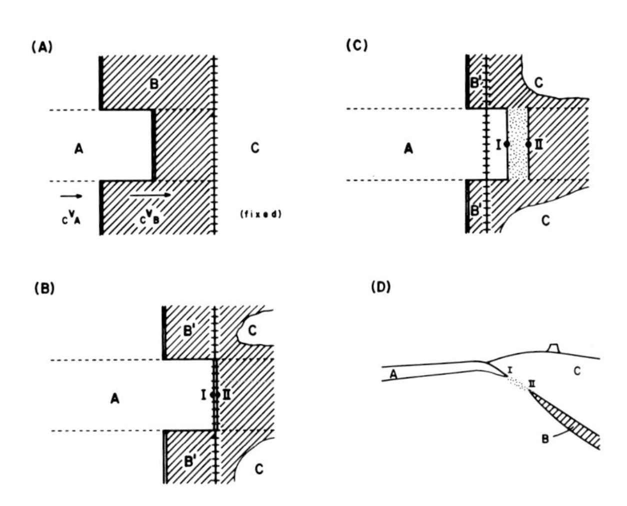

Figure 2. (A) Three-plate system, with velocities as defined in Figure 1(A). (B) Central ridge segment at trench, with underthrust portion of plate B (hachured) in cutaway view below plate C. (C) Continued subduction opens gap (stippled) between points I and II. (D) Section perpendicular to trench through points I and II for time represented in (C).

Uyeda and Miyashiro [1974] invoked essentially this process, terming it "spreading activity," to explain widespread volcanism in Japan and opening of the Japan Sea at the supposed time of subduction of the Kula-Pacific Ridge. We suggest that use of the term "spreading" in this context should subsequently be avoided, because that clearly connotes creation of oceanic lithosphere at the earth's surface. The process depicted in Figure 2d might occur instead at depths of several tens to hundreds of kilometers within the earth, and it probably has little resemblance to spreading at a ridge. It is questionable, for example, whether intrusion of asthenosphere from beneath the gap will necessarily produce widespread surface volcanism. Indeed, it is perhaps inappropriate to refer to the gap as a "ridge," because that is by definition simply the boundary between two plates [Atwater, 1970]. Thus, when the edges of the leading and trailing plates have been subducted, the ridge ceases to exist.

Model of Ridge Subduction

Topographic Considerations

The deepening of sea floor away from spreading ridges, documented for areas in all major ocean basins, is explained by thermal contraction of cooling lithosphere [Sclater et al., 1971]. As a ridge approaches a trench, progressively younger and more elevated regions of sea floor will be subducted. To an observer on the ridge crest, the approaching arc will appear to be moving "uphill," i.e., up the cooling-decay curve of the ridge flank. We suggest that some such process may occur literally.

As the ridge approaches very close to the trench, the ridge profile should be approximately maintained, because the elevation is thermally induced, in isostatic equilibrium, and most strongly expressed at the ridge crest. Relative to sea level, an approaching arc moving up the ridge flank will slowly shoal and the crest of the arc could emerge above sea level. Shoaling and subsequent sub-aerial exposure would produce significant changes in sedimentation patterns on the crest of the arc itself and in surrounding regions that received detritus from the arc. As the trailing plate is subducted, the sedimentological and topographic features should occur in reverse order as the arc moves back down the opposite flank of the ridge. The rate at which these events occurred on the uphill and downhill paths will probably differ, because the subduction rates of the leading and trailing plates will differ. We can predict, however, that in an ideal, complete rock record there would be a symmetric order of events about that point in the rock record representing the highest level of the arc.

Thermal Considerations

Many of the characteristics of subduction zones, such as volcanism and seismicity, are presumably related to contrasts in physical or chemical properties between the subducted plate and the surrounding mantle. In the usual situation, the down-going plate is colder, stiffer, and denser than the mantle that it enters (Figure 3a). As an active spreading ridge approaches a trench, however, the sea floor that is subducted is progressively younger, and therefore the plate is thinner, hotter, softer, and less dense than older, previously-subducted lithosphere. Thus, at a given depth, the physical properties of the downgoing material become increasingly like those of the surrounding mantle.

Atwater [1970] and Forsyth [1975] have noted that earthquakes are less likely to occur in hot, soft lithosphere just preceding or following a ridge into a trench than in older lithosphere; that is, seismicity may be low in an interval around the time of ridge subduction. By extension of this reasoning, we suggest that ridge subduction may cause reduced arc magmatism as well as reduced seismicity. A decrease or cessation of subduction-related magmatism in the overlying island arc as a ridge approaches may be related to one or a combination of the following mechanisms. First, if arc volcanism is due to frictional heating as a function of temperature-dependent viscosity contrast across the slip zone [Turcotte and Schubert, 1973] then a decrease in thermal contrast across the interface (Figure 3a) may lead to a decrease in the extent of melting, perhaps approaching complete cessation.

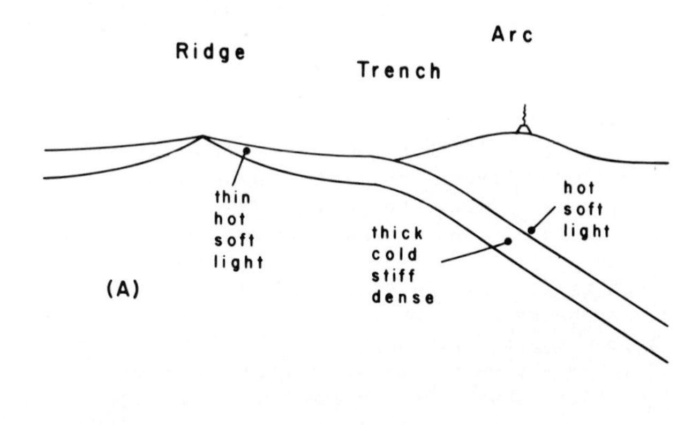

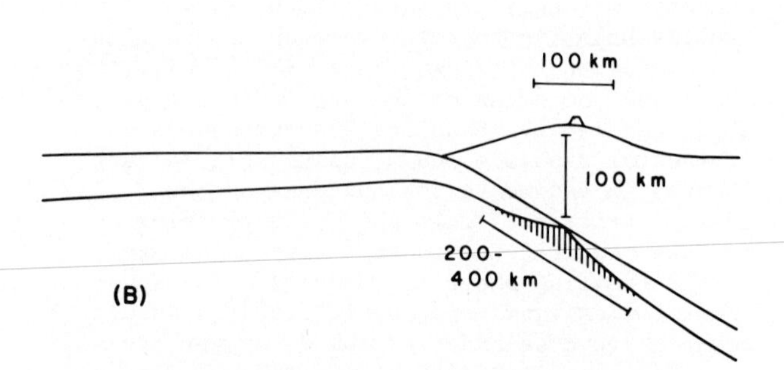

Figure 3. (A) Cross-section of ridge approaching trench, showing changes in physical properties of slab (thickness, temperature, viscosity, density) as younger crust is subducted. (B) Former ridge crest passing under arc.

Alternatively, frictional heating may be largely offset in the upper 75 to 100 km of the subduction zone by endothermic dehydration of the oceanic crust, as proposed by Anderson et al. [1976], who attributed arc volcanism to melting induced by the presence of water released from the slab. We have suggested elsewhere [Fox and DeLong, 1976, and in preparation] that the uppermost unit of oceanic crust, layer 2A, "heals" with time as fractures and other small-scale openings become filled with calcite, zeolites, and other minerals. (The initial presence of such openings and subsequent infillings is documented by D.S.D.P. samples (e.g., hole 333, Leg 37), and the progressive change is suggested by the seismic data of Houtz and Ewing [1976].) This "healing" is likely to make oceanic crust away from a ridge crest relatively impermeable. Thus, water released by dehydration may be carried down some distance in the slab (Figure 4a) until it is able to escape by flow through the reduced volume of passages, by diffusion, or by incorporation in hydrous magmas.

Where younger oceanic crust is subducted, the level of release of bound water by dehydration may become much shallower due to the increased geotherm [R. N. Anderson, personal communication, 1975]. This water may escape from the slab almost immediately (Figure 4b) because the fractures will not yet have "healed," hence the crust will be relatively permeable. Thus, approach of a ridge to a trench will result in subduction of progressively less water to positions beneath the volcanic arc, which would promote a diminution of magmatism. (The same argument applies much more strongly to any suggestion that subducted sediments may be a significant source of water, because it is well documented that ocean-floor sediment abundances decrease to virtually zero at ridge crests [Ewing and Ewing, 1967].)

A third possibility [Uyeda and Miyashiro, 1974; J. Kelleher, personal communication, 1975] is that a subducted ridge is buoyant because of its relatively high temperature and low density. Thus, as the ridge crest approached the trench, the angle of underthrusting might progressively decrease, until the subducted plate was at too shallow a depth to melt or to cause melting.

As with the topographic effects, we can also predict a certain symmetry to arc volcanism relative to the arrival of the ridge at the trench. In the case of complete cessation of melting, the ideal geological record would preserve shutdown of magmatism preceding arrival of the ridge and on-set of magmatism at a later time. This renewed magmatic activity would presumably coincide with arrival at appropriate depth in the subduction zone of oceanic crust and lithosphere sufficiently different from younger crust or from the overlying mantle to promote melting. Again, the time intervals from shut-down of volcanism to ridge arrival and from ridge arrival to on-set of volcanism need not be equal.

The above considerations suggest a seeming paradox in which introduction of hotter material (i.e., younger oceanic crust) into a subduction zone actually decreases the extent of melting. It is necessary to consider what further thermal effect may occur as the former ridge crest passes below the arc, and particularly the effect of the last wedge of hot material emplaced beneath the ridge. (This wedge is represented schematically by the hachured area in Figure 3b, with no allowance made for the gap that should occur.)

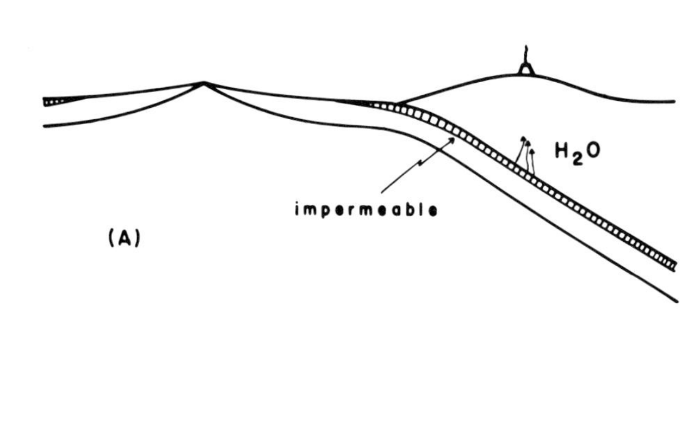

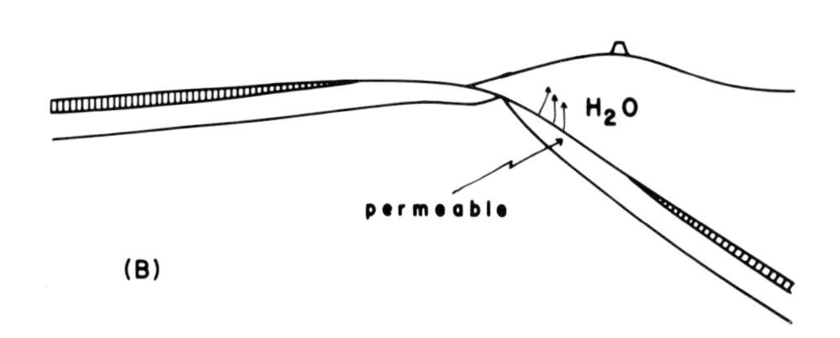

Figure 4. (A) Impermeable cap (layer 2A) prevents escape of water from slab at shallow depth. (B) Subduction of permeable young oceanic crust allows shallow escape of water.

For the purpose of predicting observable geological effects, it is adequate to examine a simple, non-rigorous case in which the wedge is viewed as a static source of heat. A single finite pulse of heat from the wedge will cause an upward deflection of isotherms, with a temperature increase, ΔT, estimated by

$$\Delta T = \frac{Q_v}{\rho C_p}, \quad (4)$$

where ρ is density, C_p is specific heat, and Q_v is heat added/unit volume (by a mechanism yet to be specified). If we use observed oceanic-ridge heat flow, Q^*, as an approximate measure of Q_v, then

$$\Delta T = \frac{Q^* t}{\rho C_p L}, \quad (5)$$

where L is the thickness of the arc overlying the slab and t is the time during which heating occurs.

The central 200-400 km of a ridge is a locus of unusually high heat flow [McKenzie, 1967]. At typical rates of subduction, this central portion of the ridge would move past the central 100 km of an arc (Figure 3b) in about 5-10 m.y. For rapid passage of the wedge (t = 5 m.y.), the temperature increases calculated from equation (5) would be 100°C for a reasonable average value of $Q^* = 6$ μcal/cm^2-sec ($\rho = 3$ g/cm^3, $C_p = 0.3$ cal/g-°C, L = 100 km). Slower passage (t = 10 m.y.) and a maximum likely value of $Q^* = 8$ μcal/cm^2-sec yields $\Delta T = 280$°C. Thus, the presence of the wedge could increase the temperature at shallow levels in the overlying arc by 100-300°C. (This range may provide only a lower limit to ΔT, because the observed Q^* at ridges is low due to heat loss by hydrothermal circulation [e.g., Sclater et al., 1976].)

There are at least three mechanisms by which temperature increases of this magnitude could be achieved: (1) If the edges of the leading and trailing plates are sufficiently buoyant that the former ridge passes at a very shallow depth under the arc (25-50 km), then an interval of 5-10 m.y. would be adequate to achieve a temperature increase of about 100°C by conductive heating [W. M.

Schwarz, personal communication, 1976]. (2) Some of the wedge of hot asthenospheric material (Figure 3b) may rise through the gap between the downgoing plates (Figure 2d), carrying heat to higher levels. (3) Similarly, water released by dehydration of young crust (Figure 4b) may also convect heat to shallow levels. In reality, all three of these mechanisms could operate simultaneously during ridge subduction.

Oxburgh and Turcotte [1971] have calculated that temperatures of 100-300°C are typical of shallow crust in island arcs. Temperature increases such as those calculated above for subduction of a former ridge should thus raise the temperature in this shallow crust to values in the approximate range 200-600°C. At a depth of a very few kilometers, the effect of the combined pressure-temperature regime on basalts and andesites of the arc should be thermal metamorphism, probably in the greenschist to epidote-amphibolite facies [Miyashiro, 1973]. We thus suggest that subduction of a ridge under the conditions outlined may produce a low-grade thermal metamorphism that could be detected in the geological record of the arc. (We are not suggesting, of course, that all alteration of rocks in present or former arcs is due to ridge subduction.)

Transient heating effects will obviously lead to a more complicated situation than we have described here. Nonetheless, preliminary numerical studies of the time-dependent thermal behavior of a ridge and an arc as they converge [W. M. Schwarz, S. E. DeLong, and R. N. Anderson, in preparation] are compatible with this largely qualitative outline of the thermal history of ridge subduction. In summary, the important points of this predicted history are: (1) arc volcanism is likely to diminish or cease during approach and subduction of the ridge; (2) heat from the former ridge crest will produce low-grade metamorphism in shallow crustal rocks of the overlying arc; and (3) arc volcanism will be reinstituted as older oceanic crust is again underthrust to appropriate depth.

Tests of the Model

Detailed tests of various aspects of the model have been made for the Aleutian arc [DeLong, Fox, and McDowell, 1976], where the Kula Ridge was subducted in the mid-Tertiary [Grow and Atwater, 1970], and for southern Chile [S. E. DeLong, J. Casey, and D. R. Spydell, in preparation], where part of the Chile Rise has been subducted as recently as about 1 m.y. ago [E. M. Herron, personal communication, 1976]. Summaries of these tests are given here.

Subduction of the Kula Ridge

A synthesis of the geology of the Aleutian Islands [DeLong et al., 1976] documents or is generally compatible with the model outlined above. A simplified geological column is shown in Figure 5, and we recognize the following sequence of events in our interpretation:

(1) Subduction-related arc magmatism greatly diminished on approach of the Kula Ridge about 45 m.y. ago (mid-Eocene). This reduction, first noticed by Marlow et al. [1973], is indicated throughout the insular arc (from Attu to Unalaska) by a conformable transition from deep marine, volcanic-rich rocks to volcanic-poor rocks.

(2) These deep marine rocks are succeeded by shallow marine rocks and subaerially derived conglomerates that indicate shoaling of the crest of the Aleutian arc in the Late Eocene to Oligocene.

(3) These shallower units are cut by an arc-wide unconformity that we attribute to subaerial exposure of the arc during maximum uplift and subduction of the most elevated, central portion of the Kula Ridge. Rocks below the unconformity are generally deformed and show greenschist mineralogy (albite + epidote + chlorite) or other indication of low-grade thermal metamorphism, as predicted. New K-Ar ages [DeLong and McDowell, 1975; L. M. Gard, Jr., personal communication, 1975; B. D. Marsh, personal communication, 1975] for igneous rocks from the pre-mid-

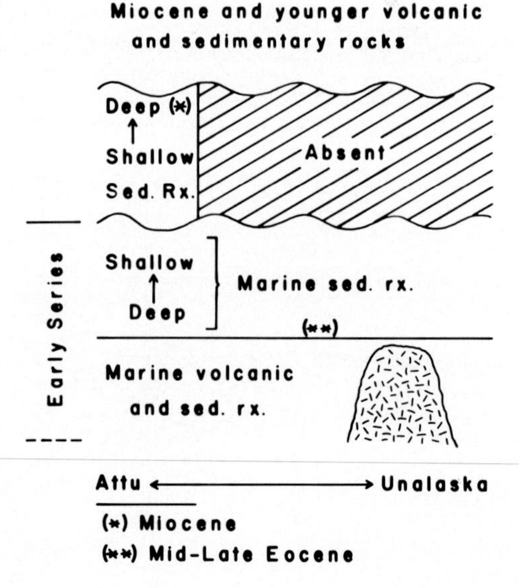

Figure 5. Simplified geological column for Aleutian Islands from Attu (173°E) to Unalaska (167°W).

Eocene, deep marine, volcanic-rich units indicate that the metamorphism occurred about 30 m.y. ago (mid-Oligocene). This is in excellent agreement with the 30 m.y. time of ridge subduction calculated by Grow and Atwater [1970] for constant-velocity Pacific-North America plate motion. Use of a simplified variable-velocity model for these two plates, based on the results of Atwater and Molnar [1973], suggests that the Kula Ridge reached the Aleutian Trench slightly earlier, at ~35 m.y. ago. These two ages (30 and 35 m.y.) are indistinguishable within the precision of presently available geological data.

(4) The arc subsided down the south flank of the Kula Ridge in the Late Oligocene to Miocene. This event is recognized only on Attu, where the unconformity is overlain by conglomerates, and these are succeeded in turn by non-volcanic, deep marine sedimentary rocks (cherts, argillites, shale) of probable Miocene age [Gates et al., 1971].

(5) Magmatism recommenced abruptly about 15 m.y. ago [Marlow et al., 1973; DeLong and McDowell, 1975], and has continued to the present throughout the arc, although with local volumetric fluctuations.

Thus, all three predicted geological effects--uplift of the arc, a magmatic hiatus, and low-grade thermal metamorphism--occurred in the Aleutians during the interval 45-15 m.y. ago, with ridge subduction at about 30-35 m.y. ago. Within the precision of the geological data, this history of ridge-trench interaction also accords remarkably well with inferred plate motions in the North Pacific [Atwater and Molnar, 1973; Moore, 1976]. Several other tests of consistency of the geological history with marine geological and geophysical data, including D.S.D.P. results, and an assessment of unresolved problems with this interpretation are given by DeLong et al. [1976].

Southern Chile

Marine magmatic anomalies indicate ridge subduction along southern Chile from less than 20 m.y. ago in the extreme south to about 1 m.y. ago just south of the present triple junction at 46°S [E. M. Herron, personal communication, 1976]. The area of Chile south of the triple junction today is largely devoid of magmatism and seismicity, in contrast to that north of the triple junction where the Chile Rise is still present offshore. Forsyth [1975] attributed the lack of seismicity to the warmer, thinner, softer, and lighter nature of the young plate now being subducted. We suggest that the lack of widespread volcanism has a similar cause, with a specific mechanism yet to be identified.

Pronounced topographic differences also occur to the north and south of the triple junction that may be related to impending and completed ridge subduction, respectively. In particular, to the south of the triple junction the South American Plate (and continent) may be attempting to move "downhill" along the cooling Antarctic Plate (the trailing plate behind the subducted portion of the Chile Rise). To the north of the triple junction the South American Plate may be moving "uphill" as the Chile Rise approaches from offshore. In north-south section, the plate profile would be a monocline.

We have recognized [DeLong, Casey, and Spydell, in preparation] three lines of evidence that are broadly consistent with this interpretation: (1) There is a progressive decrease in elevation from north to south, with a divide between areas higher and lower than 2-3 km occurring at 39°S. (2) The Chilean margin has an extensive drowned coastline south of about 42°S. (3) Analysis of drainage patterns indicates widespread stream capture and displacement of the Atlantic-Pacific watershed well east of the locus of highest elevation to the south of 42°S, suggesting westward tilting on a regional scale [Katz, 1962]. These topographic discontinuities occur a few hundred km north of the seismic and volcanic discontinuities centered about the triple junction. This discrepancy may reflect adjustment within the South American Plate to the conflicting geometric constraints imposed by side-by-side subduction of a cooling, trailing plate (Antarctic) which is following a ridge into a trench and a leading plate (Nazca) which defines a ridge that is yet to be subducted.

Acknowledgments. We thank R. N. Anderson, J. Kelleher, W. D. Means, and W. M. Schwarz for discussion of various aspects of the model presented here. Reviews by T. Atwater and R. Kay were extremely helpful.

References

Anderson, R. N., S. Uyeda, and A. Miyashiro, Geophysical and geochemical constraints at converging plate boundaries. Part I: Dehydration in the downgoing slab, Geophys. J. Roy. Astron. Soc., 44, 333, 1976.

Atwater, T., Implications of plate tectonics for the Cenozoic tectonic evolution of western North America, Geol. Soc. Amer. Bull., 81, 3513, 1970.

Atwater, T., and P. Molnar, Relative motion of the Pacific and North American plates deduced from sea-floor spreading in the Atlantic, Indian, and South Pacific Oceans, in Proceedings of the Conference on Tectonic Problems of the San Andreas Fault System, edited by R. L. Kovach and A. Nur, pp. 136-148, Stanford University Press, Palo Alto, Calif., 1973.

Bird, J. M., and J. F. Dewey, Lithosphere plate-continental margin tectonics and the evolution of the Appalachian orogen, Geol. Soc. Amer. Bull., 81, 1031, 1970.

DeLong, S. E., P. J. Fox, and F. W. McDowell, Subduction of the Kula Ridge at the Aleutian Trench, Geol. Soc. Amer. Bull., in press, 1976.

DeLong, S. E., and F. W. McDowell, K-Ar ages from the Near Islands, western Aleutian Islands, Alaska: Indication of a mid-Oligocene thermal event, Geology, 3, 691, 1975.

Dewey, J. F., Plate tectonics, Rev. Geophys. Space Phys., 13, 326, 1975.

Dewey, J. F., Ophiolite obduction, Tectonophysics, 31, 93, 1976.

Dickinson, W. R., and W. S. Snyder, Geometry of triple junctions and subducted lithosphere related to San Andreas transform activity (abstract), EOS Trans. AGU, 56, 1066, 1975.

Ewing, J., and M. Ewing, Sediment distribution on the mid-ocean ridges with respect to spreading of the sea floor, Science, 156, 1590, 1967.

Forsyth, D. W., Fault plate solutions and tectonics of the South Atlantic and Scotia Sea, J. Geophys. Res., 80, 1429, 1975.

Fox, P. J., and S. E. DeLong, Healing of oceanic layer 2A: Geophysical and geological consequences (abstract), EOS Trans. AGU, 57, 329, 1976.

Gates, O., H. A. Powers, and R. E. Wilcox, Geology of the Near Islands, Alaska, U.S. Geol. Survey Bull. 1028-U, 1971.

Grow, J. A., and T. Atwater, Mid-Tertiary tectonic transition in the Aleutian arc, Geol. Soc. Amer. Bull., 81, 3715, 1970.

Hayes, D. E., and J. R. Heirtzler, Magnetic anomalies and their relation to the Aleutian island arc, J. Geophys. Res., 73, 4637, 1968.

Hayes, D. E., and W. C. Pitman III, Magnetic lineations in the North Pacific, in Geological Investigations of the North Pacific, Geol. Soc. Amer. Mem., 126, edited by J. D. Hays, pp. 291-314, 1970.

Houtz, R., and J. Ewing, Upper crustal structure as a function of plate age, J. Geophys. Res., 81, 2490, 1976.

Katz, H. R., Fracture patterns and structural history in the sub-Andean belt of southernmost Chile, J. Geol., 70, 595, 1962.

Marlow, M. S., D. W. Scholl, E. C. Buffington, and T. R. Alpha, Tectonic history of the central Aleutian arc, Geol. Soc. Amer. Bull., 84, 1555, 1973.

McKenzie, D. P., Some remarks on heat flow and gravity anomalies, J. Geophys. Res., 72, 6261, 1967.

Miyashiro, A., Metamorphism and Metamorphic Belts, Allen & Unwin, London, 1973.

Moore, G. W., Basin development in the California border land and the Basin and Range Province, Pacific Sec. Am. Assoc. Petrol. Geol. Misc. Pub. 24, 383, 1976.

Oxburgh, E. R., and D. L. Turcotte, Origin of paired metamorphic belts and crustal dilation in island arc regions, J. Geophys. Res., 76, 1315, 1971.

Pitman, W. C. III and D. E. Hayes, Sea-floor spreading in the Gulf of Alaska, J. Geophys. Res., 73, 6571, 1968.

Pitman, W. C. III, R. L. Larson, and E. M. Herron, The Age of the Ocean Basins, Geol. Soc. America Map and Chart Ser., 6, 4 p. 1974.

Sclater, J. G., R. N. Anderson, and M. L. Bell, Elevation of ridges and evolution of the central eastern Pacific, J. Geophys. Res., 76, 7888, 1971.

Sclater, J. G., J. Crowe, and R. N. Anderson, On the reliability of oceanic heat flow averages, J. Geophys. Res., 81, 2997, 1976.

Turcotte, D. L., and G. Schubert, Frictional heating of the descending lithosphere, J. Geophys. Res., 78, 5876, 1973.

Uyeda, S., and A. Miyashiro, Plate tectonics and the Japanese Islands: A synthesis, Geol. Soc. Amer. Bull., 85, 1159, 1974.

GEOCHEMICAL CONSTRAINTS ON THE ORIGIN OF ALEUTIAN MAGMAS

R. W. Kay

Department of Geological Sciences, Cornell University, Ithaca, New York 14853

Abstract. The composition of Aleutian arc magmas contrasts with oceanic ridge magmas, backarc magmas, and intraplate magmas of the North Pacific-Bering Sea region. The differences in Pb and Sr isotope composition and large-ion lithophile (LIL) element content can be explained within the context of convergent plate interaction. Processes that have controlled Aleutian magma composition include: crystal settling, partial melting, and contamination of "oceanic" mantle and crust by admixture of a "continental component" rich in Ba, Rb and K, and radiogenic Pb and Sr. The rare earth elements (REE) and Ti are only slightly contaminated and are key elements in testing a two-stage melting-mixing model for the generation of Aleutian basalts with calc-alkaline and tholeiitic affinities. The model, which is based on a model of Ringwood (1974), postulates that partial melts are derived from the underthrust oceanic crust, transformed to quartz eclogite. These partial melts are mixed with the overlying peridotite mantle, forming a modified mantle which rises and segregates to form the erupted basalt magma and garnet-free peridotite residue. This model can explain both tholeiitic and calc-alkaline basalts of the Aleutian arc.

Introduction

Aleutian volcanic arc magmas erupt from central vent volcanoes located north of the Aleutian trench, and are intimately associated with subduction of the Pacific lithospheric plate. With the exception of Amak and Bogoslof, the volcanoes lie about 100 km above the Benioff zone. In the western Aleutians, where plate motion changes from convergent to strike-slip (see Figure 1), abundant recent volcanism has not been detected (Scholl et al., 1976).

Ringwood (1975) reviewed several processes that are widely recognized to account for the chemical characteristics of volcanic arc magmas (e.g., calc-alkaline and tholeiitic types, see Figure 2). The purpose of this paper is to evaluate the applicability of these processes to the genesis of some Aleutian basalts and andesites. Data on Pb and Sr isotope ratios and lithophile element concentrations provide important constraints on the role of sediments and sea water contamination, crystal fractionation, and partial melting in the evolution of Aleutian magmas. Several lithophile elements in two Aleutian basalts can be successfully reproduced by a specific model for volcanic arc magma genesis.

Isotopic and Trace Element Evidence of Sediment and Sea Water Contamination of Aleutian Magma Sources

Lead Isotopes

Lead isotope analyses of some Aleutian volcanic rocks are plotted in Figure 3. Also plotted are some ocean ridge and intraplate volcanic rocks from the Pacific Ocean-Bering Sea region. The major observation to be made in Figure 3 is that the values of volcanic arc and oceanic ridge-intraplate magmas do not overlap. One interpretation of the data is that the volcanic arc Pb is a mixture of two components (note the linear arrays of points), one with oceanic ridge basalt composition and the other not sampled in volcanic rocks of the region. Figure 3 shows that the missing component is richer in ^{207}Pb and resembles detrital sediment derived from continental weathering (Church, 1976). As in other volcanic arcs (Armstrong, 1971; Sun, 1973; Church, 1976), mixing proportions range from pure oceanic end members (Aleutian magnesian andesites; Kay, 1976) to large fractions of continent-derived Pb. Per cent contamination varies within a volcano. Because of the high Pb concentration in sediment relative to oceanic ridge basalt, significant Pb contamination requires only a small percentage of sediment in the mixture (Armstrong, 1971).

Pelagic sediment leads appear to have too low ^{206}Pb to be significant end members (Church, 1976). Conversely, ^{206}Pb is too high in most of the intraplate magmas. Church (1976) has come to the same conclusion for Cascade arc lavas, revising his earlier model (Church and Tilton, 1973). Hydrated oceanic crust is not a suitable source for the Pb contamination, because Pb concentrations are very low in sea water.

The site of the contamination is not easy to

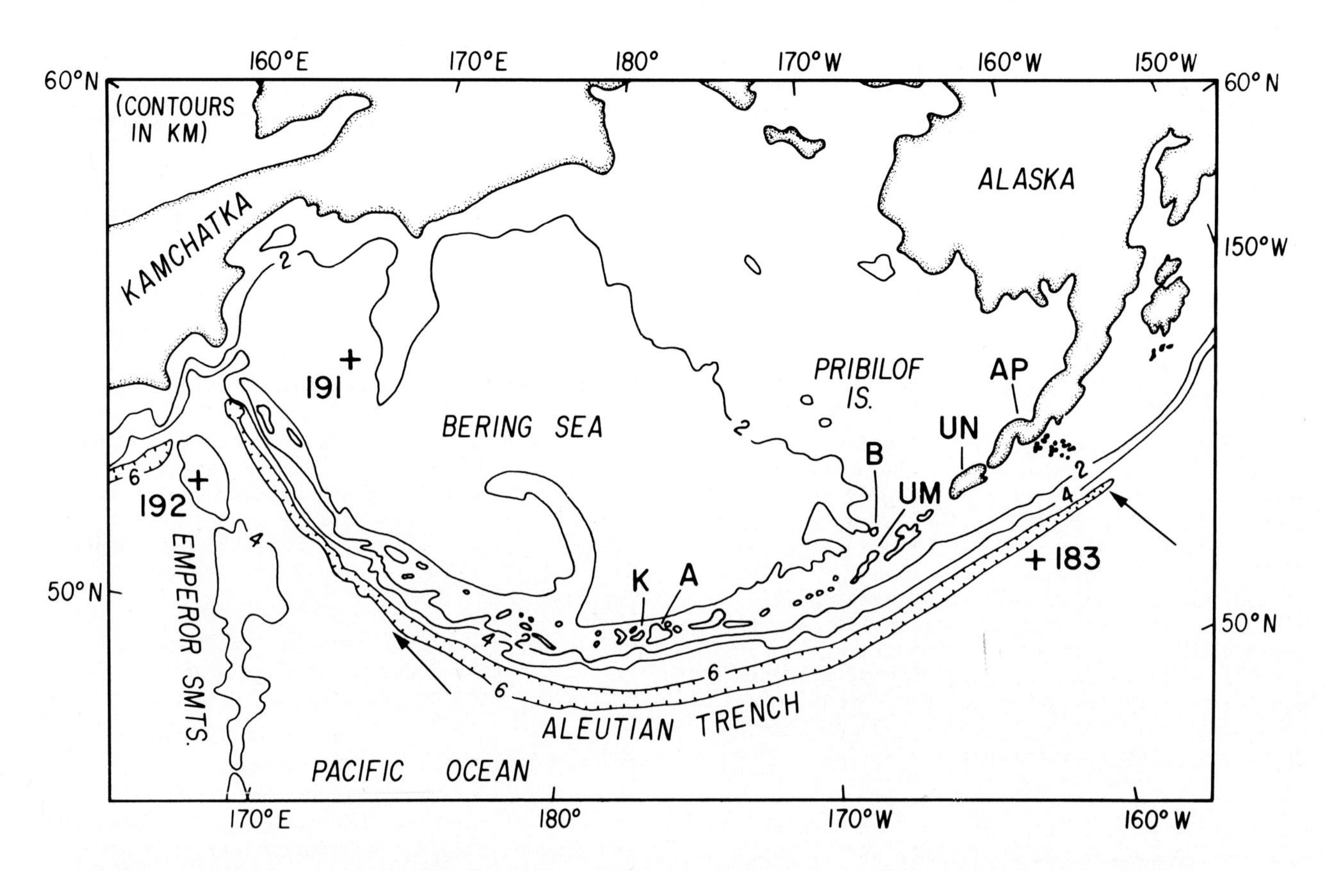

Figure 1. Location map of samples from the Aleutian region. Arrows near the Aleutian trench show present directions of relative plate motion. Samples from the Pacific Plate (Figure 6) include basalts from seamounts drilled by DSDP in holes 192 and 183 (Stuart et al., 1974). Locations of Giacomini and Hodgkins seamount basalts and basalts from the Gorda and East Pacific rises and Hawaii are off the map to the south and east. Samples from the Bering Sea basin (Figure 7) are basalts from the Pribilof islands and Kamchatka Basin (DSDP hole 191). Aleutian volcanic arc samples are basalts and andesites from active and dormant volcanoes on the Alaska Peninsula (AP) and the Aleutian islands (UN: Unimak, UM: Umnak, B: Bogoslof, A: Adak, K: Kanaga).

specify. Church (1976), citing Karig and Sharman's (1975) observation that most of the sediment reaching oceanic trenches is not subducted, prefers contamination by "underplated" sediment, rather then by sediment carried down with the underthrust lithosphere. The amount of sediment need be only a small fraction of the total underthrust column, however, and it seems hard to rule out "deep" contamination on physical grounds. Interestingly, some other arcs appear not to show Pb contamination (Oversby and Ewart, 1972; Meijer, 1974).

Strontium Isotopes

Kay et al. (1976) report an average $^{87}Sr/^{86}Sr$ value of 0.70322 $\pm$ 8 (2σ) for 24 Pleio-Pleistocene volcanic rocks from nine Aleutian volcanic centers. The values are among the lowest reported for volcanic arc rocks (see Figure 4), but are higher than values for oceanic ridge basalts from normal ridge segments (see Hofmann and Hart, 1975). Therefore, Aleutian magmas are not produced by equilibrium partial melting of the unaltered basaltic layer of the oceanic crust. Strontium isotope values of many oceanic volcanic rocks from seamounts and islands are in the range of the Aleutian values (see Hofmann and Hart, 1975) but seamounts and islands are probably only minor constituents of the oceanic crust, and in any case, Pb isotope data (see above) rule them out as a major contributor to Aleutian arc magmas.

The following observations are pertinent in evaluating explanations for Aleutian Sr isotope values:

a) The values do not change along strike of the volcanic arc, even across the transition from older to younger crust at the Bering shelf intersection (see Figures 1 and 4). This argues against crustal contamination, since the older crust should have higher ratios than younger crust;

b) The values do not correlate with Rb/Sr

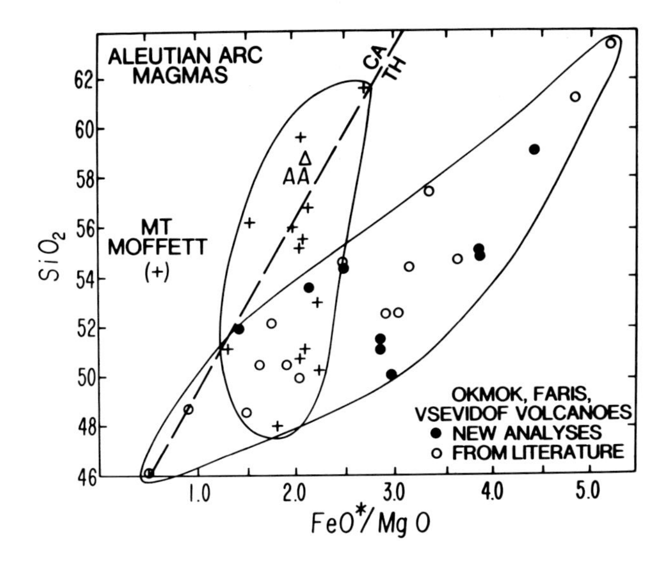

Figure 2. Iron-magnesium ratio versus silica for two volcanic rock series of the Aleutian islands. Analyses marked with filled dots are new (Table 1). The series with iron enrichment closely resembles the Japanese pigeonite series (Byers, 1961), which is referred to as tholeiitic by many workers; analyses fall in the tholeiitic (TH) field of Miyashiro (1974). Phenocrysts include plagioclase, olivine, and clinopyroxene. The series with almost constant iron-magnesium ratio (Coats, 1952) closely resembles the Japanese hypersthene series, which is referred to as calc-alkaline by many workers. The more silicic members of the series fall in Miyashiro's (1974) calc-alkaline (CA) field. Phenocrysts include plagioclase, olivine, clinopyroxene, hypersthene and hornblende. Basaltic members of the two series have different trace element abundances (see Figure 8). Point labeled AA is Chayes' (1969) average Cenozoic andesite (Table 1).

ratios (Figure 4). This argues against crustal contamination (see Taylor and Turi, 1976). Data from some other volcanic arcs do show a correlation (see Figure 4) -- with more acidic rocks having higher $^{87}Sr/^{86}Sr$ and Rb/Sr ratios. James (1975) has worked out a disequilibrium partial melting model involving phlogopite (a high Rb/Sr mica) to explain the Peruvian data on Figure 4. The Aleutian data permit no such interpretation;

c) Strontium and Pb isotope compositions correlate (see Figure 5). Ocean ridge basalts (ORB) generally plot off Figure 5 in the lower left hand corner and appear to be one component of the mixture. Lead isotope ratios (Figure 3) rule out mantle sources of intraplate magmas similar to alkali basalts from the Pribilof Islands as a second component. The second component has "continental" Pb isotopes (Figure 3) and higher Sr isotopes than are found in ORB;

d) The difference between Aleutian and oceanic ridge Sr isotope values is small. This limits the percentage of Sr contamination that can occur. There is little problem in generating the .7032 ratios of the Aleutians from oceanic crust with a .7025 ratio by addition of Sr from sea water (ratio of .709) or continental detritus (ratio of about .710: Peterman et al., 1967; Hart et al., 1970). Strontium contamination from these sources is limited to about 10% of the total Sr by mass balance. Models of oceanic crust containing extensive hydration may have values that are too high, rather than too low. The Troodos ophiolite, often used as an example of oceanic crust is an example; the lowest value reported is .7034, and many hydrated rocks have values over .7050 (Peterman et al., 1972; Chapman et al., 1975; Coleman and Peterman, 1975).

The relative importance of sediment and sea water contamination is not resolved by the present data. The correlation of $^{87}Sr/^{86}Sr$ and $^{207}Pb/^{204}Pb$ is an argument in favor of Sr contamination by sediment, but the data do not rule out variable amounts of sea water contamination.

Lithophile Elements

Normalized concentrations of a series of large ion lithophile (LIL) elements in volcanic rocks are plotted in Figures 6-8 for Pacific Ocean, Bering Sea and Aleutian arc samples. The general fractionation trend of REE in the samples is a smooth progression in abundance versus ionic radius (which increases from right to left in the figures). The range in La/Yb ratios of island arc rocks and oceanic rocks are similar.

Two observations about the La/Yb ratios of Aleutian arc rocks are relevant to models for their origin:

1) Island arc tholeiites (IAT), lavas with low La/Yb ratios (e.g., Tonga and Japan samples in Figure 8) have not been reported in the Aleutians. Compared to IAT, Aleutian tholeiites have higher contents of all elements plotted in Figure 8 as well as Sr, Ti, and Zr;

2) Aleutian tholeiites have lower La/Yb ratios than Aleutian calc-alkaline volcanic rocks; tholeiites are also lower in Sr, but are higher in Ti and heavy REE.

Note that regardless of REE fractionation pattern, all volcanic arc samples have at least a factor of three higher Ba/La, K/La and Rb/La ratios than do oceanic samples. For instance, the lowest Ba/La ratios along Aleutian lavas (average of 33, range 28 to 40 for eight tholeiites) are three times the highest alkali basalt values (Figures 7 and 8; see also: Kay and Gast, 1973). Barium/lanthanum ratios of ocean ridge basalts are even lower (Kay et al., 1970).

A mechanism for explaining the Ba/La ratio is to postulate that although the ratios in oceanic and volcanic arc sources are similar, fractionation during melting or crystallization was different in the two cases. Phases with large

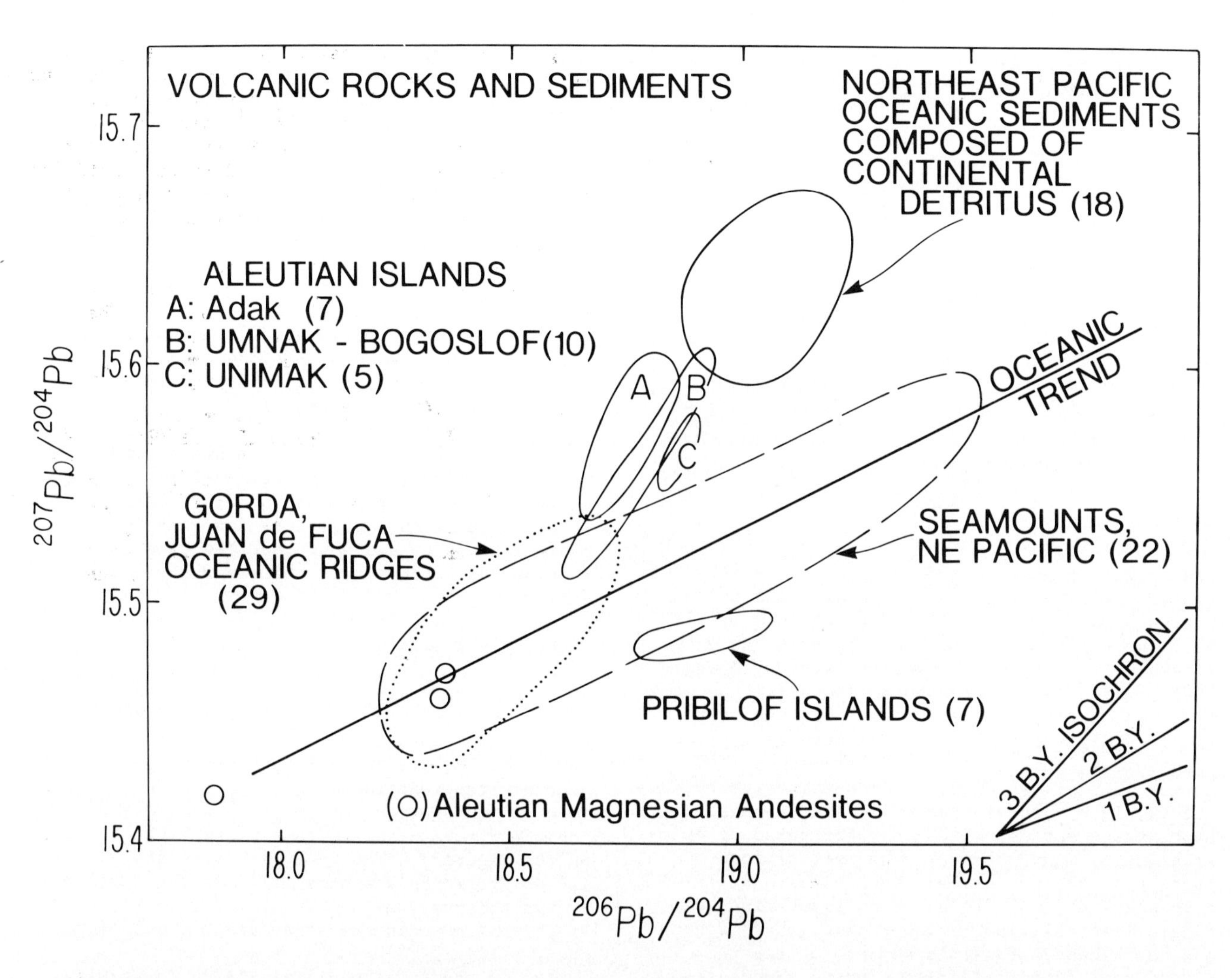

Figure 3. Ratios of two radiogenic lead isotopes (^{206}Pb, ^{207}Pb) to non-radiogenic lead (^{204}Pb) in volcanic rocks from the Aleutian arc, NE Pacific ocean (Sun, 1973; Church and Tatsumoto, 1975; Kay et al., 1976), Pribilof Islands (Kay et al., 1976), and in oceanic sediments composed of continental detritus (Church, 1973, 1976; Sun, 1973). Aleutian lead appears to be a mixture of oceanic ridge Pb and sediment Pb. Aleutian magnesian andesites are from Kay (1976).

cation sites (hornblende, phlogopite, sanidine) occur in some volcanic arc magmas, and equilibration of the melt with these phases could fractionate ratios of Ba, K, Rb to La. But fractionation would result in lower ratios in the melt relative to solid (residue or phenocrysts), and we want to generate higher ratios. Residual phases rich in light REE but not Ba, K and Rb could result in Ba/La ratios that are higher in the melt than in the starting material. However, many of these phases are among the first to melt during partial fusion (Gill, 1974). Experimental verification of residual minor phases in equilibrium with island arc tholeiites (Tofua 17, Table 1) would require K-, Ti-, P- and Zr-rich phases to be liquidus phases of melts with K = .33%, Ti =.49%, P_2O_5 = .09%, and Zr = 8 ppm, which seems unlikely.

Mixing of material with a high Ba/La ratio and source rocks or magmas with a low "oceanic" ratio is the preferred mechanism for generating high ratios in Aleutian volcanic rocks. Quantitative mass balance is not presented but the contamination hypothesis is consistent with Pb isotope evidence of sediment contamination. The high Rb, K and Ba contents (and ratios with La) of continental detrital and pelagic sediment (Church, 1973) compared to mixing end members such as ocean ridge basalt make several per cent contamination sufficient to triple the Ba/La ratios. In contrast, the low La/Yb ratios of Tongan lavas with high Ba/La show that the REE are not significantly contaminated. Note that the low Sr isotope ratios in Aleutian arc rocks (see previous section) demand high Ba/Sr ratios in the contaminant. Altered ocean ridge basalt (Frey et al., 1974) and basalts from ophiolites (old oceanic crust?, Kay and Senechal, 1976) do not often attain ratios comparable to those in the island arc rocks (e.g., Ba/La > 30, Ba/Sr ≈ 1). In summary, oceanic source ratios of Ba, K. Rb to La are probably increased by both sediment and sea water alteration of subducted ocean ridge basalts. The contamination hypothesis implies that a major portion of K, Rb and Ba in Aleutian arc rocks is recycled from oceanic sediment or sea water (Hart et al., 1970; Armstrong, 1971).

Effects of Crystal Fractionation in Aleutian Magmas

The Aleutian magmas are commonly andesite and basalt, and depending on the volcano, may follow either a tholeiitic or calc-alkaline differentiation (see Figure 2). The two trends are treated separately here.

Tholeiitic Magmas

Basalts and andesites of the tholeiitic series (48-62% SiO_2) show systematic increases in Fe/Mg ratios (Figure 2), TiO_2 and K_2O as silica increases (Byers, 1961). For Okmok volcano, Umnak Island, Byers (1961) states that "fractional crystallization was probably the dominant process in the formation of quantitatively minor andesite and rhyolite masses associated with the basaltic shield volcano."

The trace element data listed in Table 1 substantiate Byers' (1961) conclusion. Elements that are highly partitioned into the melt are higher in andesites than in basalts, but their relative abundance is the same. This can be seen by comparing the average of four basalts with the average of four andesites on Figure 8. Slightly decreasing Sr and increasing negative Eu anomalies in the differentiation series can be explained by fractionation of plagioclase. Constant Ba/La and La/Yb ratios and the positive correlation of Fe/Mg with silica argue against amphibole and garnet fractionation at high water pressures in the mantle (Ringwood, 1975). Neither amphibole nor garnet is a liquidus phase. There is no foundation for attempts to derive andesites in the tholeiitic series directly from mantle depths; all appear to have been derived from basaltic parents at crustal pressures.

Calc-Alkaline

Fractionation of Aleutian calc-alkaline magmas (Figure 2) appears to be more complex than fractionation of the tholeiitic series. The large proportion of phenocryst phases in many of these lavas (e.g., Marsh, 1976) is partly due to crystal accumulation, as evidenced by positive Eu anomalies in some plagioclase-rich basalts from Adak (Walker, 1974) and Bogoslof (Figure 8). Other phenocryst phases, also probably involved in crystal fractionation, are

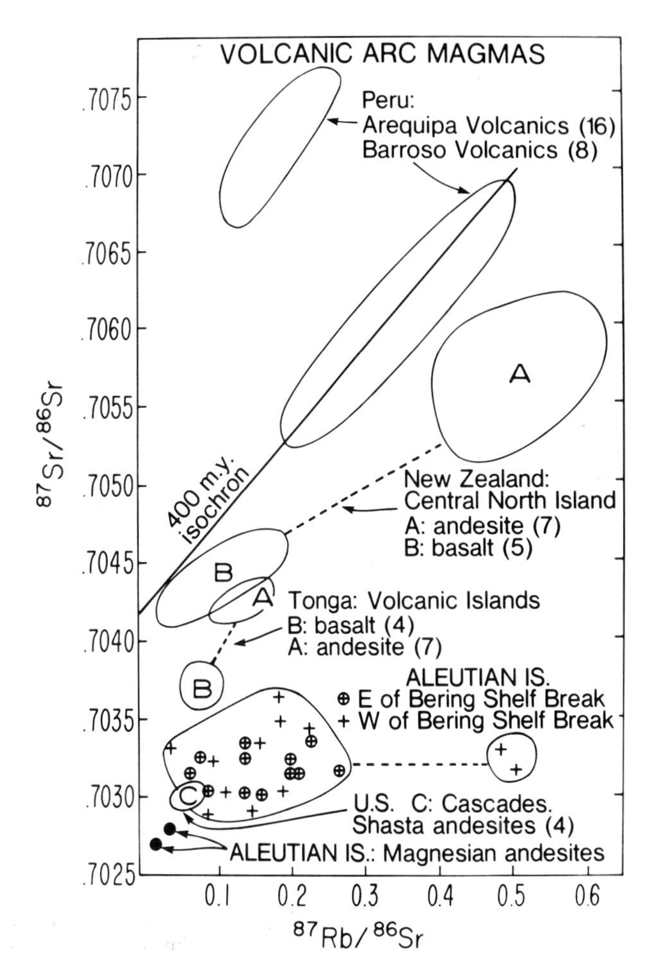

Figure 4. $^{87}Sr/^{86}Sr$ versus $^{87}Rb/^{86}Sr$ for volcanic rocks from the Aleutians (Pleio-Pleistocene) and other island arcs. Aleutian $^{87}Sr/^{86}Sr$ values (normalized to a ratio of .71022 for NBS-987) are among the lowest so far reported in any volcanic arc. Points with crosses are from "oceanic" part of the arc, points with crosses in circles are from "continental" part. The linear arrays of points found in Peru (James et al., 1976), New Zealand (Ewart and Stipp, 1968), and Tonga (Gill and Compston, 1973) are not found in the Aleutians. Aleutian magnesian andesites are from Kay (1976).

olivine, clinopyroxene, hypersthene and hornblende. Even though basalts are a common magma type (Coats, 1952), trace element abundances do not allow derivation of some Aleutian andesites from basalt by low pressure crystal fractionation (Kay, 1976). Peterman et al. (1970) and Thorpe et al. (1976) reach the same conclusion for some Cascades and Chilean andesites.

A Melting-Mixing Model for the Genesis of Aleutian Basalts

Models for the origin of volcanic arc magmas range in complexity from single stage (e.g.,

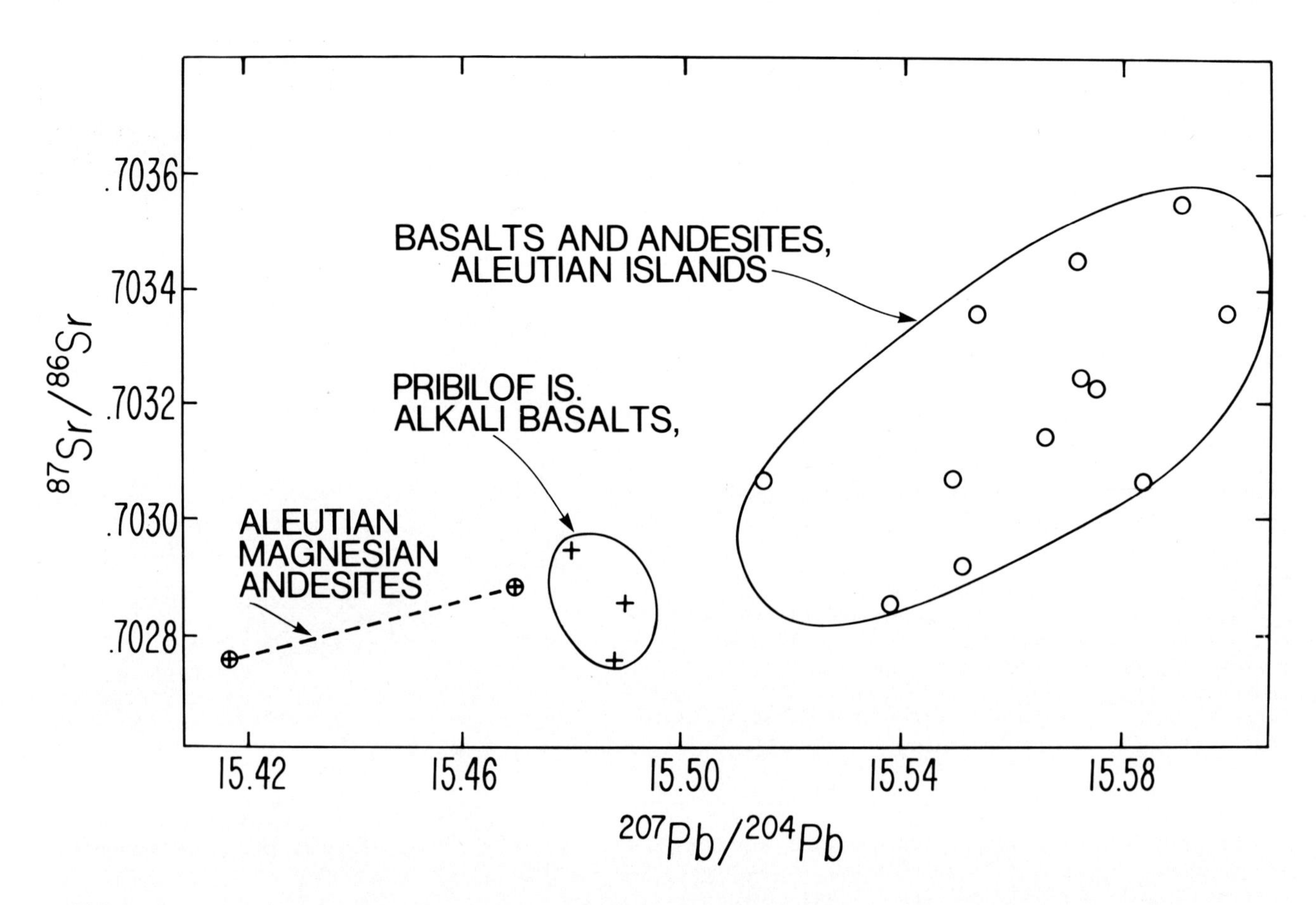

Figure 5. Ratios of radiogenic to non-radiogenic Sr and Pb in Aleutian island lavas showing the positive correlation, thought to be a mixing line (see Figure 3).

eclogite melting) to multiple stage (e.g., Ringwood, 1974). Gill (1974) carefully documented the inadequacy of simple partial melting of eclogite (subducted oceanic crust) to explain the chemistry of calc-alkaline andesites from Fiji. Ewart et al. (1973), DeLong (1974), Lopez-Escobar et al. (1976) and Thorpe et al. (1976) also reject the eclogite model. In contrast to the above workers, Marsh (1976) derives Aleutian andesites directly from the subducted oceanic crust. The multiple stage model of Ringwood (1974) receives support from calculations presented in this section, which attempt to match calculated and observed concentrations of some elements in Aleutian basalts.

Ringwood's (1974) model distinguishes two types of volcanic arc magmas. The first type is formed at shallow depth (less than 100 km) by partial melting of the (lithospheric) mantle wedge overlying an actively dehydrating subducted plate of oceanic lithosphere. These magmas, the "island arc tholeiite" series, receive no melt from the subducted plate. The second type is formed at depths greater than 100 km by mixing of partial melt from the subducted plate and overlying mantle. Both magma types often have isotopic and lithophile element ratios consistent with contamination with a "crustal" component.

This model implies that we can infer the composition of the mantle overlying the subducted plate from the chemistry of "island arc tholeiites". However, tholeiites having trace element characteristics of "island arc tholeiites" (e.g., low La/Yb ratios) have not been reported in the Aleutians. They are of widespread occurrence in volcanic arcs of the western Pacific (e.g., Tonga and Japan, see Figure 8) and in the Cascades (Warner basalt, Philpotts et al., 1971). Mantle capable of giving rise to these magmas must be similarly widespread; it is assumed to exist between the underthrust lithosphere and the volcanic arc in the Aleutians. In the calculations that follow, island arc tholeiite Tofua 17 (Tonga Island, Table 1) will be used.

If the percentage of partial melting of the mantle necessary to produce a melt of the composition of Tofua 17 is known, the composition of the mantle can be calculated. This melting percentage, treated as a variable in preliminary calculations, is controlled by the temperature at the segregation depth and the water-peridotite mixing ratio. Assuming that melting occurred when the mantle was shallow enough to be in a

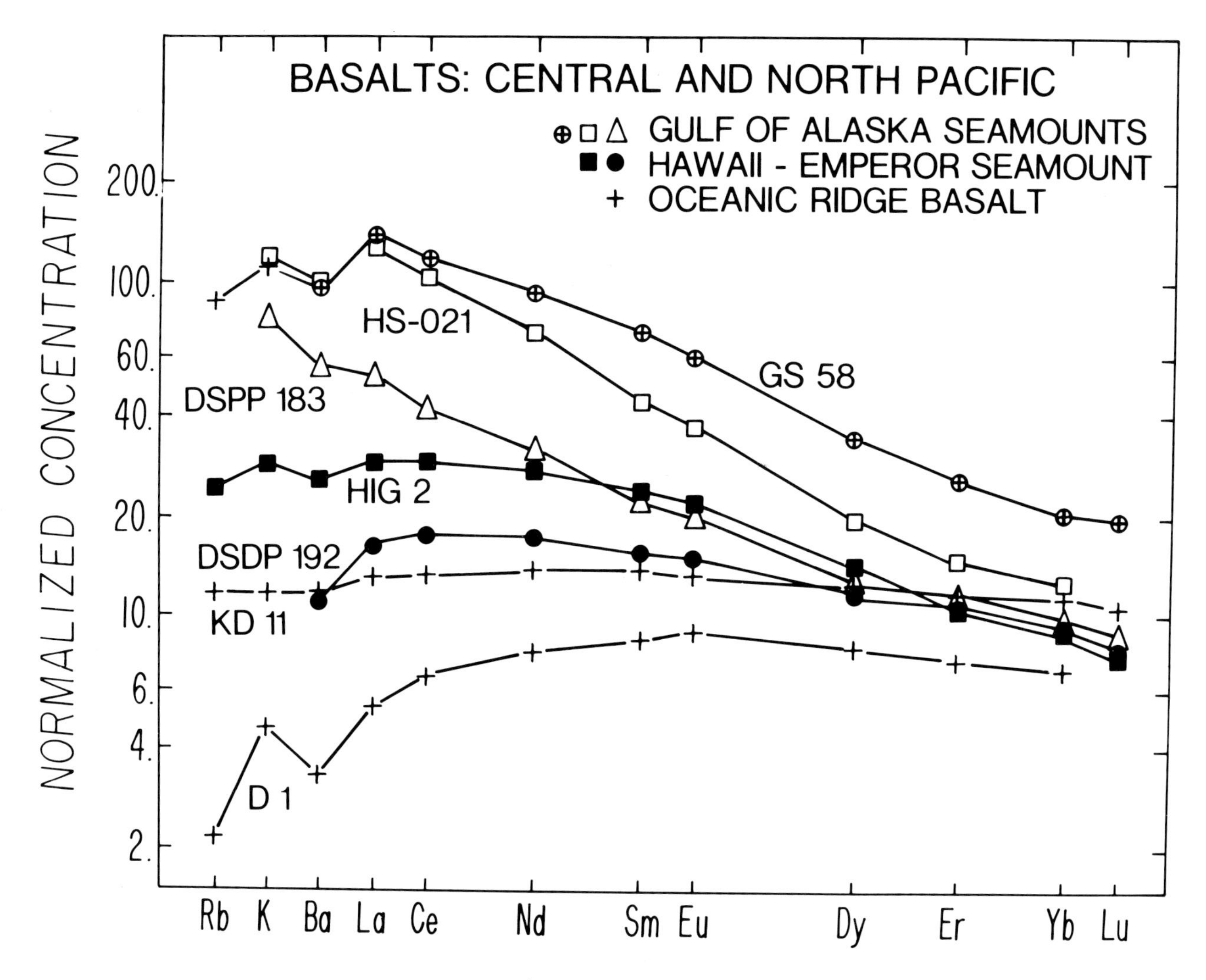

Figure 6. Rare earths, Ba, K, and Rb in basalts from the central and north Pacific ocean basins (Kay et al., 1970; Kay, 1976). Analyses are normalized to Leedy chondrite for REE, and Ba (Masuda et al., 1973) and to sample KD11 (Kay et al., 1970) for K and Rb; normalized values of Ba, K and Rb were made equal for this sample. Note the increase of La/Yb ratios from ocean ridge tholeiitic basalts to Hawaiian tholeiites to seamount alkali basalts.

garnet-free, two-pyroxene peridotite stability field, and that five per cent partial melting occurred, the mantle source of Tofua 17 would have the Ba, Sr, REE and Ti values shown in column D of Table 2 and plotted in Figure 9. This mantle composition will be used in subsequent calculations as representative of the mantle above the downgoing slab in the Aleutians.

Following Ringwood's (1974) model, partial melts of the subducted basaltic oceanic crust (transformed to quartz eclogite) mix with the overlying mantle (column D, Table 2) at depths greater than 100 km. The garnet-clinopyroxene ratio of the eclogite is not well known and is treated as a variable. It decreases with increasing ferric-ferrous ratio (Gill, 1974), but increases with depth. Using several garnet-clinopyroxene ratios and varying the percentage of melting, trace elements in numerous partial melts were calculated. Three of these melts, along with the parameters of the calculation, are listed in Table 2 and plotted in Figure 9. Melt A, from Kay (1976), closely matches the composition of some Miocene magnesian andesites from the Aleutians. Melts B and C provide good mixing end members to be combined with the overlying mantle (composition D) to produce modified mantle sources for calc-alkaline and tholeiitic basalts from the Aleutians.

Two modified mantles are plotted in Figure 10. The proportions of B and D used to calculate modified mantle E, and the proportions of C and D used to calculate modified mantle F, are shown on the bottom of Figure 10. In the model, these

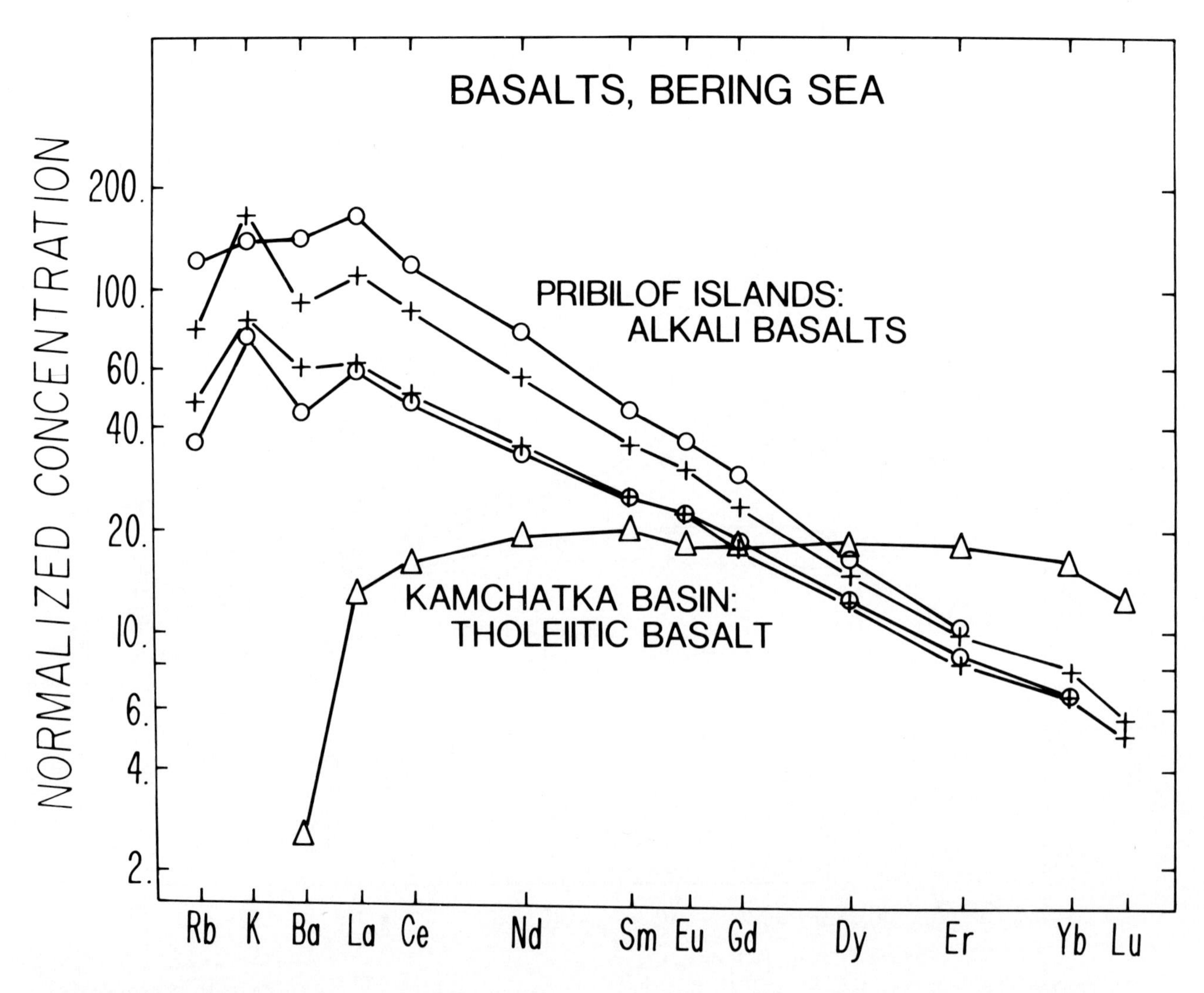

Figure 7. Rare earths, Ba, K and Rb in basalts from the Bering Sea (Kay, 1976). Analyses are normalized as in Figure 6. Trace element ratios of alkali basalts from the Pribilof islands resemble those from equivalent rocks from the Pacific ocean basin (Figure 6). The Kamchatka Basin basalt resembles oceanic ridge basalt (Figure 6).

modified mantle compositions, formed near the top of the subducted lithospheric slab, rise into the pyroxene peridotite stability field and partially melt to yield Aleutian tholeiitic basalt (UM4) and calc-alkaline basalt (SIM2). The amount of partial melting required and the observed versus calculated values for the trace elements in UM4 and SIM2 are plotted in Figure 10. The agreement is good for the REE (especially for UM4). Titanium and the Ti/Yb ratio are too high in the model for the calc-alkaline basalt (SIM2); minor phases such as rutile or ilmenite may be stable in the residue and hold back more Ti than indicated by the calculation presented here (Gill, 1974; Marsh, 1976).

Barium is too high and Sr is too low in both calculated compositions. This disagreement is not surprising, since the concentrations of these elements in Tofua 17, the "island arc tholeiite" used to calculate one of the mixing end members is heavily influenced by the proportion of "crustal component" thought to contaminate its mantle source. The proportion may be different in the Aleutian basal sources.

Summary and Conclusions

At this stage, the model presented in the previous section is largely heuristic. The results hinge on the many assumptions and the values of the parameters used. The model is consistent with Ringwood's (1974) model of island arc magmatism in that it outlines one set of conditions which do give rise to correct trace element fractionation trends for both tholeiitic and calc-alkaline volcanic rocks in the Aleutian arc.

Table 1[1].

	Aleutian Tholeiites[2]									
	UM10	UM4	UM5	UM18	SAR11	SAR24	SAR18	SAR39	AB21	AB26
La	6.50	10.1	9.26	14.2	11.0	23.2	---	---	12.3	18.3
Ce	15.1	24.3	22.2	35.4	27.3	55.3	---	---	29.3	43.8
Nd	10.3	15.4	14.8	24.3	18.8	36.1	---	---	19.7	29.6
Sm	2.60	4.05	4.06	6.81	5.05	9.28	---	---	5.31	7.78
Eu	0.834	1.23	1.30	2.02	1.59	2.51	---	---	1.51	2.12
Gd	2.66	4.17	---	---	---	10.1	---	---	6.03	8.59
Dy	2.46	4.26	5.00	7.51	6.00	10	---	---	6.09	8.61
Er	1.43	2.47	3.05	4.47	3.37	5.95	---	---	3.60	5.07
Yb	1.34	2.42	3.92	4.24	3.15	5.74	---	---	3.44	4.29
Lu	---	0.350	---	---	---	0.864	---	---	---	---
Ba	220	300	370	467	309	669	---	---	388	615
Sr	557	442	367	349	379	360*	396	382	331	290*
Rb	8.07	18.8	20.4	27.7	19.1	51	20	21.9	22	36*
K	6,276	9,489	8,224	10,521	8,402	16,300	9,760	9,921	8,528	14,100
Zr	85	140	130	180	170	260	175	190	159	237
Y	10	23	26	44	---	51	20	20	36	47
Ba/La	34	30	40	33	28	29	---	---	32	34
K/Rb	780	505	400	380	440	320	490	450	390	390
SiO_2	51.3	51.7	53.6	55.2	51.2	59.0	50.0	51.2	54.3	61.3
TiO_2	0.68	1.12	1.30	2.48	1.56	1.46	2.14	2.21	1.2	1.0
Al_2O_3	16.2	17.7	16.5	14.8	16.8	16	16.5	15.6	18	16.3
Fe_2O_3	---	---	---	---	---	---	9.41	7.35	1.7	2.1
FeO	8.3	7.8	9.4	10.3	10.9	8.7	3.40	5.18	7.8	6.0
MnO	0.15	0.15	0.21	0.25	0.26	0.28	0.24	0.31	0.18	0.20
MgO	7.45	5.45	4.46	2.72	3.76	1.96	3.97	4.06	3.8	1.6
CaO	11.32	10.12	9.17	6.86	9.28	5.19	9.82	8.56	7.9	5.6
Na_2O	2.22	3.12	3.47	4.11	3.79	4.95	3.34	4.02	3.4	4.6
K_2O	0.61	0.97	0.77	1.14	0.95	1.96	1.14	1.18	1.0	1.7
P_2O_5	0.13	0.20	0.18	0.51	0.24	0.49	0.36	0.23	0.32	0.42

	Aleutian:Calc-Alkaline[2]					Tonga[3]	Average[4] Andesite	KD11[5]
	AB23	SIM2	B1927	ADK58	KAN-1	Tofua 17		
La	10.7	14.8	7.82	11.3	5.85	1.38	---	4.90
Ce	26.3	32.3	20.0	24.2	15.35	3.73	---	12.7
Nd	16.7	17.7	15.1	13.5	10.5	3.34	---	9.72
Sm	4.13	3.81	3.76	2.86	2.64	1.19	---	3.08
Eu	1.01	1.16	1.45	0.864	0.820	0.449	---	1.13
Gd	4.50	---	3.45	2.50	0.24	1.56	---	---
Dy	4.35	2.96	3.10	2.24	2.11	1.85	---	4.81
Er	2.53	1.71	1.66	1.35	1.13	1.34	---	2.93
Yb	2.38	1.73	1.53	1.38	1.05	1.53	---	2.81
Lu	---	0.261	---	---	0.153	---	---	0.408
Ba	473	508	864	551	180	107	---	44.6
Sr	308	645	813	630	---	205	---	135
Rb	51.2	31	43	29	---	4.6	---	3.88
K	12,900	11,400	15,900	---	---	3,300	---	1,400
Zr	171	---	90	80	---	8	---	80
Y	29	15	15	20	---	---	---	---
Ba/La	44	34	110	49	31	78	---	9.1
K/Rb	250	400	370	---	---	720	---	360
SiO_2	57.4	---	48.6	---	---	53.8	58.17 ± 4.06	49.5
TiO_2	0.90	---	1.46	---	---	0.49	0.8 ± 0.35	1.22
Al_2O_3	18.2	---	18.4	---	---	14.47	17.26 ± 1.56	16.2
Fe_2O_3	2	---	---	---	---	2.71	3.07 ± 1.38	---
FeO	5	---	9.62	---	---	7.07	4.18 ± 1.62	8.4
MnO	0.10	---	0.19	---	---	0.17	---	---
MgO	4	---	4.54	---	---	7.34	3.24 ± 1.24	8.2
CaO	7.7	---	11.93	---	---	11.92	6.93 ± 1.63	12.1
Na_2O	2.9	---	2.99	---	---	1.48	3.21 ± 0.72	2.57
K_2O	1.4	---	1.75	---	---	0.33	1.61 ± 0.75	0.17
P_2O_5	0.16	---	0.17	---	---	0.09	0.21 ± 0.15	---

[1]Unless otherwise noted, major element analyses by M. Budd, trace element analysis by isotope dilution mass spectrometry, except Zr, Y, and starred Sr and Rb analyses, by x-ray fluorescence.

[2]Samples from Okmok volcano, Umnak Island (UM series), Faris volcano, Unimak Island (SAR series), Mt. Vsivdedof, Umnak Island (AB21, AB26), Mt. Recheschnoi, Umnak Island (AB23), Mt. Simeon, Unimak Island (SIM2), Bogoslof Island (B1927), Mt. Moffett, Adak Island (ADK58), and Kanaga Island (KAN-1). Sample location and description references L. Byers (1961), DeLong et al. (1975), Arculus et al. (1976), Kay et al. (1976). Analysis AB21, AB34, AB26, and ADK 58 by A. Walker (Walker, 1974).

[3]Analysis from Bauer (1970), trace element analysis from Kay and Hubbard (1976).

[4]Chayes (1969) errors are ± 1σ.

[5]Kay et al. (1970) ocean ridge basalt. Trace element analyses are new.

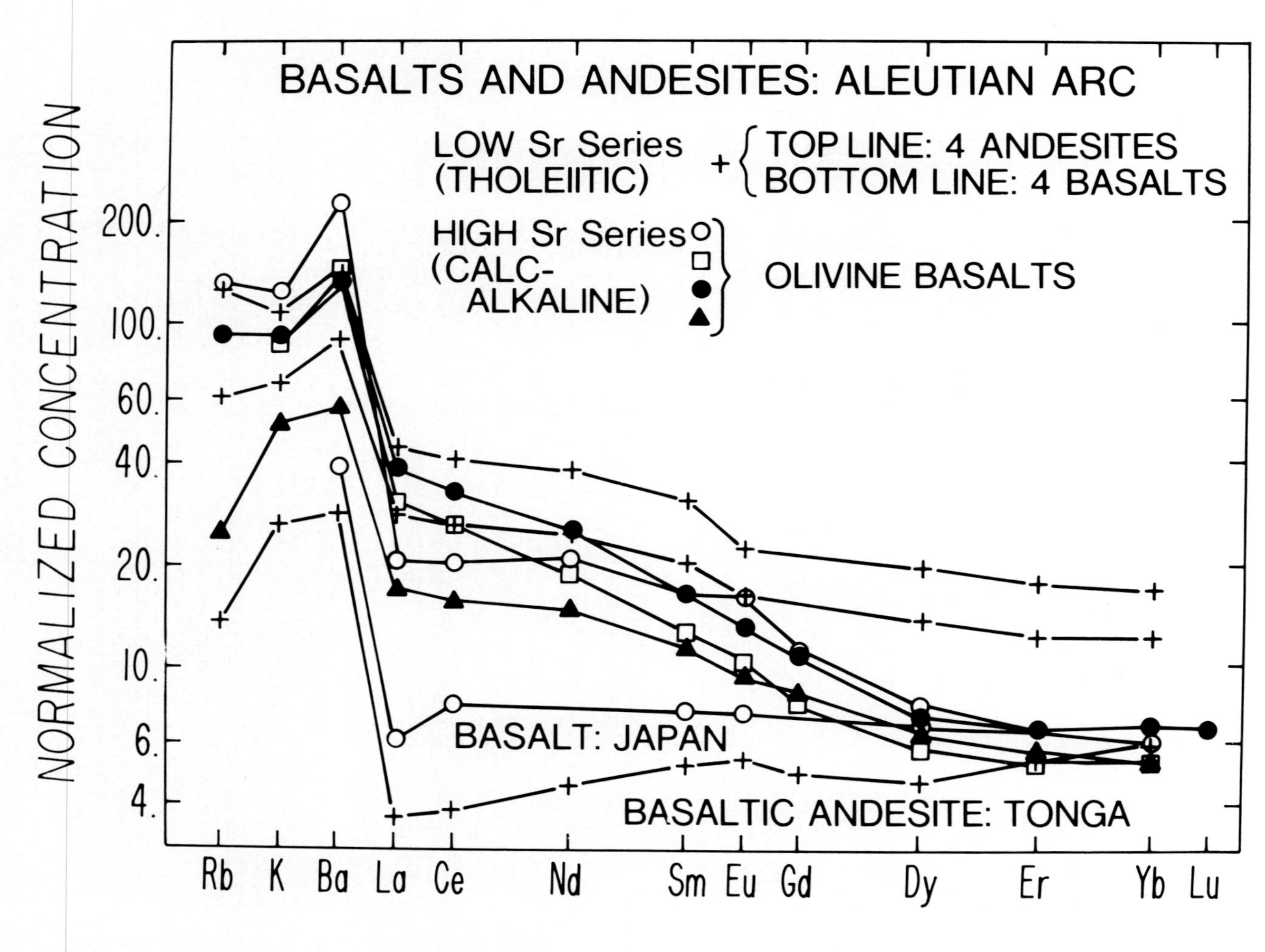

Figure 8. Rare earths, Ba, K and Rb in basalts and andesites from the Aleutian islands (Table 1) and in two basaltic rocks from the "island arc tholeiite" series in Japan (Katsui et al., 1974; sample H1 and Tonga; Table 1; sample Tofua 17). Samples are normalized as in Figures 6 and 7.

Table 2.

	D^{Cpx}_{melt}[1]	D^{Gar}_{melt}[1]	A[2,4]	B[2,4]	C[2,4]	D[2,5]	Chondrites[2]	D1[2,3]
Ba	0.001	0.001	400	240	82	5.4	3.8	12.4
Sr	0.078	0.01	1860	1215	670	13.3	15	135
La	0.05	0.001	35.7	22	10.7	0.082	0.378	1.95
Sm	0.261	0.107	9.2	7.1	5.7	0.114	0.230	1.92
Dy	0.313	1.27	3.5	5.3	4.5	0.197	0.390	2.80
Yb	0.227	4.20	0.79	1.7	1.4	0.139	0.249	1.72
Ti	0.5	0.01	1.59	1.03	0.93	0.041	0.072	0.44
(Cpx/Gar) STARTING SOLID	---	---	1	4	3	---	---	---
(Cpx/Gar) INTO MELT	---	---	1	1	1	---	---	---
%Melt	---	---	3	5	15	---	---	---

[1]Distribution coefficients used by Kay (1976). Cpx = clinopyroxene, Gar = garnet.

[2]Concentrations in ppm except Ti, in per cent.

[3]Ocean ridge basalt D1 concentrations from Kay et al. (1970).

[4]A, B, and C are three equilibrium partial melts (Shaw, 1970, equation 11) of eclogite with composition of D1, variable proportions of clinopyroxene and garnet, and equal amounts of each melting. They are plotted in Figure 9.

[5]Composition of olivine-orthopyroxene-clinopyroxene mantle from which island arc tholeiite Tofua 17 (Table 1) can be derived by 5% melting, assuming 20% model clinopyroxene, and clinopyroxene contributing 40% of the melt. Distribution coefficients for clinopyroxene are in column 1.

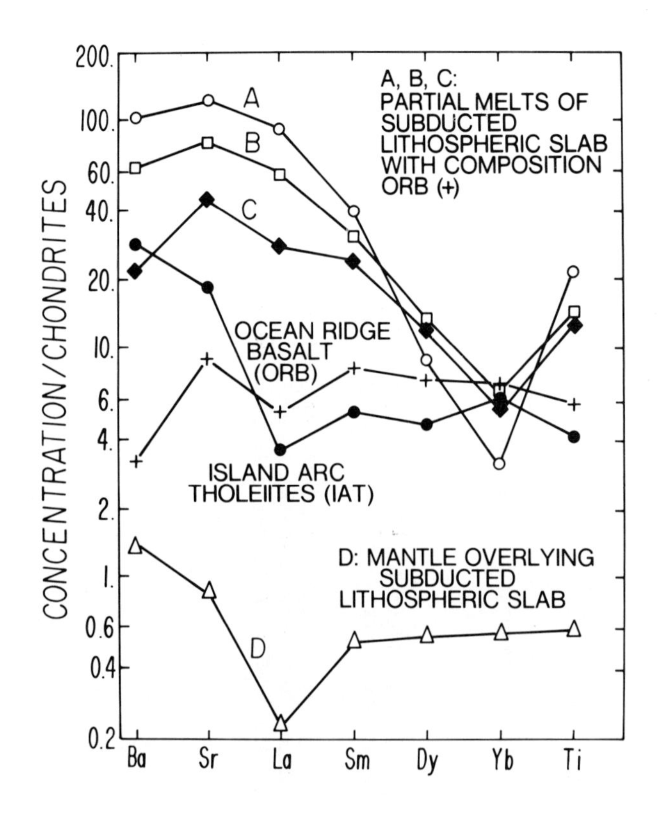

Figure 9. Chondrite-normalized mixing end members (A, B, C, D: Table 2) of Aleutian basalt model. The ocean ridge basalt (ORB) is sample D1 (Table 1); the island arc tholeiite (IAT) is sample Tofua 17 (Table 1). A, B, and C are partial melts of the ORB; D is a mantle peridotite that can yield IAT by partial fusion.

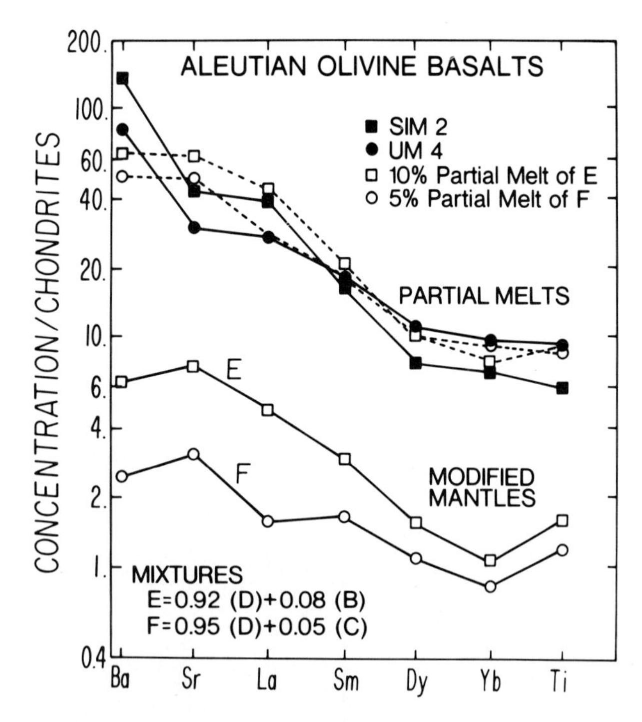

Figure 10. Chondrite-normalized composition of contaminated mantle, partial melts derived from it and two Aleutian olivine basalts. Modified mantle compositions E and F are mixtures of melts from the subducted lithospheric slab (compositions B and C) and mantle that can yield island arc tholeiites (D). The contaminated mantle peridotites rise from the Benioff zone and segregate volcanic arc magmas (5-10% partial melts) under the melting conditions of footnote 5, Table 2).

Overall, the diversity of Aleutian volcanic arc magmas is probably explained by complex melting and mixing processes within and above the subducted oceanic lithospheric plate, including sediment and sea water contamination of igneous rocks on the plate. The magmas are modified by crustal fractionation of liquidus phases.

Acknowledgments. The research was supported by NSF grants GA 16457 and DES 7420945. I thank S.S. Sun, C.-N. Lee-Hui and A. Walker for analytical assistance and discussion. J. Kaufman, J. Sells and S. Kay were of great assistance in the writing and preparation of the paper. The Department of Geophysics and Space Physics at UCLA was very accomodating during the course of the research.

References

Arculus, R.J., S.E. DeLong, R.W. Kay, C. Brooks, and S.S. Sun, The alkalic rock suite of Bogoslof, Aleutian island arc, unpublished manuscript, 1976.

Armstrong, R.L., Isotopic and chemical constraints on models of magma genesis in volcanic arcs, Earth Planet. Sci. Lett., 12, 137-142, 1971.

Bauer, G.R., The geology of Tofua Island, Tonga, Pac. Sci., 24, 333-350, 1970.

Byers, F.M., Petrology of three volcanic suites, Umnak and Bogoslof islands, Aleutians Islands, Alaska, Geol. Soc. Amer. Bull., 72, 93-128, 1961.

Chapman, H.J., E.T.C. Spooner, and J.D. Smewing, ^{87}Sr enrichment of ophiolitic rocks from Troodos, Cyprus indicates sea water interaction (abstract), Trans. Amer. Geophys. Union E⊕S, 56, 1074, 1975.

Chayes, F., The chemical composition of Cenozoic andesite, Proc. Andesite Conf., A.R. McBirney, Ed., Dept. Geol. Mineral Ind. State of Oregon Bull., 65, 1-11, 1969.

Church, S.E., Limits of sediment involvement in the genesis of orogenic volcanic rocks, Contrib. Mineral. Petrol., 39, 17-32, 1973.

Church, S.E., The Cascade Mountains revisited: A re-evaluation in light of new lead isotope data, Earth Planet. Sci. Lett., 29, 175-188, 1976.

Church, S.E., and M. Tatsumoto, Lead isotope

relations in oceanic ridge basalts from the Juan de Fuca-Gorda Ridge area, N.E. Pacific Ocean, Contrib. Mineral Petrol., 53, 253-279, 1975.

Church, S.E., and G.R. Tilton, Lead and strontium isotopic studies in the Cascade Mountains: Bearing on andesite genesis, Geol. Soc. Amer. Bull., 84, 431-454, 1973.

Coats, R., Magmatic differentiation in Tertiary and Quaternary volcanic rocks from Adak and Kanaga islands, Aleutian Islands, Alaska, Geol. Soc. Amer. Bull., 63, 485-514, 1952.

Coleman, R.G., and Z.E. Peterman, Oceanic plagiogranite, J. Geophys. Res., 90, 1099-1108, 1975.

DeLong, S.E., Distribution of Rb, Sr, and Ni in igneous rocks, central and western Aleutian Islands, Alaska, Geochim. Cosmochim. Acta, 38, 245-266, 1974.

DeLong, S.E., F.N. Hodges, and R.J. Arculus, Ultramafic and mafic inclusions, Kanaga Island, Alaska, and the occurrence of alkaline rocks in island arcs, J. Geol., 83, 721-736, 1975.

Ewart, A., and J.J. Stipp, Petrogenesis of the volcanic rocks of the central North Island, New Zealand, as indicated by a study of $^{87}Sr/^{86}Sr$ ratios, and Sr, Rb, K, U, and Th abundances, Geochim. Cosmochim. Acta, 32, 699-736, 1968.

Ewart, A., W.B. Bryan, and J.B. Gill, Mineralogy and Geochemistry of the younger volcanic islands of Tonga, S.W. Pacific, J. Petrol., 14, 429-465, 1973.

Frey, F.A., W.B. Bryan, and G. Thompson, Atlantic Ocean floor: Geochemistry and petrology of basalts from legs 2 and 3 of the Deep Sea Drilling Project, J. Geophys. Res., 79, 5507-5527, 1974.

Gill, J., Role of underthrust oceanic crust in the genesis of a Fijian calc-alkaline suite, Contrib. Mineral. Petrol., 43, 29-45, 1974.

Gill, J.B., and W. Compston, Strontium isotopes in island arc volcanic rocks, in The Western Pacific Island Arcs, Marginal Seas, Geochemistry, edited by P. Coleman, pp. 543-566, Western Australia Univ. Press, Perth, 1973.

Hart, S.R., C. Brooks, T.E. Krogh, G.L. Davis, and D. Nava, Ancient and modern volcanic rocks: A trace element model, Earth Planet. Sci. Lett., 10, 17-28, 1970.

Hofmann, A.W., and S.R. Hart, An assessment of local and regional isotopic equilibrium in a partially molten mantle, Carnegie Inst. Yearbook, 74, 195-210, 1975.

James, D.E., Strontium isotopic composition of late Cenozoic Central Andean volcanic rocks: A disequilibrium melting model, Carnegie Inst. Yearbook, 74, 250-256, 1975.

James, D.E., C. Brooks, and A. Cuyybamba, Andean Cenozoic volcanism: Magma genesis in the light of strontium isotopic composition and trace element geochemistry, Geol. Soc. Amer. Bull., 87, 892-900, 1976.

Karig, D.E., and G.F. Sharman, III, Subduction and accretion in trenches, Geol. Soc. Amer. Bull., 86, 377-389, 1975.

Katsui, Y., Y. Oba, S. Ando, S. Nishimura, Y. Masuda, H. Kurasawa, and H. Fujimaki, Petrochemistry of the Quaternary volcanic rocks of Hokkaido, North Japan, Japanese-Soviet Seminar, Tokyo Geodyn. Proj. Preprint, 1-36, 1974.

Kay, R., Aleutian volcanism in its local and regional context, Manuscript for Internat. Geol. Cong., 1976.

Kay, R.W., and P. Gast, The rare earth content and origin of alkali-rich basalts, J. Geol., 81, 653-682, 1973.

Kay, R.W., and N.J. Hubbard, Trace elements in ocean ridge basalts, Manuscript for volume to honor P. Gast, 1976.

Kay, R.W., and R.G. Senechal, The rare earth geochemistry of the Troodos ophiolite complex, J. Geophys. Res., 81, 964-970, 1976.

Kay, R.W., S.S. Sun, and C.-N. Lee Hui, Pb and Sr isotopes in volcanic rocks from the Aleutian Islands and Pribilof Islands, Alaska, in preparation, 1976.

Kay, R.W., N. Hubbard, and P. Gast, Chemical characteristics and origin of oceanic ridge volcanic rocks, J. Geophys. Res., 75, 1585-1613, 1970.

Lopez-Escobar, L., F.A. Frey, and M. Vergara, Andesites from Central-South-Chile: Trace element abundances and petrogenesis, Bull. Volcan., in press, 1976.

Marsh, B.D., Some Aleutian andesites: Their nature and source, J. Geol., 84, 27-45, 1976.

Masuda, A., N. Nakamura, and T. Tanaka, Fine structures of mutually normalized rare-earth patterns of chondrites, Geochim. Cosmochim. Acta, 37, 239-248, 1973.

Meijer, A., A study of the geochemistry of the Mariana Island Arc system and its bearing on the genesis and evolution of volcanic arc magmas, Ph.D. Dissertation, Univ. California, Santa Barbara, 214 pp., 1974.

Miyashiro, A., Volcanic rock series in island arcs and active continental margins, Amer. J. Sci., 274, 321-355, 1974.

Oversby, V.M., and A. Ewart, Lead isotopic compositions of Tonga-Kermadec volcanics and their petrogenetic significance, Contrib. Mineral. Petrol., 37, 181-210, 1972.

Peterman, Z.E., C.E. Hedge, R. Coleman, and P.D. Snavely, Jr., $^{87}Sr/^{86}Sr$ ratios in some eugeosynclinal sedimentary rocks and their bearing on the origin of granitic magma in orogenic belts, Earth Planet. Sci. Lett., 2, 433-439, 1967.

Peterman, Z.E., I.S.E. Carmichael, and A.L. Smith, $^{87}Sr/^{86}Sr$ ratios of Quaternary lavas of the Cascade Range, Northern California, Geol. Soc. Amer. Bull., 81, 311-318, 1970.

Peterman, Z.E., R.G. Coleman, and R.A. Hildreth, Sr^{87}/Sr^{86} in mafic rocks of the Troodos Massif, Cyprus, U.S. Geol. Surv. Prof. Paper 750-D, 157-161, 1972.

Philpotts, J.A., W. Martin, and C.C. Schnetzler,

Geochemical aspects of some Japanese lavas, Earth Planet. Sci. Lett., 12, 89-96, 1971.

Ringwood, A.E., The petrological evolution of island arc systems. J. Geol. Soc. London, 130, 193-204, 1974.

Ringwood, A.E., Composition and Petrology of the Earth's Mantle, McGraw-Hill, New York, 618 pp., 1975.

Shaw, D.M., Trace element fractionation during anatexis, Geochim. Cosmochim. Acta, 34, 237-243, 1970.

Scholl, D., M. Marlow, N. MacLeod and E. Buffington, Episodic Aleutian Ridge isneous activity; implications of Miocene and younger submarine volcanism west of Buldir Island (175.9°E), Geol. Soc. Amer. Bull., 87, 547-554, 1976.

Stuart, R., J. Natland, and W. Glassley, Petrology of volcanic rocks recovered on DSDP Leg 19 from the North Pacific Ocean and the Bering Sea, Init. Rep. Deep Sea Drilling Proj., 19, 615-627, 1974.

Sun, S.S., Lead isotope studies of young volcanic rocks from oceanic islands, mid-ocean ridges, and island arcs, Ph.D. Dissertation, Columbia Univ., New York, 135 pp., 1973.

Sun, S.S., and G.H. Hanson, Evolution of the mantle: Geochemical evidence from alkali basalt, Geology, 3, 297-302, 1975.

Taylor, H.P., and B. Turi, High ^{18}O igneous rocks from the Tuscan magmatic province, Italy, Contrib. Mineral. Petrol., 55, 33-54, 1976.

Thorpe, R.S., P.J. Potts, and P.W. Francis, Rare earth data and petrogenesis of andesite from the North Chilean Andes, Contrib. Mineral. Petrol., 54, 65-78, 1976.

Walker, A.T., Trace element variation in the volcanic rocks of Adak and Umnak islands of the Aleutian Arc, M.A. Thesis, Columbia Univ., New York, 39 pp., 1974.

TRENCH-VOLCANO GAP ALONG THE ALASKA-ALEUTIAN ARC: FACTS, AND SPECULATIONS ON THE ROLE OF TERRIGENOUS SEDIMENTS FOR SUBDUCTION

Klaus H. Jacob, Kazuaki Nakamura[1], and John N. Davies

Lamont-Doherty Geological Observatory of Columbia University,
Palisades, New York 10964

Abstract. Spatial variations in the present tectonic style are observed along the Alaska-Aleutian arc: changes in dip and maximum depth of the Benioff zone, widening of the trench-volcano gap, progression from oblique to normal subduction in the western and central Aleutians to almost continental collision in the Gulf of Alaska. The trench-volcano gap is 170 km wide in the central Aleutians, 300 km wide at the Alaska Peninsula, and 570 km wide in the Gulf of Alaska/Cook Inlet/Mt. McKinley region. Widening of the trench-volcano gap and related changes in dip of the Benioff zone correlate in the Gulf of Alaska with supply of large amounts of terrigenous sediments to the subduction zone. The unusual width of the gap, associated with a shallow dip of the subducted plate may be generated by at least two mechanisms: sedimentary accretion, and sediment-induced inhibition of subduction. During subduction of Pacific lithosphere, the trench in the Gulf of Alaska may have migrated 200 km seaward by accretion. The width of the remaining portion (300 to 400 km) of the gap can be explained by assuming shallow dip during initiation of Pacific-plate subduction following consumption of the Kula ridge. Possible reasons for the initial shallow dip of the Pacific plate include among others: (1) Inhibition of subduction by incorporating low-density terrigenous sediments into the descending dense oceanic lithosphere. If subducted by the oceanic conveyor belt, sediments add buoyancy to the consumed plate. (2) Suction between the buoyant continental overthrusting wedge and the sinking oceanic plate. This suction can exist as long as the nature of the contact zone prevents vertical decoupling between the two overlapping tectonic units. Reasons (1) and (2) apply only at shallow depths (< 100 km). At larger depths (> 100 km), thermally induced density differences between the subducted plate and the surrounding mantle, and mantle mobility maintained by convective heat from the asthenosphere may be more important factors controlling the dip of subducted plates. From this local study it appears that abundant sediment supply tends to inhibit or retard subduction of oceanic lithosphere.

[1] On leave from Earthquake Research Institute, University of Tokyo, Japan

Introduction

The purpose of this paper is to examine various factors other than accretion that may determine the dip of oceanic plates subducted into the upper 100 to 150 km of the mantle. Our approach is phenomenological and our interpretations are intuitive where the complexity of the problem requires simplification. We start by first examining the detailed structure of the Benioff zone using accurate seismicity data of the Alaska-Aleutian arc. Comparison of dip angles with other geological parameters guide us to single out terrigenous sediment supply as a possible factor controlling shallow subduction (0-40 km), a factor which heretofore has been ignored except for accretion. The Gulf of Alaska is a favorable place to study sediment-induced inhibition of subduction. A brief global survey of active or extinct subduction zones, and of non-subducting continental margins suggests that sediment-induced inhibition of subduction is an important tectonic principle.

Data

Seismicity and Plate Motions

Figure 1 shows the seismicity, the arc geometry, and the direction and rates of relative motion between the Pacific and North American plates. Plate motions are based on global sea-floor spreading data and seismic slip vectors as analyzed by Minster et al. (1974). Note in Figure 1 that: (1) progressing from west to east, the zone comprising shallow seismicity (near the trench) and intermediate deep

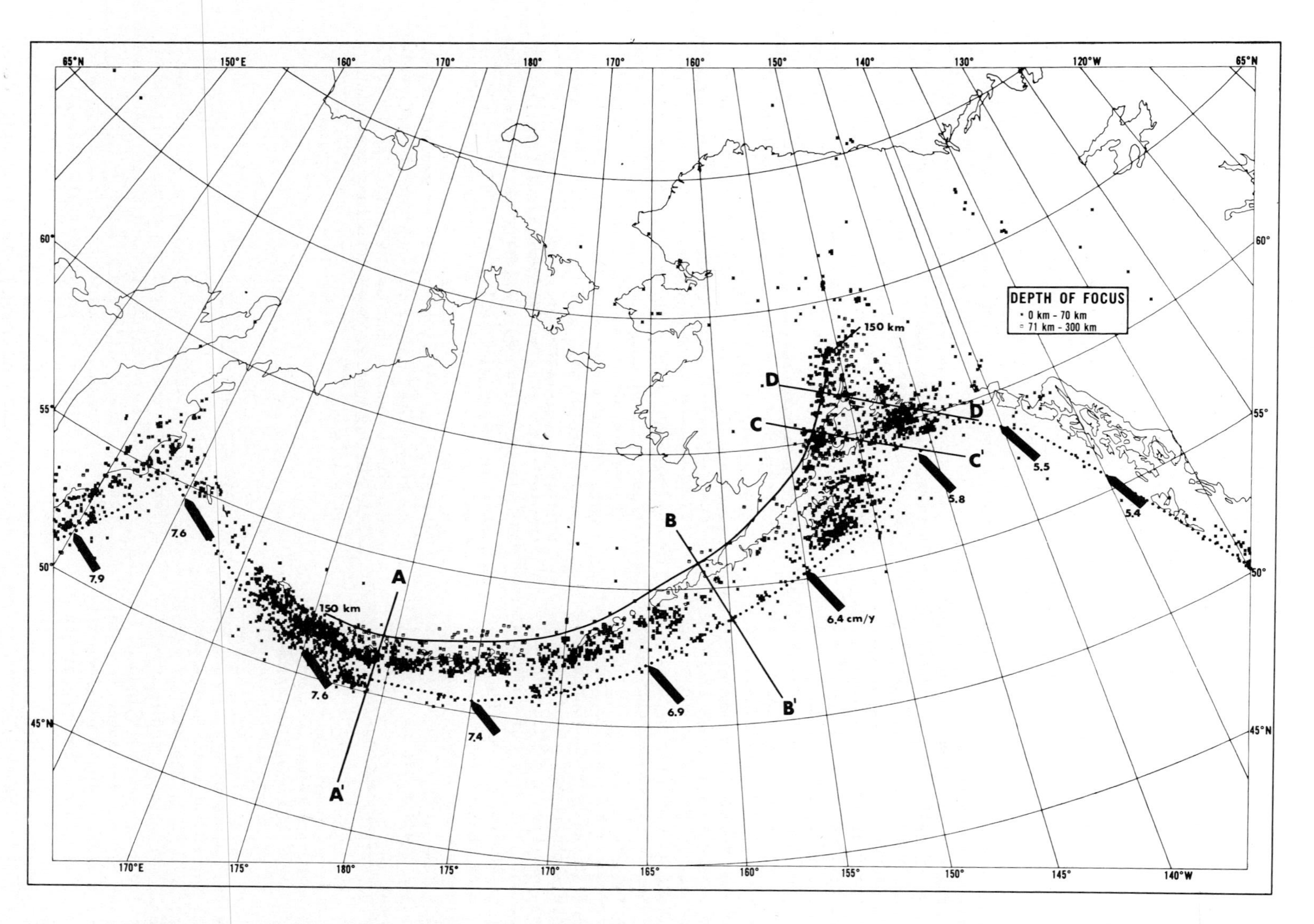

Fig. 1. Vectors of relative motion and seismicity between Pacific and North American plates after Minster et al. (1974) along the Alaska-Aleutian arc. Active zone widens towards east. Trench indicated by dotted line, 150-km depth contour of the Benioff zone by a solid line. AA' through DD' are locations of seismicity profiles in Fig. 2.

seismicity (150-km depth contour of the Benioff zone) widens by a factor of 2 to 3; (2) along the arc, marked transitions occur in the strike of the plate boundary relative to the strike of slip vectors. Right-lateral strike slip prevails off-shore British Columbia and southeast Alaska. Near the Alaska Peninsula and the central Aleutians slip vectors are almost normal to the arc in accordance with observed northwesterly thrusting solutions for seismic focal mechanisms (Stauder, 1968). Toward the western Aleutians computed slip vectors become gradually oblique to the arc and near the junction with Kamchatka, their strike virtually parallels the arc resulting in strike-slip motion (Cormier, 1975).

Cross Sections through the Benioff zone. Recently the geometry of the Benioff zone has been accurately determined at four locations along the Aleutian arc (profiles AA' to DD' in Figure 1). Refined hypocenter data were obtained from telemetered seismic networks operated at these sites by CIRES, Lamont-Doherty, USGS, and University of Alaska. Vertical cross sections through the hypocentral distributions are shown in Figure 2 and are taken from four different sources: Engdahl (1976) for profile AA', Davies et al. (1977) for BB', Lahr et al. (1974) for CC', and Davies (1975) for DD'. The four cross sections are normalized in scale, and aligned so that the locations of active volcanoes coincide laterally. That the strikes of these profiles are perpendicular to the local strike of the arc, is dictated by the choice made in the papers quoted above. For our purposes a strike parallel to the slip vectors (Figure 1) would have been more suitable. The lengths of the arc segments from which seismic events were projected were 100, 70, 300 and 90 km respectively for profiles AA' to DD'.

Profile DD' contains fewer and less accurately determined hypocenters than the other three profiles. Therefore it does not demonstrate certain characteristics of the Benioff zone as clearly as profiles AA' to CC'. Some of these features are: three different depth intervals of subduction, each comprising a range of dip angles of the Benioff zone and a characteristic aseismic wedge above the dipping plate, trenchward of the volcanic line. The three depth ranges of the Benioff zone are: (1) a shallow dipping thrust zone, from 0 to 40 km deep. It plunges from the trench beneath a wedge of accreting, thrust-faulted sediments. Seismicity close to the trench is poorly documented because of lack of seismic stations there. Alternatively, there may be actual lack of seismicity near the trench as suggested by Engdahl (1976). (2) from 40 to 100 km deep, the slab and associated Benioff zone dip at intermediate angles beneath a distinctly aseismic wedge which is bounded seaward by a clearly defined 'aseismic front' (Yoshii, 1975). The latter is located about 80 km trenchward of the volcanic front. In the second depth range the seismic zone is a well-defined narrow band often as thin as 15 km, and rarely as much as 40 km thick; (3) at depths below 100 km, the continuation of the Benioff zone plunges at steep angles into the mantle and reaches a maximum depth of about 250 km in the central Aleutians (section AA', Figure 2). An almost identical configuration of the Benioff zone with a steep dip at depths between 100 and 250 km is reported by Engdahl (1976) for the central Aleutians near Adak Island, about 300 km east of profile AA'. The onset of this steeper dip (at depths of about 100 km) coincides at its surface projection with the locus of active volcanoes. The three portions of the Benioff zone are jointed by two knee-like bends, one at a depth of 40-50 km, another at depths of about 100 to 110 km.

The aseismic wedge extends as far as 80 km from the volcanic front, toward the trench, where its leading edge has a depth of 40 km, abutting against the first knee-like bend in the Benioff zone. This leading edge of the aseismic wedge, the 'aseismic front', has been observed in other island arcs, e.g. in Japan (Yoshii, 1975). A few earthquakes are located beneath the volcanoes at shallow crustal depth. The location of Alaska-Aleutian volcanoes coincides closely with the surface projection of the 100-km depth contour of the upper surface of the Benioff zone.

The most striking differences in the cross sections AA' to DD' are associated with the highly variable distance between the trench axis and the volcanic front, referred to hereafter as trench-volcano gap. The apparent trench-volcano gap measures 180 km in section AA', and 500 km in section DD', almost a threefold increase. Most of this increase can be attributed to the shallowest portion of the dipping plate between trench and aseismic front, at depths from 0 to 40 km. The average dip of this section varies from 17 degrees in profile AA' to 5 degrees in profile DD'. This average dip of the subducted plate is obtained by implying subduction at the trench and continuation to the clear Benioff zone beginning at a depth of about 40 km, seaward of the aseismic front. This interpolation for a shallow megathrust is supported by the largest Alaska-Aleutian earthquakes, e.g. the Gulf of Alaska earthquake of 1964, M = 8.6 (Plafker, 1969).

The different angles of dip in the second portion of the Benioff zone, at depths from 40 to 100 km, contribute less prominently to widening of the trench-volcano gap as one progresses from section AA' to section BB'. The average dip of this portion measures about 44° in section AA', and 29° in section BB'.

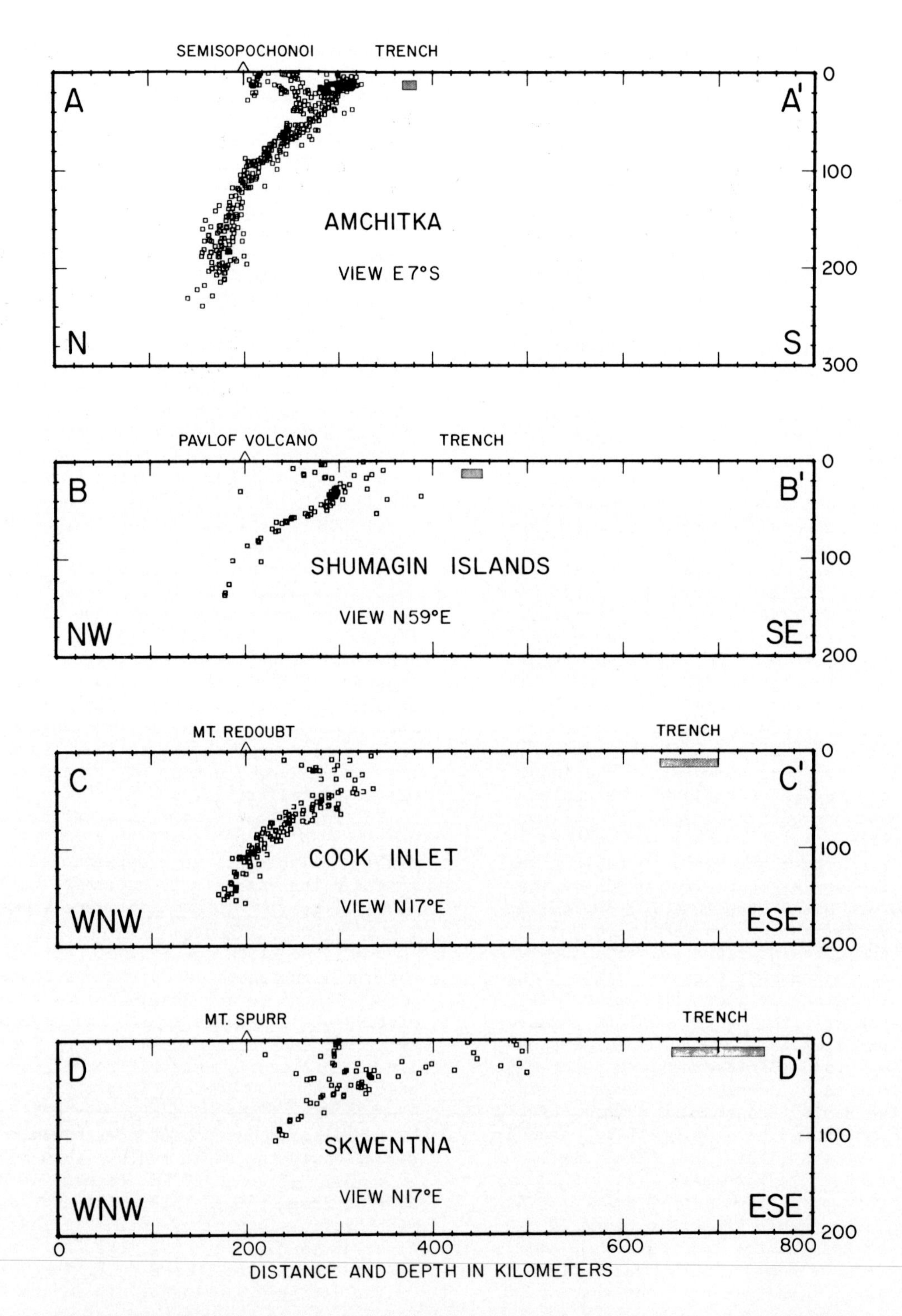

Fig. 2. Cross-sections through Benioff zone on profiles AA' through DD'. Bends in Benioff zone occur at depths of 40 km and 100 km. Aseismic wedge above 40-100 km deep parts of Benioff zone starts at 'aseismic front' 80 km trenchward of volcanic front. Distance from trench to volcanoes increases from AA' to DD'. Data sources in text.

Other details of the seismicity associated with the downgoing Pacific lithosphere and implications for the subduction process are discussed by Engdahl (1976) for the central Aleutians and by Davies et al. (1977) for the Alaska Peninsula region. Our main purpose for showing the seismicity in four cross-sections AA' to DD' (Figure 2) is to demonstrate the relationship between widening of the trench-volcano gap and flattening of the dip angle of the Benioff zone or subducted plate, particularly at depths between 0 and 40 km.

Trench-Volcano Gap

To generalize the concept of 'trench-volcano gap' for arc segments where volcanoes are absent we make use of the coincidence of volcanic front and surface projection of the 100-km depth contour of the Benioff zone. Such coincidence of volcanic front and 100-km depth contour exists in the Aleutian arc, but, allowing some scatter in depth range, is a common global pattern as well. Therefore we define the trench-volcano gap as the distance either between the trench and the volcanic front, or between the trench and the surface projection of a certain depth contour of the Benioff zone, i.e. the 100-km depth contour. The second option applies wherever volcanoes are absent. This generalized definition enables us to use the concept of trench-volcano gap along the Aleutian arc from about 176°E, near the westernmost volcano on Buldir Island, to about 140°W near the eastern end of the arc at the Wrangell Mountains. Most important, we can include the Mt. McKinley region near 150°W, in which no volcanoes are known to occur, but which has a well-developed Benioff zone (Figure 1) to depths of at least 150 km (Davies, 1975; Lahr, 1975).

We intend the trench-volcano gap to be a physically meaningful measure of the horizontal component of motion of a point on a subducting plate. Therefore, given a stationary pole of the convergent plate motion for the entire period of subduction, the trench-volcano gap should be measured along small circles about this pole. Note that this kinematically defined trench-volcano gap is distinct from the geologically defined arc-trench gap of Dickinson (1973).

Figure 3 shows the trench-volcano gaps (or trajectories) at 60 Alaska-Aleutian volcanoes assuming the recent (instantaneous) pole of rotation of the Pacific versus North American plates (Minster et al., 1974) is applicable for the entire period of Pacific plate subduction in the Aleutians and Alaska. Minor variations of pole position may have taken place during this period but they should not change the plate motions to the extent that would modify our conclusions. Table 1 contains the pertinent data used in Figure 3.

The igneous and accretionary history in the Gulf of Alaska and Alaska Peninsula regions goes back to early Jurassic time (Burk, 1965). Larson and Pitman (1972) estimate 7,000 km of combined Kula and Pacific lithosphere has been consumed beneath Eurasia and the Proto-Aleutian arc. Prior to Cenozoic time, the arc stretched from the Alaska Peninsula along the Beringian margin to the Koryak mountains in Siberia (Moore, 1972; Scholl et al., 1975). The present configuration of the Aleutian island arc was formed probably as late as 60 m.y. ago in early Cenozoic time (Scholl et al., 1975) during northerly motion of the Kula plate. According to Dickinson (1973), the width of the various segments of the present arc-trench gap should reflect the entire magmatic history in each arc segment. In our interpretation this is not necessarily so. Subduction of the Kula plate ceased with the consumption of the Kula ridge. Delong et al. (1976) propose 35 m.y. ago (early Oligocene) as the likely time of Kula-ridge consumption. A new and reoriented subduction process involving the Pacific plate started along essentially the present Alaska-Aleutian arc almost immediately after absorption of the Kula ridge. During initial subduction of Pacific lithosphere, the initial dip of the plate was probably not constrained by the former Kula plate subduction. The Pacific plate may have been detached or mechanically decoupled from the Kula plate by a lithospheric gap, a mechanically weak space occupied formerly by the Kula ridge. Therefore, a new trench-volcano gap formed at the onset of Aleutian island arc magmatism in mid-Miocene. This new trench-volcano gap was not necessarily related to the former Kula-plate subduction or pre-Oligocene magmatic history of the arc. Delong et al. (1976) date the abrupt resumption of Aleutian magmatism (related to Pacific subduction) as 15 m.y. (mid-Miocene). Hence, the width of the present trench-volcano gap should reflect igneous activity and accretion since about 15 m.y. ago.

Our goal is to examine factors which may have determined the initial dip angles of the newly subducted Pacific plate and, consequently, the variations in initial width of the mid-Miocene trench-volcano gap. We want to know how these dip variations correlate with the character of the overriding North American plate and with such related phenomena as terrigenous sediment supply. Did the leading edge of the Pacific plate start plunging everywhere along the Alaska-Aleutian arc with similar dip angles? If not, why do variations exist in the initial dip angles?

First, let us examine the configuration of the present trench-volcano gap (Figure 3) and

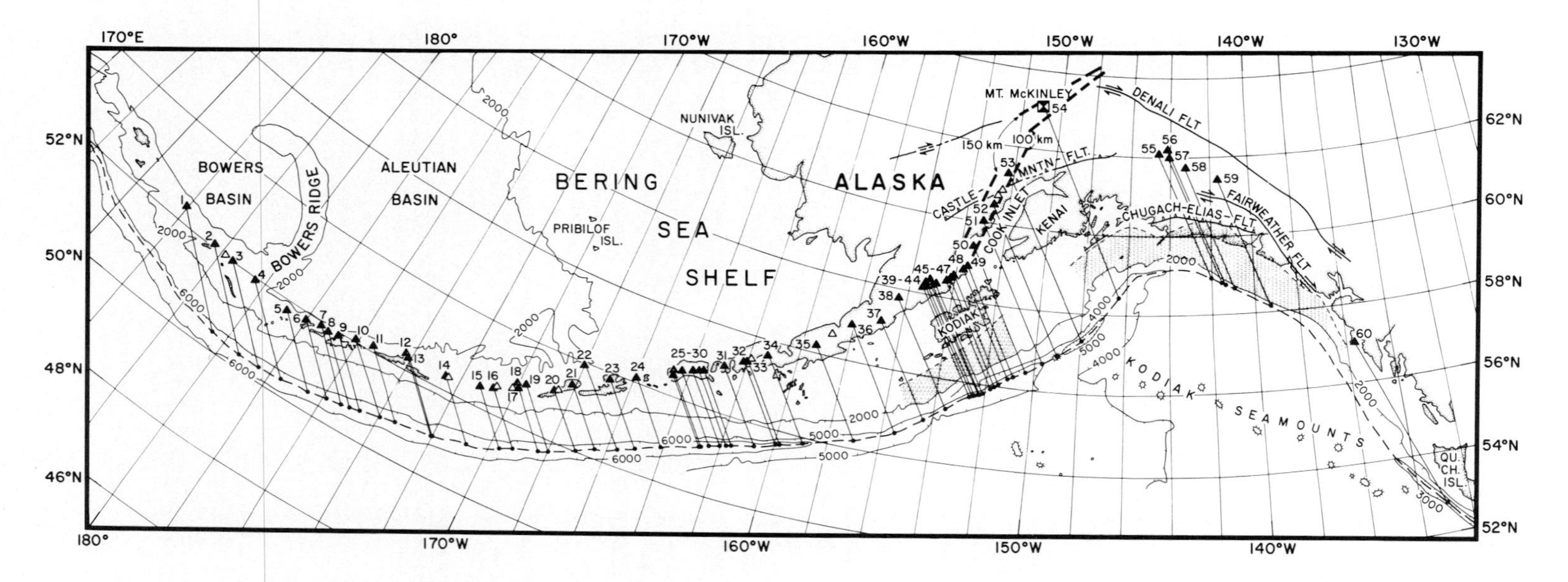

Fig. 3. Trench-volcano gap along the Alaska-Aleutian arc as defined in text. For data on individual trench-volcano trajectories, compare Table I. Shaded areas south of Kodiak Island and the Chugach - St. Elias system are interpreted as the maximum amount accreted during the last 15 m.y. Triangles represent individual volcanoes. Near the 100 and 150-km depth contours of the Benioff zone in the Mt. McKinley region volcanoes are absent. Bathymetry in meters.

TABLE 1 : Data on Alaska-Aleutian Trench-Volcano Gap for Pole and Rate of Rotation* for Pacific and North American Plates after Minster *et al.*, 1974

Volcano		Coordinates (deg)		Slip-Vector		Distance from Pole of Rotation		Length of Trench-Volcano Trajectory
#	Name	Lat.	Long.	Azimuth °N of W	Rate cm/y	deg.	km	km
1	BULD	52.317	175.767	W39.266°N	7.53	64.606	7183.87	328
2	KISK	52.100	177.600	39.222	7.50	64.069	7124.22	285
3	LISI	51.950	178.533	39.692	7.48	63.819	7096.38	278
4	CERB	51.933	179.583	40.264	7.46	63.414	7051.39	258
5	GARE	51.800	-178.800	41.103	7.42	62.862	6989.94	218
6	TAMA	51.883	-178.117	41.503	7.40	62.519	6951.89	212
7	BOBR	51.883	-177.500	41.838	7.38	62.265	6923.62	208
8	KANA	51.917	-177.167	42.030	7.37	62.102	6905.47	201
9	MOFF	51.933	-176.750	42.261	7.35	61.917	6884.58	200
10	GRSI	52.067	-176.117	42.652	7.33	61.554	6844.49	205
11	KONI	52.217	-175.133	43.240	7.30	61.030	6786.32	202
12	KORO	52.383	-174.166	43.827	7.26	60.501	6727.43	222
13	SARI	52.317	-174.050	43.863	7.26	60.499	6727.26	215
14	SEGU	52.317	-172.383	44.759	7.20	59.784	6647.76	182
15	AMUK	52.500	-171.267	45.433	7.16	59.171	6579.53	170
16	YUNA	52.650	-170.650	45.828	7.13	58.796	6537.91	168
17	CARL	52.900	-170.067	46.251	7.10	58.368	6490.23	184
18	CLEV	52.817	-169.967	46.268	7.10	58.381	6491.75	175
19	KAGA	52.967	-169.733	46.461	7.09	58.175	6468.81	182
20	VSEV	53.133	-168.700	47.092	7.04	57.607	6405.61	167
21	OKMO	53.417	-168.050	47.576	7.00	57.127	6352.25	176
22	BOGO	53.933	-168.033	47.835	6.98	56.771	6312.73	234
23	MAKU	53.867	-166.933	48.402	6.94	56.331	6263.75	190
24	AKUT	54.133	-166.000	49.045	6.89	55.741	6198.16	190
25	POGR	54.567	-164.700	49.986	6.82	54.880	6102.40	204
26	WEST	54.517	-164.650	49.986	6.82	54.890	6103.50	203
27	FISH	54.633	-164.417	50.177	6.81	54.711	6083.64	208
28	SHIS	54.750	-163.967	50.489	6.78	54.435	6052.95	208
29	ISAN	54.750	-163.733	50.618	6.77	54.330	6041.30	205
30	ROUN	54.817	-163.567	50.747	6.76	54.213	6028.30	203
31	FROS	55.083	-162.733	51.359	6.72	53.672	5968.05	218
32	PAVL	55.417	-161.900	52.017	6.67	53.090	5903.36	229
33	PSIS	55.450	-161.850	52.065	6.66	53.047	5898.60	231
34	DANA	55.617	-161.183	52.538	6.63	52.644	5853.85	245
35	VENI	56.167	-159.383	53.896	6.53	51.502	5726.84	262
36	ANIA	56.883	-158.167	55.064	6.44	50.537	5619.47	307
37	CHIG	57.133	-157.000	55.912	6.38	49.869	5545.18	285
38	PEUL	57.750	-156.350	56.732	6.32	49.233	5474.50	325
39	MART	58.150	-155.367	57.609	6.25	48.575	5401.34	325
40	MAGE	58.200	-155.250	57.717	6.24	48.496	5392.53	330
41	NOVA	58.283	-155.217	57.799	6.24	48.437	5385.95	330
42	GRIG	58.367	-155.167	57.892	6.23	48.369	5378.48	342
43	TRID	58.233	-155.117	57.821	6.24	48.419	5383.95	329
44	KATM	58.267	-154.983	57.927	6.23	48.340	5375.26	322
45	DENI	58.400	-154.367	58.398	6.20	47.994	5336.74	321
46	KUKA	58.483	-154.300	58.503	6.19	47.920	5328.55	319
47	KAGU	58.583	-154.083	58.711	6.18	47.771	5311.94	319
48	FOUR	58.767	-153.667	59.108	6.15	47.489	5280.64	321
49	DOUG	58.850	-153.567	59.234	6.14	47.402	5270.93	325
50	AUGU	59.367	-153.417	59.740	6.11	47.072	5234.23	371
51	ILIA	60.033	-153.100	60.481	6.06	46.600	5181.71	422
52	REDO	60.467	-152.750	61.065	6.03	46.235	5141.13	450
53	SPUR	61.317	-152.133	62.192	5.95	45.564	5066.53	500
54	MCKE**	63.072	-151.000	64.532	5.82	44.299	4925.88	570
55	DRUM	62.117	-144.633	67.977	5.58	41.988	4668.91	370
56	SANF	62.217	-144.117	68.436	5.55	41.726	4639.76	403
57	WRAN	62.000	-144.012	68.282	5.55	41.760	4643.59	372
58	BLAC	61.730	-143.400	68.417	5.54	41.591	4624.72	362
59	WHRI	61.450	-141.467	69.436	5.45	40.829	4540.03	385
60	EDGE	57.012	-135.767	W68.176°N	5.32	39.702	4414.73	513

* Pole Position: Lat. 50.9°N, Long. 66.3°W, Rate: 0.75 deg/My
** No Volcano (see text)

then infer the approximate mid-Miocene configuration by subtracting the possible maximum portion added via accretion since 15 m.y. ago.

It is apparent from Figure 3 that there are at least two factors controlling the length of the trench-volcano trajectories: (1) Pure arc geometry, (2) Tectonic processes dynamically affecting the dip of the slab. The trajectory L becomes longer as the angle β, between slip vector and the direction normal to the arc, increases; ($L = n/\cos\beta$; with n, width of the trench-volcano gap, normal to the strike of arc). Hence, for the westernmost trajectories (#1 through #10) and for the easternmost ones (#55 through #60) where strongly oblique slip occurs, lengthening of the trench-volcano trajectories is explicable by arc geometry alone. The strong remaining variations in trajectory length (Table 1) cannot be explained by geometrical reasons. The shortest trajectory (167 km) is #20 (Vsevidof volcano on Umnak Island), while the longest trajectory (570 km) is located in the Mt. McKinley section (#54). Both show a small deviation between slip direction and direction normal to the arc. The shortest trajectory is located where the overriding plate is oceanic lithosphere (Aleutian basin). The longest trench-volcano trajectories are located well within the continental segment of the overriding plate (Kenai Peninsula, Mt. McKinley region) implying an overall steeper dip of subducted Pacific plate beneath the oceanic section and a shallower dip beneath the continental section of the arc.

Figure 4 shows schematically the total length of trajectories along the arc in comparison with other geologic or tectonic elements, such as depth to the bottom of the Aleutian trench, or slip rates. The latter are subdivided in total slip rate and slip-rate component normal to the arc.

The trench-volcano gap increases where the trench depth decreases. The nearly constant depth (7 km) of the trench changes rather abruptly to a depth of less than 6 km near the tip of the Alaskan Peninsula where the overriding plate changes from oceanic to continental type. Trench depth decreases even more abruptly to less than 4 to 2 km where the supply of terrigenous sediments becomes abundant (Gulf of Alaska). The decrease of trench depth is mostly due to filling with terrigenous sediments (Scholl and Marlow, 1974). The source regions from which to derive the terrigenous sediments (Figure 5) increase manyfold in the Gulf of Alaska as compared to arc segments farther to the west. Partial glaciation, steep slopes, and high meteorological precipitation rates exceeding 200 mm/y further promote high sedimentation rates by rapid erosion from these rugged coastal ranges. The terrigenous sediments may be laterally distributed in the trench by currents flowing along its axis. Portions of the terrigenous sediments are deposited in small basins landward of the trench slope break (Karig and Sharman, 1975). Therefore the degree to which the trench is filled with sediments does not always reflect the potential for terrigenous sediment supply (Figure 5) that might be expected from the fluvial drainage area per unit length of coastal shore line. Off southeast Alaska and British Columbia recent sedimentary deposits transgress across the mostly transcurrent plate boundary. Because of this local sedimentary blanket the plate boundary has little bathymetric expression except due west of Queen Charlotte Island where a trough (> 3,000 m deep) developed in the sedimentation shadow behind this island (Figure 3). The trough remained here unfilled apparently because the land-derived sediments are trapped in the Hecate Strait between the island and the Canadian mainland (Figure 5). Except for such minor and local irregularities the degree of trench fill usually corresponds well with the amounts of terrigenous sediments available from nearby coastal ranges.

In summary, a comparison of Figures 3, 4, and 5 reveals that the trench-volcano trajectories along the Aleutian arc are as short as 170 km in the oceanic arc section, lengthen to about 200 to 300 km as the arc becomes continental near the tip of the Alaskan Peninsula, and measure as much as 570 km in those arc segments of the Gulf of Alaska where terrigenous sediment supply is high.

Discussion

Initial Slab Dip and Accretion

Trench-volcano trajectories consist of at least two portions: (1) the initial distance between trench and volcanic front formed at the time of early plate subduction and (2) the contribution, if any, subsequently added by accretion. The former part is dependent on the average dip of the plate during initial subduction at depths between 0 and 100 km. The latter portion depends mostly on sediment supply. A third portion of the trench-volcano trajectory exists in some arcs where the volcanic front has moved landward, which may have occurred in the Aleutians, judging from the northernmost locations of many active volcanoes within the volcanic belt. Similar landward migration of the volcanic front is known from northeast Japan (Nakamura, 1969; Matsuda and Uyeda, 1971), and in Chile after the Miocene time. These distances are generally small, less than a few tens of km, however. Therefore, we can neglect such a third contribution to the trench-volcano gap for the Aleutian arc and the period considered (15 m.y.).

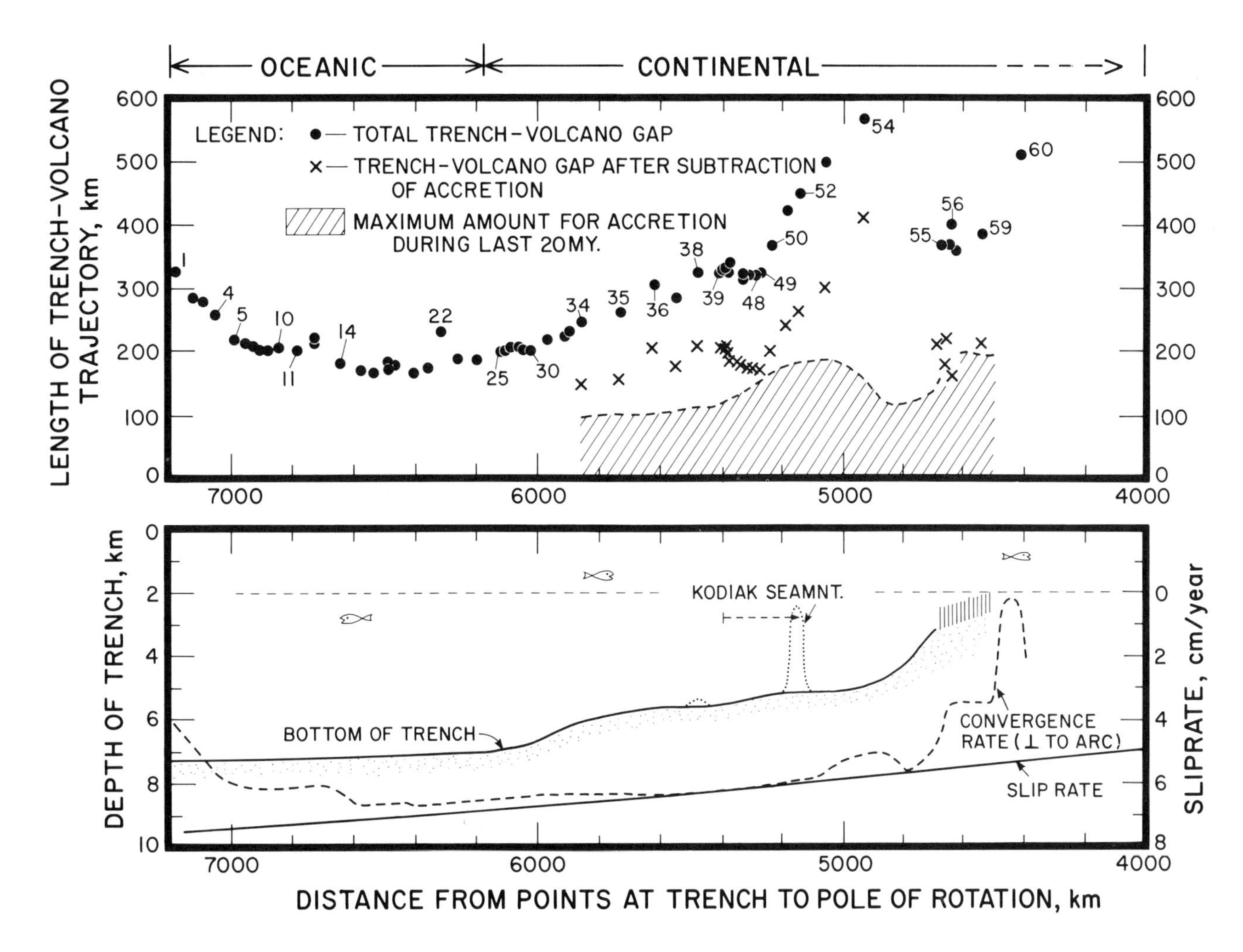

Fig. 4. Trench-volcano gap (upper frame) versus distance from pole of NA-PC relative motion (Minster et al., 1974) plotted separately for total gap, maximum portion due to accretion, and initial gap. Trench bathymetry and slip rates (lower frame). Note transition from oceanic to continental arc.

On a generalized tectonic map of the Gulf of Alaska, Plafker (1969) differentiates between early Cenozoic bedded rocks and late Cenozoic bedded rocks. The line separating the two groups of sedimentary units runs roughly parallel to the shoreline south of Kodiak and Montague Islands and straddles the shorelines along the Chugach, St. Elias, and Fairweather fault and mountain systems. We interpret the younger sequence, south of this dividing line (Figure 3), as representing the upper limit for the accretionary wedge added to the initial trench-volcano gap since mid-Miocene time. The 15 m.y. date, according to Delong et al. (1976), signifies the formation of the volcanic front related to Pacific plate subduction, and hence, of the initial trench-volcano gap. The dividing line (Figure 3) is interpreted as the northernmost possible position of the Aleutian trench at this mid-Miocene time. The actual trench position may have been considerably farther to the south. Near Kodiak the dividing line is very close to the now exposed Kodiak batholith whose age is about 60 m.y. (Moore, personal communication, 1976). Already at this early Cenozoic time the trench must have been located south of the batholith. Accepting the dividing line (Figure 3) as the northernmost possible trench position during mid-Miocene time, we find that this maximum estimate for the width of the accretionary wedge varies between 100 and 200 km (Figure 4, upper frame). After elimination of accretion, the initial trench-volcano gap shows widening in the Gulf of Alaska reaching up to 400 km in the Kenai Peninsula/Mt. McKinley segment (Figure 4). In the central Aleutians the

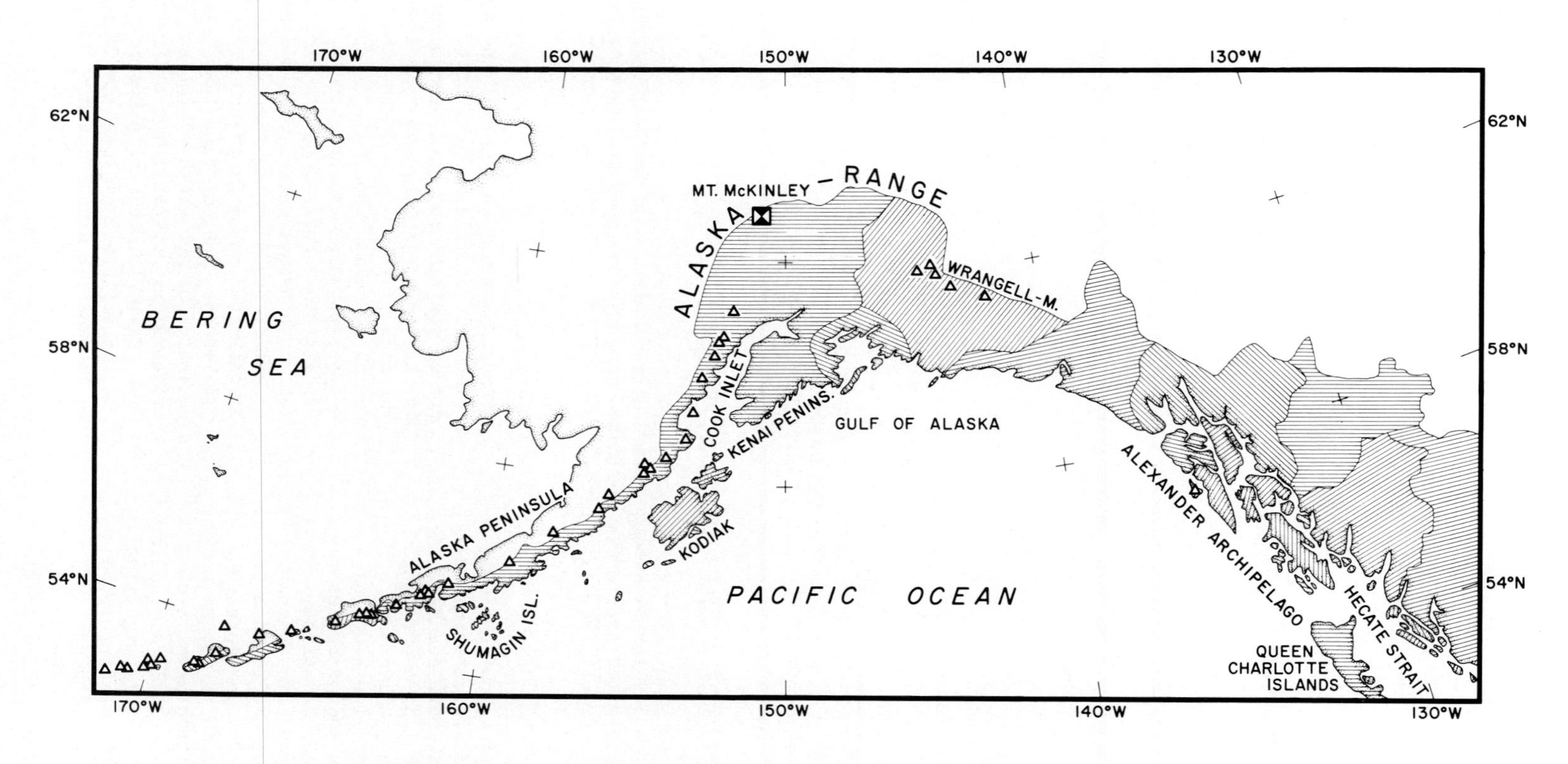

Fig. 5. Areas of potential sediment supply in the Gulf of Alaska. The subaerial drainage area abruptly increases west to east, past Kodiak Island. Shading patterns indicate major river or coastal drainage systems. Note variation in ratio of drained area/length of coastal front.

initial gap was less than 200 km wide. Clearly, the Pacific plate was subducted in the Gulf of Alaska at a much shallower dip angle than in the central Aleutians or near the tip of the Alaskan Peninsula.

Before we discuss the reasons for the shallow initial dip of the Pacific plate beneath the Kenai Peninsula segment, we test whether it is likely that the northernmost and deepest part of the Benioff zone beneath Mt. McKinley belongs to Pacific plate, rather than to a remnant piece of Kula plate. The trajectory is about 720 km long at a depth of about 150 km down-dip along the Pacific slab from the present trench to the deepest part of the Benioff zone beneath Mt. McKinley. The present convergence rate of 5.8 cm/yr implies that the deepest part of the active slab approximately 12.4 m.y. ago was at the position of the present trench near 58.5°N and 146.5°W. This datum of 12.4 m.y. is more recent than the 15 m.y. age for which Delong et al. (1976) report resumption of magmatism in the more western segments of the arc, and is much younger than the Oligocene date after which the Pacific plate may have started to underthrust, subsequent to consumption of the Kula ridge. This suggests two results: (1) The deepest portion of the Benioff zone belongs to Pacific rather than Kula plate; (2) an aseismic portion of Pacific slab extends to depths deeper than indicated by the Benioff zone which terminates presently at a depth of about 150 km beneath the Mt. McKinley range. Note, however, that this does not imply that the now deepest seismically active part of the slab has been subducting for the entire period of 12.4 m.y. because subduction took place at the mid-Miocene trench position farther north by the amount added since then via accretion. A subduction time of about 10 m.y. is therefore more probable. These different values are consistent in magnitude with an averaged time constant of about 10 m.y. found by Isacks et al. (1968) for thermally absorbing a downgoing slab into the mantle such that it no longer displays seismic activity, nor a major high-velocity/high-Q anomaly.

In comparison, we obtain a time constant for a thermal equilibrium of about 6.7 m.y. in the central Aleutians assuming 250 km for the deepest seismicity near 180° longitude, about 500 km (ignoring accretion) for the down-dip trajectory to the 250 km depth, and 7.5 cm/yr relative (oblique) slip rate. In the central Aleutians the downgoing slab reaches thermal equilibrium with the surrounding mantle faster than in the Gulf of Alaska. This is compatible with a generally steeper dip of the slab which allows the slab to reach a higher temperature environment sooner. A higher subduction rate may have a similar effect.

Speculations on Causes for Shallow Initial Dip

Two principal causes which have a tendency to produce shallow dip during initial subduction of the oceanic plate can be assumed to exist: (1) reduction of excess density of a downgoing slab, thus reducing the downward pulling forces from within the slab; (2) exterior boundary conditions acting on the slab, that prevent it from sinking steeply into the mantle, despite gravitational imbalances. Of course, a combination of these causes may actually exist.

The first category can be subdivided into (a) direct ways to reduce the average density of the downgoing slab, e.g. by incorporating a thick sedimentary layer of low density into the otherwise dense oceanic crust, and (b) indirect causes which would thermally reduce the density surplus. The latter possibility requires incorporating sediment-derived heat sources into the slab. Alternatively a thermally insulating sedimentary blanket deposited on a young and hence relatively warm oceanic lithosphere prior to subduction would prevent its cooling before entering the trench.

If a 5 km thick compacted (ρ = 2.7 g/cc) sediment layer is subducted with an aged oceanic lithosphere (100 km thick, ρ = 3.4 gg/c), it should lower the average density by 1 to 3.5% depending on thermal and rheological boundary conditions near the bottom of the oceanic lithosphere. These conditions determine the depth at which isostatic compensation is reached. The higher value of 3.5% is based on data for ocean floor elevation versus age (Sclater et al., 1976) which indicate that the upper 100 km of very young oceanic lithosphere and upper mantle has a 3.5% lower average density than a well-aged cool lithosphere of the same thickness. Very young oceanic lithosphere is considered too buoyant (Kelleher and McCann, 1976) to actively subduct, but may be passively subducted when pulled into the mantle by adjacent older lithosphere. Reduction of the average density of the lithosphere, whether by thermal expansion or by incorporation of low-density sediments, reduces the density instability by which descending slabs are presumably driven. The question is whether sufficient quantities of sediments participate in the subduction process to produce a strong enough buoyancy recognizable by a shallower dip of the leading edge of a newly subducting lithosphere. No direct data are available on how much sediment can be subducted with oceanic crust. Only indirect estimates exist (Von Huene, 1972; Karig and Sharman, 1975). They are based on assessing the sediment supply to the trench and the volume of accreted materials, and on assuming that the remainder of non-accreted materials were subducted. These estimates are probably too crude to provide the needed information.

Besides density reduction by low-density sediments, suction (Jischke, 1975) of the downgoing plate to an overriding continental wedge may also produce shallow dip of the descending plate. Jischke pointed out that a tensional (suction) force should exist across a low-viscosity boundary layer separating the dense downgoing slab from the buoyant overriding plate. This suction, he suggests, may prevent the plunging plate from vertically peeling away from the overriding plate.

The two causes of shallow dip (sediments and suction) may be difficult to distinguish at least at shallow depths (< 40 km). Conditions required for both processes occur often at the same arc segments because abundant sediment supply requires a nearby continental source. There are subduction zones along continental arcs with a minimum of terrigenous sediment supply, e.g. the trenches of arid southern Peru and northern Chile into which the Nazca plate is subducting. Sedimentary trench fill is less than 50 m at some arc segments (Scholl and Marlow, 1974). On-shore precipitation in the most arid parts is as low as 25 mm/y. The Benioff zone associated with the Nazca plate is dipping there at rather shallow angles (Barazangi and Isacks, in press, 1976) as indicated by a 300 km wide trench-volcano gap. The configuration of the South American Benioff zone differs from the ones associated with high sedimentation supply at their trenches, e.g. in the Gulf of Alaska or at the Hikurangi trench of the North Island of New Zealand (Scholl and Marlow, 1974). The South American Benioff zone has no conspicuous knee-like bend at a depth of about 40 km. In the Benioff zone of the Gulf of Alaska and New Zealand the bend is clearly present. Another peculiar feature of the Gulf of Alaska and the Hikurangi trench subduction zones is a strong negative gravity anomaly (> 100 mgal) which is present in both the Bouguer and in the isostatic residuals for New Zealand (W.I. Reilly, 1965), and at least for the free-air residuals (Figure 6) in Alaska (Watts, 1975). No such negative anomaly seaward of the volcanic front is observed above the Peru/Chile subduction zone, nor is it common in regular island arcs with ocean/ocean convergence. Island arc ridges are typically the sites of a local gravity high seaward of the volcanic front; this local high is, in most island arcs, superimposed on a broad gravity low which is most pronounced outside and close to either end of the trench-volcano gap.

In summary, two major causes for initial shallow subduction are proposed, suction and sediment supply. At shallow depths (< 100 km), the Benioff zones beneath continents are dipping at a shallower angle than beneath island arcs (even after elimination of accretion effects) probably because of vertical suction between the dense downgoing slab and the buoyant overriding continental plate. But when in addition high sedimentation supply is present at a subducting continental margin, it results in two additional unexplained features: a knee-like bend in the Benioff zone at a depth of about 40 to 50 km, and a strong negative gravity anomaly directly above it, seaward of the volcanic front.

Speculations on Causes for Flexures in the Benioff Zone

Two knee-like bends in the Alaska-Aleutian Benioff zone were identified, one at 40-50 km, another at about 100-110 km depth. Occurrence of very large earthquakes in the Gulf of Alaska region, (e.g. in 1964, M = 8.6) indicates that the downgoing plate is at least in the upper 40 km elastically coupled to the overriding plate. Great thrust earthquakes (M > 7.5) occur only rarely at depths below 40 km. Below this depth the magnitudes of the largest events are generally smaller. A major difference in the seismicity of the overriding crustal wedge occurs near the bend at 40 km depth. Minor shallow (< 20 km) seismicity is commonly observed in the accretionary wedge above the thrust zone, in the nonvolcanic outer arc. Yet seismicity is virtually absent landward of the aseismic front directly above the Benioff zone where the latter reaches depths between 40 and 100 km (Figures 2 and 6). The aseismic zone starts about 80 km trenchward of the volcanic front.

Two causes for the sudden flexure in the Benioff zone are considered: (1) dehydration of oceanic crust and (2) resumption of oceanic plate properties by shearing off subducted sediments. The temperature and pressure conditions at the depth of 40 km are probably sufficient to cause dehydration of mineral assemblages in hydrothermally altered ocean crust (P.N. Anderson, personal communication, 1976), with the remaining dehydrated mineral assemblages gaining higher densities, both for compositional and thermal reasons. Dehydration of altered oceanic crust is an attractive proposal since it also may explain the virtual aseismicity in the wedge landward of the aseismic front. The upward migration of water could promote aseismic creep in this wedge via low-temperature water-induced chemical corrosion on grain boundaries similar to that observed in lab experiments on quartz (Scholz, 1972). Alternatively, if any sediments were subducted with the oceanic crust, at least at the depth of 40 km they would be highly metamorphosed and sheared off from the downgoing slab. They could be incorporated from below into the overriding wedge such

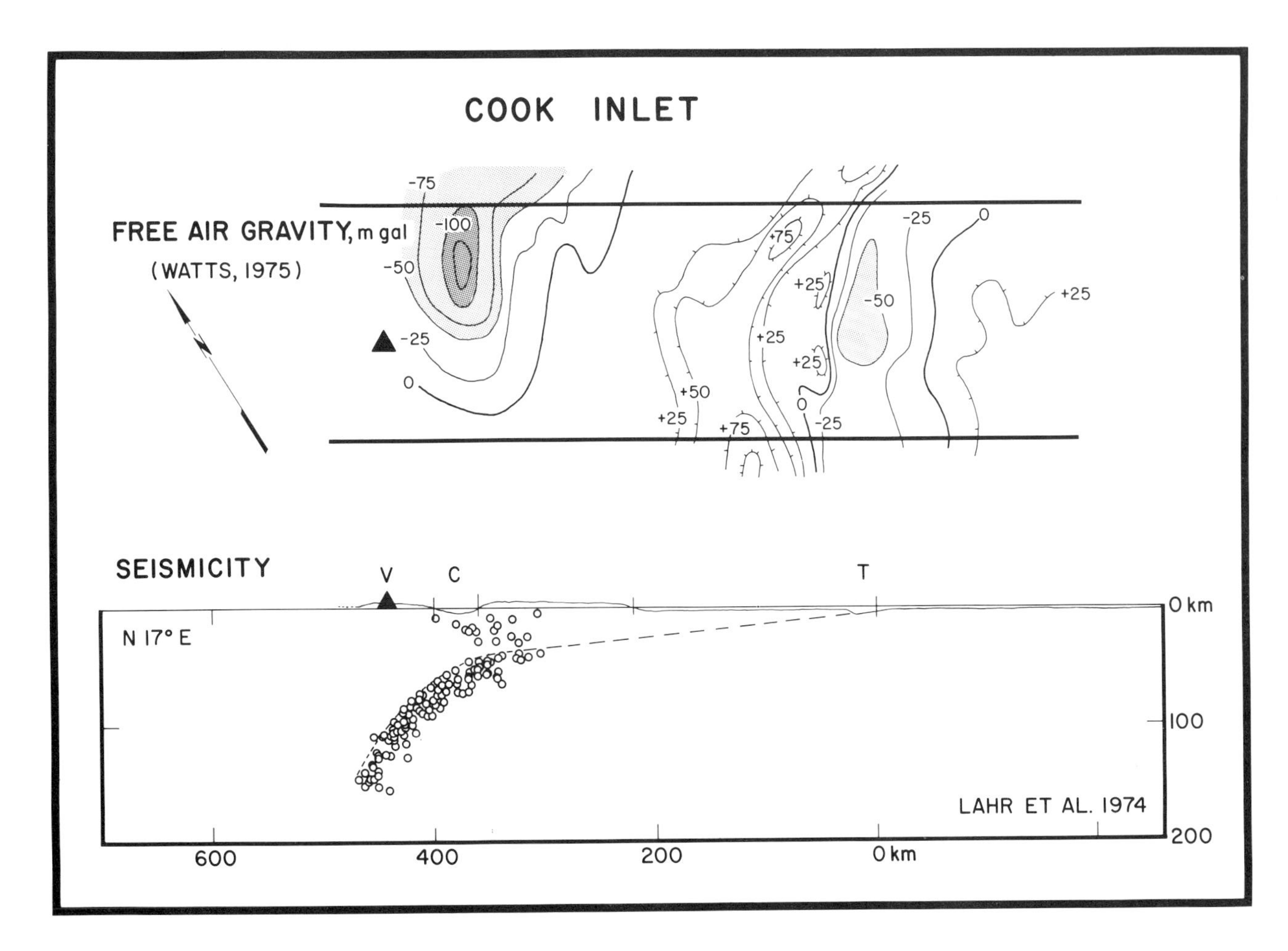

Fig. 6. Free-air gravity (Watts, 1975) and seismicity (Lahr et al., 1974) in a 300-km wide zone centered on cross-section CC' striking N73°W through trench (T), Kenai Peninsula, Cook Inlet (C) and Mt. Redoubt volcano (V). Gravity anomaly at Cook Inlet coincides with 'aseismic front' and downward flexure in Benioff zone.

that the underlying plate can resume its oceanic properties (high density) and descend from there on at a steeper angle. The more steeply dipping slab beneath the aseismic front may locally drag down the overriding wedge, surficially forming a graben-like feature (Cook Inlet). The sudden onset of the subsidence may be correlated to the sudden lowering of shear rigidity from water-induced creep, just landward of the aseismic front. The subsidence within the aseismic wedge in turn causes the gravity low discussed earlier for the Cook Inlet region (Figure 6). Similarly the negative gravity anomaly in New Zealand may have been formed.

In our proposed model, the knee-like bend in the Benioff zone at d = 40 km is due to an increase of the average density of the slab. This density increase can be produced either by shearing off the terrigenous sedimentary components, or by initiation of extensive dehydration of subducted and altered oceanic crust or by both. The steeper dip could be enhanced by decoupling the downgoing slab (for shear motion) from the overriding plate by water-induced creep in the aseismic wedge above the Benioff zone. Above and seaward of the 40 km deep bend, shear coupling between the overriding accretionary wedge and the descending plate is elastic, with accumulating stresses released by large earthquakes (that is, brittle behaviour); beneath and landward of the bend shear coupling between the overriding and sinking plates is weakened by the hydrous conditions; accumulating shear stresses between the two units may be released by creep rather than by large-scale brittle failure. The consequence would be that the observed seismicity at depths below the 40-km bend occurs mostly inside the descending slab, while seismicity trenchward of the bend at depths shallower than 40 km would be mostly thrusting motion between the overriding and the subducting plates. This relationship is demonstrated for the northeast Japan arc (Yoshii, in preparation).

The second knee-like bend, at a depth of about 100 km, coincides with the volcanic front directly above it. Anderson et al., 1976, summarized the literature and discussed the thermal and geochemical reasons why at this depth of 100 km or more the andesitic magmas apparently originate. The important mechanical implications are: the lighter volatile or hydrous components, if still present, are being removed from the descending slab. The availability of partial melts in the mechanically weak low-Q volcanic wedge above the Benioff zone further decouples (for shear motion) the descending slab from the overriding plate, such that at depths below formation of the andesitic magmas (> 100 km) the descending cold slab is sinking essentially into a hot viscously flowing asthenosphere. Whether this completely different style of subduction (compared to shallow subduction) is reached appears to be at least in some cases clearly dependent on or linked to the occurrence of andesitic volcanism above it. New seismicity data from South America (Barazangi and Isacks, in press, 1976) show that a weak knee-like bend of the Benioff zone at about 100 km occurs only in those sections where volcanoes are presently active; yet in those segments where no active volcanoes are presently observed (from 27°S to 33°S) the Benioff zone keeps advancing almost horizontally or at extremely shallow angles to depths of almost 200 km, as if suspended by suction to the overriding South American continent as discussed earlier. It is apparently crucial for initiation of the decoupling between the two overlapping plates that melts or mobile asthenospheric materials can penetrate inbetween the two overlapping plates. Only if this is the case on a large scale, then apparently the lower plate can peel away from the buoyant continental plate and descend at steep angles into the viscous asthenosphere. We suggest that the bend at 100 km depth indicates the onset of this decoupling process. In some arcs decoupling may occur at slightly larger depth where mantle mobility and convective heat supply is low, or it may occur at slightly shallower depths in actively spreading back-arc regions where mantle mobility is high and, hence, is reaching to shallower levels.

Did Sediments Terminate Kula-Plate Subduction on the Beringian Arc-System?

What created the Aleutian arc in the first place? Probably about 60 m.y. ago, the Kula plate subduction suddenly (?) jumped from the Alaska-Beringian-Koryak trench system to the Alaska-Aleutian-Kamchatka system without any apparent cause (Burk, 1965; Moore, 1972; Scholl et al., 1975). This was a major tectonic event. The Aleutian basin is different from most other Pacific marginal basins in that it appears to be a piece of remnant ocean floor. It was not created as an actively spreading marginal basin post-dating the arc. It predates the arc to which it is now the back-arc region. What caused the jump in the subduction? We can provide a possible explanation for this unique tectonic event with sediment-induced inhibition of subduction: because of the angular relationship between the Koryak and the Beringian arc segments, abundant sediments may have been laterally supplied from one coastal range and rafted to the other as trench fill. The situation may have been similar to the present one in the southeast Alaskan and British Columbian ranges which now provide sediments laterally onto the Pacific plate that are rafted into the easternmost Aleutian trench near the Yakutat Bay. The accretionary wedge in front of the Beringian-Koryak arc may have migrated southward loading the advancing Kula plate. By accretion, it flattened the primary thrust plane to such a shallow dip angle that it became easy to form a new thrust zone, amidst the Kula plate. It had probably a smaller rupture surface than the old greatly enlarged thrust plane emerging at the sediment-clogged Beringian trench. Also, the old low-angle thrust plane may have become increasingly subjected to higher friction because of the increasing sedimentary load, thus approaching almost a continental collision situation. We know that Kula subduction jumped seaward; why remains unclear; yet sedimentary clogging appears a possible cause to be considered.

Conclusions

This paper is divided into one part dealing with observable facts of subduction in the Alaska-Aleutian arc and another part with speculations on the role of sediments for subduction.

Recent accurate seismicity data show that the dip of the Benioff zone varies considerably within the Alaska-Aleutian arc. In the upper 100 km, subduction occurs at shallower dip angles in the continental part of the arc and at steeper dip angles at the oceanic island arc segments. Within the continental arc subduction flattens, particularly in the upper 40 km of the thrust zone, where the supply of terrigenous sediments from coastal ranges is abundant.

We define a trench-volcano gap based on relative plate motions, the location of the present trench axis and of the volcanic front or 100-km depth contour of the Benioff zone. We evaluate the portions of the gap due to accretion and find that to account for the present trench-volcano distance in the eastern Aleutian arc the leading edge of the Pacific plate must have been underthrusting initially at a shallower dip angle than at the central Aleutian arc.

Two major causes for the shallow dip are proposed: (1) sediment-induced inhibition of subduction and (2) suction to the overriding continental plate.

A comparison of the eastern Aleutian Benioff zone to other subduction zones with, respectively, similar and different sedimentation conditions at their trenches, e.g. in New Zealand and South America, suggests that both mechanisms are actively controlling the dip angles in the upper 100 to 150 km of a subduction zone.

Sediments may even prevent subduction. As a possible example of termination of a once ongoing subduction by sedimentary clogging, we propose that the jumping of Kula plate subduction from the Beringian to the Aleutian trench system, about 60 m.y. ago, may have been produced by dumping large amounts of sediments into the Beringian trench system.

Acknowledgements. This research was supported under Contract E(11-1) 3134 of the Energy Research and Development Agency. The paper was critically reviewed by R. Engdahl, M. Marlow, P. Richards, and A. Watts. Helpful discussions and/or suggestions for improvements were provided by R.N. Anderson, W. McCann, S. Delong, D. Forsyth, M. Langseth, R. Schweickert, L. Seeber, and T. Yoshii. We gratefully acknowledge the constructive criticisms of these individuals. We do not imply that these colleagues concur with the content or form of this paper for which the authors take full responsibility.

Lamont-Doherty Contribution No. 2407

References

Anderson, R. N., S. Uyeda, and A. Miyashiro, Geophysical and geochemical constraints at converging plate boundaries - part 1: dehydration in the downgoing slab, Geophys. J. R. astr. Soc., 44, 333-357, 1976.

Barazangi, M. and B. L. Isacks, Spatial distribution of earthquakes and subduction of the Nazca plate beneath South America, J. Geology, in press, 1976.

Burk, C. A., Geology of the Alaska peninsula - island arc and continental margin, Geol. Soc. Am. Memoir 99, 250, 1965.

Cormier, V. F., Tectonics near the junction of the Aleutian and Kuril-Kamchatka arcs and a mechanism for middle Tertiary magmatism in the Kamchatka basin, Geol. Soc. Am. Bull., 86, 443-453, 1975.

Davies, J. N., Seismological investigations of plate tectonics in south central Alaska, Ph.D. thesis, Univ. of Alaska, Fairbanks, Alaska, May 1975.

Davies, J. N., L. House, and K. H. Jacob, Morphology of the Aleutian Benioff zone and some plate tectonic implications, in preparation, 1977.

Delong, S. E., P. J. Fox, and F. W. McDowell, Subduction of the Kula ridge at the Aleutian trench, Geol. Soc. Am. Bull., in press, 1976.

Dickinson, W. R., Widths of modern arc-trench gaps proportional to past duration of igneous activity in associated magmatic arcs, J. Geophys. Res., 78, 3376-3389, 1973.

Engdahl, E. R., Seismicity and plate subduction in the central Aleutians, Preprint, AGU monograph Ewing Symposium, 1976.

Isacks, B., J. Oliver, and L. R. Sykes, Seismology and the new global tectonics, J. Geophys. Res., 73, 5855-5899, 1968.

Jischke, M. C., On the dynamics of descending lithospheric plates and slip zones, J. Geophys. Res., 80, 4809-4814, 1975.

Karig, D. E. and G. F. Sharman, III, Subduction and accretion in trenches, Geol. Soc. Am. Bull., 86, 377-389, 1975.

Kelleher, J. and W. McCann, Buoyant zones, great earthquakes, and unstable boundaries of subduction, J. Geophys. Res., 81, 4885-4896, 1976.

Lahr, J. C., Detailed seismic investigation of Pacific-North American plate interaction in southern Alaska, Ph.D. thesis, Columbia Univ., Faculty of Pure Science, New York, New York, 1975.

Lahr, J. C., R. A. Page, and J. A. Thomas, Catalog of earthquakes in south central Alaska, April-June, 1972, U.S.G.S.-N.C.E.R. Open-file Report, 1974.

Larson, R. L. and W. C. Pitman, III, World-wide correlation of Mesozoic magnetic anomalies, and its implications, Geol. Soc. Am. Bull., 83, 3645-3662, 1972.

Matsuda, T. and S. Uyeda, On the Pacific-type orogeny and its model: Extension of the paired belts concept and possible origin of marginal seas, Tectonophysics, 11, 5-27, 1971.

Minster, J. B., T. H. Jordan, P. Molnar and E. Haines, Numerical modelling of instantaneous plate tectonics, Geophys. J. R. astr. Soc., 36, 541-576, 1974.

Moore, J. C., Uplifted trench sediments: southwestern Alaska-Bering shelf edge, Science, 175, 1103-1105, 1972.

Nakamura, K., Island arc tectonics, a hypothesis, Sumposium on problems concerning "green tuffs", Annual Meeting of Geol. Soc. Japan, 31-38, 1969.

Plafker, G., Tectonics of the March 27, 1964 Alaska earthquake, U.S. Geol. Surv. Prof. Paper, 543-I, 1969.

Reilly, W. I., Gravity map of New Zealand 1:4,000,000 isostatic anomalies, 1st edit., Dept. Scient. Industr. Res., Wellington, New Zealand, 1965.

Scholl, D. W. and M. S. Marlow, Sedimentary

sequence in modern Pacific trenches and the deformed circum-Pacific eugeosyncline, (eds.) R. H. Dott and R. H. Shaver, Modern and ancient geosynclinal sedimentation, Soc. Econ. Paleo. Min., Spec. Publ. 19, 193-211, 1974.

Scholl, D. W., E. C. Buffington, and M. S. Marlow, Plate tectonics and the structural evolution of the Aleutian-Bering Sea region, In: Geology of the Bering Sea and Adjacent Regions, Geol. Soc. Am. Memoir 151, in press, 1975.

Scholz, C. H., Static fatigue of quartz, J. Geophys. Res., 77, 2104-2114, 1972.

Sclater, J. G., J. Crowe, and R. N. Anderson, On the reliability of oceanic heat flow averages, J. Geophys. Res., 81, 2997-3006, 1976.

Stauder, W., Tensional character of earthquake foci beneath the Aleutian trench with relation to sea-floor spreading, J. Geophys. Res., 73, 7693-7701, 1968.

Von Huene, R., Structure of the continental margin and tectonism at the eastern Aleutian trench, Geol. Soc. Am. Bull., 83, 3613-3626, 1972.

Watts, A. B., Gravity field of the northwest Pacific Ocean basin and its margins: Aleutian island arc-trench system, Geol. Soc. Am. Map and Chart Series, MC-1, 1975.

Yoshii, T., Aseismic front, Zisin, 28, 365-367, (in Japanese), 1975.

SEISMICITY AND PLATE SUBDUCTION IN THE CENTRAL ALEUTIANS

E. R. Engdahl

Cooperative Institute for Research in Environmental Sciences
University of Colorado/NOAA, Boulder, Colorado 80309

Abstract. Data from two high-gain, high-frequency seismograph networks in the central Aleutians are used to describe spatial variations in seismicity and focal mechanism patterns. Shallow earthquakes (0-50 km) occur the most frequently, mainly along a northward-dipping diffuse zone beneath the frontal slope of the ridge, with localized concentrations in activity. Although some normal faulting is observed in the low density crust of the ridge above the main zone of activity, these shallow events are characterized primarily by thrust mechanisms, not all uniformly consistent with the slip direction suggested by larger earthquakes and by relative plate motion calculated for the north Pacific. Only a few shallow earthquakes were detected beneath the Aleutian terrace.

Intermediate depth earthquakes, which occur as deep as 250 km, appear to separate into two groups, suggesting a change in slab properties at a depth of about 110 km, directly beneath the volcanic arc. Events between 50 and 110 km in depth are uniformly distributed along the arc in a zone no more than 20 km thick dipping steeply beneath the ridge. Focal mechanisms for events within this group suggest down-dip compression or lateral extension along the strike of the arc. Earthquakes deeper than 110 km are located in a more steeply dipping diffuse zone, beneath and behind active volcanoes, with mechanisms suggesting that the slab is neither strongly in compression or tension at depth. Except for the upper part, the wedge of crust and mantle above the Benioff zone is aseismic.

To explain these observations fluid dynamical models of the Aleutian slab require a high viscosity of about 2 x 10^{22} poise for the material adjacent to at least the deeper parts of the slab.

Introduction

With the exception of the complex Japan arc, contemporary descriptions of subduction zone seismicity are based on far field determinations of focal parameters. These descriptions are often incomplete and highly inaccurate. Intermediate and deep focus earthquake activity is usually low level and not well defined. Focal depths are poorly constrained and high velocity plates introduce considerable bias in conventional hypocenter locations [Herrin and Taggart, 1968; Engdahl et al., 1976]. Mislocation errors of up to 50 km are probably not uncommon.

Two independent opportunities to study the fine details of seismicity in the central Aleutians arose with the installation of microearthquake networks to (1) monitor explosion-related effects near Amchitka Island and (2) develop principles and techniques for earthquake prediction in the Adak Island region. Results of the Amchitka study have been previously reported [Engdahl, 1971; Engdahl, 1972]. The data were also used to study plate effects [Engdahl, 1973; Engdahl et al., 1976] and earthquake prediction methods [Kisslinger and Engdahl, 1973; 1974] in the central Aleutians. The Adak network began operation in late July, 1974, and the study of prediction techniques there is still in progress. Although installed for different purposes, these two networks have already provided a wealth of data on seismicity in an active subduction zone that has yet to be fully documented. It is the purpose of this paper to describe general characteristics of the seismicity revealed by data from these two networks.

Amchitka Data

The Amchitka network operated continuously in the configuration shown in Figure 1 from October 1970 through April 1973. The basic network consisted of eight seismic systems on four islands with matched high-frequency response for monitoring microearthquakes and explosion-related effects in the region [Engdahl, 1972]. Two levels of data, with 18-db separation, were transmitted from each site to a central facility on Amchitka for recording on 16 mm Develocorder film. The high-gain level was passed through a 2 Hz low-cut filter to reduce natural microseismic background noise. A third level, an additional 18 db down, at ASB and horizontal data from orthogonal seismometers at AMA, SSI, RAT, and ASB were also recorded.

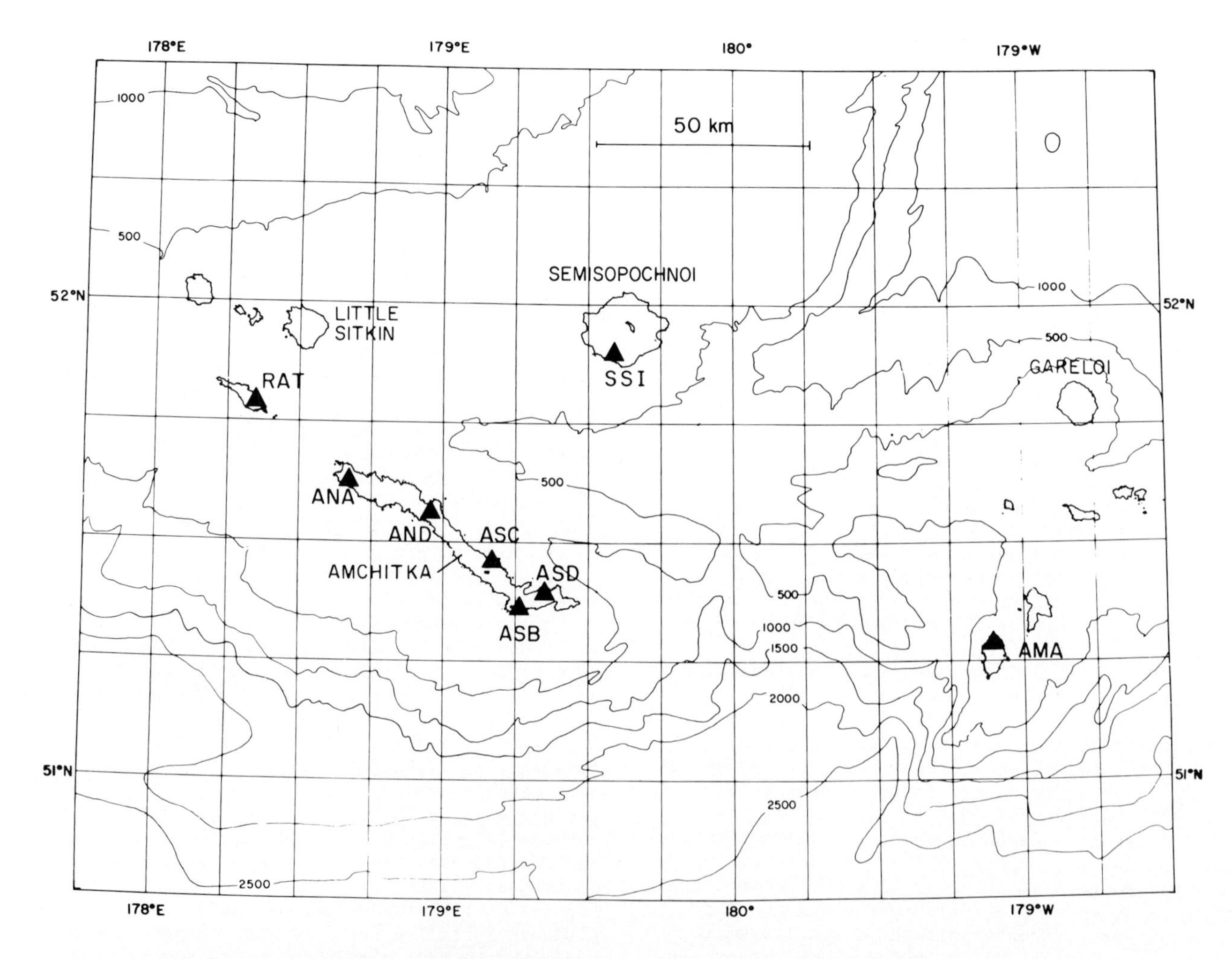

Figure 1. Amchitka seismograph network. Bathymetric contours are at 500 fathom intervals (1 fathom = 1.83 meters) in this and all subsequent figures. Volcanoes and other features mentioned in text are labelled.

P and S arrival times were read to an accuracy of $\pm$ 0.05 sec from film for all events detected by at least two or more stations of the network. The maximum amplitude in the envelope of the P-wave onset was also measured for magnitude estimation. A velocity model based on investigations by Engdahl [1972] was used with observed P and S arrival times and amplitudes to estimate hypocenters and magnitudes.

Amchitka Seismicity

Epicenters

To obtain a representative and accurate picture of the seismicity, we have plotted in Figure 2 epicenters for all events magnitude 3 and above and located with Amchitka Island stations and at least two of the off-island stations AMA, SSI, and RAT during the 2 1/2-year operational period of the network. Although the capability existed to locate earthquakes south of this map, on the Aleutian terrace, none were found and we must conclude that this region between the trench and that shown here is effectively aseismic. Earthquakes apparently occurring near the south wall of the trench were frequently detected and, although poorly located, were accompanied by large T waves which made them readily identifiable on the network [Engdahl, 1970].

Shallow earthquakes (0-50 km) occurred the most frequently, along a diffuse zone beneath the south slope of the ridge with localized concentrations in activity. Some of the larger events (∿ magnitude 5) had considerable numbers of lower magnitude aftershocks not shown here. Trends and complications in this pattern seem to correlate with bathymetric features along the arc, suggesting that these are structural features which may have expression at greater depth. A gap in shallow seismicity near 179.7°W is close to the boundary between aftershock sequences of major earthquakes occurring in the Rat Islands in 1965 and in the Delarof-Andreanof Islands in 1957 [Stauder, 1968a; 1972]. A similar gap is not evident in the deeper seismicity (> 50 km), however, a cluster of shallow earthquakes near 51.75°N and 180°W corresponds to the location of a large shallow strike-slip earthquake suggested by Stauder as a permanent boundary fault between two tectonically independent Aleutian arc segments or blocks. The region is also characterized by an apparent offset in the volcanic arc and in the bathymetry of the trench.

In this view deeper seismicity appears to be uniformly distributed along the arc between the ridge and volcanic line to the north and then to cluster spatially behind active volcanoes (Little Sitkin in the W Section, Semisopochnoi in the C Section, and Gareloi in the E Section).

The data have been conveniently separated into the three regions shown in Figure 2 for purposes of presentation.

Frontal View

Figure 3 is a view looking towards the center of curvature of the arc and projected horizontally at an arbitrary reference radius. Intense activity marks the shallow subduction zone with a pronounced gap near the boundary identified in Figure 2. The gap does not appear to extend into the deeper seismicity as earthquakes between 50 and about 110 km in depth are uniformly distributed along the arc. Earthquakes deeper than 110 km appear to cluster spatially beneath and behind the active volcanoes indicated by the triangles. On the left there is a cluster of earthquakes 110 to 150 km in depth beneath and slightly west of Little Sitkin; in the center a group 150 to 200 km in depth beneath and behind Semisopochinoi; and on the right the deepest group, not well-defined, behind Gareloi. This observation becomes more fully convincing when similar plots are made for earthquakes of all magnitudes.

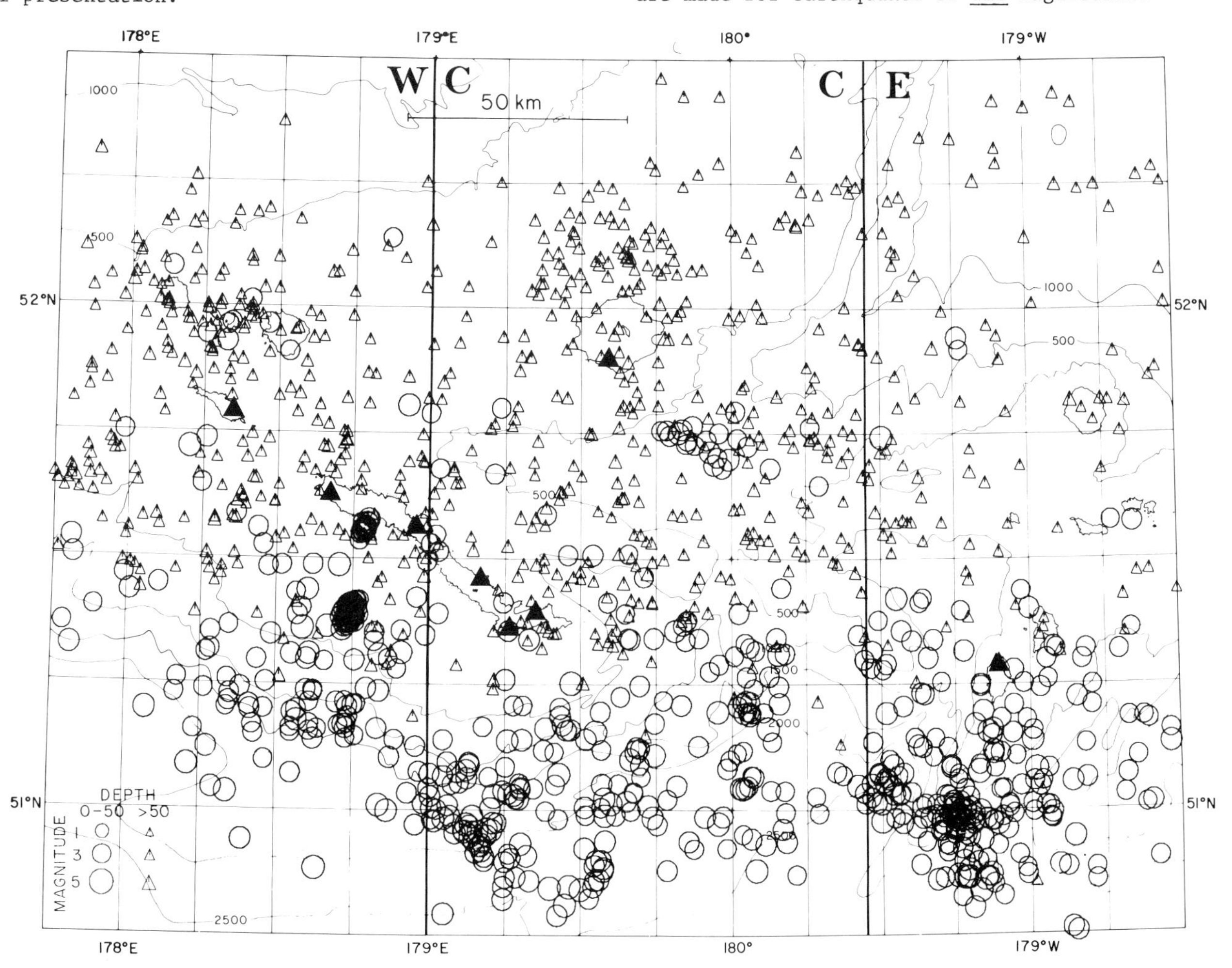

Figure 2. Epicenters for earthquakes magnitude 3 or greater and located using Amchitka Island stations and at least two of the off-island stations AMA, SSI and RAT. Event symbol size is scaled by magnitude as shown in the legend.

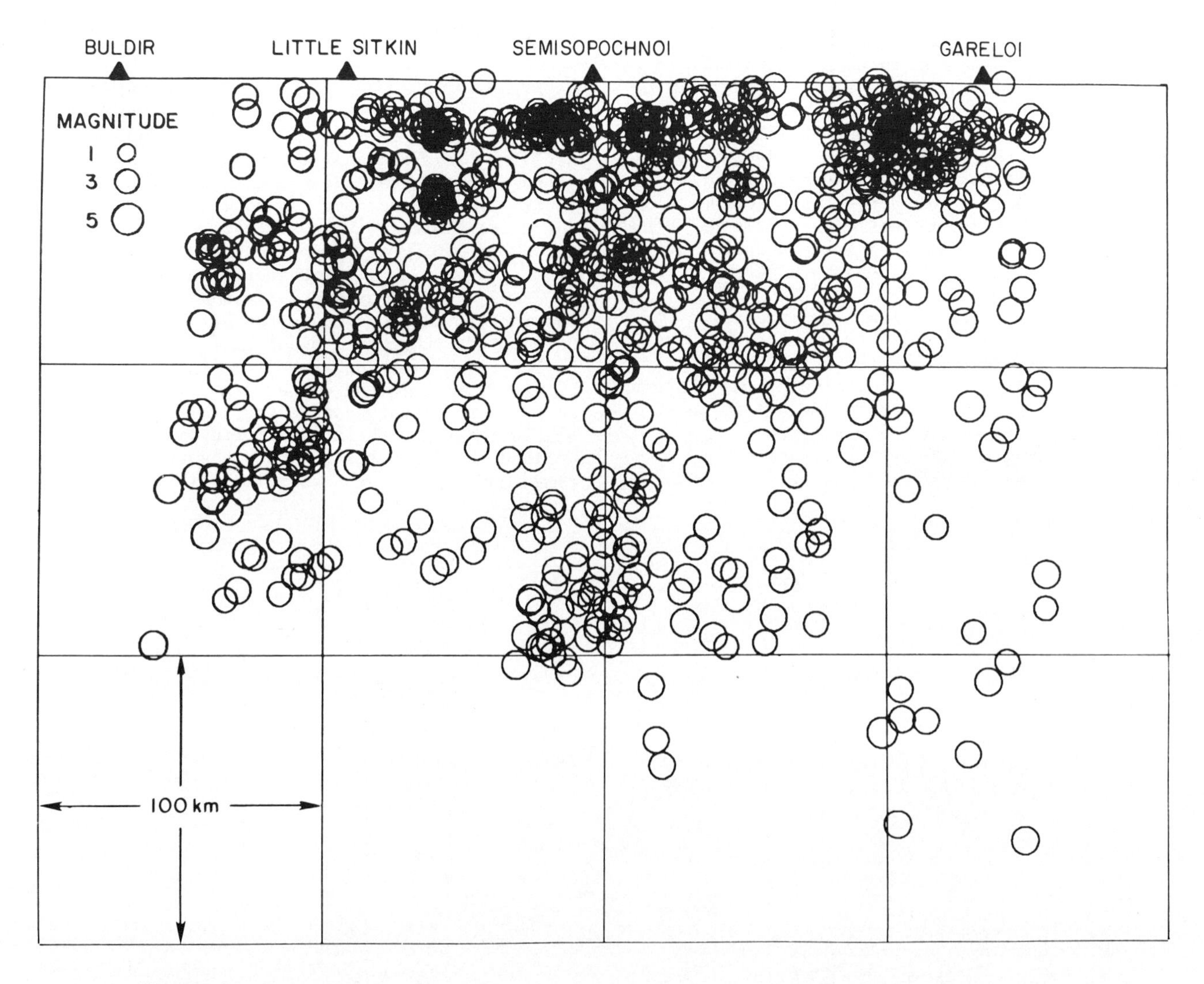

Figure 3. Frontal projection of earthquakes in Fig. 2. Vertical axis is depth and horizontal axis is arc distance at an arbitrary reference radius from the center of curvature of the arc. Active volcanoes are indicated by triangles.

Side Views

In Figure 4a all the Amchitka data previously shown are projected in section using as horizontal coordinate the distance to the center of curvature for the central Aleutians. This point at 58.392°N and 178.565°W is determined by a fit to the curvature of the trench axis at a radius of 8.1° in the central Aleutians [Engdahl et al., 1976].

In this projection the seismicity collapses into the familiar classic Benioff zone. Shallow earthquakes are now concentrated in an intense but diffuse zone of activity. The well distributed group of earthquakes between 50 and 110 km in depth in Figures 2 and 3 are now confined to a thin zone, no more than 20 km thick, dipping steeply beneath the ridge. At about 110 km the dipping seismic zone steepens and becomes more diffuse. No earthquakes were detected in the wedge above the seismic zone except for a minor amount of shallow seismicity near the ridge axis and above the main seismic zone, and along the volcanic arc.

Figures 4b-4d show the same data as Figure 4a subdivided into the W, C, and E sections defined in Figure 2. The subdivided data all show the same characteristics as the combined data set previously discussed. In Figure 4b we see localized concentrations of the minor shallow seismicity near the ridge axis above the main zone of activity and beneath the volcanic arc. We also note what appears to be a 20 km wide gap near the ridge axis at very shallow depths (0-20 km). Earthquakes in Section C are probably the best located and most of the deep seismicity falls beneath and behind Semisopochnoi volcano. Sec-

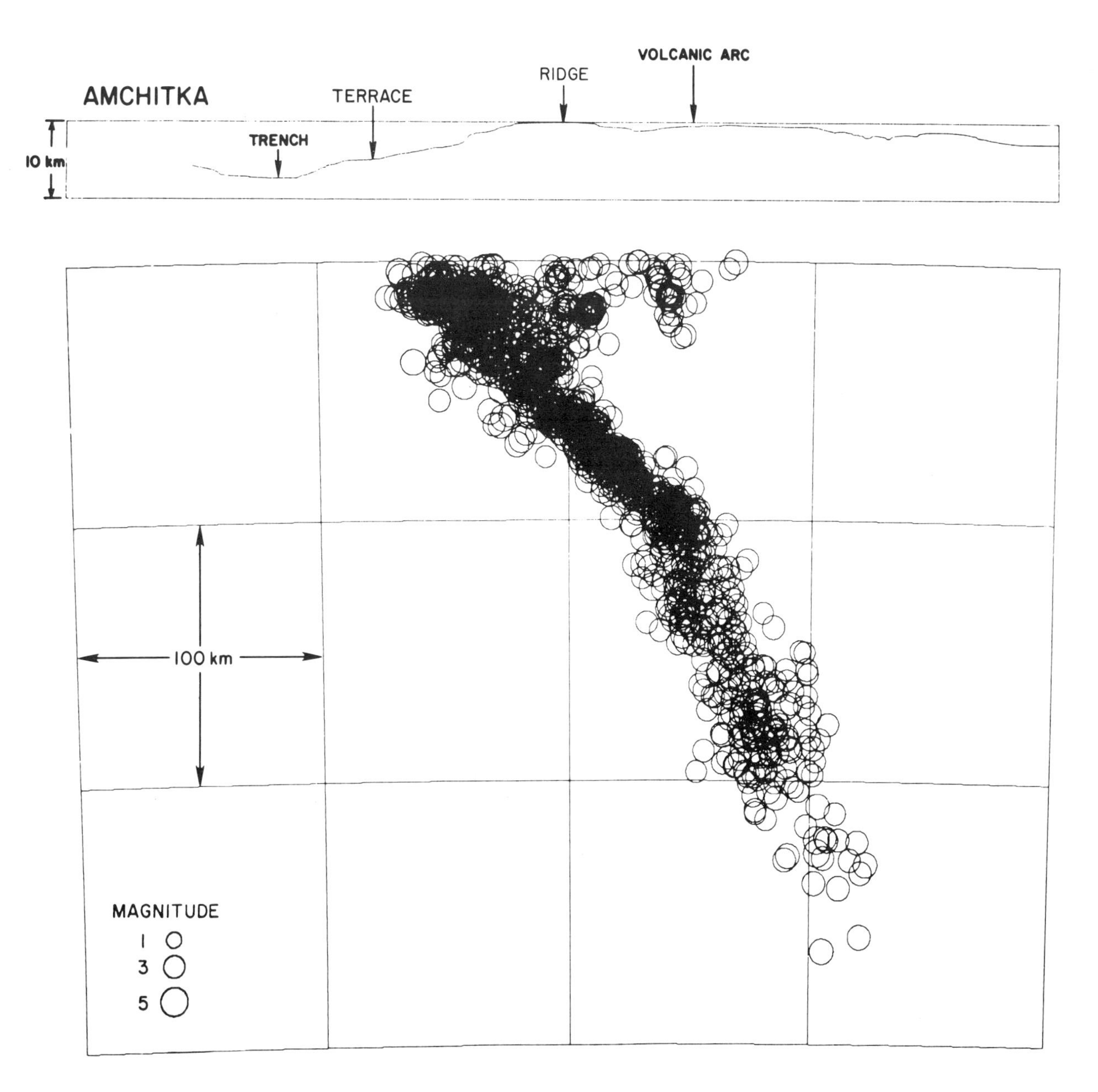

Figure 4a. Side projection of earthquakes in Fig. 2. Vertical axis is depth and horizontal axis is distance to center of curvature of the arc. Bathymetric section is normal to arc through Amchitka Island.

tion E contains the deepest events and also the most poorly located, particularly focal depths of the shallow events.

Other Amchitka Studies

Because of a concern that anomalous velocities associated with plate structure would seriously affect the interpretation of seismicity and focal mechanisms in an active island arc, an extensive investigation of plate effects in north Pacific subduction zones was made, using thermal models and seismic ray tracing [Engdahl et al., 1976]. In the Amchitka region 39 intermediate events of magnitude 4 or greater, six of which were detected teleseismically, were selected for special study. A velocity model of the plate, constructed from thermal contours, with its position constrained by seismic travel-time residuals and amplitude anomalies from the LONGSHOT explosion [Jacob, 1972; Sleep, 1973], was used to relocate all events, using only local data, and the teleseismic events, using both distant and local data. In both cases, the relocated hypocenters appear to define a thin zone, interior to the plate, corresponding to the colder, more brittle region of the slab suggested

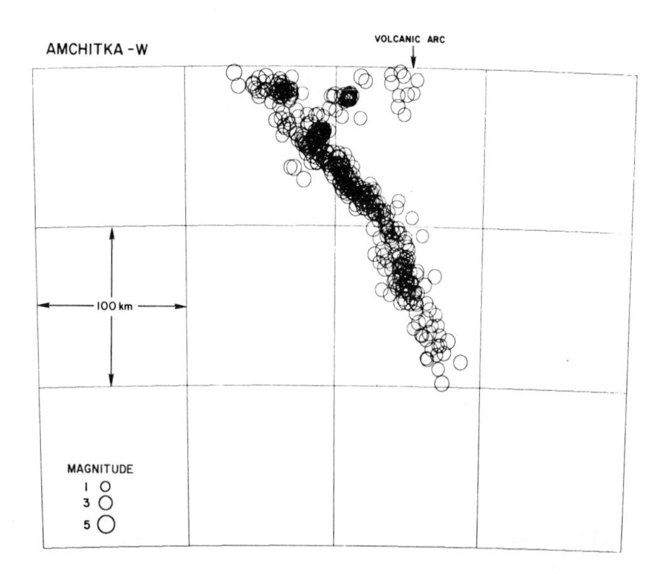

Figure 4b. Same projection as Fig. 4a with data only from W section defined in Fig. 2.

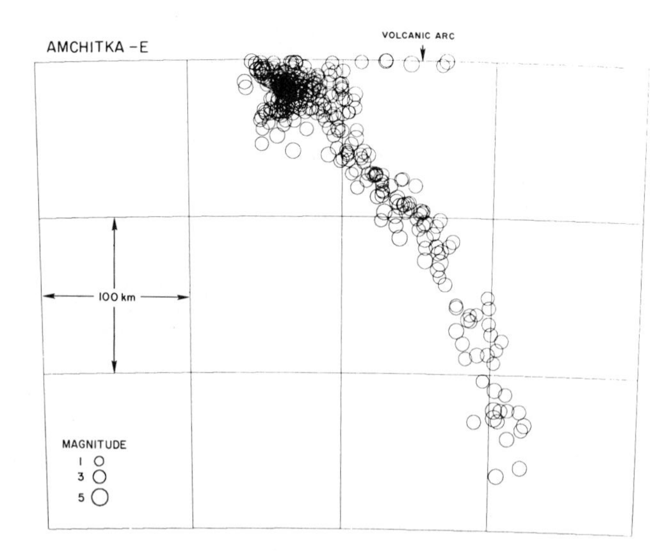

Figure 4d. Same projection as Fig. 4a with data only from E section defined in Fig. 2.

by thermal models. The thinness of this zone we believe to be model independent, but the relative location within the model is probably not as well determined, since little is known about the details of upper mantle structure in the wedge traversed by rays to the local network. In the case of teleseismic events, the locations using a plate model were close to the hypocenter distribution shown in this paper. Using only local network data, a plate model gave locations about 20 km further north and 20 km shallower. The thickness agrees well with an estimate of 11 km by Wyss [1973], from the source dimensions of large intermediate depth earthquakes in the Tonga

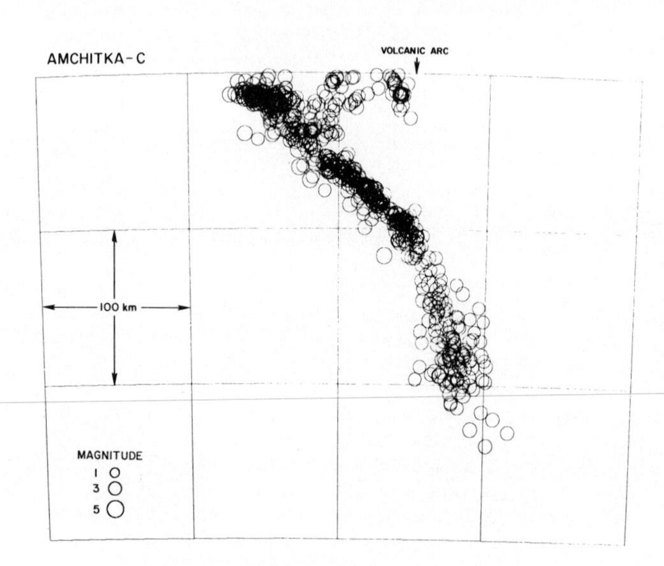

Figure 4c. Same projection as Fig. 4a with data only from C section defined in Fig. 2.

region. The systematic improvement in these event locations clearly demonstrates the need for local stations and a knowledge of lateral variations in structure near island arcs for accurate hypocenter determination. This capability will, in turn, make it possible to establish meaningful relations between active zones and particular geologic features.

We used these relocated earthquakes to obtain more consistent composite focal mechanism determinations. Intermediate events from about 70 to 120 km in depth were uniformly characterized by down-dip pressure axes or lateral extension along the strike of the arc. Among events deeper than about 140 km there was a large subset that appeared to have near vertical tension axes, although in some instances they coexisted with mechanisms of opposite sense.

In another study, two very intense concentrated source volumes were examined for systematic temporal variations in the ratio of the two body-wave velocities [Kisslinger and Engdahl, 1974]. One source lies at a depth of 20 km, well above the main activity near the ridge axis and not related to the principal tectonics of the arc in any obvious way (see Figure 4b). A composite focal mechanism shows normal faulting, with the tension axis east-west. Another source is located in the subduction zone (see Figure 4b), at a depth of about 45 km, and a composite focal-mechanism solution suggests thrust-type earthquakes, implying northwesterly plate convergence, in agreement with Stauder's [1968a] results for the 1965 series.

Studies of lateral variation in attenuation and velocity of high-frequency P and S waves near island arcs are also especially important in mapping lateral variations in upper mantle structure. Data from other arcs suggest that the presence of

volcanoes correlates well with high attenuation in the mantle. Studies by Grow [1973] and Grow and Qamar [1973], in the latter case with data from the Amchitka network, also suggest an attenuating low-density zone in the crust and mantle beneath the volcanic arc in the central Aleutians.

Adak Data

The Adak seismograph network was established on and around Adak Island, Alaska, in late July of 1974 and has been in continual operation since that time. The continued success of this field operation is due in large part to helicopter support provided by the U. S. Navy. In may of 1975 the network was expanded to the west by the addition of three stations. The network now has dimensions of about 50 by 150 km along the Aleutian arc and consists of 11 high-gain, high-frequency, two-component seismic systems on six islands (Figure 5). All the data are radio-telemetered to the U. S. National Weather Service Adak Observatory where they are recorded on 16 mm Develocorder film. An additional three channels of data from the Adak seismometers (ADK) are also recorded.

P and S arrival times are read to an accuracy of ± 0.03 sec from film for events occuring within the network. Duration of the wave coda is also measured for magnitude estimation. The earthquakes are located using the P and S arrival times and a model of the crust beneath Adak determined from seismic refraction profiles [Shor, 1964]. A velocity ratio of $\sqrt{3}$ (σ = 1/4) is assumed and station corrections by source region applied to shallow events.

Adak Seismicity

Epicenters

Epicenters for all events located with data from the network during the first 1 1/2-year

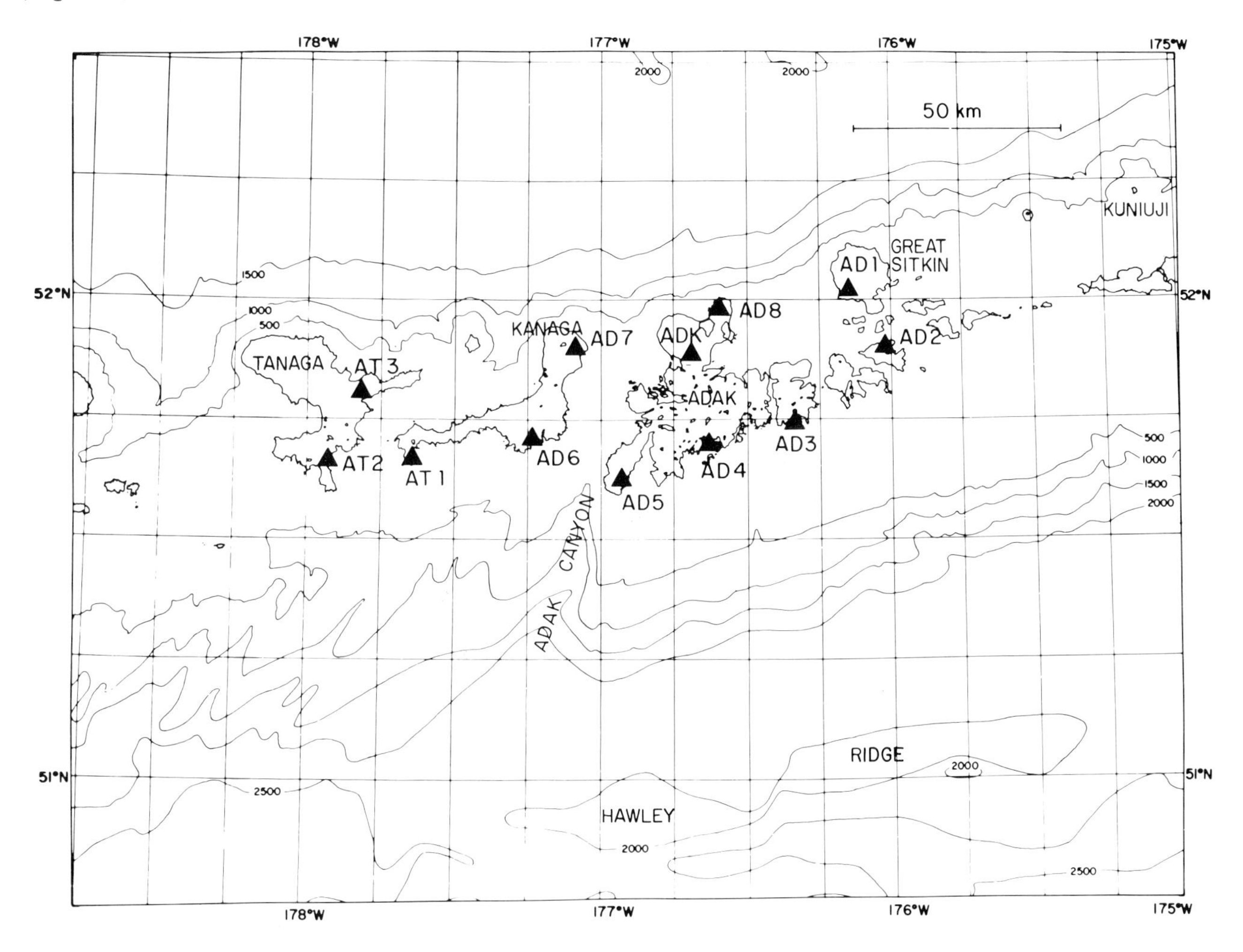

Figure 5. Adak seismograph network. AT1, AT2, AT3 were installed during the last half of the operational period reported in this paper. Volcanoes and other features mentioned in text are labelled.

period of operation are shown in Figure 6. As in the case of the Amchitka data, the capability to locate earthquakes south of this distribution between the trench and the Aleutian terrace exists, but only a few earthquakes have been found there to date. Recently, we have detected some shallow earthquake activity near Hawley Ridge (see Figure 5), but they are not plotted in this data set.

The seismicity shown in Figure 6 is consistent with the patterns described for Amchitka, except at a lower level of activity (recall in Figure 2 the events are magnitude 3 or greater). Shallow earthquakes (0-50 km) occur the most frequently, beneath the frontal part of the ridge, with localized concentrations. Complications to the pattern seem to correlate structurally with Adak Canyon, west of Adak (see Figure 5), possibly corresponding to topography in the top of the plate. Deeper seismicity is not so well defined but appears to be uniformly distributed along the arc. An exceptional feature of this region is a transition to more regular bathymetric contours east of Adak Canyon, marked on this map by the subdivision into W and E sections. In the W region the earthquakes do not seem to cluster into the kinds of localized sources found in the E section.

Frontal View

Figure 7 is similar to Figure 3, a view toward the center of curvature of the arc and projected at the same reference radius as the Amchitka data. The intense shallow zone of activity is even more evident in the Adak data where events of all magnitudes are plotted (perhaps suggesting a difference in b slope between shallow and deep seismicity). On the whole these data are probably more accurately located than the Amchitka data so that the marked upper boundary to the shallow seismicity is probably a feature common to both regions. We believe this boundary to be real and not an artifact of the location procedure or of a nearby refracting

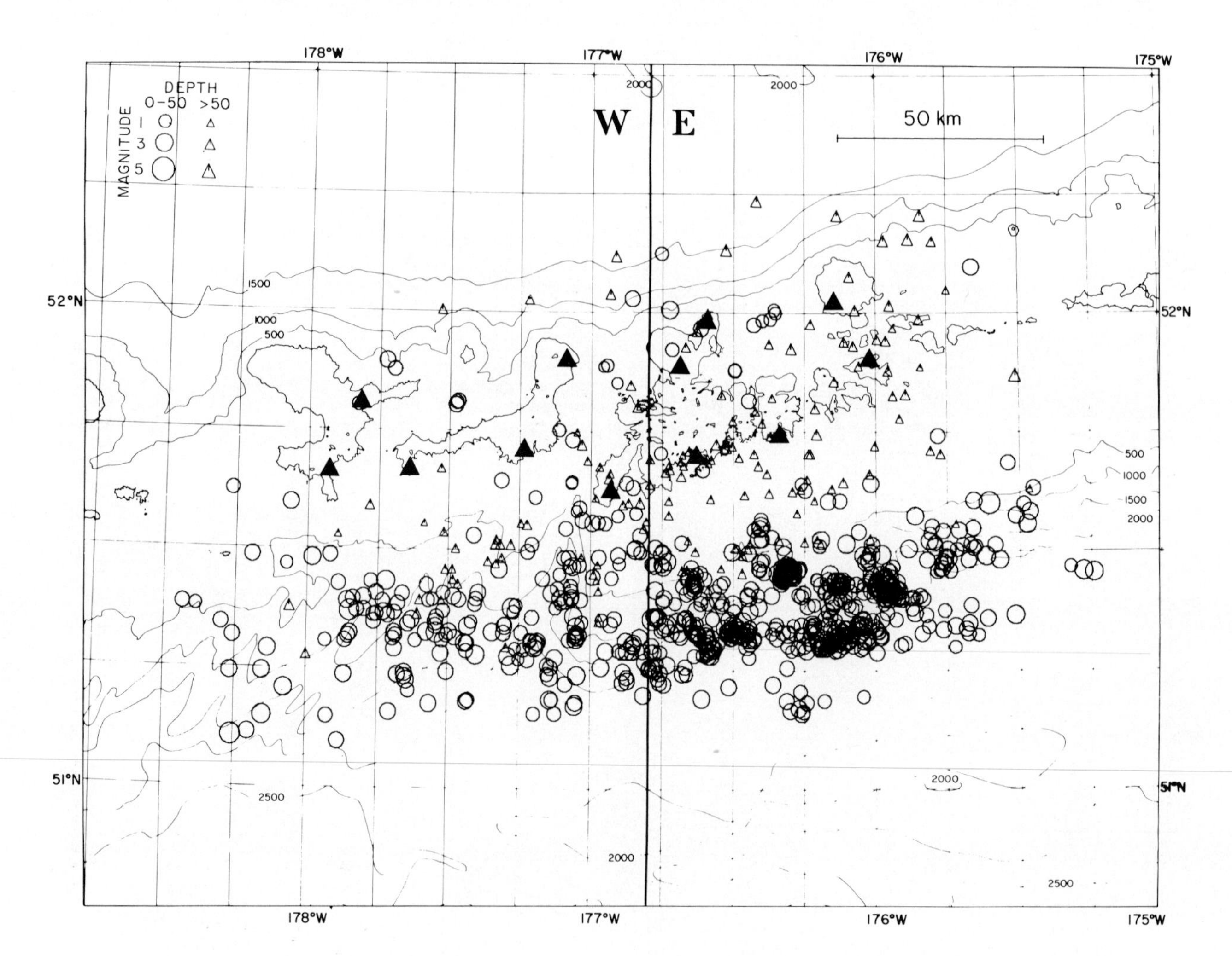

Figure 6. Epicenters for earthquakes of all magnitudes located with Adak network data.

horizon (most network arrivals are direct rays and both P and S waves are used). An interesting note is that Shor's [1964] crustal model has a layer boundary at 26 km. Most of the westerly located earthquakes were located during the second half of the operational period of the network reported in this paper, so that these figures are not true portrayals in time. Nevertheless, it appears with this limited data that earthquakes between 50 and about 110 km in depth are uniformly distributed across the arc, whereas deeper earthquakes appear to cluster beneath and below Great Sitkin (the second triangle from the right). There was a minor eruption of this volcano in February, 1974, shortly before the network was installed. We can only speculate if the present level of deep seismicity near Great Sitkin is somehow related. A large number of very shallow events near Great Sitkin were recorded by the Adak station (ADK) prior to eruption.

Side Views

In Figure 8a the data are projected in section as the distance to the center of curvature of the arc. Again we see features similar to Amchitka. Shallow earthquakes are concentrated in an intense but diffuse zone of activity. Earthquakes between 50 and 110 km in depth have collapsed into a thin zone, no more than 20 km thick, dipping steeply beneath the ridge. At about 110 km, the dipping seismic zone steepens and becomes more diffuse. Except for a minor amount of seismicity in the upper part, the wedge above the Benioff zone is aseismic.

Figures 8b and 8c show the same data subdivided into W and E sections defined in Figure 6. In the E section, where bathymetry is more regular, the shallow seismicity is not so diffuse and few events occur in the wedge above this zone. The characteristics of deep seismicity are about the

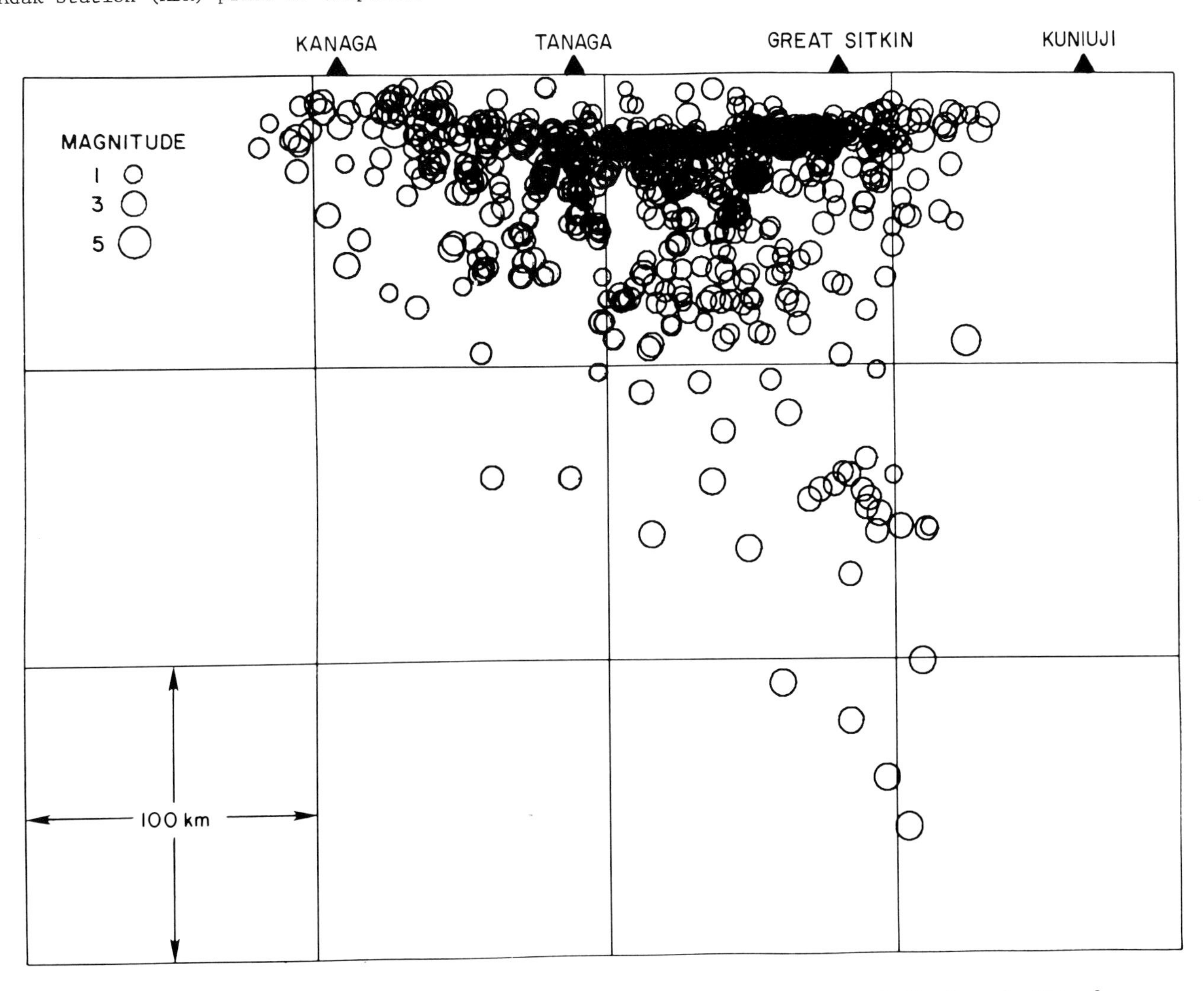

Figure 7. Frontal projection of earthquakes in Fig. 6. Axes are defined as in Fig. 3. Active volcanoes are indicated by triangles.

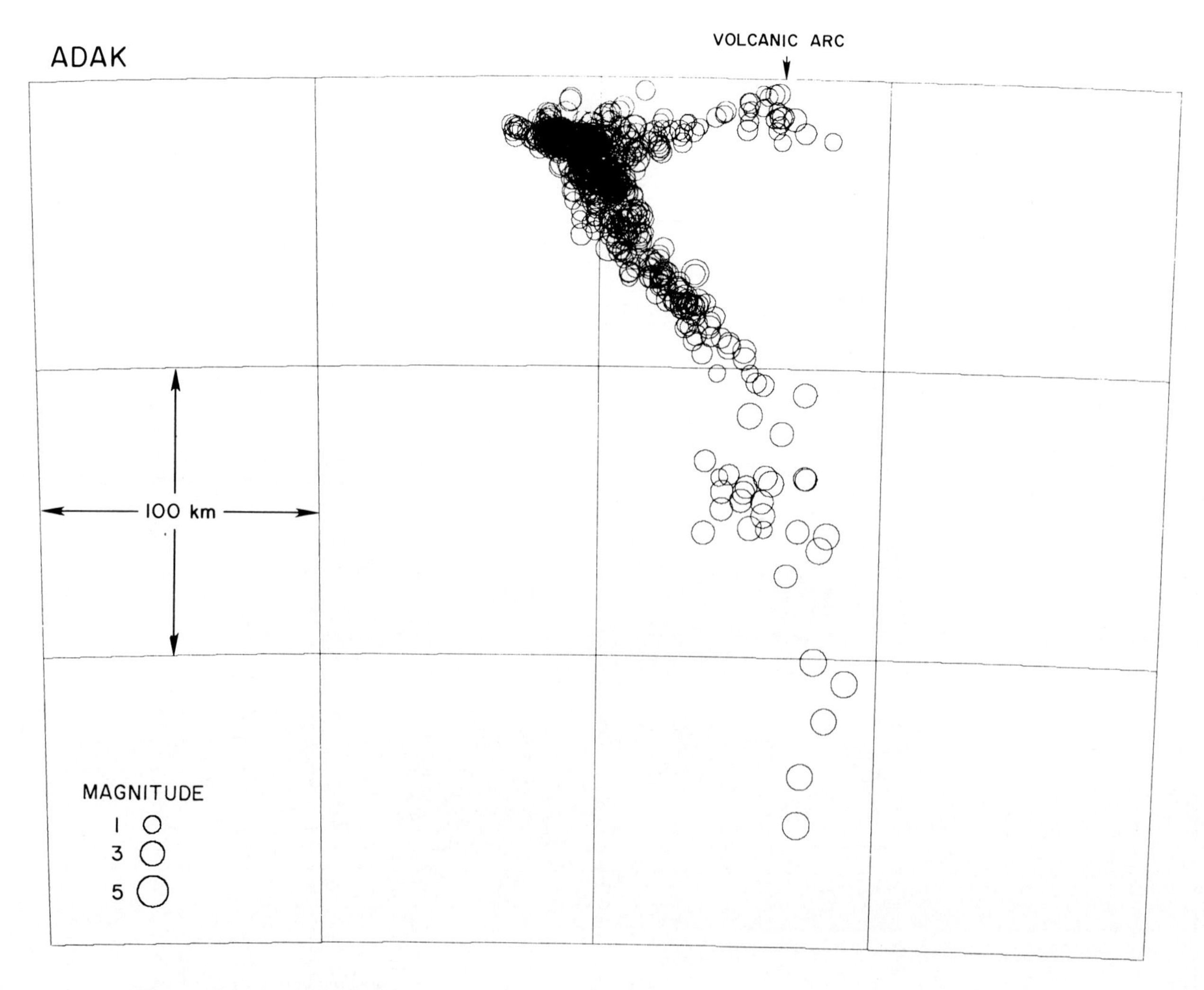

Figure 8a. Side projection of earthquakes in Fig. 6. Axes are same as Fig. 4.

same. Because the Adak network has a better capability to locate shallow earthquakes along the volcanic arc, we now see a well-defined zone of very shallow seismic activity located in a relatively continuous narrowly defined zone beneath the line of volcanoes.

Other Adak Studies

The data plotted in Figure 8a suggest about the same downward curvature of the seismic zone as observed near Amchitka, and are consistent with the location of plate Model B for the crust and mantle density structure beneath Adak proposed by Grow [1973] on the basis of gravity data.

Data on focal mechanisms are at this time limited, but there have been two recent studies. An examination of mechanisms of shallow events above the main seismic zone in Adak Canyon, corresponding to the confined lineament in Figure 8a, suggests normal faulting as the causal mechanism [R. La Forge, personal communication, 1976]. La Forge also finds the main seismic zone at depth to be characterized by thrust-type events, possibly structurally decoupled from the activity above. A special study of 35 events within 10 km of a magnitude 4.2 earthquake along the subduction zone and within a very small active source [Engdahl and Kisslinger, 1975] revealed that most of the mechanisms were also of the thrust type, but not all uniformly consistent with the slip direction suggested by larger earthquakes [Stauder, 1968b] and by relative plate motion calculated for the north Pacific [Minster et al., 1974].

Discussion

By now it must be obvious to even the most casual reader that several characteristics of the seismicity shown suggest that a sharp change in physical properties occurs at about 110 km beneath the volcanic arc and that an almost certain relationship must exist between the two features. The evidence includes: (1) progression from a

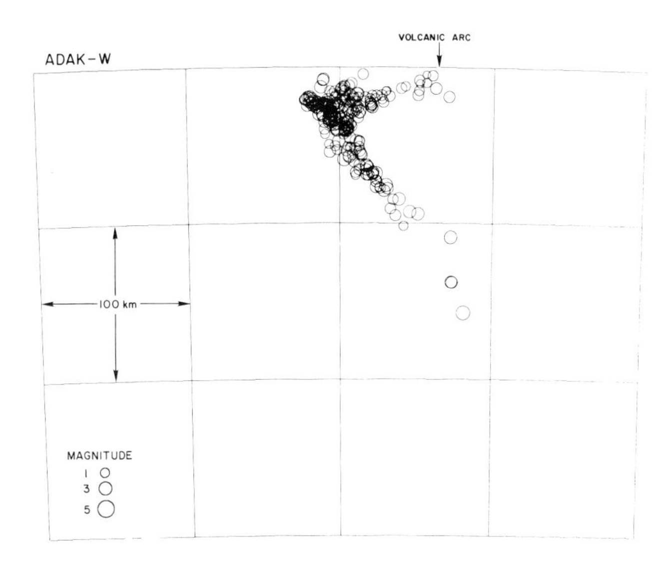

Figure 8b. Same projection as Fig. 8a with data only from W section defined by Fig. 6.

thin dipping seismic zone to a more steeply dipping diffuse deeper zone; (2) change from uniformly distributed intermediate depth activity between the ridge and volcanic arc to indications of clustering of earthquakes beneath and behind the active volcanoes; (3) change in the pattern of focal mechanisms between the two intermediate depth groups; and (4) decreased activity rate for the deepest earthquakes. We also observe that active volcanism is absent in the western Aleutians where earthquakes occur at depths less than 110 km [Engdahl, 1973]. If mechanisms such as viscous shear strain heating, phase changes, or dehydration reactions are responsible for magmatic generation and the formation of volcanoes, then we must examine these possibilities for inducing melting at depths near 110 km. Future studies of these data will center on what, if anything, can be learned about the detailed seismicity and earthquake mechanisms across this transition depth.

If intermediate depth earthquakes do occur in the colder, more brittle region of the slab, we would expect focal mechanism solutions to suggest that the descending lithosphere acts as a stress guide that aligns earthquake generating stresses along the dip of the slab [Isacks and Molnar, 1971]. Engdahl et al. [1976] found intermediate depth events of magnitude 4 or greater near Amchitka to be uniformly characterized by down-dip pressure axes or lateral extension along the strike of the arc from about 70 to 120 km in depth. Hence, in this range of depths the slab appears to be in compression, which would occur if the subduction rate were faster than the rate at which the lithosphere were sinking [Isacks and Molnar, 1971; Smith and Toksoz, 1972] or if gravitational instability (a density increase) resulted from a phase change. On the other hand, extension along the arc might be expected if lateral spreading of the plate, as it descends beneath a margin convex to it, were the operative mechanism [Stauder, 1968b; 1972]. Among events deeper than about 140 km we find a large number with near vertical tension axes but others, in some instances colocated, of an opposite sense, so that the deeper portions of the slab are both weakly in tension and compression (or little difference between maximum and minimum stress). The diffusion of hypocenters is probably due in part to mislocation errors, but it is not likely that the location capability would break down so completely at exactly 110 km. One explanation could be that more extensive fault planes are needed to accommodate these events [Billington and Isacks, 1975].

Fluid dynamical models of the Aleutian slab calculated by Sleep [1975], consistent with topography and observed gravity anomalies, erroneously presumed that the slab was in down-dip tension. In light of the seismic results found in this paper, new calculations were warranted. Engdahl and Sleep [1976] recently reported new models that generally conformed to the general characteristics suggested by the data of this paper. The observations that the slab is neither strongly in compression or tension at depth, and that intermediate focus Aleutian earthquakes are small and infrequent suggest a balance between viscous drag on the top and bottom of the slab with the negative buoyancy of the slab, about 1000 bars. To produce 500 bars of shear stress on each side of the slab, for the Aleutian velocity and geometry and a yield stress of 200 bars, a higher viscosity of about 2×10^{22} poise is required for the material adjacent to at least the deeper parts of the slab (> 200 km). The east-west orientation of the T-axis may thus possibly originate

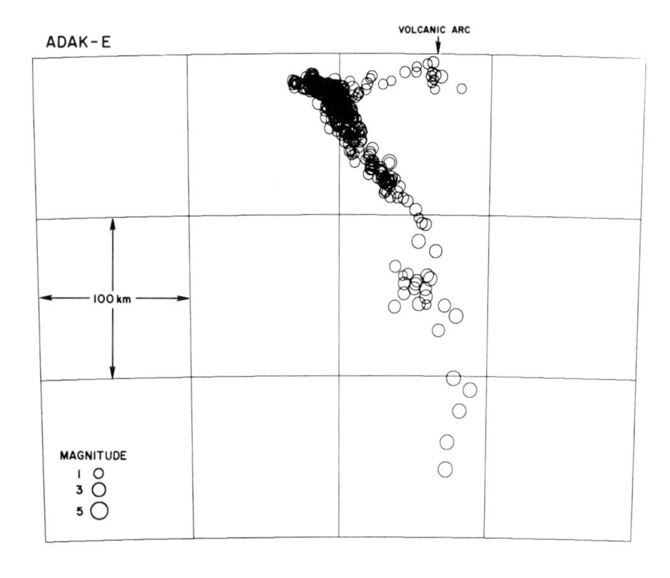

Figure 8c. Same projection as Fig. 8a with data only from E section defined in Fig. 6.

by means of viscous drag on the lower part of the slab resisting the component of horizontal movement required if the computed oblique plate movement [Minster et al., 1974] applies, suggesting a surface movement towards west relative to substrata in the central Aleutians. Normal faulting in the region above the zone of major seismicity is consistent with the tendency of the low density crust to rise and spread, producing extensional stresses, as predicted by the fluid dynamic models. Low stresses in the wedge above the Benioff zone and beneath the volcanic arc are consistent with the lack of seismicity in this zone, although aseismic flow (creep) in a highly stressed aseismic wedge is an alternative possibility.

The location and definition of a tectonic boundary between the Rat Islands and the Delarof-Andreanof Islands, near Amchitka Pass, is still not well resolved. The boundary fault suggested by Stauder [1968a] could very well be only a local feature, as suggested by the minor activity still occurring there. The gap in shallow seismicity near Amchitka Pass described in this paper actually occurs between groups A and B of recent earthquakes in the Delarof-Andreanof region reported by Stauder [1972] as an easterly progression of activity from the boundary in that block. Because only far-field observations are available, it is doubtful that we will ever be able to define the aftershock boundaries well enough to resolve these questions.

The higher seismicity near Amchitka suggests that post-earthquake effects of the major 1965 earthquake in this region are still operative. This is consistent with a model of episodic slip proposed by Anderson [1975].

Conclusions

Let us now consider a hypothetical model for plate subduction in the central Aleutians based on features of the seismicity and model calculations reported in this paper. The slab begins its descent at the trench, marked by the occurrence of normal faulting [Stauder, 1968b], and is subducted beneath a melange (the Aleutian terrace) that does not appear to be supportive of earthquake generating stresses. At the point of collision with the overriding abducted plate, there is a high stress concentration, diffuse seismicity, and major thrust earthquakes occur. The flattened upper section of this seismic zone is a result of the growth of accreted material beneath the inner trench slope [Karig et al., 1976]. Near 50 km the plate is driven or sinks into the mantle and earthquake generating stresses transfer to the colder, more brittle, internal portion of the slab.

In the central Aleutians the slab is in horizontal extension along the arc in the depth range of 70 to 120 km, suggesting that arc curvature is responsible for lateral spreading or that the slab is responding to resistance to oblique slip from higher viscosity material adjacent to its deeper parts. At about 110 km, directly beneath the volcanic arc, a dramatic change in the character of the seismicity occurs. The Benioff zone changes dip, the distribution and rate of activity changes, and there are differences in focal mechanisms. Let us suppose that the slab at this point becomes segmented, either by a lateral variation in slab properties or by tearing [Abe, 1972]. Continued penetration of slab segments into the mantle induces melting and magmatic generation. Volcanoes are formed above these slab segments and earthquakes continue to occur beneath and behind the active volcanoes as the segmented slab continues into the mantle. Eventually the slab is heated sufficiently to become assimilated into the mantle.

Future work will focus on a better definition of focal mechanism patterns within the two networks. The observed stresses can then be tested with three dimensional stress models to account for the effects of oblique slip.

Acknowledgments. I thank C. Kisslinger for a critical review of the manuscript and for many stimulating discussions, and K. Jacob for helpful review comments. This work was partially supported under USGS Contract No. 14-08-0001-14581.

References

Abe, K., Seismological evidence for a lithospheric tearing beneath the Aleutian arc, Earth Planet. Sci. Letters, 14, 428-432, 1972.

Anderson, D. L., Accelerated plate tectonics, Science, 187, 1077-1079, 1975.

Billington, D., and B. L. Isacks, Identification of fault planes associated with deep earthquakes, Geophys. Res. Let., 2, 63-66, 1975.

Engdahl, E. R., Aftershocks of the Aleutian trench earthquake of February 27, 1970, (abstract), EOS, 51, 779, 1970.

Engdahl, E. R., Explosion effects and earthquakes in the Amchitka Island region, Science, 173, 1232-1235, 1971.

Engdahl, E. R., Seismic effects of the MILROW and CANNIKIN nuclear explosions, Bull. Seism. Soc. Amer., 62, 1411-1423, 1972.

Engdahl, E. R., Relocation of intermediate depth earthquakes in the central Aleutians by seismic ray tracing, Nature Phys. Sci., 245, 23-25, 1973.

Engdahl, E. R., and C. Kisslinger, Prediction of earthquakes in an island arc using a local seismographic network, (abstract), EOS, 56, 1018-1019, 1975.

Engdahl, E. R., and N. Sleep, Seismicity and Stress beneath the central Aleutian arc, (abstract), EOS, 57, 329, 1976.

Engdahl, E. R., N. H. Sleep, and M. T. Lin, Plate effects in North Pacific subduction zones, Tectonophysics, in press, 1976.

Grow, J. A., and A. Qamar, Seismic wave attenuation beneath the central Aleutian arc, Bull. Seism. Soc. Amer., 63, 2155-2166, 1973.

Grow, J. A., Crustal and upper mantle structure of the central Aleutian arc, Bull. Geol. Soc. Amer., 84, 2169-2192, 1973.

Herrin, E., and J. Taggart, Some bias in epicenter determinations, Bull. Seism. Soc. Amer., 58, 1791-1796, 1968.

Isacks, B., and P. Molnar, Distribution of stresses in the descending lithosphere from a global survey of focal mechanism solutions of mantle earthquakes, Rev. Geophys. Space Phys., 9, 103-174, 1971.

Jacob, K., Global tectonic implications of anomalous seismic P travel times from the nuclear explosion Longshot, J. Geophys. Res., 77, 2556-2593, 1972.

Karig, D. E., J. G. Caldwell, and E. M. Parmettier, Effects of accretion on the geometry of the descending lithosphere, submitted for putlication, 1976.

Kisslinger, C., and E. R. Engdahl, The interpretation of the Wadati diagram with relaxed assumptions, Bull. Seism. Soc. Amer., 63, 1723-1736, 1973.

Kisslinger, C., and E. R. Engdahl, A test of the Semyenov prediction technique in the central Aleutian Islands, Tectonophysics, 23, 237-246, 1974.

Minster, J. B., T. H. Jordan, P. Molnar, and E. Haines, Numerical modelling of instantaneous plate tectonics, Geophys. J. Roy. astr. Soc., 36, 541-576, 1974.

Shor, G. G., Structure of the Bering Sea and the Aleutian Ridge, Marine Geol., 1, 213-219, 1964.

Sleep, N. H., Teleseismic P-wave transmission through slabs, Bull. Seism. Soc. Amer., 63, 1349-1373, 1973.

Sleep, N. H., Stress and flow beneath island arcs, Geophys. J. Roy. astr. Soc., 43, 827-857, 1975.

Smith, A., and M. N. Toksöz, Stress distribution beneath island arcs, Geophys. J. Roy. astr. Soc., 29, 289-318, 1972.

Stauder, W., Mechanism of the Rat Island earthquake sequence of February 4, 1965 with relation to island arcs and sea-floor spreading, J. Geophys. Res., 73, 3847-3858, 1968a.

Stauder, W., Tensional character of earthquake foci beneath the Aleutian trench with relation to sea-floor spreading, J. Geophys. Res., 73, 7693-7702, 1968b.

Stauder, W., Fault motion and spatially bounded character of earthquakes in Amchitka Pass and the Delarof Islands, J. Geophys. Res., 73, 2072-2080, 1972.

Wyss, M., The thickness of deep seismic zones, Nature, 242, 255-256, 1973.

POST MIOCENE TECTONICS OF THE MARGIN OF SOUTHERN CHILE

E.M. Herron,* R. Bruhn,** M. Winslow,** and L. Chuaqui+

*Lamont-Doherty Geological Observatory, Palisades, N.Y. 10964
** also Columbia University, Department of Geological Sciences
+ Instituto des Investigaciones geologicas, Santiago, Chile

Abstract. Marine and terrestrial data along the margin of Chile south of 52°S have been combined to show that deformation due to relative motion between the three crustal plates in the area (the Antarctic, Scotia and South America) is spread over a wide zone extending from the trench onto the continent. The triple junction of these plates has not been precisely located and even the nature of the junction, whether TTT or TFF, is debatable. Inspite of the apparent aseismicity of the area, evidence for present day relative motion between the three plates is found in the seismic reflection records obtained on the margin between the shelf and the floor of the Peru-Chile Trench and on the land as fault scarps that have been active since the Miocene.

Introduction

During March 1975, the R/V CONRAD surveyed the western margin of South America between Cape Horn and the Northwest Strait of Magellan. Gravity, magnetics, bathymetry and seismic profiler data were collected continuously and sediment cores, rock dredges and bottom photographs were also collected at selected sites. These data allow us to interpret the post-Miocene tectonic history of the area. In addition to the marine data, the pattern of post-Miocene faults on Tierra del Fuego and in Patagonia has also been examined. We believe that the data along the margin can be related to the terrestrial data and that a full understanding of the tectonics of this area can only be attained by incorporating both marine and terrestrial data.

Regional Tectonic Setting

Our study area is located about 1000 km south of the seismically active segment of the Peru-Chile Trench, a structural feature that has been traced along the entire western margin of South America (Hayes, 1966). The earliest sketch of the global pattern of crustal plates (Morgan, 1968) placed the western boundary of the South American Plate at the Peru-Chile Trench. A single oceanic plate (the Antarctic) abutting South America, was indicated on his sketch. In more recent studies, (Herron and Hayes, 1969; Chase, 1972; Minster and others, 1974) this oceanic plate has been divided into two plates: the Antarctic and Nazca. The Chile Ridge defines the boundary between these two plates and a triple junction between the Antarctic, Nazca, and South American plates is located at the continent margin near 46°S (Figure 1).

Both the Antarctic and Nazca Plates are converging on South America though at very different rates (Minster and others, 1974, Chase, 1972). The calculations of Minster and others (1974) predict that for the area near 45°S, the rate of convergence between the Nazca and South American Plates is 9 cm/yr, whereas the rate between the Antarctic and South American Plates at the same latitude is only 2 cm/yr. The slow convergence rate across the Antarctic/South American boundary is probably the main reason for the observed aseismicity of the western margin of South America south of 46°S. The distribution of seismic stations may also be responsible because small events that did occur in this region would not be recorded at the relatively distant stations. In spite of the apparent aseismicity of this margin, we believe that the Antarctic Plate is still converging on and/or slipping past South America along this boundary. Oceanic basement recorded on seismic profiler records plunges sharply down as the continent is approached (Ewing and others, 1969), and the large negative free-air gravity anomaly associated with the trench continues as far south as Cape Horn (Hayes, 1966).

Local Tectonic Setting

South of 50°S, the pattern of crustal plates is more difficult to determine than it is to the north, and small microplates must be con-

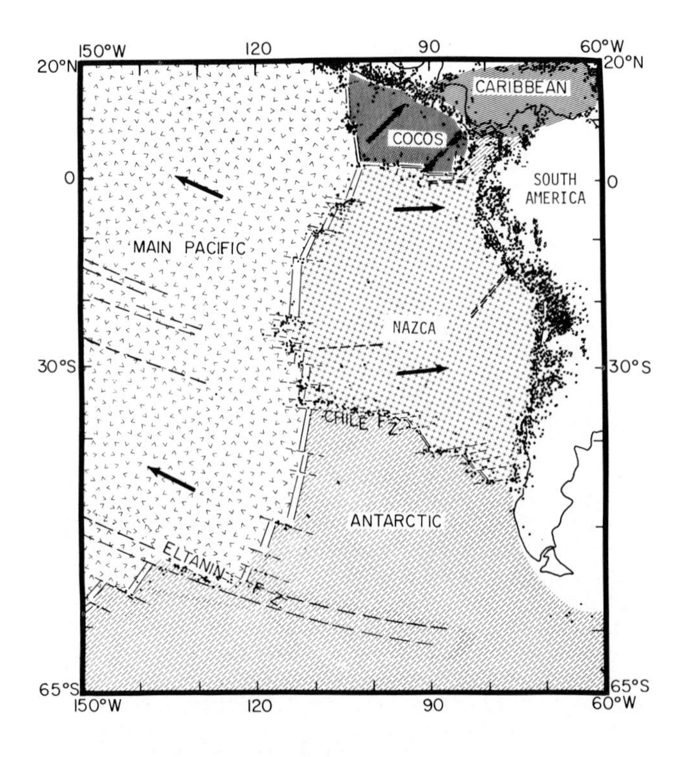

Figure 1. Major crustal plates and seismicity (Barazangi and Dorman, 1969) in the South Pacific and Scotia Sea. Arrows indicate plate motion relative to the Antarctic Plate.

sidered. In the Scotia Sea (Figure 2), Forsyth (1975) introduced a new, small plate, the Scotia Plate, on the basis of earthquake distribution and first motion studies. As shown in Figure 2, the Scotia Plate is bounded on three sides by strike-slip faults; the Malvinas Chasm, the South Scotia Ridge, and the Shackleton Fracture Zone (Barker, 1970; Forsyth, 1975). A triple junction between the Antarctic, South American and Scotia Plates must exist along the Chilean coast near 54°S. Thus in Forsyth's model, the Peru-Chile Trench in the area south of the triple junction, could reflect the relative motion between the Scotia and Antarctic plates rather than the relative motion between the South American and Antarctic Plates. For this paper, we assume that the western margin of the southern tip of South America, including Tierra del Fuego, lies on or near the Scotia-Antarctic plate boundary, and that the Shackleton Fracture Zone splays at its northern end into a series of northwest stepping faults that follow the continental margin. Finally, the pole of relative motion between South America and Antarctica as calculated by Forsyth (1975) is preferred to that calculated by Minster and others (1974).

Gravity and Bathymetry Data

In Figures 3 and 4 we present contour plots of the free-air gravity and bathymetry along the western margin of Tierra del Fuego. A negative free-air anomaly of up to -100 mgals is associated with the Trench and can be traced on Figure 3 to 68°W, 57°S. This negative anomaly is truncated by a southeast trending gravity high that parallels and may coincide with the northern extension of the Shackleton Fracture Zone (Barker and Griffiths, 1972). The bathymetry contour plot presented in Figure 4 displays the structural trench only as a broad, slightly depressed area adjacent to the continental margin. The axis of the trench deepens progressively from north to south across the study area and appears to continue to the eastern edge of the map area. The southward dip of the surface in this area probably reflects the presence of strong bottom currents flowing south into the Drake Passage and Scotia Sea (Gordon, personal communication). Between the trench floor and the continental shelf a well developed terrace has developed which we have termed the Fuegian Terrace. This terrace is present between the 1800 and 2600 meter contours and is outlined by shading in Figure 4. Mapping

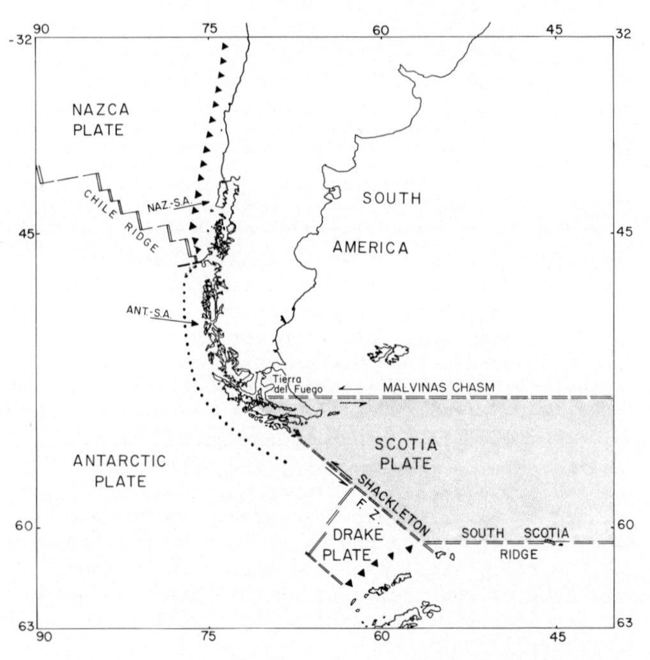

Figure 2. Plate configuration south of 30°S. (After Forsyth, 1975). Northwest limits of Scotia Plate not well defined by seismicity. Filled triangles outline seismic trenches, dots outline aseismic trench. Relative motion vectors from Forsyth (1975).

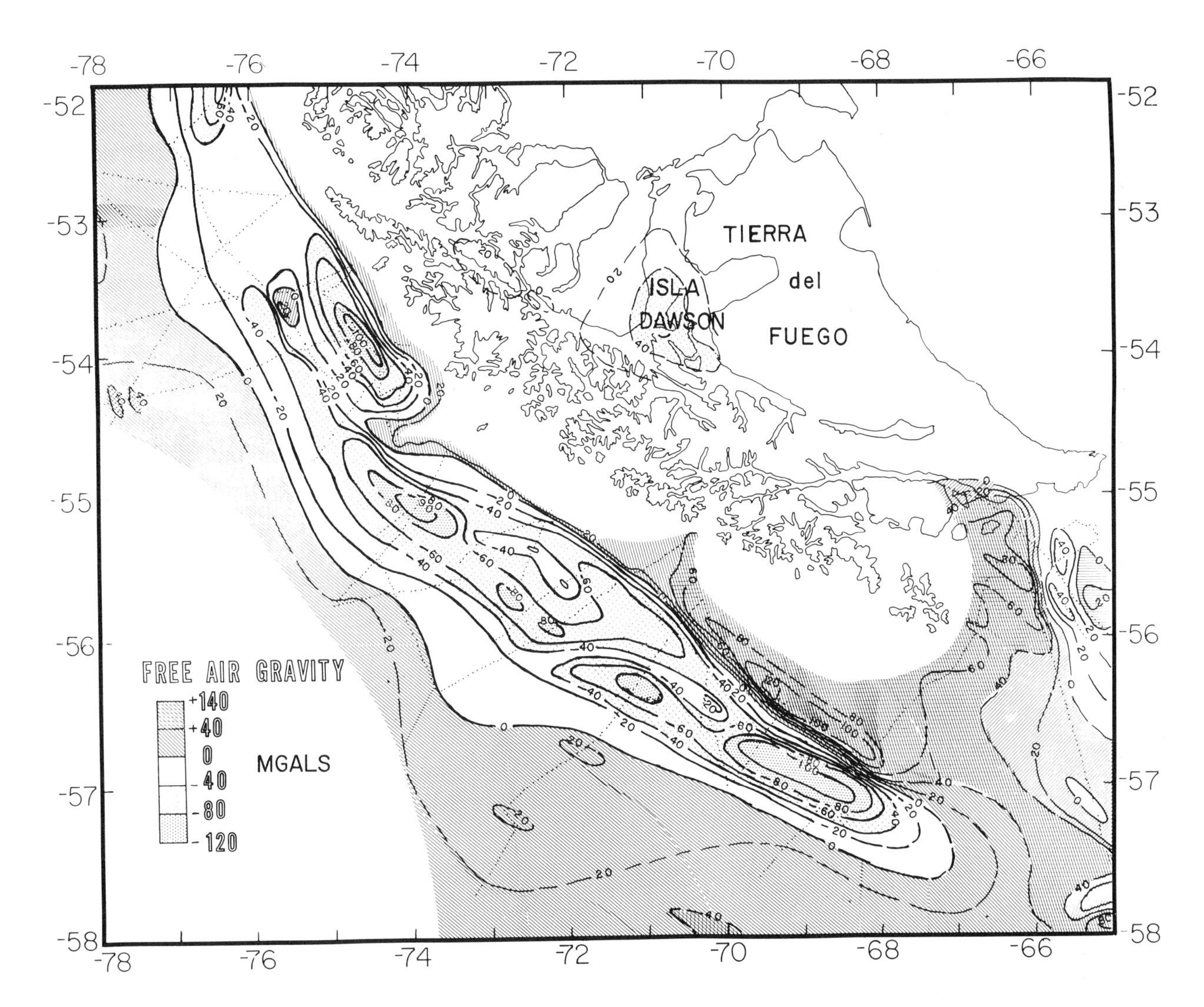

Figure 3. Free-air gravity anomaly contours along the Chilean margin of Tierra del Fuego. Control lines are indicated by the light dotted lines. Data were obtained during cruises of the R/V CONRAD and USNS ELTANIN.

of this terrace was undertaken on the basis of bathymetric chart of Mammerickx and others (1975).

Comparison of Figures 3 and 4 reveals that the axis of the gravity minimum coincides with the trench except in the area near 54°S where the largest negative anomaly is centered over the terrace. Bathymetric terraces with pronounced gravity anomalies are common features of active trenches, especially those associated with island arcs such as the trenches adjacent to the Bonin Islands or Java (Karig and Sharman, 1975). Well developed terraces are not, however, characteristic of the Peru-Chile Trench (Hayes, 1966; Hayes and Ewing, 1970), between the equator and the Northwest Strait of Magellan. Of all the terraces known, the morphology and gravity signature of the Atka Basin of the Aleutian Terrace (Grow, 1973) is most similar to that of the Fuegian Terrace. As the Atka Basin is also the best studied terrace, we will compare the structures interpreted for both these areas in a later section of this paper.

Seismic Profiler Data

In Figure 6 we present tracings of selected profiler data obtained along tracks identified in Figure 5. These records were made with a small (20 cu in air gun) sound source while

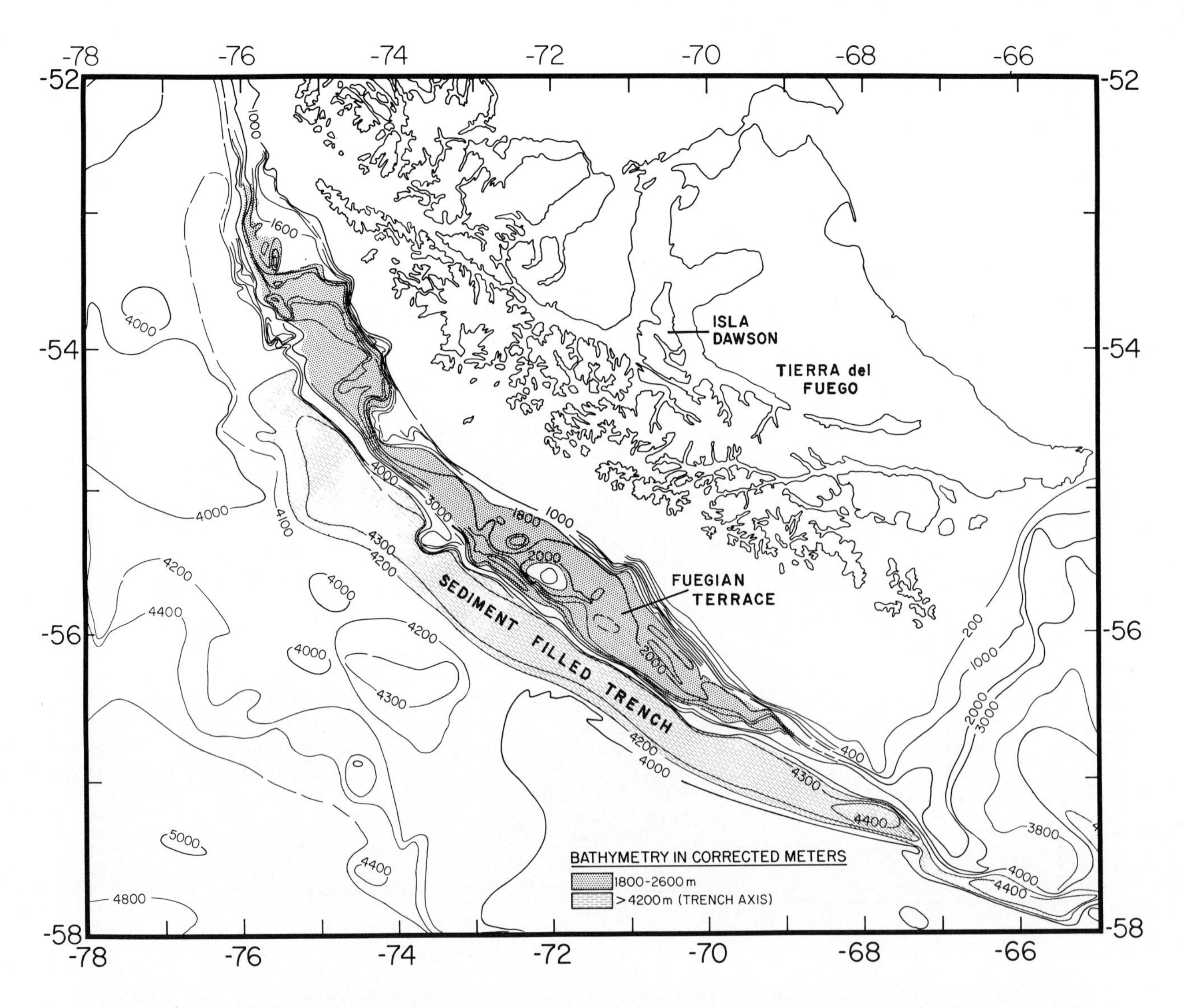

Figure 4. Bathymetric contours in corrected meters along the Chilean margin of Tierra del Fuego. Fuegian Terrace, sediment filled, and shelf are outlined by overlays.

the ship was travelling at approximately 5 knots across the margin. Thick clastic sediment fill is revealed in both the Trench and in the Fuegian Terrace. If we assume a velocity of 2 km/sec for sound propagation through these sediments, the total thickness of sediment in the trench and under the terrace revealed by the profiler exceeds 2 km. The bulk of the sediments in the area are assumed, on the basis of their acoustic signature and other studies (Scholl, 1974), to be clastic sediments of terrigenous origin.

The seismic reflection data show that along this southern extremity of the Peru-Chile Trench, oceanic basement plunges towards the continent. The turbidite fill in the trench often displays evidence of minor deformation: in Figure 6-A a thrust fault may be present near the landward wall of the trench and in Figure 6-B, possible reverse faults are present on the seaward side of the trench. These reverse faults have elevated the trench turbidites above the present level of the trench floor. In Figures 6-C and 6-D, well developed features that resemble diapirs are observed near the lower continental slope, possible evidence of recent thrust faults along this margin.

The Fuegian Terrace appears only in Figures 6-B, 6-C and 6-D. In Figure 6-B, the deformation of the sediments in the terrace suggests recurrent deformation, with the deformation apparently continuing into the youngest sediments. North of the line of Figure 6-B, the sediments in the terrace do not show similar

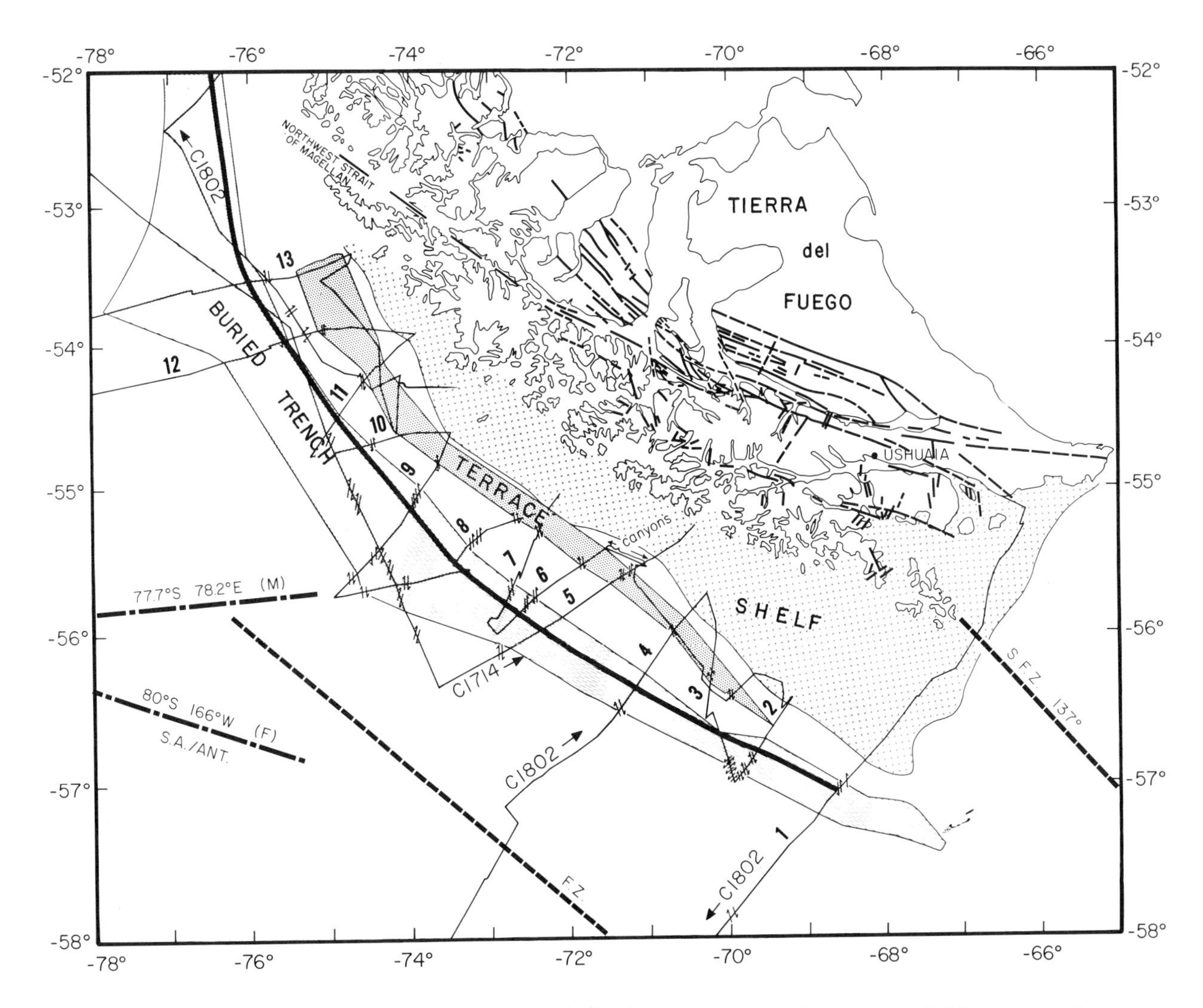

Figure 5. Summary chart of the major morphologic provinces and structural lineaments in the study area together with the location of seismic reflection lines presented in Figures 6, 8 and 9. Two major parallel fracture zones, including the Shackleton Fracture Zone, outline the assumed Scotia/Antarctic relative motion vector. Two different calculated relative motion vectors for South America/Antarctica are superimposed on the chart. M refers to Minster and others (1974) and F refers to Forsyth (1975). Small faults with observed vertical offset are indicated by the small single ended arrows over the trench and terrace. Up side is equated with north facing arrow. The faults shown on the land have all been active since the Miocene (Bruhn and others, 1976), and two main trends can be defined: a northwest trend north of 54°S, and an east-southeast trend to east trend between 54 and 55°S.

evidence of deformation, but the ridge that separates the terrace from the trench is acoustically opaque and its composition has not been determined. Only slight deformation can be observed in standard seismic profiler records such as the ones presented in Figure 6, and the absence of visible sedimentary structure in the ridge seaward of the Fuegian Terrace does not preclude the interpretation that the ridge is composed of sediment, more intensely deformed than the sediments observed in the terrace.

Structure of the Margin

In Figures 5 and 7 the extent of the terrace and its gross morphology are outlined in plan view and in profile. The shape of the terrace as outlined in Figure 5 illustrates the close agreement between the trend of the Shackleton Fracture Zone and the Northwest Strait of

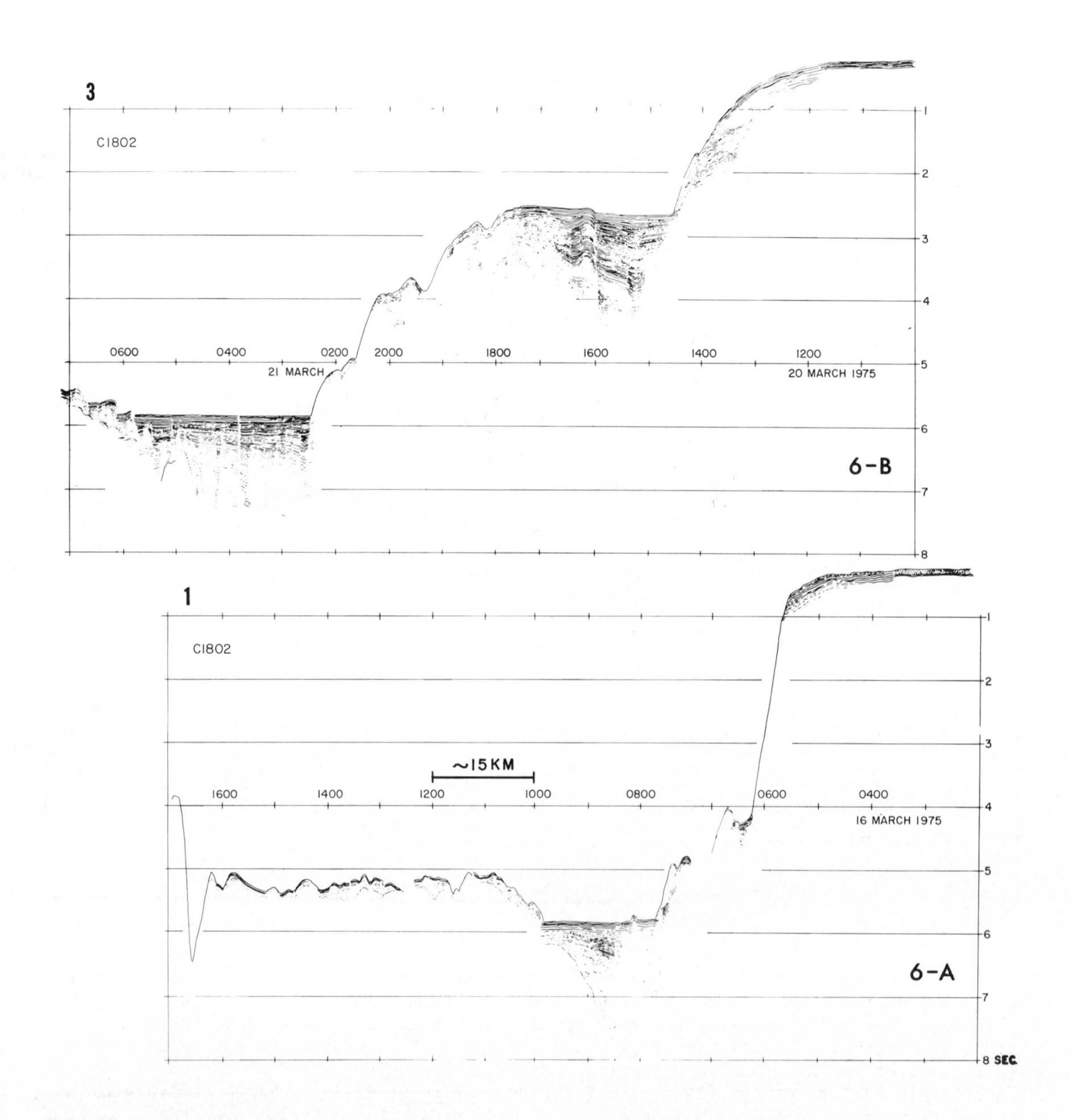

Figure 6A-6D. Tracings of selected profiler records located in Figure 5. Profiles have been selected to illustrate development of the Fuegian Terrace and progressive development of seaward deformation as one progresses from Cape Horn to the northwest. Vertical exaggeration is approximately 10X.

Magellan fault zone. This similarity in trend is most apparent in the central part of the terrace, between 54 and 56°S. The northern and southern limits of the terrace trend more northerly and the distance between the axis of the trench and the terrace decreases in these regions. Consideration of the geophysical profiles shown in Figure 7 also suggests that the history or the structure of the margin north of 54°S may be somewhat different from that to the south. Magnetic anomalies correlatable with Miocene age reversals of the earth's magnetic field are identified on the two northern profiles, nos. 12 and 13. These data are supported by more extensive magnetic studies along the margin of South America and the Antarctic Peninsula (Herron and Tucholke, 1976; Herron, in prep.) and indicate that approximately 18 m.y. ago a spreading center collided with the subduction zone. Magnetic anomalies generated at this spreading center have not been traced along the margin south of 54°S. Between 54°S and 56°S, where the terrace begins to pinch out, the terrace is developed at a depth of 2000 to 2500 m.

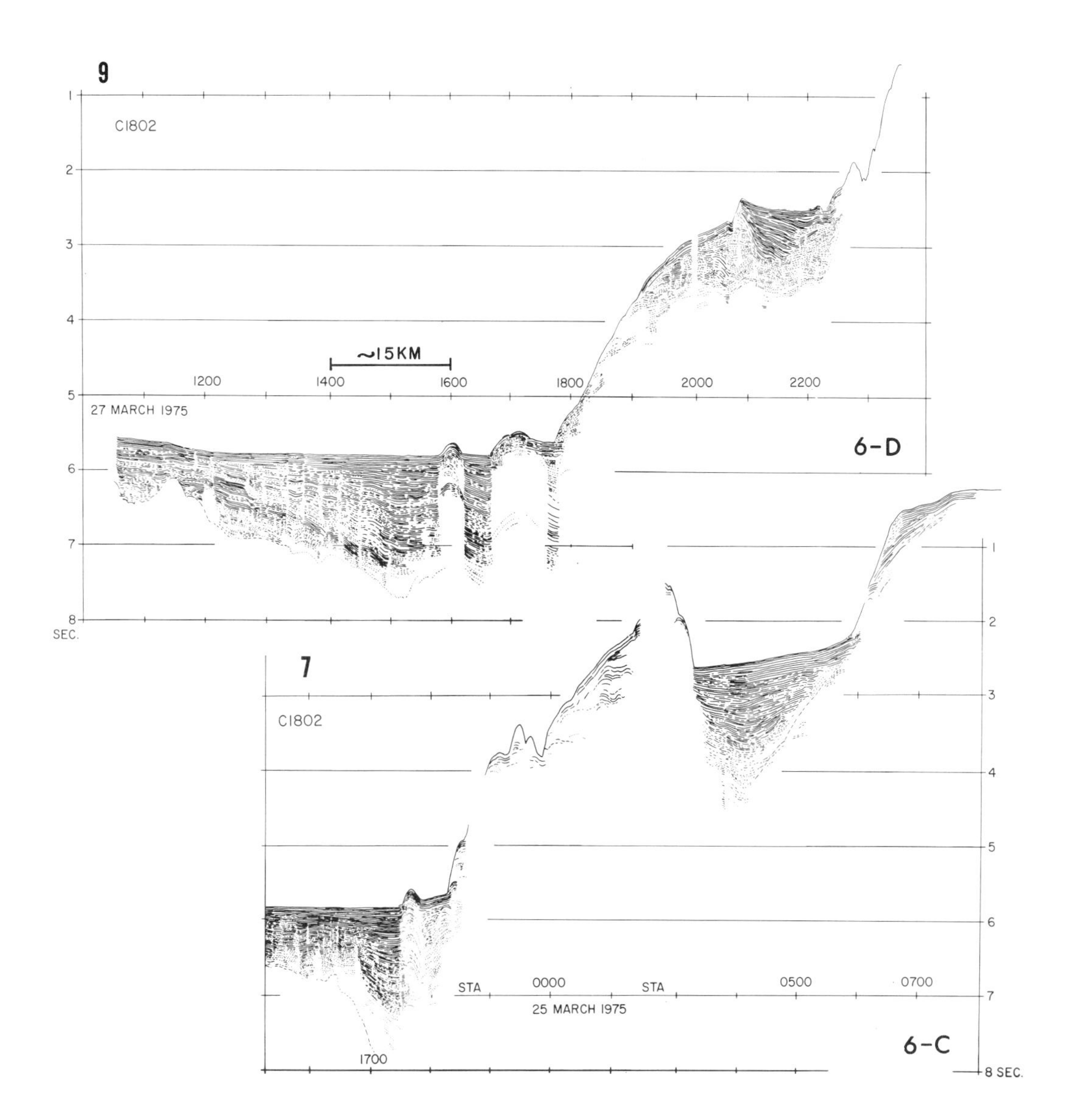

North of 54°S, the terrace is found at a depth of 2500 m and the planar surface is not well developed. (Figure 9-A and Figure 7, Profiles 12 and 13). The magnetic and gravity anomalies in Profiles 12 and 13 differ from those to the south in other aspects as well. Only on these two northern profiles do the magnetic anomalies appear to continue beyond the inner wall of the trench and only on these profiles is the largest negative free-air gravity anomaly associated with the terrace rather than with the trench.

We computed the Bouguer gravity anomaly over the trench and terrace along selected crossings near the northern and south-central parts of the terrace (Figures 5, 8 and 9). In contrast to the significant differences in the observed free-air anomalies, the Bouguer anomalies in both northern and southern areas are similar. This determination suggests that although the bathymetric depth of the terrace differs (shallower to the south) between the two areas, the underlying upper crustal structures are similar, if not identical.

We also analyzed the gravity data using a direct method of interpretation described by Tanner (1967). This technique defines the lower surface of an anomalous body when the local Bouguer anomaly, the shape of the upper surface of the anomalous body, the density contrast between the body and the presumed basement are specified. In order to use this technique across the margin, we assumed that the calculated Bouguer anomaly was composed of two main elements, a long wavelength anomaly reflecting the subducting slab of oceanic crust beneath the margin, and a short wavelength anomaly reflecting the presence of a significant thickness of low

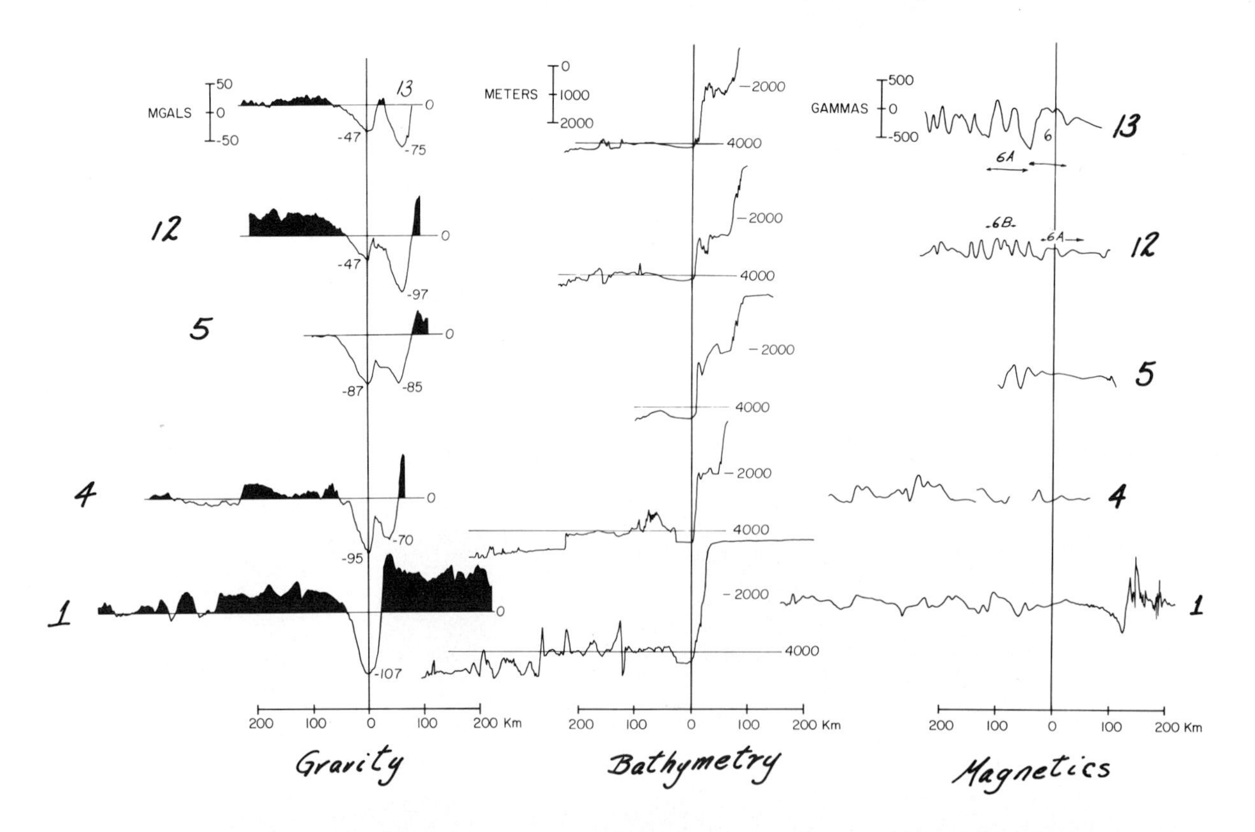

Figure 7. Projected bathymetry, free-air gravity and residual magnetics profiles across the Chilean margin of Tierra del Fuego aligned about the gravity minimum associated with the trench. Note the progressive northward development of the minimum free-air gravity anomaly associated with the Fuegian Terrace and the apparent continuation of identifiable magnetic anomalies generated by sea-floor spreading landward of the trench axis near the northern end of the terrace.

density sediment beneath the terrace. We computed the Bouguer anomaly for several profiles and selected a single straight line that we felt best fit the long wavelength trend in all the profiles. The local short wavelength anomaly was then calculated by subtracting the computed total anomaly from this assumed regional anomaly.

Figures 8 and 9 present the regional field selected for each profile and the data points on the Bouguer anomaly and topography profiles that were used in the computations. Excellent fits to the observed data and geologically reasonable depths for the bottom of the low density body were determined for density contrasts ranging between -0.4 and -0.8 gm/cc. The results show that although only 1 to 1.5 km of sedimentary deposits are revealed on the seismic profiler data, a sedimentary section as much as 6 km thick may underlie the terrace. More importantly, the process reveals a subsurface basement ridge of denser rocks beneath the outer edge of the terrace.

At the southern limit of the terrace, the outer ridge is covered by one km of sediment, (Figure 8-B), but near the Northwest Strait of Magellan, the subsurface ridge is nearly exposed (Figure 9-B). Topographic ridges that dam terrigenous sediment to form terraces are common features of many trenches, especially those associated with island arcs (Uyeda, 1974; Karig and Sharman, 1975). One interpretation for the origin of the terrace assumes that most of the clastic sediment is not subducted but instead is scraped off and folded against the continental margin. Repetition of this offscraping and folding process leads to the gradual buildup of the ridge and the possible development of a sediment filled terrace on the landward side of the ridge. Grow (1973) modelled the structure across the Aleutian Trench and Terrace. His model required a density contrast of 0.2-0.4 gm/cc between the terrace sediments and the underlying and adjacent metamorphosed trench sediments and our model requires a density contrast of at least 0.4 or 0.5 gm/cc. The free-air gravity anomaly associated with the Aleutian Terrace and the adjacent ridge is much smaller than that developed over the Fuegian Terrace and ridge, especially at the northern end of this feature. This size difference may account in part for the different density con-

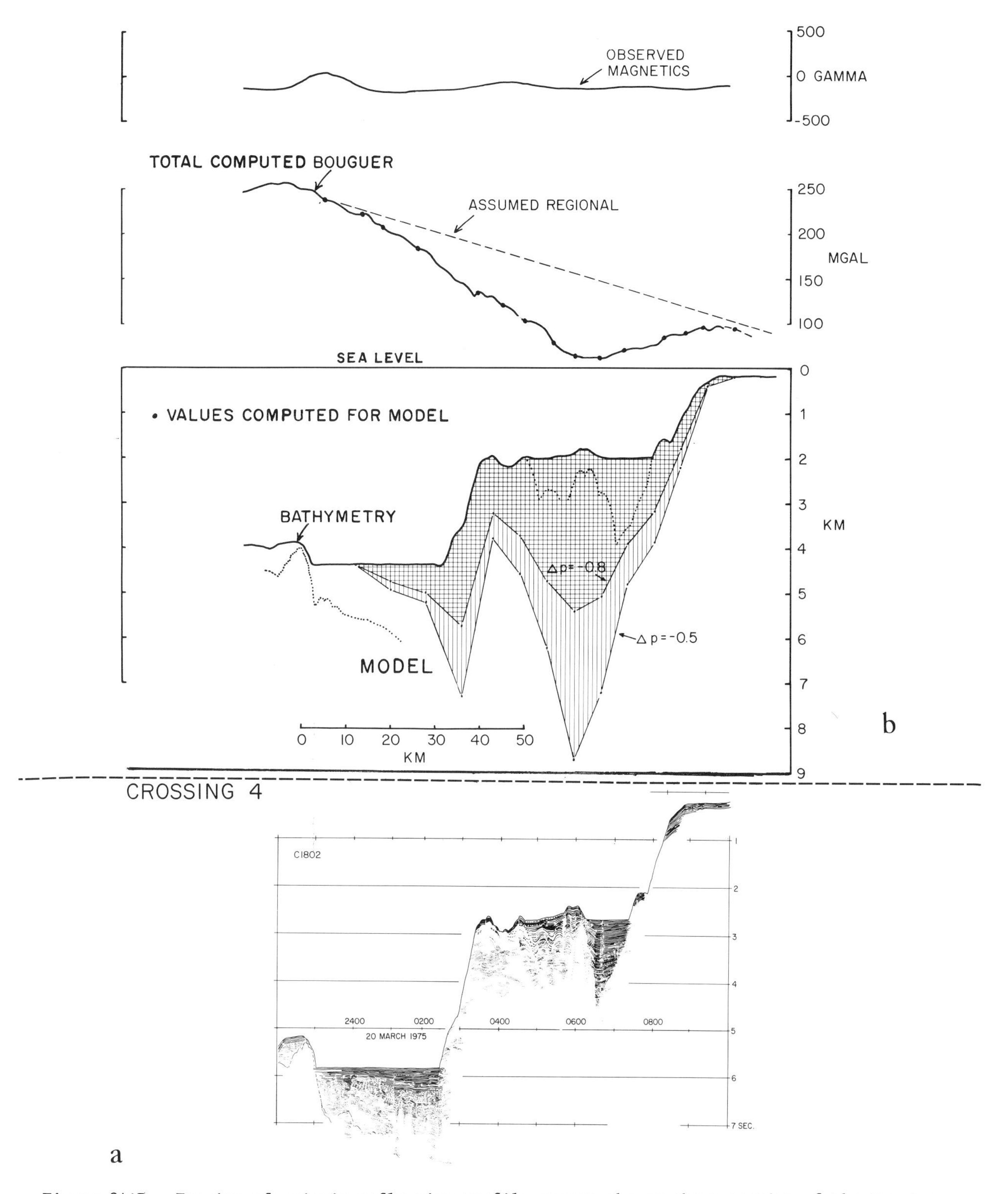

Figure 8A+B. Tracing of seismic reflection profile across the southern section of the Fuegian Terrace together with the computed Bouguer anomaly and an interpretation of the local Bouguer anomaly associated with the terrace. Two interpretations for the shape of the lower boundary of the anomalous body are shown; the lower boundary computed for a density contrast of -0.5 gm/cc and the higher boundary computed for a density contrast of -0.8 gm/cc.

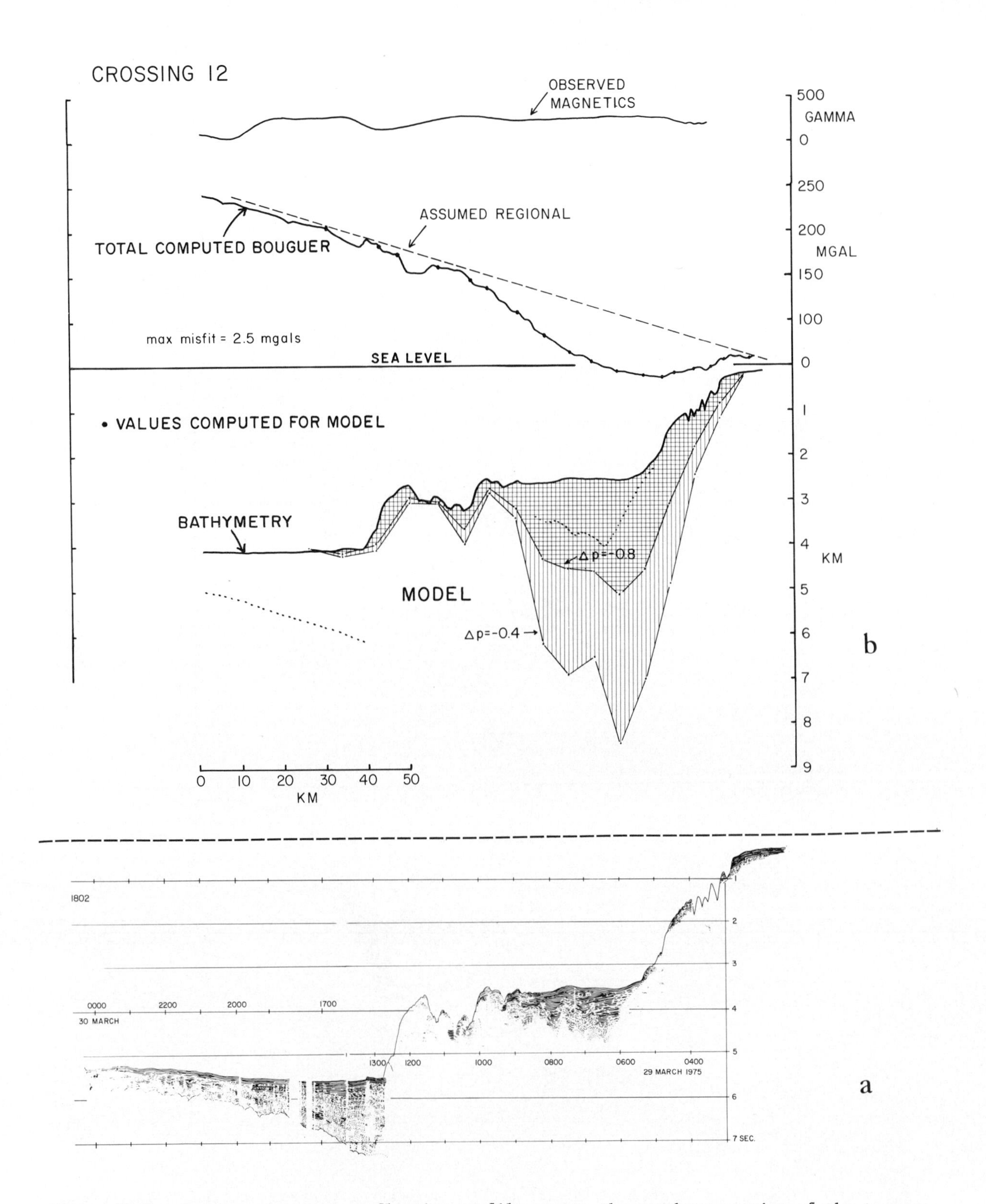

Figure 9A+B. Tracing of seismic reflection profile across the northern section of the terrace together with the computed Bouguer anomaly and an interpretation of the local Bouguer anomaly associated with the terrace. See text for discussion of this profile and compare with profile 4 shown in Figure 8.

trasts required for his model and ours. If we assume a density of 2.0 gm/cc for the terrace sediments, then the materials incorporated in the basement ridge must have a density of 2.4 to 2.5 gm/cc, and could consist either of metamorphosed trench sediments, or a melange of oceanic crust and meta-sediments, or even foundered continental crust.

As shown in Figure 7, the magnetic anomaly pattern associated with crustal rocks generated by sea-floor spreading terminates at the inner wall of the trench in the southern part of the Tierra del Fuego margin. However, in the northern area, where the hypothetical basement ridge rises close to the sea floor, the spreading type magnetic anomalies continue, but with decayed amplitude, across the inner wall and to the terrace. We computed synthetic magnetic anomalies for these areas using data from Weissel and Herron (in prep.) to guide the anomaly identification and spreading rates. Our calculations did not support the possibility that fragments of oceanic crust are near the surface along the outer part of the northern terrace. The observed anomalies caused by an increased depth to magnetized oceanic crust, or alternatively, the anomalies could be generated by weakly magnetized metamorphosed oceanic crust, continental crust or sedimentary rocks associated with the basement ridge and terrace.

Terrestrial Geology

Geological studies on land have provided pertinent data to the understanding of the Late Cenozoic tectonics of southern South America (Bruhn and others, 1976). Numerous post-Miocene offsets, with a major component of left-lateral strike-slip movement form two major, divergent fault systems. One system trends northwest-southeast, parallel to the trend of the Shackleton Fracture Zone, the other trends nearly due east, aligned with the offshore strike of the Malvinas Escarpment east of Tierra del Fuego (Katz, 1964). It is likely that many of the faults are reactivated older zones of weakness in the earth's crust. The two fault systems intersect in the vicinity of Isla Dawson (54°S, 70.4°W).

Conclusions

Data from the offshore and onshore surveys suggest that the Antarctic/South American/Scotia triple junction should be located near Isl Dawson. The triple junction has not been precisely located. It appears to be located within the continent along the margin of South America, but the geologic control exerted by pre-Miocene structures (Katz, 1973; Bruhn and Dalziel, this volume) mask and modify the response of the continental crust to post-Miocene stress patterns. We believe that the deformation associated with the relative motions between the Antarctic and Scotia Plate are spread over the continent and onto the margin. The Shackleton Fracture Zone does not, in our model, extend directly into the continent north of Cape Horn, but disintegrates into a fault zone that steps along the margin, controlling the morphology of the Fuegian Terrace and Chile Trench. TFF and TTT triple junctions are inherently unstable (McKenzie and Morgan, 1967) and this instability may aid in the development of the Shackleton Fracture Zone along the margin.

Acknowledgements. We are grateful to the Chilean authorities who supported this study, especially Dr. Eduardo Gonzalez of ENAP, Dr. Oscar Gonzalez of IIG and Captain Juan Mckay of YELCHO. The crew of CONRAD made this cruise a scientific success inspite of severe weather and mechanical problems. We thank A. Watts and D.W. Scholl for their critical review of the manuscript and their careful and constructive suggestions. The marine program and E.M. Herron were supported by NSF grants DES 71-00214 and OCE 76-01811. R.L. Bruhn and M.A. Winslow were supported by NSF grants DES 75-04076, OPP 74-21415 and IDO 72-06426 A03.

Lamont-Doherty Geological Observatory Contribution No. 2406.

References

Barazangi, M., and J. Dorman, World seismicity maps compiled from ESSA Coast and Geodetic Survey, epicenter data, 1961-1967, Seismol. Soc. Amer. Bull., 59, 369-380, 1969.

Barker, P.F., Plate tectonics in the Scotia Sea region, Nature, 228, 1293-1296, 1970.

Barker, P.F., and D.H. Griffiths, The evolution of the Scotia Ridge and Scotia Sea, Phil. Trans. Roy. Soc. A., 271, 151-183, 1972.

Bruhn, R., and I.W.D. Dalziel, The evolution and collapse of the Late Mesozoic "Rocas Verdes" marginal basin in the southern Andes, Ewing Symposium Volume, Amer. Geophys. Union (in press).

Bruhn, R., I.W.D. Dalziel, and M.A. Winslow, Lower tertiary to recent structural evolution of southernmost South America, EOS, 57, 334, abstract only, 1976.

Chase, C.G., The N-plate problem of plate tectonics, Geophys. J. Roy. astr. Soc., 29, 117, 1972.

Ewing, M., R. Houtz, and J. Ewing, South Pacific sediment distribution, J. Geophys. Res., 74, 2512, 1969.

Forsyth, D.W., Fault plane solutions and tectonics of the South Atlantic and Soctia Sea, J. Geophys. Res., 80, 1429-1443, 1975.

Grow, J.A., Crustal and upper mantle structure of the Central Aleutian Arc, Bull. Geol. Soc. Amer., 84, 2169-2192, 1973.

Hayes, D.E., A geophysical investigation of the Peru-Chile Trench, Mar. Geol., 4, 309-351, 1966.

Hayes, D.E., and M. Ewing, Pacific boundary structure, in The Sea, v. IV, Pt. II, A.E. Maxwell ed., 29-72, 1970.

Herron, E.M., and D.E. Hayes, A geophysical study of the Chile Ridge, Earth and Planetary Sci. Letters, 6, 77-83, 1969.

Herron, E.M., and B.E. Tucholke, Sea-floor magnetic patterns and basement structure in the southeastern Pacific, in Initial Reports of the Deep Sea Drilling Project, 35, U.S. Government Printing Office, Washington, D.C., 263-278, 1976.

Katz, H.R., Strukturelle verhaltnisse in den sudlichen Patagonischen Anden und deren Beziehung zur Antarktis: eine Diskussion, Geol.Rdsch, 54, 1195-1213, 1964.

Mammerickx, J., S.M. Smith, I.L. Taylor, and T.E. Chase, Bathymetry of the South Pacific - in ten sheets, Scripps Institution of Oceanography, IMR Technical Report Series, 1971-1974.

McKenzie, D.P., and W.J. Morgan, The evolution of triple junctions, Nature, 224, 125-133, 1967.

Minster, J.B., T.H. Jordan, P. Molnar, and E. Haines, Numerical modelling of instantaneous plate tectonics, Geophys. J. Roy. astro. Soc., 36, 541, 1974.

Morgan, W.J., Rises, trenches, great faults and crustal blocks, J. Geophys. Res., 73, 1959, 1968.

Scholl, D.W., Sedimentary sequences in North Pacific trenches, in The Geology of Continental Margins, C.A. Burk and C.L. Drake, eds., 493-504, 1974.

Tanner, J.G., An automated method of gravity interpretation, Geophys. J. Roy. astr. Soc., 13, 339-347, 1967.

Uyeda, S., Northwest Pacific trench margins, in The Geology of Continental Margins, C.A. Burk and C.L. Drake, eds., 473-492, 1974.

A PRELIMINARY ANALYSIS OF THE SUBDUCTION PROCESSES ALONG THE ANDIAN CONTINENTAL MARGIN, 6° to 45^oS

L. D. Kulm and W. J. Schweller

School of Oceanography, Oregon State University, Corvallis, Oregon 97331

A. Masias

Petroleus del Peru, Lima Peru

Abstract. The Nazca oceanic plate and the South American block converge rapidly to produce the complex Andean subduction zone. While this is a recognized region of subduction, there still is a general lack of information about the dominant subduction processes which are responsible for the evolution or destruction of the Andean continental margin. In this paper the subduction process is subdivided into two major components, accretion (addition of plate and trench material to the margin) and consumption (tectonic removal of pre-existing continental block) and one minor component, a transitional state between the two major processes. This study establishes a set of criteria which are related to the absence or presence of morphological and shallow structural features, that may be useful in identifying these processes within the past several million years. Using the features in the criteria, the Andean margin is divided into three physiographic provinces: province 1, 6° to 19.5^oS; province 2, 19.5^o to 27^oS; and province 3, 27^oS to 45^oS. The latter province is separated into subprovince 3A (27° to 34^oS) and 3B (34° to 45^oS). Provinces 1 and 3 typically have long prominent benches on the lower continental slope, sedimentary basins on the shelf and upper slope, and thick trench deposits that are characteristic of accretion. Province 2 has a general lack of benches on the lower slope, discontinuous upper slope basins and no shelf basins; these features are believed to be indicative of consumption. Alternatively, province 2 may be in a transitional state following a period of consumption. Short-term accretion in an overall long-term consumption process also is feasible for portions of this province. A few margin deep crustal sections were used to test our interpretations of these processes. We cannot find any correlation between the subduction processes and trench depth, crustal rupture zones or closure rate between plates in this preliminary study. There may be a significant correlation between the positions of segments of the inclined seismic zones and the physiographic provinces.

Introduction

Early geophysical studies of the Peru-Chile continental margin and trench produced the first relatively detailed information about the morphology and crustal structure of the Andean continental margin (Fisher and Raitt, 1962; Hayes, 1966; and Scholl et al., 1970). In more recent years numerous detailed studies have been conducted in this region by the IDOE Nazca Plate Project to further elucidate the structural and tectonic framework of this region (Kulm et al., 1973; Rosato, 1974; Masias, 1976; Prince and Kulm, 1975; Kulm et al., 1975; Johnson et al., 1975; Hussong et al., 1976; Kulm et al., 1976; Coulbourn and Moberly, 1976; Schweller, 1976). Although the subduction zone is fairly well defined by the spatial and temporal distribution of earthquakes (eg: Kelleher, 1972; Stauder, 1973, 1975; Barazangi and Isacks, 1976), there still is a lack of information about the dominant subduction processes present along various segments of this convergent zone.

While a synthesis of many types of data, such as oceanic and continental crustal structure, seismicity, volcanism, plutonism and ore deposits, eventually will be required to characterize the subduction processes in much greater detail, we believe it is useful at this intermediate stage in our investigations to use the morphology and shallow structure of the Andean margin to attempt to identify which of the subduction processes (accretion, consumption, etc.) may be operating along this margin. Although the structure of the downgoing Nazca Plate may exert some influence on the structure of the margin (Prince and Kulm, 1975; Schweller, 1976; Schweller and Kulm, 1976), we will concentrate this preliminary analysis on the processes occurring along continental shelf and

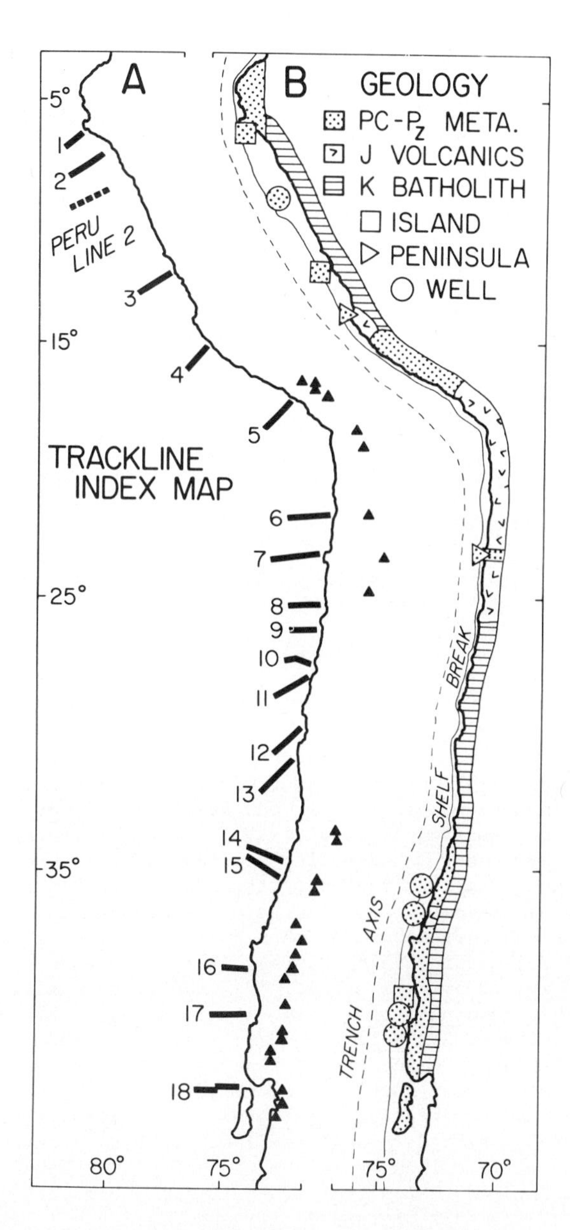

Fig. 1. A. Index map of South American coast and trackline locations of profiles shown in Figures 4 and 8. Active volcanoes shown as solid triangles. B. Geology of coastal region and continental shelf from various sources. Note symbols indicating special features (ie: island, peninsula, oil well) and the basement rocks recovered in these areas.

slope between 6° to 45° S latitude (Fig. 1A). The trench and margin are being studied independently at this stage of our investigations and both systems will be integrated in a future study.

In this paper we will utilize the most recent theories for the evolution of convergent continental margins to establish a set of criteria which will help to identify the dominant subduction processes. The structural framework outlined in the criteria would normally be expected to evolve over the past 10 million years of convergence history. We then examine the Andean margin for the features listed in the criteria in order to identify lengths of the margin that have similar morphology and shallow structure. Finally we utilize deep crustal sections, morphology and shallow structure to evaluate which process is present in various localities. Our overall purpose is to produce generalized conceptual models for the evolution of this complex Andean margin off South America.

Criteria for Recognition of Subduction Processes

In this paper we will subdivide the broadly classified subduction processes into two major component processes (accretion and consumption) and one minor component process (a transitional state between accretion and consumption) which may also be present. Figure 2 shows the various features that should characterize a given process. Each of these subduction processes is briefly defined below.

Definition of Processes

Accretion is the process whereby material from the oceanic plate and trench is added to the outer continental margin or arc by various mechanisms such as imbricate thrusting (Beck and Lehner, 1974; Seely et al., 1974; Kulm and Fowler, 1974; Karig and Sharman, 1975), combination fold-thrust sequences (Moore and Karig, 1976), or deformed sediment masses (Kulm, von Huene et al., 1973). In virtually all cases the outer margin experiences uplift of varying degrees often with a topographic ridge developing along the outermost margin and some sort of elongate depression or basin occurring behind the newly formed ridge. This basinal feature can vary in size depending upon the amount of uplift and the previous structural framework of the older accreted sediment. Sediment on the continental shelf and slope, which is carried down toward the trench by turbidity currents or slumping, will be deposited in these slope basins or the trench axis and thus be reincorporated into the margin.

Consumption is generally defined as the tectonic removal of a certain amount of preexisting material, including sediment and/or crystalline rock, from the outer continental margin. This material is presumably carried down the Benioff Zone. This causes a slow landward retreat of the leading edge of the

FEATURES	ACCRETION SHORT TERM	ACCRETION LONG TERM	TRANSITIONAL SHORT TERM	TRANSITIONAL LONG TERM	CONSUMPTION SHORT TERM	CONSUMPTION LONG TERM
BENCHES						
LOWER SLOPE	●	●	○	○	–	–
MIDDLE SLOPE	●	●	○	○	–	–
UPPER SLOPE	○	●	○	–	–	–
BASINS						
CONTINENTAL SLOPE						
LOWER SLOPE	○	○	○	○	–	–
MIDDLE SLOPE	○	●	○	●	–	–
UPPER SLOPE	○	●	○	●	○	○
CONTINENTAL SHELF	○	●	○	●	○	○
TRENCH AXIS SEDIMENT	●	●	○	○	–	–

● ESSENTIAL ○ POSSIBLE – NOT PRESENT

Fig. 2. Criteria for recognition of the subduction processes (accretion, consumption and transition) along a convergent continental margin. See text for a detailed discussion of features and processes.

continental block. Any trench sediment lying adjacent to a consuming margin no doubt would be removed along with the continental block. Any continental margin with old strata (eg: Paleozoic or Precambrian) along the leading edge of the continental block in a region with a long history of convergence is a place of possible consumption.

A transitional process may exist between the accretion and consumption processes whereby the basaltic and sedimentary material from the oceanic plate, together with the sediment from the trench, is carried down the Benioff Zone without removing significant amounts of material from the leading edge of the continental block. This should be a steady state process whereby material is neither added nor removed from the outer continental margin and may represent a delicate balance between accretion and consumption. It is difficult to imagine how such a process could exist for a long period of time and we therefore relegate it to a minor role in the overall subduction process. Evidence for transitional subduction, if it is a viable process, would be difficult to interpret in the geologic record.

Application of Criteria to Processes

A set of criteria was established to determine which of these subduction processes may be occurring along the Andean margin. We prefer to use a descriptive terminology rather than a genetic terminology at this preliminary stage of our investigation of the Andes margin and will refrain from using any terminology attributed to island arcs (see later section on Comparison of Andean Margin with Island Arcs).

The three primary features that are associated with these processes and that can be identified in the marine data available for this study are listed in Figure 2. Two of these features, structural benches and sedimentary basins, develop as a result of the subduction processes and one additional feature, trench axis sediment, is required to facilitate accretion of a significant amount of material to the outer margin. Rather thick terrigenous, oceanic plate sediments (eg: abyssal plain sequences of the Northeastern Pacific Ocean) also would provide sufficient material for large scale accretion to the margin and could be substituted for the trench sediments given in Figure 2. We recognize that this is a very simplistic approach to complex processes and that the criteria represent, in the case of the accretion process, the full range of features expected from the accretion models proposed and test cases described to date. The continental slope is divided into three regions, the lower slope, middle slope and upper slope to further define the distribution patterns of basins and benches. There is seldom more than one prominent and possibly one minor basin along a given section of continental slope in the eastern Pacific convergent zones. Benches may occur simultaneously at several levels on the continental slope and develop through the interaction of tectonism and sedimentation as described below.

We include the element of time in this set of criteria because the benches and the basins are modified by continuing tectonism and the increasing effects of continental margin sedimentation. On the basis of previous experience with the tectonic framework of convergent margins (Kulm, von Huene et al., 1973; Kulm and Fowler, 1974; Prince and Kulm, 1975), short term processes are defined as those processes that occur over periods as short as a few thousand years to as long as one million years. Long term refers to those processes that are active for several million years.

The accretion model is characterized by benches which are generally narrower on the lower slope and wider on the upper slope. These benches are structural features, 5-20 km wide, representing either imbricated thrust sheets, chaotic piles of sediment, or sediment-basalt melange that may be smoothed over by the ponding of sediments behind these tectonic barriers. Although large scale slumping has been documented along some continental margins, the authors believe that it is only of minor importance in producing bench-like features. One or two basins may be present on the continental slope with one prominent basin or several smaller interconnecting basins occupying the subsiding continental shelf. The adjacent trench has a variable amount of sedimentary fill that fluctuates with sediment supply and tectonism in the trench.

Because a consumption model has not been well defined, by other studies, we must treat it from a theoretical point of view. In this paper, the consumption model is characterized by the absence of benches or tectonic ridges which form the seaward boundaries of troughs or basins. Basins are less likely to occur on the continental slope unless the pre-existing continental block is broken into structural blocks that form long term grabens in the basement. Although a continental shelf and shelf basins may be present, in this model they probably are of limited extent if consumption has been the dominant subduction process for a long period of time. We believe that large, subsiding sedimentary basins are less likely to occur in the consumption model than in the accretion model.

A model for transitional process, as defined previously, is difficult to envision since we are not certain that it exists as a separate entity. If a steady state condition occurs after a period of accretion we would expect to find some of the attributes of the accretion process as shown in Figure 2. In this case, the transitional model may be more affected by continental margin sedimentation which has a smoothing effect upon inactive structural features such as benches. On the other hand, if a steady state condition commences after a period of consumption, there may be no visible effects on the structure of the margin other than a thickening of the continental margin deposits by various continental shelf and slope sedimentation processes, so the listed transitional criteria would not apply.

Geologic Setting of the Coastal Region and Margin

In many areas of the Andean region only the most general lithologies are given and the ages of these strata are largely inferred from their lithologies which are tentatively correlated with similar strata elsewhere. Metamorphic strata (eg: gneiss, schist, phyllite, quartzite) and granitic strata commonly occur along the coast and underlie the sedimentary deposits of a part of the adjacent continental margin (Fig. 1B). Several islands off Peru are known to contain metamorphic rocks (Masias, 1976). The few reliable absolute age determinations for these crystalline rocks all give ages in the late Precambrian to Paleozoic range. For example, radiometric age determinations give ages of 679 ± 12 m.y. and 642 ± 16 m.y. for these strata along the trend of the Andean batholith in southern Peru (Stewart et al., 1974). Radiometric dating of the plutonic and volcanic rocks of the coastal region of Chile within 200 km of the shoreline suggests a progressive eastward migration of the axis of intrusive activity since lower Jurassic (Farrar et al., 1970 and McNutt et al., 1975). The oldest plutonic rocks are exposed in the Coast Range while the youngest rocks and active volcanoes are found nearly 200 km to the east. This migration suggests that subduction has been active along this margin for more than 100 million years.

The oldest sediments overlying the metamorphic strata on the continental shelf are late Cretaceous to Pliocene deposits in the wells drilled on the continental shelf off Chile from 35° to 39°S (Mordojovich, 1974) and early and late Cenozoic deposits in wells drilled off central Peru (Masias, 1976). This indicates the Andean outer continental block has had a complex history of deposition punctuated by episodes of uplift and erosion resulting in hiatuses of various length.

Morphology and Structure of the Peru-Chile Margin

Using the conceptual models outlined in the criteria for recognition of subduction processes, we studied the morphology (continental slope benches) and shallow structure (continental shelf and slope basins; and distribution of trench axis sediment) of the Andean margin and trench between 6 ° to 45 ° S latitude. Data from previous published studies and unpublished data available through the Nazca Plate Project were used in this investigation.

Morphology

About 140 bathymetric profiles from the margin were computer processed and plotted at a vertical exaggeration of 10:1. Although most of these profiles were perpendicular to the trend of the margin, a few oblique profiles were horizontally compressed to conform to the normal profiles.

After visual inspection of the morphology of individual bathymetric profiles, it became apparent that the Peru-Chile margin could be divided into at least three distinct morphological or physiographic provinces (1, 6° - 19.5° S; 2, 19.5° - 28°S; and 3, 28° - 45 ° S) separated by fairly well defined transition zones (Fig. 3). Province 3 is separated into subprovinces 3A (28 ° -34 ° S) and 3B (34 ° -45 ° S). Examples of the variation in morphology in each of these provinces and subprovinces are given in the bathymetric profiles in Figure 4. The widths of the transition zones vary from a rather abrupt change in morphology over less than 30 km between provinces 2 and 3A to a much more gradual change in morphology over 50 to 100 km between provinces 1 and 2 .

Each bathymetric profile was examined for benches (described in criteria; Figure 2); these were plotted on a map according to their position along the continental slope and separated into lower, middle and upper slope regions for each profile (Fig. 5). Some profiles in the middle and upper slope regions have broad, flat features that approach the dimensions of marginal plateaus (20-40 km wide).

Province 1 is characterized by prominent lower slope benches that occur from 0 to 2 km above the trench axis and that frequently have a dip landward into the margin (Figs. 4 and 5). This province also has a wide upper slope plateau in two areas as shown in profiles 3 and 5; and a broad continental shelf. The average inclination of the continental slope is steepest (5.7°) at the base and flattens in the middle and upper slope regions to 3.5°.

Province 2 is characterized by relatively

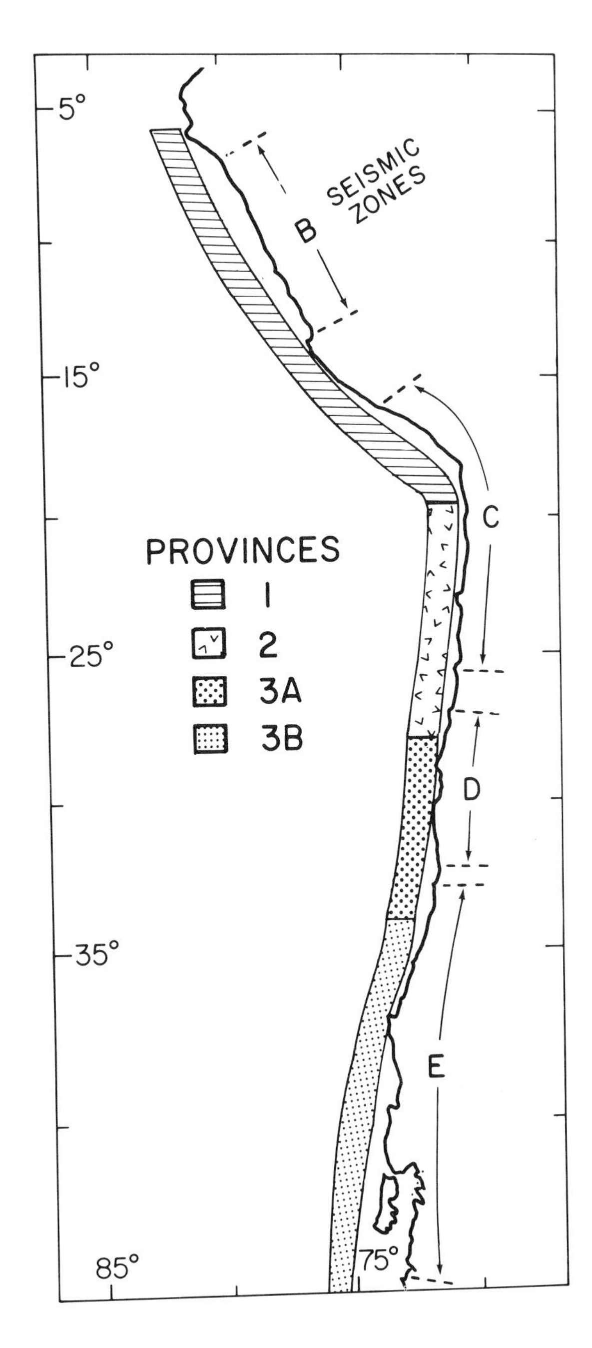

Fig. 3. Distribution of continental margin physiographic provinces (1, 2 and 3) and inclined seismic zones (B, C, D, E; Barazangi and Isacks, 1976). Province 3 is divided into subprovinces 3A and 3B.

short, narrow benches on the lower slope or by the complete absence of benches. Where present, the benches generally are flat or dip seaward. The continental shelf is only a few kilometers wide or absent. The lower slope and middle-upper slope are steeper,

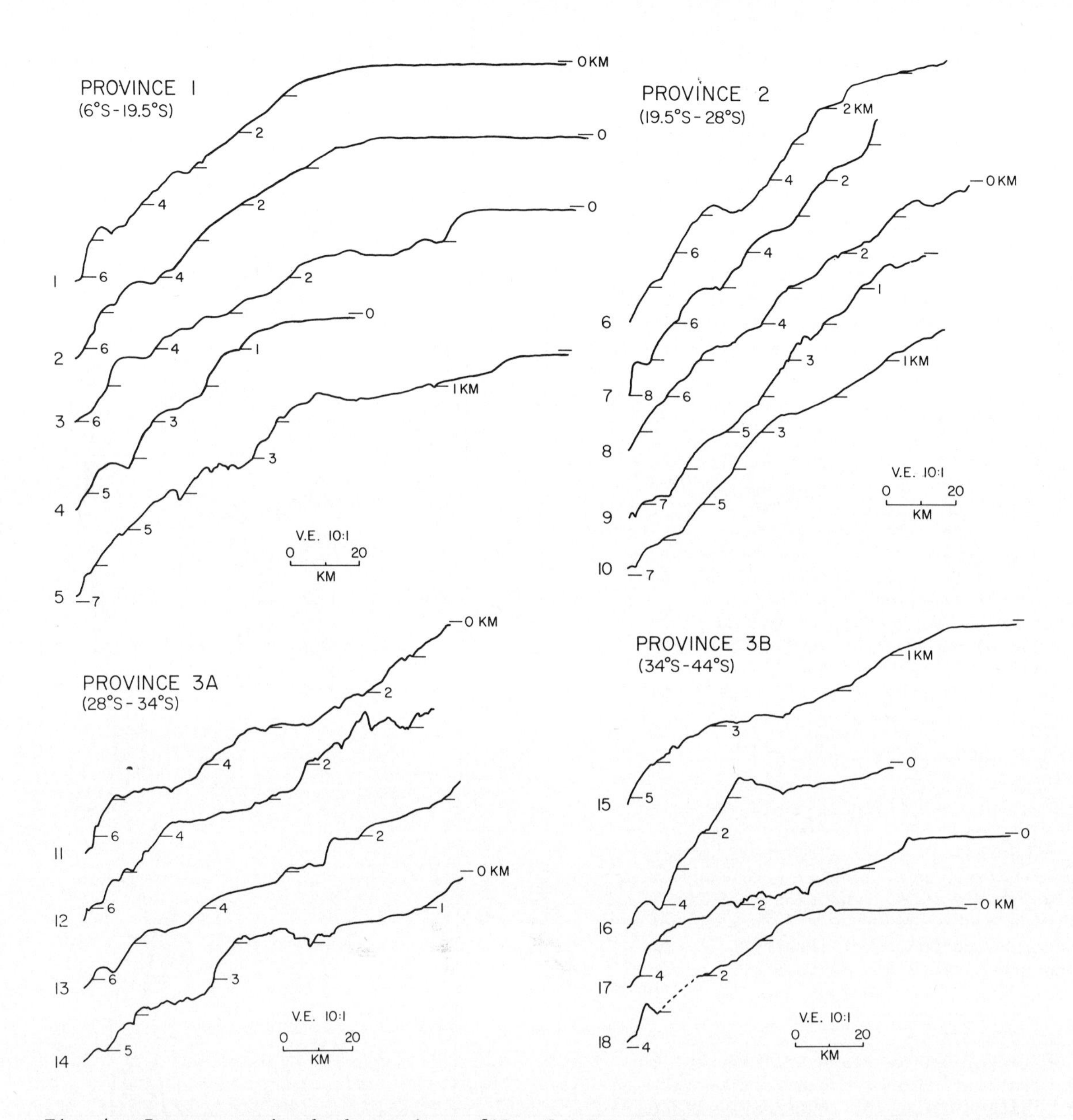

Fig. 4. Representative bathymetric profiles for Peru-Chile continental shelf and slope (6° to 45°S). Profiles are grouped according to physiographic provinces shown in Figure 3 and extend from near the shoreline to the base of the slope.

6.5° and 4.4°, respectively, when compared with either provinces 1 or 3.

Province 3 is separated into subprovince 3A and 3B and displays a much more irregular and flattened continental slope, although the lower slope is still as steep as province 1 (Fig. 4). The basic difference between subprovinces 3A and 3B in province 3 is that 3A has a prominent mid-slope plateau which generally is absent in 3B. Subprovince 3A has a mid- and upper-slope inclination of 2.8° or about the same as that of province 1. It is less dissected by canyons than province 1 and 3B and has a prominent mid-slope plateau (Fig. 4). Subprovince 3B is characterized by its very gentle dipping middle and upper slope (2.2°). There is a more variability in the morphology among the representative bathymetric profiles in province 3 and the shelf is considerably wider in subprovince 3B than in province 2.

Comparing the three provinces (Figs. 4, 5), we note that in province 2 the inclination of the continental slope is steeper on the average, particularly in the mid- and upper-slope region, than 1 and 3 and that the benches in 2 are either absent or much less pronounced than 1 and 3. Furthermore,

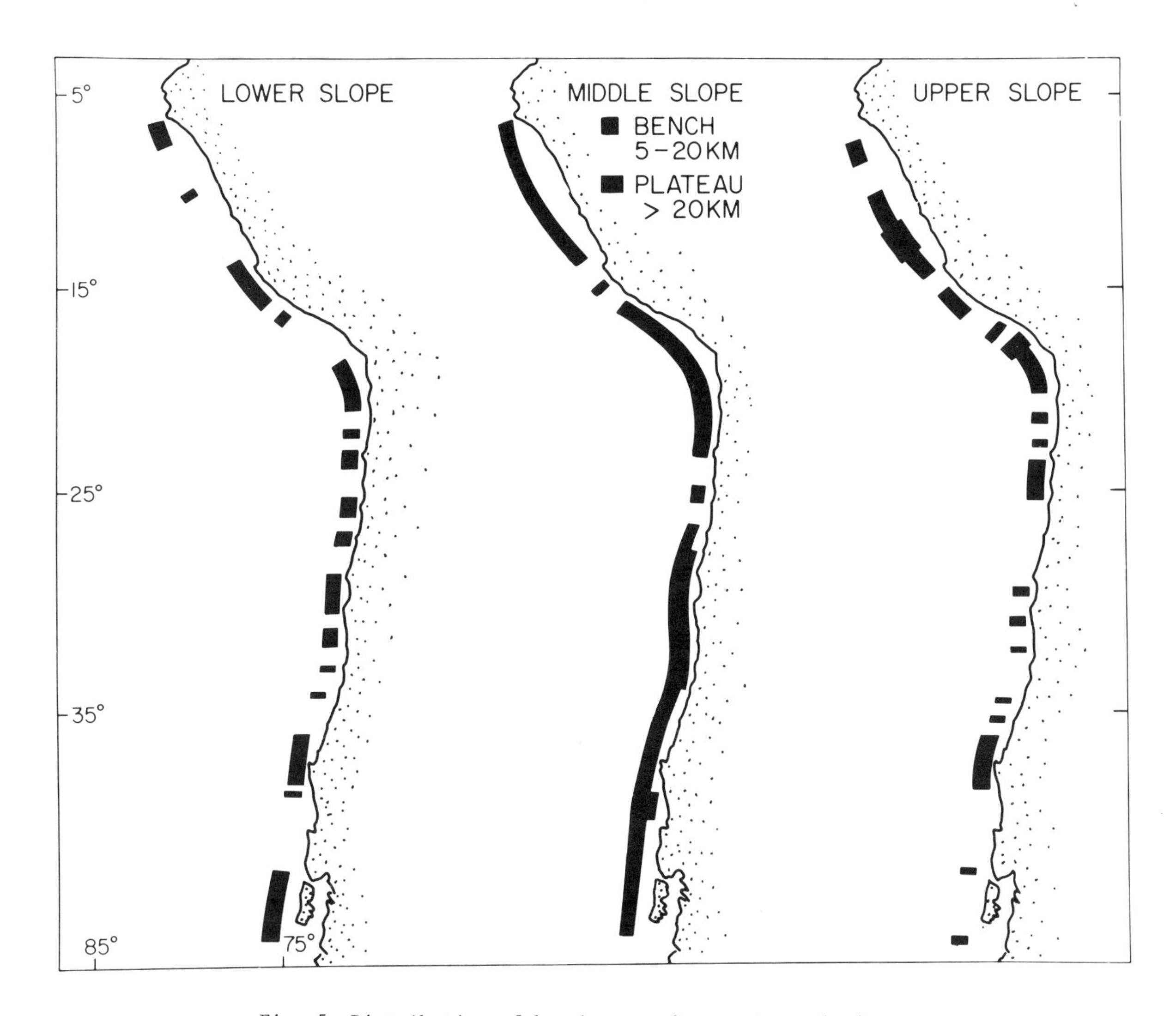

Fig. 5. Distribution of benches on the continental slope.

the continental shelf is virtually absent in 2 and 3A. The morphology of the continental slope in provinces 1 and 3A is quite similar, while 3B shows more variation and a slightly less steep upper slope. Province 2 is clearly different from province 1 and 3, especially in the middle and upper slope regions.

Because of the limits of our data, we cannot define a northern boundary for province 1 nor a southern boundary for province 3B.

Distribution of Sedimentary Basins on the Margin

Previous investigations (Hayes, 1966; Scholl et al., 1970; Mordojovich, 1974; Masias, 1976; Whitsett, 1976; Coulbourn and Moberly, 1976; Schweller, 1976) have indicated that large sedimentary basins, with a few notable exceptions, are generally lacking on the Andean margin. We have compiled the thickness and location of all continental shelf and slope basins from available data (Fig. 6). Between 6° and 45° S the prominent shelf basins are the Sechura, Salaverry and Pisco Basins (Masias, 1976) which generally contain from 1 to 3 km of sediment (Fig. 6). Two of these marine basins (Sechura and Pisco) extend into the coastal region. South of 35° S the prominent shelf basins reappear but usually do not exceed 3 to 3.5 km in thickness (Mordojovich, 1974). As shown in Figure 1B, the Mesozoic to Precambrian crystalline rocks apparently form the basement of these shelf basins with the seaward flank of each basin either these older crystalline rocks (Hussong et al., 1976) or accreted sediment melange of various ages (Kulm et al., 1976).

Upper continental slope basins frequently occur where the shelf basins are absent (Fig. 6), but they are not restricted to these most recently formed basin of latest Cenozoic deposits.

Trench Axis Sediments

As discussed in the criteria (Fig. 2), we need to know whether sediment is absent or present in the Peru-Chile Trench to test the subduction process models. Sedimentary fill varies irregularly from zero to more than two kilometers from 6° to 45°S. In Figure 7, the blank areas generally have less than 100m

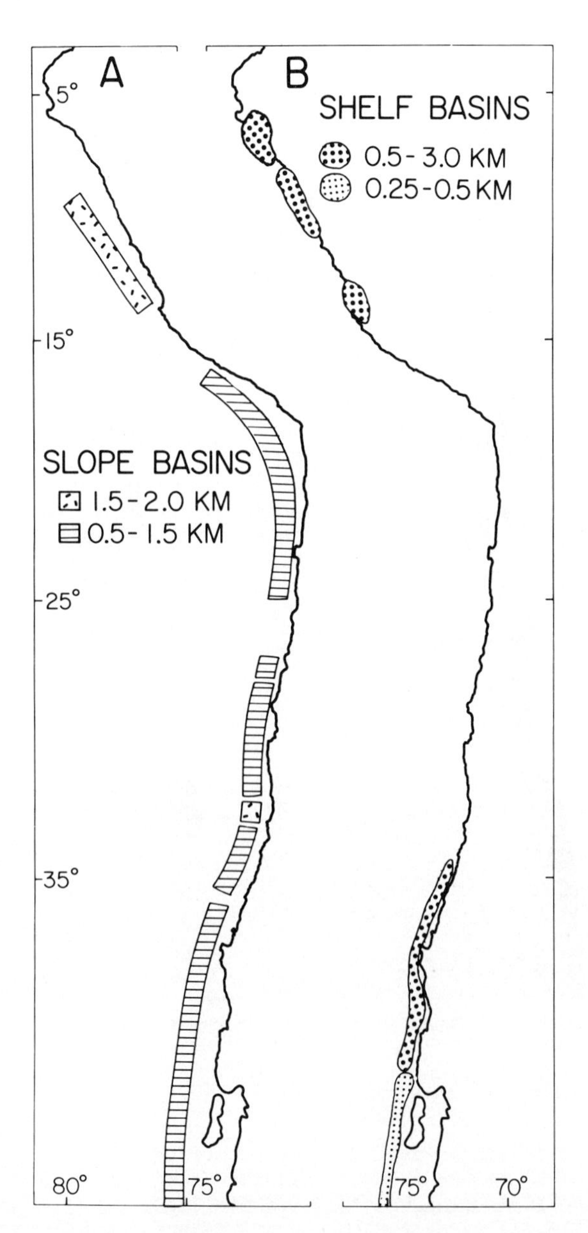

Fig. 6. A. Location and thickness of upper continental slope basins from seismic reflection profiles. These basins vary irregularly in width and shape and their locations are shown in schematic fashion. An average velocity of 2 km/sec was used to convert time sections to thickness in kilometers. B. Location and thickness of continental shelf basins. Basin length and width are displayed as accurately as the data allow. Sources of information include seismic reflection and refraction profiles, gravity data, drill hole data, and measured land sections.

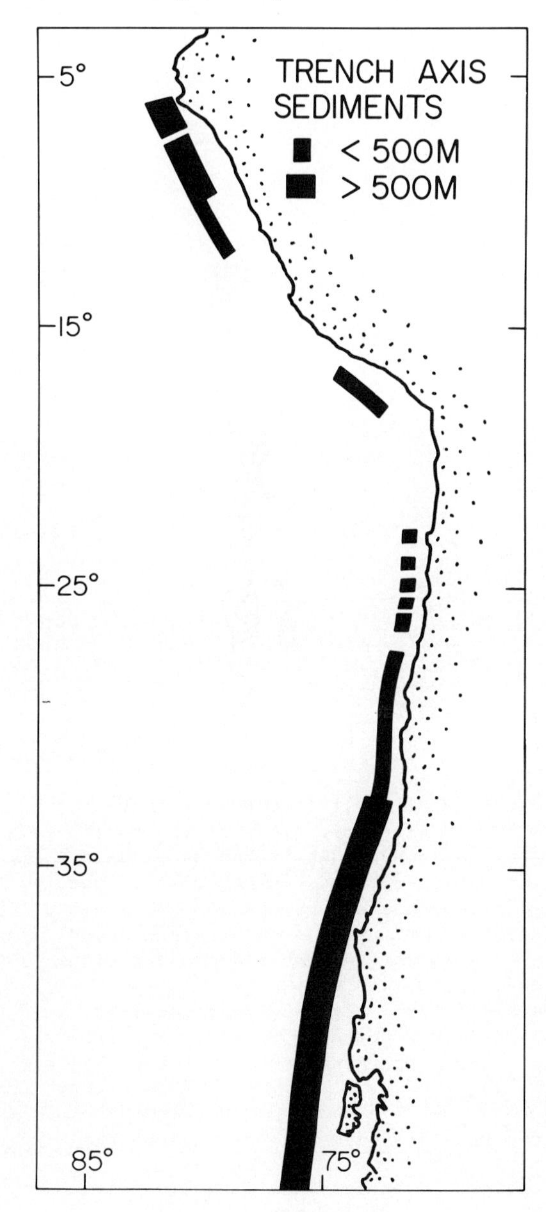

Fig. 7. Distribution and thickness of trench axis sediment. An average velocity of 1.75 km/sec was used to convert time sections to thickness in meters.

regions. They vary in width from about 10 to 45 km and may contain up to 2 km of sediment. A second, smaller slope basin may occur about mid-slope depths and represents the of sediment which is mainly pelagic and hemipelagic material on the downbending Nazca Plate. A wedge of terrigenous turbidites appears in the areas of the trench axis marked < 500 m and is well developed in the axial regions with > 500 m of sediment. South of 33° S the wedge thickens and may exceed 2 km.

Although several factors regulated the amount of sediment present in the trench at any given time, two factors, sediment supply and trench structure, are believed to be the most important ones (Schweller and Kulm, 1976a). Despite the obvious lack of fluvial sources and therefore sediment supply off northern Chile today (Scholl et al., 1970), we find a few hundred meters of turbidite fill in the trench as far north as 23°S opposite the Atacama Desert (Fig. 7). These turbidite deposits were transported northward along the trench axis from sediment sources south of 33°S (Schweller, 1976; Schweller and Kulm, 1976a). Submarine canyons found on the continental slope in province 3B (Fig. 4) serve as the conduits through which the turbidity currents carry sediment to the trench axis with a final connection to an axial deep sea channel which extends from 40°S to 30°S.

Short-lived structural barriers such as the one found at 33°S off central Chile can temporarily restrict or block turbidity current flows from reaching more distant points to the north (Schweller, 1976). Such barriers are also present from 6° to 18°S off Peru. These structural features can form and apparently disappear (ie: features are subducted) in a few thousands or tens of thousands of years (Prince and Kulm, 1975).

Crustal Sections

Province 1. A 24-channel depth section (Peru line 2) of the Peru-Chile Trench and the Peruvian continental slope is available for province 1 (Kulm et al., 1976; Figs. 8 and 9). It is characterized by five main elements which include: (1) a trench axis which consists of a large axial tholeiitic basalt ridge and two axial turbidite basins (not shown here); (2) a strongly reflecting acoustic basement deep beneath the continental slope; (3) a faintly outlined, small, one kilometer thick middle slope sedimentary basin; (4) a broad, complex two kilometer thick upper slope sedimentary basin; and (5) a thick, highly diffracting section sandwiched between the upper slope sedimentary basin and a deep reflecting surface below.

Several lines of evidence indicate that oceanic layers 1 and 2 have been rupturing along the eastern edge of the Nazca Plate (Hussong et al., 1975) and in the vicinity of the trench axis during the past 3-5,000 years and beneath the continental slope (Kulm et al., 1973; Prince et al., 1974; Prince and Kulm, 1975; Hussong et al., 1976; Kulm et al., 1976). Oceanic layer 2 extends beneath the trench sediment wedge and the lower continental slope as shown by the prominent deep reflector between kms 22 to 53 in Figures 9 and 10. This layer is offset at km 32 and becomes less distinct between kms 53 to 87. Although a prominent reflector outlines the basal part of the upper slope sedimentary basin, two prominent reflectors also occur at a depth of about 10 km.

From the velocity analysis of this reflection section (Kulm et al., 1976) and a nearby, parallel refraction section (Hussong et al., 1976), we know that a low velocity (1.8 to 3.0 km/sec) prism of sediment occurs along the lower slope (Fig. 10) from kms 26 to 45 and in the middle and upper slope sedimentary basins. A high velocity (5.7 to 6.2 km/sec) metamorphic block exists along the outer continental shelf and extends beneath the upper continental slope (Figs. 1B, 10). The remainder of the depth section is difficult to interpret because of its highly diffracting character. It could consist of a melange of sediment and/or crystalline rock. Kulm et al. (1976) favor the interpretation where the accreted section extends from kms 26 to 90 because of the basin tectonism noted along the margin (Fig. 10).

Masias (1976) and Coulbourn and Moberly (1976) have shown that the depositional centers of the upper slope basins migrate landward in an apparent response to a higher rate of uplift on the seaward flank than landward flank of these basins. This type of uplift is to be expected in an accreting regime (Seely et al., 1974; Kulm and Fowler, 1974). If this interpretation is correct, the basal portion of the upper slope basin deposits should consist of accreted material and could be as old as Mesozoic (Fig. 10). Both the accreted and basin deposits would become progressively younger in a seaward direction with late Cenozoic deposits filling the middle slope basin.

Province 2. The refraction data of Fisher and Raitt (1962) and Ocola and Meyer (1973) and the gravity data of Hayes (1966) are the only deep crustal information available for province 2. Unfortunately, only one pseudo-reversed pair of refraction profiles was obtained on the continental slope in a water depth of 3100 m (location of profile 7 in Figures 1A and 4) or over the upper slope as defined in this paper.

Figure 11 shows a thick 5.9 to 6.0 km/sec layer of intermediate density (2.86 gm/cc) beneath the coastal region which extends seaward at least as far as the upper and possibly the middle slope region. A 1.2 km thick layer of 4.7 km/sec material overlies

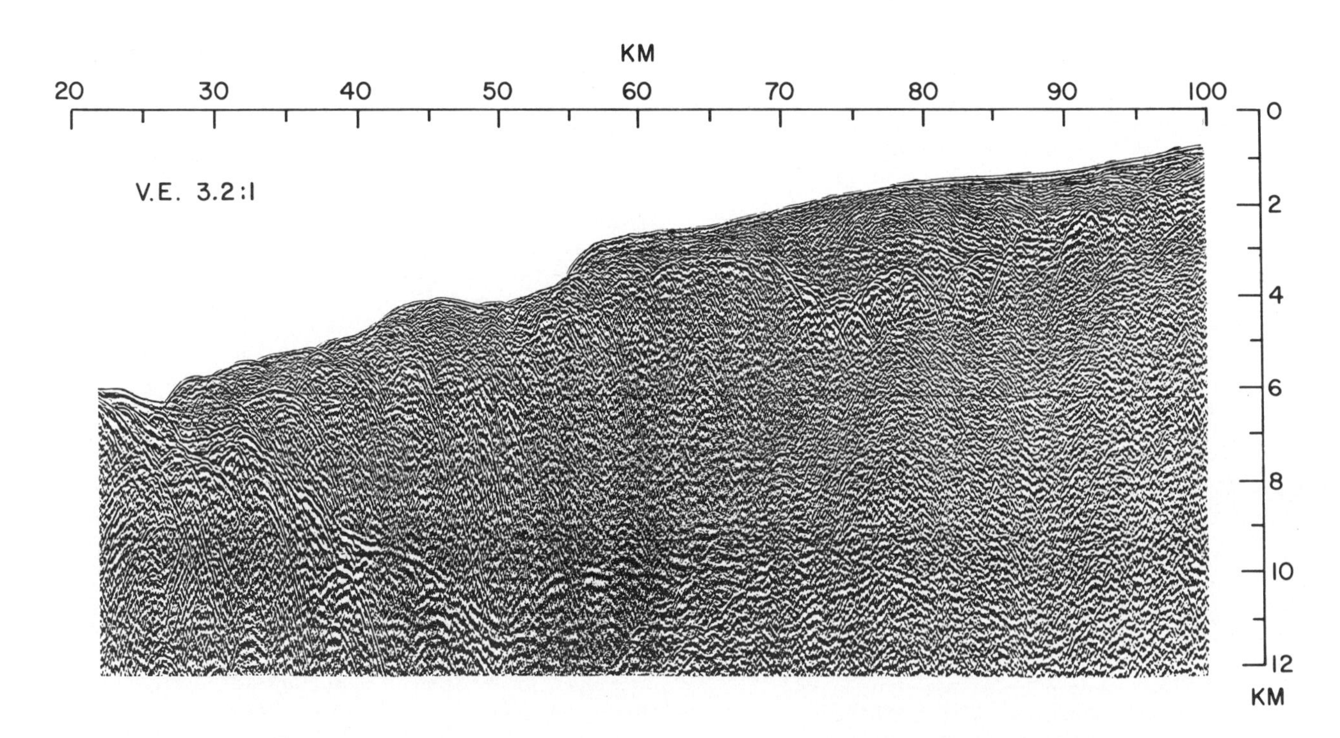

Fig. 8. Non-migrated seismic depth section of Peru line 2 (thick dashed line, Fig. 1A) across the continental slope and inner trench axis basin (modified from Kulm et al., 1976).

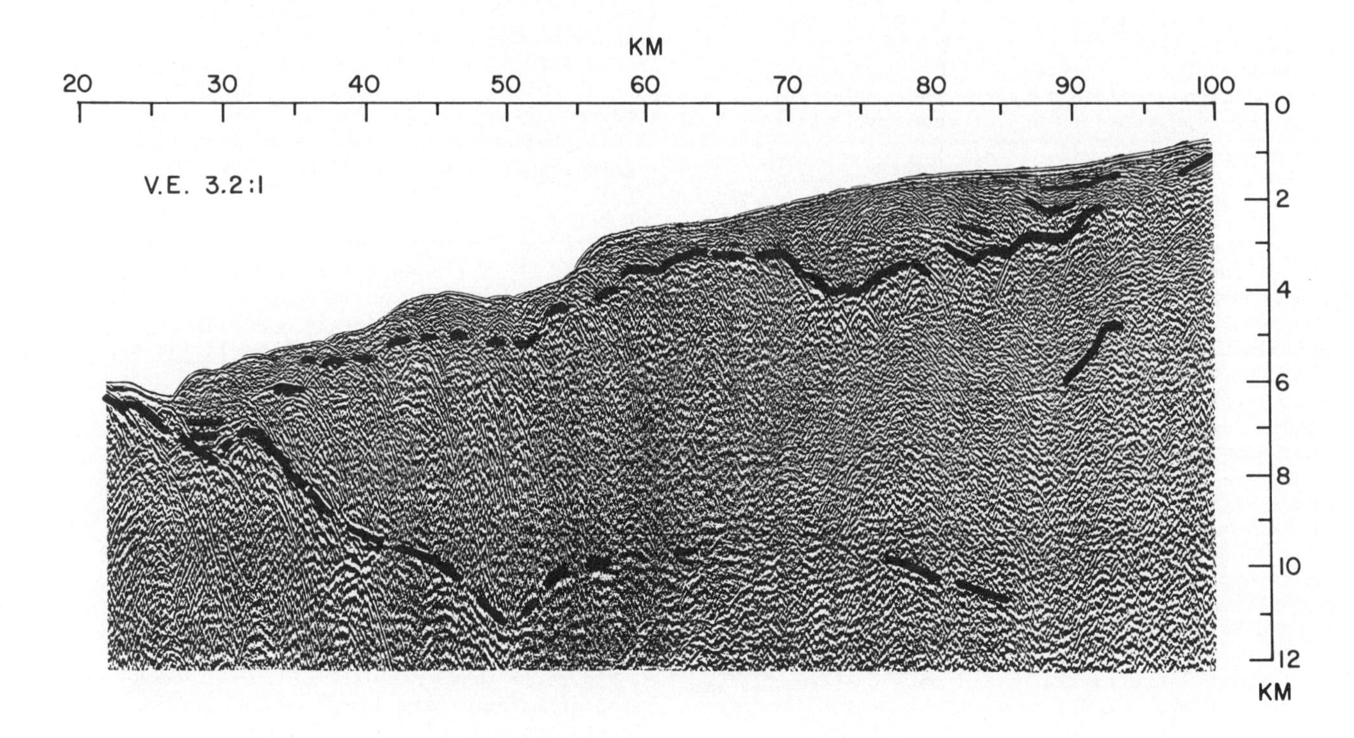

Fig. 9. Delineation of prominent reflectors in seismic depth section shown in Figure 8.

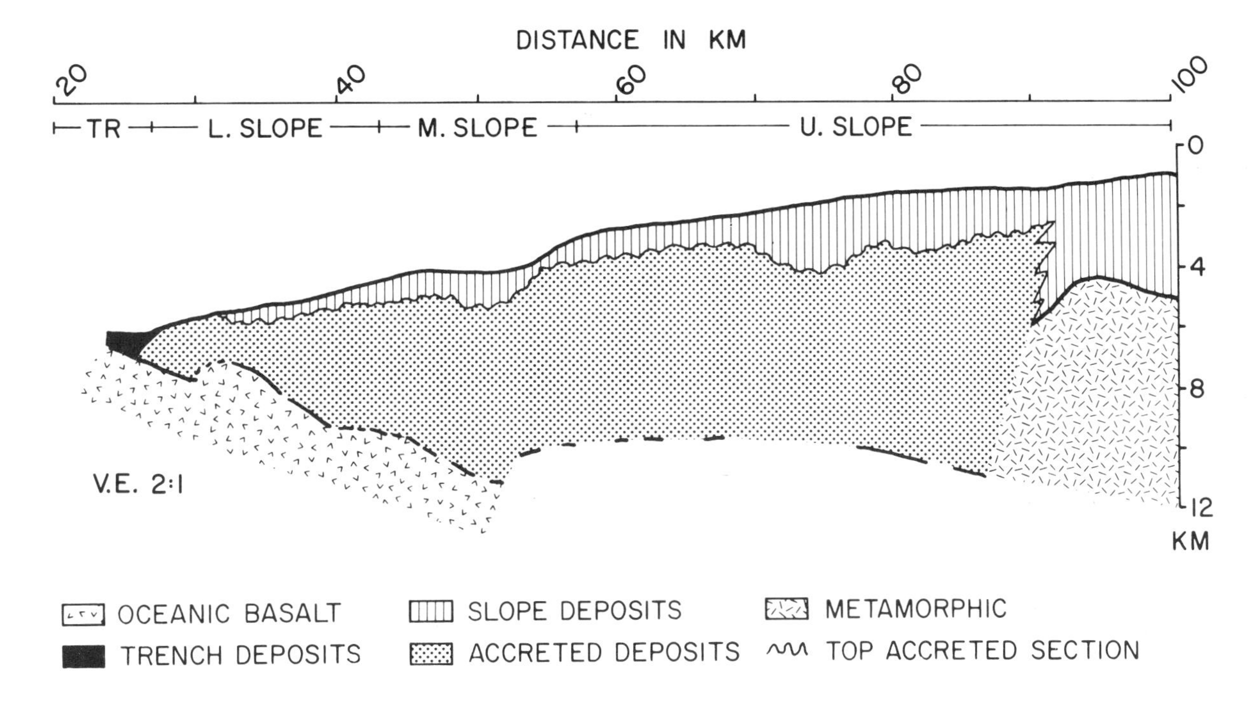

Fig. 10. Geologic interpretation of seismic depth section shown in Figures 8 and 9.

this higher velocity material and is believed to be restricted to the small intermittent basins found on the upper slope. Our seismic reflection profile in this area shows about one kilometer of sediment on the upper continental slope. The 5.5 km/sec material in the coastal region apparently is the Mesozoic coastal volcanics shown in Figure 1B. If there is an accreted sediment section along the outer margin, it is probably of very

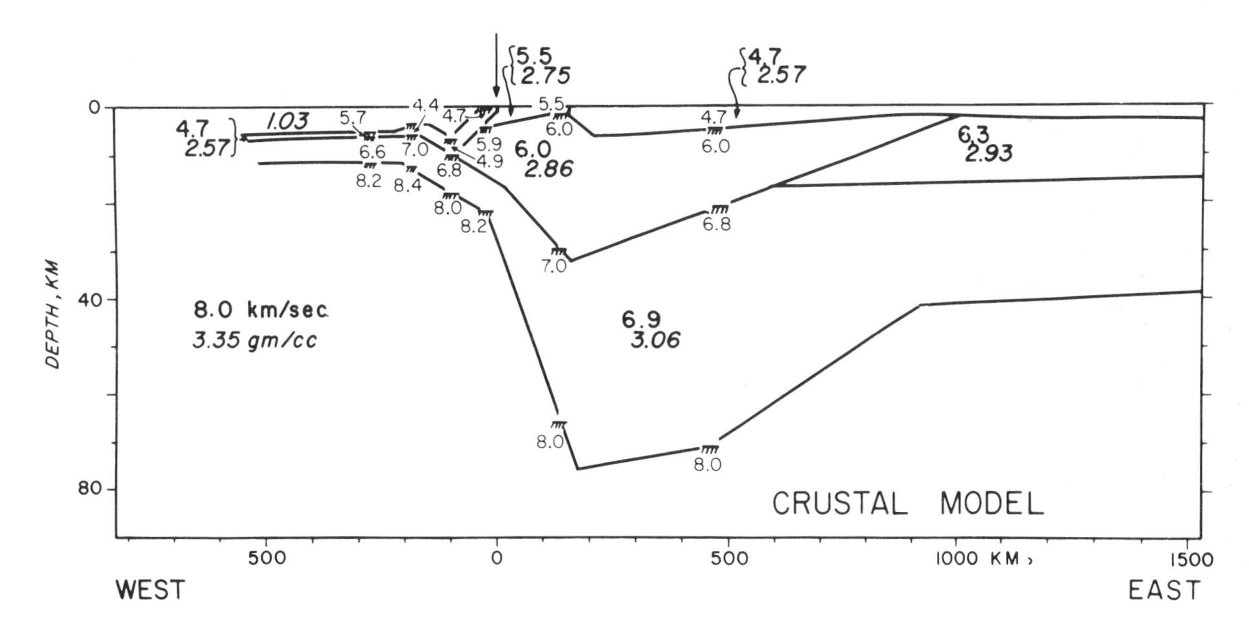

Fig. 11. Oceanic and continental crustal structure of northern Chile - southern Bolivia (modified from Ocola and Meyer, their Figure 4, 1973). Note that both refraction velocities and densities (slant numbers) are given. Coastline is indicated by arrow.

limited extent. There are no deep seismic crustal sections for province 3.

Discussion

Distribution of Dominant Subduction Processes

Based upon our preliminary analysis of the morphology and shallow structural features of the Andean continental margin, which are the features listed in the criteria (Fig. 2), and deep crustal sections, we attempt to identify the dominant subduction processes (accretion, consumption, transitional) that are present along the Andean continental margin from 6° to 45°S. We hypothesize that the physiographic provinces 1 and 3 are those regions where accretion has occurred along the outer continental margin over the long term (ie: the past 10 m.y.). Likewise, province 2 may represent a region of consumption. Alternatively, province 2 may now be in a state of transition between accretion and consumption that was initiated after a period of consumption.

Accretion Model. Province 1 and to a certain extent province 3 are characterized by several of the features usually associated with an accretionary prism along a convergent boundary. Large benches are especially prominent along the lower continental slope. A number of them have a reverse landward inclination, typical of imbricate thrust sheets (Seely et al, 1974) superimposed upon the normal seaward inclination of the continental slope. However, there is no indication of imbricate thrust sheets in the sedimentary portion of the crustal section in province 1 (Figs. 8, 9, 10), although they may be present and be highly fractured into acoustically incoherent pieces that produce the diffractions noted in the non-migrated depth section. The broken basaltic crust beneath the lower continental slope suggests that the overlying sediments are also being faulted and perhaps extensively deformed. Lithologies recovered from the lower continental slope indicate there has been some incorporation of Nazca Plate and Nazca Ridge pelagic sediments into the margin (Rosato, 1974; Kulm et al., 1974); however, no mechanism of emplacement is implied by their data. There apparently has been an ample supply of trench turbidite sediment available in provinces 1 and 3 for accretion during the past several million years while the relatively thin hemipelagic and pelagic sediments of the Nazca Plate have remained rather constant in volume over the same period of time (Kulm et al., 1976a).

The landward migrating depositional centers in a number of the upper continental slope basins in province 1 suggest that the outer margin has been tectonically active with basins forming and being uplifted. The ages of these basin deposits probably range throughout the Cenozoic and perhaps into the Mesozoic judging from the ages of adjacent continental shelf deposits (Fig. 1B; Masias, 1976). This implies that uplift has occurred over a long period of time. There may have been periods of transitional subduction (no net loss or addition of material to the margin) in the past, but they probably occurred prior to or were interspersed with times of formation and filling of the migrating upper slope basins.

Long term accretion to the outer margin is strongly implied in province 1 and suggested in province 3 for the data presented in this paper. Figure 12A is a simplified accretion model showing our interpretation of the present geotectonic setting of the Andean continental margin in provinces 1 and 3. As shown in the model, a large portion of the margin consists of Mesozoic and pre-Mesozoic volcanic or metamorphic strata. Seismic refraction work along the Andean margin (Fisher and Raitt, 1962; Hussong et al., 1976) and in the coastal region (Ocola and Meyer, 1973) indicates that the velocities of these crystalline basement rocks are apparently in the range of 5.7 to 6.2 km/sec. Much lower velocity sedimentary material (1.6 to 4.8 km/sec) overlies the basement strata on the continental shelf and blankets the shallow subsurface regions and lowermost portion of the continental slope (Fig. 10). If our velocity determinations are close to the actual values and our features of accretion are valid, a substantial portion of the outer margin in province 1 appears to be accreted as shown in Figure 12A. The material in the velocity range of 5.0 to 5.7 km/sec could be either well consolidated, cemented accreted material or crystalline strata.

Consumption Model. Province 2 is characterized by several features that we consider, from a theoretical point of view (see Criteria, Fig. 2), as best explained by a model of consumption, which produces tectonic erosion of the leading edge of the continental block over the long term (Fig. 12B). Benches are either absent or are rather small and narrow in this province. Sedimentary basins are absent on the continental shelf or small; intermittent basins occur on the upper continental slope. Many areas of the margin have little or no sediment cover on the Mesozoic and pre-Mesozoic crystalline basement. The crustal section in Figure 11 suggests that the seaward continuation of the crystalline block extends beneath the upper slope and possibly almost as far as the trench axis.

If benches are features of accretion as we propose in this paper, the small benches on the lowermost continental slope indicate that a small amount of sediment is accreted in some regions in province 2. We dredged consolidated sandstone and siltstone from the lowest bench on profile 7 (Fig. 4). A second

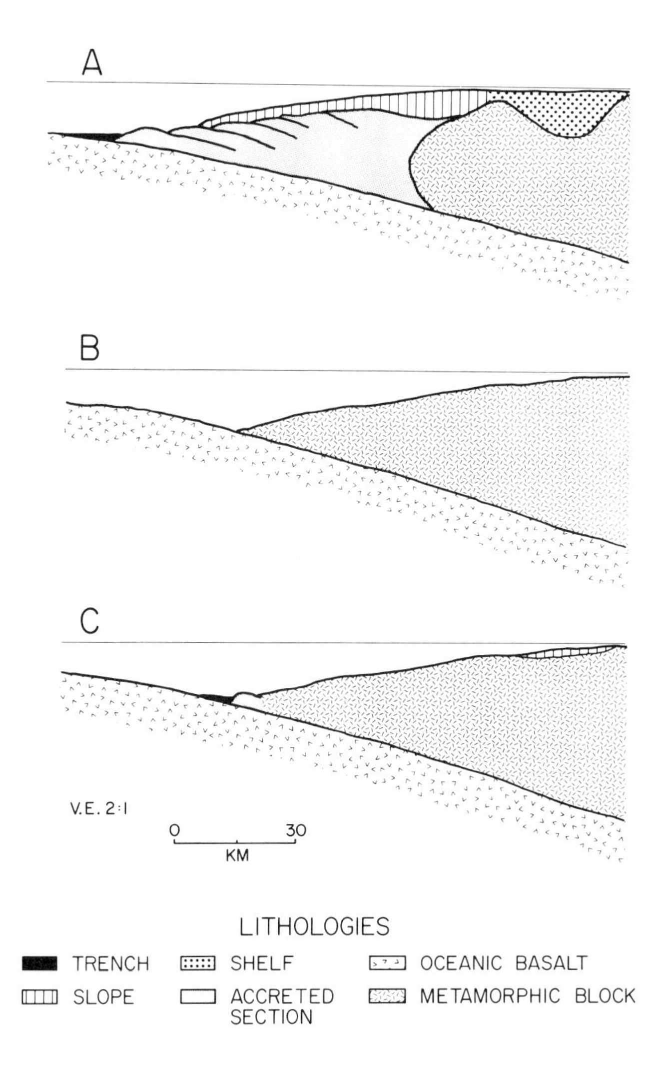

Fig. 12. Subduction process models for the Andean margin off South America. A. Long-term accretion, B. Long-term consumption, and C. Long-term consumption with short-term accretion. See text for details.

consumption model must be considered which allows the accretion of trench and oceanic plate sediments to the outer margin over the short term with the eventual loss of this material over the long term (Fig. 12C). The availability of trench sediments may be a controlling factor in the short term accretion in province 2 because of the patchy distribution of turbidites noted in this area (Fig. 7).

The thin sedimentary cover on the margin in province 2 could be due to the lack of sediment supply in the vicinity of the Atacama Desert (Scholl et al., 1970). If the supply has been minimal during much of the Cenozoic time, it is possible that province 2 is presently a region in transition between accretion and consumption and that this steady state condition followed a former period of consumption. In this case the features listed in Figure 2 under the transitional process essentially would be absent. On the other hand, if sediments have been supplied during the Cenozoic and Mesozoic Eras and if the crystalline block forms the bulk of the continental slope as we tend to believe, then province 2 must have undergone consumption over the long term. In any case, it is difficult to imagine an equilibrium condition such as the transitional process being maintained for tens of millions of years.

Transitional Subduction Model. If the transitional subduction process is occurring along the Andean margin, we cannot identify it on the basis of the criteria given and the available marine data. Although we have concentrated our discussion largely upon marine data, there is reason to believe, based upon the geologic history of the southern Peru coastal region, that a much larger accreted section should be present along the adjacent continental margin because of the uplift and erosion initiated in the late Cretaceous in this area (Masias, 1976). A period of transitional subduction, or even consumption, could explain this discrepancy.

Factors that Affect Subduction Processes

Our working hypothesis states that the morphology and structure of the Andean continental margin reflect the distribution of the dominant subduction processes in time and space. Some commonly mentioned factors that may influence these processes and thus affect the development of the margin include: (1) trench depth, (2) oceanic plate and trench axis structure, (3) trench axis sediment, (4) dip of the Benioff Zone, and (5) direction and rate of closure between the Nazca Plate and the South American block.

Trench Depth vs. Continental Slope Inclination. The inclination of the lower continental slope is relatively uniform along the Andean margin and thus apparently independent of both trench depth and subduction processes (Fig. 4). The middle and upper slope regions show more variability in inclination than the lower slope, both among individual profiles and between provinces. In general, middle and upper slope inclinations steepen with increasing trench depth. The relation between inclinations and processes are not clear from these data.

Crustal Rupture Zone. The crustal rupture zones (basaltic axial ridges) discovered within the Peru-Chile Trench axis (Kulm et al.,

1973; Prince and Kulm, 1975; Schweller, 1976) are located opposite all three provinces. Although these layer 2 features may be thrust sheets, we still cannot correlate the occurrence of such features with the presence of benches on the lower slope. It is still too early in our study of the Andean system to determine if the structure of the trench and downgoing slab has an influence on the subduction processes on the margin.

Trench Axis Sediments. According to the criteria given in Figure 2, trench axis deposits (or a thick sediment section on the plate) are required if a sizeable accretionary prism is to develop on the adjacent margin. Sediment fill is intermittent along the Peru-Chile Trench except that portion of the axis south of 27° S which has a continuous sediment fill (Hayes, 1966; Schweller and Kulm, 1976). At the present time we find no direct correlation between the occurrence of trench sediments and the occurrence of lower slope benches, although provinces 1 and 3 certainly have a larger sediment supply available than province 2.

Our data suggest that items such as axial sediments and trench structures are often out of phase with the subduction process reflected in the margin due to the rapidity of convergence and the consequent brevity of existence of axial features. Sections of the trench axis which are temporarily isolated from sediment input due to a structural barrier will become empty of terrigenous sediment in a few tens of thousands of years, while lower slope benches adjacent to these empty sections reflect dominant accretion as a somewhat longer term average.

Benioff Zone. There is some correspondence between our physiographic province boundaries and the segments of inclined seismic zones defined by Barazangi and Isacks, (1976) from the spatial distribution of precisely located hypocenters of South American earthquakes (Fig. 3). These studies utilize different data bases and were done independently of one another. Segments with shallow dips (about 10°) were found beneath northern and central Peru (B, 2° - 15° S) and beneath central Chile (D, 27° - 33° S). Steeper dips (25° - 30°) were found for segments located beneath southern Peru and northern Chile (C, 15° - 27° S) and beneath southern Chile (E, 33° - 45° S). Our province and subprovince boundaries are located at approximately 19.5° S, 28° S and 34° S latitude (Fig. 3), with the northern limit of province 1 and southern limit of province 3 not defined because they lie outside of the area of our data base.

Despite the geographic limitations of our data, there is a rather good correlation between the location of our subprovince 3A (28° - 34° S) and the adjacent segment of the shallow dipping seismic zone (27° - 33° S) (Fig. 3). On the other hand, the boundary between province 1 and 2 (19.5° S) lies within one of the steeper dipping seismic zones (15° - 27° S) or in the region where the continental block undergoes its sharpest deflection. This is also the region of broadest transition zones between seismic segments and between physiographic provinces. There is closer correspondence of transition zones in the more narrowly defined zones at 27° - 28° S and 33° - 34° S. Barazangi and Isacks (1976) conclude that the 15° to 20° S region is more complex than the other regions of the South American seismic zone, perhaps because of the concave-seaward curvature of the margin.

Direction and Rate of Closure Between Plates. The nature of closure between the plates seems to have little effect on the subduction processes along the Andean margin. Convergence rate between the two plates varies only slightly from 9 to 11 cm/yr along the margin (Minster et al., 1974). Although the angle of convergence varies from 90° to 60° mainly because of inflections in the trend of the margin the sharpest changes do not occur near our province boundaries.

In summary, neither trench depth nor crustal rupture at the trench axis seem to be directly related to the dominant subduction process reflected in the adjacent continental margin. Trench axis deposits probably have some influence on these processes (especially accretion) over the long term, but their correspondence with short term features is much less distinct. Our physiographic province boundaries correlate closely with one of the Barazangi and Isacks (1976) seismic zone boundaries, but the cause-effect relation is indeterminant. Finally, variations in the direction and rate of plate convergence along this margin apparently do not visibly influence the subduction processes. Other factors such as oceanic and continental crust thickness may be more important in determining the evolution of the Andean margin, but are beyond the scope of this study at this time.

Comparison of Andean Margin with Island Arcs

In this paper we have refrained from using any terminology that suggests a genetic relationship between island arc structure and the structure of the Andean margin. For example, the trench-slope break in the arc nomenclature is not well defined in the morphology of the Andean system, although similar gross accretionary processes may be present in both systems. We need to make a more detailed synthesis of the Andean system before we attempt a genetic classification for its morphological-structural-tectonic elements.

Conclusions

1. The broadly defined subduction process, which is associated with convergence zones, is subdivided into two major component processes, accretion and consumption; and one minor process, a transitional state between the two major processes. Accretion is the addition of oceanic plate and trench to the outer continental margin. Consumption is the tectonic removal of pre-existing material from the continental block. A transitional process or steady state condition may exist between periods of accretion and consumption whereby material is neither added nor removed from the outer margin (ie: carried down the Benioff Zone).

2. A set of criteria was established in an attempt to distinguish the dominant subduction processes along a given length of convergent margin (Fig. 2). The criteria are the presence or absence of benches and basins on the continental shelf and slope and the presence or absence of trench deposits.

3. The Andean continental margin off South America is divided into three physiographic provinces using the features given in the criteria. They are province 1 (6° to 19.5° S), province 2 (19.5° to 28° S) and province 3 (28° to 45° S); the latter province is separated into subprovinces 3A (27° to 34° S) and 3B (34° to 45° S).

4. Provinces 1 and 3 are characterized by long prominent benches, particularly along the lower continental slope, sedimentary basins, and trench axis deposits. Continental shelf sedimentary basins commonly range in thickness from 1 to 3 km and are limited in areal extent to northwestern Peru, south-central Peru (Pisco) and south-central and southern Chile. Upper continental slope basins are more extensive than shelf basins, but they generally contain less than 2 km of sediment. Province 2 is characterized by a general lack or the small size of benches on the lower slope, by intermittent small sedimentary basins on the upper continental slope, and by a complete lack of shelf basins.

5. Based upon the criteria, provinces 1 and 3 display the characteristics of an accretion subduction process whereas a consumption process is suggested for province 2. Alternatively, province 2 may be in a state of transition between accretion and consumption that was initiated after a period of consumption.

6. A limited amount of deep crustal data were used to test our interpretation of the subduction processes present along the Andean margin. The 24-channel seismic depth section in province 1 (9 ° S) shows a highly diffracting prism of low velocity sediment along the outermost margin. Geologic data from previous studies of this area suggest that at least a portion of this section may be accreted Nazca Plate sediments. Landward migration of the depositional centers of several upper slope basins indicates higher rates of uplift along the seaward rather than the landward flank of these basins. These data also argue for a long term (several million years) accretion model for province 1 and by analogy through the criteria for province 3 (Fig. 12A). In both cases, a metamorphic block forms the eastern boundary of the section. Deep crustal data available for province 2 is less conclusive in regard to the dominant subduction process (Fig. 12B), although a relatively high velocity, dense block of volcanic or metamorphic material appears to extend nearly to the trench axis. Based upon the morphology and shallow structure, it appears that the consumption process is dominant, but the transitional process (no net loss or gain along the outer margin) cannot be ruled out because of the uncertainty in the long-term sediment supply. An alternate model of short-term accretion on a long-term consumption process is used to explain the intermittent occurrence of lower slope benches in province 2 (Fig. 12C).

7. In this preliminary investigation, we find no definitive relations among trench depth, crustal rupture zones, and subduction processes. Trench axis sediments should affect the subduction processes, but often may be out of phase with the long term processes reflected in the adjacent margin. There appears to be some correlation between the inclined seismic zone segments determined by Barazangi and Isacks (1976) for South America and the positions of the physiographic province boundaries identified in this study. Province 3A and segment D show a rather close correlation if one considers the errors involved in selecting these boundaries (Fig. 3).

Acknowledgments. We thank the staff members and students of the School of Oceanography, Oregon State University and of the Hawaii Institute of Geophysics, including personnel of the R/V YAQUINA and R/V KANA KEOKI who helped with the data collection at sea. A number of bathymetric and seismic reflection profiles of the Peru-Chile Trench were kindly supplied by Roland von Huene. We thank Roger Prince for his assistance in the data reduction. This research was conducted under the auspices of the Nazca Plate Project and was supported by the National Science Foundation, International Decade of Ocean Exploration (NSF Grants GX-28675, IDO 71-04208 A07 and OCE 76-05903).

References

Barazangi, M., and B. L. Isacks, Spatial distribution of earthquakes and subduction of the Nazca Plate beneath South America, Geology, (in press), 1976.

Beck, R. H. and P. Lehner, Oceans, new frontier in exploration, Amer. Assoc. Petrol. Geol. Bull., 58, 376-395, 1974.

Coulbourn, W. T. and R. Moberly, Structural evolution of fore-arc basins off southern Peru and northern Chile, Can. Jour. Earth Sci., (in press), 1976.

Farrar, E., A. H. Clark, S. J. Haynes, G. S. Quirt, H. Conn, M. Zentilli, K-Ar evidence for the post-Paleozoic migration of granitic intrusion foci in the Andes of northern Chile, Earth Planet. Sci. Lett., 10, 60-66, 1970.

Fisher, R. L. and R. W. Raitt, Topography and structure of the Peru-Chile Trench, Deep-Sea Res., 9, 423-443, 1962.

Hayes, D. E., A geophysical investigation of the Peru-Chile Trench, Marine Geology, 4, 309-351, 1966.

Hussong,D., M. E. Odegard, and L. K. Wipperman, Compressional faulting of the oceanic crust prior to subduction in the Peru-Chile Trench, Geology, 3, 601-604, 1975.

Hussong, D. M., P. B. Edwards, S. H. Johnson, J. F. Campbell, and G. H. Sutton, Crustal structure of the Peru-Chile Trench: 8° S-12° S latitude, In: The Geophysics of the Pacific Ocean Basin and its Margin, G. H. Sutton, M. H. Manghnani, R. Moberly and E. U. McAfee (eds.), Geophy. Mono. 19, AGU, 71-86, 1976.

Johnson, S. H., G. E. Ness and K. R. Wrolstad, Shallow structures and seismic velocities of the southern Peru margin, EOS Trans. Amer. Geophy. Un., 56:443, 1975.

Karig, D. E. and G. F. Sharman, Subduction and accretion in trenches, Geol. Soc. America Bull., 86, 377-389, 1975.

Kelleher, J., Rupture zones of large South American earthquakes and some predictions, J. Geophy. Res., 77, 2087-2103, 1972.

Kulm, L. D., R. von Huene et al., Initial Reports of the Deep Sea Drilling Project, 18, Washington (U. S. Government Printing Office), 1077 p., 1973.

Kulm, L. D., K. F. Scheidegger, R. A. Prince, J. Dymond, T. C. Moore, Jr., and D. M. Hussong, Tholeiitic basalt ridge in the Peru Trench, Geology, 1, 11-14, 1973.

Kulm, L. D., and G. A. Fowler, Oregon continental margin structure and stratigraphy: A test of the imbricate thrust model, in Burk, C. A. and Drake, C. L. (eds.), The Geology of Continental Margins, New York, Springer-Verlag, 261-283, 1974.

Kulm, L. D., J. M. Resig, T. C. Moore, Jr., and V. J. Rosato, Transfer of Nazca Ridge pelagic sediments to the Peru continental margin, Geol. Soc. Amer. Bull., 85, 769-780, 1974.

Kulm, L. D., R. A. Prince, W. French, A. Masias, and S. Johnson, Evidence of imbricate thrusting in the Peru Trench and continental slope from multi-fold seismic reflection data, EOS Trans. Amer. Geophy. Un., 56,442-443,1975.

Kulm, L. D., W. French, R. A. Prince, A. Masias, and S. Johnson, Basement imbricated thrusting within the trench and continental slope off central Peru, (in preparation), 1976.

Kulm, L. D., W. J. Schweller, A. Molina-Cruz, and V. J. Rosato, Lithologic evidence for convergence of the Nazca Plate with the South American continent, in Yeats, R. S., S. R. Hart et al., 1976, Initial Reports of the Deep Sea Drilling Project, 34, Washington (U. S. Government Printing Office), 795-801, 1976a.

Masias, J. A., Morphology, shallow structure, and evolution of the Peruvian continental margin, 6° to 18° S (M.S. thesis), Corvallis, Oregon State University, 92 p., 1976.

McNutt, R. H., J. H. Crocket, A. H. Clark, J. C. Caelles, E. Farrar, S. J. Haynes, and M. Zentilli, Initial $^{87}Sr/^{86}Sr$ ratios of plutonic and volcanic rocks of the central Andes between latitudes 26° and 29° South, Earth Planet. Sci. Lett., 27, 305-313, 1975.

Minster, J. B., T. H. Jordan, P. Molnar, and E. Haines, Numerical modeling of instantaneous plate tectonics. Roy. Astron. Soc. Geophys. Jour., 36, 541-576, 1974.

Moore, J. C., and D. E. Karig, Sedimentology, structural geology, and tectonics of the Shikoku subduction zone, southwestern Japan, Geol. Soc. Amer. Bull., 87, 1259-1268, 1976.

Mordojovich, C., Geology of a part of the Pacific Margin of Chile, in Burk, C. A. and Drake, C. L. (eds.), The Geology of the Continental Margins, New York, Springer-Verlag, 591-598, 1974.

Ocola, L. C. and R. P. Meyer, Crustal structure from the Pacific basin to the Brazilian shield between 12° and 30° South latitude, Geol. Soc. Amer. Bull., 84, 3387-3404, 1973.

Prince, R. A., J. M. Resig, L. D. Kulm, and T. C. Moore, Jr., Uplifted turbidite basins on the seaward wall of the Peru Trench, Geology, 2, 607-611, 1974.

Prince, R. A., and L. D. Kulm, Crustal rupture and the initiation of imbricate thrusting in the Peru-Chile Trench, Geol. Soc. Amer. Bull., 86, 1639-1653, 1975.

Rosato, V. J., Peruvian deep-sea sediments: Evidence for continental accretion (M.S. thesis), Corvallis, Oregon State University, 93 p., 1974.

Scholl, D. W., M. N. Christensen, R. von Huene, and M. S. Marlow, Peru-Chile Trench sediments and sea-floor spreading, Geol. Soc. Amer. Bull., 81, 1339-1360, 1970.

Schweller, W. J., Chile Trench: Extensional rupture of oceanic crust and the influence of tectonics on sediment distribution (M.S. thesis), Corvallis, Oregon State University, 90 p., 1976.

Schweller, W. J., and L. D. Kulm, Extensional rupture of oceanic crust in the Chile Trench, Mar. Geol. (submitted), 1976.

Schweller, W. J., and L. D. Kulm, Depositional patterns and channelized sedimentation in active eastern Pacific Trenches, in Stanley, D. J. and Kelling, G. (eds.) Submarine Canyon and Fan Sedimentation in Time and Space, Stroudsburg, Pennsylvania, Dowden, Hutchinson and Ross, (in press) 1976.

Seeley, D. R., P. R. Vail, and G. G. Walton, Trench slope model, in Burk, C. A. and Drake, C. L. (eds.), The Geology of Continental Margins, New York, Springer-Verlag, 249-260, 1974.

Stauder, W., Mechanism and spatial distribution of Chilean earthquakes with relation to subduction of the oceanic plate, J. Geophys. Res., 78, 5033-5061, 1973.

Stauder, W., Subduction of the Nazca Plate under Peru as evidence of focal mechanisms and by seismicity, J. Geophys. Res., 90, 1053-1064, 1975.

Stewart, J. W., J. F. Evernden, and N. J. Snelling, Age determinations from Andean Peru: a reconnaissance survey, Geol. Soc. Amer. Bull., 85, 1107-1116, 1974.

Whitsett, R. A., Gravity measurements and their structural implications for the continental margin of southern Peru (Ph.D. thesis), Corvallis, Oregon State University, 82 p., 1976.

METALLOGENY OF AN ANDEAN-TYPE CONTINENTAL MARGIN IN SOUTH KOREA: IMPLICATIONS FOR OPENING OF THE JAPAN SEA

Richard H. Sillitoe

Department of Mining Geology, Royal School of Mines, Imperial College of Science and Technology, Prince Consort Road, London SW7 2BP, England

Abstract. In the Kyongsang basin of southeast Korea, a belt of copper and tungsten deposits, including a low-grade porphyry copper occurrence, breccia pipes and veins, give way northwards to lead-zinc vein deposits and, still further north, to fluorite, tungsten and molybdenum deposits. This metallogenic pattern was generated by Upper Cretaceous calc-alkaline magmatism which, by analogy with central Andean metallogeny, occurred above a shallow, northward-dipping subduction zone. The copper-tungsten and lead-zinc belts appear to be offset continuations of those generated at the same time in southwest Japan, thereby demonstrating a cumulative offset of about 250 km between southeast Korea and southwest Japan as a result of opening of the Japan Sea. In southwest Japan, molybdenum mineralization was superimposed on the lead-zinc belt in the Paleocene to mid-Eocene interval. Two groups of minor molybdenum occurrences in the coastal region of southeast Korea are believed originally to have constituted a single cluster prior to 100 km of dextral, strike-slip displacement on the Yangsan fault. This cluster is thought to be the offset extension of the Japanese molybdenum belt. The transition from an Andean-type metallogenic pattern to molybdenum mineralization may have been induced by subduction of the Kula Ridge. Therefore southward rafting of the Japanese islands and generation of oceanic crust in the Japan and Yamato basins began after the end of the Cretaceous (64 m.y. B.P.) and, if molybdenum mineralization in Japan and South Korea is correctly correlated, not until later than the mid-Eocene (46 m.y. B.P.).

Introduction

A marked contrast is apparent between our knowledge and understanding of metallogeny on the east and west sides of the Pacific ocean: Western North America and the Andes have been well studied and the latter region has become a type area for metallogeny at a convergent continental margin, whereas readily available metallogenic information for the eastern margin of mainland Asia is rather limited. In this context a brief description of the Upper Cretaceous metallogeny of South Korea should prove useful.

Unlike the case of the mainland Asian margin, metallogenic knowledge of the Japanese islands is relatively advanced. Bearing in mind that South Korea and Japan are accepted generally as having been united prior to the opening of the Japan Sea, a correlation of South Korean and southwest Japanese metallogeny should prove instructive: It can be utilized for the development of mineral exploration strategy in both South Korea and southwest Japan, and also to provide additional information concerning the distance and timing of separation of the two countries.

Regional Geological Setting

South Korea is divisible into four main morphostructural elements (Kim, 1974; Reedman and Um, 1975) (Fig. 1): The northeasterly-trending Okchon zone across the center of the peninsula that consists of unmetamorphosed Cambrian to Cretaceous systems in the northeast and largely of late Precambrian metamorphics in the southwest; the Kyonggi and Ryongnam massifs composed of Precambrian schists and granite-gneiss, 2000 $\pm$ 500 m.y. in age (Hurley et al., 1973), that are located to the north and south, respectively, of the Okchon zone; and the Kyongsang basin to the southeast of the Ryongnam massif and consisting of continental sedimentary and volcanic rocks of Cretaceous age. This paper focuses on the metallogeny of the last zone and adjacent areas.

In South Korea, major granitic intrusion (the Daebo granites) took place in the Jurassic in the Okchon zone and in the Kyonggi and Ryongnam massifs (Kim, 1971a). The Upper Cretaceous Bulkuksa granites are largely restricted to the Kyongsang basin but extend northwards as far as the Okchon zone. The known occurrences of Triassic granitic rocks in North Korea complete the definition of a southward-younging igneous suite that suggests a north-south retreat with time of a subduction system.

Precambrian trends are chiefly north-northeasterly. Some Phanerozoic deformation in pre-Triassic times has been recognized (Kim, 1974) but the major tectonic event is the Jurassic Daebo orogeny that follows a northeasterly, Sinian direction and is particularly intense in the Okchon zone, where overthrusting and isoclinal folding are widespread (Reedman et al., 1974). Ophiolites, indicative of sutured oceans, are absent. In the southeast of the Kyongsang basin, major north-northeast-trending wrench faulting post-dated the Bulkuksa intrusives (Figs. 2 and 3) and is related by Reedman and Um (1975) and the present writer to opening of the Japan Sea.

Geology of the Kyongsang Basin

In the Kyongsang basin and adjoining areas extending as far as the Okchon zone (Fig. 1), calc-alkaline intrusive and volcanic rocks cut and unconformably overlie buried Precambrian basement and younger formations in a northeast-trending belt some 400 x 200 km in area. These magmatic rocks are part of the more extensive, but now disrupted, Chukotsk-Cathasia volcanic belt of Cretaceous age along the eastern edge of the Asian continent from the Bering straits to the southern part of the East China Sea (Ustiyev, 1965). The belt includes the Fukien-Reinan massif of southern China and its northeastward continuation beneath the Yellow Sea (Wageman et al., 1970) to the Kyongsang basin, the Japan-Sea (Inner) side of southwest Japan, and the Sikhote-Alin zone of the eastern U.S.S.R.

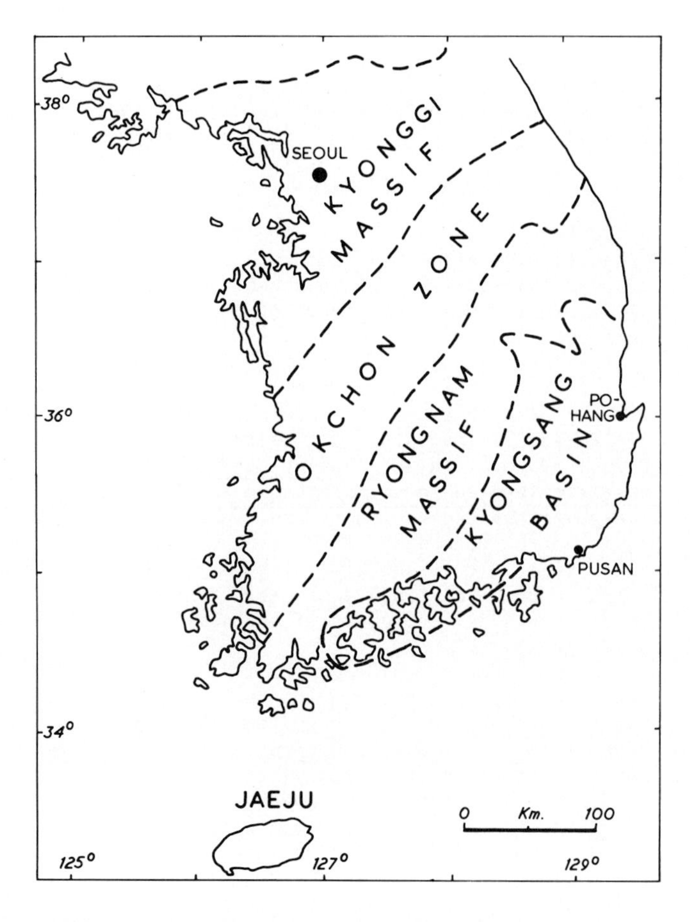

Fig. 1. Morphostructural provinces, including the Kyongsang basin, in South Korea.

Following a recent summary by Chang (1975), the stratigraphic sequence in the Kyongsang basin is assignable to the Kyongsang System which spans the Cretaceous and attains a maximum thickness of 10,000 m. The Lower Cretaceous part consists entirely of post-orogenic, molasse-type sediments, found chiefly on the west side of the basin. The Middle Cretaceous is also largely non-marine sediments, with a few volcanic and volcaniclastic intercalations. The Upper Cretaceous Yuchon Group is mainly volcanic with andesitic pyroclastics and flows below and felsic lithic tuffs and ash flows above. Most of these rocks are intruded by stocks and larger plutons of probably comagmatic intrusive rocks (Fig. 2), mainly of granodioritic and adamellitic composition but also including granites and diorites. As a result of radiometric dating, mainly by the K-Ar method, these Bulkuksa intrusives have been shown to be late Cretaceous, ranging in age from 88 to 68 m.y. (Kim, 1971a; Seo and Ju, 1971). As in southwest Japan, Lower Tertiary plutons are suspected but so far have not been proven.

The Kyongsang System strikes north-northeasterly and exhibits little structural complexity. Dips are southeasterly and the few folds are gentle. Igneous intrusion induced limited contact metamorphism but virtually no structural disturbance. The Kyongsang System has also been subjected to a regional greenschist facies metamorphism.

Metallogeny of the Kyongsang Basin and Adjoining Areas

In the southern part of the Kyongsang basin, south of lat. 36°N, copper and tungsten are the principal metals although some gold, iron and molybdenum are also present (Fig. 2). Ore deposits intimately associated with intrusive rocks include a low-grade porphyry copper occurrence at Red Hill, Dongjom (Kim and Kim, 1974), copper-tungsten-bearing breccia pipes of collapse type at Dalsung (Jordt, 1966; Won and Kim, 1966) and Ilkwang (Fletcher and Park, 1974), and a tungsten-molybdenum vein at Sannae. The Dongjom occurrence is centered on a granodiorite porphyry stock, at least 750 x 450 m at surface, emplaced in fine-grained Cretaceous clastics. Observations made on drill core show the stock to have been potassium silicate altered and subsequently sericite altered, with a tendency for sericitic alteration to be more widespread towards the stock margins. Both alteration types contain

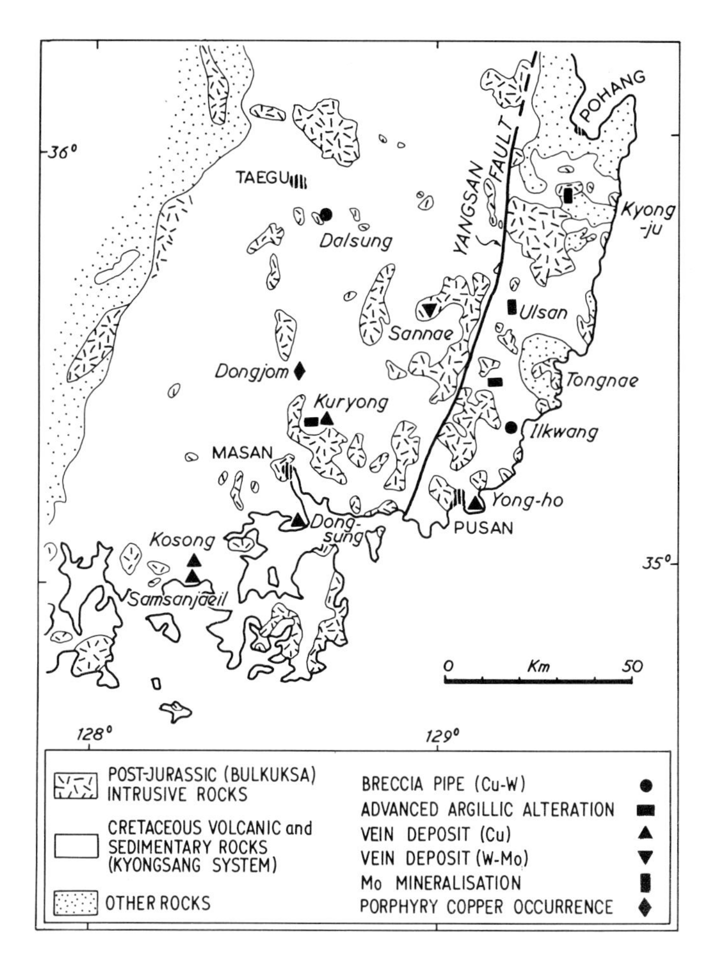

Fig. 2. Distribution of post-Jurassic intrusive (Bulkuksa) and volcano-sedimentary (Kyongsang System) rocks in the central part of the Kyongsang basin. The location and types of mineral deposits are also shown.

chalcopyrite and molybdenite but pyrite is far more abundant in the sericitic facies. Propylitic alteration constitutes a fringe in the sedimentary host rocks.

The Dalsung and Ilkwang breccia pipes are somewhat different in that the former is emplaced in andesitic rocks, although with an adamellite intrusive 1.5 km to the west, and the latter in a 1 km^2 adamellite stock. Both pipes are oval, possess sheeted contacts and are filled with angular, commonly tabular, wall-rock fragments that are sericitized and propylitized at Dalsung and sericitized (with tourmaline and garnet) at Ilkwang. Most mineralization occurs as a filling of open space between fragments, especially in a narrow, annular zone abutting the pipe contacts. It consists of chalcopyrite, scheelite and wolframite, accompanied by quartz, pyrite, arsenopyrite, pyrrhotite and bismuthinite. Specularite, calcite, siderite and K-feldspar also occur at Dalsung and tourmaline with minor galena and sphalerite at Ilkwang. The overall characteristics of these two pipes are similar to those described from Chilean examples by Sillitoe and Sawkins (1971).

The Sannae deposit is a 1-km-long vein in an adamellite pluton emplaced in flat-lying volcanics. The vein consists of quartz stringers carrying wolframite, scheelite and molybdenite, accompanied by pyrite, muscovite and K-feldspar. Apart from Sannae, most other vein deposits, such as those at Kuryong (Jordt, 1966), Dongsung (Kim, 1972), Kosong (Koo et al., 1969), Samsanjaeil and Yong-ho, are of "epithermal" type, cut volcanic or sedimentary rocks (Fig. 2), carry chalcopyrite, pyrite and in some cases bornite, and are accompanied by a propylitic gangue assemblage that includes chlorite, epidote, calcite, quartz, specularite and magnetite. Tourmaline is also a gangue phase locally, as at Yong-ho. There veins also possess an exploitable gold content and other veins in the region are copper-poor and were once worked for gold. In the extreme southeast of the Basin, two small contact-metasomatic magnetite deposits in andesitic volcanics are known (Fig. 3).

Apart from these metal deposits, a number of zones of advanced argillic alteration are found in the volcanic rocks in the southern part of the Basin and further west as far as long. 126°E. The alteration consists of a pervasive development of several of pyrophyllite, alunite, diaspore, chalcedonic silica, dickite, kaolinite, tourmaline, dumortierite, pyrite and marcasite; the first three minerals are worked at several localities. At Kuryong, the writer discovered chalcopyrite and minor molybdenite in a zone of advanced argillic alteration on the margin of which the copper-bearing vein deposits are located, and in one of the zones in the Tongnae district molybdenite was found (Fig. 2). The presence of chalcopyrite and molybdenite in these zones of advanced argillic alteration suggests that they are in some way related to the upper parts of porphyry copper systems, above generator stocks.

In the northern part of the Kyongsang basin between lats. 36° and 37°N, a concentration of lead-zinc-bearing veins occurs (Fig. 3). They are probably late Cretaceous, and certainly not earlier, in age since they cut the Kyongsang System. Intrusive rocks and related mineral deposits north of this group of lead-zinc veins are apparently pre-Cretaceous in age, except for those in the Hwanggangni district (approximately long. 128°E and lat. 37°N) (Fig. 3) that have yielded late Cretaceous ages. Mineral deposits are principally tungsten-, molybdenum- and fluorite-bearing veins and fluorite-bearing replacement deposits in limestone, but lead, zinc and copper ores also occur (Reedman et al., 1974).

At Kyongju (Jordt, 1968) and Ulsan (C. J. N. Fletcher, pers. comm., 1975), near to the east

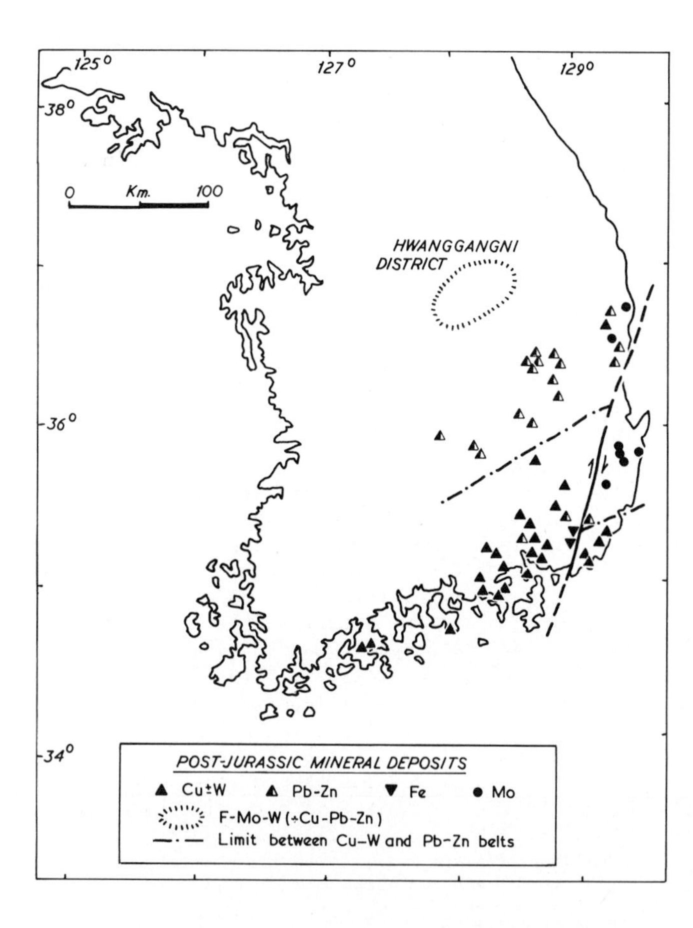

Fig. 3. Locations of mineral deposits in the Kyongsang basin and adjoining areas (after Geological and Mineral Institute of Korea, 1974) indicating the copper-tungsten and lead-zinc belts, the two groups of molybdenum occurrences believed to have been displaced by the Yangsan fault, and the fluorite-molybdenum-tungsten deposits of the Hwanggangni district.

coast south of lat. 36°N (Figs. 2 and 3), granitic to adamellitic intrusives carry very low-grade molybdenum mineralization. The Kyongju intrusive is largely a biotite granite, 9 x 5 km at surface, that is associated with leucogranite and aplite and designated as anomalous for molybdenum throughout by analysis of stream-sediment heavy mineral concentrates (Jordt, 1968). Molybdenite occurs as veinlets with or without pyrite, quartz and sericite, and locally in quartz veins, at scattered localities in leucogranite. At Ulsan, a stockwork of molybdenite-quartz veinlets is present in the southern part of a small adamellite stock. Several small molybdenum deposits are also known on the northern edge of the Basin within the lead-zinc belt (Fig. 3). They are believed to have adjoined the Kyongju-Ulsan district prior to transcurrent faulting in southeast Korea, as amplified below.

Therefore an Upper Cretaceous metal zoning is apparent in the Kyongsang basin and adjoining regions (Fig. 3), as previously observed by Kim (1971b, 1973, 1974). Deposits in the southern part of the Basin are dominated by copper with minor tungsten and, in the southernmost part, small concentrations of iron are also present. Molybdenum occurs in the extreme southeast but is believed to be present as a result of later transcurrent faulting (see below). Further north, lead and zinc occur instead of copper and at the most northerly extent of Upper Cretaceous magmatism fluorite, tungsten and molybdenum predominate.

This metallogenic pattern and the ore types involved are closely comparable to those described from the central Andes (Sillitoe, 1975). There a copper belt lies oceanwards of a belt containing copper-lead-zinc-silver and is typified by porphyry copper deposits, breccia pipes (carrying copper and tungsten) and veins. A narrow belt of contact-metasomatic magnetite deposits occurs in the western part of the copper belt, in a position comparable to that occupied by the South Korean iron deposits. The central Andean copper-lead-zinc-silver belt is bordered eastwards in Bolivia by a tin-tungsten-silver belt. The northern portion of this belt in the Cordillera Real is typified by early Mesozoic tungsten deposits with peripheral lead and zinc mineralization, in some ways comparable to the Hwanggangni district in South Korea. In Bolivia, however, fluorite is scarce, although fluorite deposits typify the landward extremity of the Mexican metallogenic province. A marked difference is provided, however, by the zones of advanced argillic alteration that are uncommon in the central Andes.

The similarities between the Upper Cretaceous metallogeny of South Korea and the Meso-Cenozoic pattern described from the central Andes prompts the suggestion that the Korean province, like the central Andes, was generated on continental basement above a shallow-dipping (20-25°) subduction zone. In Korea, subduction of the Kula plate (Larson and Chase, 1972) was clearly northwards from a trench bordering the Asian mainland in a fashion similar to the present-day central Andes. The shallow dip of the underthrust slab is corroborated by the wide expanse of the Upper Cretaceous metallogenic province in South Korea, at least 250 km inland from the present coast. This distance could be somewhat greater if the submerged continental crust south of the Peninsula and its continuation into northern Kyushu is taken into consideration, although much of this area was probably occupied by the arc-trench gap.

Metallogenic Comparison with Southwest Japan

The geological similarity between southwest Japan and southeast Korea has been realized for some considerable time (Kato, 1927). In southwest Japan, on the Japan-Sea or Inner side of Honshu, a sequence of non-marine, Lower Cretaceous sedimentary rocks is overlain by Upper Cretaceous andesitic and then more felsic calc-alkaline volcanics, including voluminous ash-flow tuffs (Ichikawa et al., 1968). Granodioritic to adamellitic intrusives, probably comagmatic with the volcanics, also occur in an east-northeasterly belt. Intrusives have been dated at 96 to 64 m.y. (Shibata and Ishihara, 1974), closely comparable to the ages (88 to 68 m.y.) obtained for the southeast Korean intrusives.

According to recent work by Ishihara (1973), Shibata and Ishihara (1974) and Shimazaki (1975), in southwest Japan a copper belt is bordered northwards by a lead-zinc belt that extends at least as far as the Japan-Sea coast. Tungsten deposits are important in the copper belt and it was in fact designated as a tungsten belt by Ishihara (1973). These ore deposits are all related to the Upper Cretaceous (96 to 64 m.y.) intrusives. Copper and lead-zinc deposits are of contact-metasomatic (skarn) and vein types. The absence of Paleozoic limestones as inliers in the Kyongsang basin precludes the presence there of outcropping contact-metasomatic deposits. The tungsten deposits are wolframite- and/or scheelite-bearing quartz veins (cf., Sannae, Kyongsang basin) and contact-metasomatic types. Therefore, the copper-tungsten and lead-zinc belts of southwest Japan appear to be the offset continuations of those defined above for southeast Korea (Fig. 4), notwithstanding the absence of known breccia pipe and porphyry deposits from the former region. Moreover, to further enhance the similarity, the volcanics in the copper-tungsten province of southwest Japan also carry zones of advanced argillic alteration (S. Ishihara, written comm., 1975). If ore deposits comparable to those in the Hwanggangni district once occurred north of the lead-zinc belt in southwest Japan, then they must now be present in continental fragments in the Japan Sea.

In southwest Japan, Ishihara (1973) and Shibata and Ishihara (1974) defined a molybdenum belt characterized by molybdenite-bearing quartz veins in leucogranite or aplite associated with adamellitic intrusives, north of the copper-tungsten belt and more or less coincident with the lead-zinc belt (Fig. 4). The molybdenum mineralization is, however, distinctly younger (65 to 46 m.y.) and related to more sodic and magnetite-rich intrusives than the earlier mineralization (Shibata and Ishihara, 1974). A well-defined molybdenum belt is absent from South Korea, a fact that led Shibata and Ishihara (1974) to conclude that further information was required before a correlation of the geology of South Korea and southwest Japan could be confirmed. Here it is suggested, however, that a correlation may in fact be made with the low-grade molybdenum mineralization in the lead-zinc

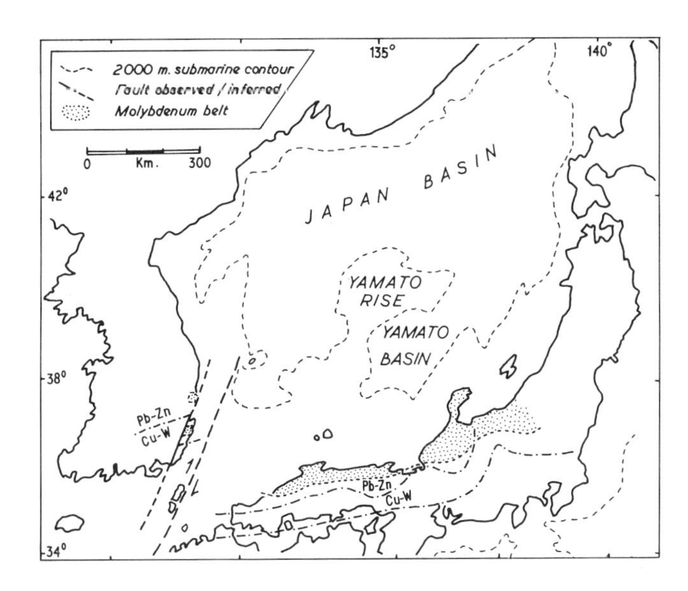

Fig. 4. Positions of copper-tungsten, lead-zinc and molybdenum belts in southeast Korea and southwest Japan, and inferred faults, including the Yangsan fault, between the two regions. Japanese data from Ishihara (1973), Shibata and Ishihara (1974) and Shimazaki (1975).

belt and in the Kyongju-Ulsan district, now further south in the copper-tungsten belt. The Kyongju mineralization bears many similarities to the molybdenum deposits in southwest Japan but chemical and radiometric evidence from the host intrusives is required before their correlation can be confirmed.

Reconstructions of South Korea and Japan

It is now widely accepted that the Korean peninsula and southwest Japan were once united and that their separation was due to the southward rafting of the Japanese arc resulting in the opening of the Japan Sea. Some marginal ocean basins, including the Japan Sea, are now generally believed to have been generated by back-arc spreading, a process of symmetrical sea-floor spreading comparable to that at ocean rises (Karig, 1971). In the Japan Sea, the generation of oceanic crust is thought to have taken place in the Japan and Yamato basins, formation of the latter splitting off the Yamato rise from Honshu as a remnant arc (Fig. 5) (Hilde and Wageman, 1973). If the now-separated copper-tungsten and lead-zinc belts in South Korea and southwest Japan are reconstructed it can be seen that back-arc spreading at these two sites in the Japan Sea has resulted in a total extension of about 250 km, between the extreme southeast of the Korean peninsula and the extreme northwest of Honshu (Fig. 4). There was clearly little overlap between the two land masses prior to separation. Separation is considered to have been on a series

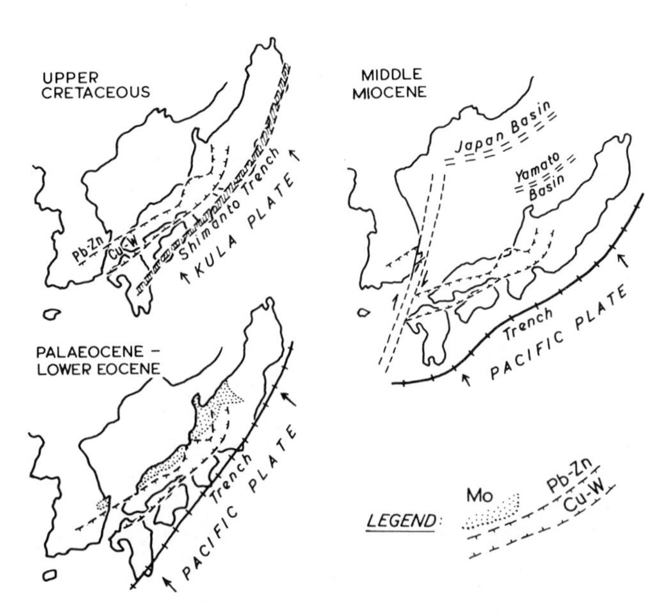

Fig. 5. Idealized reconstructions depicting the development and subsequent displacement of Upper Cretaceous-Lower Tertiary metallogenic belts in southeast Korea and southwest Japan. The Upper Cretacious time-frame shows development of the Andean-type copper-tungsten and lead-zinc belts during subduction of the Kula plate at a trench that may be now represented by the Shimanto zone. The Paleocene-Lower Eocene time-frame shows the super-position of the molybdenum belt on the earlier lead-zinc belt, perhaps induced by subduction of the Kula Ridge between the Kula and Pacific plates. The Middle Miocene time-frame assumes the region to be in essentially its present form with the metallogenic belts displaced by a dextral transcurrent (transform) fault system consequent upon spreading in the Japan and Yamato basins. The extreme southeastern sliver of South Korea, southeast of the Yangsan fault, and southwest Japan have undergone southward displacement by some 100 and 250 km, respectively. (The molybdenum belt is omitted from the Middle Miocene time-frame for the sake of simplicity.)

of north-northeast-trending faults, currently submerged beneath the Japan Sea and the Tsushima straits between Korea and Japan. One of the faults is schematized on Figs. 4 and 5, parallel to the wrench faults, including the major Yangsan fault (Figs. 2, 3, 4 and 5), in southeast Korea. The latter, considered here as part of the main fault system, underwent dextral strike-slip and eastward downthrow in the early Miocene (Reedman and Um, 1975). If the two groups of molybdenum deposits in southeast Korea were once a single cluster as suggested by their disposition with relation to the Yangsan fault (Figs. 4 and 5), then lateral dextral offset has totalled about 100 km.

This reconstruction of South Korea and southwest Japan is similar to that proposed by Hurley et al. (1973) on the basis of a comparison of Precambrian basement terrains, that based on paleomagnetic studies by Yaskawa and Nakajima (1972), and those based on general geology by Kimura (1974) and Reedman and Um (1975).

The clear separation of the copper-tungsten and lead-zinc belts in South Korea and Japan confirms that back-arc spreading began after their formation was completed 64 m.y. ago. Furthermore, if the correlation of the southeast Korean and southwest Japanese molybdenum belts is proved to be correct, then separation did not commence prior to 46 m.y. ago or the mid-Eocene on Berggren's (1969) time-scale. If back-arc spreading had begun before 46 m.y. ago, it would have been difficult to superimpose molybdenum mineralization on the older lead-zinc belt in both South Korea and Japan, as can be appreciated from Figure 5. In the mid- to late Cenozoic, the Korean peninsula was too far north to be underlain by a northward-dipping subduction zone (Fig. 5), and calc-alkaline magmatism was restricted to the Ryukyu and southern Japanese arcs. Alkaline basaltic volcanics interbedded with early Miocene sediments in the Pohang basin in coastal southeast Korea (Fig. 2) and similar volcanics as young as Quaternary in age on Jaeju island, south of the Peninsula (Fig. 1) (Reedman and Um, 1975), are believed to have been erupted through continental crust during the rifting, in regions where oceanic tholeiite was not generated.

The mid-Eocene age proposed here as the maximum age for the commencement of Japan-Sea opening accords reasonably well with the conclusions of a number of recent workers: A post-Cretaceous opening was affirmed by McElhinny (1973) on the basis of paleomagnetic data, and Murauchi (1972) and Kimura (1974) favored a beginning of opening in the Paleogene, the latter writer summarizing evidence for the Tsushima straits being open by late Paleogene times. Kaseno (1972) concluded that opening began in the early Miocene, whereas Ludwig et al. (1975) concluded that the Sea was in essentially its present form by late Oligocene times. Hilde and Wageman (1973) proposed a two-stage opening on the basis of magnetic signatures, the Japan basin forming from the late Mesozoic and the Yamato basin from the early Miocene. The conclusions reached here preclude a late Mesozoic initiation of back-arc spreading, as also favored by Uyeda and Miyashiro (1974), unless it was accomplished without significant relative displacement between South Korea and southwest Japan. It is considered more probable that at this time only basins floored by continental crust, like the Kyongsang basin and its continuation in southwest Japan, existed. Unfortunately the Deep Sea Drilling Program holes drilled on leg 31 were not deep enough to demonstrate that no pre-mid-Eocene marine sediments are present on the floor of the Japan Sea (Scientific Staff, 1973). The alkali basalts in southeast Korea may, however, correlate with the early Miocene spreading in the Yamato basin proposed by Hilde and Wageman (1973), opening of the Japan basin having taken place somewhat earlier from the mid-Eocene onwards.

Concluding Remarks

In South Korea and southwest Japan, a normal Andean-type metal zonation from copper-tungsten northwards to lead-zinc was generated during the Upper Cretaceous, with a possible fluorite-tungsten-molybdenum zone even further north. This pattern is attributed to a shallow, northward-dipping subduction zone, the outcrop of which is perhaps now represented by the Shimanto zone, a possible trench assemblage on the Outer, Pacific side of Japan (Sugimura and Uyeda, 1973) (Fig. 5). However, 64 m.y. ago this metallogenic pattern became inactive and was replaced in southwest Japan, and probably also in South Korea, by molybdenum mineralization.

From a compilation of data on the spreading history of the northwest Pacific, Uyeda and Miyashiro (1974) concluded that in late Cretaceous times an ocean rise, the Kula Ridge, intersected the trench system along the Asian mainland and was subducted. Using western North America as an example, the change from a normal, Andean pattern of mineralization, above a shallow-dipping subduction zone, to the emplacement of molybdenum deposits is attributed to the intersection of an ocean rise and a trench, whether or not the rise is subducted (Sillitoe, 1976). Off southwest Japan, Uyeda and Miyashiro (1974) tentatively dated the intersection at 80 m.y. but since the ocean-floor magnetic anomalies for the late Cretaceous have been subducted and are therefore no longer observable, a precise estimate of the timing of the event cannot be given. Might not therefore the intersection of the Kula Ridge and the trench off southwest Japan have taken place 65 to 70 m.y. ago (Fig. 5) just before, and giving rise to, the molybdenum mineralization?

Acknowledgments. This paper stems from an assignment with the Geological and Mineral Institute of Korea (GMIK) that constituted part of the first phase of a United Nations mineral exploration program. Mr. F. A. Seward, Jr. of United Nations, New York, is thanked for the invitation to participate in the program and for providing an introduction to field areas in the Kyongsang basin. Mssrs. Kim, Jong Hwan, Kim, Sun Ok, and Kim, Kil Sun, of GMIK and Dr. C. J. N. Fletcher of the Institute of Geological Sciences, U.K., are gratefully acknowledged for their help in the field and, along with Drs. K. Burke, B. F. Scales and M. J. Terman, for discussions. Written communications with Dr. S. Ishihara of the Geological Survey of Japan greatly aided the comparison with southwest Japan.

References

Berggren, W. A., Cenozoic chronostratigraphy, planktonic foraminiferal zonation and the radiometric time scale, Nature, 224, 1072-1075, 1969.

Chang, K. H., Cretaceous stratigraphy of southeast Korea, J. Geol. Soc. Korea, 11, 1-23, 1975.

Fletcher, C. J. N. and C. K. Park, The geology and origin of the Ilkwang copper- and tungsten-bearing tourmaline breccia pipe, Republic of Korea, Absts. Geol. Soc. Korea Ann. Mtg., Seoul, 9-10, 1974.

Geological and Mineral Institute of Korea, Metallogenic Map of Korea, 1:2,500,000, 1974.

Hilde, T. W. C., and J. M. Wageman, Structure and origin of the Japan Sea, The Western Pacific: Island Arcs, Marginal Seas, Geochemistry, Coleman, P. J., ed., 415-434, Univ. Western Australia Press, 1973.

Hurley, P. M., H. W. Fairbairn, W. H. Pinson, Jr., and J. H. Lee, Middle Precambrian and older apparent age values in basement gneisses of South Korea, and relations with southwest Japan, Geol. Soc. America Bull., 84, 2299-2304, 1973.

Ichikawa, K., N. Murakami, A. Hase, and K. Wadatsumi, Late Mesozoic igneous activity in the Inner side of southwest Japan, Pacific Geol., 1, 97-118, 1968.

Ishihara, S., Molybdenum and tungsten provinces in the Japanese islands and North American Cordillera: An example of asymmetrical metal zoning in Pacific type orogeny, Metallogenic Provinces and Mineral Deposits in the Southwestern Pacific, Fisher, N. H., ed., 173-189, Bur. Min. Res. Geol. Geophys. Bull. 141, Australian Govt. Publ. Service, Canberra, 1973.

Jordt, D. K., Summary report, Kyongsang andesite belt exploration, 1964-1965, Mineral Industries Engineers, Inc., Seoul, unpub. rept., p.40, 1966.

Jordt, D. K., Report of geological/geochemical investigation of the Choyang-Toktong molybdenum environment: Mineral Industries Engineers, Inc., Seoul, unpub. rept., p.13, 1968.

Karig, D. E., Origin and development of marginal basins in the western Pacific, J. Geophys. Res., 76, 2542-2561, 1971.

Kaseno, Y., On the origin of the Japan Sea basin, 24th Int. Geol. Cong., Montreal, sect. 8, 37-42, 1972.

Kato, T., The Ikuno-Akenobe metallogenic province, Japanese J. Geol. Geog., 5, 121-133, 1927.

Kim, J. T., On the geology, ore deposit and drilling summary of Dongsung copper mine (in Korean), J. Korean Inst. Min. Geol., 5, 133-144, 1972.

Kim, O. J., Study on the intrusion epochs of younger granites and their bearing to orogenies in South Korea (in Korean), J. Korean Inst. Min. Geol., 4, 1-9, 1971a.

Kim, O. J., Metallogenic epochs and provinces of South Korea, J. Geol. Soc. Korea, 7, 37-59, 1971b.

Kim, O. J., Metallogenic provinces and epochs in South Korea, Metallogenic Provinces and Mineral Deposits in the Southwestern Pacific, Fisher, N. H., ed., 211-212, Bur. Min. Res. Geol. Geophys. Bull. 141, Australian Govt. Publ. Service, Canberra, 1973.

Kim, O. J., Geology and tectonics of South Korea, United Nations ESCAP, CCOP Tech. Bull., 8, 17-37, 1974.

Kim, O. J., and K. H. Kim, A study on Red Hill copper deposits of the Dongjom mine (in Korean), J. Korean Inst. Min. Geol., 7, 157-174, 1974.

Kimura, T., The ancient continental margin of Japan, The Geology of Continental Margins, Burk, C. A. and C. L. Drake, eds., 817-829, Springer-Verlag, New York, 1974.

Koo, M. O., K. D. Kim, and J. O. Choi, Preliminary survey of Kosoung copper area, Kyongsangnamdo (in Korean), Geol. Surv. Korea Bull., 11, 47-56, 1969.

Larson, R. L., and C. G. Chase, Late Mesozoic evolution of the western Pacific ocean, Geol. Soc. America Bull., 83, 3627-3644, 1972.

Ludwig, W. J., S. Murauchi, and R. G. Houtz, Sediments and structure of the Japan Sea, Geol. Soc. America Bull., 86, 651-664, 1975.

McElhinny, H. W., Palaeomagnetism and plate tectonics of eastern Asia, The Western Pacific: Island Arcs, Marginal Seas, Geochemistry, Coleman, P. J., ed., 407-414, Univ. Western Australia Press, 1973.

Murauchi, S., Crustal structure of the Japan Sea derived by explosion seismology (in Japanese), Kagaku, 42, 367-407, 1972.

Reedman, A. J., C. J. N. Fletcher, R. B. Evans, D. R. Workman, K. S. Yoon, H. S. Rhyu, S. H. Jeong, and J. N. Park, The geology of the Hwanggangni mining district, Republic of Korea, Anglo-Korean Mineral Exploration Group, Geol. Min. Inst. Korea, Seoul, p. 118, 1974.

Reedman, A. J., and S. H. Um, The geology of Korea, Geol. Min. Inst. Korea, Seoul, p. 139, 1975.

Scientific Staff, Leg 31. Western Pacific floor, Geotimes, 18 (10), 22-25, 1973.

Seo, H. G., and S. H. Ju, Intrusive age of granitic plutons in Korean peninsula (in Korean), Geol. and Ore Deposits, 14, 31-43, 1971.

Shibata, K., and S. Ishihara, K-Ar ages of the major tungsten and molybdenum deposits in Japan, Econ. Geol., 69, 1207-1214, 1974.

Shimazaki, H., The ratios of Cu/Zn-Pb of pyrometasomatic deposits in Japan and their genetical implications, Econ. Geol., 70, 717-724, 1975.

Sillitoe, R. H., Andean mineralization: A model for the metallogeny of convergent plate margins, Metallogeny and Plate Tectonics, Strong, D. F., ed., Geol. Assoc. Canada Spec. Paper 14, in press, 1975.

Sillitoe, R. H., Evidence for molybdenum mineralization when ocean rises intersect trenches, in prep., 1976.

Sillitoe, R. H., and F. J. Sawkins, Geologic, mineralogic, and fluid inclusion studies relating to the origin of copper-bearing tourmaline breccia pipes, Chile, Econ. Geol., 66, 1028-1041, 1971.

Sugimura, A., and S. Uyeda, Island Arcs. Japan and its Environs, Developments in Geotectonics 3, Elsevier Scientific Publ. Co., Amsterdam, p.247, 1973.

Ustiyev, Ye. K., Problems of volcanism and plutonism. Volcano-plutonic formations, Int. Geol. Rev., 7, 1994-2016, 1965.

Uyeda, S., and A. Miyashiro, Plate tectonics and the Japanese islands: A synthesis, Geol. Soc. America Bull., 85, 1159-1170, 1974.

Wageman, J. M., T. W. C. Hilde, and K. O. Emery, Structural framework of East China Sea and Yellow Sea, American Assoc. Petrol. Geol. Bull., 54, 1611-1643, 1970.

Won, J. G., and K. T. Kim, On the geology and mineralization in Dalsung mine area (in Korean), J. Geol. Soc. Korea, 2, 52-68, 1966.

Yaskawa, K., and M. Nakajima, Southwest Japan migrated southward (in Japanese), Kagaku, 42, 163-165, 1972.

PETROGENESIS IN ISLAND ARC SYSTEMS

A. E. Ringwood

Research School of Earth Sciences, Australian National University, Canberra 2600

Abstract. Subduction of oceanic lithosphere beneath island arc systems is accompanied by the introduction of water into the mantle. The water is carried principally by amphibolite (derived from the mafic oceanic crust) and by bodies of serpentinite incorporated in the crust. Amphibolite transforms to eclogite plus water under subsolidus conditions at depths smaller than 100 km. The water rises into mantle pyrolite in the wedge above the Benioff zone, causing partial melting and magma formation. Experimental petrological data bearing on the nature of the partial melting process are discussed. The primary magmas produced near the Benioff zone at depths of 80-100 km consist of hydrous, tholeiitic basalts, close to silica saturation. They are not andesitic. The tholeiitic magmas fractionate as they rise, principally by olivine separation, thereby producing a spectrum of basaltic andesite to andesite magmas at shallow depths. Further shallow fractionation by separation of amphibole, pyroxene and plagioclase produces dacites and rhyolites. The differentiation trend of the volcanic suite is tholeiitic, possessing defined petrochemical characteristics, and is responsible for the tholeiitic stage of development of island arcs. Water is carried to depths greater than 100 km principally in serpentinite bodies which transform ultimately to hydroxyl chondrodite, hydroxyl clinohumite and related hydrous phases. The former serpentinite bodies maintain a high water vapour pressure in the oceanic crust (now transformed to quartz eclogite). As temperatures rise to 800-900°C, partial melting of quartz eclogite occurs between depths of 100-200 km, resulting in the production of acidic magmas possessing strongly fractionated rare-earth patterns and relatively high K/Na ratios. The acidic magmas react with overlying pyrolite, transforming some olivine to pyroxene and imprinting their trace-element characteristics. Diapirs of "modified pyrolite" rise from the Benioff zone at depths mostly of 100-150 km, and partially melt under hydrous conditions. Magmas so produced rise and fractionate to produce andesites, dacites and rhyolites possessing the calcalkaline petrochemical trends which are characteristic of mature island arc systems. The residual, refractory eclogite and peridotite in the lithosphere plates which sink below 200 km have become irreversibly differentiated. The complementary differentiate, ultimately, is the continental crust which grows through time by accretion of island arcs and by the addition of the andesitic volcanic suite. About 30-60 percent of the mantle may have passed through this process of irreversible differentiation.

Introduction

The basalt-andesite-dacite-rhyolite suite together with their plutonic equivalents and recycled sedimentary and metamorphic derivatives are the dominant constituents of island arcs. Active andesitic volcanism in island arcs is characteristically associated with oceanic trenches and regions of lithosphere subduction. Because of this association with tectonically active regions of the earth's lithosphere, it is convenient to term the basalt-andesite-dacite-rhyolite suite of island arcs, the orogenic volcanic series. Active volcanoes are usually situated in positions some 80 to 150 km vertically above Benioff zones. These relationships have suggested to many that the origin of the orogenic volcanic series is connected with processes occurring near Benioff zones.

Petrochemical Characteristics

Significant differences between fractionation behaviour of magmas in the orogenic volcanic series and of gabbroic-basaltic magmas that crystallize in the crustal environment are brought out by the FMA diagram (Fig. 1). The latter (e.g. Skaergaard) display a tholeiitic trend, characterised by iron-enrichment in the early stages. This is caused by the separation of olivines and pyroxenes possessing much lower Fe/Mg ratios than the equilibrium liquids (Table 1). On the other hand, the orogenic series often shows a different trend, occupying a band extending from the mafic side to the alkali apex and not displaying a substantial degree of iron-enrichment (Fig. 1). This is called the calcalkaline trend and is caused by the combination of two factors: (1) In the case of calcalkaline petrogenesis, partial melting in the source regions produces magmas which are initially richer in silica (at the same Fe/Mg ratio) than the parental tholeiitic magmas produced by

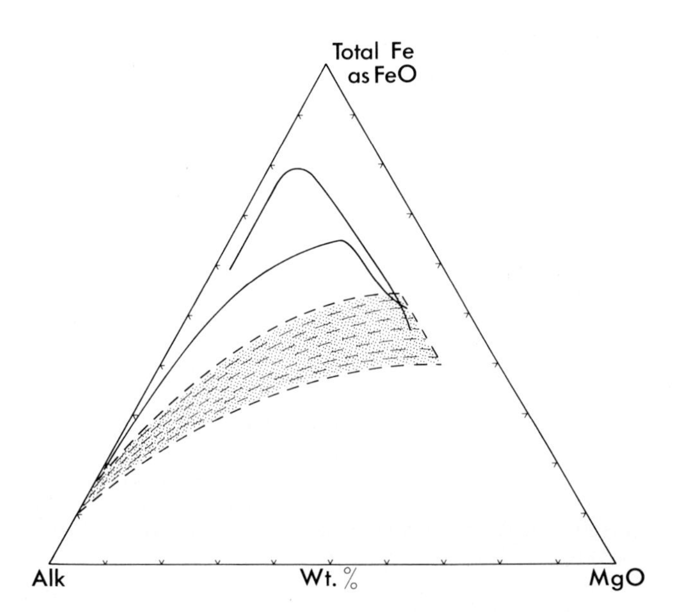

Figure 1. FMA (Total iron as FeO; MgO; Na_2O + K_2O) diagram showing characteristic tholeiitic and calcalkaline trends. The tholeiitic region (solid lines) is confined by differentiation trends observed for the Skaergaard Intrusion [Wager and Deer, 1939, upper curve] and Thingmuli volcano [Carmichael, 1964, lower curve]. The calcalkaline band (stippled) embraces the differentiation trends displayed by magmas from the Cascade, Aleutian and New Zealand calcalkaline provinces.

partial melting of mantle pyrolite. (2) Fractionation of calcalkaline magmas is influenced by phases possessing higher Fe/Mg ratios, on the average, than the olivines and pyroxenes which control the fractionation of tholeiites (Table 1).

Actually, island arc systems contain suites of rocks exhibiting both tholeiitic and calcalkaline differentiation trends and continuous gradations between the extreme trends shown on Fig. 1 may be found. Studies by Jakes and White[1969], Jakes and Gill [1970] and Gill [1970] have disclosed additional significant petrochemical trends. Tholeiitic magmas in island arcs [e.g. Tonga; Ewart et al., 1973] generally possess low abundances of incompatible elements and comparatively unfractionated rare earth patterns (relative to chondrites) which persist throughout the entire compositional range, basalt-dacite (Fig. 2). Jakes and Gill [1970] have called these the "island arc tholeiite series". On the other hand, the calcalkaline magmas generally have higher abundances of incompatible elements (e.g. K, U, Ba, REE) for a given SiO_2 content and display fractionated rare earth patterns (light REE enriched, Fig. 2). These may be called the "calcalkaline series".

Dickinson and Hatherton [1967] and Dickinson [1968] showed that when the characteristic potassium contents (for given SiO_2 contents) of orogenic-type magmas were plotted against depth to the Benioff zone, a significant trend emerged for potassium to increase with depth. The authors interpret this relationship to imply a connection between petrogenesis and physico-chemical processes occurring near the Benioff zone. Many other incompatible trace elements are known to be strongly correlated with potassium [Taylor, 1969; Gill, 1970]. The composition-depth relationship thus suggests that the K-rich calcalkaline series may be derived from greater depths than the K-poor island arc tholeiite series. It should be noted, however, that these trends are not universal and reversals have been noted, as among the volcanoes of New Guinea-New Britain [Johnson,1976].

Baker [1968] pointed to an apparent evolutionary sequence of island arc volcanism. Arcs which were believed to be relatively young, such as the South Sandwich, Mariana, Tonga and Izu islands, whilst they display a complete spectrum of compositions between basalt-andesite-rhyolite, are nevertheless dominantly composed of basalt and basaltic andesite. On the other hand, more evolved arcs (e.g. Japan, Indonesia, Kamchatka, Lesser Antilles, Aleutians) whilst also displaying a complete spectrum of compositions, contain much more abundant andesites. Subsequent studies [e.g. Nicholls and Whitford, 1976] have demonstrated that many exceptions to this evolutionary scheme exist, and that the petrochemical evolution of island arcs is in fact much more complex. Nevertheless, if viewed as expressing a broad and general trend rather than a rigorous sequential development, Baker's interpretation appears worthy of attention.

Role of Water in Development of Orogenic Volcanic Series

Melting relationships at one atmosphere are shown for a series of calcalkaline rocks in Table

Table 1. Distribution of Fe and Mg between phases crystallizing on or near liquidus of a basaltic liquid and the parental liquid.

$K = (Fe/Mg)_{crystal}/(Fe/Mg)_{liquid}$

Phase	K	Ref
Olivine	0.33	1
Clinopyroxene	0.36	1
Amphibole	1.1	1
Garnet	1.1	2

Refs. 1. Holloway and Burnham [1972], Tholeiite P = 8 kb, P_{H_2O} = 5 kb, T = 1000°C, QFM buffer .

2. T. Green and Ringwood [1968], Quartz Tholeiite, 27 kb, 1385°C.

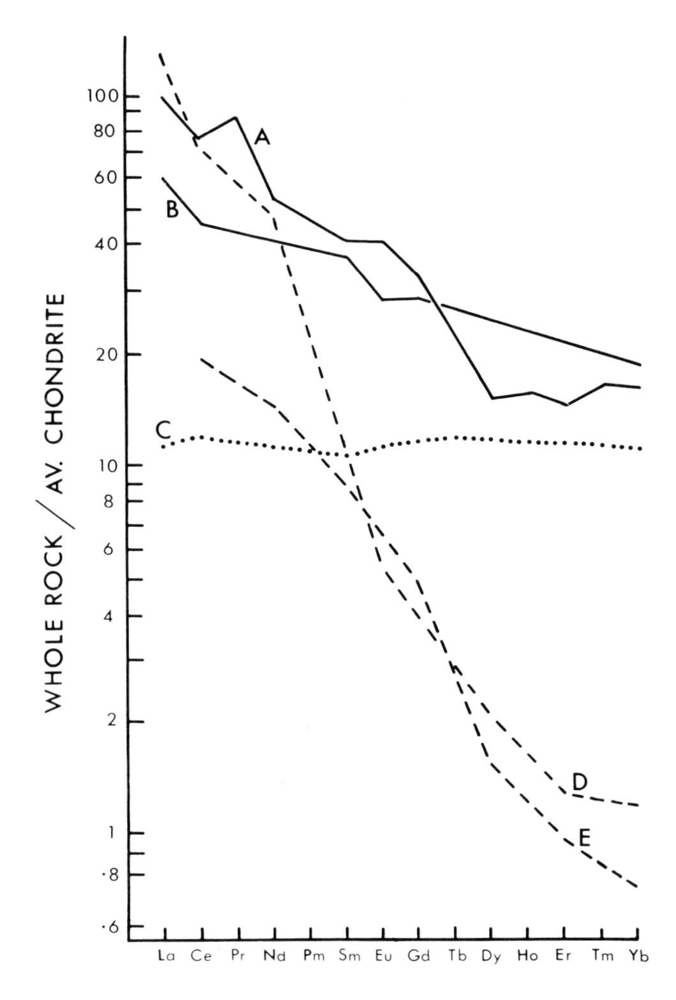

Figure 2. Chondrite - normalized rare earth patterns for two calcalkaline andesites A and B, a tholeiitic andesite C, a dacite porphyry D and the Northern Light Gneiss E.
A: Andesite 3514 from Mt. Victory, Papua [Jakes and Gill, 1970].
B: Andesite K15, Hokkaido, from Katsui et al. [1974].
C: Andesite K12, Hokkaido, from Katsui et al. [1974].
D: Dacite porphyry, Minnesota-Ontario, from Arth and Hanson [1972].
E: Northern Light Gneiss (as for D).

2. The outstanding feature is the high temperature of crystallization of plagioclase and the broad temperature interval through which plagioclase crystallizes alone. This is particularly marked in the dacite, where plagioclase appears at 1275°C and crystallizes over an interval of 100°C before being joined by pyroxene.

These relationships render it highly improbable that the orogenic volcanic series have developed from parental basaltic magma by crystallization differentiation at one atmosphere. Rocks with high normative plagioclase such as andesite and dacite effectively constitute a thermal barrier between basalt and rhyodacite. Yet analyses of aphanitic members show that a line of liquid descent exists through these compositions [e.g. Kuno, 1950].

Clearly, any hypothesis which seeks to explain the petrogenesis of the orogenic volcanic series by crystal-liquid fractionation requires a mechanism to depress the crystallization field of plagioclase relative to ferromagnesian minerals. This can be accomplished by high water pressure, high load pressure or by some combination of these factors. As we shall see, both these factors were involved.

Considerable evidence exists that water played a significant role during the evolution of the orogenic magma series. This is manifested for example by the explosive nature of andesitic volcanoes, the abundance of pyroclastics, the frequent occurrence of amphiboles and biotite, and many complex mineralogical features such as oscillatory zoning of feldspars and resorption, all of which testify to the presence of water in the original magmas [Ringwood, 1975]. Moreover, compositional data on iron-titanium oxide phases crystallizing from andesitic and dacitic magmas have indicated crystallization temperatures in the vicinity of 900-1050°C [e.g. Carmichael and Nicholls, 1967]. These are much smaller than anhydrous crystallization temperatures (Table 2) and again imply the presence of substantial pressures of water vapour during crystallization. Eggler [1972] also demonstrated the essential role of water during crystallization of a Paricutin andesite.

We shall see subsequently that magmas of the orogenic volcanic series are ultimately formed at considerable depths in the mantle. The water content of the mantle source regions of basaltic magmas is believed to be quite low, on the order of 0.1% [Ringwood, 1975]. Accordingly, it appears that the comparatively large amount of water necessary to explain aspects of the petrology of orogenic volcanic rocks has been introduced into localized regions of the mantle via some specialized process. The most plausible source of this water is to be found in subducted lithosphere plates and more specifically in the former oceanic crust. Lead isotopic considerations [Oversby and Ewart, 1972], imply that, at least in some cases, subducted oceanic sediments are unlikely to have provided most of this water. Moreover, Karig and Sharman [1975] have shown that most of the oceanic sediments are not subducted, but rather, are accreted to form a sedimentary wedge on the upper plate.

According to most current models, the oceanic crust is believed to consist mainly of a heterogeneous mixture of basalt, dolerite, gabbro and their metamorphic derivatives, mainly greenschists and amphibolites. Upon subduction into the mantle, greenschist would transform to amphibolite whilst the excess water liberated would convert most of the remaining anhydrous mafic rocks to amphibolite.

Table 2. Crystallization behaviour of a series of West Indies calcalkaline rocks.

Rock	SiO_2	Highest temperatures of crystallization of major phases
Olivine basalt 16K	47.9	Pl 1280°
Olivine basalt 20L	50.5	Pl 1280°
Olivine basalt 27V	50.5	Pl(1215°), Ol(1185°), Px(1175°)
Hypersthene andesite 19K	59.7	Pl(1240°), Px(1180°)
Hypersthene andesite 21L	60.7	Pl(1255°), Px(1180°)
Biotite dacite	69.9	Pl(1275°), Px(1180°)

Data from Brown and Schairer [1968]

Pl = plagioclase, Ol = olivine, Px = pyroxenes.

This would be one of the major "carriers" of water into the mantle. However amphibolite appears to be incapable of transporting water to depths much greater than 80 km because of its transformation at pressures around 25 kb and below 900°C into eclogite plus H_2O (Fig. 3). Since there is good evidence that water-bearing orogenic magmas may be derived from depths around 100 km and extending sometimes to beyond 150 km, some other "carrier" must be available. Data on seismic attenuation in the mantle wedge overlying Benioff zones also suggest the introduction of water into the mantle along Benioff zones at depths extending to 300 km [Barazangi and Isacks, 1971].

It appears likely that serpentinite constitutes a minor but significant constituent of the oceanic crust, being mainly introduced by rising diapirs along fracture zones [e.g. Cann, 1970]. Even if only 10 percent of serpentinite were present, this would account for a proportion of the total water in the crust comparable to that carried by amphibolite.

The behaviour of serpentinite differs drastically from amphibolite during dehydration (Fig. 3). While the transformation of amphibolite to eclogite + H_2O below 900°C is approximately isobaric at 25-30 kb, serpentinite undergoes a series of dehydration reactions up to 40 kb each of which appears to be approximately isothermal. A new series of dense hydrated magnesium silicates has been discovered [Ringwood and Major, 1967; Sclar et al., 1967; Yamamoto and Akimoto, 1974, 1975]. Some of these are stable to very high pressures and temperatures (up to 100 kb, 1200°C). The principal phases into which serpentinite may be expected to transform are "phase A" - $2Mg_2SiO_4 \cdot 3Mg(OH)_2$, hydroxyl clinohumite $4Mg_2SiO_4 \cdot Mg(OH)_2$ and hydroxyl chondrodite $2Mg_2SiO_4 \cdot Mg(OH)_2$. From the stability relationships of the latter two phases as determined by Yamamoto and Akimoto [1975], the broadest stability field is attained when the MgO/SiO_2 ratio of the bulk system is greater than two. This is substantially higher than for most serpentinites ($MgO/SiO_2 \sim 1.5$). The results of Nakamura and Kushiro [1974] show that successive dehydration of serpentine at high pressures and temperatures will be accompanied by desilication owing to the high solubility of SiO_2 in water vapour. Thus at 15 kb and at about 1300°C, water vapour in equilibrium with enstatite and forsterite contained about 20 weight percent of SiO_2. Nakamura and Kushiro pointed out that the silica thus removed from serpentinite would be transferred to the mantle wedge above the subduction zone, transforming some of the olivine to enstatite.

The stability relationships of the high pressure hydrated silicates derived from serpentinite strongly suggest that these minerals play the decisive role in carrying water into the mantle to depths greater than about 80 km. The presence of (former) serpentinite diapirs will maintain high P_{H_2O} in the 80-300 km interval throughout the oceanic crust. Moreover, water derived from these diapirs via various dehydration reactions will be introduced into the wedge overlying the Benioff zone continually throughout the 80-300 km depth interval. This may be responsible for the high seismic attenuation in this region observed by Barazangi and Isacks [1971].

Petrogenesis - General Considerations

A wide range of theories of petrogenesis of the orogenic volcanic series has been proposed in the earlier literature. These include melting of pre-existing sialic continental rocks, differentiation of basaltic magmas contaminated by crystalline sial or by sediments, hybridism, and also the high level fractional crystallization of basaltic magmas under high oxygen pressures, resulting in extensive precipitation of magnetite.

These processes have been extensively discussed in the literature [e.g. T. Green and Ringwood,

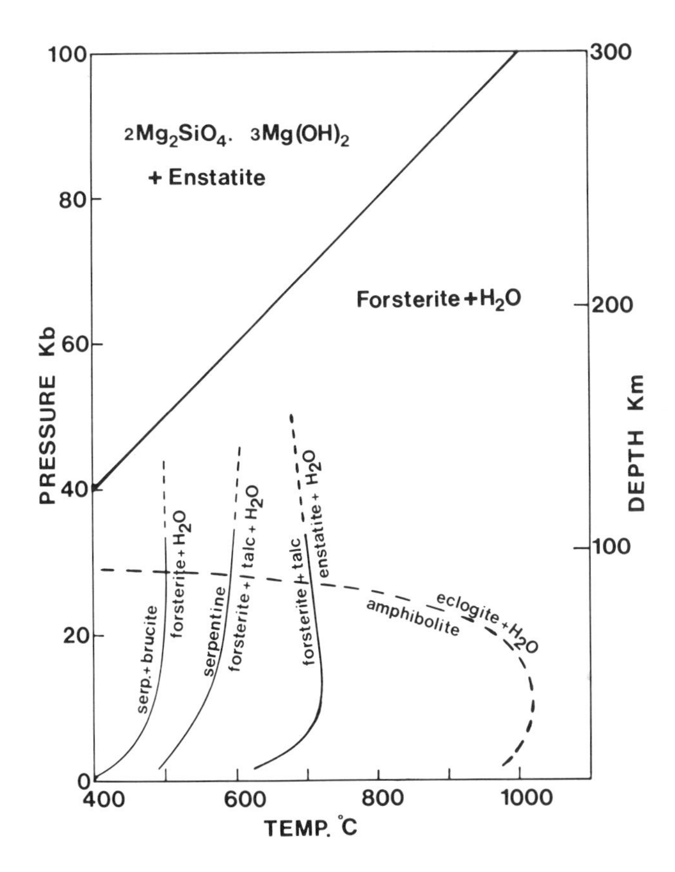

Figure 3. P,T equilibrium curves for dehydration reactions involving brucite, serpentine, talc and $2Mg_2SiO_4 \cdot 3Mg(OH)_2$, together with an estimate of the amphibolite-eclogite transformation curve. All curves for conditions of $P_{H_2O} = P_{total}$.

1968]. Generally speaking, although each of them probably plays a role in particular cases, it appears increasingly clear that they are not responsible for the ultimate petrogenesis of the entire orogenic magma series in the island arc environment. Each mechanism encounters a range of fatal difficulties when applied broadly. Not the least of these is the fact that many island arcs have developed within oceanic regions far from continents, so that pre-existing old continental rocks were not involved [Gorshkov, 1962]. This observation, combined with Sr isotope studies and with the evidence of a genetic relationship to zones of lithosphere subduction, implies that an ultimate source of these magmas must be sought in the mantle. An exception to this statement may be provided by the large extrusions of rhyolites and rhyodacites (often ignimbritic) which were emplaced upon continental crust and above subduction zones [e.g. the Andes, Pichler and Zeil, 1969; and New Zealand, Healy, 1962]. There is increasing evidence that these may have formed by remelting of the lower crust [Ewart and Stipp, 1968].

Early suggestions that orogenic magmas were derived in some way from processes occurring at the Benioff zone were made by Wilson [1954] and Coats [1962]. With the advent of plate tectonic concepts, T. Green and Ringwood [1966, 1967, 1968] undertook a detailed experimental investigation aimed at discovering how orogenic magmas might form when lithosphere was subducted into the mantle. It was demonstrated that andesitic-dacitic magmas could be formed by partial melting of the mafic oceanic crust (either as amphibolite or quartz eclogite) along the Benioff zone when lithosphere was subducted into the mantle. However, the tholeiitic magmas associated with andesites and dacites were believed to have formed not from the subducted oceanic crust, but by partial melting of pyrolite in the wedge overlying the Benioff zone. Convective instability in the wedge leading to partial melting and basalt generation might have been triggered by the uprise of orogenic magmas from the Benioff zone. Alternatively, water liberated by dehydration of subducted oceanic crust might have entered the overlying wedge, causing partial melting and producing hydrous basaltic magmas which fractionated by amphibole separation to form a range of orogenic magmas associated with hydrous high-alumina basalts.

From 1969 onward, a cascade of papers developing these themes more explicitly has appeared, and a detailed review is not practicable. The role of water liberated by dehydration of the oceanic crust has been repeatedly emphasized and a vast amount of new experimental data has been provided by several laboratories. An important new development was initiated by Kushiro and Yoder [1969] who showed that direct partial melting of pyrolite at pressures exceeding a few kilobars could produce silica-oversaturated magmas.

In the following pages, we will review recent data on key processes in orogenic magma petrogenesis and attempt to arrive at a synthesis. Generally speaking it appears that orogenic magmas are derived by partial melting under high water pressures from two principal sources: (1) subducted mafic oceanic crust and (2) the pyrolite wedge overlying the Benioff zone. The final magma compositions may reflect complex petrochemical interactions between these sources. We will commence by discussing petrogenetic processes in the pyrolite wedge.

Partial Melting of the Mantle Under Hydrous Conditions

In a study of the system MgO-SiO_2-H_2O at 20 kb, Kushiro and Yoder [1969] found that whereas $MgSiO_3$ melts congruently under dry conditions, under hydrous conditions ($P_{H_2O} = P_{Load}$), $MgSiO_3$ melted incongruently to forsterite plus a silica-oversaturated liquid (Fig. 4). It is seen that the high water vapour pressure causes the primary field of forsterite to expand through the $MgSiO_3$

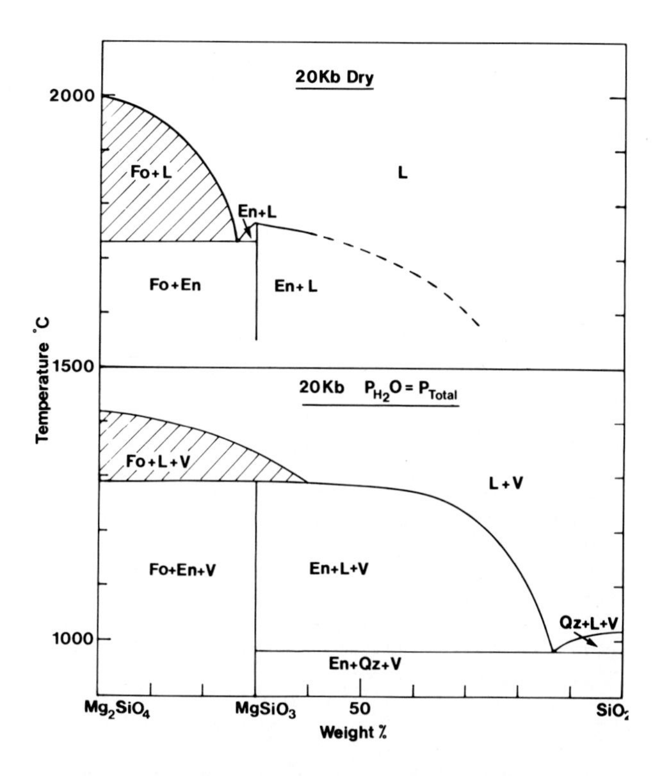

Figure 4. The system $MgO-SiO_2-H_2O$ at 20 kbs (anhydrous) and 20 kb ($P_{Load} = P_{H_2O}$). Based on Kushiro and Yoder [1969].

composition. Fractional crystallization of a composition between Mg_2SiO_4 and $MgSiO_3$ would therefore lead to the generation of a silica oversaturated liquid, providing that forsterite were withdrawn at, or before the reaction point. Likewise, partial melting accompanied by segregation of liquid from crystals could form an oversaturated liquid. Kushiro [1972] observed analogous effects in other systems under high P_{H_2O}. These results led Kushiro and Yoder [1969] to propose the hypothesis that andesitic and dacitic magmas might be produced by partial melting of pyrolite under high water pressures at depths of 70-100 km(20-30 kb). The significance of this hypothesis lay in the implication that primary andesitic and dacitic magmas might be produced directly at the Benioff zones as the result of introduction of water into the mantle.

Kushiro et al. [1972] attempted to test this hypothesis by partially melting a lherzolite nodule at high pressures in the presence of excess water vapour and measuring the composition of the liquid produced by electron-probe microanalysis. At 26 kb and 1190°C, approximately 20 percent of the charge melted to a glass in the presence of residual olivine, orthopyroxene and clinopyroxene. After correction for water content, the anhydrous glass was found to contain 68% SiO_2, 10.2% CaO, 0.6% MgO and 1.1% FeO. The result was held to confirm Kushiro's hypothesis (above).

Extensive investigations on the partial melting of peridotites and pyrolite at high pressures in the presence of excess water vapour have since been reported by several groups. Unfortunately, the results and interpretations have been plagued by the general tendency of partial melts to undergo rapid metastable crystallization during quenching, so that the glass compositions determined by microprobe analysis rarely correspond to those of the original liquids in equilibrium with residual olivine and pyroxene(s). These effects have been described and evaluated in detail by Green [1973, 1976], Cawthorn et al. [1973], Kushiro [1974], Nicholls [1974, 1976], and Nehru and Wyllie [1975] whilst Nicholls and Ringwood [1973] demonstrated that the "dacitic" liquid observed by Kushiro et al. [1972] during partial melting of peridotite had been severely modified by quench crystallization. Because of this quenching behaviour, estimates of equilibrium liquid compositions obtained by hydrous partial melting of peridotitic liquids cannot be accepted unless confirmed by independent experimental methods. The only completely rigorous manner of accomplishing this is to "reverse" the experiment. The liquid composition believed to be formed in a given partial melting experiment at a given load pressure, temperature and water pressure is first synthesized. Its crystallization behaviour is then investigated over a range of temperatures at the same load pressure and water-vapour pressure as were used in the original partial melting experiment. In order to prove that this liquid indeed represented the equilibrium melt, it must be demonstrated that the liquid phase is saturated or near-saturated with the same phases that were present in the residual refractory assemblage (olivine + pyroxene(s)) during the original partial melting experiment at the same temperature. Thus, these phases should appear on the liquidus close to the temperature at which the partial melt was produced. In the case of a reaction relationship, e.g., between olivine and liquid, the liquid is necessarily saturated with olivine at the reaction temperature even though phases other than olivine might crystallize on further cooling. Therefore, if additional olivine is added to the liquid under the same P, T, P_{H_2O} conditions as were present during partial melting, then it will persist as a phase, demonstrating olivine-saturation.

Mysen and Boettcher [1975] have recently described an extensive study of the partial melting of peridotites under hydrous conditions. On the basis of some 320 partial melting runs, they concluded that andesitic liquids are formed by partial melting of peridotite at pressures at least up to 25 kb. There are some remarkable features about this work which invite discussion. The authors state that their measurements of glass compositions represented equilibrium liquids not significantly modified by quench crystallization. This conclusion stands in sharp contrast to the behaviour of partial melts in hydrous peridotitic systems observed by five other research groups all of which recorded serious problems with quench crystallization. Moreover Mysen and Boettcher made only a single attempt to reverse one of their 320 partial melting experiments according to the principles discussed above. The attempt was unsuccessful since the liquid composition ostensibly produced by partial melting of peridotite was found to be undersaturated with olivine under identical P,T,P_{H_2O} conditions. Finally, only two glass analyses were carried out on runs above 15 kb (at 17 kb and 22kb) in which the volatile component consisted of pure water. The extremely low FeO (0.8%) in both runs and MgO (0.7%) in the 22 kb run showed that the liquids had been seriously modified during quenching. Moreover the observed abundances of these components do not remotely resemble those found in natural andesites. Mysen and Boettcher's claim to have produced andesitic liquids in equilibrium with residual olivine and pyroxene(s) at pressures at least as high as 25 kb is totally indefensible.

To circumvent the difficulties caused by quench modifications of peridotitic partial melts, Nicholls and Ringwood [1972, 1973] followed the more tedious procedure of studying the near-liquidus crystallization behaviour of a series of liquids covering a range of SiO_2 contents with the objective of determining the effects of P_{H_2O} on the primary field of crystallization of olivine in these different compositions. Results on an olivine tholeiite (46% SiO_2, 22% normative olivine) and an exactly SiO_2-saturated tholeiite (51.5% SiO_2) are shown in Fig. 5. Whereas olivine remains on the liquidus of the anhydrous olivine tholeiite up to 13 kb, it is seen that increasing P_{H_2O} extends the pressure at which olivine remains on the liquidus up to 27 kb for $P_{H_2O} = P_{total}$. In this composition, above 15 kb under dry conditions, garnet and clinopyroxenes are the near-liquidus phases. Thus, the eclogite thermal divide is indeed broken at high water pressures and the primary field of crystallization of olivine extends into more siliceous compositions as suggested by Kushiro and Yoder. The question is, how siliceous?

This question is partially answered by experiments on the saturated tholeiite (This has no free quartz or free olivine in its low pressure norm). The primary field of crystallization of olivine is extended from 3 kb (dry) to 20 kb under water saturated conditions. Thus, a magma of this composition might be generated by direct partial melting of pyrolite at depths up to 70 km (20 kb) but no deeper because the liquid could no longer have been in equilibrium with residual olivine. An important feature is that only a small amount (5-10%) of olivine is observed to crystallize from this magma in the pressure range 10-20 kb, strongly suggesting that olivine crystallization will not be capable of driving the liquid to much more siliceous compositions. This was tested by experiments on two basaltic andesites (53.7 and 56.6% SiO_2, 3.3 and 7.5% normative quartz) which show that olivine remains on the liquidus only up to pressures of 5-7 kb under water-saturated conditions. Such magmas could not be generated by direct partial melting of the mantle at depths greater than about 30 km.

There are, moreover, additional constraints on the possibility of producing primary oversaturated magmas such as quartz tholeiites and basaltic andesites by hydrous partial melting. The liquidus temperatures of water-saturated tholeiites at high pressures are greatly depressed and are smaller than 1 atm liquidus temperatures (Fig. 5). Accordingly, water-saturated magmas formed at considerable depths in the mantle are unable to reach the surface without undergoing forced crystallization as confining pressure is reduced and water escapes. In order to avoid this situation, basalts and basaltic andesites produced by partial melting in the mantle must be undersaturated in water, but this in turn requires that the depth interval under which such magmas can form in equilibrium with mantle peridotite is considerably restricted.

Detailed studies of the compositions of liquids produced by hydrous partial melting of pyrolite were also made by Green [1973] and Nicholls [1974]. Attempts were made to correct for the effects of quench crystallization on the composition of the equilibrium partial melts by means of Fe/Mg partition relationships and mass balances of phases present. In all cases, at pressures greater than 5 kb, the estimated equilibrium liquid compositions were found to be poorer in normative quartz than the directly measured compositions. Nicholls [1974] synthesized new compositions corresponding to his 'best estimates' of partial melt compositions at 5, 10 and 15 kb and studied their crystallization equilibria under similar P,T,P_{H_2O} conditions. They were all found to be substantially to seriously undersaturated in olivine, showing that they could not have formed by equilibrium partial melting of pyrolite. He then proceeded to add increasing amounts of olivine to the compositions (at 5, 10 and 15 kb) until co-saturation with olivine + pyroxene(s) ± amphibole were achieved and determined the composition of the liquid phase. The advantages of this technique over direct partial melting of peridotite is that the amount of near-liquidus crystalline phases relative to the saturated liquid is small, so that quench modification of the liquid by overgrowths on these crystalline phases is greatly reduced. Nicholls demonstrated that "andesitic" liquids containing 55-60% SiO_2 and 5-12% normative quartz could be formed only at pressures (water saturated) smaller than 10 kb. At 15 kb, liquids in equilibrium with olivine contained up to 58% SiO_2 but because of high alkali contents, they were olivine normative and thus did not resemble andesitic compositions.

A corresponding study of the partial melting of pyrolite involving estimates of melt composition corrected for quench effects, followed by 'reversals' of these compositions to find the conditions under which they could become saturated with olivine and pyroxene(s) was carried out by Green [1976] at 10 and 20 kb. He demonstrated that the partial melting product of pyrolite at 10 kb, 1100°C, $P_{H_2O} = P_{Load}$, was similar to basaltic andesite (55.5% SiO_2, 8.5% normative quartz). At 20 kb, 1100°C, the liquid was an olivine tholeiite

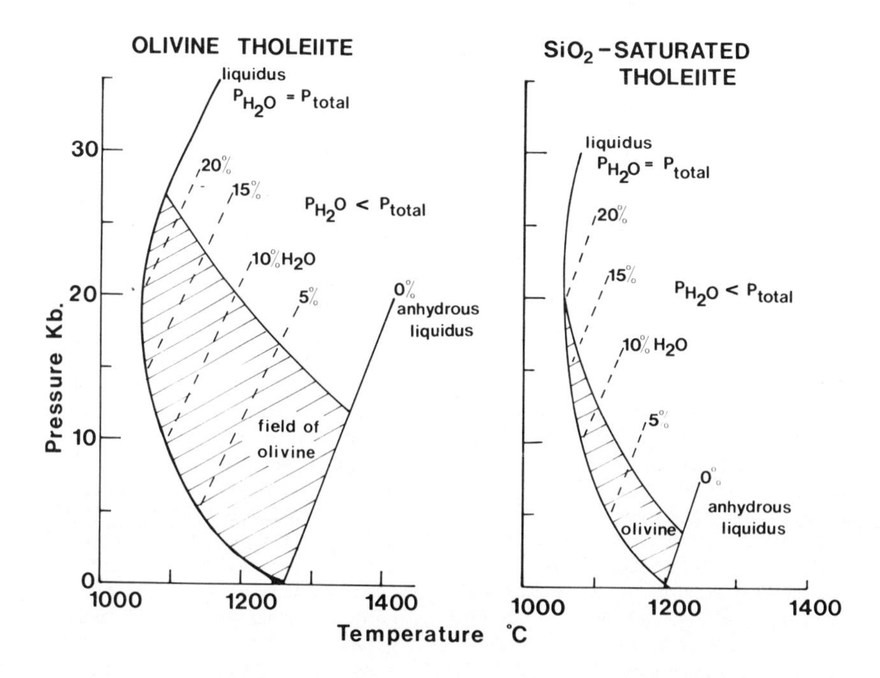

Figure 5. Liquidi of olivine tholeiite and silica-saturated tholeiite under conditions of $P_{H_2O} < P_{total}$ for a range of water contents. Shaded regions indicate fields of olivine crystallization on and near the liquidi. Based on the results of Nicholls and Ringwood [1972, 1973].

(49.8% SiO_2, 10% normative olivine).

Further studies relating to these topics have been carried out by Kushiro [1974] and Nicholls [1976]. Kushiro studied liquidus phase relationships in the system forsterite - plagioclase (An_{50} Ab_{50}) - SiO_2-H_2O at 15 kb and water-saturated conditions. Boundaries between the primary phase fields of olivine, pyroxene and amphibole were located and the composition of the piercing point Fo + En + Amph + L +V (1000°C) was determined. This composition was considered to represent a simplified model of a natural andesite. Kushiro demonstrated that the liquid in equilibrium with olivine, enstatite and amphibole at 1000°C contained 60% SiO_2 and 7 percent normative quartz. He considered this composition to be "andesitic" and to support the contention that andesitic magmas could form by partial melting of peridotite at 15 kb. The analogy is not convincing, however in view of the uncertainties in applying the results of simple to natural complex systems. The degree of oversaturation of SiO_2 (7% normative quartz) resembles that of a basaltic andesite rather than an andesite.

Nicholls repeated these experiments under identical P,T,P_{H_2O} conditions using three natural compositions differing chiefly from Kushiro's in that they contained normal quantities of FeO. Liquids in equilibrium with olivine, pyroxene(s) and amphibole ranged between weakly silica-oversaturated (4% Qz) to markedly olivine normative (12% Ol). They did not resemble natural andesites which usually contain >10% of normative quartz.

Summary

The experimental investigations reviewed above show that andesitic magmas (~60% SiO_2 and >10% normative quartz) cannot be produced by hydrous partial melting of pyrolite at pressures greater than about 10 kb. Moreover the liquids produced under these conditions contain much higher Mg/Mg+Fe ratios (at the Ni-NiO buffer) and nickel and chromium contents than natural andesites [Nicholls, 1974]. Finally, the water contents and eruption temperatures of natural andesites [Eggler, 1972] show that they were not produced under conditions of water saturation at pressures greater than 10 kb (see Fig. 5). We conclude that natural andesite magmas are not primary magmas formed by the direct partial melting of pyrolite or peridotite under hydrous conditions at the Benioff zone near depths of 80-150 km (corresponding to the commonly observed distances of volcanoes above the Benioff zone).

The experiments of Nicholls and Ringwood [1973] and Nicholls [1974] show that magmas formed by water-saturated partial melting of pyrolite at the Benioff zone at depths exceeding 70 km would possess compositions ranging between olivine tholeiite and silica saturated tholeiite. The liquidi of these magmas are depressed below the 1 atmosphere liquidus (Fig. 5), consequently upon rising, as load pressure and water pressure are reduced, these liquids would be forced to crystallize as water is exsolved. The crystallization paths lie through the field of olivine precipita-

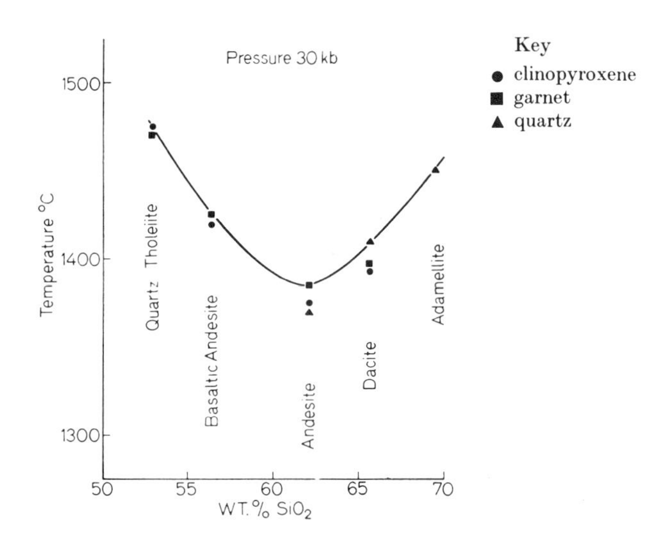

Figure 6. Extrapolated liquidus temperatures and sequence of crystallization at 30 kb for a series of calcalkaline magmas [T. Green and Ringwood, 1968].

tion, so that the residual liquid will have differentiated to a basaltic andesite, or even to an andesite composition upon ascent to within 30 km of the surface [Nicholls and Ringwood, 1973; Nicholls, 1974]. The crystallization of olivine (plus spinel) under these circumstances will produce a range of liquids varying between basalt and andesite in composition. Such liquids will show marked to strong iron enrichment (i.e., tholeiitic FMA trend) with differentiation, they will possess low abundances of incompatible elements, relatively unfractionated rare earth patterns and will be depleted in Ni, Mg and Cr relative to the abyssal tholeiites of mid-oceanic ridges. These magmas will possess all the characteristics of the tholeiitic magma suite occurring in the island arc environment.

If such magmas crystallize at shallow (e.g. 30 km) depths under closed system conditions, olivine separation will not be capable of driving residual liquid compositions to silica contents exceeding 60% SiO_2. At these compositions, olivine will be in reaction relationship to the liquid and will be replaced by amphibole at intermediate pressures and pyroxene and plagioclase at lower pressures. Further fractionation will be governed by crystallization of amphibole ± pyroxene ± plagioclase resulting in the generation of dacitic and rhyodacitic liquids. At very low pressures, extensive crystallization of plagioclase may occur, causing rhyodacites to differentiate towards rhyolite. The entire range of basalt-andesite-dacite-rhyolite magmas may thereby be generated and will be characterized by the trace element and major element abundance patterns of the island arc tholeiitic series [Nicholls and Ringwood, 1972, 1973]. The Tonga and Mariana arcs appear to provide good examples of the operation of this kind of differentiation [e.g. Ewart et al., 1973]. Nicholls and Whitford [1976] have provided a detailed and quantitative study of the operation of this kind of differentiation process, as exemplified in the Western Sunda Arc, Indonesia.

Partial Melting of Subducted Oceanic Crust

Although providing a satisfactory explanation of the petrogenesis of tholeiitic members of the orogenic magma series, hydrous partial melting of pyrolite in the wedge above the Benioff zone does not account so readily for many of the characteristics of the calcalkaline magma series. Ringwood and D. Green [1966] and T. Green and Ringwood [1966, 1967, 1968] explored the hypothesis that these were derived by partial melting of subducted oceanic crust. There are two rather compelling reasons for believing that partial melting of the mafic oceanic crust indeed occurs along Benioff zones.

1. The oceanic crust is believed to contain a substantial amount of water, mainly held in amphiboles, zoisite and serpentine. As the crust becomes heated during its descent into the mantle, high water vapour pressures are built up owing to dehydration of these minerals. Investigations of the thermal structure of the sinking slab show that the low temperatures required for partial melting of quartz eclogite "crust" under hydrous conditions [700-900°C, e.g, Hill and Boettcher, 1970; Lambert and Wyllie, 1972] are reached at quite shallow depths [e.g., Turcotte and Oxburgh, 1970]. Indeed, providing 1 or 2% of water is initially present in the oceanic crust, it is difficult to formulate models in which melting does not occur.
2. If melting and extraction of siliceous liquid did not occur, the vast volumes of oceanic crust deposited in the mantle would be transformed to quartz eclogite (5-10% Qz). On the other hand, the mafic xenoliths of mantle origin found in diamond pipes consist dominantly of bimineralic garnet-clinopyroxene eclogite and quartz eclogites are extremely rare [Mathias et al., 1970]. This rarity is inexplicable unless the low melting siliceous components are regularly extracted from the sinking oceanic crust.

Many volcanoes which erupt calc-alkaline type magmas are situated at heights of 100-150 km above the Benioff zone. The pressures at these depths would convert basalt and amphibolite in the oceanic crust into quartz eclogites. If calcalkaline magmas are indeed ultimately derived from the remelting of oceanic crust, then the melting relationships of mafic and calcalkaline rocks under eclogite-facies conditions must be of key importance.

These relationships were investigated by T. Green and Ringwood [1966, 1967, 1968] for a basalt, basaltic andesite, andesite, dacite and rhyodacite at pressures between 20 and 40 kb. Fractionation of the first three was dominated by the crystallization of garnet and pyroxene. It was demonstrated that separation of these phases would cause a parental basalt to differentiate through a basaltic andesite to an andesite composition. Further differentiation towards more acidic compositions was prevented by the very wide primary crystallization field of quartz (Fig. 6). The andesite composition is seen to occupy a thermal valley in the sequence. The results show that partial melting of a quartz eclogite (basaltic composition) at 20-40 kb would initially produce an andesite.

The inability of this mechanism to produce dacite and rhyodacite liquids suggested that some additional factor might be involved. The partial melting behaviour of calcalkaline rocks in the presence of a significant partial pressure of water (P_{H_2O} = 3 to 5 kb) was accordingly examined. A dramatic depression of the quartz field in the more acidic compositions of Fig. 6 was found. Iron-rich garnet was found to crystallize at relatively low temperatures (900-1100°C) on the liquidi of the andesite, dacite and rhyodacite compositions, instead of quartz. The importance of garnet relative to pyroxene in controlling the crystallization equilibria increased as the liquid became more acidic. The results implied that a small degree of partial melting of quartz eclogite under high pressure hydrous conditions would produce magmas generally resembling dacites and rhyodacites.

These results have been repeated in subsequent experiments in sealed capsules in which water content was controlled more rigorously than in the earlier reconnaissance work [T. Green, 1972; T. Green and Ringwood, 1972]. The essential results obtained in the earlier work have been confirmed and extended. However, it has proved difficult to define the major element compositions of the partial melts in more than general terms because of compositional modification of liquids during quenching. Stern [1974] has attempted to minimise this problem by methods based upon extrapolation of observed phase relationships. He concludes that the liquids produced by hydrous partial melting of quartz eclogite will have higher Ca/Mg+Fe ratios than typical calcalkaline volcanic rocks. A detailed experimental investigation by Harris [1976] indicates similar conclusions. Evidently, the origin of calcalkaline rocks is often more complex than envisaged in the model invoking direct partial melting of oceanic crust.

Eclogite-controlled fractionation has some important characteristics which are reflected in the chemistry of liquid differentiates. Because of the high Fe/Mg ratios of garnet (relative to pyroxenes and olivines), residual liquids were demonstrated to follow a calcalkaline rather than a tholeiitic trend [T. Green and Ringwood, 1968]. Eclogite near-liquidus pyroxenes were found to contain substantial amounts of sodium but not potassium. Liquids produced by small degrees of partial melting of quartz eclogite therefore possess a much higher K/Na ratio than their starting material. Moreover, because of strong partition of heavy rare earths in garnet as shown in detail by the experimental results of Harris and Nicholls [1976], the resultant liquids will be characterized by strong enrichment of light REE and strong depletion of heavy REE.

Petrogenetic Synthesis

We are particularly concerned to explain the existence of the tholeiitic and calcalkaline phases of volcanic evolution and the geochemical characteristics of these phases. The following idealized model is based on the proposals of Nicholls and Ringwood [1973].

The tholeiitic phase is depicted in Fig. 7. As the subducted lithosphere descends, heating along the Benioff zone is caused by viscous dissipation. Thermal models [Oxburgh and Turcotte, 1970] indicate that temperatures throughout most of the oceanic crust are unlikely to exceed 650-700°C at the stage that the crust has reached a depth of 100 km. Accordingly, transformation of amphibolite to eclogite + H_2O occurs under subsolidus conditions in the depth interval 80-100 km. [For data on the solidus of quartz eclogite with $P_{H_2O} = P_{Load}$ and the stability field of amphibole, see Hill and Boettcher, 1970, and Lambert and Wyllie, 1972]. A large amount of water produced in the slab, both by the dehydration of amphibolite and by the partial dehydration of serpentinite thereby rises into the mantle above, causing a drastic decrease in 'viscosity' and initiating the uprise of pyrolite from the Benioff zone. Partial melting occurs in the rising diapirs in the presence of high water vapour pressure, leading to the separation of hydrous tholeiitic magmas. Following earlier discussion, it is believed that these magmas fractionate by separation mainly of olivine, accompanied by pyroxene and amphibole, thereby producing the tholeiitic phase of island arc development. An essential characteristic of this phase is that there is only a small amount of transfer of silicate components (carried in the vapour phase) from the subducted oceanic crust into the tholeiitic magmas which are ultimately developed in the overlying wedge.

At greater depths most of the oceanic crust will have been converted into a quartz eclogite. However the dehydration of serpentinite and its high pressure derivatives maintains a high P_{H_2O} throughout the crust in the depth range 100-300 km. As the temperature in the crust rises above about 750°C, partial melting of quartz eclogite occurs, leading to the development of dacite-rhyodacite magmas, particularly in the 100-150 km depth interval.

Amphibolite which had not previously transformed under subsolidus conditions may also undergo partial melting between 80-100 km to produce residual eclogite plus similar magmas. The magmas will be characterised by high K/Na ratios, high abundances of incompatible elements and strongly fractionated rare earth patterns. These liquids may sometimes ascend to the surface in their original state. Arth and Hanson [1972] have described a suite of Archaean quartz diorites,

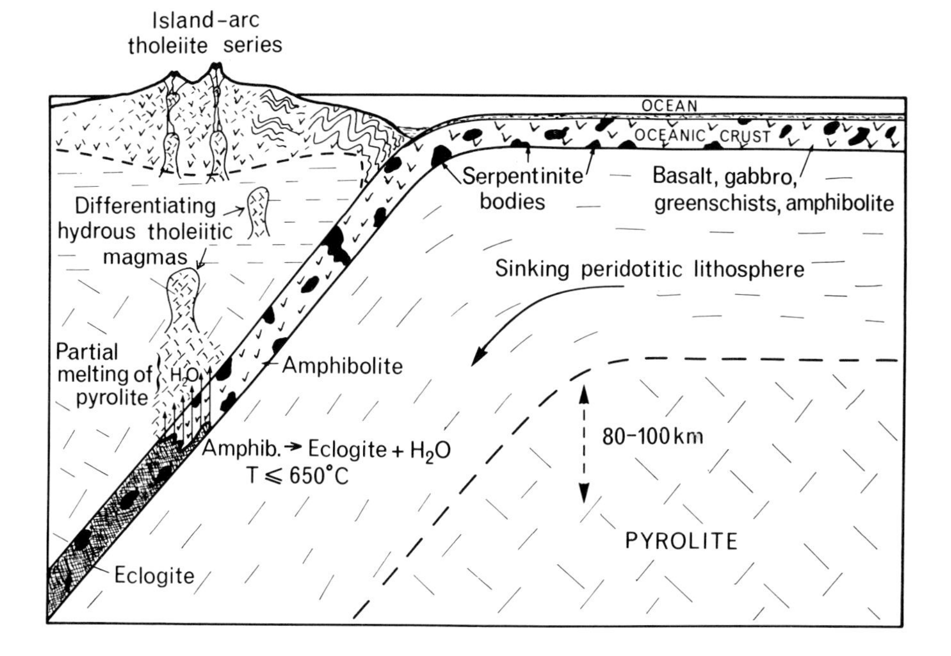

Figure 7. Early phase of development of an island-arc involving dehydration of amphibolite in subducted oceanic crust, introduction of water into the overlying wedge and generation of "island arc tholeiite" series [Nicholls and Ringwood, 1973].

tonalites and dacites which possess extremely fractionated rare earth patterns, strongly depleted in the heavy rare earths, as would be expected if these magmas had formed in equilibrium with residual eclogite (Fig. 2, curves D, E). Related rocks possess quite high Ca/Mg+Fe ratios, which is also to be expected for this kind of fractionation [Stern, 1974; Harris, 1976], see also, Goldich et al. [1972] and O'Nions and Pankhurst [1974] for further descriptions of related rocks.

Most young calcalkaline magmas do not possess the strongly depleted heavy rare earth patterns characteristic of equilibration with eclogitic residua [Gill, 1972]. Nicholls and Ringwood [1973] suggested that these had been derived from a source region formed by the reaction of dacite-rhyodacite melts derived from the former oceanic crust with overlying pyrolite or peridotite as seen in Fig. 8. The reaction would have converted some olivine to pyroxene whilst the minor elements extracted from the oceanic crust were precipitated in these zones of 'modified mantle', leading to enrichment of these elements and development of calcalkaline abundance patterns. The olivine pyroxenite bodies will possess a slightly smaller density than surrounding pyrolite. Moreover, because of the introduction of water they are likely to be highly mobile. As a result, diapirs of wet olivine pyroxenite are detached from the Benioff zone and undergo partial melting as they rise. The magmas which segregate will be characterized by high K/Na ratios, high abundances of incompatible elements and fractionated rare earth patterns.
These magmas will fractionate as they rise by the crystallization of olivine and amphibole at intermediate depths and pyroxene ± plagioclase at shallow depths. It is believed that amphibole plays a relatively more important role than olivine, perhaps because of higher alkali contents [Cawthorn and O'Hara, 1976], and is partly responsible for preventing strong iron-enrichment (Table 1, Fig. 1).

It may be seen that according to this model (Fig. 8), calcalkaline magmas are derived ultimately from two distinct sources - by partial melting of the subducted oceanic crust and by partial melting of the pyrolite wedge overlying the Benioff zone. The first of these involves a component which has been transported from a mid-oceanic ridge, perhaps thousands of kilometres away, thereby representing a process of horizontal differentiation of the mantle. The second represents a strictly vertical and local differentiation of mantle pyrolite.

Below about 150 km, the former mafic oceanic crust has become converted to a highly refractory bimineralic eclogite strongly depleted in low-melting components and incompatible elements. Likewise, the underlying peridotite which was originally formed near a mid-oceanic ridge and was complementary to the basaltic oceanic crust is also residual and refractory in nature and strongly depleted in low melting point components and incompatible elements. Thus the lithosphere which sinks into the mantle has become differentiated into two highly refractory components - eclogite and

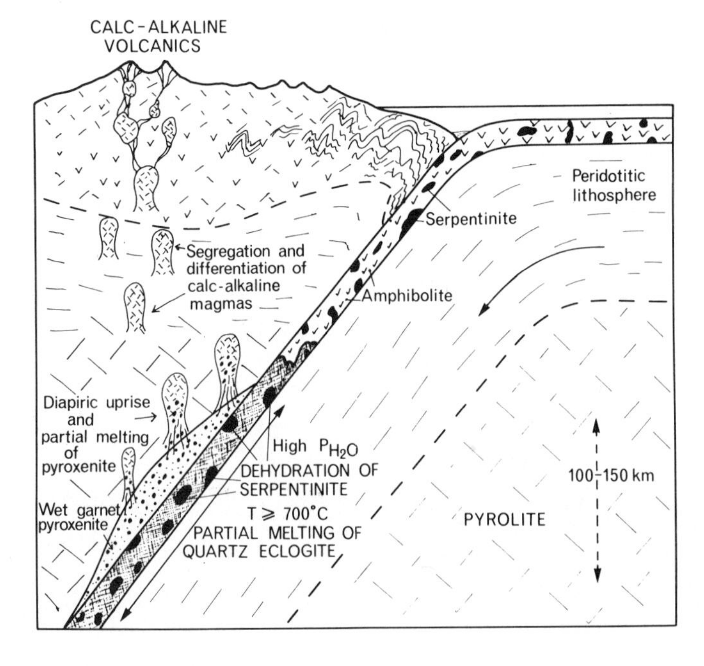

Figure 8. Mature phase if island-arc development involving partial melting of subducted oceanic crust and reaction of acidic liquids so produced with neighboring mantle above Benioff Zone to produce bodies of wet, mobile olivine pyroxenite. These rise diapirically, leading to partial melting and formation of calc-alkaline-type magmas [after Nicholls and Ringwood, 1973].

peridotite. Both these components possess higher initial melting temperatures (solidi) than pyrolite. These thermal characteristics, combined with the large size of each of these components and their depletion in incompatible elements, dictate that the sinking-lithosphere can never again serve as a source of basaltic magma in any further cycle of melting - there appears to be no obvious way in which 5 to 50 km sized blocks of refractory eclogite and peridotite can be intimately remixed at the centimetre to metre scale in the solid state in the deep mantle to reform a homogeneous pyrolite composition. Moreover, this would necessitate the reintroduction of the low melting components and incompatible elements which had been extracted from the lithosphere and added to the continental crust. A plausible model for achieving this end is not in sight.

It follows that the lithosphere has become irreversibly differentiated as emphasized by Ringwood [1969]. The complementary products of this irreversible differentiation are continental crust rocks derived ultimately via orogenic-type magmatism and the depleted, refractory sinking slab of eclogite and peridotite.

Ringwood [1975] estimated that perhaps 30 to 60 percent of the entire mantle had passed through this irreversible differentiation process.

References

Arth, J. G., and G. N. Hanson, Quartz diorites derived by partial melting of eclogite or amphibolite at mantle depths, Contrib. Min. Petrol., 37, 161-173, 1972.

Baker, P. E., Comparative volcanology and petrology of the Atlantic island arcs, Bull. Volcan., 32, 189-206, 1968.

Barazangi, M., and B. Isacks, Lateral variations of seismic-wave attenuation in the upper mantle above the inclined earthquake zone of the Tonga island arc: Deep anomaly in the upper mantle, J. Geophys. Res., 76, 8493-8516, 1971.

Brown, G. M., and J. F. Schairer, Melting relations of some calcalkaline rocks, Carnegie Inst. Washington Yearbook, 66, 460-467, 1968.

Cann, J. R., A new model for the structure of the oceanic crust, Nature Lond., 226, 928-930, 1970.

Carmichael, I. S. E., The petrology of Thingmuli, a Tertiary volcano in eastern Iceland, J. Petrol., 5, 435-460, 1964.

Carmichael, I. S. E., and J. Nicholls, Iron-titanium oxides and oxygen fugacities in volcanic rocks, J. Geophys. Res., 72, 4665-4687, 1967.

Cawthorn, R. G., C. Ford, G. Biggar, M. Bravo, and D. Clarke, Determination of the liquid composition in experimental samples: discrepancies between microprobe analysis and other methods, Earth Planet. Sci. Letters, 21, 1-5, 1973.

Cawthorn, R. G., and M. J. O'Hara, Amphibole fractionation in calc-alkaline magma genesis, Am. J. Sci., 276, 309-329, 1976.

Coats, R. R., Magma type and crustal structure in the Aleutian arc, in Crust of the Pacific Basin, pp. 92-109, Am. Geophys. Union Geophys. Monograph, 6, 1962.

Dickinson, W. R., Circum-Pacific andesite types. J. Geophys. Res., 73, 2261-2269, 1968.

Dickinson, W. R., and T. Hatherton, Andesitic volcanism and seismicity around the Pacific, Science, 157, 801-803, 1967.

Eggler, D. H., Water-saturated and undersaturated melting relationships in a Paricutin andesite and an estimate of water content in the natural magma, Contrib. Mineral Petrol., 34, 261-271, 1972.

Ewart, A., W. B. Bryan, and J. Gill, Mineralogy and geochemistry of the younger volcanic islands of Tonga, S.W. Pacific, in press, 1973.

Ewart, A., and J. J. Stipp, Petrogenesis of the volcanic rocks of the Central North Island, New Zealand, as indicated by a study of Sr^{87}/Sr^{86} ratios and Sr, Rb, K, U, and Th abundances, Geochim. Cosmochim. Acta, 32, 699-736, 1968.

Gill, J. B., Geochemistry of Viti Levu, Fiji and its evolution as an island arc, Contrib. Min. Pet., 27, 179-203, 1970.

Gill, J. B., Role of underthrust oceanic crust in the genesis of a Fijian calcalkaline suite, Contrib. Mineral. Petrol., 43, 29-45, 1974.

Goldich, S. S., G. Hanson, C. Hallford, and M.

Mudray, Early Precambrian rocks in the Saganega-Northern Light-Lake area, Minnesota-Ontario, Geol Soc. Am. Mem., 135, Part I, 151-191, 1972.
Gorshkov, G. S., Petrochemical features of volcanism in relation to the types of the earth's crust, Am. Geophys. U. Monograph, 6, 110-115, 1962.
Green, D. H., Experimental melting studies on a model upper mantle composition at high pressures under water-saturated and water-undersaturated conditions, Earth Planet. Sci. Letters, 19, 37-53, 1973.
Green, D. H., Experimental testing of 'equilibrium' partial melting of peridotite under water-saturated, high pressure conditions, Canad. Min., in press, 1976.
Green, T. H., Crystallization of calcalkaline andesite under controlled high pressure hydrous conditions, Contrib. Mineral. Petrol., 34, 150-166, 1972.
Green, T. H., and A. E. Ringwood, Origin of the calcalkaline igneous rock suite, Earth Planet. Sci. Letters, 1, 307-316, 1966.
Green, T. H., and A. E. Ringwood, Crystallization of basalt and andesite under high pressure hydrous conditions, Earth Planet. Sci. Letters, 3, 481-489, 1967.
Green, T. H., and A. E. Ringwood, Genesis of the calcalkaline igneous rock suite, Contrib. Mineral. Petrol, 18, 105-162, 1968.
Green, T. H., and A. E. Ringwood, Crystallization of garnet-bearing rhyodacite under high pressure hydrous condition, J. Geol. Soc. Australia, 19, 203-212, 1972.
Harris, K., In preparation, 1976.
Harris, K., and I. Nicholls, An experimental study of the partitioning of selected rare earth elements between garnet, clinopyroxene, amphibole and melts of andesitic and basaltic composition, in preparation, 1976.
Healy, J., Structure and volcanism in the Taupo Volcanic Zone, New Zealand, Amer. Geophys. U. Monograph, 6, 151-157, 1962.
Hill, R. E. T., and A. L. Boettcher, Water in the earth's mantle: Melting curves of basalt-water and basalt-water-carbon dioxide, Science, 167, 980-982, 1970.
Holloway, J. R., and C. W. Burnham, Melting relations of basalt with equilibrium water pressure less than total pressure, J. Petrol., 13, 1-29, 1972.
Jakes, P., and J. Gill, Rare earth elements and the island arc tholeiite series, Earth Planet. Sci. Letters, 9, 17-28, 1970.
Jakes, P., and A. J. White, Structure of the Melanesian Arcs and correlation with distribution of magma types, Tectonophysics, 8, 223-236, 1969.
Johnson, R. W., Potassium variation across the New Britain volcanic arc, Earth Planet. Sci. Letters, in press, 1976.
Karig, D. W., and G. F. Sharman, Subduction and accretion in trenches, Geol. Soc. Am. Bull., 86, 377-392, 1975.
Katsui, Y., S. Nishirmura, Y. Masuda, H. Kurasawa, and H. Fujimake, Petrochemistry of the Quaternary volcanic rocks of Hokkaido, North Japan, preprint, 1975.
Kuno, H., Petrology of Hakone volcano and the adjacent areas, Japan, Bull. Geol. Soc. Am., 61, 957-1014, 1950.
Kushiro, I., Effect of water on the composition of magmas formed at high pressure, J. Petrol., 13, 311-334, 1972.
Kushiro, I., The system Forsterite-Anorthite-Albite-Silica at 15 kb and the genesis of andesitic magmas in the upper mantle, Carnegie Inst. Wash. Yearbook, 73, 244-248, 1974.
Kushiro, I., N. Shimazu, Y. Nakamura, and S. Akimoto, Compositions of coexisting liquid and solid phases formed upon melting of natural garnet and spinel lherzolites at high pressures: A preliminary report, Earth Planet. Sci. Letters, 14, 19-25, 1972.
Kushiro, I., and H. S. Yoder, Melting of forsterite and enstatite at high pressures under hydrous conditions, Carnegie Inst. Washington Yearbook, 67, 153-158, 1969.
Lambert, I. B., and P. J. Wyllie, Melting of gabbro (quartz eclogite) with excess water to 35 kilobars with geological applications, J. Geol., 80, 693-708, 1972.
Matthias, M. J., J. Siebert, and P. Rickwood, Some aspects of the mineralogy and petrology of ultramafic xenoliths in kimberlite, Contrib. Mineral. Petrol., 26, 75-123, 1970.
Mysen, B. O., and A. L. Boettcher, Melting of a hydrous mantle II, J. Petrol., 16, 549-590, 1975.
Nakamura, Y., and I. Kushiro, Composition of the gas phase in Mg_2SiO_4-SiO_2-H_2O at 15 kb, Carnegie Inst. Washington Yearbook, 73, 255-258, 1974.
Nehru, C. E., and P. J. Wyllie, Compositions of glasses from St. Paul's peridotite partially melted at 20 kilobars, J. Geol., 83, 455-471, 1975.
Nicholls, I. A., Liquids in equilibrium with peridotitic mineral assemblages at high water pressures, Contrib. Mineral. Petrol., 45, 289-316, 1974.
Nicholls, I. A., Quartz normative liquids in equilibrium with olivine at 15 kb water pressure and their bearing on the origin of andesitic magmas, preprint, 1976.
Nicholls, I. A., and A. E. Ringwood, Production of silica-saturated magmas in island arcs, Earth Planet. Sci. Letters, 17, 243-246, 1972.
Nicholls, I. A., and A. E. Ringwood, Effect of water on olivine stability in tholeiites and the production of silica-saturated magmas in the island-arc environment, J. Geol., 81, 285-300, 1973.
Nicholls, I. A., and D. J. Whitford, Primary magmas associated with Quaternary volcanism in the Western Sunda Arc, Indonesia, preprint, 1976.
O'Nions, R. K., and R. J. Pankhurst, Rare earth element distribution in Archaean gneisses and

anorthosites, Godthab area, West Greenland, Earth Planet. Sci. Letters, 22, 328-338, 1974.

Oversby, V. M., and A. Ewart, Lead isotopic compositions of Tonga-Kermadec volcanics and their petrogenetic significance, Contrib. Mineral. Petrol., 37, 181-210, 1972.

Oxburgh, E. R., and D. L. Turcotte, The thermal structure of island arcs, Bull. Geol. Soc. Am., 81, 1665-1688, 1970.

Pichler, H., and W. Zeil, Andesites of the Chilean Andes, in Proceedings of the Andesite Conference, edited by A. R. McBirney, pp. 165-174, Bull. 65, Dept. Geology and Mineral Industries, State of Oregon, 1969.

Ringwood, A. E., Composition and evolution of the upper mantle, Am. Geophys. U. Monograph, 13, 1-17, 1969.

Ringwood, A. E., The petrological evolution of island arc systems, J. Geol. Soc. Lond., 130, 183-294, 1974.

Ringwood, A. E., Composition and Petrology of the Earth's Mantle, 618 pp., McGraw Hill, New York, 1975.

Ringwood, A. E., and D. H. Green, An experimental investigation of the gabbro-eclogite transformation and some geophysical implications, Tectonophysics, 3, 383-427, 1966.

Ringwood, A. E., and A. Major, High pressure reconnaissance investigations in the system Mg_2SiO_4-MgO-H_2O, Earth Planet. Sci. Letters, 2, 130-133, 1967.

Sclar, C. B., L. C. Carrison, and O. M. Stewart, High pressure synthesis of a new hydroxylated pyroxene in the system MgO-SiO_2-H_2O (abstract), Trans. Am. Geophys. U., 48, 226, 1967.

Stern, C. R., Melting products of olivine tholeiite basalt in subduction zones, Geology, 2, 227-230, 1974.

Taylor, S. R., Trace element chemistry of andesites and associated calcalkaline rocks, in Proceedings of the Andesite Conference, edited by A. R. McBirney, pp. 43-63, Bull. 65, Dept. Geol. and Mineral. Resource, State of Oregon, 1969.

Wager, L. R., and W. A. Deer, Geological investigations in East Greenland III. The petrology of the skaergaard Intrusion, Kangerdlugssaq, East Greenland, Med. om Gronland, 105, 1-352, 1939.

Wilson, J. T., The development and structure of the crust, in The Earth As a Planet, edited by G. P. Kuiper, pp. 138-214, Chicago Univ. Press, 1954.

Yamamoto, K., and S. Akimoto, High pressure and high temperature investigations in the system MgO-SiO_2-H_2O, J. Solid State Chem., 9, 187-195, 1974.

Yamamoto, K., and S. Akimoto, High pressure and high temperature investigation of the phase diagram in the system MgO-SiO_2-H_2O, preprint, 1975.

ISLAND ARC MODELS AND THE COMPOSITION OF THE CONTINENTAL CRUST

Stuart Ross Taylor

Research School of Earth Sciences, Australian National University, Canberra

Abstract. A popular model provides for continental growth through the addition of material derived from island-arc volcanism, ultimately derived via multi-stage melting processes from the upper mantle. Assuming that this model provides for crustal accretion at present, how far back in time is it valid? The model implies that the total crustal composition is equivalent to that of the overall average island-arc volcanic rocks. Constraints on overall crustal composition are limited, but one index is provided by the rare-earth element (REE) distribution in sedimentary rocks. In these, the REE patterns are sufficiently uniform to suggest that they represent an effective sampling of the upper crust which is exposed to weathering processes. A first-order distinction appears between the REE patterns in Archean greenstone belt sedimentary rocks, and those of later periods indicating that two different mechanisms of crustal growth or evolution may be involved. The Archean sedimentary REE patterns are rather variable. Relative to chondritic patterns (interpreted as parallel to upper mantle patterns deduced from the ocean-ridge basalt data) the Archean sediments are enriched in light REE. They show no Eu anomaly and resemble island-arc volcanic rock patterns. Taylor and Jakeš (1974) suggested from this evidence that the Archean continental crust was dominantly composed of island-arc volcanics, which had not undergone any intracrustal melting. However, the dominant rocks in Archean greenstone terrains appear to be tholeiitic basalts and felsic volcanic and intrusive rocks. Mixtures of these two types also produce REE patterns resembling those of the Archean sediments. Post-Archean sedimentary rocks show a distinctly different REE pattern. The LREE/HREE ratio is nearly constant (ΣLREE/HREE = 9.7 $\pm$ 1.8) and a negative Eu anomaly (relative to chondrites) of constant magnitude (Eu/Eu* = 0.67 $\pm$ 0.05) is present. This pattern is typical of granodioritic rocks. If the total crust has the composition of island-arc volcanic rocks, such a REE pattern could be produced via intracrustal melting, from an island-arc type REE pattern. The presence of plagioclase as a residual phase in the lower crust can produce Eu depletion while the LREE enrichment is consistent with equilibration with clinopyroxene. The absence of HREE enrichment or depletion indicates that garnet is not a major residual phase and the overall REE pattern is consistent with partial melting at pressures below 10 kb, hence implying intracrustal melting. There is no evidence for an early anorthositic crust. A table of upper crustal and total crustal abundances for major and trace elements is provided. Lower crustal rocks should display a positive Eu anomaly, relative to upper crustal rocks or to chondrites.

Crustal Compositions

Has the composition of the continental crust been constant, or has it changed with time? How can we estimate the composition? This is not a simple task, and all estimates tend to be biased toward upper crustal compositions. Because of this, some confusion has arisen over estimates of the composition of the continental crust. The values of Vinogradov (1962) and Taylor (1964a), for example, refer to the total crust, extending to the Mohorovičić discontinuity at a depth of about 40 km. On the other hand, the estimates of Eade and Fahrig (1971, 1973) and Shaw et al. (1976), for example, refer only to the upper crust, exposed to weathering and accessible to surface sampling.

As a result of extensive work, it is reasonably well established that the upper crustal composition is close to that of granodiorite with 65-70% SiO_2, 10-15 ppm Th, 3-4 ppm U and 2-3% K (e.g. Poldervaart, 1955; Eade and Fahrig, 1971; Shaw et al. 1976). It is well known from the heat flow data that such a composition cannot extend to a depth of more than 10-20 km, so that the lower crust must be depleted in these heat producing elements. This has led to the hypothesis that elements migrate vertically within the crust, by metamorphic processes, or by partial melting, leading to the observed upper crustal composition (e.g. Fyfe, 1973 a,b). Thus whole crust compositions contain higher values for elements such as Cr, V, and Sc and lower values for K, Rb, LREE, Th, U etc. than do tables of upper crustal compositions.

A very large number of models have been proposed to account for the origin of the continental crust. The most spectacular of these are the cosmic hypotheses. Donn et al. (1965) suggested that the continents represent material accreted to the earth in the final stages of formation, so that the crust is the product of the infall of large sialic meteorites of appropriate composition.

Gilvarry (1961) suggested from the depth-diameter relationships of the major ocean basins that they represented the sites of large impacts (by analogy with the lunar maria). The high standing continental masses represented the outthrown rim debris. These and other variants suppose an early primordial crust, resulting from primordial fractionation process. Most of these catastrophic models have not survived. A principal objection is that the crust appears to have grown throughout geological time, and that chemical or isotopic evidence of the reworked remnants of a primordial crust is nowhere visible. This concept arises chiefly from consideration of the Sr isotope data, which has been summarised most recently by Moorbath (1975).

Whatever the particular model, or composition adopted, it is clear that there has been extensive concentration of lithophile elements in the crust compared to the composition of the whole earth. For almost any compositional model, it turns out that about half at least of the total earth abundances of elements like Rb and Ba now reside in the continental crust, a volume about 0.4% of the earth (Taylor, 1964b).

How do such highly fractionated compositions arise? A current popular model for the continuing growth of the crust is that it results from the accretion of the eruptive products of island-arc volcanic systems (Taylor and White, 1965; Taylor 1967; Ringwood, 1969; Jakeš and White, 1971). The most voluminous material being erupted from the mantle is probably represented by the mid-ocean ridge basalts, but it is improbable that the continents have such a composition. The island-arc volcanics, derived by multistage processes from the mantle, possess both a suitable composition and an adequate volume (Dickinson and Luth, 1971).

For these reasons, the chemical aspects of island arc volcanic rocks have been studied for several years at Canberra (e.g. Taylor and White, 1965; Taylor, 1968; Taylor et al. 1969; Gill, 1970; Gorton, 1974; Whitford, 1975; Jakeš and White, 1971, 1972). The chemistry of sedimentary rocks has also been studied as an independent approach to this problem (Nance and Taylor, 1976).

Based on these considerations, data are presented in Table 1 for the composition of the upper continental crust (UC) exposed to weathering and accessible to sampling. This composition represents the average of the upper 10 to 20 km of the crust. An estimate of the composition of the whole crust, (CC) extending to the Mohorovicic discontinuity is also given in Table 1. This is based on the assumption that the whole crust has been derived by accretion from the mantle by island arc volcanism.

Upper Crustal REE Patterns

Figure 1 shows the REE patterns in Australian post-Archean shales of various ages normalized to chondrites (Nance and Taylor, 1976a). The individual REE are normalised to the abundances in chondritic meteorites. This allows a useful and simple comparison of relative abundances and patterns. The rationale for using chondritic abundances is that, to a first approximation the abundances in the various classes of chondrites are subparallel (although there are differences in detail: e.g. Masuda et al. 1973) and so are taken to be representative of the relative abundances in the early solar nebula.

All the samples show parallel patterns. The light REE are enriched relative to the heavy REE. A measure of this is the La/Yb ratio, which is constant for these patterns almost within analytical error, as is the depletion in europium ($Eu/Eu^* = 0.67 \pm 0.5$; $La/Yb = 13.6 \pm 2$). The average of these patterns is very close to that of the composite sample of North American shales (Haskin et al. 1968). This pattern also holds for differing types of sedimentary rocks. The patterns are similar for greywackes, subgreywackes, limestones and quartzites (Nance and Taylor, 1976a). Although the abundances decrease, the relative patterns remain parallel. This result, similar to that of other studies (e.g. Wildeman and Haskin, 1973) is interpreted to indicate that the mixing process during sedimentation is remarkably efficient and produces a uniform product from the very diverse patterns in igneous rocks. It is thus interpreted as the average pattern for the upper continental crust, exposed to weathering processes. This then gives a reference point for the composition of the upper crust. This uniformity holds for the Australian record at least as far back as the Mt. Isa sediments deposited about 1500 million years ago (Nance and Taylor, 1976a).

Archean Sedimentary Rock REE Patterns

Figure 2 shows data for Archean sedimentary rocks normalised to the post-Archean average Australian sedimentary rock (PAAS). This shows that the Kalgoorlie Black Flag sediments, about 2600 million years old have similar patterns to the Wind River greywackes from Wyoming, dated at 2700 m.y. and to the Fig Tree group from the Barberton Mountain Land, South Africa, dated at about 3400 m.y. (Wildeman and Condie, 1973; Wildeman and Haskin, 1973), and, most significantly to the REE patterns from sedimentary enclaves in the Amitsoq gneisses (Mason and McGregor, 1976). These are older than 3700 m.y. Compared with post-Archean sedimentary rocks these Archean sedimentary rocks have lower total

TABLE 1. Average composition of the upper crust (UC) compared with the overall continental crust (CC).

	UC %	CC %
SiO_2	66.0	58.0
TiO_2	0.6	0.8
Al_2O_3	16.0	18.0
FeO*	4.5	7.5
MgO	2.3	3.5
CaO	3.5	7.5
Na_2O	3.8	3.5
K_2O	3.3	1.5
Σ	100.0	100.3
	ppm	ppm
Rb	110	50
Pb	15	7
Ba	700	350
Sr	350	400

	UC ppm	CC ppm
La	38	19
Ce	80	38
Pr	8.9	4.3
Nd	32	16
Sm	5.6	3.7
Eu	1.1	1.1
Gd	4.7	4.2
Tb	0.77	0.64
Dy	4.4	3.7
Ho	1.0	0.82
Er	2.9	2.3
Tm	0.5	0.4
Yb	2.8	2.2
Lu	0.4	0.3

	UC ppm	CC ppm
Th	10.5	2.5
U	2.5	1
Zr	240	100
Hf	5.8	2.2
Nb	25	11
Cr	35	55
V	60	175
Sc	10	30
Ni	15	20
Co	10	25
Cu	25	60
Zn	52	-

*All Fe expressed as FeO.

Data for upper crustal (UC) composition adapted from the following sources: Major elements Eade and Fahrig (1971) Canadian Shield data; REE from Nance and Taylor (1976a) average post-Archean Australian sedimentary rocks. Most other trace element data from Shaw et al. (1976) values for Canadian Shield rocks, with some values from average granodiorite of South Eastern Australia from Kolbe and Taylor (1966). Data for overall continental crust (CC) composition is based on the assumption that the crust is derived from island-arc type volcanism. The overall composition is derived basically from the estimates by Taylor (1968) for the average andesite, modified by later estimates of Jakeš and White (1971), Jakeš (1973) and Whitford (1975).

abundances of rare earths, less relative enrichment of the light REE and are not depleted in Eu relative to chondrites.

Two possibilities exist to account for these differences: (a) The Archean sedimentary rock data is biased toward sampling from the greenstone belts or (b) the data represent an effective overall sampling of the Archean crust, exposed to weathering, as is the case for post-Archean time.

Support for the first point comes from the observation that there are more local variations in the REE patterns in the Archean. There is more scatter in the La/Yb ratios, for example, and Eu sometimes shows an enrichment relative to chondrites (Nance and Taylor, 1976b; Wildeman and Haskin, 1973). Some of the Archean sediments clearly show evidence of a local volcanogenic origin. Examples include the greywackes from the Vermillion district, Minnesota, described by Ojakangas (1972) and some samples of the Kalgoorlie Black Flag shales (Nance and Taylor, 1976b).

The second point of view is supported by the widespread occurrence of the distinctive Archean REE patterns, from Australia, South Africa, Wyoming and Greenland. The similarity in REE patterns (Fig. 2) from such widely dispersed areas, spanning a time interval of 1000 million years, argues that this is not an isolated phenomenon, and could indicate that sedimentary processes in the Archean were carrying out an effective sampling of the crust exposed to weathering.

Sedimentary rocks with typical post-Archean REE patterns have yet to be identified in Archean terrains. Rocks similar to younger granodiorites (e.g. Amitsoq gneisses, Mason and McGregor, 1976) occur in Archean terrains, but apparently did not make a significant contribution to the REE patterns of presently exposed Archean sediments.

It is clear that much more work is required to elucidate the nature of the Archean crust. The evidence at present suggests that the Archean upper crust, exposed to weathering, has a different composition from the post-Archean upper crust and was dominated by volcanogenic sediments. Whether this was a surficial effect or true for the whole thickness of the Archean crust, cannot be decided from the present data.

Archean Igneous Rocks

What REE patterns resemble the Archean sedimentary rock pattern? They show a striking parallelism to those of recent average island arc volcanic rocks. This raises the question about the average composition of volcanic rocks being erupted at island arcs (or plate boundaries). The relative percentages of island arc tholeiites, andesites, high-K andesites and dacites erupted in such regions is still a matter of debate; (e.g. Jakeš

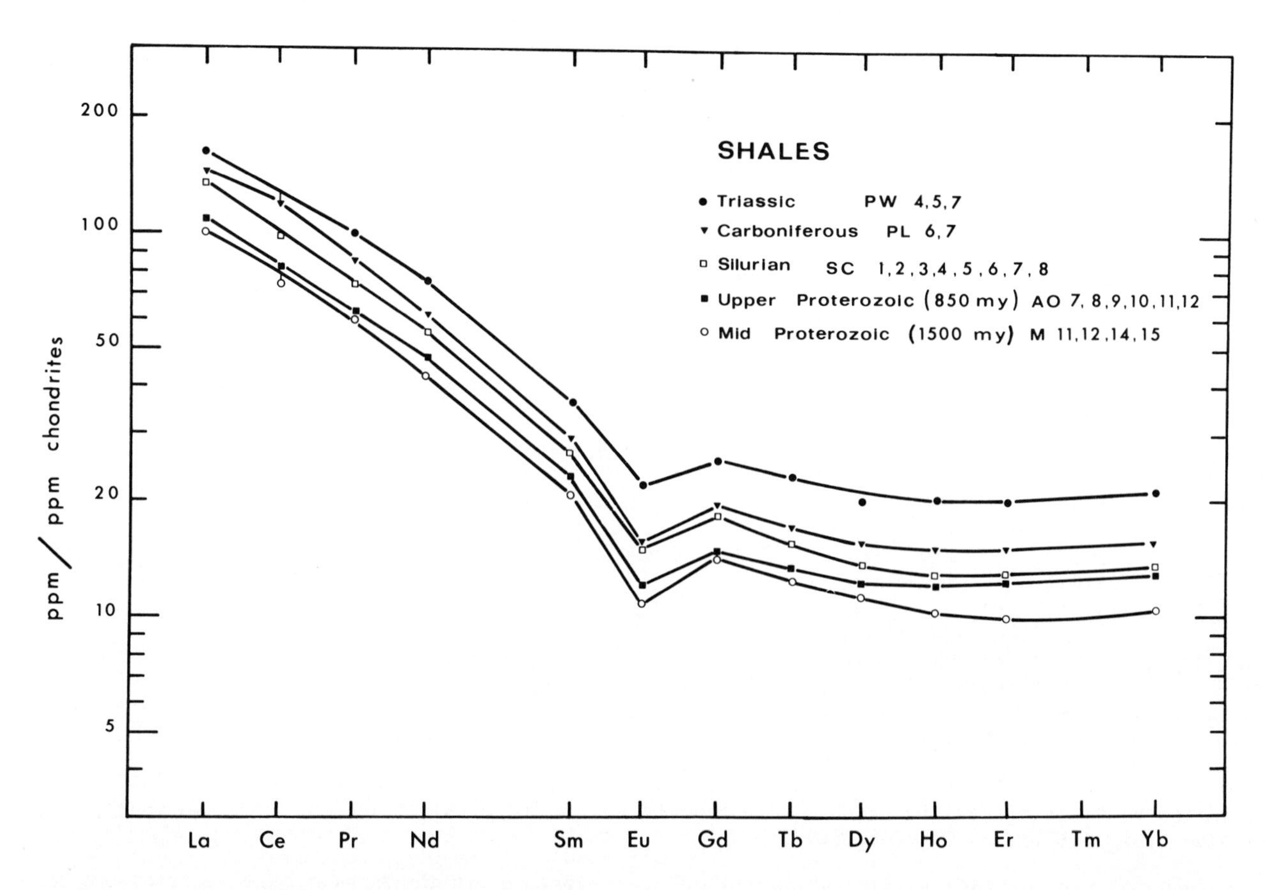

Fig. 1. REE patterns, normalised to chondrites for averages of post-Archean Australian Shale samples. Data from Nance and Taylor (1976a). Note uniformity of slope of patterns and of Eu depletion. There is an increase in total REE in the younger samples.

and White, 1971, Jakeš, 1973). The solution to the overall spatial and temporal element abundance pattern is a non-trivial matter. If the island-arc accretion model is correct for continental growth, then finding the appropriate input composition from plate boundary volcanism should solve the composition problem for the overall continental crust.

An estimate of this overall composition is given in Table 1. From these data, the REE composition of modern island arc volcanic rocks is parallel to that of the Archean sediments and shows the same features (fig. 3). Relative to post-Archean sediments, they are depleted in light REE and enriched in Eu.

Jakeš and Taylor (1974) proposed the following model to account for these observations. It is assumed that thecontinental crust is accreted from island arc volcanic rocks. Intracrustal melting of the composition produces an upper crust dominated by granodiorites, with light REE enrichment, and a negative Eu anomaly, due to the retention of divalent Eu in plagioclase in the lower crust. Divalent Eu is about 20% larger than trivalent Eu^{3+}, is very close in ionic radius to Sr^{2+}, and accordingly enters lattice sites which accommodate Sr. This proposal suggests that intracrustal evolution in the Archean was less developed than in later epochs and that the overall crust exposed to weathering in the Archean had an unfractionated composition.

It is assumed that the whole earth REE pattern is parallel to that of chondrites. Support for this assumption comes from Sm-Nd isotope systematics (De Paolo and Wasserburg, 1976). Mid-ocean ridge basalts (MORB) appear to be derived from sources either with REE parallel to chondritic patterns, or with light REE depletion. Europium anomalies relative to chondrites are negligible (Bence and Taylor, in press). The production of the europium anomaly observed in upper crustal rocks must therefore be of intracrustal origin.

Are there any recent analogues of the Archean crustal REE patterns? The Devonian greywackes of the Baldwin Formation in the Tamworth Trough, are volcanogenic sediments (Chappell, 1968) with REE patterns which are not distinguishable from island arc REE patterns (Nance and Taylor, 1976b). Some indeed show a positive Eu enrichment relative to chondrites, attributable to local concentrations of plagioclase. As noted

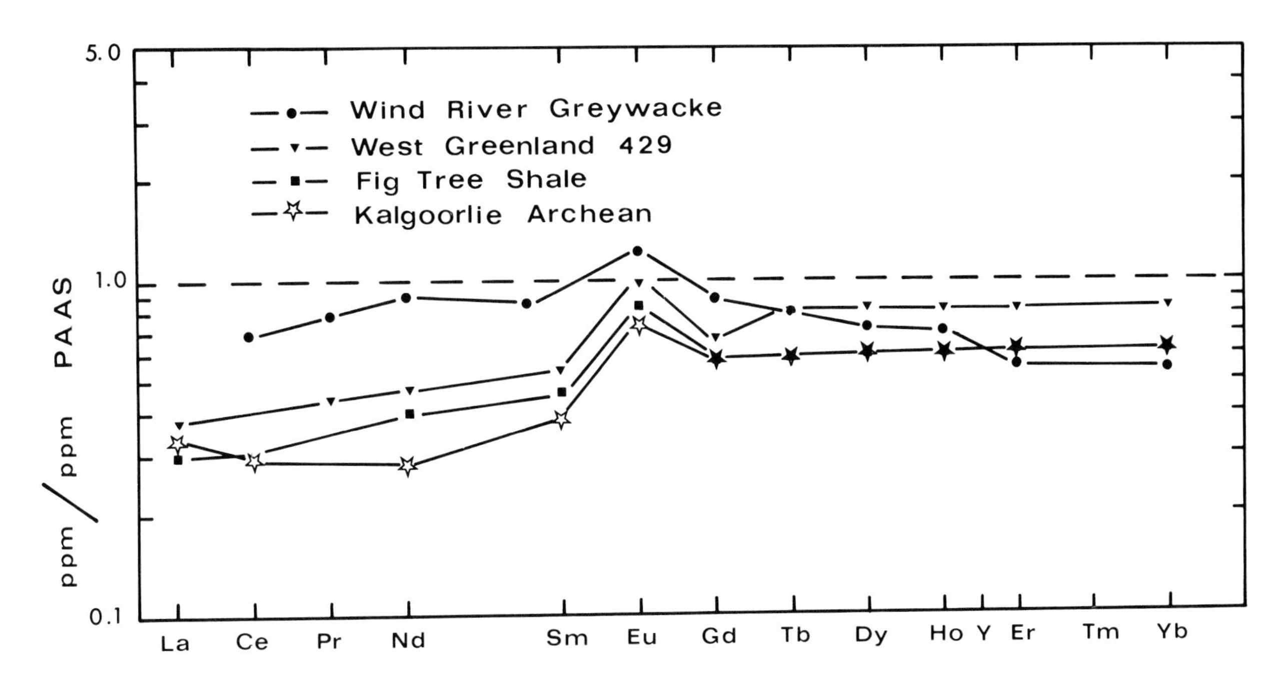

Fig. 2. REE patterns, normalised to post-Archean Australian average shale (PAAS). Note depletion of light REE and positive Eu anomaly, relative to PAAS. Data from Wildeman and Condie (1973), Wildeman and Haskin (1973), Nance and Taylor (1976b) and Mason and McGregor (1976).

earlier such enrichments occur occasionally in the Archean sediments, but are probably due to the same cause. Another possibility is that early reducing conditions in the atmosphere might reduce Eu^{3+} to Eu^{2+}. This explanation might hold for chemically deposited sediments which could directly reflect their precipitation from sea water. The concentration of REE in sea

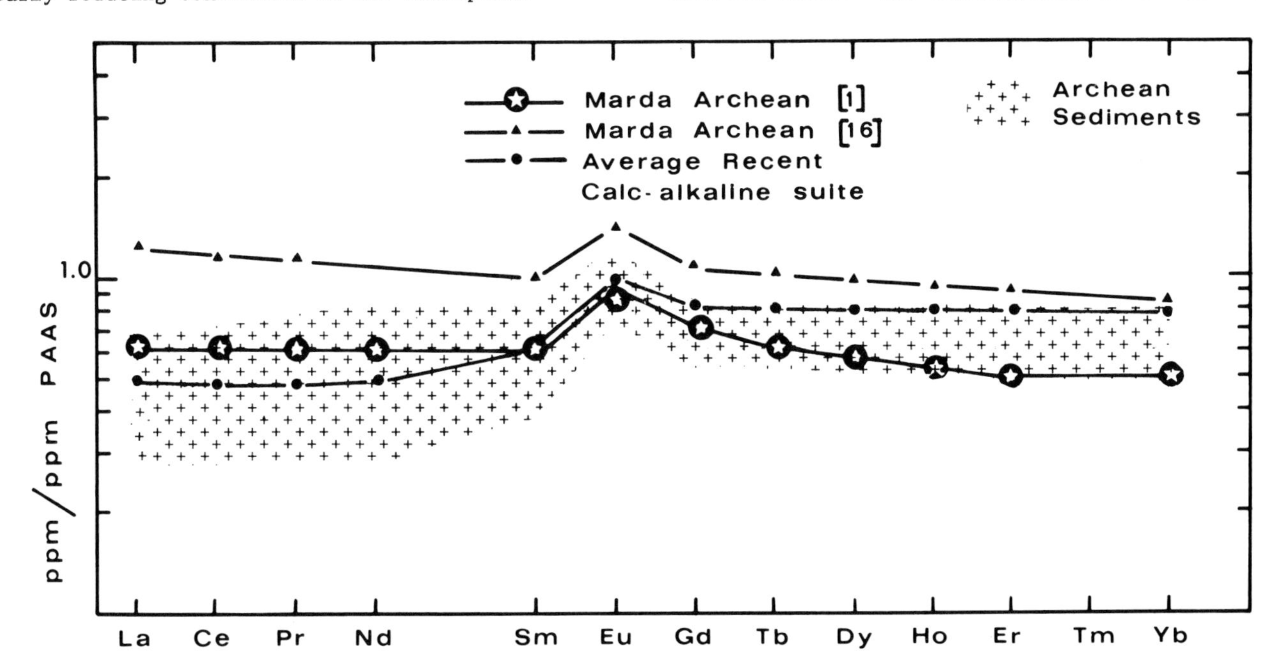

Fig. 3. REE patterns, normalised to post-Archean Australian average shale (PAAS) for Archean sediments from Figure 2, two Marda Archean andesites (Hallberg and Taylor, unpub. data) and the average recent calc-alkaline suite (Table 1).

water is so low, the residence time so short (400 years) (Piper, 1974) that I suggest that these effects will be swamped by the REE carried in the clastic components.

The Jakeš-Taylor Archean model depends critically on the assumption that island arc type volcanic rocks were produced in the Archean. What is the evidence for island arc type volcanic rocks in the Archean? Barager and Goodwin (1968) and Goodwin (1973) have provided evidence for the andesitic rocks in Canadian Archean terrains. The Marda volcanics, from Western Australia, studied by Hallberg et al. (1976) are Archean andesites, 2635 million years old. They are identical in composition to late Cenozoic andesites from Papua (Smith, in preparation) Fig. 4 shows the REE data and Figure 5 shows an element by element comparison. Curiously both are high-K suites and show rather high Ni concentrations (50-100 ppm) compared with modern island arc rocks. Nevertheless, andesitic rocks appear to be less common in Archean terrains than in modern island arc regions. Whether this observation is due to the accidents of exposure of such ancient rocks, or because andesitic rocks are more easily eroded, or represents a real effect, is unknown.

Is it possible that the REE Archean pattern could be developed in other ways? Mixing of Archean tholeiitic and felsic rocks, (the bimodal Archean suite of Barker and Peterman, 1974), can produce similar REE patterns to those observed in Archean sedimentary rocks (Nance and Taylor, 1976b; Arth and Hanson, 1975).

Can we decide between the island arc type model and the bimodal tholeiite-felsic model for the Archean crust, on the basis of the sedimentary rock data? Most elements are similar in abundance levels, provided a 50/50 mixing model is used. An increase in felsic component would deplete the heavy REE. An increase in the tholeiitic component should raise the content of elements such as nickel and chromium. Tholeiitic rocks contain more of these elements than do andesitic rocks (Taylor et al. 1969). This test is complicated by the apparently higher nickel contents of Archean andesites (e.g. Barager and Goodwin, 1969; Hallberg et al. 1976) compared with more recent examples. Whether this is a sampling artifact is not known.

Nevertheless, if the Archean greenstone belts supply 50% tholeiite to sediments, then the latter might be expected to be intrinsically higher in nickel than those of younger eras. Dachin, (1967) working at Cape Town, found high values for Ni and Cr in the Fig Tree group (Fig. 6). This led to speculation that the early crust might have a large mafic or ultramafic com-

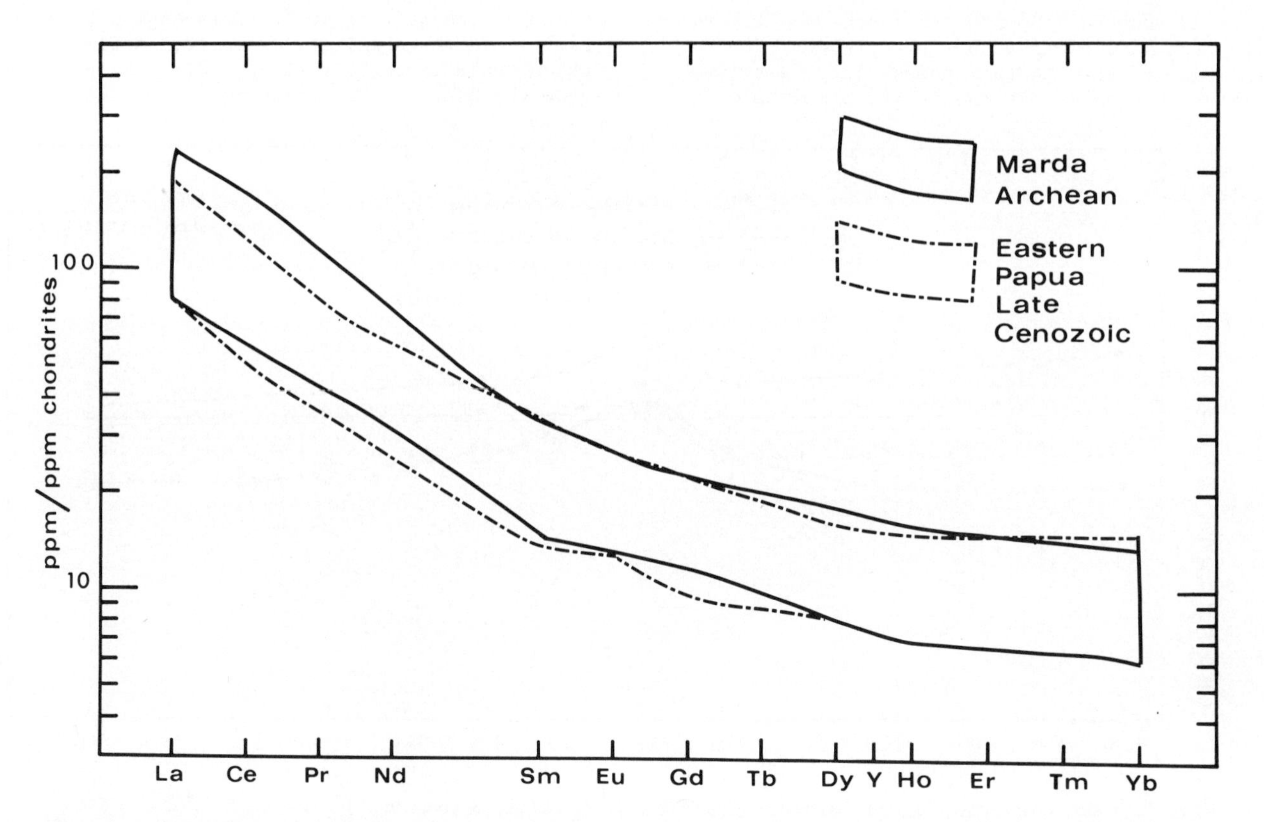

Fig. 4. Similarity between chondrite-normalised REE patterns for Marda Archean andesites (Hallberg and Taylor, unpub. data) and andesites from Eastern Papua (I.E. Smith pers. comm.).

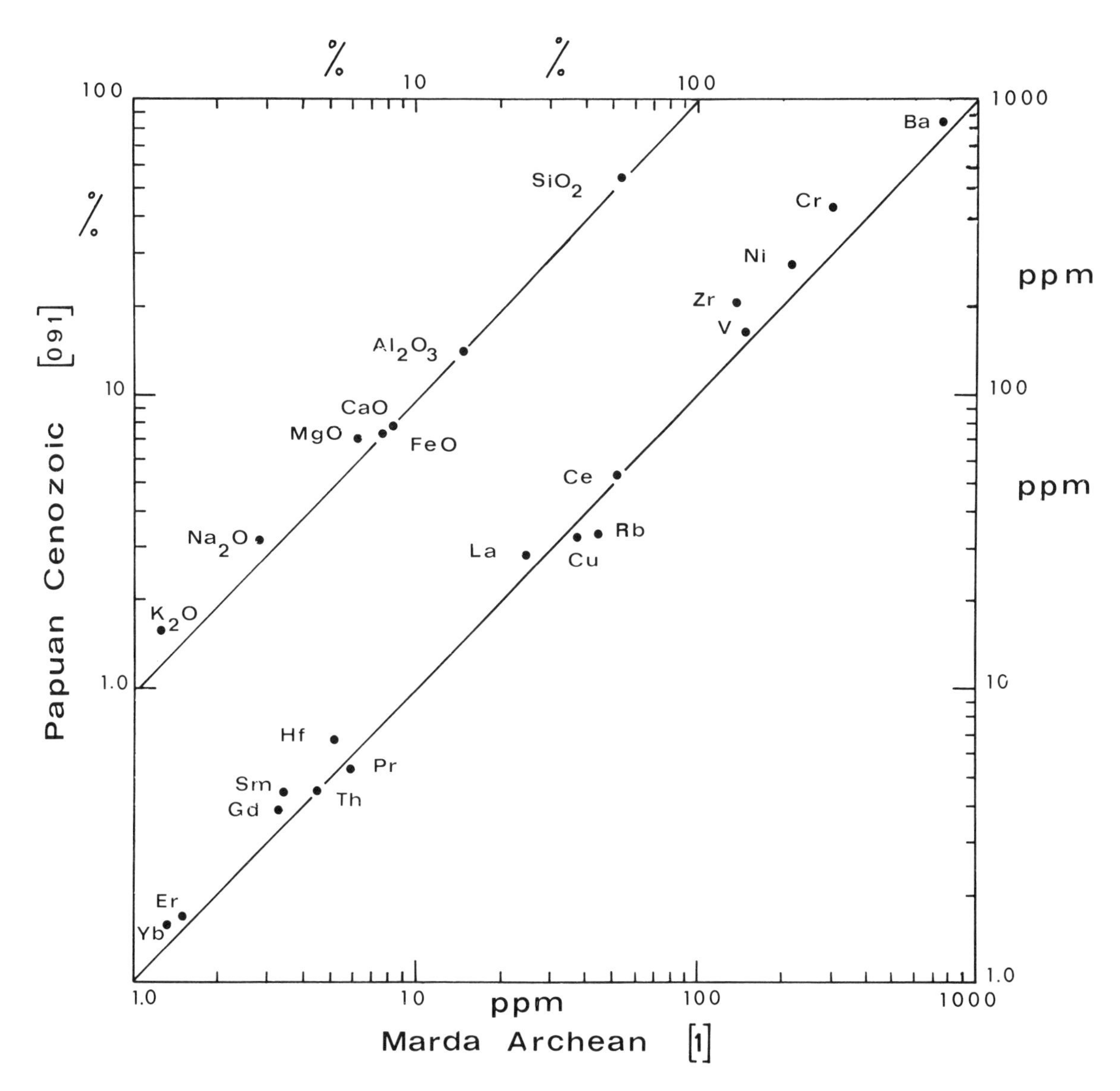

Fig. 5. Comparison of major and trace element data from a Marda Archean andesite (Hallberg et al. 1976, Hallberg and Taylor, unpub. data) and an andesite from Eastern Papua (I.E. Smith, pers. comm.). Points lying on the 45° diagonal line indicate equality of composition.

ponent. Condie et al. (1970) found similar high Ni contents in a more extended study of the Fig Tree group. The Ni and Cr values in the Kalgoorlie sediments show much scatter but the data fall among, and cannot be distinguished from those for post-Archean sedimentary rocks (Nance and Taylor, 1976a). The volcanogenic greywackes from Tamworth show similar values. Data for the Wind River greywackes, for Ni only (Condie et al. 1970) shows comparable values. Only the Fig Tree data is aberrant. But the Fig Tree group is older (3400 m.y. compared to the 26-2700 m.y. data from Kalgoorlie).

Could this effect be due to a secular trend? This concept was tested by analyzing the samples from the sedimentary enclaves from the Amitsoq gneisses. These are pre-3700 m.y. in age. The data shown in Figure 6 cannot be distinguished from more recent Ni and Cr values in sedimentary rocks, and indicate that the earliest Archean crust is not distinctively different for these elements. The high Ni and Cr contents in the Fig Tree shales are presumably derived from a local ultramafic source.

Summary and Discussion

If the sedimentary rock REE data provide an adequate sampling of the Archean crust, then by analogy with the compositions of modern island arcs and volcanogenic sediments (e.g. Baldwin greywackes), the overall composition of the

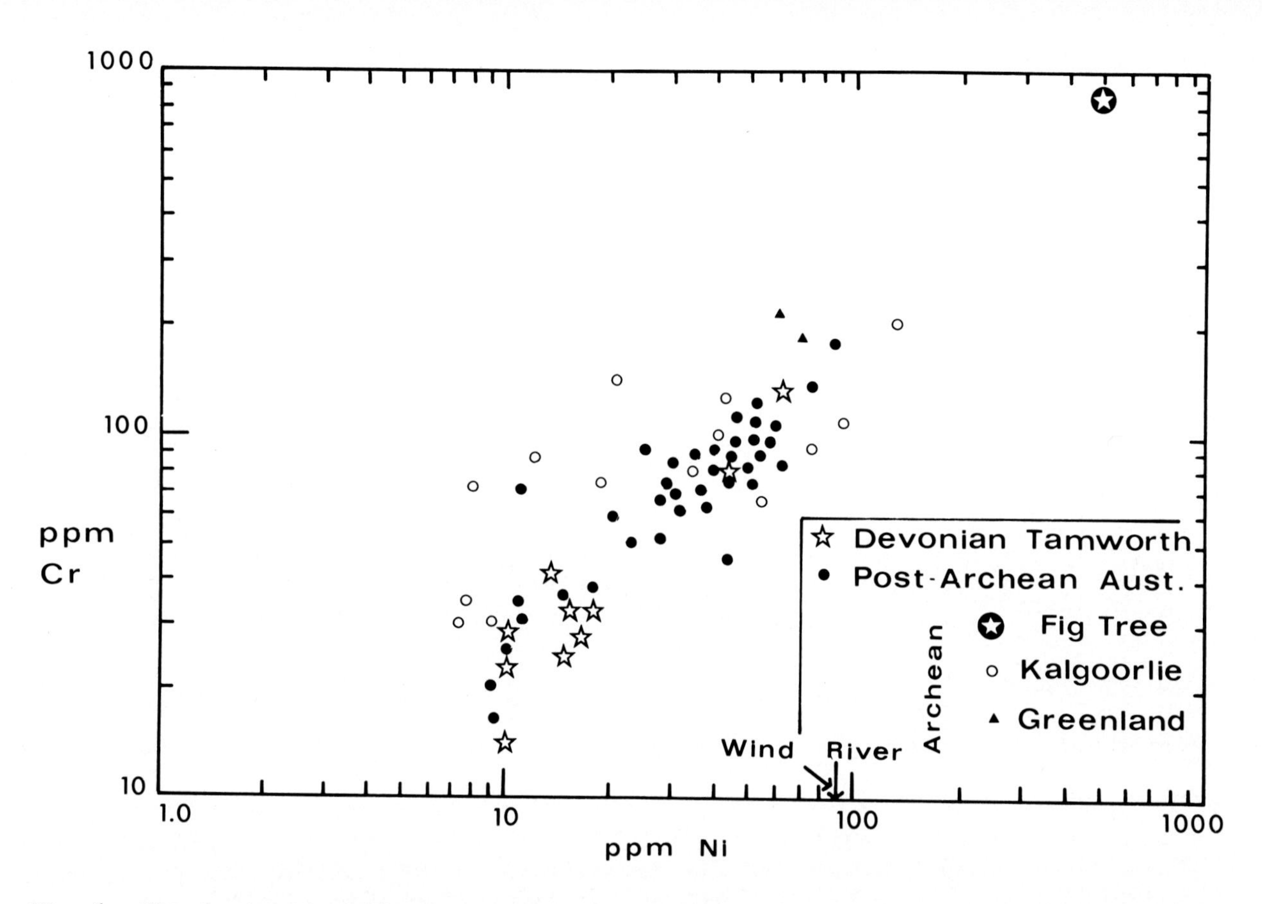

Fig. 6. The abundances of Ni and Cr in Archean and post-Archean sedimentary rocks. Data for Fig Tree Group from Danchin (1967) and for Wind River greywackes from Condie et al. (1970). All other data are from this laboratory (Method XRF, Analyst M. Kaye).

Archean crust exposed to weathering was similar to that of average present day island arc volcanic rocks (Jakeš and Taylor, 1974). Such a composition might be derived either from Archean calc-alkaline volcanism similar to present-day activity, or it might be derived from mixing of equal amounts of the tholeiitic and felsic components of the bimodal Archean volcanic rocks (Barker and Peterman, 1973).

It is unclear to what extent the sampling of Archean sedimentary rocks is biased toward locally derived volcanogenic sediments in greenstone belts. The wide distribution in space and time of the distinctive Archean REE patterns, argues for a thorough sampling of the upper crust.

An Early Anorthositic Crust? The interpretation that the Eu excess in Archean sediments might reflect an europium spike from an early anorthositic crust does not appear to be viable. Excess Eu can be due to local accumulation of feldspar, as in the Devonian Baldwin greywackes. An early anorthositic crust, based on analogy with the moon, should be revealed by the survival of very primitive strontium isotope ratios. The high Ca and Al contents required to produce such a crust would lead to the extensive occurrence of garnet in the upper mantle. This would lead to distinctive LREE enriched, HREE depleted crustal REE patterns. None of these effects appear to be present.

Lower Crustal Composition

Based on the model for total crustal and upper crustal composition (Table 1), it is possible to

TABLE 2. Predicted major element and normative composition of lower crustal rocks, based on assumed total crust of island arc volcanic composition (Table 1, CC) with upper one third crust of granodiorite composition (Table 1, UC).

	Wt %	CIPW	Norm
SiO_2	54.0	Quartz	1.5
TiO_2	0.9	Orthoclase	3.5
Al_2O_3	19.0	Albite	28
FeO*	9.0	Anorthite	35
MgO	4.1	Diopside	10
CaO	9.5	Hypersthene	20
Na_2O	3.35	Ilmenite	1.7
K_2O	0.6	*All Fe expressed as FeO	

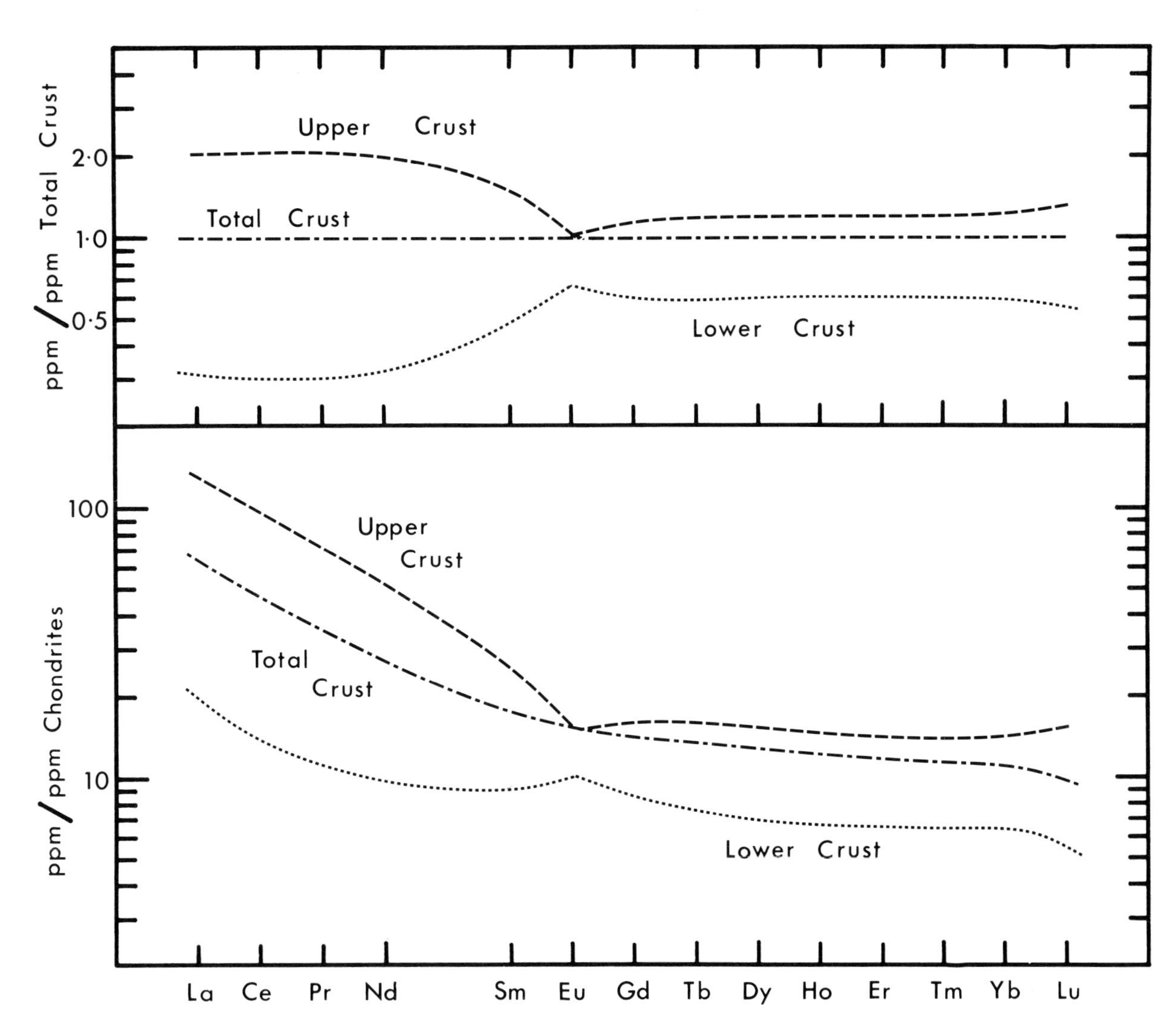

Fig. 7. Proposed REE patterns for the whole continental crust, compared with the REE patterns for the upper crust and the lower crust. The data for the whole crust, from Table 1, are based on the assumption that the continental crust has the composition of average island arc volcanic rocks. The values for the upper crust are assumed to be represented by the average sedimentary rock pattern (data from Table 1). The lower crustal composition is calculated from these data, assuming that it comprises two thirds of the total crust (see Table 2).

make some prediction about lower crustal compositions. Such estimates are heavily dependent on the thickness of upper crustal granodioritic material, relative to that of the residual lower crust. Nevertheless, it is clear that lower crustal material in this model should be characterised by a positive europium anomaly (Fig.7) relative to chondritic abundances, and depletion of large lithophile elements (e.g. K, Rb, Cs, Ba, Th, U). Based on the compositions in Table 1, residual lower crustal compositions, following extraction of 25-50% upper crustal granodiorite composition, contain normative mineralogies dominated by plagioclase.

Table 2 provides an estimate for the lower crust, assuming that the upper 1/3 of the continental crust (13 km) is composed of granodiorite. The lower crustal composition has high Ca and Al and the normative mineralogy is dominated by plagioclase. Such a composition would account for the Eu depletion observed in upper crustal rocks.

Acknowledgements. I wish to thank the Director and staff of the Lamont-Doherty Geological Observatory for their invitation to attend the meeting and for hospitality during the Ewing Symposium. I thank Joseph G. Arth and Gilbert N. Hanson for perceptive reviews of this paper.

REFERENCES

Arth, J.G. and G.N. Hanson, Geochemistry and origin of the early Precambrian crust of northeastern Minnesota. Geochim. Cosmochim. Acta, 39, 325-362, 1975.

Baragar, W.R.A. and A.H. Goodwin, Andesites and Archean volcanism of the Canadian shield. Proc. Andesite Conf., Oregon Dept. Geol. Mineral Ind., Bull. 65, 121-142, 1969.

Barker, F. and Z.E. Peterman, Bimodal tholeiitic-dacitic magmatism and the early Precambrian crust. Precambrian Res., 1, 1-12, 1974.

Bence, A.E. and S.R. Taylor, Petrogenesis of Mid-Atlantic Ridge basalts at DSDP Leg 37 sites 332A and B from major and trace element geochemistry. Initial Reports DSDP Vol. 37 (in press), 1976.

Chappell, B.W., Volcanic greywackes from the Upper Devonian Baldwin Formation, Tamworth-Barraba District, New South Wales, J. Geol. Soc. Aust. 15, 87-102, 1968.

Condie, K.C., J.E. Macke, and T.O. Reimer, Petrology and geochemistry of Early Precambrian graywackes from the Fig Tree Group, South Africa. Bull. Geol. Soc. Amer., 81, 2759-2776, 1970.

Danchin, R.V., Chromium and Nickel in the Fig Tree Shale from South Africa. Science, 158, 261-262, 1967.

De Paolo, D.J. and G.J. Wasserburg, Nd isotopic variations and petrogenetic models. Geophys. Res. Lett. (in press), 1976.

Dickinson, W.R. and W.C. Luth, A model for the plate tectonic evolution of mantle layers. Science, 174, 400-404, 1971.

Donn, W.L., B.D. Donn, and W.G. Valentine, On the early history of the earth. Bull. Geol. Soc. Amer. 76, 287-306, 1965.

Eade, K.E. and W.F. Fahrig, Chemical evolutionary trends of continental plates - a preliminary study of the Canadian shield. Bull. Geol. Surv. Can. 179, 51 pp, 1971.

Eade, K.E. and W.F. Fahrig, Regional lithological and temporal variation in the abundances of some trace elements in the Canadian shield. Geol. Surv. Can. Paper, 72-46, 46 pp, 1973.

Fyfe, W.S., The granulite facies, partial melting and the Archaean crust. Phil. Trans. R. Soc. Lond., A, 273, 457-62, 1973a.

Fyfe, W.S., The generation of batholiths. Tectonophys., 17, 273-83, 1975b.

Gill, J.B., Geochemistry of Viti Levu, Fiji and its evolution as an island arc. Contrib. Mineral. Petrol. 27, 179-203, 1970.

Gilvarry, J.J., The origin of ocean basins and continents. Nature, 190, 1048-1053, 1961.

Goodwin, A.M., Plate tectonics and evolution of Precambrian crust in Implications of Continental Drift to the Earth Sciences (Eds. D.H. Tarling and S.K. Runcorn) Academic Press, 1047-1069, 1973.

Gorton, M.P., Geochemistry and geochronology of the New Hebrides, Ph.D. Thesis, Australian National University, 1974.

Hallberg, J.A., C. Johnson, and S.M. Bye, The Archean Marda igneous complex, Western Australia. Precambrian Res. 3, 111-136, 1976.

Haskin, L.A., M.A. Haskin, F.A. Fry and T.R. Wildeman, Relative and absolute terrestrial abundances of the rare earths. In Origin and Distribution of the Elements, (editor L.H. Ahrens), pp. 889-912. Pergamon Press, 1968.

Jakeš, P., Geochemistry of continental growth. In Implications of Continental Drift to the Earth Sciences, (editors D.H. Tarling and S.K. Runcorn), Vol. 2, pp. 991-1001. Academic Press, 1973.

Jakeš, P. and S.R. Taylor, Excess europium content in Precambrian sedimentary rocks and continental evolution. Geochim. Cosmochim. Acta, 38, 739-745, 1974.

Jakeš, P. and A.J.R. White, Composition of island arcs and continental growth. Earth Planet. Sci. Lett. 12, 224-230, 1971.

Jakeš, P. and A.J.R. White, Major and trace element abundances in volcanic rocks of orogenic areas. Bull. Geol. Soc. Amer. 83, 29-40, 1972.

Kolbe, P. and S.R. Taylor, Geochemical investigation of the granitic rocks of the Snowy Mountains area, New South Wales, J. Geol. Soc. Aust., 13, 1-25, 1966.

Mason, B. and V.R. McGregor, Geochemistry of metabasaltic and metasedimentary enclaves in the Amitsoq gneisses of the Godthaab region, West Greenland. Precambrian Res. (in press), 1976.

Masuda, A., N. Nakamura, and T. Tanaka, Fine structures of mutually normalised rare-earth patterns of chondrites. Geochim. Cosmochim. Acta, 37, 239-248, 1973.

Moorbath, S. The geological significance of Early Precambrian rocks. Proc. Geol. Ass., 86, 259-279, 1975.

Nance, W.B. and S.R. Taylor, Rare earth element patterns and crustal evolution. I: Australian post-Archean sedimentary rocks. Geochim. Cosmochim. Acta (in press), 1976a.

Nance, W.B. and S.R. Taylor, Rare earth element patterns and crustal evolution, II: Archean sedimentary rocks from Kalgoorlie, Australia, Geochim. Cosmochim. Acta (in press), 1976b.

Ojakangas, R.W., Archean volcanogenic graywackes of the Vermillion District, Northeastern Minnesota. Bull. Geol. Soc. Amer., 83, 429-442, 1972.

Piper, D.Z., Rare earth elements in the sedimentary cycle: A summary, Chem. Geol. 14, 285-304, 1974.

Poldervaart, A., Chemistry of the Earth's Crust, in The Crust of the Earth (Ed. A. Poldervaart). Geol. Soc. Amer. Spec. Paper 62, 119-144, 1955.

Ringwood, A.E., Composition and evolution of the upper mantle, Amer. Geophys. Union Monog. 13, 1-17, 1969.

Shaw, D.M., J. Dostal and R.R. Keays, Additional estimates of continental surface Precambrian shield composition in Canada. Geochim. Cosmochim. Acta, 40, 73-83, 1976.

Taylor, S.R., Abundances of chemical elements in the continental crust: a new table. Geochim. Cosmochim. Acta, 28, 1273-1285, 1964a.

Taylor, S.R., Trace element abundances and the chondritic earth model. Geochim. Cosmochim. Acta, 28, 1989-1998, 1964b.

Taylor, S.R., The origin and growth of continents. Tectonophysics, 4, 17-34, 1967.

Taylor, S.R., Geochemistry of andesites. In Origin and Distribution of the Elements, (editor L.H. Ahrens), pp. 559-585. Pergamon Press, 1968.

Taylor, S.R., A.C. Capp, A.L. Graham and D.H. Blake, Trace element abundances in andesites II. Saipan, Bougainville and Fiji. Contrib. Mineral. Petrol. 23, 1-26, 1969.

Taylor, S.R. and A.J.R. White, Geochemistry of andesites and the growth of continents. Nature, 208, 271-273, 1965.

Whitford, D.J., Geochemistry and petrology of volcanic rocks from the Sunda Arc, Indonesia Ph.D. Thesis. Australian National University, 1975.

Wildeman, T.R. and K.R. Condie, Rare earths in Archean graywackes from Wyoming and from the Fig Tree Group, South Africa. Geochim. Cosmochim. Acta, 37, 439-453, 1973.

Wildeman, T.R. and L.A. Haskin, Rare earths in Precambrian sediments. Geochim. Cosmochim. Acta 27, 419-438, 1973.

Vinogradov, A.P., Average contents of chemical elements in the principal types of igneous rocks of the Earth's crust. Geochemistry, 1962, 641-664 (English trans.), 1962.

CENOZOIC EXPLOSIVE VOLCANISM RELATED TO EAST AND SOUTHEAST ASIAN ARCS*

Dragoslav Ninkovich

Lamont-Doherty Geological Observatory of Columbia University
Palisades, New York 10964

William L. Donn

Lamont-Doherty Geological Observatory of Columbia University,
Palisades, New York 10964

Department of Earth and Planetary Sciences
of the City College of New York, New York 10031

Abstract. A study of the history of Cenozoic explosive volcanism has been made using Deep Sea Drilling Project (DSDP) and piston core data, all from the Indian Ocean off Indonesia and the Western Pacific Ocean. Data from piston cores, which only penetrate Recent and Pleistocene sediments, are used primarily to establish the geographic limits of the respective ash-layer zones. DSDP sites usually yield sparse core recovery but the drilling penetrates much deeper, often through the entire Cenozoic Era and thus permit some assay of the volcanic history. Some DSDP sites are on plates that have moved toward the subduction zone; others are on plates fixed relative to the volcanic arc. Within the limits of the sparse DSDP core information, data from the fixed zones do not indicate a significant change in the frequency of occurrence of ash layers. Data from sites on moving plates show an increase in the number of ash layers in the late Pliocene-Pleistocene record. But this frequency increase corresponds to the time that the drilling localities entered the ash-layer zone and also moved close to the volcanic sources. As such, the higher frequency of ash layers in these regions cannot be taken to indicate a trend in the frequency of explosive volcanism.

Introduction

The purpose of this report is to present an interpretation of data bearing on the history of explosive volcanism during the Cenozoic Era. To do this we selected piston core and DSDP data from the northeast Indian Ocean and the northwest Pacific. All of the marine areas involved are downwind from the volcanic sources extending from western Indonesia through all of the island arcs of southeast and east Asia into Kamchatka (Figures 1 and 3). All of the Lamont piston cores from these areas were examined carefull for ash layers. Although continuous data retrieval is possible from these cores, penetration is limited to the Pleistocene. The value of these limited data is that they permit us to establish the extent of the zone of ash layers downwind from the sources. The exact limit of such a zone may of course vary in time depending on wind strength and degree of explosive activity.

To investigate explosive volcanism prior to the Pleistocene we utilize information obtained from DSDP sites both by direct analysis of cores and of published DSDP Reports. Unfortunately, core retrieval from DSDP drill holes is usually very sparse as will be shown below. This leads to difficulties in interpretation. Also, all interpretation of history based on DSDP data, extending over some 60 m.y., must include consideration of the sea floor motion relative to the volcanic sources, a motion of the order of 100 km per m.y. Clearly, older volcanic material has been lost through subduction. Further, old non-ash-layer bearing sediments have been carried into the ash-layer zone. This can lead to serious misinterpretations about an absence of volcanism corresponding to the ages of the sediments involved.

Data from Indonesia

Brief Summary of the Volcanic History of Indonesia

The present volcanic Indonesian Arc, apart from Sumatra, was formed in early Tertiary (Van Bemmelen, 1949; Westerveld, 1952). The history of Sumatra extends back to Paleozoic. The complete Cenozoic tectonic and magmatic history can be divided into two major phases: the first extended into Early Miocene and the second began in Late Miocene and lasted until the present. Near-

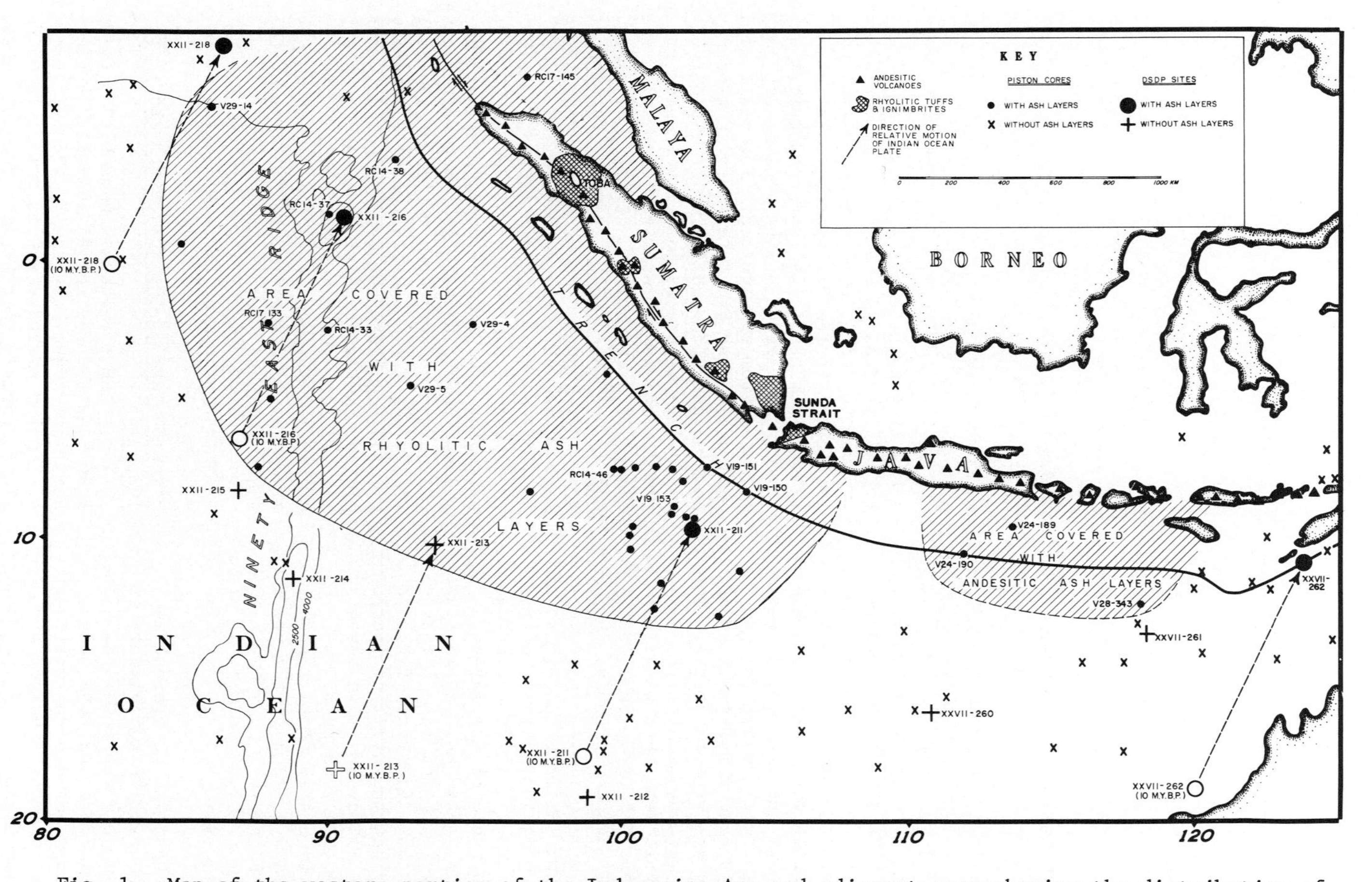

Fig. 1. Map of the western portion of the Indonesian Arc and adjacent seas showing the distribution of rhyolitic tuffs and ignimbrites and andesitic volcanoes on the islands and related ash deposits in deep sea sediments. Small dots show locations of piston cores containing ash layers. Locations without ash layers are shown by Xs. Large dots show locations of Deep Sea Drilling Sites (DSDP) from which coring sections containing ash layers have been obtained. Those without such layers are indicated with a + mark. The shaded region defines the areas of Pleistocene ash layers on the basis of piston core data. The open circles show locations of the DSDP ash layer sites about 10 m.y. ago as relocated on the basis of sea floor motion.

ly complete submergence of the arc occurred during Middle Miocene time. Both phases were characterized by intense magmatic activity except for the submergence period when almost no activity occurred. Magmatism in the early phase was limited to intrusive activity until Early Miocene when extensive andesitic volcanism became active until the submergence.

The second magmatic phase in Late Miocene began with granitic intrusion related to the uplift of sediments of Middle Miocene age. The associated explosive volcanism in the eastern section of the arc (Figure 1) began and continued as andesitic eruptions. In westernmost Java and in Sumatra, volcanism began as rhyolitic eruptions and as inferred from deep-sea cores, continued until Late Pleistocene. Eruptions from the area of the Sunda Strait (between Sumatra and Java) had a somewhat different time history from those in north Sumatra. The former started in Late Miocene and lasted until Pleistocene (Van Bemmelen, 1949) and the latter was manifest as a single eruption about 70,000 years ago (Ninkovich et al., 1971). The rhyolitic eruptions in Sumatra and westernmost Java were followed by recent andesitic volcanic activity. The recent andesitic eruptions were much less violent than the preceding rhyolitic activity judged from their limited extent as observed in cores V19-150 and 151 from the trench south of the Sunda Strait (Figure 1). A large volume of unpublished data from Lamont marine cores indicate that the very violent rhyolitic ash-producing activity ceased about 70,000 years ago and was continued by quieter, less explosive andesitic eruptions. Because south Sumatra displays a series of many layers of rhyolitic tuffs beginning in Late Miocene compared to less numerous occurrences in north Sumatra it appears that south Sumatra experienced more frequent explosive volcanism. This becomes important in core data interpretation.

The Marine Data from Indonesia

The location of all of the Lamont piston cores taken off Indonesia are shown in Figure 1 together with the sites of the Deep-Sea Drilling Project (Von der Borch et al., 1974; Heirtzler et al, 1974). All of these piston cores have been examined for the occurrence of volcanic ash layers. All core locations shown as small black dots contain such layers; those shown by Xs are barren. In the shaded zone to the west of Sumatra the ash is rhyolitic with the addition of andesitic ash near the top of the two trench cores, V19-150 and 151. Cores from the shaded zone to the southeast of Java contain only andesitic ash.

The distribution of the ash zones appears to be related primarily to the tropospheric trade winds which prevail in the area. The composition of the ash in each of the shaded zones matches that observed on the islands and indicates that the known volcanic activity on the respective islands is the source of the adjacent marine ash layers. All of the piston core recoveries are limited to the Pleistocene.

To reach ages older than Pleistocene, DSDP site observations are used. The four DSDP sites bearing ash layers are shown with heavy dots in Figure 1. Barren sites are indicated with large plus symbols. Interpretation of information from these sites must include consideration of the motion of the Indian Ocean floor relative to the Indonesian arc because sediment and ash is continuously destroyed by subduction as plate motion carries this material to the trench bordering the arc. The dashed lines between the drilling sites and the open circles in Figure 1 indicate the amount of plate motion (Minster et al., 1974) that occurred since the Late Miocene explosive volcanic phase began (about 10 m.y. ago). Note that the only site that would have been in the ash layer zone continuously is Site 216. Site 211 would have entered the ash layer zone more recently. The other two sites are essentially out of the ash layer zones. As noted previously, the past width of the ash layer zone may have been slightly different.

The core recovery information from the four ash-bearing DSDP sites (Von der Borch et al., 1974; Heirtzler et al., 1974) is shown in Figure 2, in which stratigraphic depth is plotted against absolute age established by Berggren and Van Couvering, 1974. Sites 216 (A) and 211 (B) have short bottom extensions to the Cretaceous basement which is not included here. Continuous core recovery was obtained for Site 262 (D) which includes the entire Pleistocene and Late Pliocene. The other three sites were cored discontinuously as indicated by the stratigraphic breaks. Occurrences of ash layers in the recovered cores are shown by black dots on the curves. The vertical extent of the dashed boxes indicates the time uncertainty for the absolute ages of the curves. We will consider Sites 216 and 211 together because of their position in the ash-layer zone.

Site 216 (Figure 2A), the only site continuously in the ash-layer zone since the beginning of the rhyolitic explosive phase beginning in Late Miocene, shows two ash layers in Pleistocene, one in Pliocene and one in Late Miocene, interpolated to be about 7 m.y. old. From 9 m.y. ago to the Cretaceous basement no ash layers are found. No increasing frequency of volcanism can be inferred from this site for it would be a remarkable coincidence to expect that the small sections cored were taken from the only two ash layers in Pliocene and Late Miocene. We would, on the contrary, expect as many ash layers in the missing sections as in the cored sections. Also, the absence of ash layers prior to 9 m.y. is easily explained. The site had not yet entered the ash-layer zone. Site 211 shows five ash layers in Pleistocene and one in the Late Pliocene with none earlier. Again, there is no reason not to expect that other ash layers occurred in the missing section between Late

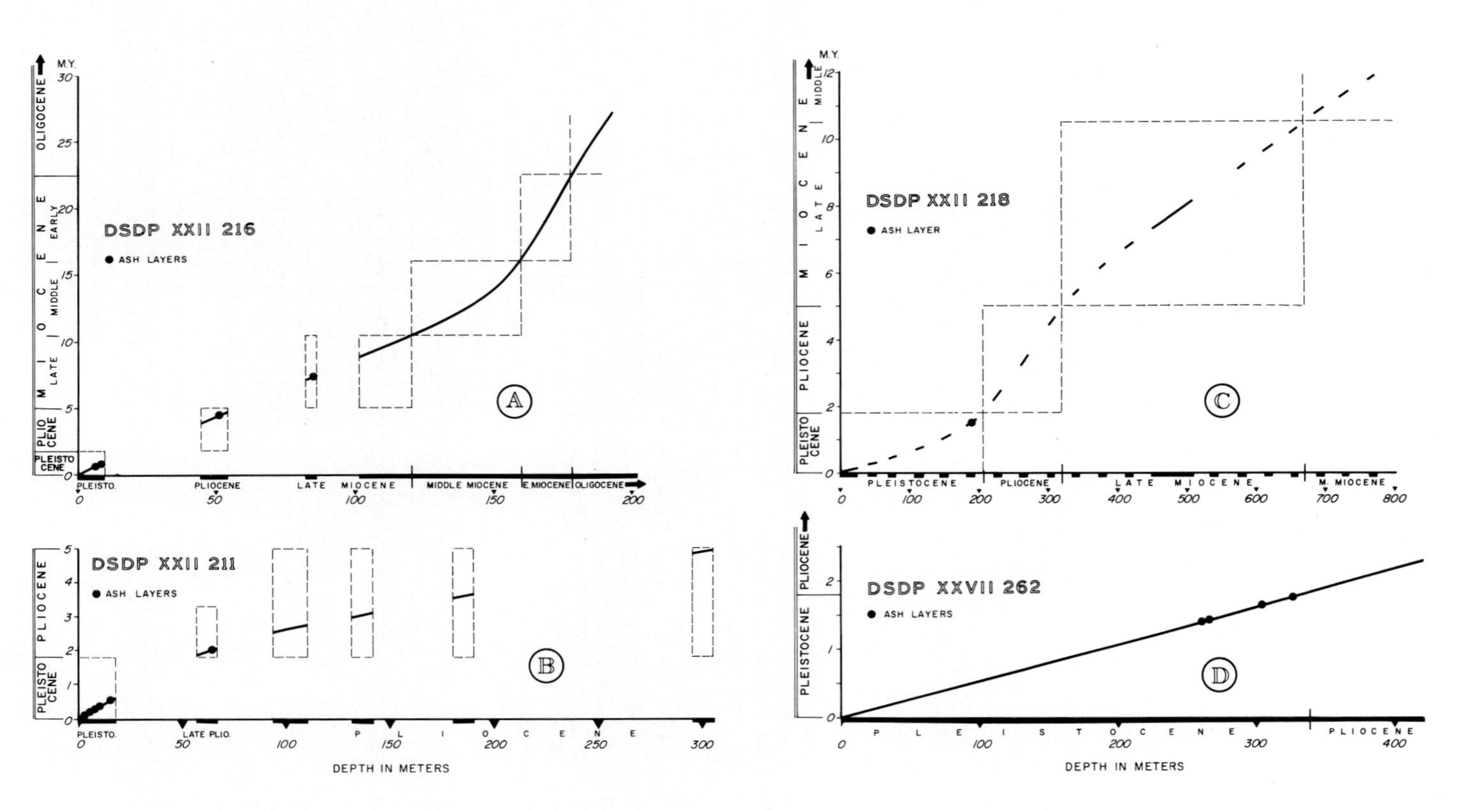

Fig. 2. Diagrams showing amount of core recovery and ash layer distribution (black dots) for the four DSDP sites in Fig. 1. The boxes show the uncertainty in the ages for the sediment recovered.

Pleistocene and Late Pliocene. Also, we cannot say whether any ash layers occurred prior to the Late Pliocene event and the underlying barren zone (∿100 m depth). But the lack of ash layers below the 100 m depth is explainable because the site would not yet have entered the ash-layer zone. Of these two sites, only 216 is really representative of the explosive volcanic history since Late Miocene. Site 211 gives only the later part of the history. We note that the number of ash layers for the Pleistocene of Site 211 is greater than that for 216 within the same ash-layer zone. This appears to be a consequence of the motion of Site 211 close to the source of most frequent explosive volcanism located, as noted earlier, in south Sumatra.

Site 218 is just beyond the northern edge of the Pleistocene ash-layer zone (Figure 1). Core recovery data for this site is shown in Figure 2C which indicates that semicontinuous recovery was obtained to Middle Miocene, the limit of drilling. Only one ash layer occurs at the beginning of Pleistocene. From the site motion in Figure 1 we would expect that any ash that does occur would be only in the upper part of the record. Enough core data exists above and below this ash layer to indicate that no other significant number of ash falls occurred at this site for the time interval represented. This is also what would be expected from the motion of the site relative to the ash-layer zone.

Site 262 (Figure 1) which was drilled in the Java Trench outside of an ash-layer zone was cored continuously through Late Pliocene. As with the other three sites, the motion since Late Miocene is shown by the broken line from the open circle to the present location. Four ash layers occur in Pleistocene (Figure 2D). The data from this core are not particularly instructive about the volcanic history of the region because: (1) the drilling operation did not penetrate below Late Pliocene and (2) the section above the ash layers, when the site presumably reached the trench, shows thick sediments with nannofossils that include Cretaceous and Cenozoic forms (Heirtzler et al., 1974). This indicates rapid sedimentation from turbidity currents and obscures the Pleistocene chronology of the ash layers.

Most of the sites located outside the ash-layer zone (plus marks in Figure 1) have excellent core recovery for the Cenozoic. But none of these sites contains ash layers. This emphasizes the importance of considering the position of the DSDP sites relative to ash-layer zones in making historical interpretations.

Kennett and Thunell (1975) also studied the DSDP data. Their count of ash layers from the same sites for Indonesia shows an increase in the number of layers beginning 3 to 4 million years ago and they interpret this as evidence for a greatly increased rate of volcanism. Although we have similar results in Figure 2, it is our contention that this does not represent an increase in volcanism but is a consequence of the sea floor plate motion relative to the ash-layer zone. Unfortunately, they did not have all of the pertinent data on the areal distribution of ash layers as available from our piston cores.

Support is given our conclusion by the work of Vallier and Kidd (1976) who give a detailed volcanogenic analysis of all of the DSDP site core material from the entire Indian Ocean. Rather than recording only ash layers they studied the distribution of volcanogenic material throughout the entire sediment series. The distribution of silicic glass shards in the sediments shows essentially continual production since the beginning of Late Miocene. Although they analyzed all sites in the Indian Ocean a particularly detailed analysis of glass shards was made for Site 213 (Figure 1) which contains no distinctive ash layers despite the continuous, disseminated shard content. It seems clear that the absence of discrete ash layers is a consequence of the location of the site relative to the ash-layer zone since Late Miocene. A small amount of ash, not adequate to accumulate a distinct layer, is always airborne to great distances. It is this material that constitutes the disseminated material. Vallier and Kidd (1976) conclude from the detailed study of Site 213 and others in the Indian Ocean that pulses of explosive volcanism occurred rather uniformly in the last 10 m.y. No evidence of a Late Cenozoic increase in such activity shows in their data.

Data from the Northwest Pacific

In examining the northwest Pacific region we recognize three areas with somewhat different histories: the Bering Sea north of the Western Aleutians; the chain of arcs from Kamchatka through the Kurils and Japan; and the seas adjacent to the Marianas Arc. A large number of piston cores and DSDP data exist here and all of the ash could have fallen into the sea so that a good marine record is expected.

The Cenozoic volcanic history of the Japanese, Kuril and Kamchatka arcs is very similar to that of Sumatra (Minato et al., 1965; Gorshkov, 1970). A "Green tuff" phase of activity which occurred in Early Miocene was followed by a profound submergence of the Japan and Kuril arcs in Middle Miocene. The Late Miocene emergence was associated with rhyolitic explosive activity which continued into Pleistocene. This in turn was followed by a more recent andesitic volcanic phase which is currently active.

All available piston cores from the Lamont collection from these regions were analyzed for the presence of volcanic ash layers. Locations are shown in Figure 3 which also indicates those DSDP sites that contained ash layers. The shaded area shows the distribution of volcanic ash layers in piston cores of Pleistocene age (Horn et al.,

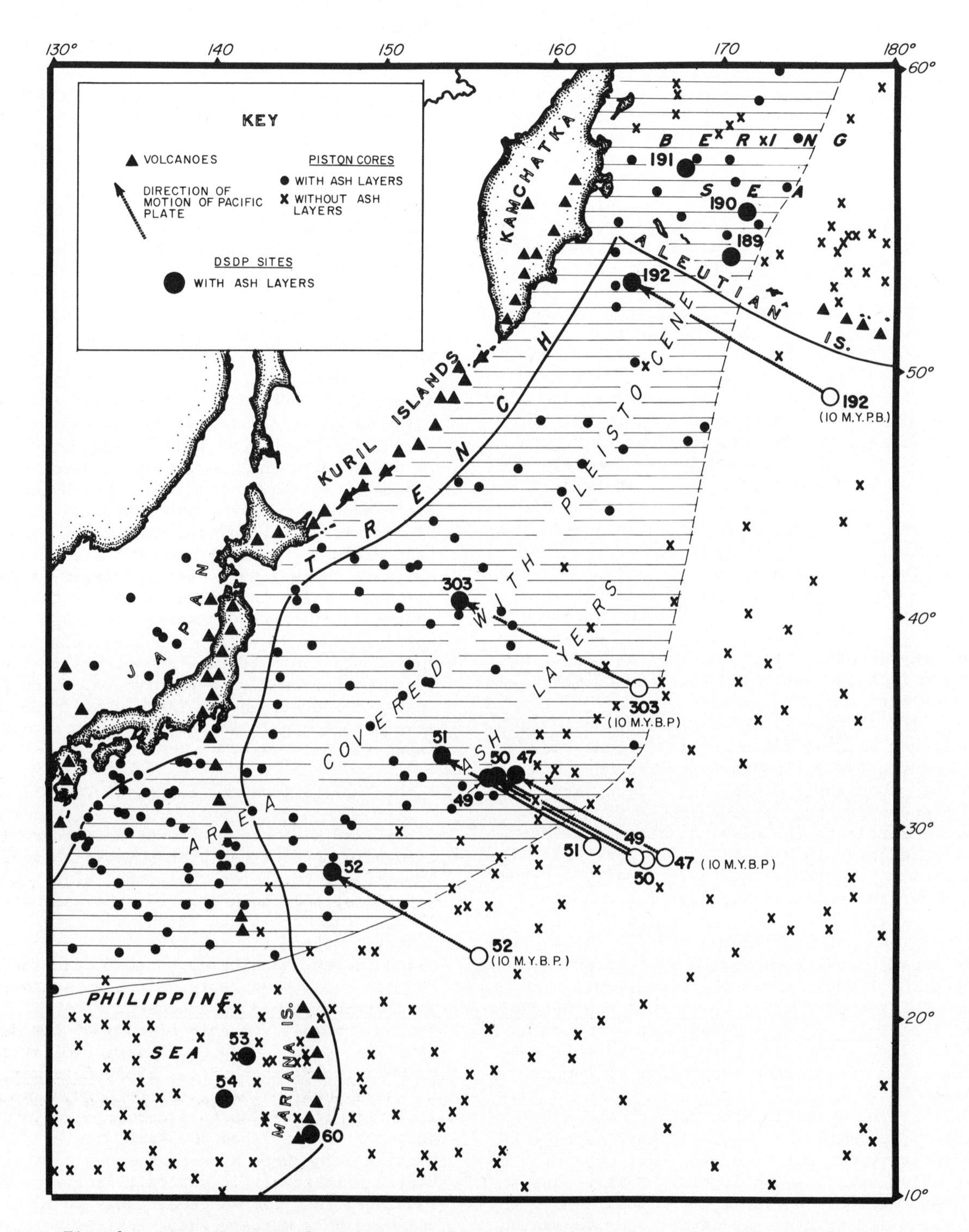

Fig. 3. Distribution of ash layers for the northwest Pacific. All symbols are used as in Fig. 1. Note that north of the Aleutians and in the Philippine Sea, the DSDP sites are stationary relative to the arcs bearing the volcanic ash sources.

1969; Hays and Ninkovich, 1970). The ash indicated in this region has been carried by the westerlies from the adjacent arcs. The Marianas (which lie in the trade winds) have no neighboring ash-layer zone. This is expected because the geological history of the Marianas (Cloud et al., 1956; Tracey et al., 1964) shows that the intense explosive volcanism of the region was limited to

Oligocene and Miocene. We will examine first the data from DSDP sites for the Marianas (Heezen et al., 1971) and Bering Sea (Scholl et al., 1973) as these sites were stationary relative to adjacent arcs during Cenozoic. Then we will consider data from the other sites from the northwest Pacific plate (Heezen et al., 1971) which has been in motion relative to the adjacent arcs.

The Marianas Region

As noted, no Pleistocene ash-layer zone lies adjacent to the Marianas. Two DSDP sites (Figure 3) with ash layers exist to the west of the Marianas, in the Philippine Sea and one to the east, between the arc and the trench. Data from these sites, compiled from the DSDP Reports (Heezen et al., 1971), are shown in Figure 4. Note that core recovery is very sparse at these sites. For Site 53, apart from a 20 m sediment section barren of ash on top and a thin barren section on bottom, all of the cored sediment described contains numerous ash layers with microfossils that indicate ages. The ash deposits do not occur above Late Miocene and reach at least Early Oligocene. Although only one of the three sites (Site 53) has recovery from the top of the hole, the absence of volcanic ash in the upper 20 m correlates with a similar absence in the records of the piston cores in this region.

Sites 54 and 60 (Figure 4) have an even poorer percentage of core recovery and at both sites all recovery contained sediments interbedded with numerous ash layers of Middle Miocene age. From these data we can only conclude that explosive volcanism in the Marianas region extended from Early Oligocene to Late Miocene. No significant conclusions about volcanic frequency can be drawn from these data.

The tectonic history of the Marianas region (Karig, 1971) shows that these sites are located on portions of the sea floor that have been essentially stable relative to the Marianas throughout the Cenozoic, unlike the DSDP site locations described for the Indian Ocean. The composite of these sparse data gives a very rough, discontinuous history of most of the Cenozoic and, for the Marianas at least, an absence of Late Cenozoic explosive volcanism seems evident.

The Bering Sea

The three sites 189, 190 and 191, in the Bering Sea (Figure 3) are located on a plate that has been stationary relative to Kamchatka (Morgan, 1968; Le Pichon, 1968) and like those for the Marianas, could give a fairly complete Cenozoic record of explosive volcanism in the neighboring arc.

Figure 5 shows the core recovery information for the three sites in the Bering Sea. Note again how poor actual core recovery is for these sites. Of the three, Site 190 has the largest percent of core information and penetrates sediment of Late Miocene. The number of ash layers in the sediment of each recovered section is indicated with adjacent numbers. Ash layers are present in all but one thin section. The number of ash layers is simply proportional to the thickness of the recovered section and indicates nothing, in this case, about the frequency of explosive volcanism. If anything, uniform activity is indicated from Late Miocene to the present time. Recovery at Sites 189 and 191 is so poor that little interpretation can be made, but the data generally agree with the above conclusion.

The Northwest Pacific Plate

The Pacific Ocean plate is moving toward Asia and is subducted along the trenches marginal to the chains of arcs in the Western Pacific (Figure 3). The amount of plate motion (Minster et al., 1974) is shown by the broken arrows that connect present DSDP sites that bear ash layers with their locations 10 m.y. ago. We would not expect ash layers at these sites until the site vectors entered the ash-layer zone shown by the shaded region developed from the ash distribution in piston cores penetrating Recent and Pleistocene sediments.

Logs of the core recovery data for the DSDP sites east of Japan are shown in Figure 6. Of the five sites east of Japan, Site 47 has nearly continuous recovery of the Cenozoic geologic record. Only two ash layers are found, one in Early Pleistocene and the second in Late Pliocene. According to the site vector in Figure 3, no older ash is expected because the site lay outside the ash-layer zone.

Core data from the adjacent Sites 49 and 50 (Figure 6) show Pleistocene sediment lying on Cretaceous sediment. Although a zone of intermixed ash and sediment is present in the Pleistocene of each core, no historical interpretation can be made. At Site 51, core recovery is very scanty and a thin sediment section of Early Pleistocene and Late Pliocene is totally intermixed with sediment. Again, no interpretations can be made. For Site 303, four thin core sections were recovered above Cretaceous (Larson et al., 1975). One zone of intermixed ash and sediment is present in the recovered part of the Pleistocene and one ash layer is present in Late Miocene. The entire site vector for the last 10 m.y. lies within the ash-layer zone so that ash older than that in Sites 47, 49, 50 and 51 is expected. All of the data from sites east of Japan are essentially useless in interpreting the history of explosive Cenozoic volcanism, but do not indicate any Late Cenozoic increase in such activity. We should note that we disregarded Site 52 shown in Fig. 3 despite its good core recovery and numerous ash zones because the record from Miocene to Recent is reported (Heezen et al., 1971) to be thoroughly mixed.

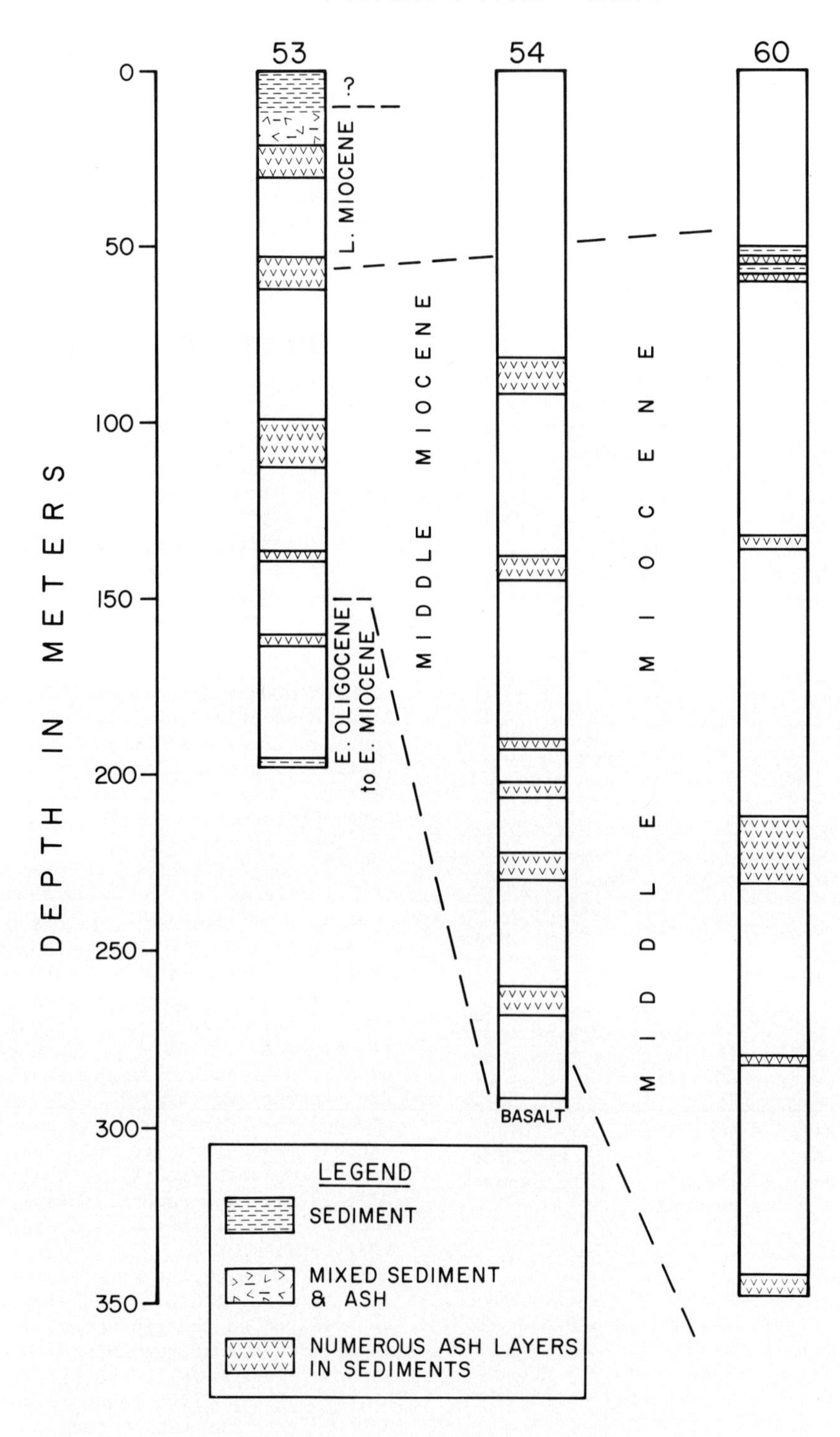

Fig. 4. Columns showing the lithology in the relatively small core recovery zones for DSDP sites in the Marianas area.

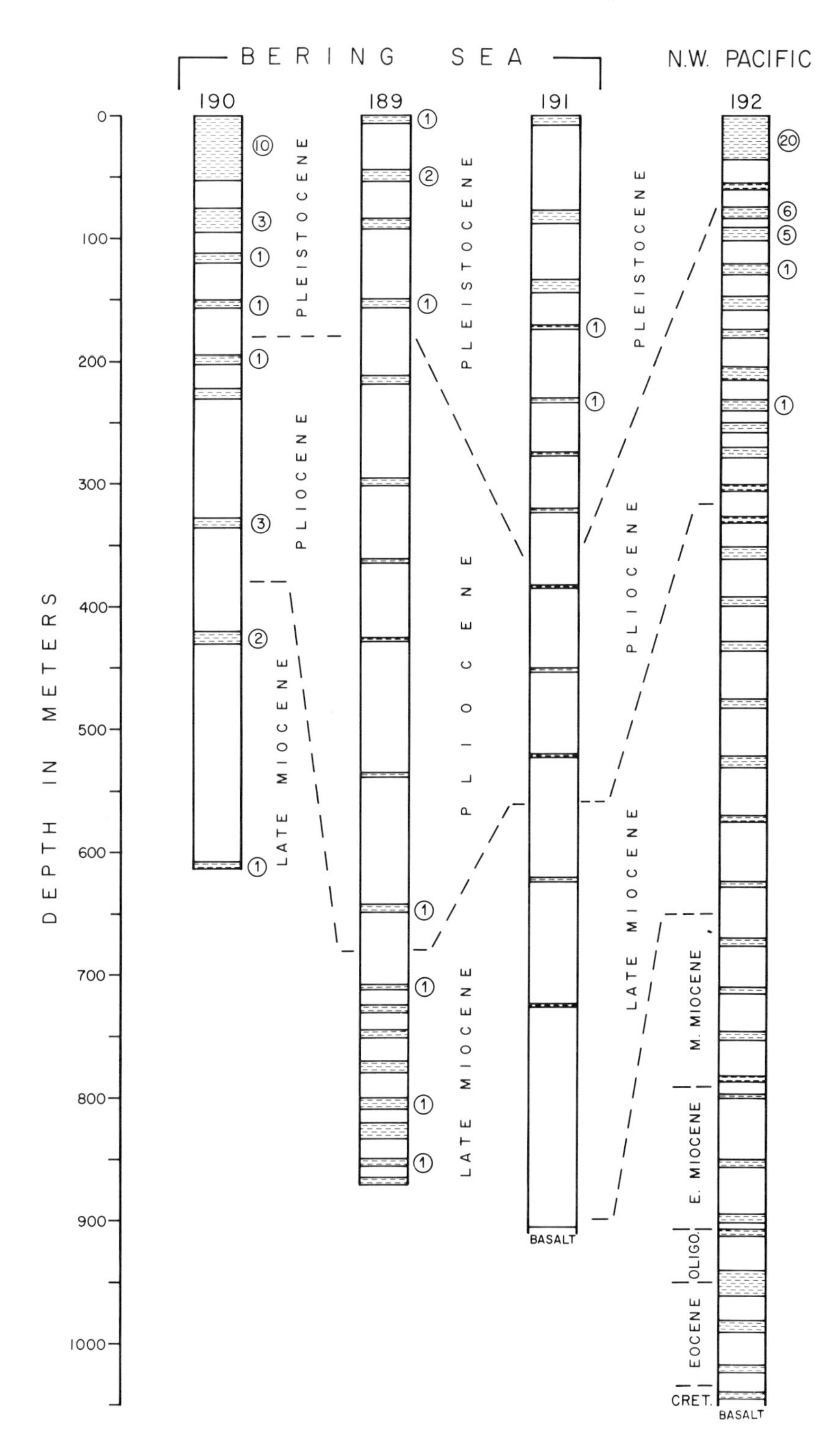

Fig. 5. Columns showing vertical distribution of sediment zones recovered in cores from DSDP sites east of Kamchatka. Circled numbers give the number of ash layers in the recovered sediment sections.

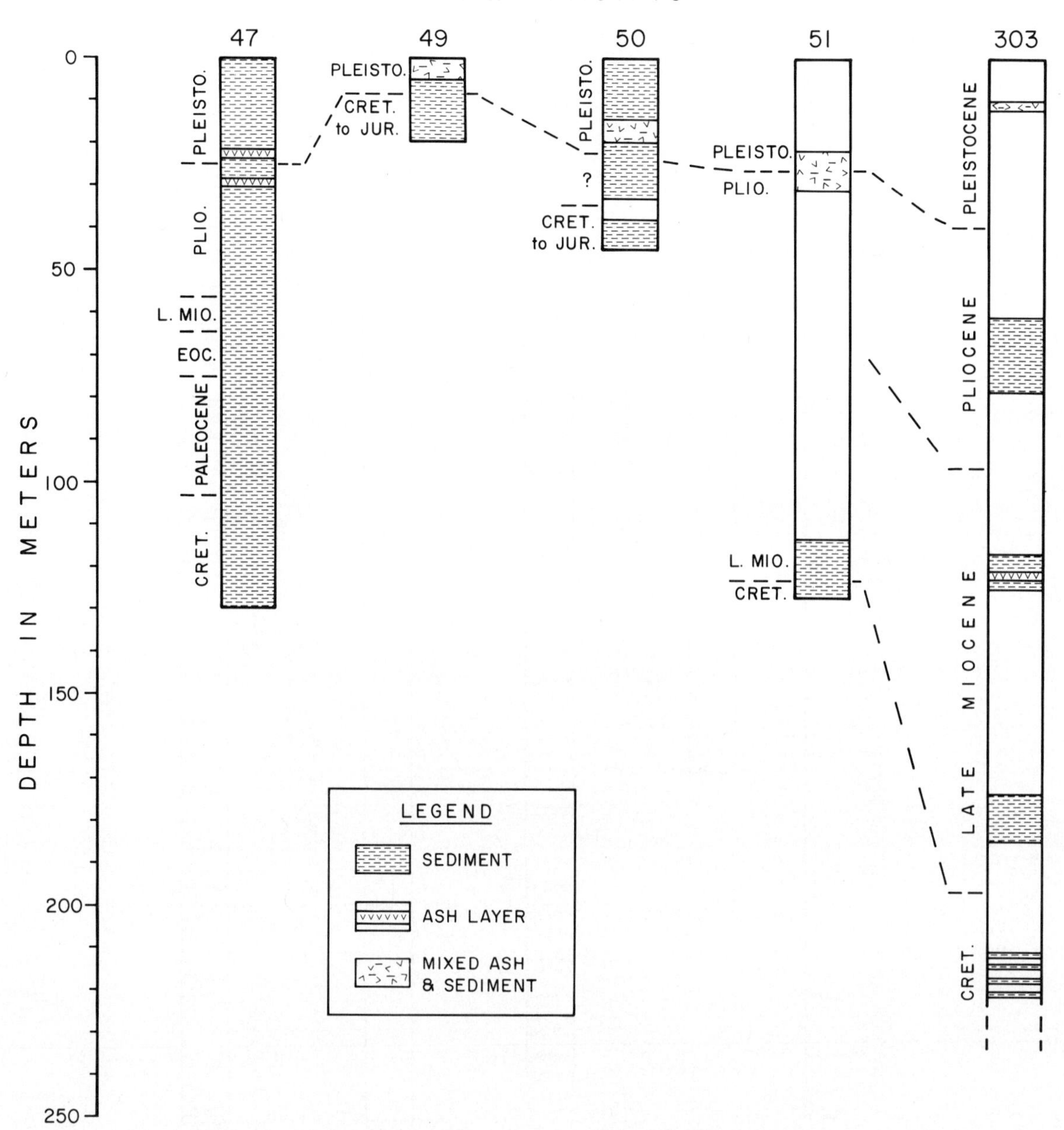

Fig. 6. Columns showing lithology and core recovery for the DSDP sites east of Japan.

Site 192, far to the north, but still on the Pacific plate (Fig. 3), lies close to the sites described for the Bering Sea. Owing to a scarcity of piston cores (particularly with ash layer) for many degrees immediately south of the Aleutians, the eastern limit of the ash-layer zone cannot be located too precisely in these areas. Apparently the site vector entered the ash-layer zone after about 5 m.y. ago. As with the Bering Sea sites, the ash in Site 192 probably came from Kamchatka. But, unlike these sites it does not contain ash older than Middle Pliocene (Figure 5) because of the plate motion. Following Middle Pliocene, there appears to be a numerical increase in ash layers for this site. This is reasonable because the site moved so close to the volcanic sources and could have received ash from relatively small eruptions. This increase in the number of ash layers as the Site 192 approaches the volcanic source is just like the case of Site 211 off south Sumatra as compared to Site 216 off north Sumatra.

Conclusions

From the data and analysis given we may conclude that: (1) the amount of data available from DSDP sites is very low, leading to a dangerous potential for serious errors in interpretation; (2) interpretation that is made must be consistent with the data of sea floor plate motion; (3) the data from Recent and Pleistocene piston cores provide a valuable reference for the time when ocean plates entered the marine ash-layer zone marginal to island arcs; and (4) the best conclusions we can reach on the basis of information from Indonesia and the Western Pacific area are (a) the increase in the number of ash layers in the deep sea cores of Late Cenozoic is explainable by the motion of the sea floor and the data sites, relative to the volcanic ash-layer zones as well as motion toward the eruptive sources of volcanogenic sediments; (b) approximate uniformity of explosive volcanism is indicated by data from sites that were essentially motionless relative to the volcanic source as is the case with the western Bering Sea and the Marianas region.

Clearly the present interpretation of the history of Cenozoic explosive volcanism is seriously limited by the sparseness of data from core material older than the Pleistocene. More continuous core recovery is needed to establish a more complete history of volcanism for each arc.

---

* Lamont-Doherty Geological Observatory Contribution No. 2410.

Acknowledgments. Drs. W. Alvarez, B.C. Heezen and R.L. Larson read the manuscript and offered many valuable comments and suggestions. The study was supported by the National Science Foundation Grant OCE72-01707, and by grants from the U.S. Steel Foundation and NASA Institute for Space Studies.

References

Berggren, W.A., and J.A. Van Couvering, The Late Neogene biostratigraphy, geochronology and paleoclimatology of the last 15 million years in marine and continental sequences, Paleogeogr., Paleoclim. and Paleoecol., 16, 216, 1974.

Cloud, P.E., R.G. Schmidt, and H.W. Burke, Geology of Saipan, Mariana Islands, U.S. Geological Surv. Prof. Paper 280-A, 126, 1956.

Gorshkov, G.S., Volcanism and the Upper Mantle, Plenum Press, New York-London, 1970.

Hays, J.D., and D. Ninkovich, North Pacific deep-sea ash chronology and age of present aleutian underthrusting, Geol. Soc. Am. Mem. 126, 263-290, 1970.

Heezen, B.C., et al., Site data, in Initial Reports of the Deep-Sea Drilling Project, 6, edited by A.G. Fisher et al., 17-629, U.S. Government Printing Office, 1971.

Heirtzler, J.R., et al., Site data, in Initial Reports of the Deep-Sea Drilling Project, 27, edited by J.J. Veevers et al., 15-335, U.S. Government Printing Office, 1974.

Horn, D.R., M.N. Delach and B.M. Horn, Distribution of volcanic ash layers and turbidites in the North Pacific, Geol. Soc. Am. Bull. 80, 1715-1724, 1969.

Karig, D.E., Structural history of the Mariana Island arc system, Geol. Soc. Am. Bull. 82, 323-344, 1971.

Kennett, J.P. and R.C. Thunell, Global Increase in Quaternary explosive volcanism, Science, 187, 497-503, 1975.

Larson, R.L., et al., Site data, in Initial Reports of the Deep-Sea Drilling Project, 32, edited by R.L. Larson et al., 17-43, U.S. Government Printing Office, 1975.

Le Pichon, X., Sea floor spreading and continental drift, J. Geophys. Res., 73, 3661-3697, 1968.

Minato, M., M. Gorai, and M. Hunahashi, The Geologic Development of the Japanese Islands, Tsukiji Shokam, Ltd., Tokyo, 1965.

Minster, J.B., T.H. Jordan, P. Molnar, and E. Haines, Numerical modelling of instantaneous plate tectonics, Geophys. J. R. Astr. Soc. 36, 541, 1974.

Morgan, W.J., Rises, trenches, great faults, and crustal blocks, J. Geophys. Res., 73, 1959-1982, 1968.

Ninkovich, D., J.D. Hays, and A.A. Abdel-Monem, Late Cenozoic volcanism and tectonics of Sumatra (abstract), Geol. Soc. Am., Annual Meeting, 3, 661, 1971.

Scholl, D.W., et al., Site data, in Initial Reports of the Deep-Sea Drilling Project, 19, edited by J.S. Creager, U.S. Government Printing Office, 325-553, 1973.

Tracey, J.L., S.O. Schlanger, J.T. Stark, D.B. Doan, and H.G. May, General geology of Guam, U.S. Geological Surv., Prof. Paper 403-A, 104, 1964.

Vallier, T.L., and R.B. Kidd, Volcanogenic sediments in the Indian Ocean, Geol. Soc. Am. Mem., in press.

Van Bemmelen, R.W., The Geology of Indonesia, Netherlands Government Printing Office, The Hague, 1949.

Von der Borch, C.C., et al., Site data, in Initial Reports of the Deep-Sea Drilling Project, 22, edited by C.C. Von der Borch et al., 13-348, U.S. Government Printing Office, 1974.

Westerveld, J., Quaternary volcanism on Sumatra, Geol. Soc. Am. Bull., 63, 561, 1952.

COMMENTS ON CENOZOIC EXPLOSIVE VOLCANISM RELATED TO EAST AND SOUTHEAST ASIAN ARCS

James P. Kennett and Robert C. Thunell

Graduate School of Oceanography, University of Rhode Island
Kingston, Rhode Island 02881

The paper by Ninkovich and Donn (1977) requires comment as it bears strongly on our earlier discussion of Cenozoic volcanism (Kennett and Thunell, 1975). Major conclusions by Ninkovich and Donn (1977), based on the distribution of volcanic ash in western Pacific and Indonesian region DSDP sites of Neogene age and Quaternary piston cores, include the following. The upward increase of volcanic ash in Neogene deep-sea drill cores is primarily a reflection of sea-floor plate motion towards volcanic sources. The recovered marine record is insufficiently continuous to allow determinations of differences in rates of explosive volcanism through time. The sites which have been essentially motionless relative to potential volcanic sources during the Neogene record uniformity of explosive volcanism. The validity of our observed increase in Quaternary explosive volcanism and our belief of its potential paleoclimatic significance is thus questioned.

The major conclusions of our previous investigation (Kennett and Thunell, 1975) are as follows. DSDP sequences as a whole represent a valuable series of observations to infer the distribution of volcanic ash and associated changing rates of explosive volcanism. The distribution of volcanic ash reported in DSDP sites throughout the oceans is consistent with a much higher rate of explosive volcanism during the last 2 million years. Volcanism on a global basis and in various regions has not been uniform throughout the Neogene, but rather shows maxima in activity occurring during the Quaternary; at various times during the Pliocene; and during the Middle Miocene. These episodes are separated by intervals of much lower inferred intensity, although variability exists from region to region. The distribution of volcanic ash in deep-sea sequences is clearly the result of a number of factors other than the volcanism itself. These include the effects of plate motion, wind direction, wind speed, and diagenetic alteration of the ash, but these are unlikely to be the primary controls of the observed distribution. Increased explosive volcanism during approximately the last 2 million years coincides with that episode of the Cenozoic marked by major and rapidly fluctuating climate.

These are obviously strongly conflicting opinions and therefore, further discussion is justified.

The idea that plate motion has had an effect on the distribution of volcanic detritus in Neogene sequences was clearly demonstrated by Heezen et al. (1973) in drill sites southwest of Japan and recognized by us (Kennett and Thunell, 1975) and Stewart (1975) as having an important effect. The amount of influence which such plate motion has had on the conspicuous increase in Quaternary volcanic ash in deep-sea sites throughout the world observed by Kennett and Thunell (1975) is, however, minimal as we now demonstrate. The paleopositions of all ash bearing DSDP sites have been determined for 5, 10 and 15 million years ago, using the age of marine magnetic anomalies compiled by Pitman et al. (1974), and the clockwise rotation of Pacific sites around a pole at 67°N., and 59°W. at a rate of 0.83° per m.y. (Minster et al., 1974). The DSDP sites as a whole are distributed over a wide range of tectonic regimes, with many located landward of island arc areas, within marginal basin complexes (such as the southwest and western Pacific, Panama Basin, and Caribbean) or in enclosed seas (such as the Mediterranean and Gulf of Mexico). A number of sites are on oceanic crust moving normal to potential source regions of volcanic ash. We have constructed a series of histograms in which the number of ash layers at these DSDP sites are plotted within the framework of the Neogene planktonic foraminiferal zones (Figure 1). The data from those sites that are moving normal to volcanic sources have been eliminated from the compilation, so that the resulting pattern (Figure 1) can be compared with the initial histogram containing all data (Figure 1). Inspection shows that the amplitudes of certain peaks in Figure 1 are changed slightly; the very conspicuous Quaternary peak is still very dominant. Thus, no changes are required in the original interpretation of the

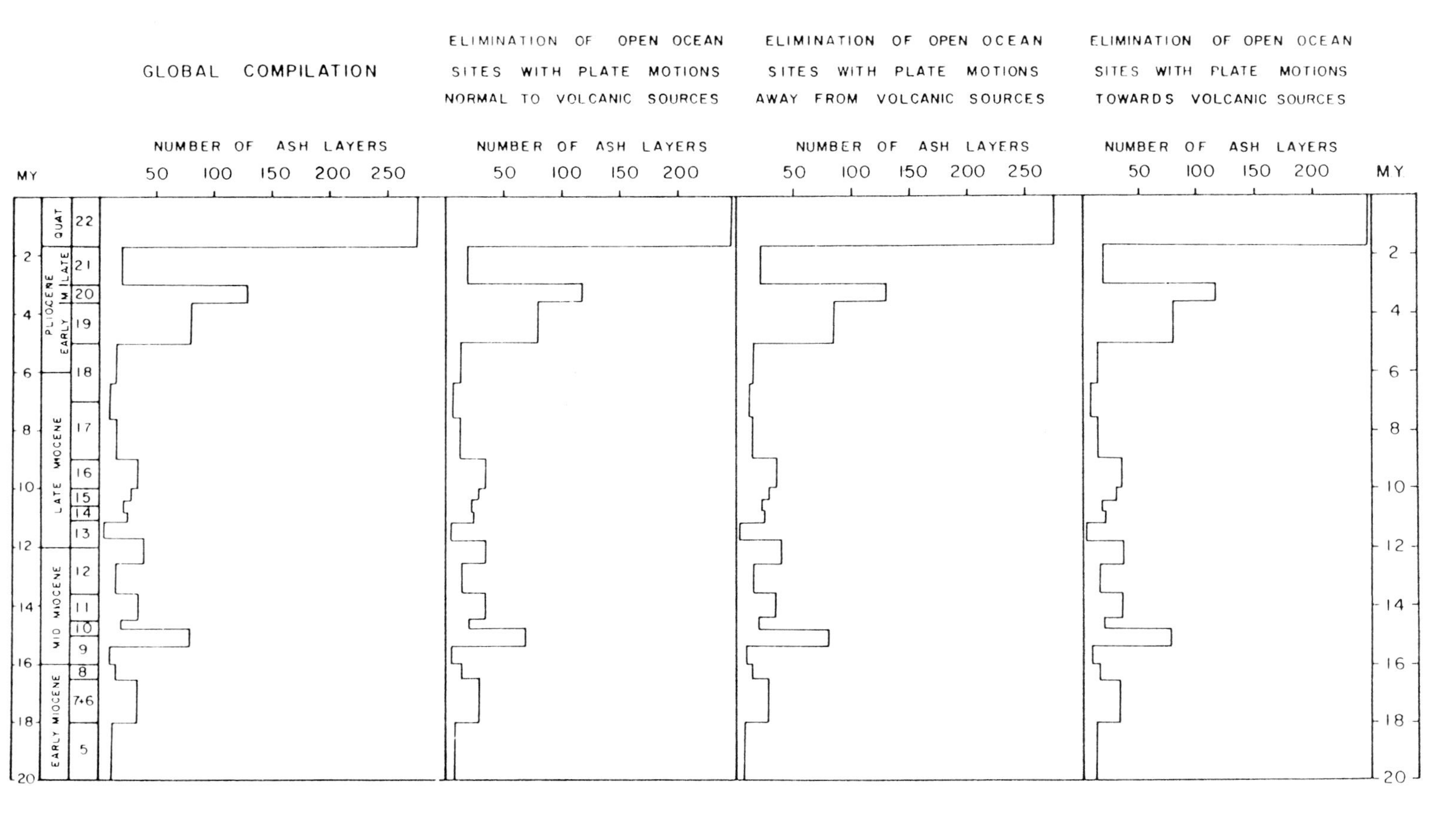

Fig. 1. Comparison of histograms representing the number of ash horizons in various deep-sea sedimentary sections (DSDP)(Legs 1 - 31) demonstrating the effect of plate motions on the stratigraphy of volcanic ash in the Neogene. A, represents a global compilation of all such recorded ash horizons plotted within the chronological framework of the N foraminiferal zonations. In B, all sites are eliminated that clearly show plate motion normal to potential volcanic ash sources such as in certain areas of subduction. These sites include 47, 49, 50, 51, 52, 60, 79, 80, 81, 194, 195, 199, 211, 212, 213, and 216. C, represents a compilation excluding sites that have undergone motion away from potential volcanic sources and D, towards potential volcanic sources.

sequences, and we conclude that plate motions are very clearly not the primary cause of the observed ash distributions in DSDP sequences.

For the sites south of Indonesia which are under the influence of northward plate motion towards the volcanic sources, the sequence we described for the last 10 m.y. is similar to that of Ninkovich and Donn (1977). The source of volcanic ash in the different oceanic regions we differentiated may be derived from various sources. The assumption that the Middle Miocene ash peak in the Indonesian section is derived from the occurrance of ash in Site 212 south of Indonesia is incorrect; it is based on ash in Site 292 near Luzon, Philippine Islands. They prefer, however, to interpret the curve as a result of seafloor motion moving the sites within a zone of ash distribution, the limits of which are defined by piston core observations, but they have not included some relevant data. Ninkovich and Donn (1977) do not seem to be consistent in what they consider to be an ash layer or an ash zone. For example, in Site 211 they refer to 5 Quaternary ash layers and yet in the Leg 22 Initial Report 2 layers and 3 zones are distinguished. In contrast, they disregard Early Pliocene ash zones (25% to 75% ash) in Sites 212 and 213. For example, Site 212 has two important levels of ash described by smear slide analysis in the initial report (von der Borch et al., 1974) as being 75% ash. The levels are dated as Early Pliocene age and therefore, deposition would have occurred when the paleoposition of this site was well removed from the Quaternary ash layer zone. These horizons of abundant ash clearly represent major volcanic episodes that created distribution patterns that cannot simply be related to plate motion history as required in the Ninkovich and Donn model (1977). Ninkovich and Donn (1977) have unfortunately attempted to explain the ash distribution entirely in terms of plate motion history. A single Quaternary layer in Site 218 is interpreted to be the result of movement of this site into the inferred edge of the Quaternary ash zone (see their Figure 1). The edge of this ash zone can be equally well drawn to show that Site 218 was potentially within range of volcanism throughout the last 10 m.y. Thus, based on the piston core data presented, the Quaternary volcanic ash in this sequence can just as readily be explained to reflect wider dispersal of ash perhaps resulting from greater volcanic explosivity.

Ninkovich and Donn (1977) question the validity of our conclusions because of poor core recovery. We recognize that caution must be used in interpretations where the record is poor, but on the other hand, they seem unaware that some normalization is required and was used (Kennett and Thunell, 1975) to assist in compensating for this effect, even in sites where core recovery is relatively high. While stressing the poor core recovery of many of the DSDP sites, they also failed to include the best cored sites in this region (DSDP Leg 31). Detailed analyses of the volcanogenic debris were made on these sites by Donnelly (1975). Furthermore, despite their call for restraint in interpreting the data, they have concluded after unexplained data rejection that explosive volcanism has been relatively uniform during the Late Cenozoic. A major criticism of their paper relates to this point involving discussion of details of western Pacific volcanic history which they have not considered.

Site 292 near Luzon, Philippine Islands, has excellent core recovery throughout an almost continuous well-dated stratigraphic sequence from the Early Oligocene to the Quaternary. In addition, this site has not been under the influence of any significant plate motion (Karig et al, 1975). The quantitative distribution of total volcanogenic debris, including volcanic ash has been studied by Donnelly (1975) and clearly shows that the tempo of explosive volcanism of this region has changed greatly during the Neogene and Oligocene. Although diagenetic alteration of volcanic glass has definitely occurred at some levels (Donnelly, 1975), the volcanic glass is relatively more abundant during periods of high eruptive activity (Donnelly, 1975) and strong similarity exists between the curve representing total volcanogenic debris in Site 292 and a curve we generated based on volcanic glass for the Philippine Basin sites (Kennett et al., in press). Thus, the result of our more general approach is in agreement with more detailed studies. Donnelly (1975) interprets the curve of volcanogenic abundance for Site 292 (10, pg. 586) to indicate that the volcanic activity of Luzon in the Oligocene was moderate, rose to a maximum near the Oligocene - Miocene boundary, and decreased in the earliest Miocene. A hiatus prevents evaluation of the Early Miocene, but the Middle Miocene is inferred to be a period of moderate volcanic activity, diminishing towards the Late Miocene and increasing to a probable maximum during the Late Miocene (12 to 7 m.y. B.P.). Activity began to increase significantly again during the latest Cenozoic at about 3 m.y. ago and reached a maximum during the Recent.

The distribution of volcanogenic material in Site 296, located to the southeast of Japan is important for the study of rates of volcanism of southern Japan and the Ryukyu Islands (Donnelly, 1975). This site has excellent core recovery for most of the Neogene and has not been under the influence of plate motion. This section has levels of volcanic ash in the intervals representing the last 4 million years and the late Early to early Middle Miocene. Unfortunately a hiatus cuts out some of the Middle Miocene at a time when important activity is known in the Japanese Islands (Sugimura, personal communication, in Donnelly, 1975), in contrast to Ninkovich and Donn's summary (1977) of the history for Japanese volcanism which states that the Middle Miocene was a period of submergence. This site was also examined in detail by Donnelly (1975) for total volcanogenic debris and again shows that the tempo of volcanic activity of Japan changed substantially throughout the Neogene. Donnelly (1975) interprets the curves for Site 296 to indicate a peak in volcanic activity in the late Early Miocene to Middle Miocene declining towards the Late Miocene, a short-lived less significant peak about 9 m.y. ago; declining activity through most of the Late Miocene and sharply increased activity occurring about 5 m.y. ago. This latest episode can possibly be divided into early and late active phases and a middle somewhat less active phase (Donnelly, 1975). The sequence mirrors ash distributions in Site 297, which is 300 km southeast of Kyushu, Japan. In this sequence the greatest concentrations of volcanic ash clearly occur in the latest Early to Middle Miocene and within the Quaternary. These results, as well as the distribution of the volcanic ash, agree with the independent analyses of Sugimura and Uyeda (1973), which showed that the most important pulses of volcanism in Japan were during the early Miocene and the Quaternary, and with the analyses of Karig (1975) who has concluded from geophysical and DSDP sedimentological evidence that the Philippine Sea has undergone tectonically related volcanic pulses along the northern and eastern margins (including the Mariana Islands). This region is marked by highly active volcanism during the last 5 m.y., minimal activity from the late Middle Miocene to the Early Pliocene and important Early Neogene volcanism.

In the extreme N.W. Pacific region adjacent to Kamchatka and the Aleutian Islands, Ninkovich and Donn (1977) consider that a marked increase in the number of ash layers within the Quaternary of Site 192 reflects the movement of this site towards Kamchatka. Although the relative motion of this site is not well known (Creager et al., 1973), the site was also clearly within range of volcanic debris from the Aleutians throughout the Late Cenozoic. Thus a dual volcanic source is probable and the stratigraphic record of ash also partly reflects rates of change of volcanism in the Aleutians. Ninkovich and Donn (1977) show no Quaternary ash in piston cores from the western part of the Aleutian Basin and use this distribution as a basis for their model. They failed, however, to note the Early and Middle Quaternary ash layers in Site 188 located in the same region. Thus, ash distribution in Quaternary piston cores may not necessarily represent a useful standard of reference for ash distribution through even the entire Quaternary if there have been marked changes in volcanic explosivity. Large changes in the rates of volcanism have already been recorded in Late Cenozoic volcanic sequences of Kamchatka (Erlich and Melekestev, 1972) and in the Aleutian region (Scholl et al., 1976).

For the Indonesian region, Ninkovich and Donn (1977) prefer to interpret our curve (Kennett and Thunell, 1976) only in terms of plate motion. The authors refer to the early work of van Bemmerlen (1949) concerning the volcanic history of Indonesia. Van Bemmerlen, however, is one of the strongest advocates of episodic volcanism, as presented by his work in Indonesia (van Bemmerlen, 1949; 1961). Van Bemmerlen (1961) refers to the intense tectonic and volcanic activity during the Middle Miocene of Indonesia in contrast to the statement of Ninkovich and Donn (1977) that this was a time of submergence. This changing activity is at least partly responsible for the peaks of Neogene activity shown in our work (Kennett and Thunell, 1975). An important related study is that of Vallier and Kidd (in press) who semiquantitatively examined the distribution of volcanic ash (smear slide analyses) in Site 213 to the south of Indonesia. They show traces of volcanic ash throughout most of the section separated by strong maxima during the latest Miocene, within the Early Pliocene, and in the Early Quaternary. The maxima are the same peaks discerned by us (Kennett and Thunell, 1975). Despite this, Ninkovich and Donn (1977) interpret the curve for Site 213 (Vallier and Kidd, in press) to indicate that glass shards have shown "essentially continued production since the beginning of the Late Miocene." The presence of trace amounts of volcanic debris, including volcanic ash, in DSDP sequences is common (Vallier and Kidd, in press) and of very little value in interpretation. We recognize that in the sites south of Indonesia, the absence of Middle Miocene volcanic ash and the greater amplitude of the Quaternary peak are partly related to plate motions, but we also believe that the record contains valuable information on the rates of explosive volcanic activity for this region.

In summary, we cannot agree with Ninkovich and Donn's (1977) conclusions that the existing DSDP data indicates relative uniformity of explosive volcanism during the Middle and Late Cenozoic in the western Pacific and Indonesian regions, or that plate motions is the primary factor controlling the observed distribution of volcanic ash in DSDP sequences of Middle and Late Cenozoic age. On the other hand, ash in the DSDP cores can reflect changing rates in explosive volcanism through the Neogene, although the computed relative amplitudes of the episodes can be expected to change with future adjustments in chronology, future improvement in core recovery, and related normalization procedures, and with better knowledge of the effect of diagenetic alteration of volcanic ash layers.

References

Creager, J. S., D. W. Scholl, et al., Site 192 Initial Reports of the Deep-Sea Drilling Project, (Government Printing Office, Washington, D. C.), 19, 464-465, 1973.

Donnelly, T. W., in D. E. Karig, J. C. Ingle, et al., Neogene Explosive Volcanic Activity of the Western Pacific, Site 292 and 296, DSDP Leg 31, Initial Reports of the Deep-Sea Drilling Project, (Government Printing Office, Washington, D. C.), 31, 577-597, 1975.

Erlich, E. N. and I. V. Melekestsev, Quaternary Explosive Volcanism of Kamchatka, Modern Geology, 3, 183, 1972.

Heezen, B. C. et al., The Post Jurassic Sedimentary Sequence of the Pacific Plate; A Kinematic Interpretation of Diachronous Deposits, in Initial Reports of the Deep-Sea Drilling Project, (Government Printing Office, Washington, D. C.), 20, 725-738, 1973.

Karig, D. E., J. C. Ingle, Jr., et al., Site 292 Initial Reports of the Deep-Sea Drilling Project, (Government Printing Office, Washington, D. C.), 31, 67-79, 1975.

Karig, D. E., in D. E. Karig, J. C. Ingle, et al., Basin Genesis in the Philippine Sea, Initial Reports of the Deep-Sea Drilling Project, (Government Printing Office, Washington, D.C.), 31, 857-879, 1975.

Kennett, J. P., and R. C. Thunell, Global Increase in Quaternary Explosive Volcanism, Science, 187, 497, 1975.

Kennett, J. P., A. R. McBirney, J. F. Sutter, and R. C. Thunell, Episodes of Cenozoic Volcanism in the Circum-Pacific, Jour. Volcanology and Geothermal Res., in press.

Minster, J. B., T. H. Jordan, P. Molnar, E. Haines, Numerical Modeling of Instant Plate Tectonics, Royal Astron. Soc. Geophys. Jour., 36, 541, 1974.

Ninkovich, D., and W. L. Donn, Cenozoic Explosive Volcanism Related to East and Southeast Asian Arcs, Ewing Memorial Volume, 1977.

Pitman, W. C., R. L. Larson, and E. M. Herron, Geol. Soc. Am. Map and Chart Series, 6, 1974.

Scholl, D. W., M. S. Marlow, N. S. MacLeod, and E. C. Buffington, Episodic Aleutian Ridge igneous activity: Implications of Miocene and younger submarine volcanism west of Buldir Island, Geol. Soc. Am. Bull., 87, 547, 1976.

Stewart, R. J., Late Cenozoic Explosive Eruptions in the Aleutian Kuril Island Arcs, Nature, 258, 505, 1975.

Sugimura, A. and S. Uyenda, Island Arcs: Japan and Its Environs., New York, 1973.

Vallier, T. L. and R. B. Kidd, Volcanogenic Sediments in the Indian Ocean, Geol. Soc. Am. Bull., in press.

van Bemmelen, R. W., The Geology of Indonesia, (Netherlands Printing Office, The Haque), 1949.

van Bemmelen, R. W., Volcanology and Geology of Ignimbrites in Indonesia, North Italy, and the U.S.A., Geologie en Mijnbouw, 40, 399-411, 1961.

von der Borch, C. C., J. G. Sclater, et al., Site 212 Initial Reports of the Deep-Sea Drilling Project, (Government Printing Office, Washington, D. C.), 22, 48, 1974.

REPLY

Dragoslav Ninkovich and William L. Donn

To refute all of the comments of Kennett and Thunell would require considerable space. We will, therefore, reply only to some of their more cogent comments but note in passing, that Kennett and Thunell discuss volcanicity in terms of climatic consequences. We did not touch the subject of climate in our paper. As the subject has been brought up, we do not understand their correlation between stated increased volcanism during the past 2 million years and marked and rapid fluctuations of climate. The known glacial-interglacial cycles began about 700,000 years ago and the known cooling that culminated in glaciation began in the early part of Cenozoic.

Kennett and Thunell attack our interpretation of the importance of Indonesian plate motion on volcanic frequency partly on the basis of our interpretation of a strong Miocene ash layer peak in Site 212 which they ascribe to Site 292. However, we do not discuss such a peak at either of these sites anywhere in our paper and regard such observations as irrelevant to the Indian Ocean plate motion we discussed, particularly as Site 292 is east of Luzon on the Philippine Sea plate. They also err in stating that we do not distinguish between ash layers and ash zones. All of the ash layers reported for the Indian Ocean were either examined directly by the senior author, or examined at his request by DSDP personnel. All were found to be pure ash layers. We disregarded other reported ash zones because when examined they were found to represent mixtures of sediment and ash rather than ash layers. We also disregarded similar zones in piston cores in drawing our ash-layer bounderies.

The real substance of the discussion of Kennett and Thunell is found in their Fig. 1 and the related discussion. They purport to show that the same frequency of volcanicity exists for all sites with global distribution as for the same sites when those sites with motion either toward or away from potential volcanic sources are eliminated. This, of course, refers also to elimination of sites moving relative to subduction zones.

To consider their Fig. 1A, we understand that the numbers of ash layers are not absolute numbers but are arrived at by what they describe as "normalization." However, we have shown, and they have admitted that core recovery is as low as 10 percent. How then can one normalize missing data. Presumably their normalization technique really involves considerable interpolation to fill in missing data and is a very hazardous procedure. Also, if one normalizes vertically one should also do it horizontally to avoid coloring results by counting the same ash layer many times. For example, in our Figure 3, it would seem that Sites 47, 49, 50 and 51 which are very close to each other must show the same Plio-Pleistocene ash layers which they received after entering the ash-layer zone. Thus, these sites would show four times the number of young ash layers than would be shown by single sites having older ash, such as Site 303 which was actually **in** the ash-layer zone for 10 million years.

Kennett and Thunell state that their Figure 1B eliminates all sites with motion clearly toward the sources and related subduction zones, but the list of sites in their figure caption does not include many of the sites we show in our Figures 1 and 3 which are obviously on moving plates and which contribute to the increase of late Cenozoic ash layers. These are sites 218, 262, 192 and 303. Also, five of the sites that are eliminated, 60, 79, 80, 81 and 199 are not applicable. Site 60 as shown in our Figure 3 is between the Mariana Islands and the trench and is not on a moving plate. Site 199 is far to the east of the Marianas in the trade wind zone and hence not in any late Cenozoic ash layer zone. Also, it contains mixed

Eocene to Pliocene sediment only the lower part of which is ash bearing. Sites 79, 80 and 81 are on the west flank of the East Pacific Rise about on the equator and are moving toward the Japanese Island Arc about 10,000 km away. The volcanic material described for those sites consists of andestic and basaltic pumice fragments in early and middle Miocene sediment. Probably these Miocene fragments floated in when the sites were closer to a Central American source. The four sites do not meet the stated criterion of Kennett and Thunell as being examples clearly showing plate motion toward a potential volcanic source.

In place of these four inapplicable sites, it seems much more appropriate to include the sites which are now clearly close to and moving toward the potential sources such as those omitted from our Figure 1 and 3 as well as those in the north and northeast Pacific which are moving toward Alaska and the Aleutians. Work in progress by Ninkovich, Cadet, Burkle and Hammond (1977) show that the sites moving toward these north Pacific sources indicate that the same progressive increase in ash layers exists from late Miocene to the Present as is shown by the western Pacific and Indonesian sites, but sites in the Bering Sea, on a stationary plate do not show this progression.

The Pacific region is certainly the most productive of all volcanic areas. Sites from at least half of the Pacific source regions show that only those sites on plates moving toward the sources show the time-increase in numbers of ash layers. We do not understand how the proper elimination of sites close to and moving toward the sources will not materially change the apparent global trend in number of ash layers. Certainly their results are strongly colored by the lack of elimination of so many sites known to be on moving plates.

As regards their Figure 1C, no statement is given of which sites have been eliminated so that little discussion can be made. However, from the title, this diagram eliminates only sites moving away but includes all of the sites which are moving toward the sources. Hence, this presentation must show the same increase in ash layers as Figure 1A.

Figure 1D has the same title as Fig. 1B and seems to represent the information we discussed previously. In general, we find it rather curious that the diagrams are so nearly identical in view of their apparently different numerical bases.

We see nothing in the attack on our work by Kennett and Thunell which in any way requires altering our major premise that interpretation of the history of explosive volcanism must include the effect of plate motion. Current data available are very sparse and present serious hazards in the study of the frequency of volcanism. We suggested that if anything, the data from stationary plates can be interpreted to indicate relative uniformity of the frequency of volcanicity in Cenozoic, but this suggestion is based on data from only two localities, the East Philippine and Bering Seas. Better core recovery from a global distribution of stationary sites is necessary before any firm conclusion can be drawn.

References

Ninkovich, D., Cadet, J.P., Burkle, L. and Hammond, S. "Le volcanisme explosive de l'arc des Aléoutiennes: répartition spatio-temporelle des cendres dans des sédiment marins; implications géodynamiques", (Abstract). Cinquiéme Réunion Annuelle des Sciences de la Terre, Rennes, France, April 19-22, 1977.

PETROLOGIC AND GEOCHEMICAL CHARACTERISTICS OF MARGINAL BASIN BASALTS

James W. Hawkins, Jr.

Geological Research Division
Scripps Institution of Oceanography
University of California, San Diego
La Jolla, California 92093

Abstract. Marginal basins (shallow seas underlain by oceanic crust) may have multiple origins, but at least some have properties indicating an origin by processes like those which form oceanic lithosphere at spreading ridges. The Lau, Mariana, Woodlark and North Fiji Basins and the Scotia Sea have geophysical and bathymetric features which support an origin by lithosphere dilation. Petrologic and geochemical data for tholeiitic rocks dredged from these basins resemble oceanic ridge tholeiites (ORB) in major and minor elements, isotope ratios, and in modal and normative composition. The most extensive data are for Lau Basin basalts (LBB); these show a range in chemistry which, in least altered rocks, can largely be attributed to low-pressure fractional crystallization. The compositional range overlaps ORB. Samples from the Mariana, Woodlark and North Fiji Basins resemble the more strongly fractionated LBB and ORB samples in terms of solidification index vs. CaO/Al_2O_3 , FeO^*/MgO and in alkali content. The LBB samples include both PL (plagioclase) and OL(olivine) tholeiites; the two types are about equally represented in a suite of 31 analyzed samples. Vitrophyres have microphenocrysts of En_{48} Fs_{13} Wo_{39} with .25% Ti, 1.6% Al and ∿ 40 ppm K. Chemical data for least altered LBB are (wt. percent) TiO_2, 0.4 to 1.9; K_2O, 0.03 to 0.39; P_2O_5, 0.01 to 0.16; FeO^*/MgO, 0.8 to 1.6: K/Rb 750-1400, Ba/Sr 0.03 to 0.2, 150-670 ppm Cr, 60-280 ppm Ni and 35-280 ppm Sr. REE data normalized to chondrites show light REE depleted, flat or slightly light REE enriched patterns depending on extent of fractionation. $^{87}Sr/^{86}Sr$ for four samples ranges from .7020 to .7035, one Sr-poor sample has .7051. Oxygen isotope data, $\delta\, ^{18}O$ = 5.7, are typical of ORB. The data for LBB indicate that the most probable origin is by fractional melting of a peridotitic mantle and emplacement in slowly spreading lithosphere (half-rate 1-2 cm/yr). Melting of subducted oceanic crust or a genetic relationship to "island arc tholeiites" is unlikely. North Fiji Plateau samples also were emplaced in slowly spreading lithosphere. Data are limited but indicate that the tholeiites are slightly more fractionated than LBB although still retaining relatively high Mg/Mg+Fe. All samples are moderately altered; chemical data are (wt. percent) TiO_2, 1.0 to 1.6; K_2O, .08 to .58; P_2O_5, 0.05 to 0.12; Cr, 210 to 370 ppm; Ni, 60 to 160 ppm; Sr, 80 to 255 ppm. Vitrophyres have microphenocrysts of bytownite, Fo_{86} and rare cpx, (En_{49} Fs_{11} Wo_{40} with 0.2% Ti, 1.5% Al and ∿ 25 ppm K). These data are similar to reported analyses of Woodlark and Mariana basin samples and to differentiated LBB. The minor variations in marginal basin basalt chemistry suggest control by differences in depth of melt separation, extent of mantle melting or subsequent fractional crystallization. In general, it appears that these basalts evolve in a manner similar to ORB; the range in chemistry may be related to differences in thermal gradients under the basins.

Introduction

The western Pacific Ocean basin is ringed by a festoon-like arrangement of island arcs and trenches. In the southwestern part of the basin the arc-trench systems are nested in a roughly concentric arrangement. The shallow sea areas between the arcs, or arcs and the continent, include areas known to be underlain by oceanic crust. Some of these oceanic basins are believed to be areas in which new oceanic crust is actively forming today; in others there is evidence that there are trapped fragments of older oceanic crust. Wegener (1929) was perhaps one of the first to suggest that "the island arcs, and particularly the eastern Asiatic ones, are marginal chains which were detached from the continental mass, when the latter drifted westwards, and remained fast in the old sea floor which was solidified to great depths. Between the arcs and the continental margin later still-liquid areas of sea floor were exposed as windows". Wegener's suggestion that the origin of the shallow seas of the southwestern Pacific was related to crustal (lithosphere) dilation has been followed and embellished by a number of workers (eg. Karig,

1970, Packham and Falvey, 1971, Moberly, 1972, Sclater et al., 1972, Hawkins, 1974, 1976) who support the concept that at least some of these shallow seas are essentially young ocean basins developing by tectonic and petrologic processes similar to those that form oceanic lithosphere at oceanic ridge spreading centers. Karig 1970 proposed a nomenclature for components of the arc-trench region in which the term inter-arc basin is used for the shallow sea which lies landward of the active volcanic arc. Many workers use the term marginal basin more or less synonymously for this area and in this paper I use it for the area of shallow sea lying between an active volcanic arc and an inactive volcanic arc or a continental block. As a further clarification, this paper will only address the geologic features of those Pacific marginal basins which have high heat flow, active seismicity, thin sediment cover and for which there is evidence that relatively unaltered oceanic-ridge type basalt forms the sea floor. Pacific marginal basins which meet these criteria are: Lau Basin, Mariana Basin, North Fiji Plateau and Woodlark Basin. The Scotia Sea (Barker, 1970; Tarney, this volume) is an example from the Atlantic Basin which appears to be similar in all essential respects. Other Pacific marginal basins may actually have the same origin but are excluded from this discussion either because they are buried by thick sediment or because petrologic data are lacking.

Marginal basins are of obvious importance as loci of generation of new oceanic crust by fractional melting of the mantle. They also are important in understanding the evolution of orogenic belts and in understanding the origin of ophiolites which commonly are interpreted as fragments of oceanic crust. One of the objectives of this paper is to present a geochemical and mineralogic description of those marginal basins for which we have data and to emphasize the similarity between marginal basin basalts (MBB) and oceanic ridge basalts (ORB). This similarity poses a problem in any attempt to distinguish between MBB and ORB in the study of ophiolites. The data presented here are largely from the Lau Basin for which there are extensive geophysical, geochemical and petrologic data (Hawkins, 1974, 1976). Some petrologic data are available from the Mariana Basin and N. Fiji Plateau and there is one sample from the Woodlark Basin. The geophysical data for these basins give strong support for their extensional origin. Because of the limited number of samples from other basins, the petrologic characteristics of the Lau Basin dominate the data and any attempt to compare ophiolitic material with "typical" marginal basin basalt as described in this paper should recognize this possible bias. In the following section I will summarize some of the morphologic and geophysical data for the "active" Pacific basins to help establish the similarity between them. In another section I will show the petrologic comparisons between the basin basalts and ORB.

Geology of Marginal Basins

The active marginal basins lie on the concave or "landward" side of volcanic arc-trench systems (Fig. 1). Volcanic arc magmatism is believed to be caused by fractional melting of subducted oceanic lithosphere, and/or mantle, beneath the arcs. The arc-trench system itself is in a zone of net lithosphere shortening but the marginal basin obviously requires a dilation of the lithosphere. The Lau Basin is located immediately above the Tonga Trench Benioff Zone (which is about 250 km beneath the basin) but the N. Fiji Plateau has a different configuration. A well-defined Benioff Zone extends down to 350 km but some N. Fiji plateau areas presumed to be undergoing extension lie 300 km east of the apparent termination of the New Hebrides Trench Benioff Zone.

Active marginal basins are characterized by shallow depths (e.g. 2.25 to 2.5 km, Fig. 1) and this helps to distinguish them from fragments of old and deeper sea floor trapped or isolated behind island arcs. Active basins typically have 1-2 km of relief, rough topography and numerous short ridges and seamounts (Hawkins, 1974).

High heat flow is characteristic of the active basins as well as some basins having a thick sediment filling. The high heat flow of the N. Fiji Plateau led Sclater and Menard (1967) to suggest that it was underlain by upwelling hot mantle material. Further support for this idea is the zone of anomalously high seismic wave attenuation in the upper mantle beneath the Fiji Plateau (Barazangi et al, 1974). The Lau Basin (Sclater et al., 1972, Hawkins, 1974) Mariana Basin (Anderson 1975) and Woodlark Basin (MacDonald et al., 1972) all have heat flow similar to that considered typical of spreading oceanic ridges.

If marginal basins develop by a sea floor spreading mechanism, then we should find sequential magnetic anomalies symmetrically arranged about central bathymetric highs. Luyendyk et al. (1973) presented an interpretation of Woodlark Basin data which, they claim, shows symmetry in some profiles spaced about 30 km apart in the eastern part of the basin. Detailed studies of the Lau Basin (Hawkins 1974, Lawver et al., 1976) and the Mariana Basin (Karig, 1971) show that these basins lack the large scale, symmetrical anomalies typical of spreading ocean ridges. In part of the Lau Basin, the data suggest symmetric patterns several tens of kilometers long which are offset by transform faults for distances of tens of kilometers. Large areas of the basin are dominated by seamounts and no consistent striped pattern can be mapped with certainty. North Fiji Plateau data also suggest the presence of linear anomalies but our detailed surveys show that these symmetric linear patterns are difficult to trace for more than a few tens of kilometers (Hawkins and Batiza, 1975; Hartzell, 1975).

The intrusion mechanism which operates in marginal basins may be different in detail from that which controls the linear magnetic stripe pattern

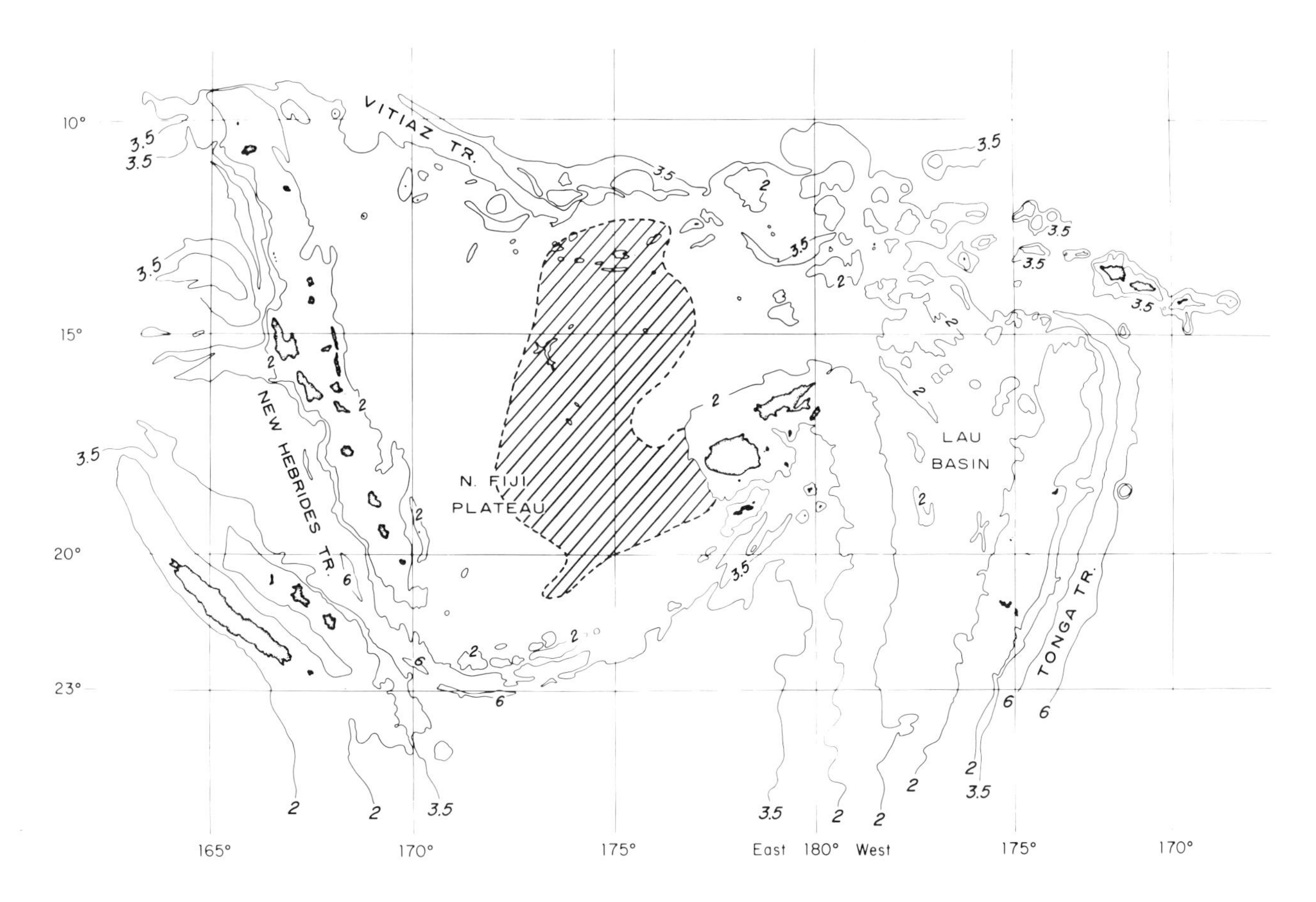

Figure 1. Chart showing relationships between Tonga and New Hebrides trench-arc systems and the N. Fiji Plateau and Lau Basin marginal seas. Depth intervals are in kilometers. The diagonal lined area is the approximate extent of a region of rough topography, high heat flow and ridge-trough topography, with depths between 2.5 and 3 km.

formed with the crust at mid-ocean ridges. If the intrusion mechanism differs, then it is possible that the petrology of the crustal rocks may be different. For example, if seamounts are one of the fundamental expressions of magmatism in marginal basins, and predominate over linear vent systems which would form symmetric magnetic stripes, then marginal basins may have a significantly greater proportion of strongly fractionated basalts, or alkali basalts, than "normal" oceanic crust.

Petrologic Data

Tholeiitic basalts have been dredged from the four western Pacific marginal basins described above and their compositions appear to be like that of ocean ridge basalts (ORB) (Table 1). Karig (1970) predicted that the Lau Basin would not be underlain by silicic volcanic rocks in spite of its tectonic setting and speculated that basalt would be present. Karig's prediction has been verified by the extensive surveys of the Lau Basin. Further, it has been shown that the chemistry of the Lau Basin basalt (LBB) requires a mantle source and precludes contamination by sialic material as might be expected if continental or arc rocks were under the basalt floor of the basin (Hawkins, 1974, 1976). The data for the North Fiji Plateau, Mariana Basin and Woodlark Basin are very much more limited but they most closely resemble LBB and ORB samples and it seems likely that all four basins follow a common tectonic-petrologic evolutionary style.

In this section I will first discuss the mineralogy and geochemistry of LBB and show how they resemble our estimate of the composition of ORB. Because of the fundamental similarities in a number of basalt types, it will be essential to show that LBB can be distinguished from basalt types considered to be typical of island arcs, oceanic islands and linear volcanic chains (i.e. "hot spot traces"). The Lau Basin samples will be used to model marginal basin basalt compositions because there are more data for them than for all of the other basins combined. The range in composition of LBB, both in terms of fractionation effects and alteration, is sufficient to overlap

TABLE 1. Basalt, Average analyses (wt%)

	1	2	3	4	5	6	7	8	9
SiO_2	48.8	48.6	49.21±0.74	48.6	47.7	49.0	48.9	49.6	49.5
TiO_2	1.0	1.0	1.39±0.28	1.1	2.7	0.8	1.5	1.5	1.2
Al_2O_3	16.2	16.4	15.81±1.50	15.4	13.0	18.6	16.3	16.6	15.5
Fe_2O_3	1.6	1.6	2.21±0.74	4.7	6.1	5.3	2.9	2.3	3.9
FeO	7.2	7.0	7.19±1.25	6.2	7.8	4.8	7.0	6.5	6.2
MnO	0.2	0.2	0.16±0.03	0.2	-	0.2	0.2	0.1	0.1
MgO	9.3	9.5	8.53±1.98	7.8	6.3	5.1	6.7	6.8	6.7
CaO	12.8	12.2	11.14±0.78	12.2	8.5	9.0	12.4	11.4	11.3
Na_2O	2.2	2.3	2.71±0.19	2.1	2.5	3.0	2.2	3.2	2.7
K_2O	0.12	0.20	0.26±0.17	0.1	0.87	0.58	0.24	0.42	0.3
H_2O^+	-	-	-	1.3	3.8	2.5	0.7	0.9	1.4
H_2O^-	-	-	-	-	-	0.6	0.4	0.7	0.1
P_2O_5	0.07	0.09	0.15±0.04	0.09	0.21	0.14	0.1	0.2	0.1
TOTAL	99.49	99.09	98.76	99.79	99.48	99.62	99.44	100.22	99.9
FeO*/MgO	.93	.89	1.10	1.33	2.10	1.88	1.43	1.26	1.45
DI	23.6	21.2	19.7	2.25	16.0	6.7	22.2	20.9	21.5
OL	10.8	12.1	7.3	-	-	-	-	7.0	-
HY	7.6	7.2	11.4	15.1	13.8	13.6	14.7	5.3	13.3
Q	-	-	-	1.7	5.0	3.1	1.4	-	1.5
Or	0.7	1.2	1.2	0.6	5.4	3.6	0.9	2.5	1.8
Average Analyses (ppm)									
V	224	233	289±73	275	525	275	283	-	330
Cr	459	520	296±80	263	38	35	241	230	300
Co	67	71	-	48	65	30	71	40	45
Ni	199	226	123±56	101	50	20	117	70	90
Cu	101	93	87±26	123	140	70	121	-	90
Zn	62	-	122±21	-	-	-	74	-	-
Rb	< 1	∿ 1	1	.67	-	4	1.5	4.5	-
Sr	97	142	123±46	111	265	250	134	190	150
Zr	-	-	100±42	65	120	50	-	-	-
Ba	< 2	-	12±8	11	45	95	-	40	-
Pb	2	3	-	-	-	∿ 3	2	-	1.6

1. ave of 6 least altered, least fractionated basalts, Lau Basin (Hawkins, 1976)
2. ave of 14 least and moderately altered, least fractionated basalts, Lau Basin (Hawkins, 1976)
3. ave of 33 basalts, MAR (Melson and Thompson, 1971)
4. ave of 12 basalts, Caribbean, DSDP sites 146, 150, 153 (Donelly et al, 1973)
5. ave of 7 basalts, Indian Ocean, Ninety East Ridge, DSDP site 216 (Thompson et al, 1975)
6. ave of 7 island arc tholeiites (Ewart and Bryan, 1972, Jakes and Gill, 1970)
7. ave of 9, least and moderately altered, fractionated basalts, Lau Basin (Hawkins, 1976)
8. ave of 6 basalts, Mariana Basin (Hart et al, 1972)
9. ave of 10 basalts, North Fiji Plateau (Hawkins unpub. data)

the compositions reported for samples from other basins. Thus, it seems likely that any supposed differences between the compositions of marginal basin basalts and "average" ORB are more likely reflections of the limited sample data available and of fractionation - alteration effects. When essentially fresh unfractionated marginal basin samples are compared with equally unmodified ORB, the similarity is quite apparent.

Rocks dredged from the Lau Basin are largely pillow basalt fragments; many are very fresh and have clear sideromelane in vitrophyre rims without palagonite or manganese crusts. Rock textures include vitrophyric, variolitic and fine grained basaltic; in addition, there are diabasic-textured samples which are either dike-sill fragments or interior parts of pillows. One dredge collection consisted of greenstone, layered anorthositic gabbro and moderately sheared gabbro (Table 2). Collectively, these samples are believed to represent fragments of oceanic seismic layer 2 (pillows, feeder dikes and sills) and possibly parts of seismic layer 3 (layered anorthositic gabbro).

The vitrophyric pillow rinds have microphenocrysts of olivine and plagioclase in pale tan sideromelane. The olivine microphenocrysts range in composition from Fo_{80} to Fo_{87}; grains have slight normal zoning of about 1 or 2 percent Fo. Nickel content ranges from 900 to 1700 ppm and shows positive correlation with Mg. The most Mg-rich olivine (Fo_{85-87}) is in the least differentiated rocks ($FeO^*/MgO < 1.0$). Olivine also forms phenocrysts in pillow interiors and in the variolitic zone of pillows. Euhedral chromite is common as an inclusion in both phenocrysts and microphenocrysts of olivine. Although euhedral olivine is very common, especially as phenocrysts, skeletal forms are also common and many microphenocrysts have highly irregular shapes due to glass inclusions surrounded by negative crystal faces.

Plagioclase microlites are common in the pillow rinds and indicate that it too must have been a liquidus mineral along with olivine and chromite. Hopper-shaped or swallow-tail forms similar to ORB microlites shown by Bryan (1972) are common in the vitrophyre but most of the plagioclase in the variolitic and coarser grained samples is euhedral. Plagioclase ranges from An_{65} to An_{85} depending on the extent to which the parent liquid was differentiated. Slight normal zoning of a few percent An is common in the samples. The K content of all plagioclase crystals is low and most samples have less than 0.3% Or. A typical calcic labradorite sample ($An_{65.8}$ $Ab_{34.1}$) has 0.06% K_2O, 0.84% FeO^* and 0.38% MgO. The low K and high Fe + Mg seems to be typical of plagioclase from sea floor tholeiitic rocks. The groundmass is largely either glass or opaque minerals, plumose microlites of clinopyroxene and glass.

Normative data plotted in a plagioclase - pyroxene - olivine ternary diagram (PL-PX-OL) show that the samples lie close to the presumed cotectic (Shido et al 1971). Some samples lie on the Pl side, some on the Ol side, but all of the least fractionated relatively unaltered samples lie close to this field boundary (Fig. 2). This in an additional aspect of their chemistry which resembles ORB samples. Clinopyroxene is present mainly as ophitic to sub-ophitic plates enclosing plagioclase in diabasic rocks or as a plumose quench material in the groundmass of variolitic textured and fine grained basalt. It is very rare as a microphenocryst and no grains larger than 1 mm were found. Microprobe analyses of the pyroxene are shown in Fig. 3 and summarized in Table 3. The microphenocrysts (0.1 to 0.25 mm) are low Cr, low Ti, low Al, augite. They are zoned from Mg-rich cores to Fe-rich rims but oscillatory zoning is present. Ti decreases and Al increases from core to rim, Ca increases with Fe, K and Na are very low. The samples are silica saturated, sub-alkalic pyroxenes.

The host rock for the Lau Basin pyroxenes was dredged from a 250 meter high seamount built on the edge of a steep scarp which may be the trace of a fault. The rock sample is quartz normative tholeiite and is considered to be only moderately altered, close to a liquid composition, but fractionated (FeO^*/MgO is 1.72). Low-pressure fractionation of olivine (Fo_{86}) from an OL-normative tholeiite could account for its composition. The dominant microphenocrysts in the glass rind of this sample are olivine (Fo_{80}) and Ca-Plagioclase but clinopyroxene is more abundant in the inner parts of the pillow fragment. This sample appears to be fractionated more strongly than many of the OL-tholeiites as it has higher V content (355 ppm) lower Cr (50 ppm) and lower Ni (60 ppm) than the more "typical" Ol-rich LBB.

The sub-ophitic clinopyroxene of the diabasic and micro-diabasic textured rocks are generally the same composition as the microphenocrysts except that they tend to have higher Al and Ti and have a higher proportion of Wo and Fs. Some are salite, most are augite.

The Lau Basin samples include rocks which have been modified by varied amounts of fractionation and alteration. The original liquid chemistry would be obscured if the analyses were averaged together. The data were separated by element ratios and abundances to eliminate the effects of these modifications. "Least fractionated" basalts are those in which FeO^*/MgO is less than 1.0. Samples termed "least altered" have $Fe''' / Fe'' < 0.4$, $K_2O < 0.2\%$ and $H_2O+ < 0.6\%$. Moderately altered basalts have Fe''' /Fe'' as high as 0.9, up to 0.6% K_2O and up to 1.1% H_2O+. Three average analyses are shown in Table 1. They are: the average least fractionated, least altered, phenocryst-free LBB; the average least fractionated, phenocryst free LBB, based on least altered and moderately altered samples; and least altered, phenocryst-poor, fractionated LBB ($FeO^*/MgO > 1.0$). These average analyses closely resemble the published estimates of the composition of ORB (Table 1). It will be

Table 2. - Gabbro, altered basalt and ophiolite rocks, average analyses (wt%)

	1	2	3	4	5	6
SiO_2	48.8	49.6	46.7	51.1	50.4	48.9
TiO_2	0.6	1.5	0.2	1.3	1.4	1.0
Al_2O_3	16.8	15.7	18.2	17.2	14.6	15.6
Fe_2O_3	2.7	-	2.1	4.1	3.4	2.5
Fe O	4.5	7.6	5.0	5.3	6.3	5.4
MnO	0.1	0.2	0.1	0.2	0.2	0.2
MgO	8.4	9.3	9.2	4.6	6.2	8.4
CaO	12.9	11.2	14.8	10.0	7.7	11.7
Na_2O	2.2	2.7	1.1	2.9	3.8	2.3
K_2O	0.20	0.08	0.25	0.87	0.59	0.25
H_2O^+	1.9	2.2	1.7	1.2	3.1	2.2
H_2O^-	0.5	-	0.4	0.9	1.5	-
P_2O_5	0.06	0.03	0.05	0.2	0.2	0.2
TOTAL	99.6	100.11	99.8	99.87	99.39	98.65

1. ave of 5 feldspathic gabbros, Lau Basin (Hawkins, 1976)
2. ave of MAR gabbros (Miyashiro, 1970)
3. ave of 8 Al-rich gabbros, Franciscan fm. ophiolite, (Bailey & Blake, 1974)
4. ave of 5 altered and fractionated basalts (Lau Basin, Hawkins, 1976)
5. ave of 21 greenstones and spilites, Franciscan fm. ophiolite (Bailey and Blake, 1974)
6. ave of 8 samples, Nicoya complex, Costa Rica (Dengo, 1962; Henningsen and Weyl, 1967)

shown in a later section that there is a good comparison between LBB data and the limited data available for other marginal basins. It is important to note that the comparisons are not as good if all data for an area are indiscriminately lumped together but similarities are quite apparent when comparison is made between phenocryst-free samples having similar degrees of fractionation and alteration. The trace element comparisons (Table 1) duplicate the major element similarities.

Selected normative data for the average LBB analyses are in Table 1 and Fig.4 shows all of the LBB samples plotted on the NE-DI-OL-HY-Q diagram. The LBB data points largely fall in the field for ocean ridge basalts (Kay et al, 1970). Some of the samples from the least altered, least fractionated group (Table 1) have high Mg/Mg + Fe" (e.g. 0.65 to 0.69), high Ni (120-280 ppm, mean value 215 ppm) and high Cr (300-670 ppm, mean value 480 ppm) suggesting that they are "primitive" liquids (complete major and trace element data for these samples are published in Hawkins, 1976). These "primitive" rocks are essentially liquid compositions as they are phenocryst-free. Their chemistry seems to be a strong argument in support of their origin as a fractional melt of peridotitic mantle material.

Excess He^3 relative to atmospheric He^3/He^4 in Lau Basin vitrophyres was interpreted by Lupton and Craig , (1975) as indicating a mantle source. Strontium isotope data (Hart, 1971, Montigny, 1975) are compatible with a mantle source although they are slightly more radiogenic than "typical" ORB. Five samples have been analyzed and gave values of 0.7020, 0.7030, 0.7033, 0.7034 and 0.0751. The lowest value is for an anorthositic gabbro, the high value is for an Ol-tholeiite (123-95-1) with only 35 ppm Sr. Minor seawater (or sediment) contamination could explain its composition or, alternatively, the source may have been more radiogenic than for the other samples. Rare earth element data (Montigny, 1975; Gill, 1974 as cited in Hawkins, 1976) show a

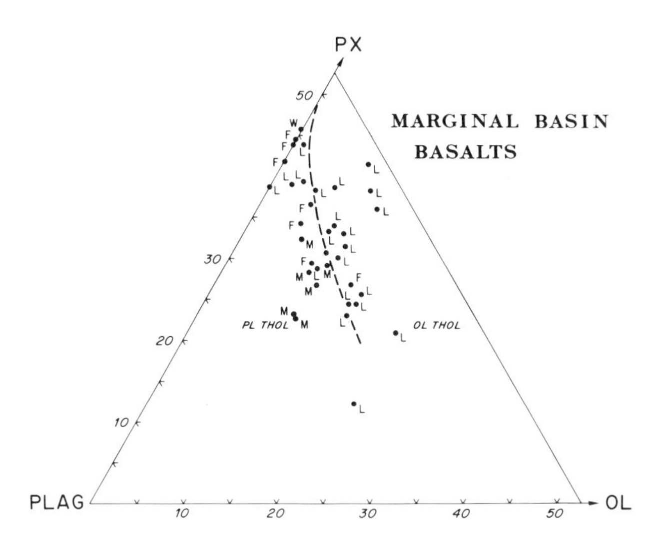

Figure 2. Marginal basin basalt normative data plotted in Pl-Ol-PX system. The dashed line is the presumed cotectic separating OL-from PL-tholeiites (Shido et al, 1971). Symbols for data points are: L = Lau Basin least altered and moderately altered tholeiites, F = North Fiji Plateau, M = Mariana Basin, W = Woodlark Basin.

flat pattern like ORB for least fractionated samples and overall enrichment with relative light RE enrichment for fractionated samples (see Fig 5) one sample (123-95-1) has marked relative depletion in light RE; it is from a presumed leaky transform fault in the Lau Basin. A sample, with a similar pattern, from the Troodos Complex (Kay and Senechal, 1976) is interpreted as a basalt emplaced on a transform fault (Moores and Vine, 1971 and E. Moores, oral commun. 1976). The most enriched pattern shown, 103-1, is from a strongly fractionated basalt or basaltic andesite from the leaky fransform fault. It is not typical of the most widespread "Primitive" LBB.

The major and trace element data, element ratios, Sr isotope ratio, He^3/He^4 and REE patterns all point to a magma type similar to the magmas leaking out of the mantle at oceanic ridges. The chemistry and mineralogy of the samples plus oxygen isotope data ($\delta\, O^{18}$ mean value 5.7 for unaltered basalt) presented by Pineau et al (1976) make it unlikely that Lau Basin basalt is merely a veneer overlying continental or island arc rocks. Basalts forming new sea floor at mid-ocean ridges are interpreted as fractional melts of mantle material which have separated from residual mantle solid phases at depths of 30 to 50 km. Similar melting and fractionation processes are inferred for the Lau Basin.

The data for samples from the North Fiji Plateau, Mariana Basin and Woodlark Basin are very limited and it is not possible to estimate meaningful average compositions of the basaltic crust of these marginal basins. Hart et al (1972) presented analyses for six samples; all of them show effects of fractionation (FeO^*/MgO ranges from 1.2 to 1.5) and none fit the criteria used for LBB to separate out least altered basalts. Nevertheless, their chemistry is typical of oceanic ridge tholeiites and they are HY and OL normative; they compare well with average ocean ridge basalt. Norms calculated from the reported analyses with Fe'''/Fe'' adjusted to 0.2 give essentially the same results except that one sample then shows a small amount of normative NE. The Mariana Basin rare earth element data show a flat, light depleted, pattern with an overall REE concentration enriched by 2 to 4 x ORB values. The strontium isotope ratios, and element ratios such as K/Rb and K/Ba, are like ORB (Hart et al, 1972).

A direct comparison between the Mariana Basin samples and ORB indicates some minor differences, mainly in alkali metals and degree of fractionation. However, the data presented by Hart et al. (1972) show that there are important similarities and the Mariana Basin samples clearly show similarities to some of the Lau Basin samples.

In addition to their higher alkali metal concentration, one of the major distinctive features of Mariana Basin basalts is that Ni abundances are lower than either the Lau Basin or ORB samples. A partial explanation for this Ni may be related to the degree of fractionation which these samples have experienced as shown by their high FeO^*/MgO (1.13 to 1.52) and by their MgO content (ave of 6 samples is 6.8% as compared with least altered LBB which has 9%). Hart et al

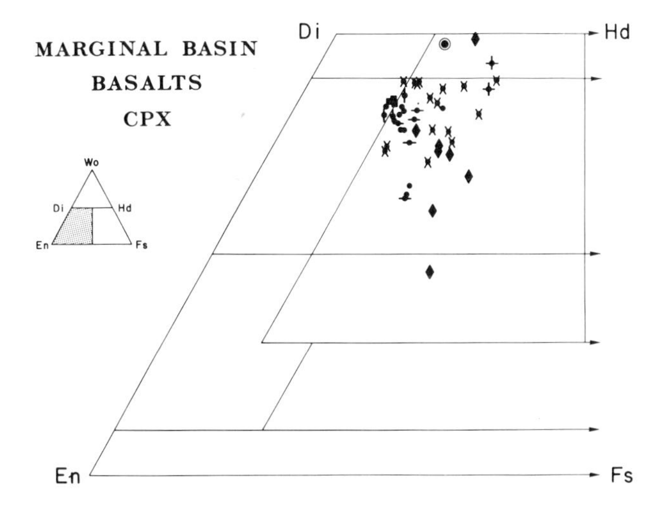

Figure 3. Microprobe data for clinopyroxene in marginal basin basalts plotted in pyroxene quadrilateral. Symbols are: ● = N. Fiji Plateau, microphenocryst, ◆ = N. Fiji Plateau, Diabase; ✦ and ■ = Lau Basin, diabase; -●- Lau Basin, microphenocryst; ⧫ = Mariana Basin, microphenocrysts; ⌘ = Mariana Basin, diabase; ⊙ Lau Basin gabbros. Data points lie in field delineated by clinopyroxene from DSDP sites 319 and 321 (Bunch and La Borde, 1976)

Table 3. Clinopyroxene Data

CLINOPYROXENE MICROPHENOCRYSTS IN VITROPHYRE

		Mole %			Wt%		Atoms per 6-O	
		Wo	En	Fs	Ti	Al	Al^{IV}	Al^{VI}
1. ANT-225-1	core	39.4	48.7	12.0	.191	1.257	.072	.030
	inner zone	37.6	48.7	13.7	.280	1.806	.122	.024
	outer zone	42.0	45.4	12.6	.291	1.630	.097	.036
	rim	40.7	46.5	12.8	.461	2.333	.129	.061
2. ANT-225-1	core	31.3	52.5	16.2	.289	1.940	.092	.066
	rim	31.9	51.9	16.3	.269	1.465	.039	.082
3. ERDC 14-12	core	40.8	48.5	10.9	.221	1.782	.092	.053
	intermed. zone	41.1	47.7	11.2	.201	0.970	.063	.017
	rim	41.6	47.5	11.0	.251	1.661	.097	.037
4. ERDC 14-12	core	32.7	51.3	16.0	.289	2.195	.138	.036
5. T-11-D-4	core	41.0	48.6	10.4	.262	1.184	.065	.031
	core	43.8	45.8	10.5	.332	1.933	.120	.039

1. Lau Basin, 0.25 mm Microphenocryst
2. Lau Basin, 0.10 mm microphenocryst
3. N. Fiji Plateau, 0.6 mm microphenocryst
4. N. Fiji Plateau, 0.05 mm microphenocryst
5. Mariana Basin 0.5 mm microphenocryst

SUB-OPHITIC CLINOPYROXENE IN DIABASE

	Mole %			Wt %		Atoms per 6-O	
	Wo	En	Fs	Ti	Al	Al^{IV}	Al^{VI}
1. 7-TOW 86-6	44.7	37.1	18.3	.736	2.240	.142	.042
7-TOW 86-6	47.6	35.3	17.1	1.056	3.658	.223	.082
2. 7-TOW 103-6	43.1	47.4	9.5	.393	1.720	.099	.041
7-TOW 103-6	42.8	47.5	9.7	.332	1.534	.067	.059
3. T-11-D-8	45.2	38.9	15.9	.818	2.749	.156	.074
4. ERDC 17-3	39.6	47.2	13.1	.381	1.654	.091	.042
ERDC 17-3	30.6	49.4	20.0	.317	.919	.055	.018
5. T-11-D-13	38.6	44.0	17.5	.475	1.449	.083	.040
T-11-D-13	37.0	51.5	11.6	.023	.819	.025	.042
6. T-13-D-2	43.5	43.8	12.7	.545	2.028	.103	.067
T-13-D-2	43.4	44.0	12.6	.826	2.978	.150	.099

1. Lau Basin
2. Lau Basin
3. Mariana Basin
4. Fiji Plateau
5. Mariana Basin
6. Mariana Basin

All analyses done on polished thin sections with electron probe microanalyzer

(1972) discuss the problems posed by the trace element composition and suggest that the differences between ORB and Mariana Basin samples is best explained by differences in relative degree of partial melting of the mantle sources.

Another significant aspect of the Mariana Basin samples is that they have clinopyroxene phenocrysts. Microprobe data reported by Hart indicate a composition of Wo_{42} En_{49} Fs_9 with .47% Ti, 1.74% Al and less than 50 ppm K. With exception of Ti, it compares closely with North Fiji Plateau and Lau Basin microphenocrysts.

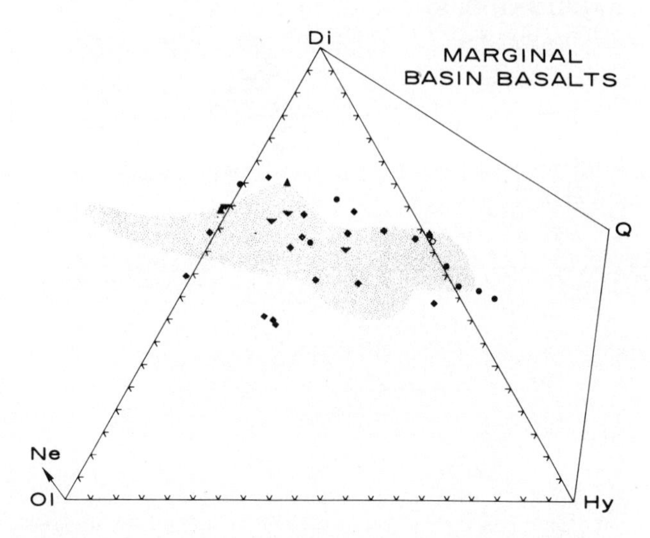

Figure 4. Marginal basin basalt normative data plotted in NE-OL-HY-DI-Q diagram. Field for ocean ridge basalts is shown in shaded pattern (Kay et al, 1970). Symbols shown are: ● and ▲ = two separate N. Fiji Plateau basalt collections; ◆ = Lau Basin, phenocryst free least and moderately altered basalts; ○ = Woodlark Basin vitrophyre, ▼ = Mariana Basin basalts.

Additional microprobe data for tholeiitic samples collected on SIO TASADAY expedition (1974) are in Table 3 and Fig. 3. The TASADAY microphenocrysts are essentially the same as those analyzed by Hart; the microdiabase clinopyroxene is closer to salite ($Wo_{45}En_{39}Fs_{16}$ with .82% Ti and 2.75% Al). The relative enrichment of the sub-ophitic pyroxene in Ca, Fe, Ti and Al with respect to the microphenocrysts is also seen in Lau Basin and N. Fiji Plateau diabase samples.

The North Fiji Plateau and Woodlark Basin samples are from areas which are not yet known well enough to defend strongly as actively spreading marginal basins. Geophysical data support the interpretation that they are active and the petrologic data in turn offer further support for this interpretation. A microprobe analysis of a vitrophyre sample (Luyendyk et al 1973) is the only petrologic evidence for Woodlark Basin basalts but it has the major element characteristics of the moderately fractionated marginal basin rocks (Fe^*/MgO = 1.16) and compares well with the estimate of an average abyssal Ol-tholeiite of similar FeO^*/MgO (Shido et al 1971).

Chemical Comparisons With Ocean Ridge Basalt

The data in table 1 are averages based on the analyses available for marginal basin basalts and

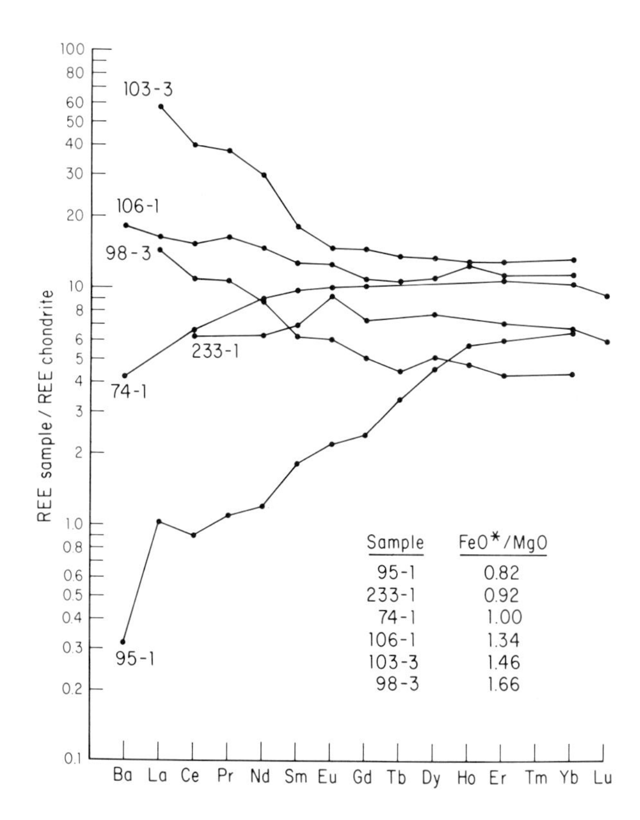

Figure 5. Barium and rare earth element concentrations normalized to chondritic abundances for Lau Basin rocks. Sample 74-1 is typical of least fractionated tholeiites which are the most abundant rock type in the basin, 233-1 is a feldspathic gabbro. 95-1 is a moderately fractionated basalt sample from a seamount. 103-3 is a strongly fractionated basaltic andesite, 98-3 is hypersthene dacite from a seamount

on samples from ocean basins. The Lau Basin Data (col. 1 and 2) have been screened so that one can see the composition of unmodified basalt liquids which are considered to be the magma type initially erupted. These averages, plus the estimates for altered and fractionated LBB (col. 7) are intended to represent the possible range in composition of rock types which might be seen if Lau Basin type crust were exposed in an ophiolite suite. Table 2 includes more limited data bearing on the probable composition of Lau Basin gabbroic rocks (col. 1) and data for MAR (col. 2) and Franciscan fm gabbros (col.3).

The close similarity in major and minor element chemistry between Lau Basin basalt samples and presumed average ORB (Table 1, col. 3) is apparent. Caribbean (DSDP) basalt data (col. 4) also look like LBB. The Cr, Ni and Mg of the LBB tend to be a little higher than the ORB and Caribbean averages because LBB is less fractionated. However, some individual LBB samples make a close match (Hawkins, 1976) and the moderately fractionated LBB also match closely. The Lau Basin samples differ from other oceanic basin basalts. One of the more critical points to consider is whether or not they are related to island arc tholeiites. The island arc tholeiite data (Table 1, col. 6) show significant differences in Ti,Al, alkali metals, FeO*/MgO and especially in Cr, Ni, Ni/Co, Sr and K/Rb. The two basalt types may be formed at essentially the same time in close proximity but it appears that they are geochemically distinct. An average analysis for DSDP samples from the Ninety East Ridge in the Indian Ocean (col. 5) is also distinctly different from LBB and shows that Ninety East Ridge samples are more strongly fractionated (lower in Ca, Al, Mg and higher in Fe, Ti and K) than any of the LBB samples. If the Ninety East Ridge is indeed a hot spot trace, than LBB samples are different from rocks formed by processes related to hot spots. Ocean island tholeiites (eg. Iceland and Hawaii) are geochemically distinct from ORB and thus also different from LBB.

In summary - ORB is the class of basalt which is geochemically most like LBB and the Lau Basin is underlain by oceanic crust. The data for Lau Basin gabbros is based on only one dredge haul but they too look like oceanic ridge samples. (Table 2, col. 1,2). Samples from the other marginal basins (Table 1, col. 3, 9) do not show as close a match to ORB averages and do not compare as well to the "primitive" LBB. They do look quite comparable to the fractionated LBB and through this similarity it seems that it is justified to say that they too are rock varieties closely related to ORB. The other marginal basin samples are too limited in number to make meaningful estimates of, for example, average North Fiji Plateau basalt. However, the chemical and mineralogic characteristics seem sufficiently similar to postulate that when more data for unaltered and unfractionated samples are available the averages for these basins will be like LBB and ORB values.

Chemical Comparisons With Ophiolites

The ophiolite rock series comprises rock types typical of oceanic crust (e.g. Bailey and Blake, 1974, Kidd, this volume).The rock types, textures, mineralogy and chemistry of marginal basin basalts and gabbros are so similar to those of oceanic crust that it is necessary to consider that some ophiolite units may contain fragments of marginal basin crust. Table 2 summarizes data for greenstone and spilite samples and for feldspathic gabbros from the Franciscan fm., California and the Nicoya complex, Costa Rica. The greenstone and spilite data (col. 5,6) do not necessarily represent an average ophiolite basalt but they do show that in a qualitative sense that there is not a great difference between them and altered rocks of the Lau Basin (col. 4). The gabbro data should be used with some caution considering the compositional variation likely in cumulate textured rocks, and the effects due to low temperature alteration, but perhaps the comparison is still reasonably good. A number of workers have suggested that some ophiolites may have been derived from marginal basins - for example, the ophiolites of the Coast Range (Bailey and Blake, 1974) and Baie Verte, Newfoundland (Kidd, this volume). There are important implications to the understanding of the evolution of orogenic belts and to the interpretation of paleo-tectonic configurations of arc-trench systems if marginal basin crust remnants form parts of ophiolites. From the available data on marginal basin basalts, it appears that it probably is not possible to distinguish Lau Basin-type basalt from ORB. The more fractionated basalts such as those from the North Fiji Plateau also have their counterparts in "normal" oceanic crust.

The prevalence of point source volcanism associated with seamounts seem to be typical of young active marginal basins. This style of volcanism raises problems with the interpretation of magnetic lineations in marginal basins but it may be a key to aid in recognizing marginal basin crust in ophiolites. An abundance of strongly fractionated basaltic rocks, pillow basalts mixed with volcaniclastic breccia, coralline limestone and reef debris, and minor silica and alkali-rich differentiates are all typical rock types found on seamounts. The sedimentary filling of marginal basins may be very restricted especially on the seaward side where the volcanic arc is the only sediment source. The thin sediment ponds of the Lau Basin adjacent to the Tonga Arc illustrate this point (Hawkins 1974). It seems, however, that there is no unique indicator of marginal basin crust; collectively some of these suggestions may aid in their recognition.

Summary

The crust of the Lau Basin consists of basaltic and gabbroic rocks which have textural, mineralogic and geochemical properties like those of oceanic ridge basalt. Rock samples which appear to be derived from phenocryst-free magmas, and are unmodified by either fractionation or alteration effects, were probably formed from liquids which separated from residual refractory phases at depths of about 30 km. Some samples, especially those in seamounts, have been modified by low pressure fractionation of olivine or plagioclase or both. The parental magmas were derived by fractional melting of upper mantle peridotite; there is no evidence of sialic contamination. North Fiji Plateau, Mariana Basin and Woodlark Basin samples are similar to fractionated Lau Basin or ocean-ridge basalt and a common magma source and evolution is envisioned for each. Marginal basin crust may be an important component of ophiolite belts but, in view of the overall similarity to oceanic crust, it may be difficult to identify by petrologic or geochemical characteristics. The abundance of seamounts relative to linear ridges in some marginal basins may offer a possible clue to identifying the origin of some ophiolite occurrences.

Acknowledgments. This work was supported by NSF Grants GA-30315, Ga-33227 and DES 74-17299

References

Anderson, R. N., Heat flow in the Mariana Marginal Basin, J. Geophys. Res., 80, 4043-4078, 1975.

Bailey, E. H. and M. C. Blake, Jr., Major chemical characteristics of Mesozoic Coast Range ophiolite in California, Jour. Res., U. S. Geol. Surv., 2, 637-656, 1974.

Barazangi, M., B. Isacks, J. DuBois and G. Pascal, Seismic wave attenuation in the upper mantle beneath the southwest Pacific, Tectonophysics, 24, 1-12, 1974.

Barker, P. F., Plate tectonics of the Scotia Sea region, Nature, 228, 1293-1296, 1970.

Bryan, W., Morphology of crystals in submarine basalts, J. Geophys. Res., 77, 5812-5819, 1972.

Bunch, T. E. and R. LaBorde, Mineralogy and compositions of selected basalts from DSDP Leg 34 in Initial Reports of the Deep Sea Drilling Project, R. Yeats and S. Hart, editors, XXXIV, 263-275, U.S. Gov. Printing Office, Wash. D. C., 1976.

Dengo, G., Tectonic-igneous sequence in Costa Rica, in Buddington Vol. Geol. Soc. Amer., A. Engel, H. James and B. Leonard, editors, 133-161, 1962.

Donnelly, T., R. Kay and J. Rogers, Chemical petrology of Caribbean basalts and dolerites; Leg 15 DSDP, EOS, Trans. Am. Geophys. Union, 54, 1002-1004, 1974.

Ewart, A. and W. Bryan, Petrography and geochemistry of the igneous rocks from Eua, Tongan Islands, Geol. Soc. Amer. Bull., 83, 3281-3298, 1972.

Hart, S., Dredge basalts: some geochemical aspects, Trans. Am. Geophys. Union, 52, 376, 1971.

Hart, S. R., W. Glassley and D. Karig, Basalts and seafloor spreading behind the Mariana Island arc, Earth Planet. Sci. Lett., 15, 12-18, 1972.

Hartzell, S., Geophysical study of the Fiji Plateau near 15°30'S, 173°30'E, EOS, Trans. Am. Geophys. Union, 56, 1073, 1975.

Hawkins, J. W., Geology of the Lau Basin, a marginal sea behind the Tonga Arc, in: Geology of Continental Margins, C. Burk and C. Drake, editors, Springer-Verlag, Berlin, 505-520, 1974.

Hawkins, J. W. and R. Batiza, Tholeiitic basalt from an active spreading center on the North Fiji Plateau near 15°30'S, 173°30'E, EOS, Trans. Am. Geophys. Union, 56, 1078, 1975.

Hawkins, J. W., Petrology and geochemistry of basaltic rocks of the Lau Basin, Earth Planet. Sci. Lett. 28, 283-297, 1976.

Hawkins, J. W., Petrology of Lau Basin basalts: Oceanic crust of a marginal basin, EOS. Trans. Amer. Geophys. Union, 57, 410, 1976.

Henningsen, D. and R. Weyl, Ozeanische Kruste im Nicoya-Komplex von Costa Rica, Geolog. Rund., 57, 33-47

Jakes, P. and J. Gill, Rare earth elements and the island arc tholeiitic series, Earth Planet. Sci. Lett., 9, 17-28, 1970.

Karig, D. E., Ridges and basins of the Tonga-Kermadec Island arc system, J. Geophys. Res., 75, 239-254, 1970.

Karig, D. E., Structural history of the Mariana Island arc system, Geol. Soc. Amer. Bull., 82, 323-344, 1971.

Kay, R. W., N. Hubbard and P. Gast, Chemical characteristics and origin of oceanic ridge volcanic rocks, J. Geophys. Res., 75, 239-254, 1970.

Kay, R. W. and R. G. Senechal, The rare earth geochemistry of the Troodos ophiolite complex. J. Geophys. Res., 81, 964-969, 1976.

Kidd, W. S. F., The Baie Verte lineament, Newfoundland: Ophiolite complex floor and mafic volcanic fill of a small Ordovician marginal basin (this volume). 1976.

Lawver, L., J. Hawkins and J. Sclater. Magnetic anomalies and crustal dilation in the Lau Basin, Earth Planet. Sci. Lett., 28, 1976 (in press).

Lupton, J. E. and H. Craig, Excess ^{3}He in oceanic basalts: evidence for terrestrial primordial helium, Earth Planet. Sci. Lett., 26, 133-139, 1975.

Luyendyk, B., K. MacDonald and W. Bryan, Rifting history of the Woodlark Basin in the southwest Pacific, Geol. Soc. Amer. Bull., 84, 1125-1134, 1973.

MacDonald, K., B. Luyendyk and R. von Herzen, Heat flow and plate boundaries in Melanesia, J. Geophys. Res., 78, 2537-2546, 1973.

Melson, W. and G. Thompson, Petrology of a transform fault zone and adjacent ridge segments, Phil. Trans. Roy. Soc. London, Ser. A, 268, 423-442, 1971.

Moberly, R., Origin of lithosphere behind island arcs, with reference to the western Pacific, Geol. Soc. Amer. Mem. 132, 35-55, 1972.

Montigny, R., Geochimie comparee des corteges de roches oceaniques et ophiolitiques - problems de leur genese, These, l'Université de Paris, 288, p., 1975.

Miyashiro, A., F. Shido and M. Ewing, Crystallization and differentiation in abyssal tholeiites and gabbros from mid-ocean ridges, Earth Planet. Sci. Lett., 7, 361-365 1970.

Moores, E. M., and F. J. Vine, The Troodos Massif, Cyprus and other ophiolites as oceanic crust: evaluation and implications, Phil. Trans. Roy. Soc. London, Ser. A, 268, 443-466, 197.

Packham, G. and D. Falvey, An hypothesis for the formation of marginal seas of the western Pacific, Tectonophysics, 11, 79-109, 1971.

Pineau, F., M. Javoy, J. Hawkins and H. Craig, Oxygen isotope variations in marginal basin and ocean ridge basalts, Earth Planet. Sci. Lett., 28, 299-307, 1976.

Sclater, J., and H. Menard, Topography and heat flow of the Fiji Plateau, Nature, 216, 991-993, 1967.

Sclater, J., J. Hawkins, J. Mammerickx and C. Chase, Crustal extension between the Tonga and Lau ridges: petrologic and geophysical evidence, Geol. Soc. Amer. Bull., 83, 505-518, 1972.

Shido, F., A. Miyashiro and M. Ewing, Crystallization of abyssal tholeiites, Contrib. Mineral. Petrol., 31, 251-266, 1971.

Tarney, J., Geochemistry of island arc and marginal basin volcanics in the Scotia Arc. (this volume) 1976.

Thompson, G., W. Bryan and F. Frey, Petrology and geochemistry of basalts and related rocks from DSDP leg 22, Sites 214 and 216, Ninety East Ridge, Indian Ocean, EOS, Trans. Am. Geophys. Union, 54, 1019-1021, 1973.

Wegener, A., The origin of continents and oceans, 1929, (translated by J. Biram), Dover Publ., N. Y., 246 p., 1966.

GEOCHEMISTRY OF VOLCANIC ROCKS FROM THE ISLAND ARCS AND MARGINAL BASINS OF THE SCOTIA ARC REGION

John Tarney, Andrew D. Saunders and Stephen D. Weaver

Department of Geological Sciences, University of Birmingham, England

Abstract. The main petrological and geochemical features of the igneous rocks from three island arc/marginal basin systems in the Scotia Arc are discussed and compared. (a) Fast active spreading behind the primitive intraoceanic S. Sandwich arc has been underway for 8 m.y. The back-arc basalts are slightly enriched in large ion lithophile (LIL) elements and are more radiogenic than normal mid ocean ridge (MOR) basalts. (b) Volcanism in the S. Shetland Is. has been of low-K calc-alkaline type since the Jurassic. On cessation of active spreading in Drake Passage ca 4 m.y. ago, a small marginal basin began to open up Bransfield Strait, behind the arc. Recent volcanism in Bransfield Strait has characteristics transitional between calc-alkaline and MOR tholeiite, and may be related to mantle diapirism behind the arc. The off-axis volcano of Penguin Is. is mildly alkaline. (c) On the continental margin of S. Chile, a narrow marginal basin developed behind an active continental-based calc-alkaline andesitic arc in the Jurassic, but the basin was closed and uplifted in the Cretaceous, preserving the marginal basin floor as a pillow lava-sheeted dyke-gabbro complex. Although the complex is affected by low-grade metamorphism, the fresher rocks have a geochemistry which is transitional between MOR and continental tholeiites. Whereas basalts in marginal basins with a long history of back-arc spreading are essentially similar to MOR basalts, magmas generated during the early stages of back-arc spreading seem to have more LIL-enriched characteristics, particularly where spreading was initiated along a continental margin. This may reflect some vertical LIL-element heterogeneity in the mantle rather than variations in partial melting conditions. The LIL-depleted mantle source for MOR basalts may be deep rather than shallow.

Introduction

The Scotia Sea, bounded by the extended loop of the Scotia Arc linking the Antarctic Peninsula with S. Chile, is at present situated near the junction of two major plates, the S. American and the Antarctic. During the Mesozoic and early Tertiary there was subduction of S.E. Pacific ocean lithosphere under S. Chile and the Antarctic Peninsula, but this segment of the S.E. Pacific is now coupled with the Antarctic Plate. Marine geophysical studies in the Scotia Sea (Barker, 1972; Barker and Griffiths, 1972) have revealed a complex pattern of magnetic anomalies which are linked to various phases of sea floor spreading since the mid-Tertiary. This resulted in the formation of a number of microplates, some of them no doubt quite short-lived.

In at least three situations in the area subduction has been associated with some form of back-arc spreading. It is the purpose of this paper to summarise available geochemical data on the igneous rocks produced as a result of back-arc spreading in relation to the geochemistry of the associated island arc volcanics. In the first situation, that of the East Scotia Sea (Fig. 1), relatively fast back-arc spreading behind the primitive intraoceanic S. Sandwich island arc has been underway for almost 8 m.y. In the second situation, bordering the Antarctic Peninsula, back-arc spreading may have been initiated relatively recently behind the continental-based S. Shetland volcanic arc, giving rise to the extensional feature of Bransfield Strait. The active or recently active volcanoes of Deception Is., Bridgeman Is. and Penguin Is. lie close to what may be the axis of back-arc spreading. In the third situation, in southern Chile, a small marginal basin opened up behind a continental-based arc in the Late Jurassic, linked to subduction of Pacific Ocean floor. But by the mid-Cretaceous the back-arc spreading had ceased, the basin was closed, and the oceanic floor uplifted and preserved as an ophiolite complex.

The three examples of back-arc spreading are of course not related in time, nor even perhaps by equivalent mechanisms. They are however relatively youthful features; in the case of the S. Chile fossil marginal basin it was an episode of back-arc spreading that was abruptly ended not long after it had got underway. They do therefore provide an insight into the type of magmatism associated with the initial stages of back-arc activity. In the equivalent early stage of development of mid-ocean ridges it is possible to argue that, in the case of the E. African Rift-Red

Sea system for instance, magmatism changes from alkaline to ocean floor tholeiite type with time (Gass, 1970). Since back-arc spreading is closely associated with subduction, one might expect a corresponding transition from island arc tholeiite or calc-alkaline magmatism to oceanic tholeiite with time, particularly if models such as that of Karig (1971) are correct in suggesting that mantle diapirs split the volcanic arc. Finally, it is important to establish whether or not there are any significant geochemical differences between mid-ocean ridge and marginal basin basalts because of suggestions (e.g. Dewey, 1976) that many ophiolite complexes could represent obducted marginal basin rather than oceanic lithosphere.

The three marginal basin examples will be described separately and then compared in the final discussion.

The South Sandwich Arc and the S. Sandwich Spreading Centre

At the easternmost extremity of the Scotia Arc, the S. Atlantic section of the S. American plate is subducting at a relatively high rate (ca 8 cm yr^{-1}) below the Scotia Sea. Approximately 80 km above the subducting plate lie the volcanic islands of the S. Sandwich Arc, which are at present erupting magmas of the island arc tholeiite series. The chemical characteristics of these magmas are very similar to the Tongan suite (Ewart et al., 1972) in having relatively low LIL element abundances, variable light-RE depleted rare earth patterns with both positive and negative europium anomalies, and rather uniform $^{87}Sr/^{86}Sr$ ratios of about 0.704 (Baker, 1976; Hawkesworth et al., 1976).

Marine geophysical investigations by Barker (1972) and Barker and Griffiths (1972) have established that there is rapid spreading (ca 4 cm yr^{-1} half-rate) behind the arc some 440 km west of the trench (Figs. 1 and 2). Well defined magnetic anomalies indicate that spreading has been underway for approximately 8 m.y. It would appear from the distribution of magnetic anomalies that the arc itself could be resting on lithosphere generated during the spreading episode, unless there was asymmetric spreading or a jump in the axis of spreading during the initial stages of back-arc activity. In the first case there is an implication that the initial sinking and subduction of the S. American plate under the relatively young oceanic lithosphere of the mid-Scotia Sea was accompanied by spreading immediately west of the trench, and that the volcanic arc developed later on this newly generated lithosphere. Alternatively, following a Karig (1971) model, the present volcanic arc might be superimposed on an older buried frontal arc with the remnant arc being positioned some 500 km to the west. The tectonic configuration at present is that of a small D-shaped plate (the Sandwich

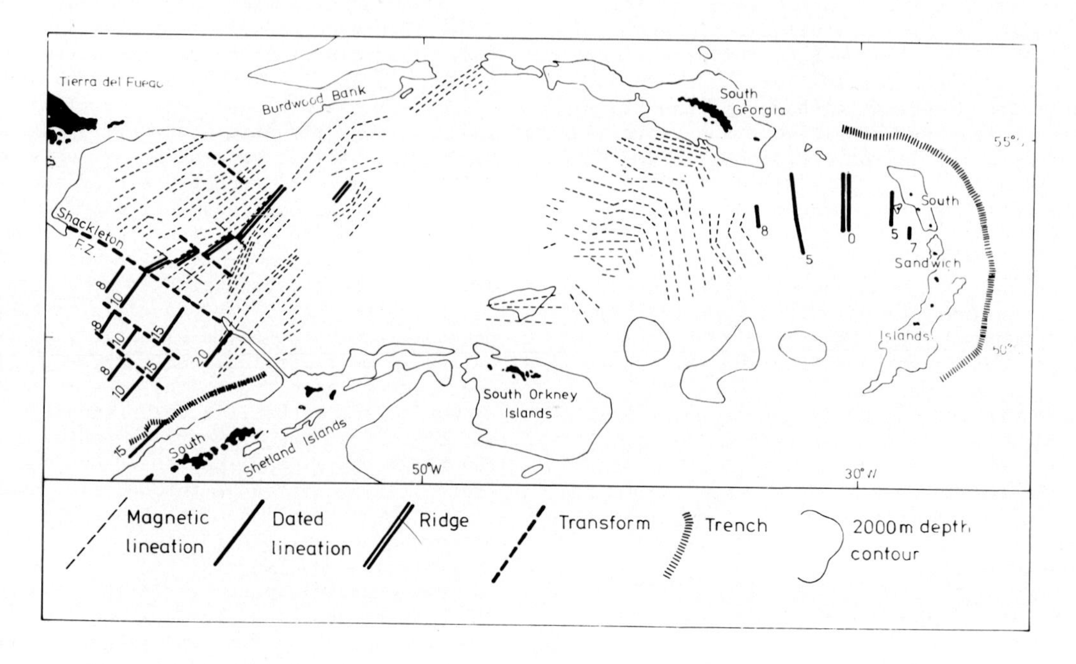

Figure 1. Map of the Scotia Arc showing pattern and ages of magnetic lineations in the Scotia Sea (after Barker and Griffiths, 1972). Dredge hauls 20, 22, 23 and 24 were located at points along the S. Sandwich spreading centre near 30°W while dredge hauls 17, 16 and 12 were located west of the spreading centre, progressively nearer the point of inception of spreading 8 m.y. ago.

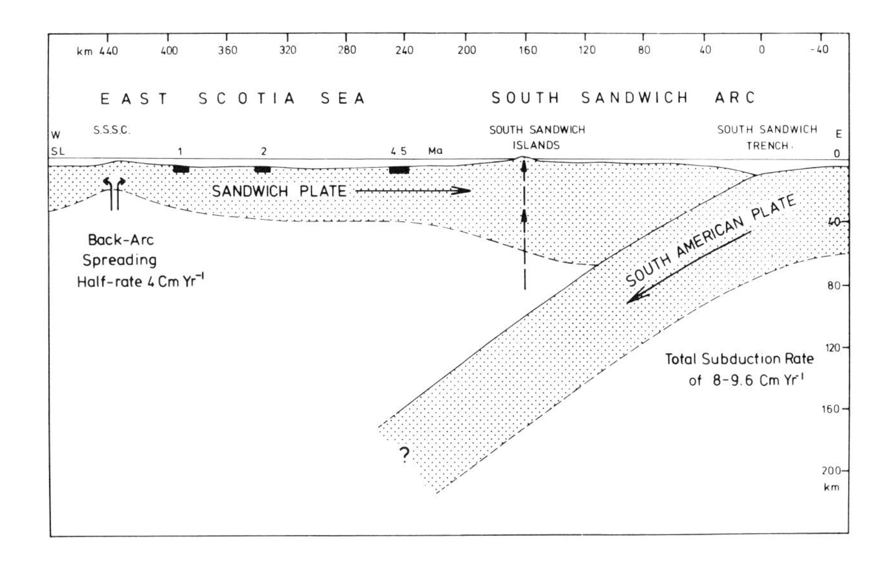

Figure 2. True scale section across the S. Sandwich Arc and the S. Sandwich back arc spreading centre (based on Barker, 1972 and Forsyth, 1975).

Plate) moving eastwards and growing by accretion at the back-arc spreading centre at a rate of 8 cm yr^{-1}, and moreover almost totally enclosed by the S. American-Antarctic Plates.

In an attempt to clarify the situation and to examine the geochemistry of basalts formed by back-arc spreading, dredging was carried out by RRS Shackleton to recover basalts from various points along the axis of the spreading centre and also at intervals westwards along the charted magnetic anomaly tracks to the point of commencement of spreading. Dredge hauls along the spreading axis yielded adequate pillow basalt fragments, but unfortunately the recovery from the older scarps was small in proportion to the glacial debris, and only obvious pillow basalts were analysed. Those basalts recovered are fresh. The majority carried phenocrysts of olivine and/or plagioclase, with little groundmass alteration.

Average analyses of basalts from the axis of the spreading centre and from older scarps are shown in Table 1. The basalts range from quartz-normative to olivine-normative tholeiites. Although the major element chemistry of these basalts is broadly similar to that of other mid-ocean ridge tholeiites, they are significantly enriched in some lithophile elements (K, Rb, Ba, Ce, La, P) and are rather poorer in Ni. It is possible to rule out sea water alteration or contamination as a cause of these higher LIL element abundances for several reasons (Saunders and Tarney, 1976). On the one hand Sr-isotope ratios are uniform within each dredge haul, and little difference in trace element chemistry is observed between the centres and glassy margins of pillows. On the other hand there is a strong degree of covariance between various lithophile elements in samples from different dredge hauls. This can be illustrated (Fig. 3) by plotting various lithophile elements against Zr, an incompatible element with very low crystal-melt distribution coefficients for most igneous minerals. These variations would appear to be mostly dependent upon the degree of partial melting because the compositional variations can be related only by appealing to some clinopyroxene fractionation, yet clinopyroxene is not a phenocryst phase in any of the basalts (with the exception of dredge 24).

The strong geochemical coherence between Zr, Ti, Sr and P in the Scotia Rise basalts may indicate that these elements are located in one mantle mineral phase (probably clinopyroxene). On the other hand the fact that K, Rb and Ba show a similar distribution for each dredge haul, but with obvious differences between dredge hauls, suggests that these elements may be located in another mineral phase (?phlogopite) and that the relative proportions of these two phases may vary in the mantle source. There is no apparent correlation with the petrological character of the basalts (whether quartz-normative or olivine-normative) and hence with differing P, T or pH_2O conditions during partial melting. Instead this would seem to indicate some degree of mantle inhomogeneity.

Rare-earth patterns for Scotia back-arc basalts (Fig. 4) are slightly light-RE enriched compared with normal MOR basalts, and lack Eu anomalies. The overall RE abundances correlate with other LIL-element abundances (i.e. those samples richer in REE are also richer in Zr, Sr, P, Ti, etc.) and appear to be largely a function of degree of partial melting.

Strontium isotope ratios for the same samples are higher than those for normal MOR basalts, but

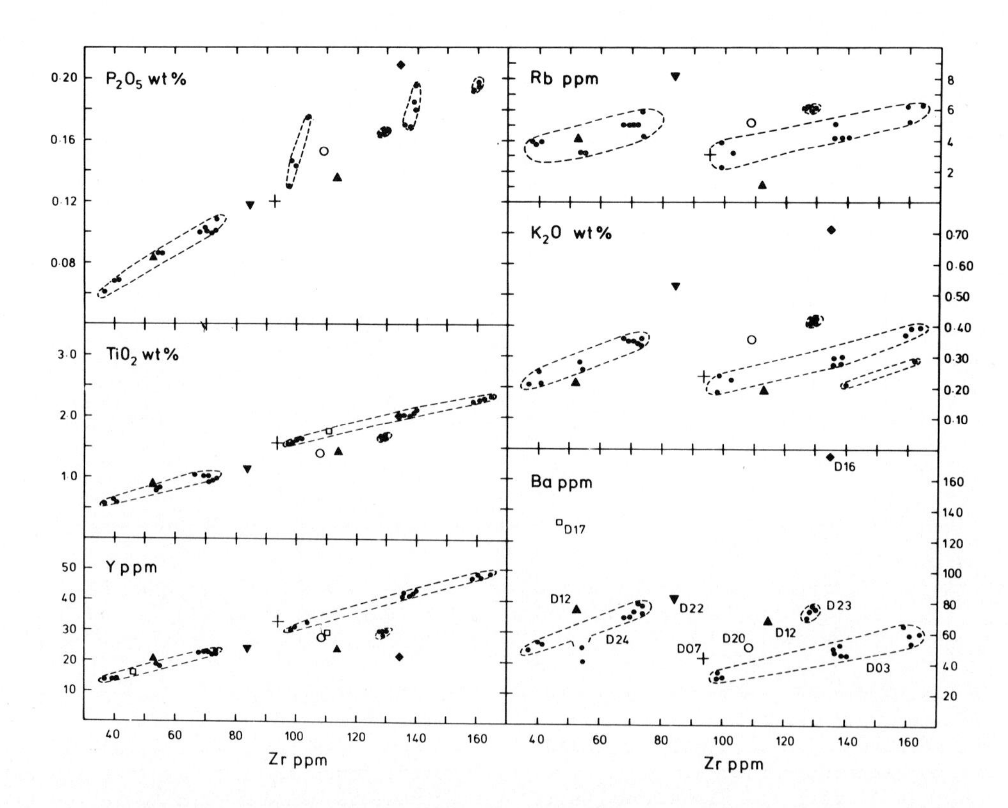

Figure 3. Incompatible element variations in Scotia Sea basalts. Zirconium is used as a fractionation index. Key to dredge hauls given in Ba v. Zr plot. Dashed lines enclose range of samples from each dredge haul. D20-D24 from back arc spreading centre. D12-D17 were located west of the spreading axis (see text). Data on basalts from two dredge hauls (D03 and D07) in the W. Scotia Sea are included for comparison.

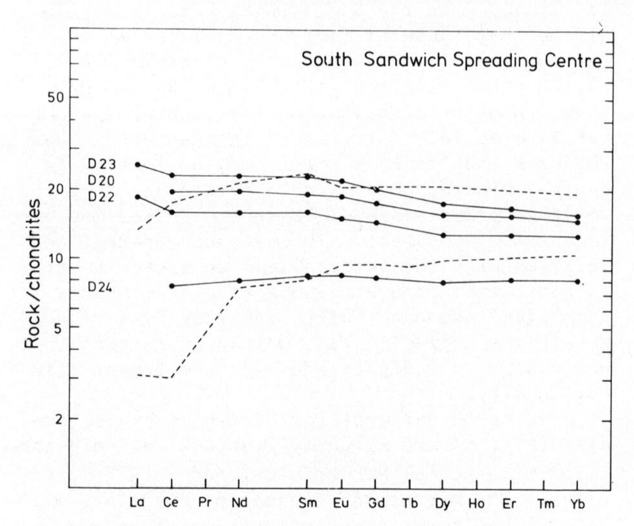

Figure 4. Chondrite-normalised REE patterns for basalts from the S. Sandwich spreading centre. Dashed lines indicate range for normal MOR basalts.

are uniform within each dredge. There is only an approximate correlation of $^{87}Sr/^{86}Sr$ with Rb/Sr ratio, suggesting that the present Rb/Sr ratios, which are higher than those in normal MOR basalts, may reflect only relatively recent mobility of Rb in the source region of the back-arc basalts.

Some of the basalts recovered from the older scarps in the east Scotia Sea (dredge hauls 12, 16 and 17) are rather more enriched in LIL elements compared with those from the spreading axis, but there is no obvious systematic variation with distance from the spreading centre, at least with the small number of samples recovered. It is possible of course that these more LIL-enriched samples may represent the products of off-axis volcanic activity.

One of the samples from Dredge 12 (i.e. located close to the point of inception of spreading) has some geochemical characteristics of basalts of the island arc tholeiite (IAT) series: low Zr, Ti, Ni and P_2O_5. While this might be taken as evidence for the presence of a remnant island arc before the present episode of back-arc spreading was initiated, there are other geochemical characteristics (high Cr, low Fe/Mg ratio) which do not conform with those of the

Table 1 Analyses of Marginal Basin and Associated Island Arc Basic Volcanics

	1	2	3	4	5	6	7	8	9	10	11	12	13	14	15	16
N	44	1	4	11	1	1	1	1	1	1	1	20	1	22	34	6
SiO_2	50.40	50.69	50.24	52.77	47.70	49.50	49.36	50.22	51.50	51.41	48.87	53.94	51.64	48.55	50.54	48.72
TiO_2	1.42	1.13	1.64	0.85	1.75	2.06	0.74	0.56	0.98	0.73	1.23	0.65	1.57	1.79	1.44	1.56
Al_2O_3	14.62	15.75	14.41	15.21	11.55	13.00	14.06	19.59	15.35	16.47	15.44	18.54	15.62	11.10	11.76	12.47
tFe_2O_3	8.81	8.19	9.74	9.98	11.74	14.50	10.85	10.60	9.70	10.72	10.59	7.29	9.50	14.25	13.24	11.86
MnO	0.17	0.17	0.18	0.18	0.24	0.25	0.21	0.17	0.19	0.23	0.19	0.13	0.18	0.22	0.23	0.21
MgO	8.13	8.08	7.99	6.28	9.80	8.28	9.16	5.60	5.83	6.64	9.07	5.19	6.43	8.66	7.89	8.89
CaO	11.10	10.85	11.01	10.57	12.18	8.51	11.16	12.53	10.82	9.80	10.10	9.73	9.83	11.35	9.82	7.01
Na_2O	3.11	2.39	3.44	2.29	1.19	3.26	1.29	1.67	2.52	3.24	3.74	3.59	4.16	1.55	1.97	4.26
K_2O	0.36	0.53	0.42	0.30	0.69	0.72	0.21	0.11	0.45	0.51	0.53	0.56	0.31	0.39	0.23	0.33
P_2O_5	0.15	0.12	0.17	0.09	0.26	0.21	0.08	0.04	0.25	0.15	0.30	0.06	0.21	0.14	0.24	0.20
Trace elements in p.p.m.																
Cr	263	196	269	171	137	296	129	24	178	81	508	68	139	152	115	248
Ni	72	66	67	32	21	124	11	11	52	23	163	26	35	30	27	79
Cu	76	68	88	120	77	-	26	74	-	-	-	-	-	-	-	-
Zn	67	69	69	70	103	111	76	-	-	-	82	63	76	82	70	105
Ga	16	13	16	13	17	20	14	13	-	-	22	20	22	17	18	14
Rb	5	8	6	4	20	12	4	2	6	5	5	12	3	12	8	3
Sr	195	195	214	148	414	337	143	153	361	514	550	332	342	140	191	121
Y	28	24	29	20	16	22	21	11	25	14	12	10	28	21	32	29
Zr	109	84	129	60	47	135	53	41	105	60	80	71	156	53	122	130
Nb	4	3	8	1.5	2	18	3	2	5	3	2	1	2	1	3	6
Ba	51	83	74	63	131	176	74	46	144	183	186	110	114	96	92	163
La	6	7	9	3	5	12	7	-	11	8	10	3	8	4	9	6
Ce	12	14	18	7	17	22	13	3	21	21	26	10	23	10	22	17
Pb	3	4	3	2	5	4	5	-	8	5	7	5	6	3	2	3
Th	1	1	1	1	1	3	1	-	1	1	3	2	1	1	2	1
Fe*/Mg	1.23	1.18	1.41	1.87	1.70	2.13	1.37	2.19	1.93	1.87	1.33	1.59	1.68	1.91	1.95	1.55
K/Rb	654	537	581	619	286	511	396	522	612	864	936	380	830	270	239	913
Rb/Sr	0.03	0.04	0.03	0.03	0.05	0.04	0.03	0.01	0.02	0.01	0.01	0.04	0.01	0.09	0.04	0.02
Ba/Rb	10.2	10.1	12.8	14.3	6.5	15.0	16.8	23.0	23.6	37.4	35.5	8.9	36.8	8.0	11.5	53
Ba/Sr	0.26	0.43	0.36	0.42	0.32	0.52	0.52	0.30	0.40	0.36	0.34	0.33	0.33	0.69	0.48	1.4
Zr/Nb	27	28	16	40	23	7.5	18	20	21	20	40	71	78	53	41	22
$^{87}Sr/^{86}Sr$	.7028	.7032	.7030	.7032	-	-	-	.7038	-	-	-	-	-	-	-	-

N = no. of analyses in means.

S. SANDWICH. Nos. 1-4 from spreading axis (dredge hauls 20, 22, 23 and 24). Nos. 5-7 from progressively older scarps west of spreading axis (dredge hauls 17, 16 and 12). No. 8 Island arc tholeiite from Bristol Is. (after Baker, 1976; Hawkesworth et al., 1976).

S. SHETLAND. No. 9 Calc-alkali basalt, Byers Peninsula (Mesozoic). No. 10 calc-alkali basalt, Fildes Peninsula (Tertiary). No. 11 Penguin Is. alkali basalt. No. 12 Bridgeman Is. basaltic andesite. No. 13 Deception Is. basalt (all Recent).

SARMIENTO, S. CHILE. No. 14, gabbros. No. 15 sheeted dykes. No. 16 pillow lavas.

island arc tholeiite series. In fact the suite of samples from Dredge 24, on the spreading axis itself, has even closer similarities to the IAT series in that Zr, Ti, Ni, P and REE levels are relatively low, the basalts are quite silica-rich (51.5-53.5% SiO_2) and, as a result of olivine and pyroxene fractionation, there is a fair range of Fe/Mg ratios. However, Cr levels are higher than in most arc tholeiites and $^{87}Sr/^{86}Sr$ ratios much lower than in any arc tholeiite.

In summary, basalts from the S. Sandwich back-arc spreading centre are more LIL-element enriched, have more light-RE enriched rare-earth patterns and have higher $^{87}Sr/^{86}Sr$ ratios than normal MOR basalts. However they are within the range encompassed by MOR basalts from Iceland (O'Nions et al., 1976) and some other areas along the mid-Atlantic ridge (e.g. 45°N, Erlank and Kable, 1976). Some of the basalts, both at the present spreading axis, and those generated 8 m.y. ago at the inception of spreading, have geochemical characteristics transitional towards arc tholeiites. However, considering the observed systematic chemical variations and the Sr-isotope differences between the back-arc basalts (Saunders and Tarney, 1976) and the adjacent S. Sandwich arc tholeiite volcanics (Baker, 1976, Hawkesworth et al., 1976) it seems unlikely that the chemistry of the arc tholeiite series could be duplicated exactly by further fractionation of Scotia Sea basalt magmas. Finally, there is no positive evidence, admittedly on the basis of limited sample recovery, of any major change in the composition of the basalts generated during the back-arc spreading episode.

The South Shetlands Island Arc and Bransfield Strait Marginal Basin

The South Shetland Island Arc, with a volcanic history extending back into the Mesozoic, is separated from the Antarctic Peninsula by the long narrow trough of Bransfield Strait (Fig. 5). The marine seismic investigations by Ashcroft (1972) supplemented by the gravity data of Davey (1971) and earlier geological investigations (bibliography in Ashcroft, 1972 and Baker et al., 1975) have established that the arc is based on 15 km thick continental crust. This is confirmed by the presence of quartzite and high-grade gneiss blocks in the volcanics and the fact that 3 km of U. Palaeozoic sediments (equivalent to the Trinity Peninsula Series of the Antarctic Peninsula) are exposed on Livingston Island.

About 100-120 km northwest of the arc is a 5 km deep trench, the site of subduction of 15-20 m.y. old oceanic crust generated at the spreading centre in the West Scotia Sea south of Cape Horn

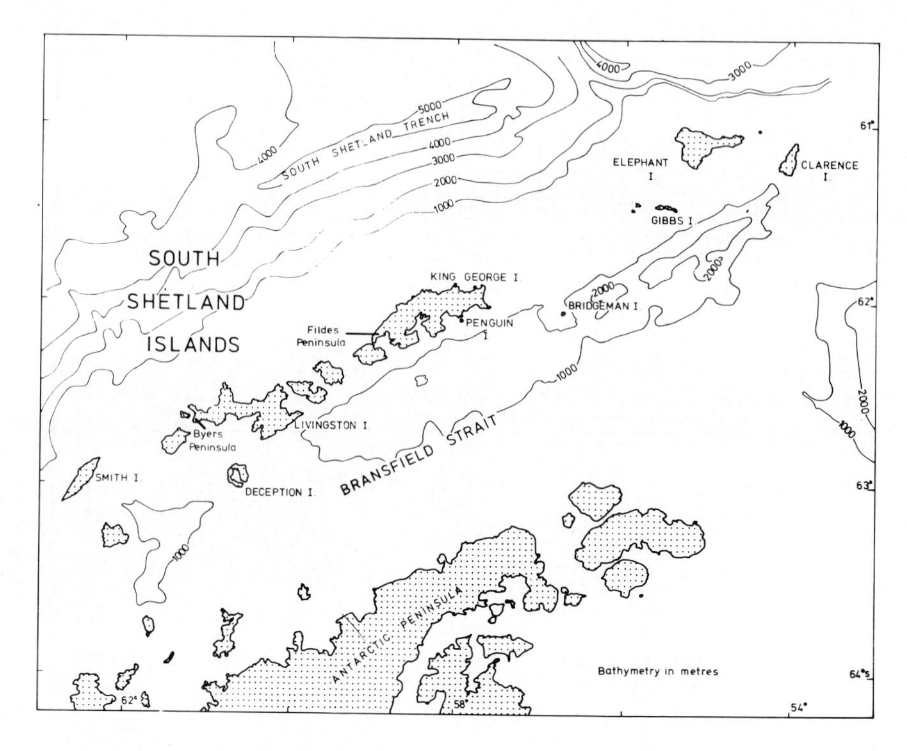

Figure 5. The S. Shetland island arc, separated from the Antarctic Peninsula by the extensional trough of Bransfield Strait. The active and recent volcanoes of Deception Is. and Bridgeman Is. lie along the axis of the trough while Penguin Is. lies just north of the axis.

(Griffiths and Barker, 1971). Low density sediments up to 6 km thick occur in the arc-trench gap.

Bransfield Strait (Fig. 6) is a graben feature at least 400 km long, 65 km wide and 4 km deep, partly filled with low density sediments 1-2 km thick. Major normal faults downthrowing towards the Strait occur along the south-east coast of the Shetland Islands, their presence being supported by both geological and seismic data. The mantle lies at a depth of only 14 km below the axial trough, but is of abnormally low velocity (7.6-7.7 km sec^{-1}). An unusually thick basaltic crust is suggested by the fact that rocks of 6.5-6.9 km sec^{-1} velocity occur only 5 to 6 km below the deep central trough and can be traced for 250 km along strike.

There seems little doubt that Bransfield Strait is a back-arc extensional feature and that its floor is oceanic in character. However it also seems to be a fairly young feature, probably less than 3-4 m.y. old. Two recent volcanoes, Deception Is. and Bridgeman Is., are located some 200 km apart along the axis of the trough and there are other bathymetric features interpreted as submarine volcanoes along the same axis (Ashcroft, 1972). Recent volcanic activity has also been recorded on Penguin Is. (just SE of King George Is.) some 20 km NW of the axis. It would seem that as spreading ceased, or slowed down, in Drake Passage during the last few million years (Barker and Griffiths, 1972), continuing subduction under the S. Shetland arc was accompanied by extension in Bransfield Strait and movement of the S. Shetland Is. northwestwards away from the Antarctic Peninsula. At the same time the locus of volcanic activity moved southeastwards to be centred over the axis of back-arc spreading (Deception and Bridgeman) with minor off-axis activity (Penguin). The situation thus provides an unusual opportunity to examine the geochemical nature of magmas produced at the inception of back-arc spreading. However with the cessation of subduction at the S. Shetland trench it is not possible to predict that back-arc spreading will necessarily continue in Bransfield Strait.

The volcanic history of the S. Shetlands, and the active volcano of Deception in particular, has been the subject of a number of investigations (Hawkes, 1961 a, b; Baker et al., 1975). To provide a more comprehensive geochemical picture we have, within the last year, sampled more extensively the volcanics of the region. The results will be published in full elsewhere, but the following summarises the more important features bearing upon the mechanism of back-arc spreading.

A series of dominantly basalt, basaltic andesite and andesite lavas with occasional dacites and rhyodacites characterises the S. Shetland arc from the Jurassic to the Late Tertiary. Plutonic intrusions ("Andean Intrusive Suite") of Late Cretaceous age are also present. Basalts

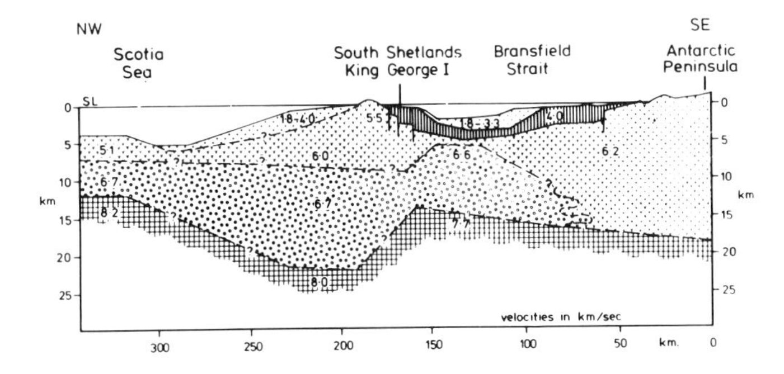

Figure 6. Crustal structure of the S. Shetland Is. and Bransfield Strait (after Ashcroft, 1972).

are less conspicuous amongst the Jurassic volcanics, but are more important components of the Tertiary lava sequence. On the other hand more salic lavas are commoner in the Jurassic sequence. Aphyric basalts and andesites occur, but the majority of lavas have phenocrysts of plagioclase and olivine with or without clinopyroxene, hypersthene, hornblende and (rarely) titanomagnetite.

Chemically (Table 1) the lavas have low K_2O and Rb contents and fairly high K/Rb and low Rb/Sr ratios similar to arc tholeiites of primitive island arcs (Jakes and Gill, 1970). However, elements such as Sr, Ba, Zr and Cr are higher than in arc tholeiites, and rare-earth patterns are light-RE enriched, unlike any so far reported for members of the island arc tholeiite series. The lavas would therefore be better regarded as members of a low-K, high alumina calc-alkaline volcanic series. The Tertiary lavas have in fact rather lower K and Rb contents than those erupted earlier in the Jurassic.

With the extensional opening of Bransfield Strait there is a change in the character of the volcanics. Of the two volcanoes lying along the spreading axis, Bridgeman Is. is largely made up of high-alumina basalts and baslatic andesites rich in plagioclase phenocrysts but with minor clinopyroxene and olivine phenocrysts. Deception Is. however, horseshoe shaped as a result of caldera collapse and breaching by the sea, has a longer history of crystal fractionation, displaying a range of rock types from basalt (50% SiO_2) to rhyodacite (70% SiO_2). The basalts and basaltic andesites may be aphyric, but most lavas have plagioclase phenocrysts, with additional olivine and clinopyroxene in the more basic lavas, hypersthene in the intermediate and fayalitic olivine in the salic lavas.

The lavas of the off-axis volcano of Penguin Is. however are mildly alkaline (up to 5% <u>ne</u>) olivine basalts with phenocrysts of olivine, minor spinel and clinopyroxene.

The strong mineral fractionation observed at Deception produces considerable enrichment of incompatible elements such as K, Rb, Ba, Zr, Nb, Pb and Th in the salic volcanics compared with their values in the basalts, but Sr values fall

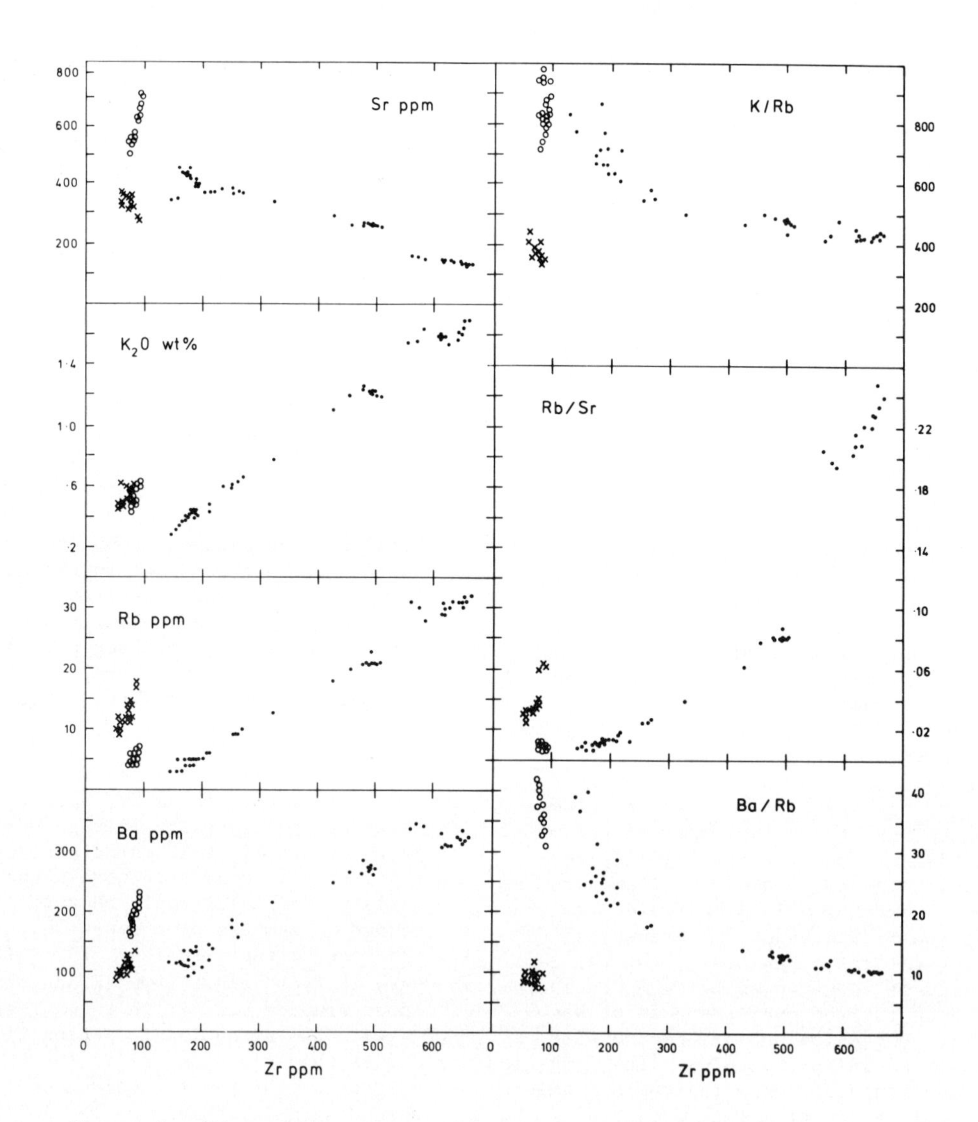

Figure 7. Plots of incompatible elements and element ratios against zirconium for Deception (filled circles), Penguin (open circles) and Bridgeman (crosses) lavas.

as a result of plagioclase fractionation. There is a similar, but smaller, degree of enrichment in the 'less incompatible' elements, Na, Ce, La and Y. The highly incompatible element, Zr, can be used as an indicator of the degree of fractionation since it is not contained in any significant quantity in the phenocrystic minerals. Plots against Zr produce smoother trends for incompatible elements than those against SiO_2 or Fe/Mg which are more useful in dealing with cumulates. At the same time it allows meaningful comparisons to be drawn with the Bridgeman and Penguin magmas.

Although it is not possible to reproduce more than a small fraction of the data here, most such Zr-normalised plots link the Deception volcanics as much with the Penguin alkali olivine basalts as with the Bridgeman lavas, suggesting that Deception and Penguin parental magmas have been generated from similar mantle sources. Plots of K_2O, Rb, Sr, Ba, K/Rb, Rb/Sr and Ba/Rb against Zr (Fig. 7) emphasise this relationship and demonstrate the low Rb/Sr ratios of the Deception-Penguin lavas. $^{87}Sr/^{86}Sr$ ratios as low as 0.703 have been recorded for some Deception lavas (Baker et al., 1975). Bridgeman lavas have much higher Rb/Sr and lower K/Rb and Ba/Rb ratios, suggesting that the mantle source may have been geochemically slightly different, at least with respect to these elements, at the time of magma generation.

There would seem to be no way of generating the Deception and Penguin magmas by fusion of the subducting Scotia Sea oceanic crust, since Rb/Sr ratios are lower and K/Rb and Ba/Rb ratios are higher in the volcanics than in the subduct-

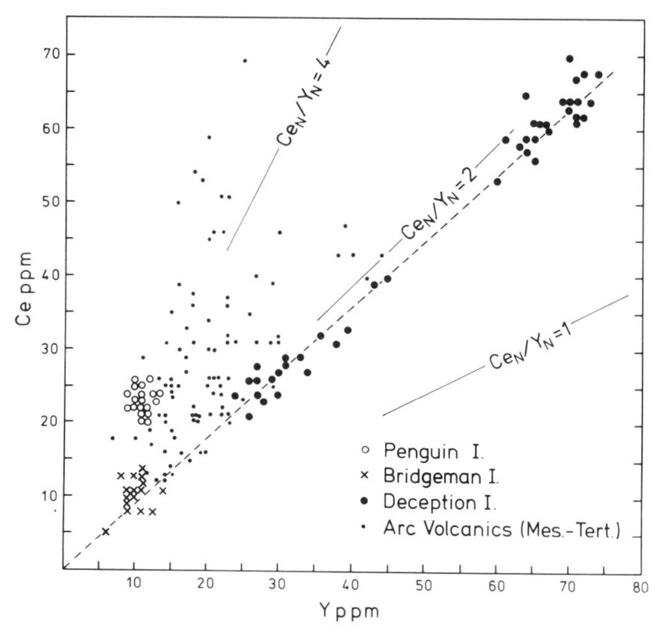

Figure 8. Ce v. Y plots for Deception, Penguin and Bridgeman lavas and for the Jurassic and Tertiary calc-alkaline volcanics of the S. Shetlands arc.

ing oceanic crust. This is also confirmed by the high Cr and Ni contents of the Deception, and especially the Penguin basalts, which suggests mantle derivation. The association of alkaline Penguin and calc-alkaline Deception magmas geographically is not without precedence: the association is not uncommon in other island arcs (c.f. Arculus, 1976). However the off-axis Penguin magma seems to have been generated at greater depth. This is indicated by the fact that whereas basic magmas from both volcanoes have similar Ce contents, Y levels are much lower in the Penguin basalts, suggesting that garnet may have been a stable residual phase during magma generation, holding back Y (and possibly some Zr too). Ce/Y ratios are in fact lower in the Deception (and Bridgeman) volcanics than in any of the earlier S. Shetlands calc-alkaline volcanics. If Y behaviour reflects that of the heavy REE, this suggests that chondrite normalised REE patterns are less fractionated than for the older volcanics, though not as flat as those of MOR basalts. Shallower level melting for the production of the Deception and Bridgeman magma types would be consistent with a model invoking the uprise of a hot mantle diapir to account for the back-arc extension in Bransfield Strait. There is a strong linear relationship between Ce and Y throughout the whole series of Deception lavas, demonstrating that, if Y follows the heavy REE, rare-earth patterns are essentially unaffected by extensive crystal fractionation.

In summary, the Deception and Bridgeman magmas appear to have been generated through relatively shallow level melting of mantle which has some geochemical features (low K, Rb, Rb/Sr, $^{87}Sr/^{86}Sr$; high K/Rb, Ba/Rb) characteristic of the source for MOR basalts. Yet the differentiation trends are broadly calc-alkaline, but with some Fe/Mg enrichment. The geochemistry is thus transitional between calc-alkaline and mid-ocean ridge. This would be consistent with a model of mantle diapirism splitting the volcanic arc during the initial stages of back-arc spreading. It would appear however that Ba and Sr levels are still high, and that further spreading would be necessary before basalts with true MOR geochemistry were to be generated in Bransfield Strait.

Mesozoic Marginal Basin Floor Ophiolites from S. Chile

The tectonic setting of the 'Rocas Verdes' marginal basin ophiolites from southernmost Chile has been described by Dalziel and co-workers (c.f. Dalziel et al., 1974; Bruhn and Dalziel, this vol.). Briefly, just before the opening of the S. Atlantic, extension behind a Jurassic continental-based island arc caused rifting and the development of a narrow marginal basin composed of oceanic crust. Extension ceased and the basin was closed and uplifted in the mid-Cretaceous, preserving the marginal basin floor as an ophiolite complex composed of gabbros, sheeted dykes and pillow lavas, with minor plagiogranite. The present outcrop pattern suggests that the marginal basin may have been as much as 100 km wide originally near Cape Horn, narrowing to less than 30 km northwards (near 51°S), and probably over 1000 km in length. The original tectonic situation, particularly in the northern area, may have been similar to the S. Shetlands arc and Bransfield Strait. Moreover the thick mafic crust in the latter, if typical of small marginal basins, may explain the absence of an ultramafic component in the Chilean ophiolites.

As with most other ophiolite complexes, the rocks have suffered low-grade (zeolite- to greenschist facies) metamorphism. The metamorphic affects however appear unrelated to the localised deformation which occurred during basin closure, or to the intrusion of the batholiths of the Cordillera, but are more directly linked to the hydrothermal activity associated with the spreading episode itself (Stern et al., 1976; Saunders et al., 1976). Many of the rocks analysed have in fact suffered low grade metamorphism, but some only to a very limited extent. Nevertheless the geochemical effects accompanying the hydrothermal activity can be allowed for in discussing the primary chemistry.

Mean analyses of gabbros, sheeted dykes and pillow lavas from the Sarmiento Complex are presented in Table 1, and are based on the range of 70 samples analysed and discussed by Saunders et al. (1976). Most of the lavas and dykes fall in the MOR basalt field on Pearce and Cann (1973) discrimination diagrams, but REE patterns are light-RE enriched (La_N/Yb_N=2) compared with MOR basalts or even Scotia Sea basalts. The levels of Zr, Y, Sr and Ti are thus comparable with those in MOR basalts. However the gabbros have lower values of incompatible elements such as Zr, Ce, La, Sr and P_2O_5, as may be expected if the gabbros are partly cumulates. There is a wide range of Fe/Mg ratios (0.8 to 4.6) in the dykes and gabbros, and there is a reasonable degree of correlation of Cr, Ni, Zr and TiO_2 with Fe/Mg ratio, indicating considerable crystal fractionation or variable degrees of partial melting (or both) during development of the complex. The fractionation trend is tholeiitic, there being no silica enrichment with increasing Fe/Mg ratio (except in the late stage plagiogranites). Most of the rocks are quartz- rather than olivine-normative. This, coupled with the low alumina contents and relatively flat rare-earth patterns would seem to indicate relatively shallow-level melting. The plagiogranites are rich in Zr, La, Ce and Y, thus inviting comparison with the dacitic Deception lavas, and suggesting moreover that the plagiogranites may be late stage differentiates of the mafic rocks.

There is a much poorer degree of correlation of incompatible elements such as K, Rb and Ba with Zr or with Fe/Mg ratio, although they correlate fairly well with each other. Values for K_2O, Rb and Ba are higher and much more uniform in the fresher dykes and gabbros, where K/Rb ratios are very much lower than in MOR basalts. The amphibolised and chloritised dykes and gabbros have much lower K and Rb values and K/Rb ratios are much higher, suggesting loss of Rb and K during the hydrothermal alteration; this is not unexpected since minerals such as chlorite, hornblende and epidote hold little K or Rb in their structures. More extreme effects are seen in the pillow lavas which are spilitised and have high Na_2O contents.

We would regard the K_2O, Rb and Ba contents of the fresher dykes and gabbros (ca 0.5% K_2O, 13 p.p.m. Rb, 90 p.p.m. Ba) as more typical of the initial magmas. This, together with the more fractionated REE patterns, would imply a mantle source rather more enriched in these elements compared with that for MOR basalts. Finally, we note that the geochemical characteristics of the magmas are transitional towards continental tholeiites rather than calc-alkaline magmas. This would be more consistent with the development of the marginal basin slightly behind rather than within an active pre-existing arc.

Discussion

Comparison of the tectonic settings of the three different marginal basin situations and the geochemistry of the volcanic products suggests that there may be no single uniform mechanism responsible for back-arc spreading. Whereas the geochemistry of the Bransfield Strait volcanics would be consistent with the initial stages of splitting of the calc-alkaline volcanic arc, the back-arc products in S. Chile and the Scotia Sea are essentially tholeiitic. Much may depend of course on the relative activity of the mantle diapir producing the back-arc extension. Slow uprise might produce transitional characteristics, as in Bransfield Strait, whilst rapid active diapirism may account for the essentially tholeiitic volcanism in the Scotia Sea and S. Chile.

None of the marginal basin igneous products is as geochemically depleted in lithophile elements as MOR basalt, but each shows some transitional characteristics. Furthermore, although the tectonic situations are not exactly equivalent, there appears to be an increase in the 'depleted' characteristics of the marginal basin volcanics in going from Bransfield Strait, through S. Chile to the Scotia Sea (i.e. with increasing stages of opening of the back-arc basins). Basalts from marginal basins with a longer history of back-arc spreading, such as those behind the Mariana (Hart et al., 1972) and Tongan (Hawkins, 1976) arcs are much closer to MOR basalts in their geochemistry. Note however that in S. Chile and Bransfield Strait the basaltic magmas were generated in sub-continental lithosphere.

Geochemical and isotopic variations in basalts along the mid-Atlantic ridge have been linked with the uprise of LIL-element enriched deep mantle plumes(e.g. Schilling, 1973). While the Zr-normalised plots of the marginal basin rocks also suggest that differences in lithophile element abundances and ratios are partly a function of mantle inhomogeneity, we feel that the three cases of back-arc spreading examined here would be just as compatible with a LIL-enriched mantle source which is shallow rather than deep. For instance, in the initial stages of back-arc spreading the influence of a deep mantle plume would seem to be precluded by the presence of the subducting slab. Yet it is at this stage in their development that marginal basin products seem to display more LIL-element enriched characteristics.

Acknowledgements. Geochemical studies in the Scotia Arc were supported by the Natural Environment Research Council, U.K. We thank the British Antarctic Survey, the Royal Navy, D. H. Griffiths and P. Barker for logistic support; B.A.S., I.W.D. Dalziel, M. J. de Wit and C. R. Stern for donating samples; R. K. O'Nions and R. J. Pankhurst for carrying out Sr isotope and REE determinations; G. L. Hendry and N. Donnellan for their help with XRF analysis and S. E. Delong, I. Ridley and R. Bruhn for their comments on the manuscript.

REFERENCES

Arculus, R. J., Geology and geochemistry of the alkali-basalt-andesite association of Grenada, Lesser Antilles island arc, Geol. Soc. Amer. Bull., 87, 612-624, 1976.

Ashcroft, W. A., Crustal structure of the South Shetland Islands and Bransfield Strait, Brit. Antarct. Surv. Sci. Rept., 66, 1-43, 1972.
Baker, P. E., The South Sandwich Islands: II. Petrology and Geochemistry, Brit. Antarct. Surv. Sci. Rept., in press, 1976.
Baker, P. E., I. McReath, M. R. Harvey, M. J. Roobol, and T. G. Davies, The geology of the South Shetland Islands: V. Volcanic evolution of Deception Island, Brit. Antarct. Surv. Sci. Rept., 78, 1-81, 1975.
Barker, P. F., A spreading centre in the east Scotia Sea, Earth Planet. Sci. Lett., 15, 123-132, 1972.
Barker, P. F., and D. H. Griffiths, The evolution of the Scotia Ridge and Scotia Sea, Phil. Trans. Roy. Soc. Lond., A271, 151-183, 1972.
Bruhn, R. L., and I. W. D. Dalziel, Destruction of the Early Cretaceous marginal basin in the Andes of Tierra del Fuego (this volume).
Dalziel, I. W. D., M. J. de Wit, and K. F. Palmer, Fossil marginal basin in the southern Andes, Nature, Lond., 250, 291-294, 1974.
Davey, F. J., Marine gravity measurements in Bransfield Strait and adjacent areas, in Antarctic Geology and Geophysics, edited by R. J. Adie, pp. 39-45, Universitetsforlaget, Oslo, Norway, 1971.
Dewey, J. F., Ophiolite obduction, Tectonophysics, 31, 93-120, 1976.
Erlank, A. J., and E. J. D. Kable, The significance of incompatible elements in Mid-Atlantic Ridge basalts from 45°N with particular reference to Zr/Nb, Contrib. Mineral. Petrol., 54, 281-291, 1976.
Ewart, A., W. B. Bryan, and J. B. Gill, Mineralogy and geochemistry of the younger volcanic islands of Tonga, S.W. Pacific, J. Petrol., 14, 429-465, 1973.
Forsyth, D.W., Fault plane solutions and tectonics of the South Atlantic and Scotia Sea, J. Geophys. Res., 80, 1429-1443, 1975.
Gass, I. G., Tectonic and magmatic evolution of the Afro-Arabian dome, in African Magmatism and Tectonics, edited by T. N. Clifford and I. G. Gass, Oliver & Boyd, Edinburgh, Scotland, 1970.
Griffiths, D. H., and P. F. Barker, Review of Marine Geophysical Investigations in the Scotia Sea, in Antarctic Geology and Geophysics, edited by R. J. Adie, pp. 3-11, Universitetsforlaget, Oslo, Norway, 1971.
Hart, S. R., W. E. Glassley, and D. E. Karig, Basalts and sea floor spreading behind the Mariana island arc, Earth Planet. Sci. Lett., 15, 12-18, 1972.
Hawkes, D. D., The Geology of the South Shetland Islands, I. The petrology of King George Island, Sci. Rept. Falkland Is. Dependencies Survey, 26, 1-28, 1961a.
Hawkes, D. D., The Geology of the South Shetland Islands, II. The geology and petrology of Deception Island, Sci. Rept. Falkland Is. Dependencies Survey, 27, 1-43, 1961b.
Hawkesworth, C. J., R. K. O'Nions, and R. J. Pankhurst, A geochemical study of island-arc and back-arc tholeiites from the Scotia Sea, Earth Planet. Sci. Lett., in press, 1976.
Hawkins, J.W., Petrology and geochemistry of basaltic rocks of the Lau Basin, Earth Planet. Sci. Lett., 28, 283-298, 1976.
Jakes, P., and J. Gill, Rare earth elements and the island arc tholeiite series, Earth Planet. Sci. Lett., 9, 17-28, 1970.
Karig, D. E., Origin and development of marginal basins in the western Pacific, J. Geophys. Res., 76, 2542-2561, 1971.
O'Nions, R. K., R. J. Pankhurst, and K. Gronvold, Nature and development of magma sources beneath Iceland and the Reykjanes Ridge, J. Petrol., in press, 1976.
Pearce, J. A., and J. Cann, Tectonic setting of basic volcanic rocks determined using trace element analysis, Earth Planet. Sci. Lett., 19, 290-300, 1973.
Saunders, A. D., and J. Tarney, Geochemistry of basalts from the back-arc spreading centre in the Scotia Sea, in preparation, 1976.
Saunders, A. D., J. Tarney, C. R. Stern, and I. W. D. Dalziel, Geochemistry of marginal basin floor mafic igneous rocks from S. Chile, in preparation, 1976.
Schilling, J. G., Icelandic mantle plume: Geochemical study of the Reykjanes Ridge, Nature, Lond., 242, 565-567, 1973.
Smellie, J. L., and P. D. Clarkson, Evidence for pre-Jurassic subduction in western Antarctica, Nature, Lond., 258, 701-702, 1975.
Stern, C. R., M. J. de Wit, and J. R. Lawrence, Igneous and metamorphic processes associated with the formation of Chilean ophiolites and their implication for ocean floor metamorphism, seismic layering and magnetism, J. Geophys. Res., in press, 1976.

FORMATION AND EVOLUTION OF MARGINAL BASINS AND CONTINENTAL PLATEAUS

M. Nafi Toksöz and Peter Bird[1]

Department of Earth and Planetary Sciences
Massachusetts Institute of Technology, Cambridge, Massachusetts 02139

[1]Now at Department of Earth and Space Sciences
University of California, Los Angeles

Abstract. A necessary consequence of the subduction of oceanic lithosphere is an induced convective circulation in the wedge above the slab. This may play an important role in the formation and evolution of marginal basins. We present here both laboratory models and numerical calculations of the induced convection to demonstrate their geological and tectonic effects. These results show that the maximum heating effect of induced convection corresponds in space and time to the initiation of spreading in marginal basins behind island arcs.

A classification of marginal basins as undeveloped, active spreading, mature, and inactive is proposed. The time interval of 20-40 m.y. between initiation of subduction and secondary spreading is consistent with a gradual warming and weakening of the overriding lithosphere by convection. Calculations using a realistic olivine flow law in the upper mantle show that stress induced by convection is minor; but tension exerted on the lithosphere by the downgoing slab probably plays a role in the initiation of secondary spreading.

Using evidence from elevation, volcanism, and seismic attenuation, a similar convective heating can be shown where lithosphere subducts under the margin of a continent. Secondary spreading does not occur in this case because of the excessive time required to weaken the thicker continental lithosphere. Still, heating may reach the base of the crust and may cause partial melting. This heating is a major factor in the evolution of continental plateaus such as the Altiplano and Tibet.

Introduction

The occurrence of marginal basins behind island arcs has been a subject of great interest. The origin of these basins is related to the processes along the margins of converging plates and to subduction of the oceanic lithosphere (Vening Meinesz, 1951; Hasebe et al., 1970; Karig, 1970, 1971; Oxburgh and Turcotte, 1971; Sleep and Toksöz, 1971; Andrews and Sleep, 1973). Many of the geological and geophysical features of these basins imply an extensional origin and a spreading mechanism, possibly driven by the heating of the basin floor from below.

In the mechanism we propose for the formation and spreading of the marginal basins, the convective flow induced in the asthenosphere by the subducting lithosphere plays an important role. The mechanism is illustrated in Fig. 1 in a schematic way. The subducting lithosphere drags part of the low viscosity asthenosphere until the flow is deflected by an increase of viscosity and/or density (most likely at the 350 km-deep phase boundary). This generates a flow pattern that brings hot asthenospheric material to the base of the lithosphere under the basin, behind the island arc. A combination of the upwelling of material, heating of the lithosphere, and flow-induced tension initiates rifting, and causes the spreading of the marginal basins.

In this paper we evaluate the validity and consequences of this mechanism on the basis of theoretical calculations, laboratory experiments and field data. First, we summarize briefly some general characteristics of these basins. Next, we show the generation and the effects of the induced convection. In the section, "Induced Convection Beneath Continental Plateaus", we evaluate the consequences of the subduction under a continental lithospheric plate and the development of continental plateaus such as Tibet, the Iranian Plateau and the Altiplano instead of marginal basins.

Some Geological and Geophysical Characteristics of Marginal Basins

Geological and geophysical data - including fault scarps, heat flow, bathymetry, petrology and chemistry of basalts, sediment distribution, seismicity, magnetic anomalies, velocities and

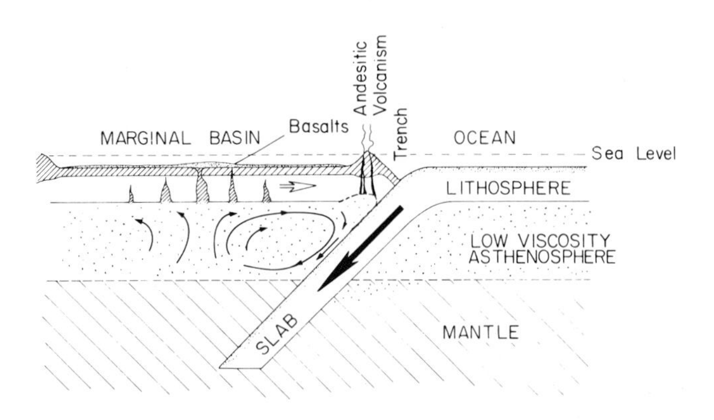

Fig. 1. Schematic diagram of convection induced by a downgoing slab and its heating of the overlying lithosphere.

attenuation of seismic waves - suggest evidence of past or active spreading under marginal basins of the western Pacific. Table 1 summarizes the most critical geophysical data for the basins.

Geology, topography, heat flow and magnetics data are discussed by Yasui and Kishii (1967), Watanabe et al. (1970), Karig (1970, 1971a), Sclater (1972), Sclater et al. (1972a,b), Uyeda and Ben-Avraham (1972), Barker (1972), Anderson (1975), Cooper et al. (1976a,b) and Hawkins (1976). Seismic wave attenuation is a good measure of the temperature regime and the degree of partial melting of the lithosphere and asthenosphere under the basins. Shear wave (S_n) attenuation, and the attenuation of surface reflected P waves (pP) have been investigated, and strong attenuation of these waves has been found under basins characterized by high heat flow (Molnar and Oliver, 1969; Barazangi and Isacks, 1971; Oliver et al., 1973; Barazangi et al., 1975).

In heat flow, seismic wave attenuation and general spreading properties, basins demonstrate certain characteristics. Karig (1971a) divided the basins into three categories: active, inactive with high heat flow, and inactive with normal heat flow. For reasons that will be clear later in this paper, we would like to classify basins as undeveloped, active, mature and inactive. This latter classification reflects the origin and the evolutionary history of the marginal basins. In our terminology, active basins are the same as in Karig's and mature basins correspond to "inactive with high heat flow." We separate Karig's "inactive with normal heat flow" basins into two categories: undeveloped (those that have not yet reached the spreading stage) and inactive (old basins that have long passed the spreading stage). In Fig. 2 and Table 1, basins that fall into each category are identified.

Undeveloped basins are those with typical basin morphologies, but without high heat flow or other evidence of spreading. Generally, they are behind young island arcs where subduction has not progressed long enough and deep enough. Examples of this may be the Aleutian and Caribbean subduction zones. The Bering Sea behind the Aleutian arc has a complicated structure and history, which may have included entrapment of an oceanic plate and the Late Cretaceous-Eocene subduction of the Kula plate prior to the development of the Aleutians (Cooper et al., 1976a,b; Schor, 1964). The current subduction under the Aleutians is only of Late Pliocene age and it has not yet produced spreading. At present, the Aleutian basin immediately behind the arc is probably being heated.

Active basins are the ones that display clear characteristics of active spreading centers with an axial topographic high, generally parallel to the trend of the trench and island arc, high heat flow and correlatable magnetic anomalies. These have shallow earthquakes with source mechanisms having the tension axis in the direction of spreading. Good examples of these are the Lau-Havre and Mariana Basins (Sleep and Toksöz, 1971; Karig, 1970, 1971a,b; Sclater et al., 1972a; Anderson, 1975; Hawkins, 1976) and the East Scotia Sea (Barker, 1972).

Mature basins (Sea of Japan, Okhotsk Basin, Parece Vela Basin, North Fiji Basin among others) are regions where there is a broad heat flow anomaly and evidence of at least past spreading. Mature basins are the same as "Inactive with high heat flow" in Karig's classification. These basins may still be spreading but the topography, magnetic anomalies, seismicity and heat flow do not characterize well-defined active centers. Heat flow and topography data shown in Fig. 3 demonstrate the difference between the active Mariana Basin and the mature Okhotsk Basin and Sea of Japan. Attenuation of seismic waves is high under these basins (Molnar and Oliver, 1969; Barazangi et al., 1975) indicating elevated temperature.

Active basins may eventually become mature basins as more sea floor is created by spreading and as the convection heats a broad region under the basin floor. Even after active spreading ceases, thermal effects (high heat flow) persist for a long time and give a broader anomaly.

Inactive basins (South Fiji Basin, West Philippine Basin) are generally old (mid or early Tertiary in age). They have normal or nearly normal heat flows and some evidence of past spreading (Karig, 1971a). If spreading indeed occurred in the past, then these basins represent the last stage in the basin evolution. After convection and spreading stop, a mature basin loses its thermal energy with time and heat flow values return to normal. Lithosphere and asthenosphere temperatures would probably be normal under these basins. Although seismic measurements are not abundant, the available

Table 1

Evolutionary Classification of Marginal Basins

Basin	Type	Age of Basin	Subduction Age m.y.	Subduction Rate (cm/yr)	Depth of Benioff Zone (km)	Average Elevation (km)	Heat Flow Range HFU	Heat Flow Ave. HFU	Attenuation of seismic waves Sn	Attenuation of seismic waves pP
Aleutian*	Undeveloped*	--	3	6	260	-4.7	0.9-1.3	1.1	Normal	Normal
Mariana	Active	Pliocene	50	9	680	-4.0	0.1-8.3	2.5	High	High
Lau-Havre	Active	Pliocene	45	8	660	-2.6	0.5-3.8	2.2	High	High
East Scotia Sea	Active	Upper Miocene	--	3	170	-3.2				
Fiji Plateau	Mature	Upper Miocene	39	9	640	-2.9	1.4-10	2.9	High	High
Sea of Japan	Mature	Miocene	26	9	610	-4.2	1.0-4.0	2.2	High	High
Okhotsk	Mature	Mid-Tertiary	65-85	8	610	-4.2	1.5-3.0	2.3	--	High
Parece Vela	Mature	Mid-Tertiary	--	--	--	-4.6			High	
South Fiji	Inactive	Early to Mid-Tertiary	--	--		-4.6	0.5-1.7	1.1	--	Normal
West Philippine	Inactive	Early Tertiary	--	--		-5.8			Normal	--

*This refers to the Aleutian episode of subduction and not to the early complicated history of the Bering Sea.

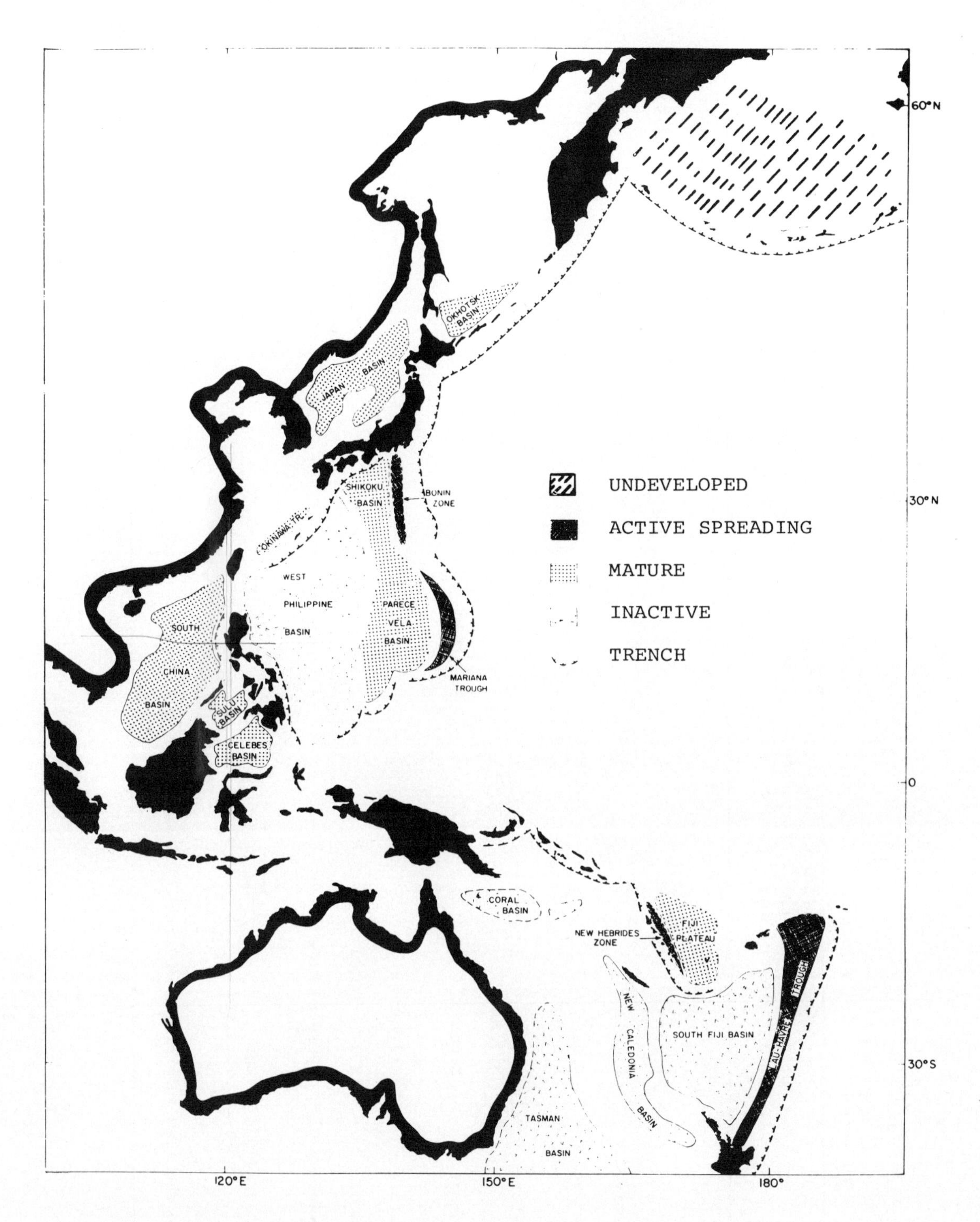

Fig. 2a. Marginal basins of the western Pacific, classified by evolutionary stage. Modified from Karig (1971a).

data (S_n data for West Philippine Basin and pP data for South Fiji Basin) do not show high attenuation (Molnar and Oliver, 1969; Barazangi et al., 1975). While the aging hypothesis could explain these features, other models of formation, including the trapping of old ocean floor, are also consistent with these data (Uyeda and Ben-Avraham, 1972).

In the following section, the four stages of the evolution of oceanic marginal basins are dis-

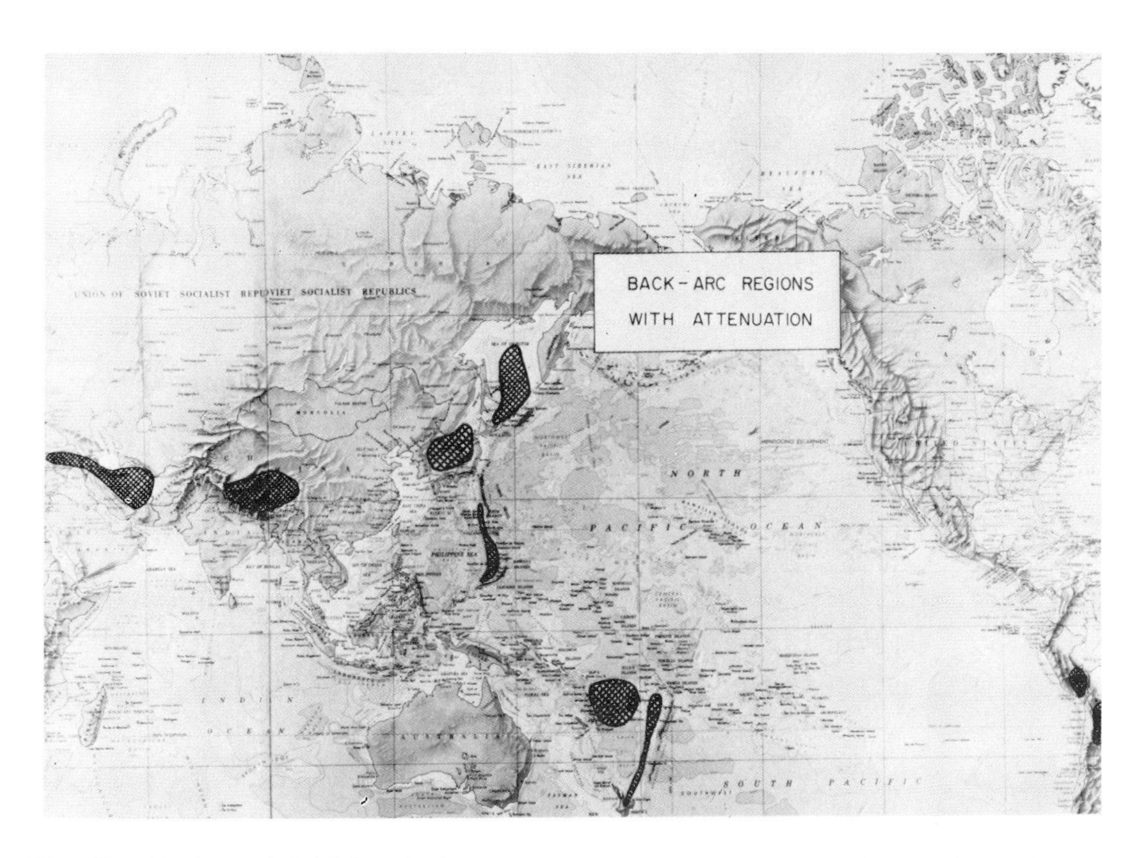

Fig. 2b. Regions with high seismic attenuation at shallow depths above a downgoing slab.

cussed in the light of theoretical calculations.

Formation and Evolution of Marginal Basins

The formation of marginal basins is related to the subduction of oceanic lithosphere. The high heat flow values observed for the active and mature marginal basins (Table 1, Fig. 3) require the upwelling of hot material to heat the lithosphere beneath the basins.

Two mechanisms have been proposed for this upwelling: (1) diapiric upwelling of the hot asthenospheric material heated by slab subduction (Hasebe et al., 1970; Karig, 1971a; Oxburgh and Turcotte, 1971), and (2) convection induced in the asthenosphere by the subducting lithosphere (McKenzie, 1969; Sleep and Toksöz, 1971). These two mechanisms differ in one basic aspect; the first requires heating by the subducting slab to initiate the upwelling, and the second induces convective motion in the asthenosphere by the viscous drag of the subducting lithosphere. As illustrated in Fig. 1, the material that rises by convection eventually undergoes partial melting and gives rise to the upwelling of basalts observed at the margin floor.

The convection process and the evolution of marginal basins are schematically illustrated in four stages in Fig. 4. The first stage (A) represents the initiation of the subduction of the lithosphere. The second stage (B) that may occur approximately at 3 m.y. at 8 cm/yr subduction rate shows the initiation of convection induced in the asthenosphere. Although there is island arc volcanism at this stage, convection has not progressed long enough to contribute to the heating of the basin floor. Undeveloped basins correspond to this stage. The third stage (C), which may correspond to approximately 10 m.y., shows well-developed convection and heating of the lithosphere above the upwelling cell. The fourth stage (D) shows the spreading of the basin floor under the combined action of heating, regional stress, upwelling, and drag forces exerted by the convecting asthenosphere. This active stage may occur anywhere between 10 to 50 m.y. after the initiation of the subduction for subduction rates shown in Table 1.

As the spreading process continues, the basin floor widens. This gives rise to mature basins. As heat is transported close to the surface by convection, causing the high heat flow, the asthenosphere under the basin floor starts to cool. In less than 100 m.y. the basin will become inactive, having lost its thermal energy.

A geometrical consequence of the spreading of a marginal basin is that, as new lithosphere is added behind the island arc, the distance between the spreading center and trench increases. Eventually a new, active marginal basin may develop next to the subduction zone, while

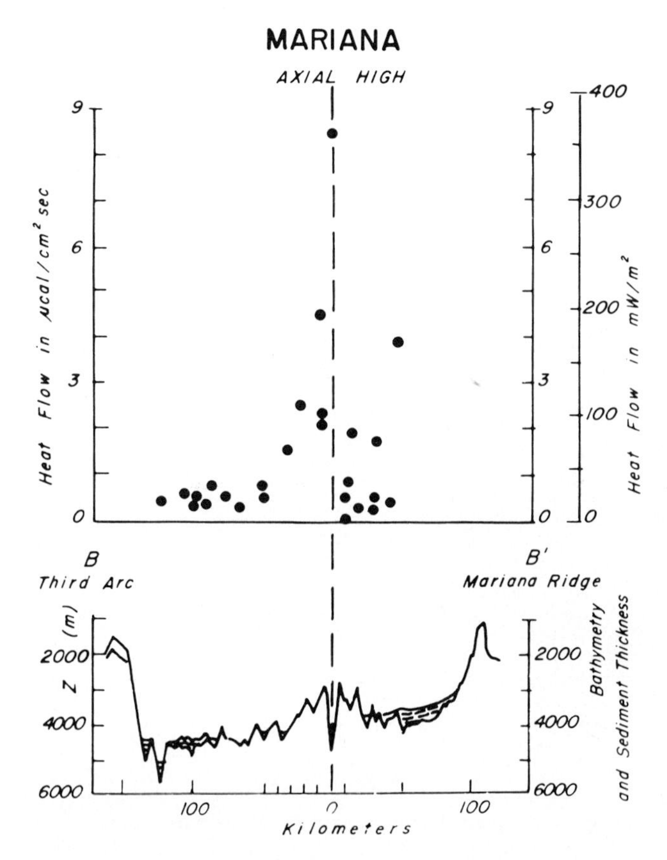

Fig. 3a. Heat flow and topography in the active Mariana Basin. Marianas data reproduced from Anderson (1975).

the old basin cools and eventually becomes inactive. The Mariana and Parece Vela Basins may represent such a pair (Karig, 1971a). The Lau-Havre and South Fiji Basins could be another example. The oceanward migration of the trench during the formation of multiple basins requires some disruption of the subducting lithosphere, or else its dip would become more shallow than we observe. If this takes place by bending of the slab at depth no surface effects would be visible. However, the lithosphere might also break in front of the trench. This would form a new subduction zone, and trap a fragment of ocean floor which would subsequently be subjected to heating. The solution of this problem must come from the geology of the island arcs and of the submerged ridges behind them.

Laboratory Measurements

A set of laboratory experiments was carried out by Patton (1976) to study the flow induced in a fluid in a confined medium by the motion of a moving boundary. The laboratory set-up was designed to model the simple geometry similar to Fig. 1. Two dimensional geometry was represented by a rectangular trough with rigid side walls and top and bottom boundaries. The down-going slab was represented by a moving boundary whose inclination could be adjusted to different angles of dip. The fluid was a silicon oil (10 stoke viscosity). There were no heat sources in the system. The velocity of the moving wall was 6 cm/min. Aluminum powder was suspended in the fluid to reflect a beam of light in order to mark the stream lines.

The photographs of the flow induced in the field by the motions of a moving wall at three different angles of inclination are shown in Fig. 5. In each case the induced flow and stream lines are clearly visible. The dip angles of the moving wall ("descending slab") in Patton's experiments ranged from 17° to 65° and in each case flow was induced in the fluid. These simple experiments support the concepts of induced flow as well as the theoretical solutions for this problem (McKenzie, 1969; Sleep and

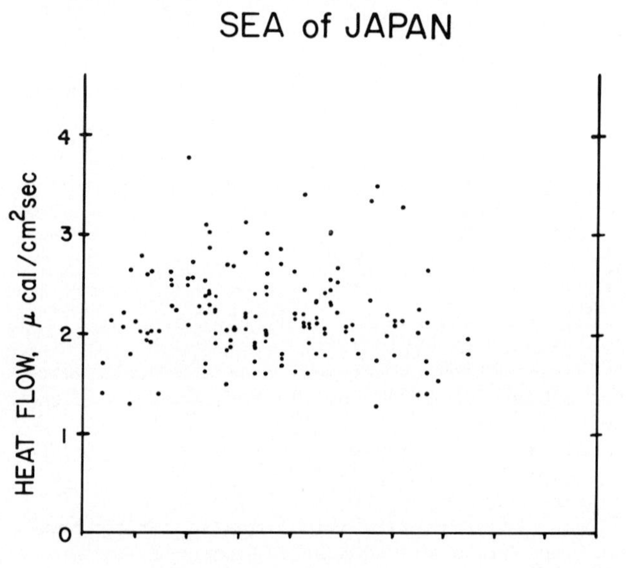

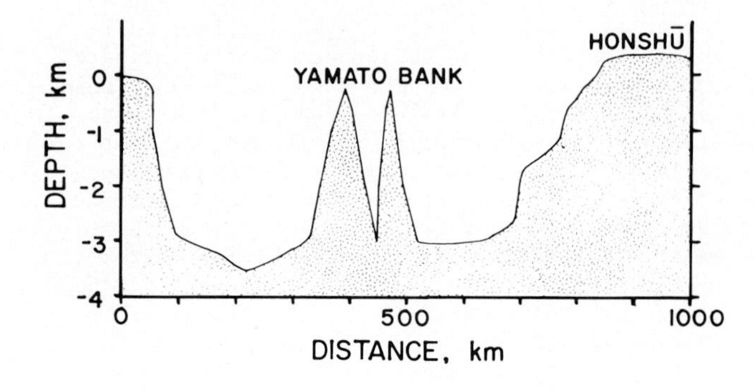

Fig. 3b. Heat flow and topography in the mature Japan Sea, projected to a NW-SE profile (SE to the right).

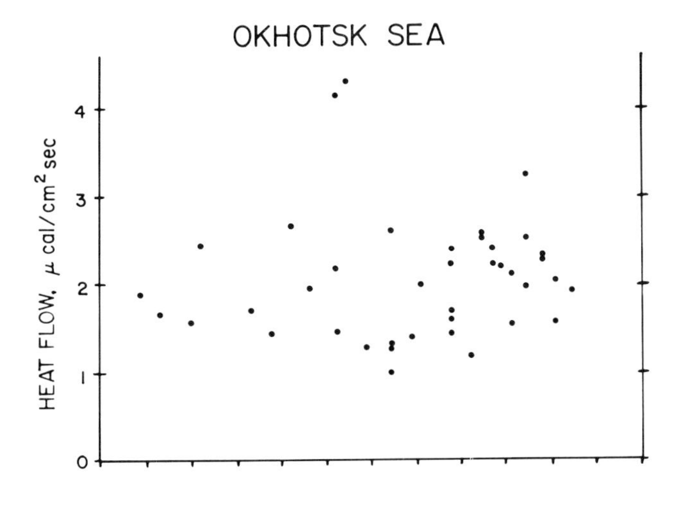

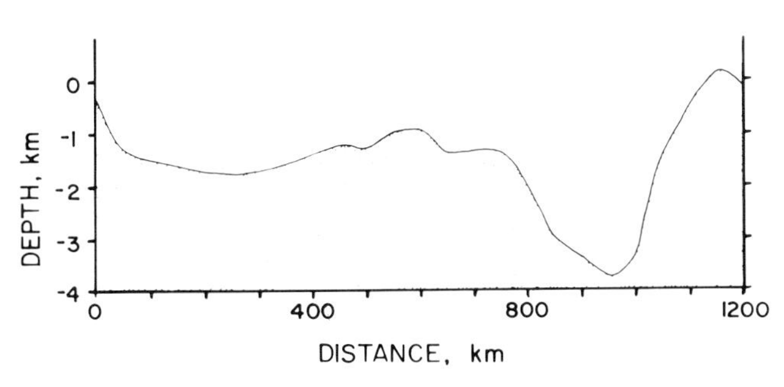

Fig. 3c. Heat flow and topography in the Okhotsk Basin, projected to a NW-SE profile (SE to the right).

Toksöz, 1971; Moffatt, 1964; Pan and Acrivos, 1961).

Results of more detailed numerical experiments are described in the next section.

Numerical Simulation of Induced Convection

The problem of induced flow in the asthenosphere in the subduction zones has been treated by different techniques. Analytical solutions can be obtained for the flow field under steady state conditions (Moffatt, 1964; Pan and Acrivos, 1961; McKenzie, 1969; Sleep and Toksöz, 1971). These solutions explain the laboratory results described in the previous section.

To understand the role of induced convection in marginal basin formation and evolution it is necessary to solve the time-dependent problem. Andrews and Sleep (1973) used a finite difference solution with variable viscosity. Here we will use a different approach. Comparison of the results from both of these calculations demonstrates the importance of some parameters that affect the spreading process.

Our time-dependent flow and temperature calculations were done by combining two programs, a finite element program to solve the Navier-Stokes equation and a finite difference program to solve the heat equation. The finite difference program is a modified version of the down-going slab program used in earlier calculations by Minear and Toksöz (1970) and Toksöz et al. (1971). The translation of the material above the slab has been obtained by the finite element program and used as an input to the finite difference temperature calculations at each step. These temperatures are used to calculate the next step in the flow field, and this process is repeated until the desired time is reached.

The specifics of the calculations of the flow field (finite element program) are described in Appendix A. A strain rate dependent viscosity (non-linear creep) is assumed (eq. A-9) and parameters for olivine given by Kohlstedt and Goetze (1974) are used to determine the viscosity as a function of strain rate, temperature and pressure. Thermal buoyancy forces are neglected. These are discussed in greater detail in the Appendix.

The calculation was begun at the initiation of subduction with a convective geotherm of Sleep (1973) which yields a surface heat flow of 45 mW/m^2 from temperatures of 1350°C at 100 km and 1700°C at 400 km. Because we did not want to excite any artificial transient convection

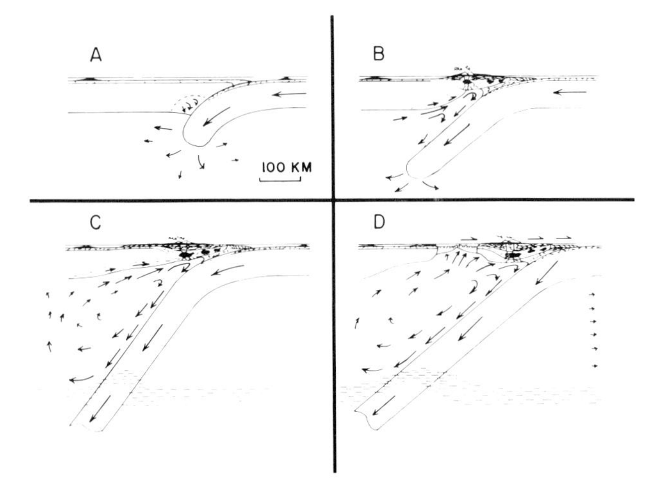

Fig. 4. Four stages in the creation of secondary spreading: (a) Initiation of a subduction zone. (b) Island arc volcanism begins almost immediately, perhaps because of mechanical erosion of island arc lithosphere and the subduction of volatiles. (c) As the slab passes through the olivine-spinel phase change (dashed lines) the pattern of induced convection is established. Convective overturn in the asthenosphere carries heat to the overlying plate. (d) The lithosphere fails because of heating and tension and secondary spreading is initialized.

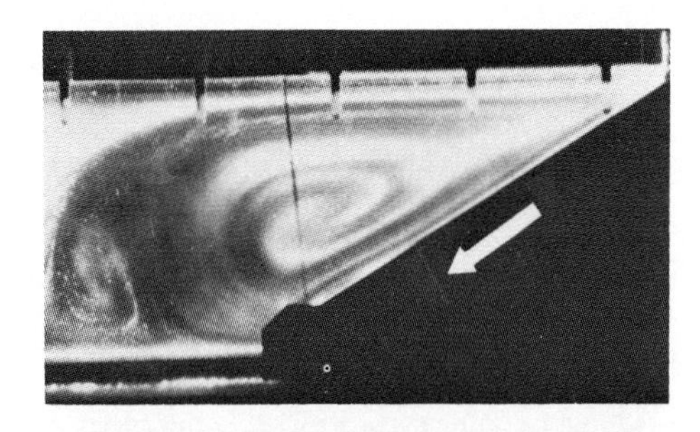

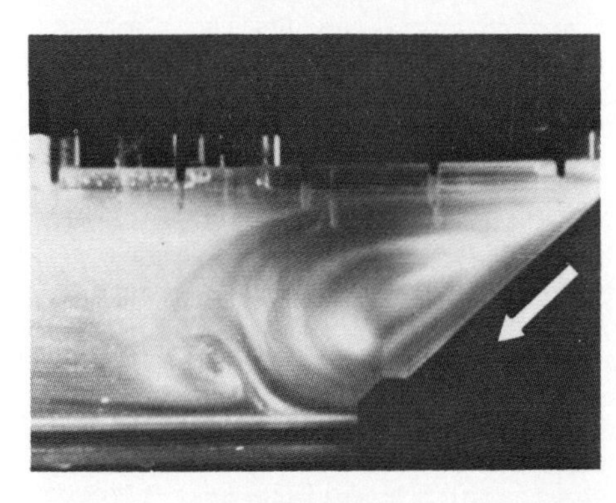

Fig. 5. A laboratory experiment on induced convection in a viscous fluid. Top and bottom boundaries are fixed, while the right boundary moves with a constant velocity, simulating a downgoing slab. Aluminum particles in the fluid cause streaks in the time exposure, showing streamlines. Three different slab dips are illustrated: 32°, 45°, and 65°. (After Patton, 1976)

that might result from an incorrect initial condition, the thermal buoyancy term of (A-1) was suppressed in this preliminary calculation. The slab subducts at an angle of 45° and a velocity of 8 cm/year, which is typical of the western boundaries of the Pacific plate. Adiabatic heating and phase changes were as specified by Toksöz et al. (1971); the grid had a 10 km vertical by 20 km horizontal increment; and frictional heating equivalent to a stress of 500 bars was applied at the top of the slab down to 90 km depth. Conductivity at all depths was obtained from the empirical formula of Schatz and Simmons (1972).

Beginning at t = 7.8 m.y., when the slab penetrated to 375 km, velocities were calculated and used to translate material above the slab. These velocities were updated every 5 m.y. The grid of elements used is shown in Fig. 6. It includes all of the lithosphere and asthenosphere above the slab except the island arc (0-100 km from the trench) in which rock types, temperatures, and stress are not well understood. The upper part of the right boundary is fixed against horizontal motion, while the slanting boundary is constrained to move along with the slab. Thus the island arc is the reference point in the velocity solution. The top and bottom boundaries are constrained to allow horizontal slip but not vertical movement. The physical justification for this is the treatment of Schubert and Turcotte (1971), which demonstrates that material with a viscosity below 2×10^{22} poise (the asthenosphere) is not likely to penetrate the 350 km phase change. The left boundary is entirely free, and does not restrict the flow.

Skipping over several steps, we present results for the time t = 27.8 m.y. in Figs. 7 and 8. Despite the nonlinear flow law and the temperature effect on viscosity, the velocity vectors are remarkably similar to the results presented earlier for a uniform viscous fluid in a laboratory tank (Fig. 5). Although the bulk of the lithosphere does not move, a stream of hot material with a velocity equal to the slab velocity is directed against its base at a point 150-200 km behind the trench.

Stresses have not been plotted because they are very small. Except in the extreme right corner, where the boundary conditions create vertical tension between slab and lithosphere, the stress

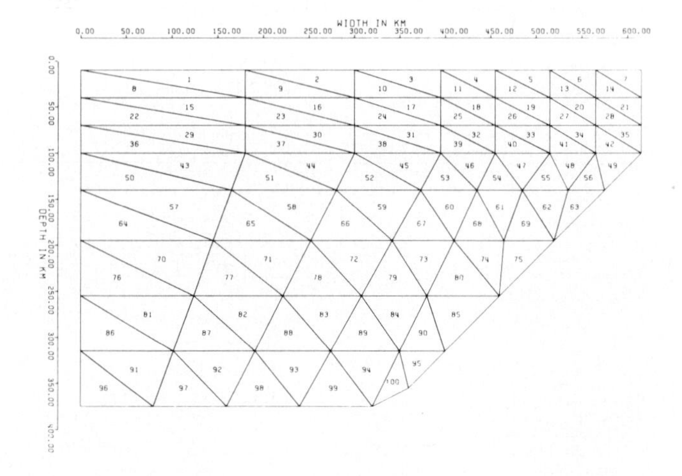

Fig. 6. Grid of 100 triangular finite elements used to represent the asthenosphere and lithosphere above a subducting slab, down to the 350-km phase change. The island arc, which would be at the upper right, is not included in the grid.

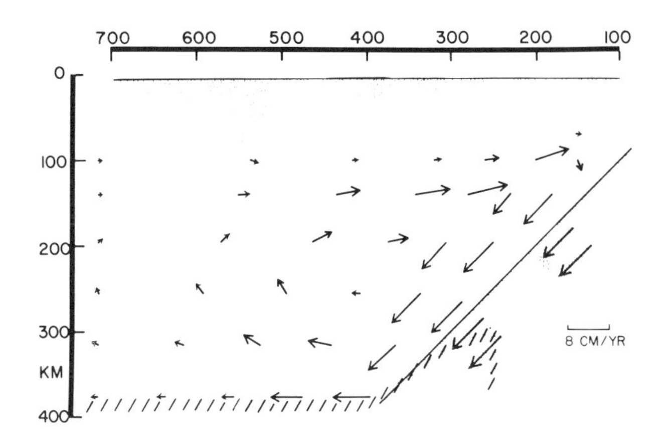

Fig. 7. Velocity vectors of induced flow at time t = 27.8 m.y. after initiation of subduction. Lithosphere is shown by shading and the phase change by diagonal slashes.

level is generally only 0.5-2.0 bars. We conclude that the stress exerted on the lithosphere by induced convection is not the direct cause of secondary marginal basin spreading, as the maximum stress created in the lithosphere is 25 bars, and even this may be an overestimate if the asthenosphere contains significant H_2O or partial melts. Instead the flow is important because it thermally weakens the lithosphere, allowing it to fail in response to tension exerted on it by the slab or by gravitational spreading away from the heated and uplifted region.

The temperatures at the latest time calculated (32.8 m.y.) are shown in Fig. 9. Throughout the circulating region the lithosphere is warmed and thinned, but this process is fastest at a point about 180 km behind the trench. This would be the logical point for the initiation of secondary spreading, although in nature a slightly greater separation is observed. Measuring from the trench to the first normal fault scarp, one obtains distances of about 170 to 280 km in the Tonga, South Sandwich Islands, and Marianas regions. The temperature increase due to induced convection at this point is shown in Fig. 10. A 200°C temperature increase at 80 km depth is achieved in 18 m.y., and temperatures continue to rise more slowly after that time. We suspect that the slowdown of thermal erosion may be linked to our neglect of thermal buoyancy, which allows the convecting material in the wedge to rotate within closed streamlines (Fig. 8) and become chilled by its contact with the slab. More vigorous heating might also be obtained if we allowed some of the cold material sinking with the slab to pass through the phase change, and replaced it with fresh hot asthenosphere from the left. These effects will be investigated in our future work.

Clearly some important effect has been omitted from this calculation since secondary spreading was not induced within the appropriate time. This effect might be a large amount of initial tension exerted on the lithosphere by the subducting slab. A model calculated by Andrews and Sleep (1973) and reproduced as Fig. 11 shows that if this tension is as great as 1 kb and if the lithosphere viscosity is 10^{24} poise, secondary spreading could be produced in as little as 10 million years. Another possibility is that the temperature increases obtained in this calculation may cause extensive partial melting. If such melting were sufficiently developed for the liquid and crystalline phases to separate, then diapiric upwellings into the lithosphere might greatly speed up the transfer of heat. If the mechanical strength of the lithosphere were reduced by such an intrusive process, then only a small initial tension from the downgoing slab would be required to produce secondary spreading.

Induced Convection Beneath Continental Plateaus

Located in the overriding plate above three of the world's major subduction zones are continental plateaus whose elevation and magmatism are not well understood. Of these, the smallest is the south-central Iranian Plateau, with an average elevation of about 1.1 km over 2×10^5 km^2. The Altiplano, actually a downfaulted block between the two cordillera of the Andes, still has a 3.8 km elevation for 6×10^5 km^2. And the highest plateau in the world is in Tibet, above the Tethyan-Himalayan subduction zone, 5.1 km high over a 7×10^5 km^2 area. While two of these regions have undergone continental

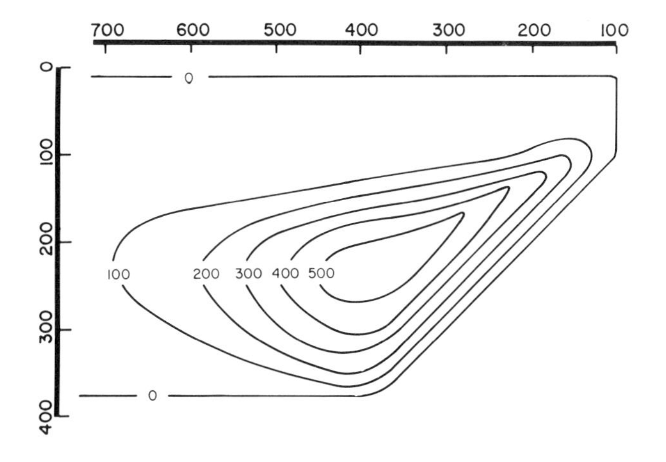

Fig. 8. Streamlines of induced flow at t = 27.8 m.y. in units of kilometer-centimeters/year. The zero contour is fixed as a boundary condition, as is the slope of the stream function along the right boundary next to the slab. Axes labeled in kilometers.

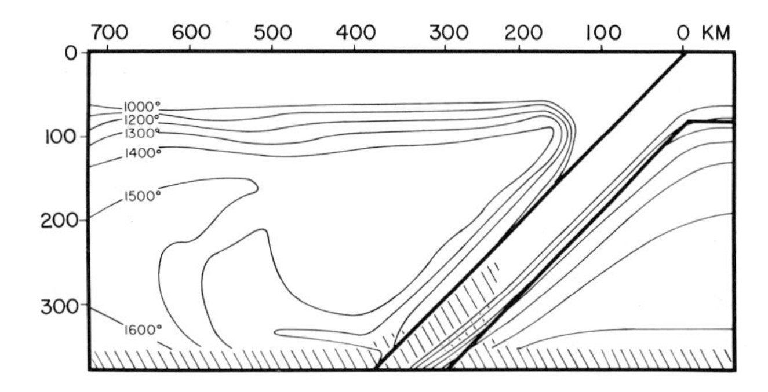

Fig. 9. Temperatures in °C after 32.8 m.y. Only temperatures about 1000° are contoured. Material to the right of the slab was not translated, and gives approximately the initial geotherm. Phase change at 300-375 km shown by hachures. The shallow plagioclase-garnet peridotite phase change was included in the calculation but omitted from the figure for clarity. Note general warming of the lithosphere, steep temperature gradients in the corner, and slow cooling of the convecting asthenosphere.

collisions after the closing of portions of the Tethys Ocean, it is unlikely that they formed entirely as a product of those collisions. The Zagros Mts. adjacent to the Iranian Plateau are no older than Pliocene (Haynes and McQuillan, 1974) and may be only Pleistocene (Bird et al., 1975). In India, the well-known drainage pattern with major rivers cutting the rising Himalayas indicates that Tibet was already elevated at the time of that collision (Holmes, 1965). Furthermore, there has been no marine sedimentation in Tibet since the Cretaceous.

What gravity data there is tends to show that these regional uplifts are isostatically rather than tectonically supported. In Iran, Bouguer anomaly values are -125 to -150 milligals (Wilcox et al., 1972); in the Altiplano -300 to -400 milligals (James, 1971); and in Tibet are found some of the lowest ever recorded: -550 milligals (Ambolt, 1937). It is not known whether the low density roots take the form of thick crust or whether they are deeper. Seismic refraction has been attempted only in the Altiplano (Tatel and Tuve, 1958), and there no mantle refractions were recorded.

Extensive volcanism is another characteristic of these plateaus. Some of the Neogene eruptions in Iran fall in an arc northeast of and parallel to the continental suture, as expected, but others have occurred to the north in the Elburz Mts. or to the east in the Lut block where they have no clear association with plate boundaries (Stöcklin and Nabavi, 1973). Many kilometers of Mesozoic and Cenozoic volcanics have been deposited on the Altiplano (Newell, 1949). And widespread calc-alkaline volcanism over large fractions of Tibet has occurred within the last 10 m.y. (Kidd, 1975). Obviously these magmas are not derived from downgoing slabs, as no such slabs of lithosphere presently exist beneath Iran or Tibet. Also, in cases where oceanic slabs subduct beneath oceanic plates, the island arc volcanism is found in narrow bands not more than 200 km behind the trench. It appears likely that these magmas have a shallower source.

Seismological studies of attenuation support the inference of anomalous heat within the continental lithosphere. Molnar and Oliver (1969) studied the phase S_n which propagates below the Moho and found that it travels sporadically across the Altiplano, poorly across Iran, and not at all across central Tibet. Barazangi et al. (1975) used the reflected phase pP to show that there is strong attenuation at some depth less than 200 km in the Altiplano.

The best waves for the study of attenuation are surface waves, because the frequency-dependence of their attenuation can be used to deduce the depth of the attenuating zone. A rough rule of thumb is that the period of the Rayleigh wave most strongly affected (in seconds) is comparable to the depth to the attenuating zone in miles. Figure 12 shows three long period records from a nuclear explosion at Lop Nor to the north of Tibet. The record at Kabul has a broad spectrum such as is expected from a surface source (Tsai and Aki, 1970). Yet when the waves pass through Tibet, and arrive at New Delhi and Shillong, most of the long period energy has been absorbed. Preliminary analysis

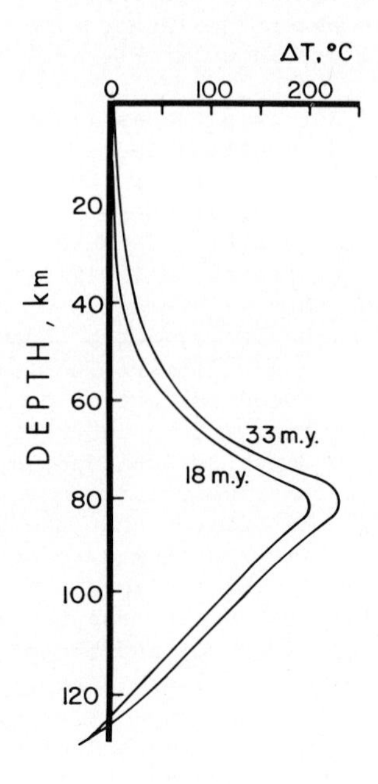

Fig. 10. Profile of excess temperature produced by induced convection at the point x = 180 km behind the trench. Two different times after the beginning of subduction are shown.

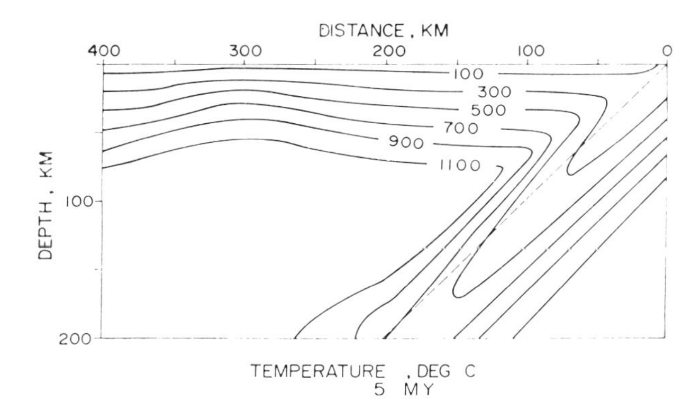

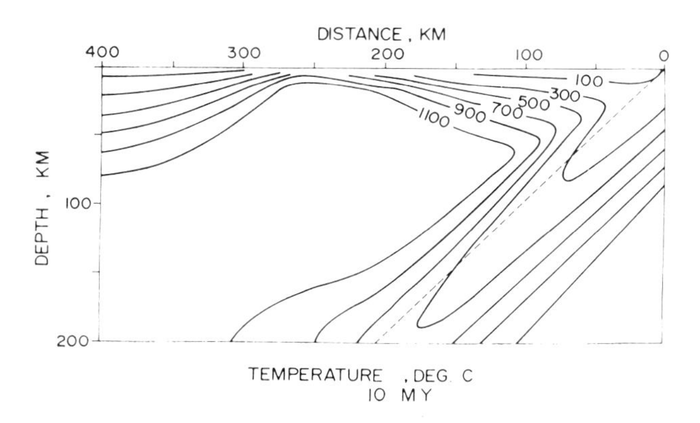

Fig. 11. A model calculated by finite-difference techniques with a temperature-dependent viscosity by Andrews and Sleep (1973). Temperature fields shown at 5 and 10 m.y.; at the latter time secondary spreading is well under way. In this model large stresses created by the slab and induced convection combine to produce spreading.

of these and other Rayleigh waves across Tibet indicates extremely high attenuation at a depth of 70 km (Bird and Toksöz, 1976).

Another example is shown in Fig. 13, where Rayleigh waves from a nuclear explosion in Novaya Zemlya have passed through central Iran. Compared to Tabriz and Mashhad, the Shiraz record contains very little long period energy. The attenuation appears at such short periods (40 sec) that it cannot be ascribed to the asthenosphere unless the asthenosphere is considered to be very shallow under Iran. Fig. 14 shows Q^{-1} as a function of period from South America, using records from stations on opposite sides of the Altiplano and three earthquakes in the South Sandwich Islands. The peak in attenuation at 30-40 seconds again requires a model with a hot layer at 40-70 km depth. This is either in the uppermost mantle, or, according to the model of James (1971), in the lower half of the crust.

Such extensive attenuation and volcanism require large-scale partial melting of the lower crust and upper mantle, far more than can be explained through occasional intrusions into a normal lithosphere. Instead, it appears that the lithosphere in these regions is relatively thin. This explanation has the virtue of explaining the gravity anomalies discussed above without the need for extravagantly thick crust. If the density contrast between average lithosphere and the hot asthenosphere is .09 gm/cc, as required to explain the topography of mid-ocean ridges, then the replacement of the lower 100 km of a continental lithosphere by the hotter asthenosphere would create a gravity anomaly of -370 milligals and an uplift of 3.3 km. This can explain everything except the height of the Tibetan Plateau, which would still require a thick 55 km crust.

Another feature of the proposed mechanism of heating lithosphere from below is that it takes a long time to become apparent. In the Paleogene Alpine subduction zone, a narrow portion of the Tethys Ocean - less than 500 km wide - was consumed beneath the Italian continental block (Dewey et al., 1973). Yet no uplift, attenuation, or volcanism are found, and the Po valley is in fact a deep depression. But in the Himalayan-Tibetan subduction zone, which consumed some 8,000 km of lithosphere over a period of 200 m.y. (Dietz and Holden, 1970), all of these characteristics are fully developed. This slow evolution argues for a steady rather than a catastrophic mechanism.

All of this evidence is consistent with the

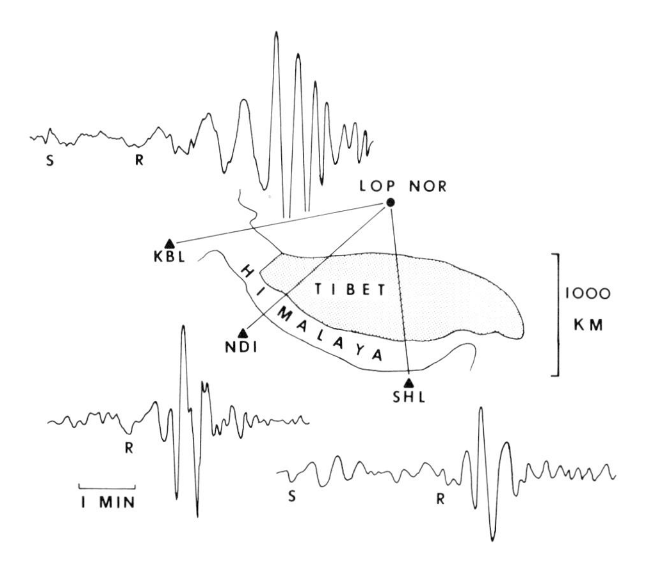

Fig. 12. Rayleigh waves (LPZ component) from a nuclear explosion at Lop Nor, Sinkiang, recorded at three WWSSN stations. Note attenuation of long-period portion of the wave on paths that cross Tibet.

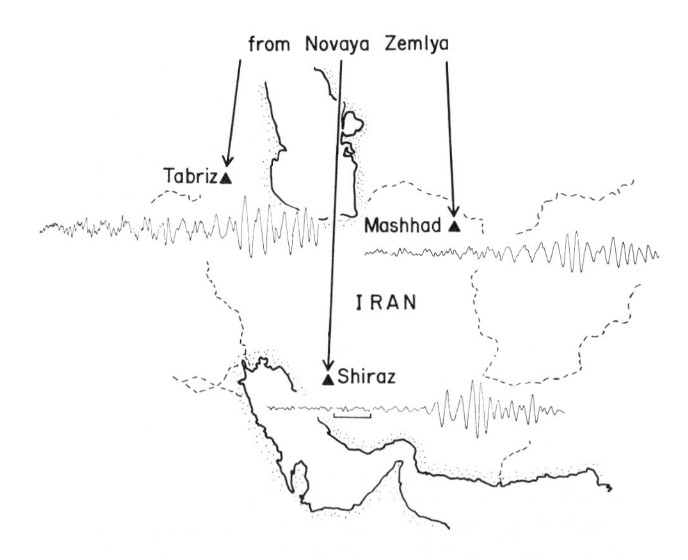

Fig. 13. Three Rayleigh waves (LPZ component of WWSSN stations) in Iran from a nuclear explosion in Novaya Zemlya. Long-period portion of the wave is attenuated while passing through central Iran. Bar shows length of one minute.

hypothesis that continental lithosphere adjacent to subduction zones is gradually heated and thinned from below, as a result of heat transferred to it by induced convection. As shown in Fig. 15 (and Fig. 9 previously), this heating takes place over a band extending at least 700 km back from the trench. The fact that secondary spreading is not induced in continental lithosphere is readily explained by its greater initial thickness, greater than 120 km (Toksöz et al., 1967). Or it may be that the spreading is only very slow and diffuse, and is represented by normal faulting like that of the Basin and Range province or the margins of the Altiplano. It will be of interest to learn, as the tectonics of Tibet are eventually unravelled, whether an earlier tensional phase preceded its present shortening in the collision with India.

Conclusions

1. Subducting lithosphere induces convection in the asthenosphere above the slab. This is demonstrated by both laboratory experiments and numerical modelling. This convection heats the lithosphere from underneath by bringing up hotter asthenospheric material.

2. For the secondary spreading to start in a marginal basin requires about 20-40 m.y. of subduction. Secondary spreading is not caused by sudden tensional rifting of the cold lithosphere but by the combined actions of gradual heating, weakening, and rifting of the lithosphere.

3. Four stages of evolution (and four types of oceanic basins) can be identified on the basis of the proposed mechanism.

a) Underdeveloped: Subduction has penetrated and started the induced convection. However, not sufficient time has elapsed to heat, weaken and rift the lithosphere under the basin.

b) Active Spreading: The heated lithosphere starts rifting and a well defined spreading center develops.

c) Mature Stage: The spreading has widened the basin floor. The asthenosphere has been cooling because of the convection and rise of the hot material toward the surface. Spreading may have slowed down or it may become less organized and more complicated.

d) Inactive Stage: After subduction ceases under a basin, and the basin material has lost its thermal energy, the temperatures return to normal oceanic geotherm and the basin floor to a normal oceanic crust.

4. The spreading of the basin floor moves the island arc and trench oceanward, decreasing the dip of the downgoing slab. The behavior of the induced convection at this stage will depend on whether it can draw on a large mass of fresh, hot asthenosphere from a wide region. If it can, it will move with the island arc away from the ridge and begin warming a new, future basin location. If the flow is trapped, its gradual cooling will produce greater and greater resistance to subduction, and might cause the formation of a new subduction zone on the oceanward side.

5. Subduction of an oceanic slab under a continental lithosphere produces continental equivalents of marginal basins. These are the high plateaus such as Tibet, the Iranian Plateau and possibly the Altiplano. Seismic attenuation in

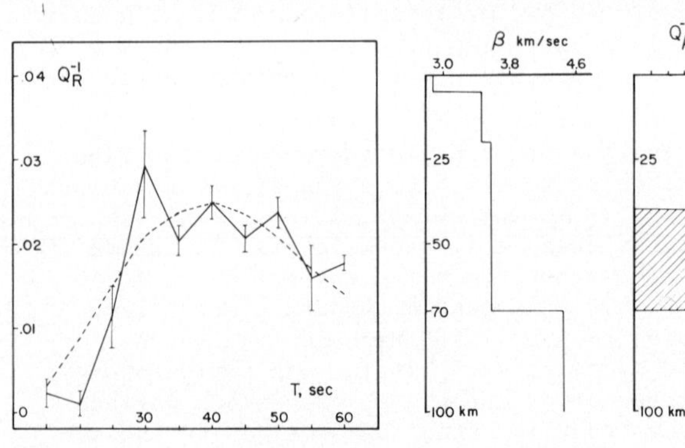

Fig. 14. Attenuation of Rayleigh waves versus period obtained from 3 pairs of records at stations LPA and ARE on opposite side of the Altiplano in the Andes. Model attenuation curve shown was calculated from the Andean velocity model of James (1971) in center, assuming a strong attenuating layer in the lower crust (right). The contribution of eastern South America to this attenuation is assumed to be small.

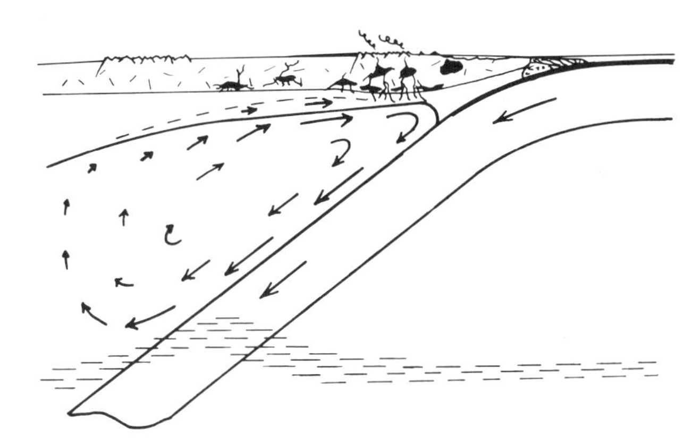

Fig. 15. Schematic diagram of induced convection beneath a continent. The thick lithosphere requires warming over many millions of years before effects are visible. At late stages, lower crust may be melted and intrusions from partial melting in the asthenosphere may contribute to crustal thickening.

these areas is indistinguishable from oceanic marginal basins. Because of the greater thickness of continental lithosphere, these areas require a longer time to develop.

Appendix

A two-dimensional time-dependent model of the induced flow was calculated by combining a finite-difference program to solve the heat equation and a finite-element program to solve the Navier-Stokes equation. In this appendix we describe the finite element calculations.

We seek to solve the modified Navier-Stokes equations

$$-\frac{\partial P}{\partial x} + 2\frac{\partial}{\partial x}\left(\mu\frac{\partial u}{\partial x}\right) + \frac{\partial}{\partial z}\left(\mu\frac{\partial u}{\partial z} + \mu\frac{\partial w}{\partial x}\right) = 0$$

$$b_z - \frac{\partial P}{\partial z} + 2\frac{\partial}{\partial z}\left(\mu\frac{\partial w}{\partial z}\right) + \frac{\partial}{\partial x}\left(\mu\frac{\partial u}{\partial z} + \mu\frac{\partial w}{\partial x}\right) = 0 \qquad \text{(A-1)}$$

subject to the incompressibility condition

$$\frac{\partial u}{\partial x} + \frac{\partial w}{\partial z} = 0 \qquad \text{(A-2)}$$

where u, w are velocities in x and z directions; μ is viscosity; P is pressure; and b_z is the gravitational body force. Incompressibility is assured by expressing u and w in terms of a stream function ϕ:

$$u = \frac{\partial \phi}{\partial z} \quad ; \quad w = -\frac{\partial \phi}{\partial x} \qquad \text{(A-3)}$$

A suitable element for this purpose is the triangular element described by Zienkiewicz (1971) which has a cubic polynomial for the stream function, a quadratic variation of velocities and a linear variation of strain. The polynomial is determined in the interior of the element by the 3 variables at each corner node; ϕ, u, and w. When the ith variable is non-zero at one node, it implies a non-zero value of $\phi_i(x,z)$ in all the adjacent elements.

We place the equilibrium equation (A-1) in a weak form by multiplying both sides by an arbitrary velocity variation $\delta\vec{v}(x,z)$ and integrating over the domain. If the variation is restricted to satisfy (A-2) the pressure terms will drop out:

$$\iint_A [2\mu u,_x(\delta u),_x + \mu(u,_z + w,_x)(\delta u),_z$$

$$+ 2\mu w,_z(\delta w),_z + \mu(u,_z + w,_x)(\delta w),_x \qquad \text{(A-4)}$$

$$- b_z\delta w]dA - \oint_S [p_x\delta u + p_z\delta w]ds = 0$$

where p_x, p_z are surface forces and commas indicate differentiation. The approximation is that instead of requiring (A-4) for all variations $\delta\vec{v}$, we consider only the finite-dimensional space of velocity variations spanned by the ϕ_i (nodal functions). This results in a set of simultaneous linear equations:

$$\underset{\sim}{K}\vec{x} = \vec{F}, \text{ where} \qquad \text{(A-5)}$$

$$\vec{x} = \{\phi_1, u_1, w_1, \ldots \phi_N, u_N, w_N\} \qquad \text{(A-6)}$$

$$\vec{F}_i = -\iint_A b_z\frac{\partial \phi_i}{\partial x}\,dA + \oint_S [p_x\frac{\partial \phi_i}{\partial z} - p_z\frac{\partial \phi_i}{\partial x}]ds \qquad \text{(A-7)}$$

$$\underset{\sim}{K}_{ij} = \iint_A [4\mu(\phi_i),_{x,z}(\phi_j),_{x,z} + \mu((\phi_i),_{z,z}$$

$$- (\phi_i),_{x,x})\cdot((\phi_j)_{z,z} - (\phi_j),_{x,x})]dA = \underset{\sim}{K}_{ji} \qquad \text{(A-8)}$$

A similar reduction to linear equations is not possible in the case of non-linear creep (strain rate dependent viscosity). Yet a large number of laboratory experiments summarized by Kohlstedt and Goetze (1974) indicate that mantle flow occurs by dislocation creep in olivine, with a variable viscosity

$$\mu = A(\dot{e})^{-2/3}\exp\left(\frac{B+Cz}{T}\right) \qquad \text{(A-9)}$$

From their results we derive A= 6.82×10^4 dynes/cm^2 and B= 21,000°K for μ in poise if $\dot{e}$ is the maximum engineering strain rate in inverse seconds. By making the assumption that the activation volume for diffusion is proportional to the pressure derivative of the melting temperature (Weertman and Weertman, 1975) and using the forsterite melting data of Davis and England (1964) we estimate the pressure effect C = 16°K/km. This flow law gives an upper limit on mantle strength down to the 375-km phase change, as the weakening effects of water and partial melting have not been considered. If the temperature at 200 km is about 1400°C and the strain rate about 10^{-14}/sec, this relation (A-9) gives $\mu = 2.8 \times 10^{20}$ poise.

We have incorporated this nonlinearity into the solution by iteration. Using the "initial-stress" method (Zienkiewicz, 1971) the difference between calculated stresses and the stresses obtained using (A-9) at each point in each element is incorporated into the load vector $\vec{F}$ for the next iteration and redistributed. The number of iterations was limited to 30 to avoid accumulating errors in stress equilibrium, and in this time convergence of stresses to within 10% of the flow law was achieved at 77% of the integration points. Thus the solution satisfies stress equilibrium at all points, although it deviates in some regions from the assumed flow law.

Acknowledgements. We thank Howard Patton for making available his work on laboratory modelling of convection. Nezihi Canitez supplied seismograms demonstrating the attenuation in Iran, and Richard Buck assisted with collection and interpretation of the Altiplano attenuation data. This work was supported by the Advanced Research Projects Agency and was monitored by the Air Force Office of Scientific Research under Contract No. F44620-75-C-0064.

References

Ambolt, N., Relative Schwerkrafts - Bestimmungen mit Pendeln in Zentralasien, in Reports from the Expedition to the North-western Provinces of China under the Leadership of Dr. Suen Hedin, Stockholm, 1937.

Anderson, R.N., Heat flow in the Mariana marginal basin, J. Geophys. Res., 80, 4043-4048, 1975.

Andrews, D.J. and N.H. Sleep, Numerical modelling of tectonic flow behind island arcs, Geophys. J. Roy. Astron. Soc., 38, 237-251, 1973.

Barazangi, M. and B. Isacks, Lateral variation of seismic wave attenuation in the upper mantle, J. Geophys. Res., 76, 8493, 1971.

Barazangi, M., W. Pennington, and B. Isacks, Global study of seismic wave attenuation in the upper mantle behind island arcs using pP waves, J. Geophys. Res., 80, 1079-1092, 1975.

Barker, P.F., A spreading centre in the East Scotia Sea, Earth Planet. Sci. Lett., 15, 123-132, 1972.

Bird, P. and M.N. Toksöz, Strong attenuation of Rayleigh waves in Tibet, submitted to Nature, 1976.

Bird, P., M.N. Toksöz, and N.H. Sleep, Thermal and mechanical models of continent-continent convergence zones, J. Geophys. Res., 80, 4405-4416, 1975.

Cooper, A.K., M.S. Marlow, and D.W. Scholl, The Bering Sea - A multifarious marginal basin, this volume, 1976a.

Cooper, A.K., D.W. Scholl and M.S. Marlow, A plate tectonic model for evolution of the eastern Bering Sea basin, Geol. Soc. Am. Bull., in press, 1976b.

Davis, B.T.C. and J.L. England, The melting of forsterite up to 50 kilobars, J. Geophys. Res., 69, 1113-1116, 1964.

Dewey, J.F., W.C. Pitman III, W.B.F. Ryan, and J. Bonin, Plate tectonics and the evolution of the Alpine system, Geol. Soc. Am. Bull., 84, 3137-3180, 1973.

Dietz, R.S. and J.C. Holden, Reconstruction of Pangaea: Breakup and dispersion of continents, Permian to present, J. Geophys. Res., 75, 4939-4956, 1970.

Hasebe, K., N. Fujii, and S. Uyeda, Thermal processes under island arcs, Tectonophysics, 10, 335-355, 1970.

Hawkins, J.W., Jr., Petrologic and geochemical characteristics of marginal basin basalts, this volume, 1976.

Haynes, S.J. and H. McQuillan, Evolution of the Zagros suture zone, southern Iran, Geol. Soc. Am. Bull., 85, 739-744, 1974.

Holmes, A., Principles of Physical Geology, Ronald Press, New York, 1288 pages, 1965.

James, D.E., Andean crustal and upper mantle structure, J. Geophys. Res., 76, 3246-3271, 1971.

Karig, D.E., Ridges and basins of the Tonga-Kermadec island arc system, J. Geophys. Res., 75, 239-254, 1970.

Karig, D.E., Origin and development of marginal basins in the Western Pacific, J. Geophys. Res., 76, 2542-2561, 1971a.

Karig, D.E., Structural history of the Mariana island arc system, Geol. Soc. Am. Bull., 82, 323-344, 1971b.

Kidd, W.S.F., Widespread late Neogene and Quaternary calc-alkaline vulcanism on the Tibetan Plateau, EOS Trans. Am. Geophys. Un., 56, 453, 1975.

Kohlstedt, D.L. and C. Goetze, Low-stress high-temperature creep in olivine single crystals, J. Geophys. Res., 79, 2045-2051, 1974.

McKenzie, D.P., Speculations on the consequences and causes of plate motions, Geophys. J. Roy. Astron. Soc., 18, 1-32, 1969.

Minear, J.W. and M.N. Toksöz, Thermal regime of a downgoing slab and new global tectonics, J. Geophys. Res., 75, 1397-1419, 1970.

Moffatt, H., Viscous and resistive eddies near a sharp corner, J. Fluid Mech., 18, 1-18, 1964.
Molnar, P. and J. Oliver, Lateral variations in attenuation in the upper mantle and discontinuities in the lithosphere, J. Geophys. Res., 74, 2648-2682, 1969.
Newell, N.D., Geology of the Lake Titicaca region, Peru and Bolivia, Geol. Soc. Am. Mem. 36, 111 p., 1949.
Oliver, J., B. Isacks, M. Barazangi, and W. Mitronovas, Dynamics of the downgoing lithosphere, Tectonophysics, 19, 133-147, 1973.
Oxburgh, E.R. and D.L. Turcotte, Origin of paired metamorphic belts and crustal dilation in island arc regions, J. Geophys. Res., 76, 1315-1326, 1971.
Pan, F. and A. Acrivos, Steady flows in rectangular cavities, J. Fluid Mech., 28, 643-655, 1967.
Patton, H., Asthenospheric convection induced by the downgoing slab: laboratory experiments, in preparation, 1976.
Schatz, J.F. and G. Simmons, Thermal conductivity of earth materials at high temperatures, J. Geophys. Res., 77, 6966-6983, 1972.
Schor, G.G., Jr., Structure of the Bering Sea and the Aleutian Ridge, Marine Geology, 1, 213-219, 1964.
Schubert, G. and D.L. Turcotte, Phase changes and mantle convection, J. Geophys. Res., 76, 1424-1432, 1971.
Sclater, J.G., Heat flow and elevation of the marginal basins of the Western Pacific, J. Geophys. Res., 77, 5705-5719, 1972.
Sclater, J.F., J.W. Hawkins, J. Mammerickx, and C.G. Chase, Crustal extension between the Tonga and Lau Ridges: Petrologic and geophysical evidence, Geol. Soc. Am. Bull., 83, 505-518, 1972a.
Sclater, J.G., U.G. Ritter, and F.S. Dixon, Heat flow in the southwestern Pacific, J. Geophys. Res., 77, 5697-5704, 1972b.
Sleep, N.H., Deep structure and geophysical processes beneath island arcs, Ph.D. thesis, M.I.T., Cambridge, Mass., 1973.
Sleep, N.H. and M.N. Toksöz, Evolution of marginal basins, Nature, 33, 548-550, 1971.
Stöcklin, J. and M.H. Navabi, Tectonic map of Iran, Geological Survey of Iran, Teheran, 1973.
Tatel, H. and M. Tuve, Seismic studies in the Andes, EOS Trans. Am. Geophys. Un., 39, 580, 1958.
Toksöz, M.N., M.A. Chinnery, and D.L. Anderson, Inhomogeneities in the earth's mantle, Geophys. J. Roy. Astron. Soc., 13, 31-59, 1967.
Toksöz, M.N., J.W. Minear, and B.R. Julian, Temperature field and geophysical effects of a downgoing slab, J. Geophys. Res., 76, 1113-1138, 1971.
Tsai, Y.-B., and K. Aki, Precise focal depth determination from amplitude spectra of surface waves, J. Geophys. Res., 75, 5729-5743, 1970.
Uyeda, S. and Z. Ben-Avraham, Origin and development of the Philippine Sea, Nature Phys. Sci., 240, 176-178, 1972.
Vening Meinesz, F.A., A third arc in many island arc areas, Koninkl. Ned. Akad., Wetenschap. Proc. B., 54 (s), 432-442, 1951.
Watanabe, T., D. Epp, S. Uyeda, M. Langseth, and M. Yasui, Heat flow in the Philippine Sea, Tectonophysics, 10, 205-224, 1970.
Weertman, J. and J.R. Weertman, High temperature creep of rock and mantle viscosity, Ann. Rev. Earth Plan. Sci., 3, 293-315, 1975.
Wilcox, L.E., W.J. Rothermel, and J.T. Voss, A geophysical geoid of Eurasia, EOS Trans. Am. Geophys. Un., 53, 343, 1972.
Yasui, M. and T. Kishii, Heat flow in the seas of Japan and Okhotsk, Oceanogr. Mag., 19, 87, 1967.
Zienkiewicz, O.C., The Finite Element Method in Engineering Science, McGraw-Hill, London, 521p, 1971.

DESTRUCTION OF THE EARLY CRETACEOUS MARGINAL BASIN IN THE ANDES OF TIERRA DEL FUEGO

Ronald L. Bruhn and Ian W.D. Dalziel

Lamont-Doherty Geological Observatory of Columbia University, Palisades, New York 10964

Abstract. The upper part of an ophiolite suite (gabbro, sheeted dikes, pillow lavas, sedimentary cover) preserved in the Andean Cordillera from 51°S latitude to Cape Horn represents the floor of a back-arc marginal basin. The basin formed near the Jurassic-Cretaceous boundary by spreading within a volcanic chain along the Pacific margin of the South American plate. Starting in the middle Cretaceous the marginal basin was destroyed when the frontal arc and basin floor were uplifted relative to the remnant arc and translated towards the interior of the plate. These processes resulted in the initiation of the present southern Andean Cordillera. Extremely heterogeneous deformation of the rocks in the Cordillera occurred at this time. The most intense deformation occurred along the boundary between the marginal basin and the remnant arc, within a regional zone of progressive shear. Although all the rock units are essentially autochthonous, the deformation process represented the initial stage of obduction of the basin floor ophiolites. If the process had continued the ophiolites would have been tectonically emplaced over the remnant arc onto the continental margin. A significant strike-slip component of deformation in the middle Cretaceous was localized along several discrete shear belts, some of which continued to be at least intermittently active through the Cenozoic.

Introduction

It is now widely accepted that ophiolite complexes are rocks which once formed the floor of present day or ancient ocean basins and were subsequently emplaced tectonically within or upon a continent or island arc. Variation in the composition, thickness and structure of the ophiolites and their various components has nonetheless led to uncertainty as to their exact origin. In particular, it is frequently uncertain whether such a complex represents part of the floor of a major ocean basin, or whether it is a segment of the floor of a small back-arc or marginal ocean basin such as occur behind some island arc-trench systems (Karig, 1971).

Processes have been proposed by which the crust of a major ocean basin could have been uplifted and preserved on land today. These include off-scraping of the downgoing oceanic crust in an arc-trench gap subduction complex, and uplift and detachment (obduction) of oceanic crust intervening between two colliding continents, two island arcs, or a continent and an island arc (Dewey and Bird, 1970; Coleman, 1971; Dewey, 1976). Some ophiolite complexes clearly do occur in such tectonic settings. However, it is perhaps easier to envisage the preservation of the floor of a narrow marginal ocean basin by its being trapped and uplifted between the frontal and remnant arcs, and there has been speculation that a large proportion of ancient ophiolite complexes, including Archaean greenstone belts, may have originally been the floor of marginal basins (Dewey, 1976; Tarney et al., 1976).

In some cases there are specific reasons for believing that an ophiolite complex, even a clearly obducted allochthonous one, represents marginal basin floor (Dewey, 1976). Frequently, however, there is insufficient evidence to discriminate between a major ocean basin and a marginal basin as the source of an ophiolite complex.

The purpose of this paper is to present a general account of the initial uplift and deformation in the middle Cretaceous of the floor of the marginal basin that existed along the Pacific margin of the South American plate during the Early Cretaceous. The geotectonic setting of this basin, including the polarity of the arc, is well established, and most elements of the island arc-marginal basin-remnant arc system are now well exposed. Hence this study gives an insight into the tectonic processes involved in the preservation of a marginal basin ophiolite complex within a continental margin orogen.

Basin Formation

The evolution of the marginal basin is discussed only briefly here as it has been described elsewhere (Dalziel et al., 1974a; Dalziel et al., 1974b; Tarney et al., 1976). The present distribution of rock types within the Cordillera of southern South America and on South Georgia

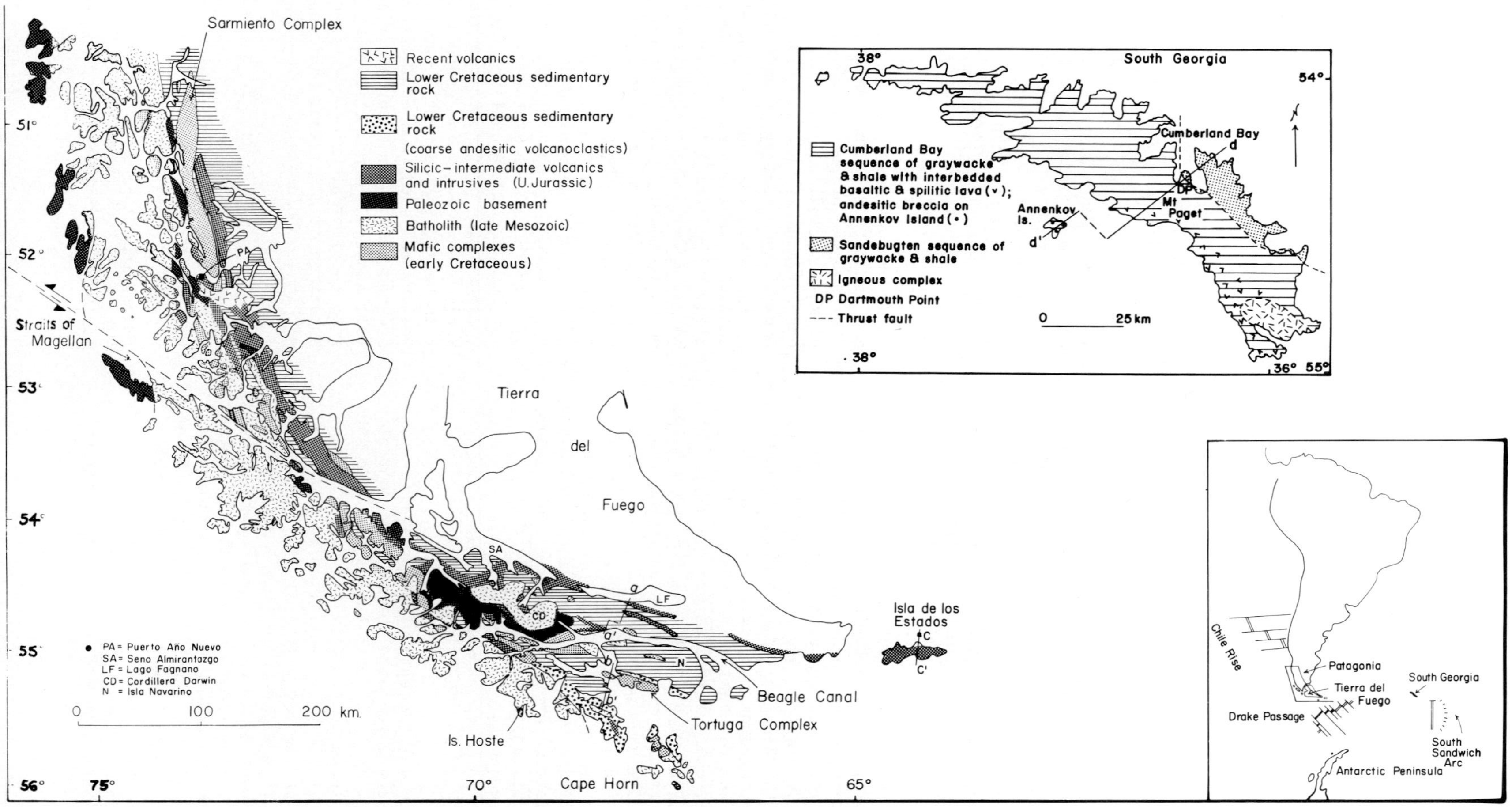

Fig. 1. Geologic map of the southern Andes south of 50°S latitude and South Georgia Island (inset). Geographic locations of the area mapped in South America and on South Georgia Island are indicated in small insert at lower right corner.

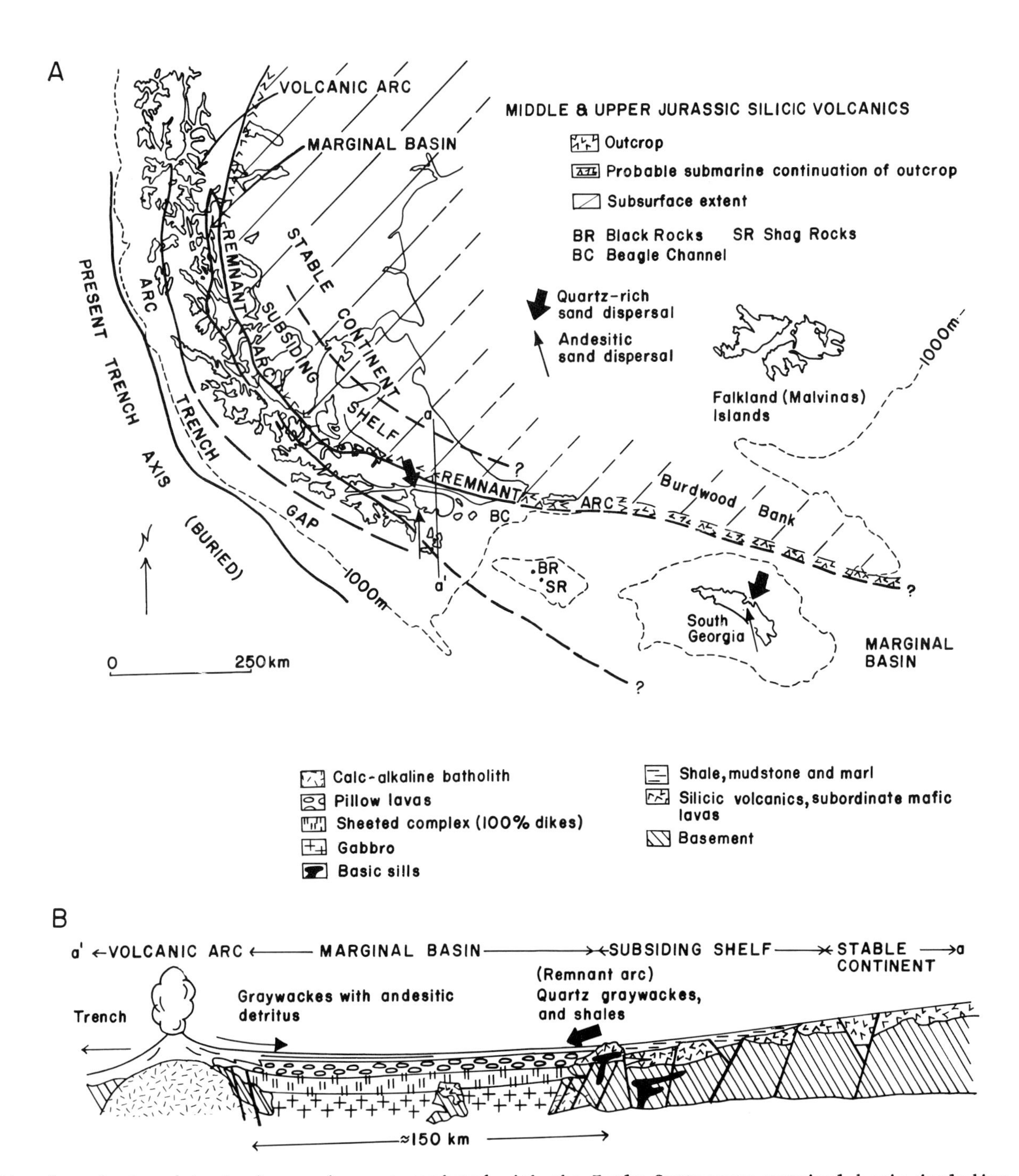

Fig. 2a. Restored tectonic provinces associated with the Early Cretaceous marginal basin including arc-trench gap, frontal volcanic arc, marginal basin, remnant arc and shelf and platform. Note wide dispersal of the silicic volcanics associated both with arc volcanism and rifting during formation of the marginal basin.

Fig. 2b. Restored, diagramatic section across the volcanic arc, marginal basin, subsiding shelf and onto the stable continent. The cross section is based on restoring rock associations exposed along line a-a′ in Fig. 2a.

Island is shown on Figure 1. A reconstruction of the tectonic setting up to Albian time is shown on Figure 2. The building of the present Andean Cordillera south of 50°S latitude was initiated in the Middle Jurassic, following an apparent hiatus in subduction during part of the Triassic and Jurassic Periods while the pre-existing continental basement was uplifted and eroded. A vast quantity of dominantly silicic volcanic rocks were extruded over the basement beginning in the Middle Jurassic while granitic plutons were intruded contemporaneously at depth (Halpern, 1973). Initially, extrusion was dominantly subaerial, although the volcanics contain intercalated shallow marine shale and volcaniclastic sedimentary rocks formed as the result of erosion and subsequent subaqueous deposition within the volcanic terrane. In the Late Jurassic an extensive volcano-tectonic rift zone formed with regional extensional faulting and subsidence within the active volcanic field. At this time a deep, fault bounded trough developed along the Pacific edge of the rift zone. During the latest Jurassic and Early Cretaceous this trough was the site of extensive mafic intrusion. The trough formed a narrow marginal basin (Katz, 1973; Dalziel et al., 1974a). The floor of the basin consists of gabbro, sheeted dikes and pillow lava; hence it represents only the upper part of a complete ophiolite complex. As only a thickness of approximately 1 km of gabbro is exposed above sea level, the absence of ultramafic rock is probably due to the level of exposure and we feel justified in referring to the complex as ophiolitic.

The Middle and Upper Jurassic silicic volcanics apparently represent the products of renewed subduction along the continental margin and/or within-plate extension during formation of the marginal basin (R. L. Bruhn, C. R. Stern and M. J. de Wit, unpublished data). Regardless of their petrogenesis, that part of these rocks which crop out along the Pacific margin, within the present Cordillera, formed part of the main continental margin volcanic chain prior to marginal basin formation.

During the Early Cretaceous the marginal basin was the site of deposition of volcaniclastic flysch derived dominantly from the andesitic volcanics of the active frontal arc on the Pacific side, and to a lesser extent from the mainly silicic volcanics of the inactive volcanic field (remnant arc) on the Atlantic side (Dalziel et al., 1975a; Winn, 1975). Relicts of the continental crust intruded by the mafic complex occur as partially remelted xenoliths and as large screens and blocks. The present total width of the intrusive part of the complex (gabbro-sheeted dikes) varies from a mere 2-3 km in the north to an extrapolated 200 km on the continental shelf near Cape Horn (Fig. 1). South Georgia Island, presently located at the eastern end of the north Scotia ridge, is interpreted as an off-faulted part of the floor and infilling of the basin. The South Georgia platform was once situated adjacent to Burdwood Bank, hence the basin may have been even wider than its extrapolated width in Tierra del Fuego (Dalziel et al., 1975a). We should emphasize that these are present widths which neglect possible subsequent tectonic shortening. However, as we shall discuss below, we do not believe that such shortening was significant.

The rifted continental shelf on the rear side of the marginal basin subsided gently during the Late Jurassic and Early Cretaceous. Between the Kimmeridgian and Albian approximately 1.5 km of black shale and marl of a comparatively shallow water facies (depth $\leq$ 500 m; Natland et al., 1974) accumulated on the shelf. Contemporaneously the volcaniclastic graywackes of the marginal basin infill were being deposited at greater, perhaps bathyal, depths (Winn, 1975).

Destruction of the Basin

Regional Synthesis

Sometime between the Aptian and Coniacian the volcaniclastic infilling of the marginal basin was deformed. The youngest fossils in the sedimentary deposits infilling the marginal basin are the Neocomian to Aptian fauna of Annenkov Island off the southwest coast of South Georgia (Wilckens, 1947; Casey, 1961). Granitic plutons dated (using Rb-Sr whole rock-mineral isochrons) at 80-90 Ma cut the folded sedimentary rocks of the Yahgan Formation in southern Tierra del Fuego (Halpern and Rex, 1972; Halpern, 1973). This major, initial phase of penetrative deformation and uplift in the southern Andes can apparently be bracketed more exactly as having occurred between the middle Albian and the Coniacian based on the following evidence. In the late Albian, marine water depth over the shelf along the Atlantic side of the remnant arc deepened from 500 m to 1 to 2 km according to Natland et al. (1974). The first flysch deposits in the present Andean foothills, located along the continental side of the marginal basin, formed in the Cenomanian (Katz, 1964; Scott, 1966; Natland et al., 1974). These deposits contain detritus eroded from the frontal arc, marginal basin and remnant arc terranes. There is therefore unequivocal evidence that penetrative deformation of the frontal arc, marginal basin and remnant arc rocks in the middle Cretaceous was accompanied by orogenic uplift, which marked the initiation of the present Cordillera (Palmer and Dalziel, 1973; Dalziel and Palmer, in press).

Deformation and uplift has continued, perhaps episodically, until the present. However, folding during the latest Cretaceous and Cenozoic has been concentrated along the Atlantic side of the Cordillera, in the Andean foothills, while the effects of Cenozoic tectonics in the main Cordillera have been confined to brittle faulting and uplift (Katz, 1973; R. L. Bruhn, unpublished data). In this paper we are concerned solely with the major, middle Cretaceous uplift and penetrative deformation which resulted in destruction of the Early Cretaceous marginal basin.

By bracketing destruction of the marginal basin between middle Albian and Coniacian, we do not imply that deformation was a singular, synchronous pulse distributed along the entire length of the marginal basin. In fact, we shall argue that deformation was progressive and clearly neither synchronous nor homogeneous with respect to either space or time.

The tectonic processes active in initiating formation of the southern part of the Andean Cordillera along the Pacific margin of the South American plate in the middle Cretaceous included uplift, horizontal shortening, and horizontal shearing. These three processes occurred contemporaneously and were directly interrelated. Uplift of the frontal arc, the marginal basin floor, and the remnant arc is manifested respectively by the appearance of granodioritic, ophiolitic and rhyolitic detritus in Upper Cretaceous conglomerates deposited in the deepening tectonic Magellan basin behind the remnant arc. The total uplift must have amounted to 5 km or more given that the gabbros of the marginal basin floor are now at the same level as the shallow water marine sedimentary rocks directly overlying the remnant arc volcanics (Dalziel et al., 1974a). The occurrence of ophiolitic detritus in the Upper Cretaceous conglomerates indicates that most of the uplift had occurred by that time. Crustal shortening is obvious from the folds and cleavage, particularly in the marginal basin sedimentary infill and the remnant arc rocks. These structures trend parallel to the Cordillera, curving around the Patagonian "orocline." Locally, the folds and cleavage tighten and are deflected into discrete, left-lateral shear zones such as the northwest end of the Straits of Magellan and the northwest Beagle Channel (see de Wit, in press; Fig. 1).

The present width of the marginal basin floor within which the ophiolitic complexes were emplaced varies from less than 5 km around 52°S latitude where the Andes are still north-south trending, to around 200 km near Cape Horn. Deformation in the north region seems to have involved mainly uplift and horizontal shortening strains, perhaps due to the narrow width of the basin floor and straight north-south plate boundary. The ophiolitic rocks there (the Sarmiento complex, Fig. 1) are now at the same level as the remnant arc rocks. Strain was concentrated along the contacts between the ophiolites and the frontal and remnant arcs and along narrow shears within the ophiolite complex (Dalziel et al., 1974a). The greatest differential movement, uplift of the basin floor, apparently occurred along the contact with the remnant arc. Otherwise, the ophiolites are autochthonous because they are sandwiched between the Patagonian batholith (roots of the frontal arc) on the Pacific side and the remnant arc volcanics on the Atlantic side. Certainly the ophiolites have been neither subducted nor obducted (see Dalziel et al., 1974a).

The structural situation further south, in Tierra del Fuego and on South Georgia Island, where the marginal basin floor was much wider, is shown in Figure 3. It differs from that in the north in that more important components of horizontal translation towards the interior of the South American plate occurred contemporaneously with uplift and horizontal shortening strains. Also, left-lateral strike-slip occurred along discrete shear belts (see de Wit, in press). Strain was concentrated along the marginal basin floor - remnant arc contact which was a zone of major, regional progressive simple shear deformation. The continental basement within this zone was reactivated, taking on an Andean (middle Cretaceous) structural fabric. In contrast, the frontal arc terrane and marginal basin ophiolites were penetratively deformed only along narrow shears (Fig. 4). Strain due to differential uplift and horizontal translation was therefore concentrated along the rear wall of the marginal basin.

The penetrative style of deformation seen in the southern Andes dies out north of the ophiolite complex (at 49°S latitude, see Dalziel et al., 1975b). Closure of the marginal basin (frontal arc - continent collision) was clearly critical for such penetrative deformation.

Arc-trench Gap

One of the surprising aspects of the late Mesozoic-Cenozoic geology in the southern Andes is the absence, or near absence, of an arc-trench gap terrane. The linearity of the western boundary of the batholith with the pre-Middle Jurassic basement from 50°S to 53°S, offset only by an apparent strike-slip displacement of 80 km along the Straits of Magellan lineament (Fig. 1), does suggest that this line marks the western edge of the late Mesozoic arc terrane. Hence the preserved arc-trench gap is very narrow. No post-Paleozoic sedimentary deposits are present at this latitude, and it does not appear that the basement rocks are extensively affected by deformation associated with late Mesozoic subduction. Options for the "disposal" of a wider late Mesozoic arc-trench gap terrane are limited. North of 52°S latitude there is no evidence of its having been removed by strike-slip faulting as is happening to the Franciscan terrane in western North America. Late Cenozoic strike-slip motion apparently has occurred south of 52°S latitude and may have affected the arc-trench gap terrane (Herron et al., this volume). Tectonic erosion (by subduction) and uplift followed by subaerial erosion are the two other possibilities for removal of the arc-trench gap terrane. Given the deep level at which the frontal arc and marginal basin terranes are now exposed, uplift together with subaerial erosion constitutes the more attractive possibility to us.

Frontal Arc

As mentioned in previous publications (Dalziel

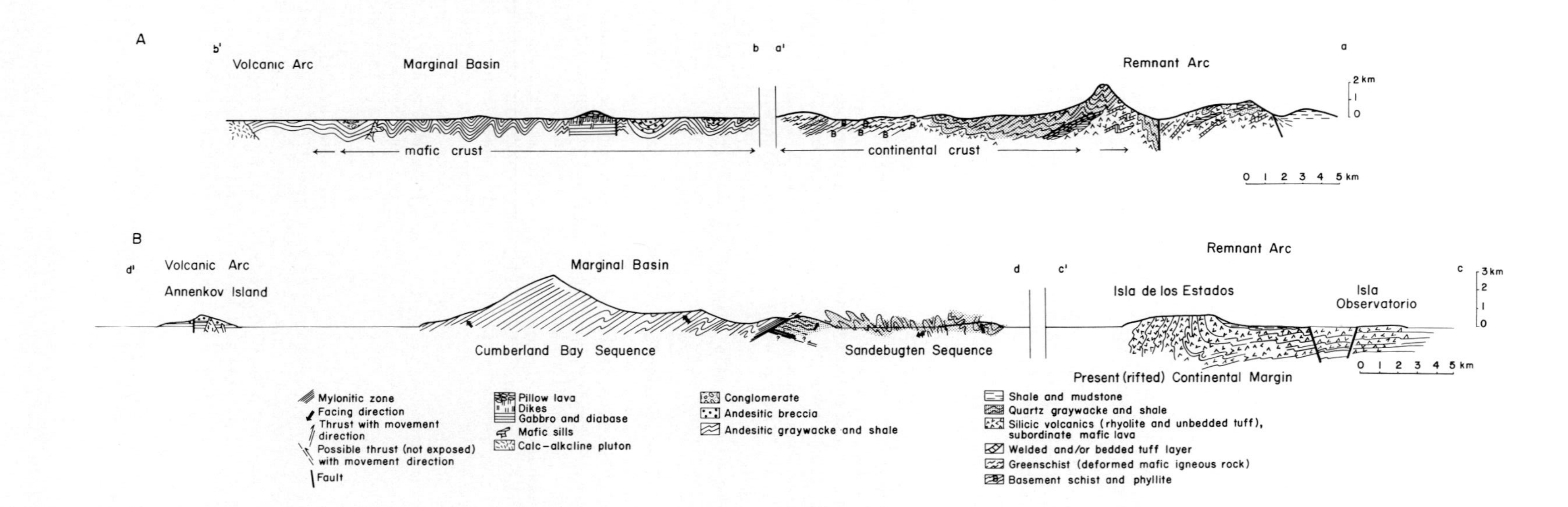

Fig. 3a. Composite structure section illustrating distribution of rock types and nature of middle Cretaceous structures in the Cordillera of Tierra del Fuego. Location of profile lines indicated on Fig. 1.
Fig. 3b. Structure sections as in 3a across Isla de los Estados and Isla Observatorio (c-c′) and South Georgia Island (d-d′). Location of profile lines indicated on Fig. 1.

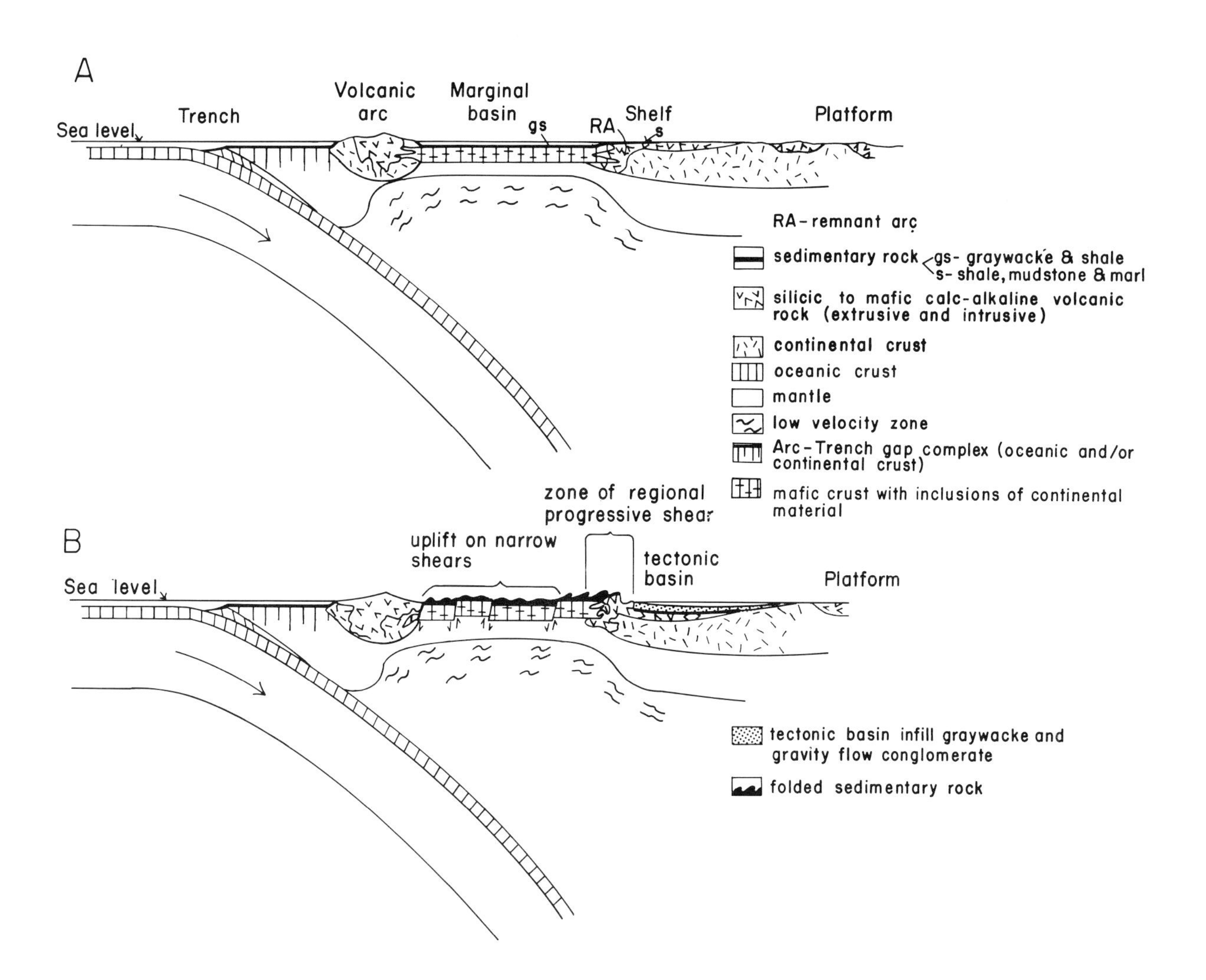

Fig. 4a. Cartoon depicting relation between the trench, volcanic arc, subducting lithospheric slab, marginal basin and adjacent continent. Note the upwarp of the low velocity zone in the back-arc region. Evidence for the upwarp is cited in the text.
Fig. 4b. Cartoon illustrating the location and type of uplift and deformation during the middle Cretaceous within the frontal arc, marginal basin and remnant arc in southern South America. Note uplift of marginal basin contemporaneous with subsidence of the tectonic basin (shelf of 4a) and intervening area of progressive shear deformation.

and Cortés, 1972; Dalziel and Elliot, 1973; and Suarez, 1976), the amount of deformation in the frontal arc terrane during the middle Cretaceous was limited. West of the marginal basin between 51°S and 52°S latitude and near Cape Horn (56°S), the effects of strain on the volcanic and volcaniclastic deposits are limited to open folding and the imposition of a sporadically developed, steeply dipping slaty cleavage trending parallel to the Cordillera. On Annekov Island, near the frontal arc, deformation is clearly less pronounced than in the marginal basin rocks to the northeast on the island of South Georgia itself (Dalziel et al., 1975a; Suarez, 1976; Fig. 3b).

Marginal Basin

Across the marginal basin the deformation is extremely heterogeneous. The intrusive part of the ophiolite complex and the pile of pillow lavas are mostly undeformed. Only locally along shear zones do the mafic rocks show the effects of strain. The volcaniclastic infilling of the basin on the other hand, while little deformed near the arc terrane becomes very intensely folded towards the continental side of the basin (Fig. 3a). It appears as if the sediments must have been very ductile and rich in pore fluid judging from the tectonic structures and the presence of syntectonic sedimentary dikes in the sedimentary rocks of Navarino Island (R. L. Bruhn, unpublished data). In southern Tierra del Fuego the volcaniclastic sedimentary rocks are affected by rather irregular folding while exposed, unfaulted blocks of mafic crust are mostly undeformed. If the nature of deformation in the mafic crust underlying the Yahgan sedimentary rocks is similar to that in the exposed blocks of mafic crust, then a décollement must have formed at depth between the sedimentary rocks and the mafic crust of the marginal basin floor. However, in spite of the irregular nature of folding, vergence is dominantly towards the continent (Fig. 3a).

On South Georgia similar north to northeasterly vergence is the rule in the andesitic detritus rich Cumberland Bay rocks, while the Sandebugten sequence, which was derived from the siliceous remnant arc, mainly verges south to southwest (Fig. 3b). The dominantly Atlanticward vergence led one of us initially to think in terms of the basin having closed by subduction of the floor towards the Pacific (Dalziel et al., 1974a). The situation now appears more complex. The differing vergence in the Cumberland Bay and Sandebugten sequences may reflect their originally having been deposited in different sites, the former sequence behind the active frontal volcanic arc and the latter sequence along the rear edge of the marginal basin. The parallelism of structures in the two sequences, including the fact that the mineral alignment lineations in the main slaty cleavages of the sequences occupy essentially the same vertical plane although plunging in different directions, certainly indicates deformation at the same time and within the same broad basin.

Polyphase deformation of the sedimentary infill of the basin is apparent in some places in southern Tierra del Fuego and in South Georgia particularly near the zone where the Cumberland Bay sequence is thrust north and east over the Sandebugten sequence. While these more complex structures can be interpreted in terms of distinct episodes of deformation, it seems more likely to us that they reflect continuing movement and progressive strain along certain zones within the marginal basin. For example, it is demonstrable in Tierra del Fuego that the post-deformational plutons dated at 80-90 Ma by Halpern (1973) cut Yahgan rocks previously affected by at least two sets of structures. Also, the parallelism of later crenulation cleavages and associated tight asymmetric folds in both the Cumberland Bay and the Sandebugten sequences of South Georgia argue for their development during progressive strain.

Remnant Arc

Deformation is concentrated along the rear or continental wall of the marginal basin. As mentioned above, substantial uplift occurred here together with translation of the marginal basin floor towards the stable continent. Detailed study of the rocks of the remnant arc on Isla de los Estados, off the eastern tip of Tierra del Fuego, has revealed a complex structural history (Dalziel and Palmer, in press; Fig. 3b). Horizontal shortening at right angles to the continental margin first produced a strong slaty cleavage axial planar to small-scale folds in well-bedded sedimentary and volcanic deposits. Overturning of the folds towards the north and southward dip of the cleavage indicate a rotational component of the strain in keeping with uplift of the Pacific margin even at this stage. Subsequent development of a large buckle-fold with a south-dipping axial surface folded the slaty cleavage. Buckling was accompanied by renewed cleavage development only in the highly-strained, overturned, south-dipping limb of the fold. The latest structures are flat-lying crenulation cleavages deforming both earlier sets.

Like the structures of the marginal basin infilling, these structures could be regarded as the result of deformation of several unrelated episodes. However, the first two sets of structures share an east-west strike, a northerly vergence, and a south-plunging maximum finite extension direction within the cleavages. It therefore seems more reasonable to consider the polyphase structures as the result of progressive deformation, with early layer-parallel shortening at right angles to the continental margin, and subsequent large-scale buckling. Both apparently reflect compression across the Pacific margin (Dalziel and Palmer, in press).

This structural situation appears to exist along the entire remnant arc. On Burdwood Bank, to the east of Tierra del Fuego, material with a seismic velocity comparable to that established for the silicic volcanics of the remnant arc appears to crop out along the southern margin of the bank (Ludwig et al., 1968; Dalziel et al., 1974b; Fig. 2). In the west the volcanics occupy the central part of the Cordillera in Tierra del Fuego (Fig. 1).

In Tierra del Fuego the volcanics are folded into several major anticlinoria and synclinoria (Fig. 3a). The structures of the remnant arc are extremely heterogeneous due to the original complex topography and facies relations within the volcanic terrane. For example, the location of individual anticlines is often controlled by the presence of irregular rhyolite domes that form their cores. Also, the along strike persistence and cross sectional geometry of folds is extremely variable. This variability is directly related to the heterogenous lateral and vertical distribution of igneous and sedimentary rock types which developed as the result of the igneous, sedimentary and tectonic processes active in the remnant arc terrane prior to and during formation of the marginal basin (R. L. Bruhn, unpublished data).

It is in the zone of very intense deformation that the pre-Jurassic continental basement was reactivated. It displays a strong, new Andean (i.e. mid-Cretaceous) foliation. Ophiolitic rocks of the marginal basin floor, which intrude and locally overlie the remnant arc volcanics and basement, have undergone polyphase, penetrative deformation, in contrast to the relatively undeformed ophiolites within the central part of the basin to the south (Fig. 3a).

Mapping in central Tierra del Fuego has established that strain within the basement, the remnant arc volcanics and the sedimentary cover had a strong constrictional component at the lowest stratigraphic levels where the axes of the main Andean folds were rotated toward alignment with the southerly dipping extension lineation, the maximum finite extension direction (R. L. Bruhn, unpublished data). In contrast, at higher stratigraphic levels within the shales and graywackes flattening strains predominated and shortening was in part taken up along localized ductile and brittle thrusts. The development of the L-S tectonic fabrics and the sense of rotational strain indicated by the dominantly northward fold vergence implies that a major component of regional, progressive simple shear was involved in the middle Cretaceous deformation along the rear edge of the marginal basin. We wish to emphasize, however, that this deformation was far from homogeneous presumably due to the effect of paleotopography and of the heterogeneous interfacing of rocks of widely varying lithology within the back wall of the marginal basin. Development of discrete zones of left-lateral horizontal shear further complicated the nature of the deformation. It is not surprising, therefore, that in some areas folds verge away from the continent (Fig. 3b), and that in restricted areas cleavage and fold axes within the upper Mesozoic cover locally rotate abruptly out of the regional trend (Dalziel et al., 1975a; R. L. Bruhn, unpublished data).

Discussion

There is strong geologic evidence that an oceanic marginal basin in a geotectonic setting comparable to that of the Japan Sea or of Bransfield Strait, opened along the southwestern margin of the South American plate in the latest Jurassic. It existed until Albian time, and was subsequently destroyed by uplift and deformation.

The plate configuration in the southeastern Pacific region during the middle Cretaceous is uncertain. It is possible, however, that an increase in the subduction rate, causing an increase in horizontal compressive stress on the western margin of the South American plate, may have been responsible for closure of the marginal basin (Dalziel et al., 1974a). The period of time during which the basin was uplifted and deformed occurred during the interval of high spreading rates reported by Larson and Pitman (1972).

The important thing about the method by which the basin was destroyed is that its floor is still essentially autochthonous. Certainly the basin did not close by subduction of the mafic floor, which appears to be close to its original level with respect to the frontal arc and is uplifted with respect to the remnant arc. This uplift, especially together with the translation towards the continent in Tierra del Fuego, does seem to constitute aborted obduction. In particular, we wish to emphasize the contemporaneous development of a tectonic basin immediately behind the uplifting marginal basin floor, with an intervening zone of regional, progressive simple shear deformation within the remnant arc terrane (Fig. 4b). If the process had continued, one could envisage the emplacement of klippen of ophiolitic material on top of the tectonic Magallanes Basin sediments, perhaps analogous to the situation in the western Appalachians (Dewey and Bird, 1970). It is interesting to observe that the klippen would probably have consisted of relatively undeformed ophiolitic rocks with a tectonic fabric due to emplacement confined to the region near their base. Such relations are reported in ophiolitic thrust sheets in the Alps (Moores, 1969), the Newfoundland Appalachians (Williams, 1971), and the western cordillera of the United States (Armstrong and Dick, 1974; also see Dewey, 1976).

It is interesting to speculate as to why the denser, mafic floor of the marginal basin should have been uplifted and translated toward the remnant arc with its underlying continental crust. We can imagine that there was a tendency for the plate margin, which was the site of active calc-alkaline volcanism within the frontal arc, to be uplifted relative to the plate interior in response to the mechanical stresses of subduction. Also, the thermal regime behind the active volcanic arc may have inhibited subduction of the floor of a narrow marginal basin during basin closure. Regions of high heat flow occur in present marginal basins behind active volcanic arcs. This thermal anomaly exists even in marginal basins which are not currently spreading, such as the Sea of Japan (Mckenzie and Sclater, 1968; Lee, 1970; Karig, 1971). The presence of a high temperature mantle and/or partial mantle melts at shallow depth beneath such basins is inferred from heat flow data (Karig, 1971) and is further supported by the high attenuation and low velocities of seismic waves which propagate through the back-arc region (Barazangi et al., 1975). We suggest that an upwarp of the base of the low velocity zone probably existed beneath the marginal basin floor in the southern Andes prior to and even during the middle Cretaceous deformation (Fig. 4a). This would have resulted in 1) a decrease in the relative average density difference between an equivalent lithospheric column within the marginal basin and adjacent continent and 2) a concomitant shallowing of the depth of the brittle-ductile transition of the rocks forming the mafic marginal basin crust and the continental crust along the back edge of the marginal basin. Such a situation is conducive to emplacement of overthrust slices of crystalline crust, and may even explain obduction (Armstrong and Dick, 1974). We note that the mafic crust of narrow marginal basins may be thicker than typical oceanic crust (Tarney et al., this volume) and also, as is the case in the southern Andes, may contain large blocks of continental basement (Dalziel et al., 1974a). These features may further reduce the average density contrast between the marginal basin lithosphere and adjacent continental lithosphere.

A further possibility is that the original normal fault zone on the continental side of the marginal basin, downthrowing and dipping towards the Pacific (Fig. 2), was later reactivated and utilized as a thrust zone during compression. Beach (1976) has argued convincingly that fluids play a major role in the activation and maintenance of deformation within major crustal shear zones which penetrate to the mantle. If the general configuration of isotherms mimics that of the low velocity zone beneath a marginal basin as depicted in Figures 4a and 4b, then one might expect thermal gradients which would permit upper mantle fluids to migrate toward the rear edge of the basin, facilitating deformation within a large zone of progressive shear.

Finally, we wish to comment on the deductions which can and cannot be made from the structures in the marginal basin and remnant arc terranes. Firstly, the dominant vergence of the rocks of the marginal basin infilling might lead to an erroneous conclusion that the floor of the basin had been subducted beneath the frontal arc if it were not possible to see the essential relationship of the arc and basin terranes. Secondly, the inhomogeneity of the structures, and the extent to which the uplift, the strike-slip motion, and the horizontal shortening are localized in discrete zones makes it impossible to deduce the relative motion at the western boundary of the South American plate from the structures. We suggest that this is always going to be an extremely dubious deduction to make across what is an essentially decoupled boundary.

Lamont-Doherty Geological Observatory Contribution No. 2405.

Acknowledgments. No study covering 1,000 kilometers of a cordillera can be accomplished without extensive co-operation. The efforts and ideas of our co-workers Maarten J. de Wit, Keith F. Palmer, Margaret Winslow at Lamont-Doherty and of Robert H. Dott Jr. and R. D. Winn Jr. of the University of Wisconsin were particularly important. We thank Dr. Mort Turner and others at the Office of Polar Programs of the National Science Foundation, and Dr. Eduardo Gonzalez P. and others at the Departamento de Exploraciones, Empresa Nacional del Petroleo, Chile, for their continuing help and encouragement. The work was financially supported by the Offices of Polar Programs, Earth Science and International Decade of Oceanographic Exploration of the National Science Foundation, and logistically supported by the Office of Polar Programs and the Empresa Nacional del Petroleo. We thank R. H. Dott Jr., D. H. Elliot, J. T. Engelder, A. B. Watts and R. D. Winn Jr. for critically reviewing the manuscript.

References

Armstrong, R. L., and H.J.B. Dick, A model for the development of thin overthrust sheets of crystalline rock, Geology, 2, 35-40, 1974.

Barazangi, M., W. Pennington, and B. Isacks, Global study of seismic wave attenuation in the upper mantle behind island arcs using pP waves, J. Geophys. Res., 80, 1079-1092, 1975.

Beach, A., The interrelations of fluid transport, deformation, geochemistry and heat flow in early Proterozoic shear zones in the Lewisian complex, Phil. Trans. R. Soc. Lond. A., 280, 569-604, 1976.

Casey, R., A monograph of the Ammonoidea of the Lower Greensand, Part 2, London Palaeontogr. Soc., 45-118, 1961.

Coleman, R. G., Plate tectonic emplacement of upper-mantle peridotites along continental edges, J. Geophys. Res., 76, 1212-1222, 1971.

Dalziel, I.W.D., R. Caminos, K. F. Palmer, F. Nullo, and R. Casanova, South Extremity of Andes: Geology of Isla de los Estados, Argentine Tierra del Fuego, Amer. Assoc. Petro. Geol. Bull., 58, 2502-2512, 1974b.

Dalziel, I.W.D., and R. Cortés, The tectonic style of the southernmost Andes and the Antarctandes, Proc. 24th Session, International Geological Cong., Montreal, 3, 316-327, 1972.

Dalziel, I.W.D., M. J. de Wit, and K. F. Palmer, Fossil marginal basin in the southern Andes, Nature, 250, 291-294, 1974a.

Dalziel, I.W.D., M. J. de Wit, and W. I. Ridley, Structure and petrology of the Scotia Arc and and the Patagonian Andes: R/V Hero cruise 75-4, Antarctic J. of U.S., X, 307-310, 1975b.

Dalziel, I.W.D., R. H. Dott, R. D. Winn, and R. L. Bruhn, Tectonic relations of South Georgia Island to the southernmost Andes, Geol. Soc. Amer. Bull., 86, 1034-1040, 1975a.

Dalziel, I.W.D., and D. H. Elliot, The Scotia Arc and Antarctic margin, in Ocean Basins and Margins, Vol. 1, The South Atlantic, edited by A.E.M. Nairn and F. G. Stehli, pp. 171-245, Plenum Press, New York, 1973.

Dalziel, I.W.D., and K. F. Palmer, Progressive deformation and orogenic uplift at the southernmost extremity of the Andes, Geol. Soc. Amer. Bull., in press, 1976.

Dewey, J. F., Ophiolite obduction, Tectonophysics, 31, 93-120, 1976.

Dewey, J. F., and J. M. Bird, Mountain belts and the new global tectonics, J. Geophys. Res., 75, 2625-2647, 1970.

de Wit, M. J., The Evolution of the Scotia Arc as a key to the Reconstruction of Southwestern Gondwanaland, Tectonophysics, in press, 1976.

Halpern, M., Regional geochronology of Chile south of 50° latitude, Geol. Soc. Amer. Bull., 84, 2407-2421, 1973.

Halpern, M., and D. C. Rex, Time of folding of the Yahgan Formation and age of the Tekenika Beds, southern Chile, South America, Geol. Soc. Amer. Bull., 83, 1831-1886, 1972.

Herron, E. M., A. B. Watts, R. L. Bruhn, M. A. Winslow, W. C. Pitman, III, and L. Chuaqui, Neotectonics of the Chile margin south of the Straits of Magellan, this volume.

Karig, D. E., Origin and development of marginal basins in the western Pacific, J. Geophys. Res., 76, 2542-2561, 1971.

Katz, H. R., Some new concepts on geosynclinal development and mountain building at the southern end of South America, Proc. 22nd Int. Geol. Cong., New Delhi, 4, 241-255, 1964.

Katz, H. R., Contrasts in tectonic evolution of orogenic belts in the southeast Pacific, J. Roy. Soc. New Zealand, 3, 333-362, 1973.

Larson, R. L., and W. C. Pitman, III, Worldwide correlation of Mesozoic magnetic anomalies, and its implications, Geol. Soc. Amer. Bull., 83, 3645-3662, 1972.

Lee, W.H.K., On the global variations of terrestrial heat-flow, Physics Earth and Planetary Interiors, 2, 332-341, 1970.

Ludwig, W. J., J. I. Ewing, and M. Ewing, Structure of Argentine continental margin, Am. Assoc. Petro. Geol. Bull., 52, 2337-2368, 1968.

McKenzie, D. P., and J. G. Sclater, Heat flow inside the island arcs of the northwestern Pacific, J. Geophys. Res., 73, 3173-3179, 1968.

Moores, E. M., Petrology and structure of the Vourinos ophiolitic complex of northern Greece, Geol. Soc. Amer. Spec. Paper 118, 74, 1969.

Natland, M. L., E. Gonzalez, A. Canon, and M. Ernst, A system of stages for correlation of Magallanes Basin sediments, Geol. Soc. Amer. Mem. 139, 1-126, 1974.

Palmer, K. F., and I.W.D. Dalziel, Structural studies in the Scotia Arc, Isla Grande, Tierra del Fuego, Antarctic J. of the U.S., VIII, 11-14, 1973.

Scott, K. M., Sedimentology and dispersal pattern of a Cretaceous flysch sequence, Patagonian Andes, southern Chile, Amer. Assoc. Petro. Geol. Bull., 50, 72-107, 1966.

Suarez, M., Plate tectonic model for southern Antarctic Peninsula and its relation to southern Andes, Geology, 4, 211-214, 1976.

Tarney, J., I.W.D. Dalziel, and M. J. de Wit, Marginal basin "Rocas Verdes" complex from southern Chile: a model for Archaean greenstone belt formation, in Early History of the Earth, edited by B. F. Windley, pp. 131-147, Wiley, London, England, 1976.

Tarney, J., A. D. Saunders, and S. D. Weaver, Geochemistry of island arc and marginal basin volcanics in the Scotia Arc, this volume.

Wilckens, O., Paläontologische und geologische Ergebrisse der Reise von Kohl-Larsen (1928-29) nach Süd-Georgian, Abhandlungen Senckenbergischen Naturf. Gesell., 474, 1-66, 1947.

Williams, H., Mafic ultramafic complexes in western Newfoundland Appalachians and the evidence for their transportation, Geol. Assoc. Canada Proc., 24, 9-25, 1971.

Winn, R., Late Mesozoic flysch of Tierra del Fuego and South Georgia Island: A sedimentological approach to lithosphere plate restoration, unpublished Ph.D. thesis, University of Wisconsin, Madison, Wisconsin, 160 p., 1975.

THE BAIE VERTE LINEAMENT, NEWFOUNDLAND: OPHIOLITE COMPLEX FLOOR AND MAFIC VOLCANIC FILL OF A SMALL ORDOVICIAN MARGINAL BASIN

W.S.F. Kidd

Department of Geological Sciences
State University of New York at Albany, Albany, New York 12222

Abstract. The Baie Verte Lineament, located in the Burlington Peninsula, northwest Newfoundland, is a steeply-dipping linear belt 90 km long by 1 to 5 km wide containing a variably disrupted ophiolite suite and other mainly mafic volcaniclastic and volcanic rocks that originally overlay the ophiolite complex. The ophiolite complex closely resembles others interpreted as samples of oceanic crust and upper mantle, although the thickness of the gabbro layer is less than 1 km. The overlying mafic volcaniclastic sediments show evidence of deposition close to (and locally at) the base of a steep scarp bounding the east side of the basin. Minor silicic tuffs are found near the top of the preserved sequence. Mafic pillow lavas comprise from 0 to 80% of the sediment section whose total preserved thickness before deformation is estimated as 5 km. Lavas within the sediment section have a chemistry closely resembling samples dredged from the Marianas Basin, while the pillow lavas and dykes of the ophiolite complex are of a distinctly different composition and include some komatiites.

The rocks of the Lineament are tectonically bounded on both sides by polyphase-deformed metasediments, metavolcanics and metaplutonic rocks, generally of coarse-grained epidote-amphibolite metamorphic facies. These rocks were metamorphosed and deformed prior to formation of the ophiolite complex and supplied some clearly identifiable detritus to the lowest sediments to prove it. They behaved as rigid bounding blocks to the Lineament. The local and regional structural, stratigraphic and detrital derivation evidence shows that the Lineament is best interpreted as the remains of a small marginal basin. Its width was probably not more than 50 km (and probably less), after spreading in it ceased. It lay in the basement to an Early Ordovician island arc, on the side remote from the trench, on the American side of the Appalachian-Caledonian (Iapetus) Ocean. The basin remained undeformed until Middle Devonian time, when it closed as a consequence of the Acadian continental collision that shut the Northern Appalachian Ocean. Closure of the Baie Verte basin did not, it appears, involve subduction, but occurred by initial formation of a large syncline with subsequent moderate-angle overturning and eastward thrusting, followed by oversteepening of the main zone of the Lineament due to continued convergence of the two rigid bounding blocks. This behavior, rather than subduction, was probably due to the small width of the basin. The best present-day analogues to the Baie Verte basin can be found in the small basins in the rear of the southern New Hebrides arc. It is suggested that the rocks exposed in the Baie Verte Lineament provide useful information about parts of present-day marginal basins that are mostly inaccessible to direct observation.

Introduction

Wilson's [1966] suggestion that the Appalachian-Caledonian orogenic belt was the site of opening and subsequent closing of a major ocean basin has been very fruitful both in the reinterpretation of its geology in terms of the effects of plate tectonics [Dewey, 1969; Bird and Dewey, 1970; Dewey and Bird, 1971; Stevens, 1970; Church and Stevens, 1971] and for the interpretation of the geological corollaries of plate tectonics in general [Dewey and Bird, 1970, 1971]. Although the quantitative plate motions responsible for the geology of Paleozoic and older orogenic belts will almost certainly never be known, this situation is also true for many Mesozoic and Tertiary belts [Dewey et al., 1973]. Therefore, even though the relative plate motions responsible for parts of orogenic belts are unlikely to be more precisely definable than having been convergent or divergent, useful information about the geological effects of plate tectonics may be obtained from any orogenic belt if the rocks concerned are well exposed and were not severely disrupted and deformed during final closure of the ocean. In particular, information may be obtained on features not readily accessible to sampling in the same tectonic environment near present plate and continental margins.

The northwestern Newfoundland Appalachians have superb coastal exposures in a section across strike and are in an area that was relatively little deformed during the Acadian collision that closed this sector of the ocean [Dewey and Kidd, 1974]. They possess three sharply defined belts of rocks containing well-preserved ophiolite complexes (Fig. 1) that are of interest both as samples of oceanic crust and mantle and for the nature of their type of origin and mode of emplacement. This paper deals with one of these ophiolitic belts, the Baie Verte Lineament (Fig. 1). This is a steeply-dipping, linear belt 90 km long by 1 to 5 km wide containing a variably disrupted ophiolite suite and other mostly mafic volcaniclastic and volcanic rocks that originally overlay the ophiolite complex. Several more or less plausible and comprehensive evolutionary schemes, seen in terms of the geologic corollaries of plate tectonics, have been proposed for the development of the northwest Newfoundland Appalachians. Although these schemes differ considerably in various respects, most workers now appear to agree with the view of Dewey and Bird [1971] that the ophiolite complexes and associated rocks were formed as the crust and fill to one or more early Ordovician marginal basins. There is disagreement, however, over the site or sites of origin of the basins with their ophiolite complex floor, and hence on the 'root zone' for the now allochthonous sheets of the Bay of Islands and Hare Bay ophiolite complexes (Fig. 1). Evidence detailed below bears on this problem and is discussed in a regional context; however, the main purpose of this paper is to illustrate the evolution of a particular marginal basin and to draw some conclusions from it about marginal basins in general.

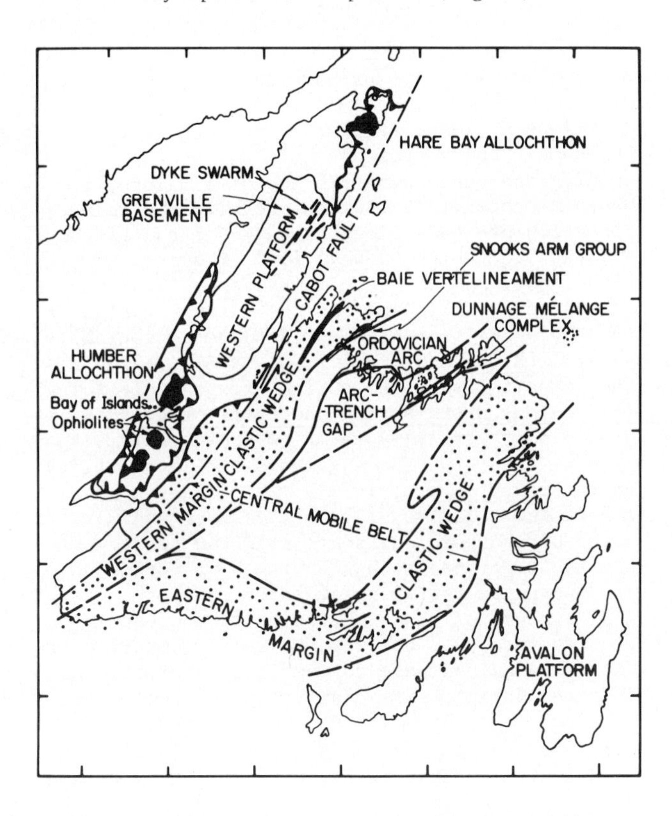

Fig. 1. Major tectonic zones of the Newfoundland Appalachians (black - ophiolites and related rocks; stippled - medium T/P metamorphic belts).

The rocks adjoining the western side and the northern part of the eastern side of the Baie Verte Lineament are polyphase deformed, thoroughly recrystallized, generally epidote-amphibolite facies metasedimentary and metavolcanic schists, the Fleur de Lys Supergroup [Church, 1969], part of the western margin clastic wedge of the Newfoundland Appalachians (Fig. 1). These rocks, which include some gneissic granitic basement to the west [deWit, 1974] are lithologically correlated with the lower part of the autochthonous Cambro-Ordovician miogeocline of the Western Platform, and with the Cambrian and older part of the allochthonous sedimentary rocks of the Humber Arm and Hare Bay allochthons (Fig. 1). The latter and most of the Fleur de Lys Supergroup are interpreted as early graben fill and overlying continental rise sediments developed on this margin of the opening Appalachian ocean [Dewey, 1969; Stevens, 1970; Bird and Dewey, 1970]. Along most of its eastern side, the Baie Verte Lineament is bounded by an extensive granodiorite, the Burlington Granodiorite. This intruded rocks of the Fleur de Lys Supergroup early in, or prior to, their complex deformation [Kidd, 1974], and therefore belongs with the Fleur de Lys structural assemblage. The rocks it intrudes are mafic metavolcaniclastic and metavolcanic schists, which pass to the east conformably and transitionally upwards into an assemblage of silicic metavolcanic and metavolcaniclastic schists and subjacent plutonic rocks, the Grand Cove Group and Cape Brule porphyry [Church, 1969]. A suggestion [DeGrace et al., 1975] that the Grand Cove Group is equivalent to the late Ordovician or younger Cape St. John Group (also silicic volcanics and related rocks) has not yet, in the author's view, been satisfactorily demonstrated. The ophiolitic Baie Verte Lineament thus lies within a zone of complexly deformed and medium-grade metamorphic rocks while itself containing rocks that, although moderately to strongly deformed, are not metamorphosed beyond greenschist facies. The Baie Verte Group has not been directly dated, but is presumed to be Arenigian (early Ordovician) in age due to its very close lithological and stratigraphic resemblance to the fossil-dated Snooks Arm Group. The latter is found 30 km from the Baie Verte Lineament on the east cost of the Burlington Peninsula (Fig. 1). During the early and middle Ordovician, the rock assemblages found to the southeast of the Burlington Peninsula in the Notre Dame Bay region are interpreted [Dewey and Bird, 1971] as representing a volcanic island arc succeded to the southeast by an arc-trench gap, in turn adjoined by an extensive melange (Dunnage) interpreted as a trench-wall assemblage (Fig. 1). The spreading age of the Bay of Island ophiolites is probably Tremadocian (earliest Ordovician), and their obduction into their present allochthonous position, in an exogeosyncline developed on the previous carbonate miogeocline, occurred during medial Ordovician time [Stevens, 1970].

Several conflicting hypotheses have been proposed for the occurrence of the Baie Verte Lineament, a belt of ophiolite-bearing low-grade rocks within a higher grade metamorphic terrain. Church and Stevens [1971] suggested that it is a downfaulted part of a once more extensive obducted ophiolite sheet of which the Bay of Islands ophiolites are also a remnant; Neale and Kennedy [1967] suggested that it is the root zone for obduction of the Bay of Islands ophiolites; while Dewey and Bird [1971], and Kidd [1974] suggested that it is the remains of a marginal basin in the sense of Karig [1971], developed by spreading to the immediate rear of an active island arc. They suggested that the Bay of Islands complex also originated in a marginal basin, but to the west of, and separated by a piece of remnant arc from the Baie Verte basin. On this hypothesis the Baie Verte Lineament cannot have been the root zone for the obduction of the Bay of Islands complex. Evidence detailed below shows that only the latter hypothesis is tenable. The stratigraphic sequence, the structure, and the chemistry of the basalts all show some interesting and distinctive features that may aid in understanding some features of present day marginal basins and assist in the recognition of similar features in orogenic belts.

Stratigraphy and Sediment Provenance

The ophiolite complex is best exposed and preserved in the Mings Bight area at the northern end of the Lineament. Although it occurs in two inverted and one upright thrust slices the ophiolitic plutonic rocks within each slice are internally undeformed. The slices are cut by some high-angle normal faults, but a reconstruction can be made by matching distinct lithological boundaries across these faults. The thicknesses shown for the ophiolite complex units (Fig. 2) are therefore minimum thicknesses for most of the units. The rocks of the ophiolite complex consist of the full sequence of units seen in, and closely resemble, others interpreted as oceanic crust and upper mantle. The minimum thicknesses obtained are comparable to other complexes with the exception of the gabbro unit, which is apparently less than 1 km thick. In this aspect the complex resembles that of Betts Cove, the intact complex found at the base of the Snooks Arm Group [Upadhyay et al., 1971].

The structurally inverted contact between the top of the pillow lavas of the ophiolite complex and the stratigraphically overlying sediment section is fully exposed on the west coast of Mings Bight. About 300 m thickness of little-deformed sediments are preserved between the ophiolite complex pillow lavas stratigraphically below, and a further section of about 1 km (probably at least 2 km before deformation) of pillow lava above (Fig. 2). A bed of maroon chert 30 cm thick lies directly on the ophiolite

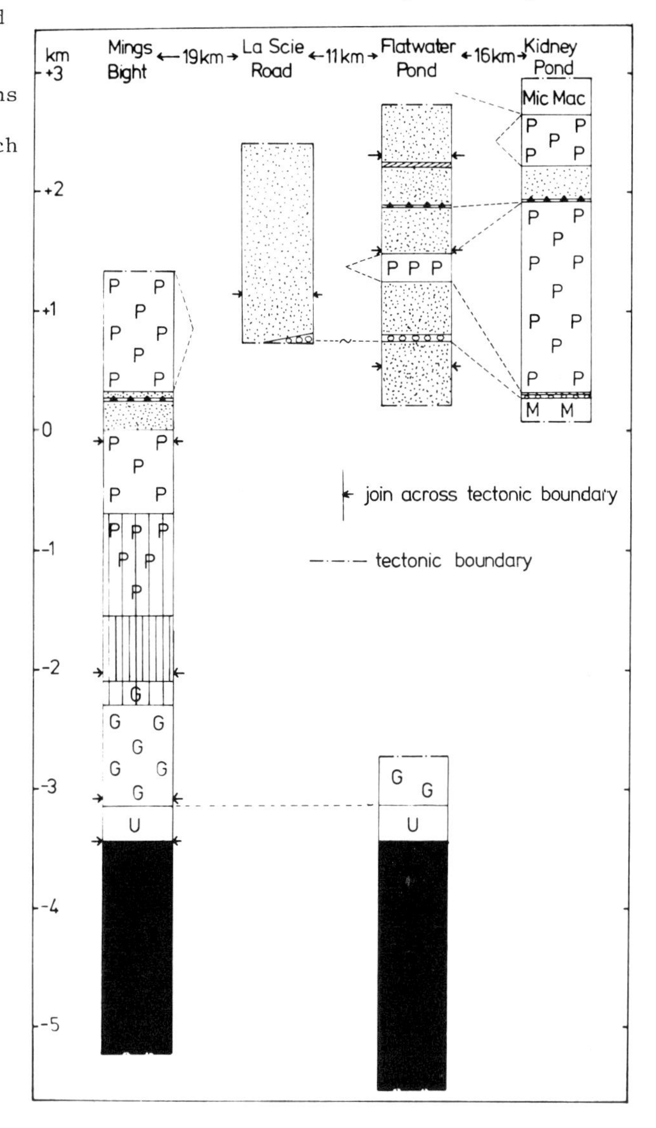

Fig. 2. Stratigraphic sections of the Baie Verte Lineament (black - non-cumulate harzburgite; U - ultramafic cumulates; G - gabbro; vertical lines - diabase dike complex; P - pillow lava; wide-spaced vertical lines with either G or P - diabase dikes with screens of gabbro or pillow lava; stipple - mafic volcaniclastics; black triangles - clinopyroxene-rich volcaniclastics; circles - black slate conglomerate; M - gabbro megabreccia; oblique lines - pink silicic tuff). Thicknesses shown for sediment and upper volcanic units are for present deformed state; these should be at least doubled for original thicknesses. Ophiolite complex rocks are internally undeformed.

complex pillow lava; however, the remainder of the section consists entirely of sandy, silty and conglomeratic mafic volcaniclastic sediments with minor interbedded mudstone to cherty mudstone laminae in places. Some of the beds are graded, but most are not; the conglomeratic beds mainly show large, mostly distinctly rounded cobbles to pebbles dispersed and supported in a sandy to silty matrix. Small-scale channeling is seen at the base of some sandy beds and larger channels can be seen in some places filled by the conglomeratic beds. These sediments thus show evidence of deposition mainly by a mud- or sand-flow (grain flow) mechanism although a minority may be turbidite deposited. None show any definite evidence of being directly deposited tuffs, settled through water or otherwise. The occurrence of medium-scale channels and this mode of deposition suggest that the depositional environment was near the base of a relatively steep submarine slope [Stanley and Unrug, 1970]. Rare slump folds consistently show a paleoslope from the northeast/east when the beds are unfolded by a strike rotation. The identifiable clastic components of the sandy and silty beds are albitised plagioclase grains, most commonly rounded, suggesting a shallow water transit; the mafic components appear to have been much finer grained and have mostly been recrystallized and obliterated in the low greenschist facies alteration affecting the rocks. Quartz is universally and conspicuously absent. Near the base of the section, the conglomeratic clasts are mainly large flakes of yellowish cherty mudstone very similar to beds within the section; upward, mafic lava clasts predominate, accompanied by rare ophiolite gabbro and diabase clasts. Near the top of the section, there is a sequence of conglomeratic beds dominantly containing very porphyritic mafic lava clasts with large oscillatory zoned phenocrysts of (altered) clinopyroxene, which also occur abundantly in the coarse sandy matrix. Such a rock type suggests derivation from an island arc, and a very similar rock is illustrated by Mitchell [1966] from Neogene arc deposits in the New Hebrides. The overlying lavas, which are in large part strongly deformed, contain two thick sills of abundantly and coarsely porphyritic plagioclase porphyry dolerite; wide feeder dikes of identical aspect are found cutting all members of the ophiolite suite including the non-cumulate harzburgite. This demonstrates that the lavas above the sediments are not, in the strict sense, part of the ophiolite suite.

Such a thick section consisting entirely of mafic volcaniclastic rocks is not expected to immediately overlie oceanic crust formed in a major ocean, except possibly near a "hot spot"-type volcanic chain originating very near the spreading axis. However, the upper lavas should then show alkaline affinities, which they do not, and the progressive erosion of the inactive part of the migrating volcanic chain should result in a rapid return to pelagic sedimentation, which is not seen. Sections developed near fracture zones should mainly show scree breccias and pelagic sediments without sandy mafic deposits.

Inland, the stratigraphic succession is, in part, different. Although the lower part of the ophiolite complex stratigraphy is preserved in one place and large ultramafic bodies are found discontinuously along the western side of the Lineament, the contact between the ophiolite complex and the overlying mafic sediments and volcanics has everywhere been removed by tectonic sliding. However, the nature of the exposed basal sediments suggests that they were deposited not far above the ophiolite complex base. The section forms an eastward-facing homocline across the Lineament. The sediments and lavas in the inland area are deformed and appropriate lithologies possess a steep cleavage. Estimates of shortening from rare suitable lithologies show that original stratigraphic thicknesses were probably at least double present thicknesses and therefore that the total thickness of 2.5 km in the southern (Kidney Pond) section (Fig. 2) was probably at least 5 km before deformation. The two southern sections show (Fig. 2), above the two basal units, sandy to silty and conglomeratic mafic volcaniclastic rocks, essentially identical to those in the Mings Bight section, intercalated on a large scale with pillow lava containing a relatively few thin horizons of mafic sediments. The pillow lava comprises about 80% of the southern section (Kidney Pond) and mapping shows that it thins northward, being entirely replaced by volcaniclastic rocks in the La Scie Road section. Dolerite sills are common throughout the Kidney Pond section and also become less common northward. Two distinctive horizons near the base and top of the stratigraphic sequence in the Flatwater and Kidney Pond sections provide control for the demonstration of this facies change. The upper of these horizons consists of conglomeratic mafic sediment containing abundant large zoned clinopyroxene phenocrysts in the matrix and in the clasts; this horizon is very similar to that in the Mings Bight section, but they cannot be proven to be equivalents. Above this horizon in the Flatwater Pond section, there are fairly abundant beds of pink, silicic, albite-phenocryst-bearing tuffs. Both these lithologies suggest the presence of a nearby island arc. Rare slump structures in the lower part of the sediment section all indicate an east to west paleoslope when rotated to horizontal about strike.

The lowest two units in the Kidney Pond section show some distinctive characteristics. The upper of the two, a black slate matrix pebbly conglomerate, has been equated with a mélange [St. Julien et al., 1976]. The term is inappropriate for this unit that is contained within a regular stratigraphic sequence, that nowhere exceeds 60 meters in thickness, that does not show any signs of disruptive post-depositional deformation, and that possesses a restricted range of clast types.

This conglomerate is merely a submarine mudflow deposit, just as the conglomeratic mafic volcaniclastics above are sandflow deposits; it differs only in the distinctive nature of its matrix and clast assemblage. This assemblage consists of 1) argillaceous clasts similar to the matrix; 2) clasts of distinctive ophiolite-derived gabbro, diabase and basalt; and 3) exotic, quartz-bearing clasts. Most of the clasts, because they are somewhat rounded, were probably worked through shallow water before incorporation in the mudflow. This observation also applies to the mafic volcanic clasts elsewhere in the succession. The restricted assemblage of ophiolitic clasts is noteworthy; apart from a few rare and small clinopyroxenite clasts, there are no ultramafic clasts present and no chromite has been detected. These ophiolite-derived clasts form the bulk (90% +) of the total assemblage. Most are pebble to cobble size, but one block of ophiolite gabbro 40 X 20 meters in section is present. The exotic clasts never comprise more than 10% of the assemblage and are usually much less abundant. The most common type are of granodiorite, ranging in size up to 1 meter across. These, as has previously been pointed out [Church, 1969; Neale and Kennedy, 1967] very closely resemble the Burlington Granodiorite, which is a large syn- or pre-kinematic intrusion into the metamorphic terrain and which immediately adjoins the eastern side of the Baie Verte Lineament. These clasts confirm the paleoslop direction of provenance, and demonstrate that the Baie Verte Lineament rocks, including the conformabl underlying ophiolite complex base, are not allochthonous, on a large scale, with respect to the eastern bounding block. Other rare clasts within this conglomerate include pieces of silicic tuff with a strong pre-depositional foliation; these range in size from pebbles to a slab 1 meter across. They can also be matched with rocks (probably Grand Cove Group) in the metamorphic block to the east of the Baie Verte Lineament. Also matchable with the Grand Cove rocks to the east is one large block (1 X 2 meters exposed) of silicic meta-siltstone possessing a well-developed, pre-depositional muscovite schistosity axial surface to tight folds, both refolded by open angular folds that are also predepositional. In addition, a boulder conglomerate, which probably occupies a large channel, and is on strike but not directly connected with the black slate conglomerate, occurs towards the La Scie Road section. It mostly contains quartzite and quartz pebble conglomerate boulders, the latter very closely resembling metasediments found in the area previously mapped as Cape Brule Porphyry, but which are probably in the Grand Cove Group. It also contains some small serpentinized and chloritized chromite-bearing ultramafic clasts.

The lower of the two lower units in the Kidney Pond section is a megabreccia. This consists entirely of large (10-100 meters across), closely packed slabs of ophiolite gabbro, with or without diabase dikes. Thin accumulations of silty mafic volcaniclastic and mudstone beds conformable to the section above the unit occur between blocks showing that it accumulated block by block as a submarine scree breccia, not as a disrupted tectonic melange or olistostrome. Several characteristics of some of the gabbros, distinct from normal ophiolite gabbros, suggest that they were deformed on a fault zone before incorporation in the megabreccia. Gabbro blocks containing diabase dikes parallel to a sheet jointing have their long dimensions parallel to the sheet jointing and lie flat in the bedding orientation so that the dike segments are also coplanar with bedding. Toward Flatwater Pond this unit is replaced along strike by a unit of sandy and silty mafic volcaniclastics with minor argillaceous laminae and very rare calcareous turbidites. In places this unit also contains large blocks, for example one of cumulate clinopyroxenite and one of recrystallized limestone both about 10 meters across. Areas of ophiolite gabbro at its base may or may not be accumulations of blocks or large single blocks several hundred meters across.

The clast types and occurrences are more fully documented and discussed in Kidd [1974].

Composition of the Basalts

Three separate units of basalt are present in the area; 1) the ophiolite complex pillow lavas and dikes; 2) the upper pillow lavas above the sediment section in Mings Bight; and 3) the upper pillow lavas in the Kidney and Flatwater Pond sections. Major element analyses of mafic lavas in low greenschist facies have been shown to give highly variable and hence unreliable results by Smith [1968]; this variability affects all of the major elements and is not merely spilitization affecting Na and K. Whether the variability (in some cases extreme) found in the major element analyses of the Baie Verte lavas is due to such migration during alteration and low greenschist facies metamorphism is an open question, but appears likely from the less mobile minor element data. An average of 6 of the more consistent analyses of the ophiolite lavas and one analysis of the Kidney Pond section upper pillow lavas are given for information in Table IA. These analyses appear to be a reasonably normal spilitized basalts, but a few within the ophiolite lavas have the characteristics of basaltic komatiites (Table IA, Sample T2). This is of interest because of the more common occurrence of such rocks in Archean greenstone belt sequences.

The less mobile minor element analyses (Table IB) show much more consistency than the major element analyses and give a clear discrimination between the ophiolite lavas and dikes, and the upper pillow lavas in the Kidney Pond section. The upper pillow lavas in the Mings Bight section appear to be very similar to the ophiolite lavas.

TABLE I

Selected Analyses of Baie Verte Lineament Basalts

Major Elements	Ophiolite Pillow Lavas (average of 6 samples)	Sample T2	Upper Pillow Lavas Kidney Pond Section
SiO_2	49.58	47.11	48.24
TiO_2	0.46	0.40	1.46
Al_2O_3	14.16	12.63	14.58
FeO*	8.52	8.94	11.93
MgO	11.59	12.80	7.79
CaO	9.30	13.81	9.37
Na_2O	3.07	0.95	3.55
K_2O	0.04	0.34	0.31

Minor Elements (ppm)	Number of Samples	Ti	Zr	Y	Nb	Sr	Cr
Mings Bight							
Ophiolite Dikes	8	2240	18	12	0.5	105	115
Ophiolite Pillow Lavas	11	2790	21	12	0.4	93	418
Upper Pillow Lavas	5	1900	19	8	0.6	101	572
Kidney Pond							
Upper Pillow Lavas	5	8760	104	29	4.4	190	nd
Marianas Trough[a]	6	8940	101	25	6	186	231
Niua Fo'ou[b]	14	8760	131	36	nd	160	282
Average Ocean Ridge Basalt[c]	33	8340	100	43	nd	123	296

a - Hart et al. 1972 b - Reay et al. 1974 c - Melson and Thompson 1971

XRF analyses by E. Nesbit and R.G.W. Kidd at the University of East Anglia

For comparison, averages of analyses of the same minor elements for samples obtained from the active Marianas Trough marginal basin [Hart et al., 1972], from the volcano Niua Fo'ou in the northern Lau active marginal basin [Reay et al., 1974] and ocean ridge basalts [Melson and Thompson, 1971] are given. It can be seen that the ophiolite lavas and dikes from the Baie Verte Lineament are quite distinct from the present active marginal basin samples and from ocean ridge basalts, but that the Kidney Pond section upper pillow lavas are very similar to all three types given for comparison, especially to the Mariana Trough samples.

Plotting of Ti, Y, Zr, and Sr values on Pearce and Cann's [1973] revised discrimination diagrams reveals the unfortunate fact that the ophiolite lavas and dikes and the Mings Bight upper pillow lavas plot in the island arc low-K tholeiite field while the upper lavas within the Kidney Pond sediment section plot in the ocean ridge basalt field. A minor but mappable horizon of dark pillow lava within the Kidney Pond upper lavas yields analyses with alkaline affinities (not shown in Table I) and that plot in the alkaline (hot-spot) field on Pearce and Cann's diagrams. It is suggested that the environment of formation of rocks from marginal basins may not always be correctly discriminated by Pearce and Cann's method.

Structure and Deformation

The structure of the Baie Verte Lineament is a tightly folded syncline severely disrupted by thrust faults, which developed synchronously with the severe horizontal shortening deformation responsible for the single steep penetrative cleavage. It can be argued from a sequence of thrust faults and cleavage development in the thrust zones exposed in the Mings Bight area (Fig. 3), that the initial stages of development involved a syncline with an axial surface dipping moderately west. This was disrupted by moderately west-dipping thrust faults as it tightened and cleavage developed. Subsequent oversteepening of the rocks caught in the main part of the Lineament occurred, together with high-angle thrust-faulting (tectonic sliding) on both east and west boundaries of this steep zone, which is essentially the whole of the width of the Baie Verte Lineament south of the Mings Bight area. During all this deformation, the bounding blocks to the Lineament (the western and eastern Fleur de Lys metamorphic terrain) behaved in an essentially rigid manner. Very little evidence of local deformation or retrograde metamorphism can be seen in most places even within 100 meters of the tectonic contacts on either side of the Baie Verte Lineament, both in the mostly epidote-amphibolite facies schists on the western, and northern part of the eastern, sides, and in the Burlington Granodiorite on the eastern side. The only exceptions to this statement are in a zone of thin thrust slivers of retrograded and deformed Burlington Granodiorite on the east side of the Lineament north from Flatwater Pond, and very close to local sympathetic thrust zones in the Fleur de Lys schists on the eastern margin in the Mings Bight area (Fig. 3).

The structural situation in the Advocate mine area at the west of the Mings Bight section appears to be more complex, but the affinities of the ophiolitic rocks in this area are not yet entirely clear [Bursnall and deWit, 1975]. Some belong to the older, polyphase deformed and metamorphosed Fleur De Lys terrain while others, although complexly deformed, may belong to the Baie Verte assemblage, perhaps having been deformed on a fault zone, possibly a transform fault, and

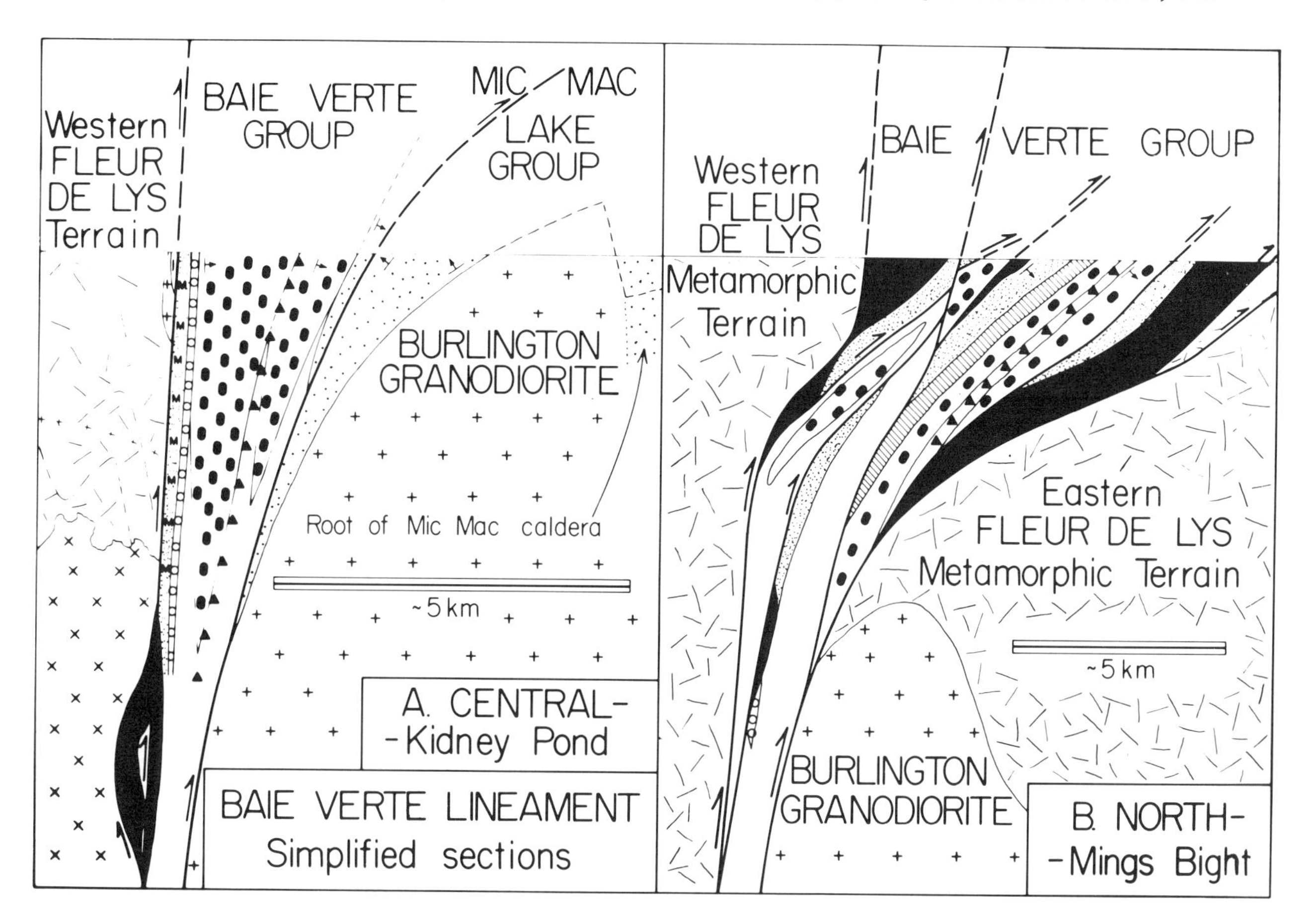

Fig. 3. Simplified and extrapolated cross-sections of the Baie Verte Lineament. (a) Bounding blocks: random dashes - schists of the Fleur de Lys terrain; + - Burlington Granodiorite; x - post-kinematic granite in the Fleur de Lys. (b) Baie Verte Lineament: black - ultramafic rocks; fine dots - ophiolite gabbro; lines - sheeted dikes; black ovals - pillow lava; M - gabbro megabreccia; circles - black slate matrix conglomerate; black triangles - mafic volcaniclastics containing some beds with abundant clinopyroxene-porphyry clasts; blank - mafic volcaniclastics; coarse dots - Mic Mac Lake Group; arrows with bar - younging direction.

perhaps the same fault zone responsible for the gabbro megabreccia and the highly strained gabbros found in places within it.

Two partially simplified and extrapolated cross-sections of the Baie Verte Lineament are shown in Fig. 3. In the Mings Bight area, two somewhat disrupted, inverted thrust slices overlie a slice containing a partial ophiolite section that is upright, defining a disrupted synclinal structure. In the Kidney Pond area, the structure is also a disrupted syncline, although another group of rocks is also involved besides the ophiolitic Baie Verte Group. This is the early Devonian Mic Mac Lake Group, a subaerial bimodal basalt-rhyolite sequence containing much proximally-derived alluvial fan sediment. In the section in the Kidney Pond area, a thin strip of east-facing Mic Mac Lake Group rocks rests with very slight, unfaulted, angular unconformity on the east-facing homocline of Baie Verte Group volcanics and sediments (Fig. 3, Fig. 2). This same contact has been suggested to be a fault [Church, 1969], and conformable with the Baie Verte Group overlying west-facing Mic Mac Lake Group [Neale and Kennedy, 1967]. Neither of these hypotheses is correct. Most of the Mic Mac Lake Group in this area faces west from a spectacular unconformity on the Burlington Granodiorite. The west and east facing sections of the Mic Mac Lake Group are separated by a major thrust fault (tectonic slide zone), which has been removed in most places by a late high angle fault with considerable westward downthrow (for clarity, not shown on Fig. 3). In the Flatwater Pond area, the east-facing Baie Verte is separated from a narrow belt of west-facing Mic Mac Lake Group by this same thrust fault, with thin thrust slices of Burlington Granodiorite intervening in places. As suggested by Neale and Kennedy [1967], the Mic Mac Lake Group shares the same single regional cleavage as is found in the Baie Verte Group. The overall structure in this area is therefore also a thrust-modified syncline, just as it is in the Mings Bight area.

Considering the east-moving thrusts, the relatively intact nature of the Baie Verte sequence, the very low angle unconformity with which the Mic Mac Lake Group (early Devonian) overlies the Baie Verte Group (probably Arenigian), and their shared single cleavage, there is no evidence of significant regional deformation of the ophiolitic Baie Verte Group until after early Devonian times. It is suggested that there is no evidence of subduction having occurred during the deformation of this belt, although some of the oceanic mantle underlying the Baie Verte Lineament must have been displaced downwards during the convergence of the two rigid bounding blocks and the resulting compressive, fundamentally horizontally directed, deformation. In addition, all the post-early Devonian thrust deformation is towards the east, and it is therefore in the wrong direction, let alone the wrong age, for the Baie Verte Lineament to be a source or root zone for the Bay of Islands ophiolites. The source along White Bay and Deer Lake thrust belt (at the western side of the Grand Lake Group schists) proposed by Dewey and Bird [1971] seems a more likely proposition.

Reconstruction and Present Day Analogues

Assuming the conclusion that there was no subduction involved during the deformation of the Baie Verte Lineament is correct, then the width of the oceanic crust that floored the Baie Verte basin can be estimated by unstacking the thrust sheets and unfolding the syncline, allowing a generous estimate for the width of material at depth and that eroded. Such an estimate for the Mings Bight section (Fig. 3) gives a width of about 30 km; it is extremely difficult to imagine that it could be more than 50 km. An approximately scale cross-section reconstructing the Baie Verte basin given a 30 km width of oceanic crust is shown as Fig. 4. This incorporates the arc volcanic source for the mafic volcaniclastic rocks and silicic tuffs, the fault zone and scarp source for the megabreccia and black slate conglomerate, and indicates the renewed vulcanism in the basin responsible for the upper pillow lavas. Data from the Mid-Atlantic Ridge [Ballard et al., 1975] shows that pillow lavas do not flow great distances and therefore that the source must have been within the basin, and is not likely to be overflow from the arc. This is also likely to be true because of the chemistry of the upper lavas, which resemble ocean ridge basalts.

The overall length of the Baie Verte Lineament can also be approximately determined. The Canadian 1:63,360 aeromagnetic maps show that the pronounced anomalies associated with the ultramafic bodies and the weak trend along the Lineament do not continue to the NNE beyond the seaward end of Burlington Peninsula. To the SSW, the Lineament proper is cut off by the Green Bay Fault, which has 20 km dextral displacement [Upadhyay et al., 1971]. However, a small ultramafic body occurs about this distance to the west of the appearent end of the Lineament near Sandy Lake [Neale and Nash, 1963] and mafic rocks like those in the Lineament are found on and around Glover Island in Grand Lake [Riley, 1957]. However, no further occurrences are found south of this area. The Baie Verte Lineament proper is about 90 km long; with the extension proposed above, it becomes 220 km long. St. Julien et al., [1976] speculate that the Baie Verte Lineament is the same structure as that around Thetford in Quebec, a distance of about 1500 km. This seems unlikely to be true, because of the absence of any direct evidence that there is a continuous structure between the two localities, because the Baie Verte Lineament was not deformed until post-early Devonian times, whereas the Quebec structure developed in early to medial Ordovician

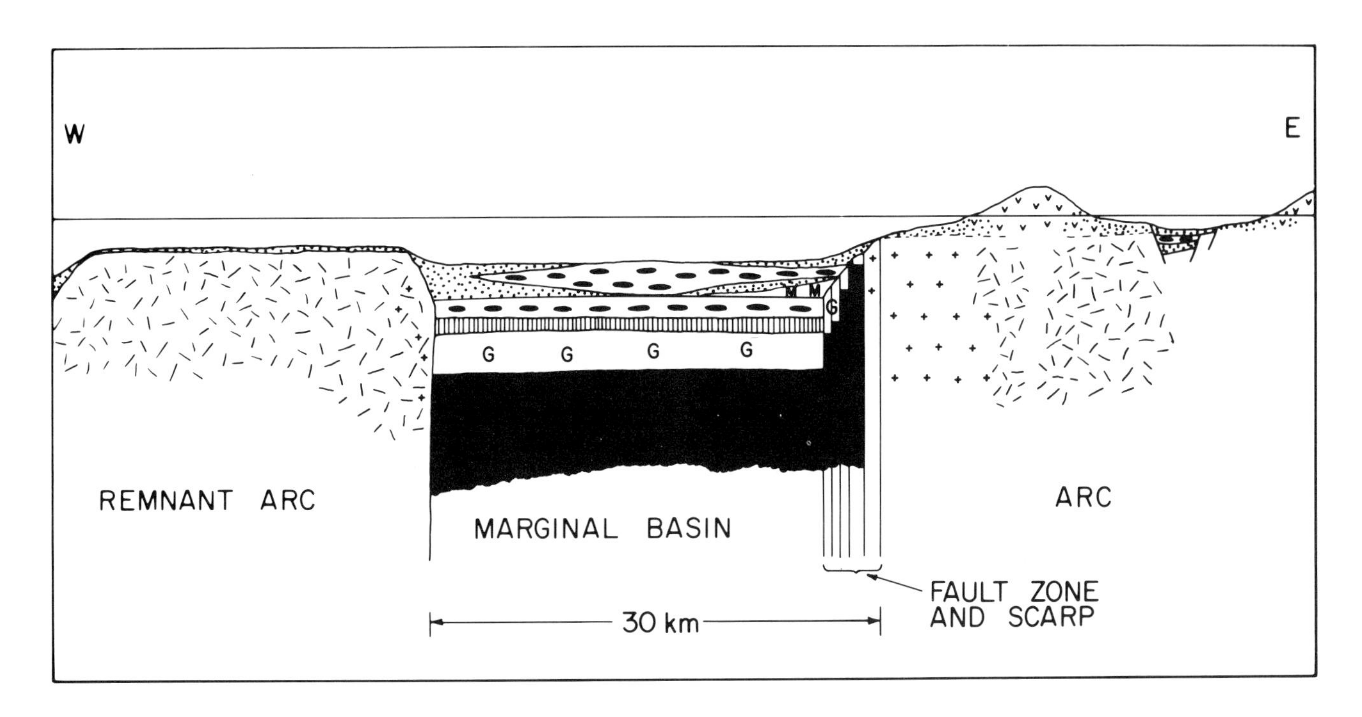

Fig. 4. Reconstructed cross-section (approx. to scale, V=H) of the Baie Verte marginal basin in Ordovician time. Ornament as for Fig. 3 except: fine dots - mafic volcaniclastics; G - ophiolite gabbro, V - arc volcanics.

time, and because of the evidence presented here and in Dewey and Bird [1971] that the western Newfoundland ophiolitic belts developed in separate marginal basins. Thus a present day analogue for the basin floored by oceanic crust that became the Baie Verte Lineament must have dimensions about 30-50 km wide and 100-200 km long and be developed behind an active island arc. Such a process is represented today in areas like the Marianas Trough and Lau-Havre Trough [Karig, 1971], but these basins are about ten times too wide and long to be precise analogues to the Baie Verte basin. However, precise analogues for size can be found behind the southern half of the New Hebrides arc (Fig. 5) [Karig and Mammerickx, 1972]. These basins have been demonstrated to be relatively free of sediment given the available supply from the arc and are clearly active extensional structures [Karig and Mammerickx, op. cit.]. The narrower ones have not yet been demonstrated to be floored by oceanic crust, but wider basin in the central part of the arc is highly likely to be so floored. To further the analogy with the Baie Verte sequence, this basin has the volcano Aoba sited within it. This [Warden, 1970] currently erupts plagioclase-phyric basalt, perhaps resembling a possible source similar to that which supplied the porphyritic dikes and sills in the Mings Bight section. It has also erupted ankaramitic lavas, and may perhaps be similar to the source of the coarse clinopyroxene porphyry conglomeratic horizons in the Baie Verte. Most of the other volcanoes in the New Hebrides erupt plagiophyric lavas [Mitchell and Warden, 1971]; this resemblance to the source of the volcaniclastics in the Baie Verte is perhaps also noteworthy. Karig and Mammerickx (op. cit.) emphasized the en echelon nature of the smaller New Hebrides basins oblique to the general trend of the arc. This obliquity to the arc is also true of the Baie Verte Lineament (Fig. 1). The Baie Verte structure is also not an isolated example; the Snooks Arm Group 30 km to the east has a very similar sequence of rocks conformably resting on an ophiolite complex base [Upadhyay et al., 1971]. These are also in a fault-founded syncline, and are rather less deformed than the Baie Verte Group. This has also been proposed to have formed in a small marginal basin [Upadhyay et al., 1971].

Discussion and Conclusions

Regional Geology

The fact that the sediments immediately and conformably above the ophiolite complex base to the Baie Verte Group contain clasts of material closely matching rocks in the eastern Burlington Peninsula shows that it and its ophiolite base are not allochthonous with respect to that area. The Snooks Arm Group is also not allochthonous with respect to the same area for the same reason [Church, 1969; Dewey and Bird, 1971]. The fact that clasts of this material in the Baie Verte

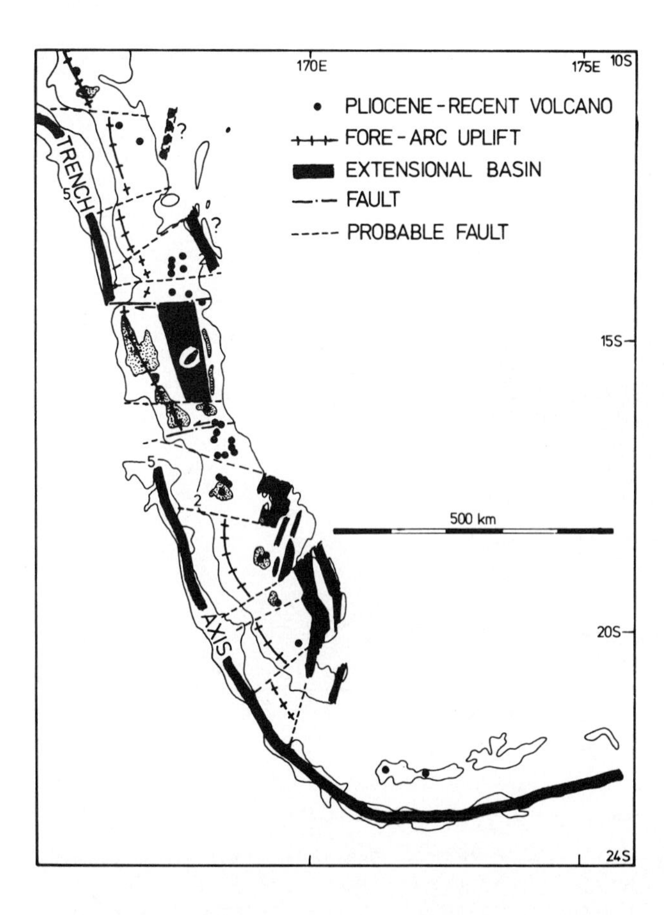

Fig. 5. Sketch map of the New Hebrides island arc. Slightly modified from Karig and Mammerickx (1972). 2 and 5 km. submarine contours shown.

Group show significant predepositional deformation fabrics demonstrates that they had undergone that compressive deformation and metamorphism not only prior to the deposition of the sediments but also prior to the formation of the ophiolite complex (oceanic crust and mantle) base. The latter must have developed in a structural environment of regional brittle tension, not the plastic, compressive behavior represented by the pervasive deformation of the metamorphic rocks. The Baie Verte Group did not undergo any regional penetrative deformation until after the deposition of the early Devonian Mic Mac Lake Group, and therefore does not show any detectable effects of the medial Ordovician obduction of the Bay of Islands ophiolites and Humber Arm allochthon. In addition, all the thrusting in the Baie Verte Lineament moved eastward. The rocks of the Fleur de Lys metamorphic terrain that form the two blocks bounding the Lineament were not internally penetratively deformed in any significant way during the compressive, horizontally directed, deformation that affected the rocks of the Baie Verte Lineament.

General

The occurrence of a coherent ophiolite complex base to the sequence in the Baie Verte Lineament (and the Snooks Arm Basin) shows that, even in such small marginal basins, the spreading results in the successive injection of coplanar dikes into the accreting oceanic crust. The fact that these dikes are coplanar and do not penetrate the cumulate and non-cumulate ultramafic rocks shows that this process is coherent at least on the scale of a few kilometers, and, therefore, that at least short segments of spreading ridge should be present in actively distending marginal basins. Whether these segments are connected in a single, orderly ridge-transform boundary in large marginal basins is not predictable from this data, but in narrow New Hebrides-type basins there is not likely to be room for more than one such boundary. There are two sets of orientations of ophiolite diabase dikes at right angles to one another (each is in a different thrust sheet) in the Mings Bight ophiolite complex. The restored original orientations of these dikes in the Baie Verte basin are near parallel and at a high angle to its length. The dikes in the Snooks Arm basin are at a high angle to its long dimension and those in the allochthonous Bay of Islands ophiolites are oblique to the long axis of the allochthon and the presumed trend of the basin they originally lay in. These data suggest that spreading axes in marginal basins are rarely parallel to their long dimension and also that radical changes of spreading direction and consequent ridge axis jumping may commonly occur.

The chemistry, particularly of some of the minor elements, of the mafic lavas of the ophiolite complex does not resemble that of lavas dredged from the larger, active marginal basins like the Marianas Trough [Hart et al., 1972] and the Lau-Havre Trough [Hawkins, 1976], and presumed to be representative of their upper oceanic crust. Perhaps smaller basins like those in the New Hebrides will yield lavas more similar to those of the Baie Verte ophiolites; their different compositions might be due to their emplacement in a narrower basin with thicker adjacent lithosphere. However, the occurrence of lavas, very closely similar to those from the Marianas Trough, within the sediments filling the Baie Verte Basin suggests an alternative possibility. If vulcanism occurs in marginal basins not only at the spreading ridge segment axes, forming the upper oceanic crust, but also in a relatively widespread manner in other parts of the basins, then some hitherto puzzling features of active marginal basins might be at least partially explicable. The relatively rough topography within the basins, and the difficulty or impossibility of identifying coherent or symmetrically disposed spreading magnetic anomaly patterns could both be, at least in part, accounted for. It can be pointed out that in a narrow basin with a high rate of volcaniclastic sediment supply, such off-axis volcanics might be separated from the oceanic crustal lavas by a significant, seismically detectable, thickness of sediments, but that in wider basins with a low rate of pelagic sedimentation such volcanics would likely be indistinguishable by such means from the oceanic crustal lavas below them.

It is suggested that the closure of the Baie Verte basin did not involve subduction because of the small width of oceanic crust in it, just as in the case of the Cretaceous marginal basin of Southern Chile [Dalziel et al., 1974].

The occurrence of basalts resembling basaltic komatiites within the ophiolite lavas suggests that many of the narrow greenstone belts in Archean terrains may be the closed remains of marginal basins, rather than cores of island arcs, a suggestion made on this and other grounds by Burke et al. [1976]. The Baie Verte Lineament in many respects closely resembles such Archean greenstone belts, particularly in its steep, pinched structure and its steep penetrative cleavage indicating severe horizontal compression and shortening.

It is not clear whether such a zone of small, en echelon extensional basins as is found in the New Hebrides represent the initial state in the formation of a large marginal basin or whether they represent the rear-arc distentional process occurring at a slower rate than that forming large marginal basins. It could perhaps be suggested that it is confusing to apply the term marginal basin to structures of such smaller size. However, the fundamental process and effects seem to be the same, merely being applied on a different scale. Their possible role in producing narrow ophiolitic lineaments in orogenic belts that may not directly mark ancient subduction zone sites or sutures of major consequence may perhaps be more important than previously realized. Small basins closing without subduction, like the Baie Verte Lineament, appear to preserve more of their contents on the site of closure than are preserved at sites where larger oceanic-floored basins, for example the Bay of Islands basin, or the main Appalachian ocean in Newfoundland, are proposed to have vanished by subduction.

Acknowledgments. I thank E. Nesbit and R. G. W. Kidd for the chemical analyses. Field work during 1969-71 was partly funded by a British NERC Research Studentship. I thank my bank manager for allowing a delay in the repayment of the remainder of those field expenses. Field work during 1974 was supported by a grant from the Council on Research of SUNY Albany.

References

Ballard, R.D., W.B. Bryan, J.R. Heirtzler, G. Keller, J.G. Moore and Tj. VanAndel. Manned submersible observations in the FAMOUS area: Mid-Atlantic ridge. Science, 190, 103-108, 1975.

Bird, J.M., and J.F. Dewey. Lithosphere plate-continental margin tectonics and the evolution of the Appalachian orogen. Geol. Soc. Amer. Bull. 81, 1031-1060, 1970.

Burke, K., J.F. Dewey and W.S.F. Kidd. Dominance of horizontal movements, arc and microcontinental collisions during the later permobile regime, in Early History of the Earth (edited by B.F. Windley) 113-129, Wiley, London 1976.

Bursnall, J.T. and M.J. deWit. Timing and development of the orthotectonic zone in the Appalachian Orogen of Northwest Newfoundland: Can. Jour. Earth Sci., 12, 1712-1722, 1975.

Church, W.R. Metamorphic rocks of the Burlington Peninsula and adjoining areas of Newfoundland, and their bearing on continental drift in the North Atlantic, in North Atlantic-geology and continental drift, edited by M. Kay, 212-233, Amer. Assoc. Petrol. Geologists Mem. 12, 1969.

Church, W.R., and R.K. Stevens. Early Palaeozoic ophiolite complexes of the Newfoundland Appalachians as mantle-oceanic crust sequences. J. Geophys. Res. 76, 1460-1466, 1971.

Dalziel, I.W.D., M.J. deWit, and K.F. Palmer. Fossil marginal basin in the southern Andes. Nature, 250, 291-294, 1974

DeGrace, J.R., B.F. Kean, E. Hsu, and D.M. Besaw. Geology of the Nippers Harbour area, Newfoundland. Newfoundland Dept. of Mines and Energy Preliminary Report 788, 59 pp. 1975.

Dewey, J.F. Evolution of the Appalachian-Caledonian orogen. Nature, 222, 124-129, 1969.

Dewey, J.F., and J.M. Bird. Mountain belts and the new global tectonics. J. Geophys. Res., 75, 2625-2647, 1970.

Dewey, J.F., and J.M. Bird. Origin and emplacement of the ophiolite suite: Appalachian ophiolites in New foundland. J. Geophys. Res. 76, 3179-3206, 1971.

Dewey, J.F., W.C. Pitman III, W.B.F. Ryan, and J. Bonnin. Plate tectonics and the evolution of the Alpine system. Geol. Soc. Amer. Bull. 84, 3137-3180, 1973.

Dewey, J.F., and W.S.F. Kidd. Continental collisions in the Appalachian/Caledonian orogenic belt: variations related to complete and incomplete suturing. Geology, 2, 543-546, 1974.

deWit, M.J. On the origin and deformation of the Fleur de Lys metaconglomerate, Appalachian fold belt, northwest Newfoundland. Can. J. Earth. Sci. 11, 1168-1180, 1974.

Hart, S.R., W.E. Glassley, and D.E. Karig. Basalts and sea floor spreading behind the Mariana island arc. Earth. Planet. Sci. Lett. 15, 12-18, 1972.

Hawkins, J.W. Petrology and geochemistry of basaltic rocks of the Lau Basin. Earth.

Planet. Sci. Lett. 28, 283-297, 1976.
Karig, D.E. Origin and development of marginal basins in the western Pacific. J. Geophys. Res. 76, 2542-2561, 1971.
Karig. D.E., and J. Mammerickx. Tectonic framework of the New Hebrides island arc. Mar. Geol. 12, 187-205, 1972.
Kidd, W.S.F. Evolution of the Baie Verte Lineament, Burlington Peninsula, Newfoundland. Unpublished Ph.D. thesis, University of Cambridge, 1974.
Melson, W.G., and C. Thompson. Petrology of a transform fault zone and adjacent ridge segments: Phil. Trans. Roy. Soc. London, A, 268, 423-441, 1971.
Mitchell, A.H.G. Geology of South Malekula. New Hebrides Condominium Geol. Surv. Rept. 3, 28-34, 1966.
Mitchell, A.H.G., and A.J. Warden. Geological evolution of the New Hebrides island arc. J. Geol. Soc. London, 127, 501-529, 1971.
Neale, E.R.W., and W.A. Nash. Sandy Lake (E½) Newfoundland. Geol. Surv. Canada Paper 62-28, 40pp., 1963.
Neale, E.R.W., and M.J. Kennedy. Relationship of Fleur de Lys group to younger groups of the Burlington Peninsula, Newfoundland. Geol. Assoc. Canada, Spec. Paper 4, 139-169, 1967.
Pearce, J.A., and J.R. Cann. Tectonic setting of basic volcanic rocks determined using trace element analyses. Earth. Planet. Sci. Lett., 19, 290-300, 1973.
Reay, A, J.M. Rooke, R.C. Wallace, and P. Whelan. Lavas from Niuafo'ou Island, Tonga, resemble ocean floor basalts. Geology, 2, 605-606, 1974.
Riley, G.C. Red Indian Lake (W½) Newfoundland. Geol. Surv. Canada Map 8-1957, 1957.
St. Julien, P., C. Hubert, and H. Williams. The Baie Verte-Brompton Line and its possible tectonic significance in the northern Appalachians. Geol. Soc. Amer. Abstracts with Programs 8, 259-260, 1976.
Smith, R.E. Redistribution of major elements in the alteration of some basic lavas during burial metamorphism. J. Petrol., 9, 191-219, 1968.
Stanley, D.J. and R. Unrug. Submarine channel deposits, fluxoturbidites and other indicators of slope and base of slope environments in modern and ancient marine basins, in Recognition of ancient sedimentary environments, edited by J.K. Rigby and W.K. Hamblin, 287-340. Soc. Econ. Palaeontologists and Mineralogists Spec. Publ. 16, 1972.
Stevens, R.K. Cambro-Ordovician flysch sedimentation and tectonics in west Newfoundland and their possible bearing on a Proto-Atlantic ocean. Geol. Assoc. Canada Spec. Paper 7, 165-177, 1970.
Upadhyay, H.D., J.F. Dewey, and E.R.W. Neale. The Betts Cove ophiolite complex, Newfoundland: Appalachian oceanic crust and mantle. Geol. Assoc. Canada Proc. 24, 27-34, 1971.
Warden, A.J. Evolution of Aoba caldera volcano, New Hebrides. Bull. Volc., 34, 107-140, 1970.
Wilson, J.T. Did the Atlantic close and then reopen? Nature, 211, 676-681, 1966.

TECTONIC EVOLUTION OF THE SOUTH FIJI MARGINAL BASIN

A. B. Watts and J. K. Weissel

Lamont-Doherty Geological Observatory of Columbia University, Palisades, New York 10964

F. J. Davey

Department of Scientific and Industrial Research, P.O. Box 8005, Wellington, New Zealand

Abstract. All available magnetic anomaly data have been used to clarify the magnetic lineation patterns in the South Fiji basin. The new data show that a RRR triple junction was active during the mid-Cenozoic and that oceanic crust was generated from the three spreading centers at a half rate of about 3 cm/yr. The three plate system began at about anomaly 12 time (∿ 35 m.y. B.P.) and spreading at the three ridges ceased just after anomaly 7A time (∿ 28 m.y. B.P.). At present only the northeast spreading center is well preserved as a basement ridge which coincides with the previously mapped Bounty "channel". The southwest plate of the system appears to be missing probably due to subduction at an island arc presently preserved between New Caledonia and Northland, New Zealand. These results show that the South Fiji basin has a history more complicated than the simple model of rifting of an island arc and subsequent generation of new crust between the rifted fragments.

Introduction

The South Fiji basin is one of a series of basins underlain by oceanic crust (Shor et al., 1971) which occur behind island arc-trench systems in the southwest Pacific. Karig (1970) described the South Fiji basin as an inactive marginal basin formed by crustal extension behind the Tonga-Kermadec arc during the Early to Middle Tertiary. The Three Kings rise and Loyalty islands ridge were interpreted by Karig (1970) as remnant arcs, isolated by the rifting process. JOIDES sites 205 and 285 (Scientific Staff, 1972, 1973) indicate that the minimum age of the basin is Middle Oligocene to Early Miocene, in general agreement with the age suggested by Karig.

As in the world's major ocean basins, magnetic lineations provide the fundamental tool with which to study the evolution of marginal basins. Foreman (1973) examined magnetic anomaly profiles obtained during the JOIDES site survey by R/V Kana Keoki and concluded that N-S trending magnetic lineations exist in the east-central part of the basin. However, he was not able to correlate the lineations with the geomagnetic time scale. Weissel and Watts (1975) using more data than was available to Foreman (1973) confirmed the existence of the N-S lineations and identified them as anomalies 8 through 12 (∿ 29 to 35 m.y. B.P.), in general agreement with the age of the basin inferred from JOIDES sites. Weissel and Watts (1975) also tentatively mapped E-W trending lineations in the western part of the basin but were unable to identify them or relate them to the N-S trending lineations.

The purpose of this paper is to present a new compilation of magnetic anomaly data which has increased our understanding of the tectonic evolution of the South Fiji basin. We have used these data to determine the details of the magnetic lineation patterns which exist in the basin.

Magnetic Anomaly Data

This study utilizes total magnetic field intensity data collected mainly on cruises of New Zealand and U.S. research vessels between 1963 and 1975. The principal data sources are Lamont-Doherty Geological Observatory (R/V Robert D. Conrad cruises 9, 12; R/V Vema cruises 28, 32; U.S.N.S. Eltanin cruises 29, 31, 40), Scripps Institution of Oceanography (R/V Argo and R/V Horizon, NOVA Expedition; Chase et al., 1972), Hawaii Institute of Geophysics (R/V Kana Keoki cruises 3, 6; Foreman, 1973; Malahoff, personal communication), Department of Scientific and Industrial Research, Wellington (R.N.Z.F.A. Tui; Hochstein and Reilly, 1967) and the Deep Sea Drilling Project (R/V Glomar Challenger cruises 21, 30; Smith, personal communication). The total magnetic field data are presented as magnetic anomalies along ships' tracks (Fig. 1) and as selected projected profiles across the basin (Figs. 2 to 7).

Three well-lineated magnetic patterns occur in the basin north of latitude 27°S: N-S trending lineations in the east-central part of the basin; NE-SW trending lineations in the northeast part

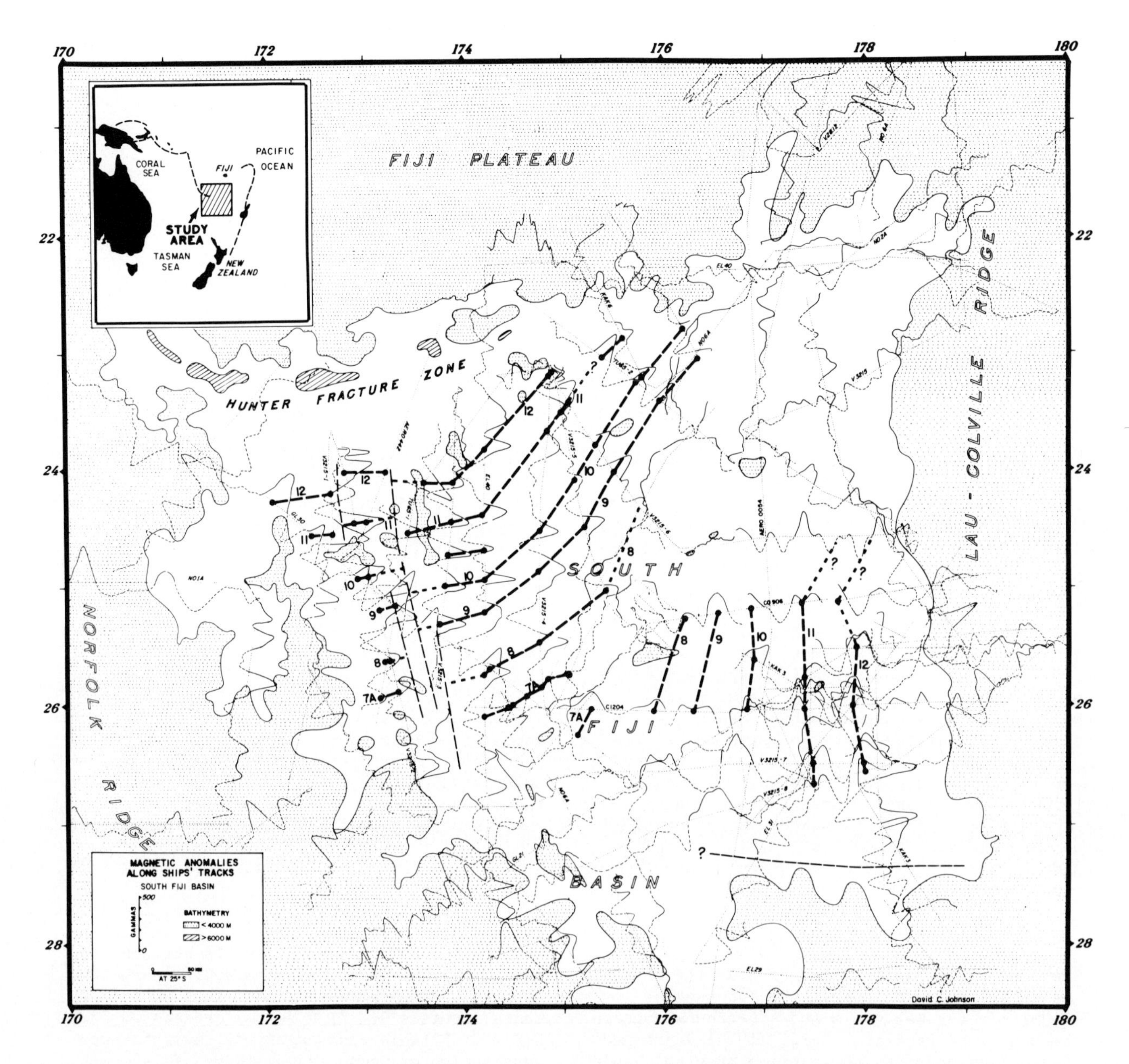

Fig. 1. Magnetic anomalies along ships' tracks in the South Fiji basin. The sources of the data are discussed in the text. Heavy dots indicate where correlations have been made on ships' tracks and aeromagnetic profiles. We have not shown magnetics data for the aeromagnetic profiles because of their poor navigation control. Thin dashed lines indicate inferred fracture zones.

of the basin; E-W trending lineations in the western part of the basin (Fig. 1). We have not been able to map and recognize magnetic lineations in the extreme northeast or in the south part of the basin between latitude 27°S and Northland, New Zealand.

We suggest that the three lineation patterns comprise a sequence of magnetic anomalies 7A through 12 (∿ 28 to 35 m.y. B.P.). In the N-S pattern anomaly 12 occurs near the base of the Lau-Colville ridge and anomaly 7A near the center of the basin. The main difference with the earlier study of Weissel and Watts (1975) is that anomaly 7A has now been identified on profiles C1204 and GL30 and anomalies 8 and 9 have been re-identified on profile C0906. We have also been able to extend the N-S lineations 70 km further south to about latitude 26.6°S. In the NE-SW pattern anomaly 12 occurs south of the Hunter Fracture Zone and anomaly 8 near the cen-

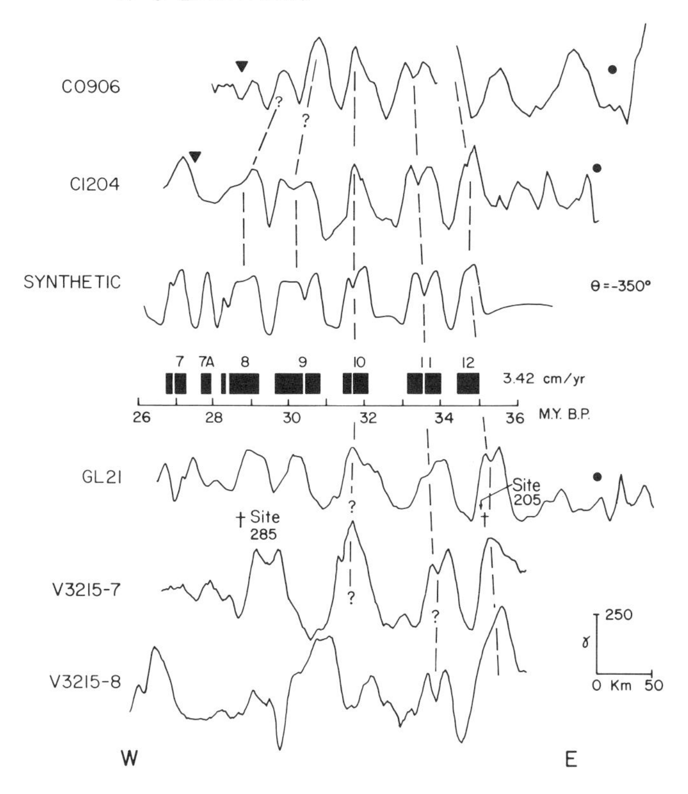

Fig. 2. Selected observed magnetic anomaly profiles across the N-S lineations compared to a computed profile based on the geomagnetic time scale (Heirtzler et al., 1968) and model parameters summarized in Table 1. The observed profiles have been projected at azimuths of 82° to 90°. The filled triangles indicate the location of the Bounty ridge (Fig. 7) and the filled circles the base of the Lau-Colville ridge. The position of JOIDES sites 205 (Scientific Staff, 1972) and 285 (Scientific Staff, 1973) are shown by daggers.

ter of the basin. This pattern cannot be extended into the extreme northeast part of the basin using the present data set. In the E-W pattern anomaly 12 occurs south of the Hunter Fracture Zone and anomaly 7A near the northern end of the Three Kings rise. This pattern is disrupted by N-S trending fracture zones (Fig. 1). Due mainly to the paucity of data the E-W lineations cannot be extended to the west of about longitude 172°E.

Ages inferred from the magnetic lineations suggest the basin formed during the Oligocene. Site 205 (Scientific Staff, 1972) was located on an anomaly which we identify as anomaly 12. The oldest sediments recovered from this site are of late Middle Oligocene age (∿ 29 to 30 m.y. B.P.; Packham, personal communication) in general agreement with the inferred age of anomaly 12 (∿ 35 m.y. B.P.). Site 285 (Scientific Staff, 1973) is located on the inferred extension of anomaly 8 and although basement was not reached at this site the drilling results are not inconsistent with the magnetics data.

To examine the validity of the magnetic anomaly identifications (Fig. 1) synthetic profiles based on the geomagnetic time scale were computed and compared to selected profiles across each lineation pattern (Figs. 2, 3, 4; Table 1). The agreement between observed profiles across the N-S lineations and the computed profile is good for anomalies 11 and 12 but rather poor for anomalies 7 through 9 (Fig. 2). Anomaly 10 agrees more closely with observed profiles from the world's major ocean basins than with the block model (Weissel and Watts, 1975). The match between observed profiles across the NE-SW lineations and the computed profile is particularly good for anomalies 9 through 12 (Fig. 3). Although anomaly 12 cannot be identified on profile TU65-2 it can be recognized on profiles V3215-5 and V3215-6. The agreement between observed profiles across the E-W lineations and the computed profile is reasonable considering the disrupted nature of the lineation pattern (Fig. 4). Magnetic effects of shallow basement morphology obscure the effects of geomagnetic reversals and regions along the observed profiles

NE-SW LINEATIONS

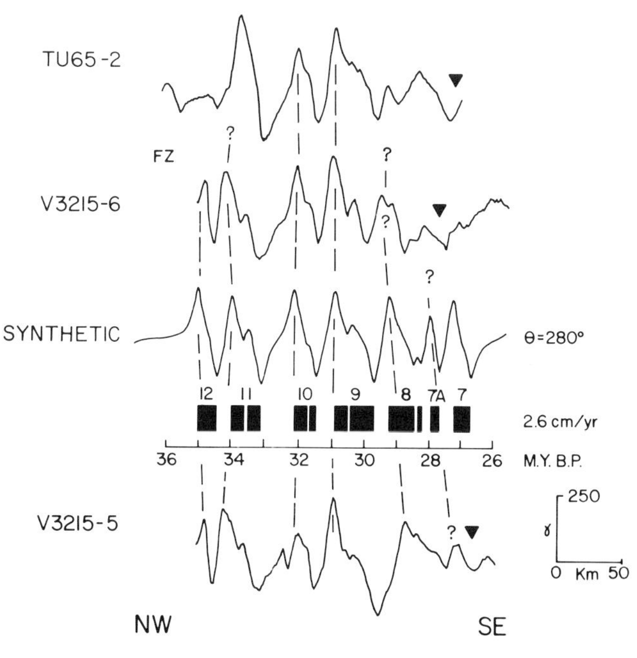

Fig. 3. Selected observed magnetic anomaly profiles across the NE-SW lineations compared to a computed profile. The observed profiles have been projected at an azimuth of 123°. The filled triangle indicates the location of the Bounty ridge (Fig. 7).

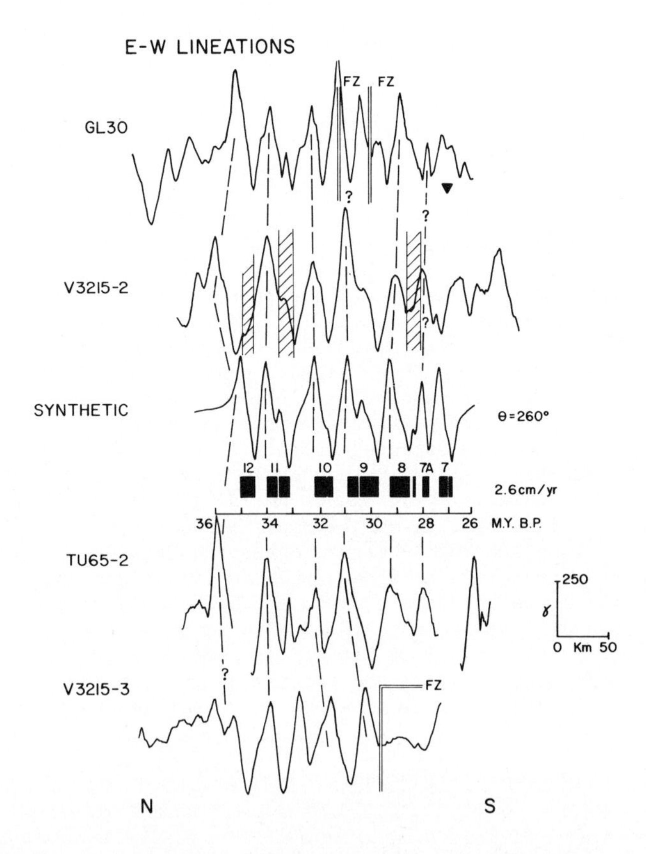

Fig. 4. Selected observed magnetic anomaly profiles across the E-W lineations compared to a computed profile. The observed profiles have been projected at an azimuth of 178°. The filled triangle on the GL30 profile indicates the location of the Bounty ridge (Fig. 8). Shaded areas indicate portions of the profile along which seismic reflection profiles indicate basement morphology changes by more than 1 km of relief. The position of inferred fracture zones (Fig. 1) are shown by double lines.

where this may occur are shown by shading in Figure 4. A higher value of the remanent magnetization is required to match amplitudes of anomalies in the E-W pattern than is required for the other patterns (Table 1).

The synthetic profiles (Figs. 2, 3, 4) are based on an average value of the skewness parameter Θ (Schouten and McCamy, 1972) determined from the shapes of observed anomalies within each lineation pattern. Phase shifting of these anomalies, however, establishes a scatter about each average Θ value of about ± 15°. Deskewed anomalies which illustrate this scatter are shown for a profile from the NE-SW lineation pattern in Figure 5.

We believe that the model studies (Figs. 2, 3, 4) support our suggestion that anomalies 7A through 12 occur in the basin. The inferred crustal ages based on these identifications are in agreement with ages inferred from JOIDES sites 205 and 285 (Scientific Staff, 1972, 1973) as well as with ages inferred from the global empirical age versus depth curve of Sclater et al. (1971).

Tectonic Fabric

Basement morphology provides a complementary way to study the tectonic fabric of the basin. As a starting point we have constructed a new bathymetric map of the South Fiji basin (Fig. 6). This map is based on earlier compilations of Mammerickx et al. (1971) and Packham and Terrill (1975) along with the recent EL40 and V3215 data. The main difference between this map and previous maps is in the detail of the topographic fabric of the deep parts of the basin. The deep smooth topography in the northeast part of the basin corresponds to a uniform cover of sediments (average thickness of about 0.5 km) derived mainly from the Lau-Colville ridge. However, the sedi-

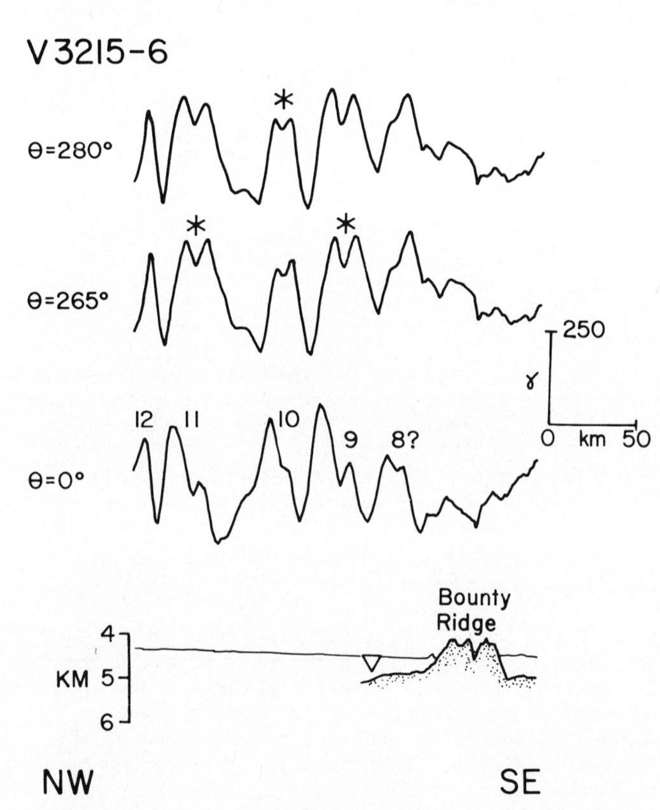

Fig. 5. V3515-6 magnetic anomaly profile (Figs. 1 and 3) phase-shifted by 280° and 265° (top two profiles). Asterisks denote the prominent anomalies which have been brought into symmetry. Anomalies 11 and 9 require values of Θ = 265° while anomaly 10 requires Θ = 280°. The bottom profile is a line drawing of the corresponding V3215-6 seismic reflection profile. The peak of the unfilled triangle indicates expected basement depth based on the global empirical depth versus age curve of Sclater et al. (1971).

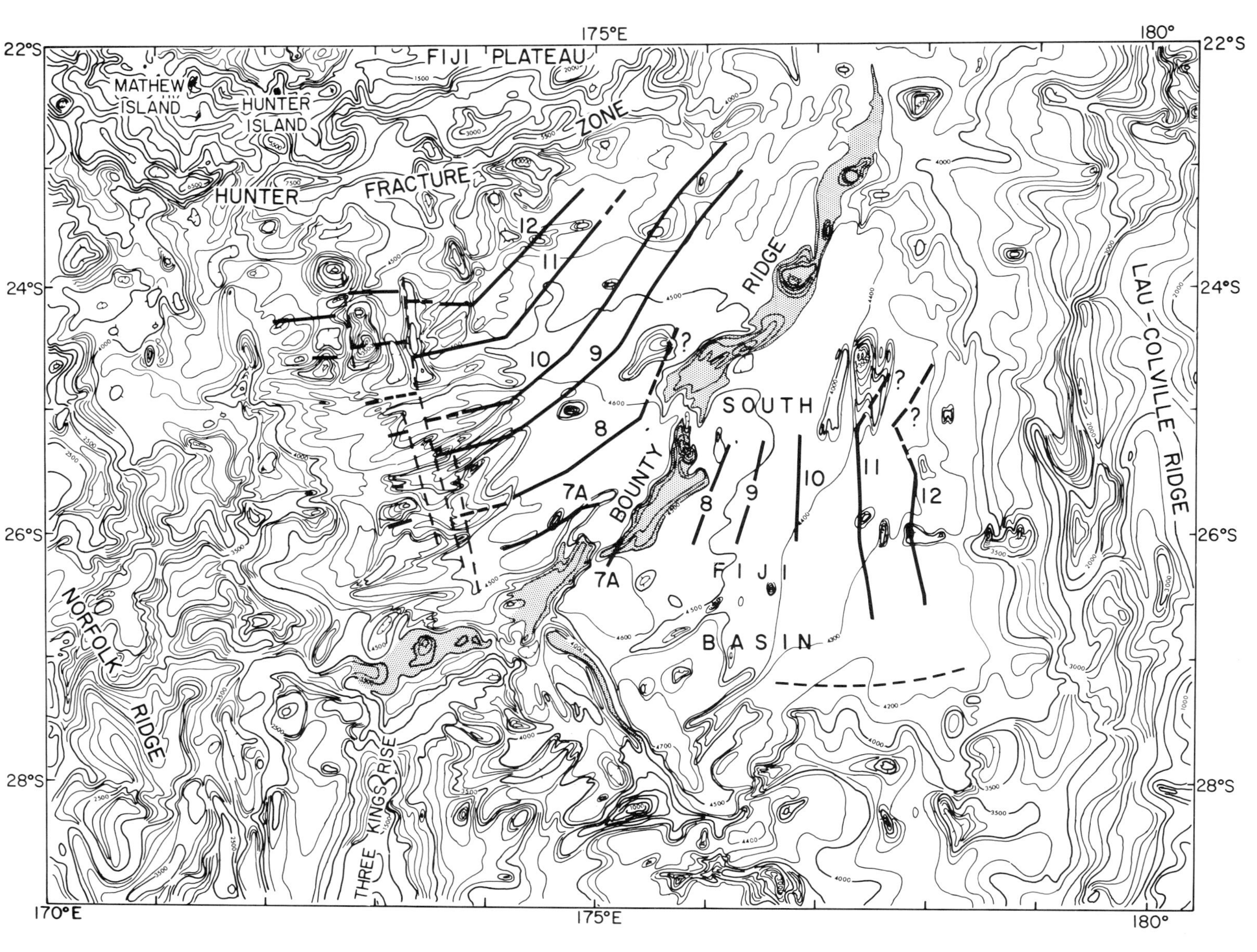

Fig. 6. Bathymetric map of South Fiji basin contoured at 100 m intervals. Data sources for the map are discussed in the text. Magnetic lineations and identifications are based on Figure 1. The crest of the Bounty ridge and other ridges which may be part of the inferred three plate system are shown by stippling.

TABLE 1.
Magnetic Reversal Block Model Parameters

	Lineations		
	N-S	NE-SW	E-W
Azimuth	82°	123°	178°
Remanent magnetization (emu)	0.005	0.005	0.007
Inc. of present field	-53°	-46°	-48°
Dec. of present field	14.7°	12.7°	12.5°
Average Θ	-350°	280°	260°
Half spreading rate (cm/yr)	3.42	2.60	2.60
Layer top (km)	4.9	4.9	4.5
Layer bottom (km)	5.4	5.4	5.0

ment cover does not obscure a NE-SW trending basement ridge which is stippled in Figure 6. This ridge corresponds closely in position to the previously mapped Bounty "channel" (Mammerickx et al., 1971) and we refer to this feature as the Bounty ridge (Fig. 6). The magnetic lineation patterns show that the Bounty ridge coincides with the region of youngest crust. We therefore suggest that the Bounty ridge was the spreading center which generated the northeast part of the basin between anomaly 12 and 7A time.

Three magnetic anomaly profiles and line drawings of seismic reflection profiles which cross the Bounty Ridge are shown in Figure 7. These profiles show that magnetic anomalies are repeated about the Bounty ridge and the basement morphology of the ridge is consistent with it being an extinct spreading center. Details of the basement morphology and its relationship to the magnetic anomalies are shown in Figure 8.

The sediment cover in the western part of the basin is thinner than elsewhere and basement morphology shows a fabric which is similar to the magnetic lineation pattern. The basement morphology shows E-W trends which are consistent with ridge flank morphology and N-S trends which reflect prominent fracture zone ridges and troughs. Both the magnetic and morphologic fabrics indicate that the crust in the western part of the basin was generated at an E-W spreading center. The crest of a possible spreading center is stippled in Figure 6 and can be seen as a basement ridge near the northern end of the Three Kings rise (see also Weissel and Watts, 1975, Fig. 4).

Another ridge is mapped (Fig. 6) trending SSE from the junction of the Bounty ridge and the suggested E-W ridge. Prominent ridges which are orthogonal to this trend are observed east of this ridge and may represent fracture zones associated with a third spreading system.

Discussion

Based on our study of magnetic anomaly, seismic reflection and topographic data we suggest that oceanic crust in the South Fiji basin was generated in the Oligocene from three spreading centers which met at a ridge-ridge-ridge (RRR) triple junction. The evidence for the inferred three plate system is derived from the ages amd geometry of magnetic lineation patterns and morphologic evidence for three extinct spreading centers. The inferred rates and directions of spreading from the three spreading centers are not incon-

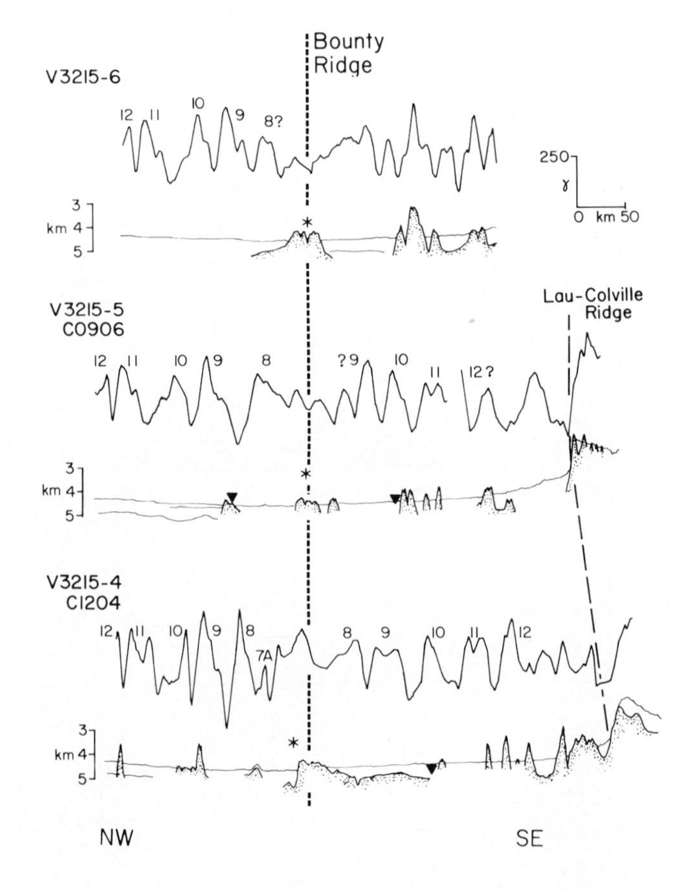

Fig. 7. Selected magnetic anomaly profiles and line drawings of seismic reflection profiles across the Bounty ridge. The profiles have been projected at an azimuth of 123°, normal to the trend of the ridge. Age identification of the individual magnetic anomalies is based on Figures 1, 2, 3 and 4. The asterisk denotes the inferred center of spreading. The peak of the solid triangles indicates expected basement depth based on the empirical depth versus age curve.

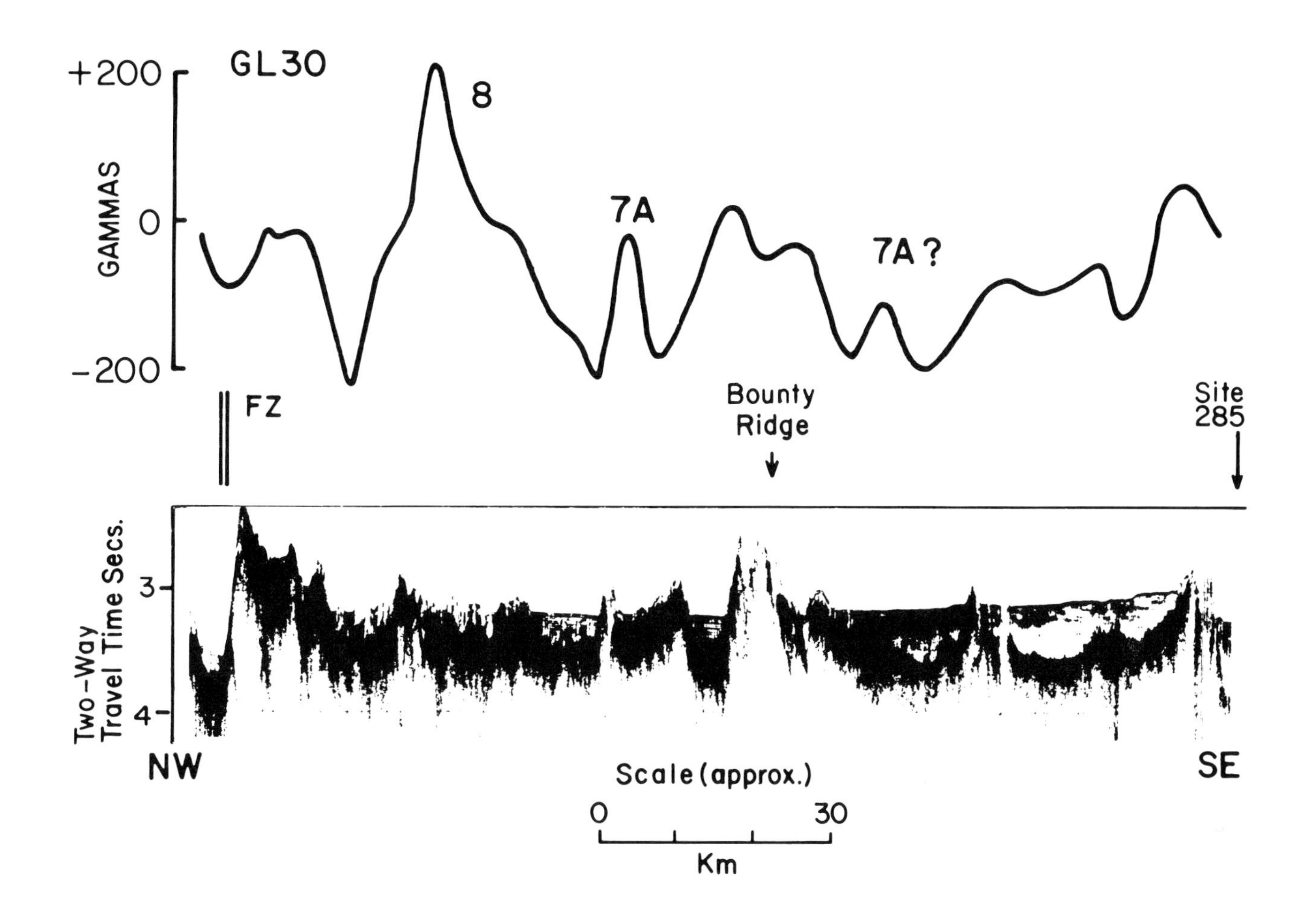

Fig. 8. Seismic reflection profile obtained by R/V Glomar Challenger (Leg 30) across the Bounty ridge (Packham and Andrews, 1975). Identification of individual magnetic anomalies is based on Figure 1.

sistent with plate tectonic models of a triple junction (C. G. Chase, personal communication). We recognize, however, that the Bounty ridge is the only well documented of these spreading centers. The extinct triple junction is located at about latitude 26.7°S, longitude 174.5°E, northeast of the northern end of the Three Kings rise (Fig. 6).

The southwest plate of the inferred three plate system appears to be mostly missing, probable due to subduction at an island arc presently represented by the complicated series of en echelon ridges and troughs between New Caledonia and Northland, New Zealand. Brothers and Blake (1973) have summarized gèological evidence for Late Eocene and Oligocene subduction-related tectonics in New Caledonia. Geological mapping in Northland led Brothers (1974) to suggest a Late Oligocene northeast dipping subduction zone which reversed polarity in the Early Miocene. Thus Late Oligocene through Miocene subduction zones in the complicated region between New Caledonia and Northland could have consumed part of the newly formed South Fiji basin. A representative seismic reflection profile shows that the ridges and troughs of this complicated region closely resemble island arc morphology (Fig. 9).

The presence of an extinct RRR triple junction suggests that the tectonic evolution of the basin is more complicated than simple two-limb spreading between an active and remnant island arc. There is no obvious remnant arc that was once contiguous with the Lau-Colville ridge (Fig. 10) when subduction was active along its eastern margin.

Lamont-Doherty Geological Observatory
Contribution No. 2412

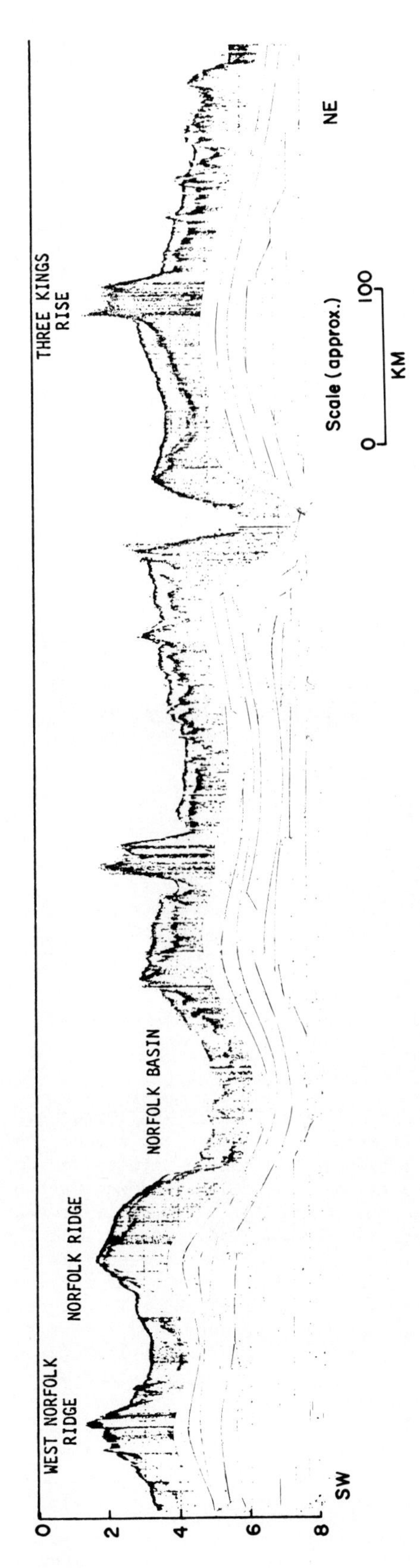

Fig. 9. Representative seismic reflection profile obtained on R/V Glomar Challenger (Leg 21) across the series of ridges and troughs between the Three Kings rise and West Norfolk ridge (Burns and Andrews, 1973; Fig. 10).

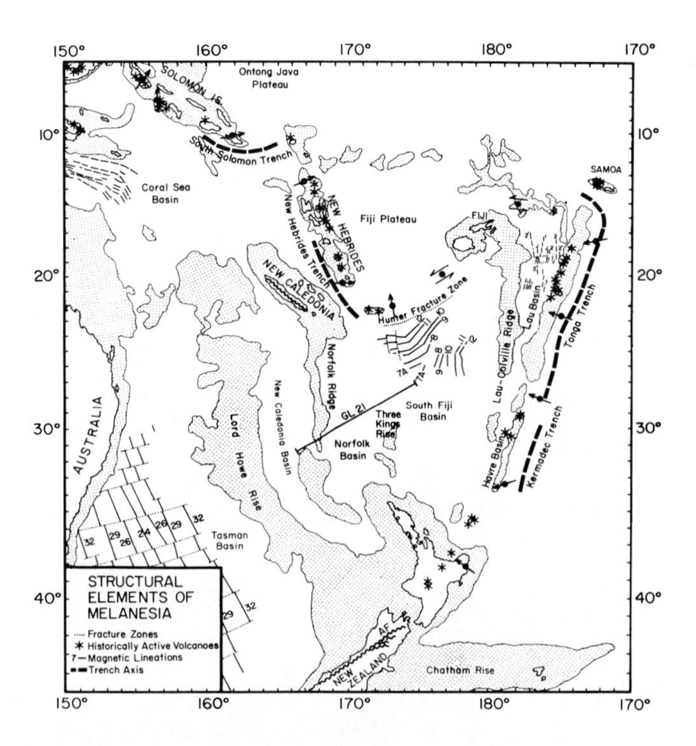

Fig. 10. Structural elements of Melanesia. Solid circles denote shallow earthquakes with thrust-fault or strike-slip focal mechanisms which are interpreted as representing motion between two plates (Johnson and Molnar, 1972). The inferred direction of relative motion is indicated by solid arrows. Magnetic lineations are based on Falvey (1972) in the Coral Sea basin, Hayes and Ringis (1973) in the Tasman basin, Weissel (this volume) in the Lau basin and Figure 1 in the South Fiji basin. AP = Alpine Fault.

Acknowledgments. We are grateful for the help of Captain H. C. Kohler and the officers, crew and scientists on board R/V Vema cruise 32, leg 15, during which most of the data used in this study was obtained. This research was supported by National Science Foundation grants DES 71-00214 and OPP 74-02238 and Office of Naval Research contract N00014-75-C-0210.

References

Brothers, R. N., Kaikoura Orogeny in Northland, New Zealand. N. Z. Jl. Geol. Geophys., 17 (1): 1-18, 1974.

Brothers, R. N. and M. C. Blake, Jr., Tertiary plate tectonics and high-pressure metamorphism in New Caledonia. Tectonophysics, 17, 337-358, 1973.

Burns, R. E. and J. E. Andrews, Regional aspects of deep sea drilling in the Southwest Pacific. In: Burns, R. E., Andrews, J. E., et al., Initial Reports of the Deep Sea Drilling Project, v. 21, Washington, D. C., 897-906, 1973.

Chase, T. E., S. M. Smith, D. A. Newhouse, W. L. Crocker, M. Schoenbechler, L. Hydock, and U. Ritter, Track charts of SIO Bathymetric Data and track charts of SIO Magnetic Data in the Pacific Ocean, IMR TR-25 Sea Grant Publication No. 7, Scripps Inst. of Ocean-

Falvey, D. A., On the origin of marginal plateaux, Ph.D. thesis, University of New South Wales, Australia, 1972.

Foreman, J. A., A structural and tectonic study of the Lau-Havre - South Fiji Basin Region, M.Sc. thesis, Univ. of Hawaii, 1973.

Hayes, D. E. and J. Ringis, Seafloor spreading in the Tasman Sea, Nature, Lond., v. 243, p. 454-458, 1973.

Heirtzler, G. O., G. O. Dickson, E. M. Herron, W. C. Pitman, and X. Le Pichon, Marine magnetic anomalies, geomagnetic field reversals and motions of the ocean floor and continents, J. Geophys. Res. 73, 2119-2136, 1968.

Hochstein, M. P. and W. I. Reilly, Magnetic measurements in the south-west Pacific Ocean, N.Z. Jl. Geol. Geophysics., 10, 1527-1562, 1967.

Johnson, T. and P. Molnar, Focal mechanisms and plate tectonics of the southwest Pacific, Jour. Geophys. Res., v. 77, p. 5000-5032, 1972.

Karig, D. E., Ridges and basins of the Tonga-Kermadec Island Arc. System, Jour Geophys. Res., 75, 239-254, 1970.

Mammerickx, J., T. E. Chase, S. M. Smith, and I. L. Taylor, Bathymetry of the South Pacific, (Chart No. 11 and 12): Scripps Inst. of Oceanography, 1971.

Packham, G. H. and J. E. Andrews, Results of Leg 30 and the geologic history of the southwest Pacific arc and marginal sea complex, In: Andrews, J. E., G. H. Packham et al., Initial Reports of Deep Sea Drilling Project, v. 30, Washington, D. C., 691-705, 1975.

Packham, G. H. and Terrill, A., Submarine geology of the South Fiji Basin, In: Andrews, J. E., G. H. Packham et al., Initial Reports of the Deep Sea Drilling Project, v. 30, Washington, D. C., 617-633, 1975.

Schouten, H. and K. McCamy, Filtering marine magnetic anomalies, Jour. Geophys. Res., 77, 7089-7099, 1972.

Scientific Staff, Deep Sea Drilling Project, Leg 21, Geotimes 14, 14-17, 1972.

Scientific Staff, Deep Sea Drilling Project, Leg 30, Geotimes 18, 19-21, 1973.

Sclater, J. G., R. N. Anderson and M. L. Bell, The elevation of ridges and the evolution of the eastern Central Pacific, Jour. Geophys. Res., 76, 7888-7915, 1971.

Shor, G. G., H. K. Kirk and H. W. Menard, Crusta structure of the Melanesian area, Jour. Geophys. Res., 76, 2562-2586, 1971.

Weissel J. K. and A. B. Watts, Tectonic complexities in the South Fiji marginal basin, Earth and Planet Sci. Lett., 28, 121-126, 1975.

Weissel, J. K., Evolution of the Lau basin by the growth of small plates, Ewing Symposium Volume, AGU Monograph Series, 1976.

EVOLUTION OF THE LAU BASIN BY THE GROWTH OF SMALL PLATES

Jeffrey K. Weissel

Lamont-Doherty Geological Observatory of Columbia University, Palisades, New York 10964

Abstract. Zones of shallow seismicity indicate that plate boundaries are presently active within the Lau basin. Some of the seismicity is associated with a previously recognized strike-slip feature, the Peggy ridge. North-south trending magnetic lineations comprising anomalies 1 through 2' (0 to 3.5 m.y.B.P.) are mapped over much of the basin. The central anomalies are associated with morphologic axial rifts and with a north-south belt of epicenters extending from the southeast end of the Peggy ridge to about 20°S. The lineation pattern is bounded obliquely on the north by the Peggy ridge and a postulated second strike-slip feature (the Roger fracture zone) which trends northeast from the southeast end of the Peggy ridge. Within a plate tectonics framework, the two strike-slip boundaries and the spreading ridge define a ridge-fault-fault (RFF) triple junction which has probably been active for the last 3.5 m.y. During this time the Lau basin evolved by the interaction of a number of small rigid plates instead of by the usual concept for marginal basin genesis of crustal accretion at a simple two-limb system.

Introduction

The Lau basin is a small marginal basin located behind a major convergent plate boundary of the southwest Pacific, the Tonga island arc-trench system (Figure 1). On the basis of shallow (2-3 km) and rough acoustic basement, high but variable heat flow, thin sediment cover, and geology of nearby islands, Karig [1970, 1971] suggested that the Lau basin has formed by crustal extension processes since the Late Miocene. He proposed that prior to 5 m.y.B.P. the Lau ridge and the Tonga ridge were contiguous and constituted an active island arc. A young age has been supported by JOIDES site 203 in the southern part of the basin (Figure 1) which penetrated Middle Pliocene sediments without reaching igneous basement [Burns and Andrews, 1973]. Shallow seismicity patterns within the basin suggest that some forms of plate interaction are occurring at the present time [Figure 1, Barazangi and Isacks, 1971]. Petrologic studies of dredged rocks have shown that the Lau basin is floored by typical oceanic tholeiites [Hawkins, 1974, 1976, this volume]. The freshness of the basaltic samples further attests to the youth of the basin.

The major problem encountered hitherto in deciphering the detailed tectonic history of the Lau basin is that magnetic lineations which reflect crustal accretion and seafloor spreading processes have proved difficult to map. Previous studies of marine magnetics data concluded that, although the correlatability of magnetic anomalies is poor, limited magnetic evidence exists for slow spreading (<2 cm/yr half-rate) from short ridge segments aligned N 40° E perpendicular to the trend of the Peggy ridge [Sclater et al., 1972; Lawver et al., in press]. Lawver et al. [in press] also suggested that a north-south magnetic lineation fabric may be present over other restricted parts of the basin.

This paper reconsiders the available magnetic anomaly data, and north-south seafloor spreading magnetic lineations comprising anomalies 1 through 2' (0 to 3.5 m.y.B.P.) are mapped south of the Peggy ridge. The geometry of the magnetic lineation pattern and other geologic and geophysical information suggest a model for the evolution of the Lau basin which may be compatible with plate tectonic concepts.

Magnetic Anomaly Data

Figure 2 is a compilation of most available marine magnetics data plotted as magnetic anomalies along ships' tracks. Data have been collected by Lamont-Doherty Geological Observatory, Department of Scientific and Industrial Research, New Zealand [Hochstein and Reilly, 1967], Scripps Institution of Oceanography [Sclater et al., 1972; Chase et al., 1972; Lawver et al., in press] and the University of Hawaii [Foreman, 1973]. Selected magnetics and morphologic profiles are shown in Figure 3 and two magnetics profiles are compared to a standard magnetic block model in Figure 4. All profiles are located on Figure 2.

The magnetic lineation pattern in the Lau basin generally trends north-south and shows several distinctive features. Seafloor spreading-type magnetic lineations comprising anomalies 1 through 2' (0 to 3.5 m.y.B.P.) are recognized on

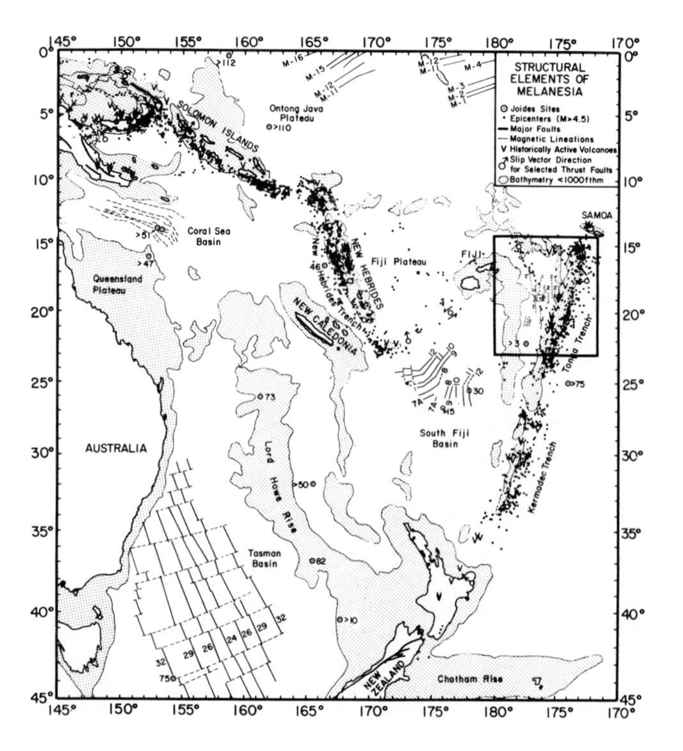

Figure 1. Structural elements map of Melanesia showing the study area (boxed).

both sides of a diffuse band of seismicity that extends south of the Peggy ridge along about 176°W to about 20°S. No magnetic anomalies have been identified in the northern part of the basin between the Peggy ridge and the boundary between the Lau basin and the Pacific plate. The mapped lineations are almost parallel to the trend of the Lau ridge which is the remnant arc isolated from the active Tonga arc by the growth of the Lau basin. Except for the east flank J (Jaramillo) anomaly lineation between the two fracture zones, the lineations generally trend obliquely to the volcanic axis of the active Tonga arc (Figures 1 and 2). Magnetic 'noise' generated by areas of rough and shallow basement (e.g., Figure 3, profile 2) and apparent ridge crest jumps during the evolution of the basin have complicated the normal seafloor spreading pattern of magnetic lineations. Evidence for a ridge jump is seen between the Peggy ridge and the first fracture zone at about 18.7°S where anomaly 2' occurs immediately west of the central anomaly (Figures 2; 3, profiles 1 and 2) and anomaly 2 is repeated on the east flank (Figure 2).

The most lineated and well-mapped anomaly is the west flank anomaly 2' lineation which corresponds to the geomagnetic reversals of the Gauss normal epoch. This lineation can be followed along 177°W south from the Peggy ridge to about 18.7°S, a distance of about 200 km. The width of anomaly 2' decreases southward (Figures 2 and 3) and the lineation is obliquely truncated on the north by the seismically active Peggy ridge and on the south by an inferred fracture zone. On the east flank, however, anomaly 2' is difficult to recognize except between 18° and 18.5°S (Figure 2).

Proposed central anomalies generally coincide with the belt of seismic activity trending N-S along 176°W. The central anomalies are also generally associated with morphologic 'troughs' about 15 km wide and up to 1 km deep which may be expressions of axial rifts (Figures 3 and 4). On some morphologic profiles across the basin (Figure 3, profiles 1 and 4) basement depths increase with distance from the inferred ridge axis as expected from simple cooling models for oceanic lithosphere [Sclater and Francheteau, 1970]. From an overall consideration of crustal depths and ages, the Lau basin appears to be a few hundred meters shallower than the global empirical depth-age curve of Sclater et al. [1971].

Anomaly 2 lineation on the east flank between 17° and 19°S shows a southward narrowing similar to the west flank 2' and the lineation is slightly disrupted along its length. Presumably because of westward ridge jumps north of 18.7°S, the anomaly 2s that were originally part of the west flank in this part of the basin now lie east of the central anomalies (Figure 2). South of 18.7°S, anomaly 2 on the west flank can be followed south across at least one fracture zone to about 20°S. Some of these west flank anomaly 2s have unusually large amplitudes (Figures 3 and 4) and a second positive anomaly not consistently seen over oceanic crust elsewhere sometimes occurs west of anomaly 2 (Figure 4).

The only part of the basin where a repeated anomaly pattern can be reliably discerned is between the two inferred fracture zones of Figure 2. Two magnetic profiles from this area are compared to a constant spreading rate block model in Figure 4. The constant half rate of 3.8 cm/yr was determined from the distance between anomaly 2 either side of the inferred axis. The observed broad anomaly shapes (Figures 3 and 4) are consistent with this high spreading rate. Even though the observed anomalies correlate with the synthetics one for one, a constant spreading rate does not satisfy anomalies younger than anomaly 2. Either asymmetric spreading with more crust accreted on the west or eastward jumping of the ridge crest has occurred in this part of the basin during the past 1.8 m.y. A reasonable skewness match is achieved for anomalies J to 2' on the east and 2 to 2' on the west supporting the general north-south strike of the lineation pattern.

In the area of the profiles shown in Figure 4, the ridge axis occurs east of the middle of the basin and only 170 km from the volcanic axis of the Tonga island arc. This spreading geometry implies that only crust of anomaly 2' and younger ages could have been generated from a single ridge in a continuous, approximately symmetric

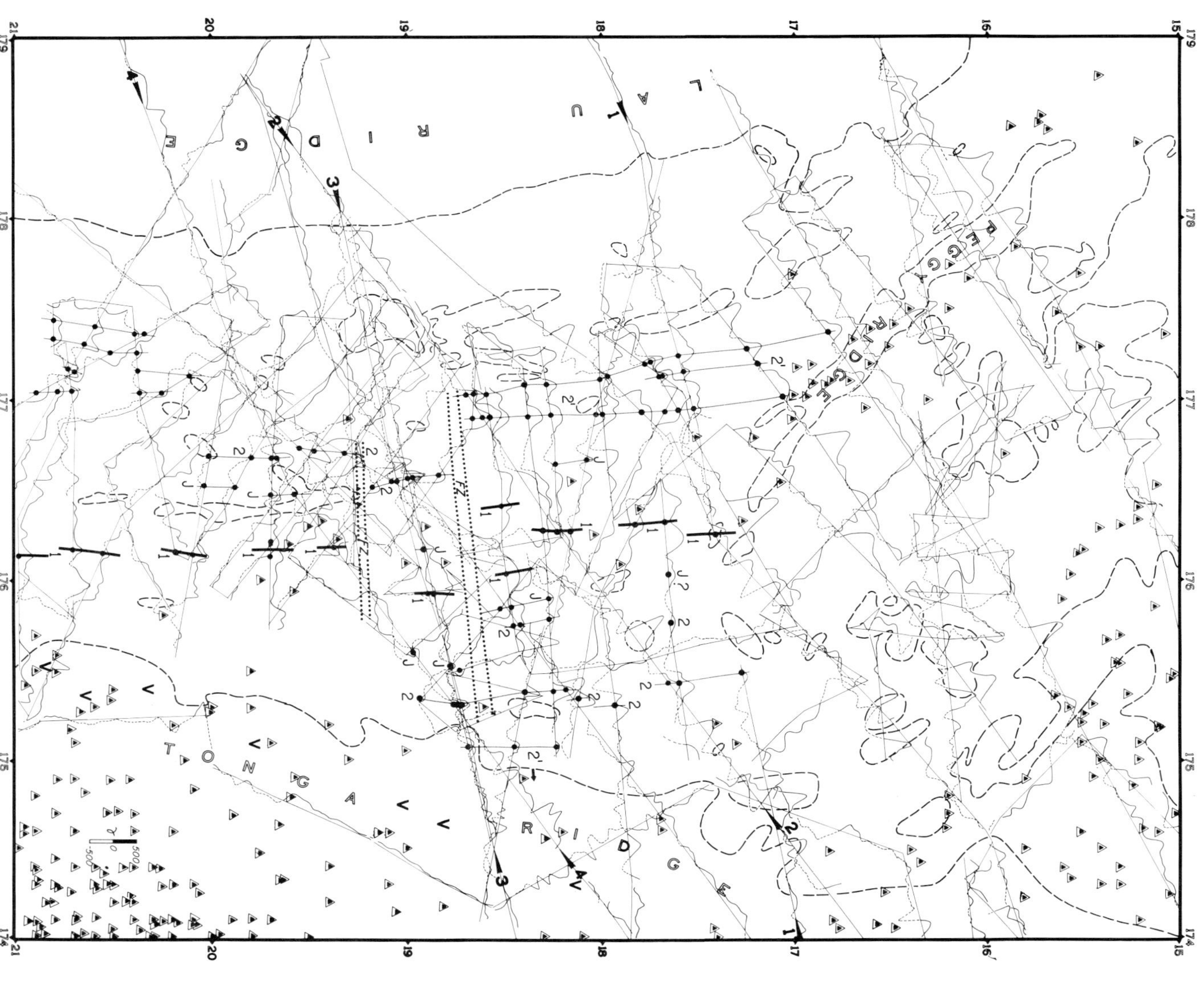

Figure 2. Magnetic anomalies (light solid and broken lines) plotted along ships' tracks, and mapped lineations (light dotted lines) in the Lau basin. Solid dots indicate the locations of identified magnetic anomalies on each track. Enclosed triangles indicate shallow earthquakes whose locations are determined by 10 or more stations. Large Vs represent volcanic centers on the Tonga island arc. 2000 m isobaths are shown by heavy broken lines.

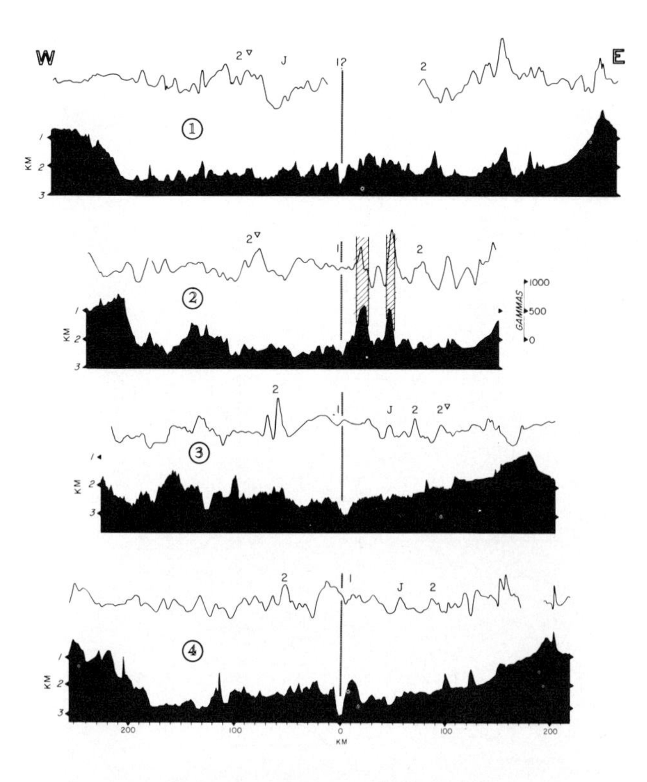

Figure 3. Magnetic anomaly and morphologic profiles across the Lau basin. These profiles are projected on to east-west azimuths. See Figure 2 for locations.

manner. A subsidiary basin structure which is developed along the western margin of the basin (Figure 3; Figure 4 between 170 and 230 km west of the ridge crest) may have been the location of crustal accretion between the time of initial separation of the Lau and Tonga ridges at 5 to 6 m.y.B.P. [Gill, personal communication] and anomaly 2' time (3.5 m.y.B.P.). The subsidiary basin is up to 80 km wide and is bounded on the east by discontinuous ridges which are shown as bathymetric highs in Figure 2.

Proposed Tectonic History

Karig [1970, 1971] proposed that the Tonga island arc and the Lau ridge began to separate in the Late Miocene and the Lau basin has been growing since then. Radiometric dates from volcanic associations on the Lau islands indicate that the Lau ridge became isolated from the sources of andesitic volcanism by 5 to 6 m.y.B.P. [Gill and Gorton, 1973; Gill, personal communication]. Geochemical evidence thus constrains the age of the Lau basin to less than about 6 m.y. In Figures 5, 6, and 8, the development of the Lau basin is viewed at two times in the past, 4 and 1.8 m.y.B.P., and at the present time. As mentioned above, crustal accretion may have first occurred in a subsidiary basin presently along the western margin of the Lau basin. The first phase of extension took place between the time of cessation of andesitic volcanism on the Lau ridge (5 to 6 m.y.B.P.) and the start of the Gauss epoch (3.5 m.y.B.P.). The situation at 4 m.y.B.P. is shown schematically in Figure 5.

The presently observed tectonic fabric, which is shown in Figure 8, is based on magnetic anomaly, bathymetric, and seismological information. The Lau basin at 1.8 m.y.B.P. (Figure 6) is an extrapolation backward in time of the present tectonic fabric. The observed fabric suggests to me that the evolution of the Lau basin can be described by a plate tectonics model. I suggest that about 3.5 m.y. ago the accreting plate boundary jumped east from the subsidiary basin and the Lau basin has since evolved by the interaction of a number of small plates. The geometric relationship between the lineation pattern and the Peggy ridge shows that the along-strike length of the accreting plate margin has decreased during the past 3.5 m.y. This suggests that the history of much of the basin can be described by the evolution of a ridge-fault-fault (RFF) triple junction ('T' in Figure 8). The three plates (A, B, and C, Figure 6) which meet at the RFF triple junction are separated by the Peggy ridge strike-slip boundary, the observed ridge axis, and a postulated strike-slip boundary (termed the Roger fracture zone) which trends northeast from the inferred triple junction.

The part of the basin north of the proposed strike-slip plate boundaries has been generated by one or more additional accreting plate boundaries; and two of these boundaries are shown schematically in Figure 8. Although no magnetic lineation fabrics have been associated with the proposed ridges, the presence of accreting plate boundaries has been inferred from the attenuation of shallow seismic waves [Barazangi and Isacks, 1971]. An intense band of shallow seismicity trending east-west at about 15°S defines a broad zone of interaction between the main Pacific plate and the northern margin of the Lau basin [Johnson and Molnar, 1972]. P_n and S_n phases emanating from events within this zone are inefficiently propagated to Samoa, Fiji, and Tonga [Barazangi and Isacks, 1971] suggesting that accreting plate margins exist in the northern part of the Lau basin as required by the tectonic model. The bathymetry in Figure 8 and Hawkins [1974] shows that crustal depths are shallow north of the inferred strike-slip boundaries, implying a young age for the crust. Although much of the northern area exhibits high morphologic relief, the greatest depths occur just north of the triple junction, consistent with the expected location of the oldest crust generated from ridge segments such as those shown schematically in Figure 8.

A major problem with the evolutionary scheme shown in Figures 5, 6, and 8, lies in explaining the differences between the seismically active

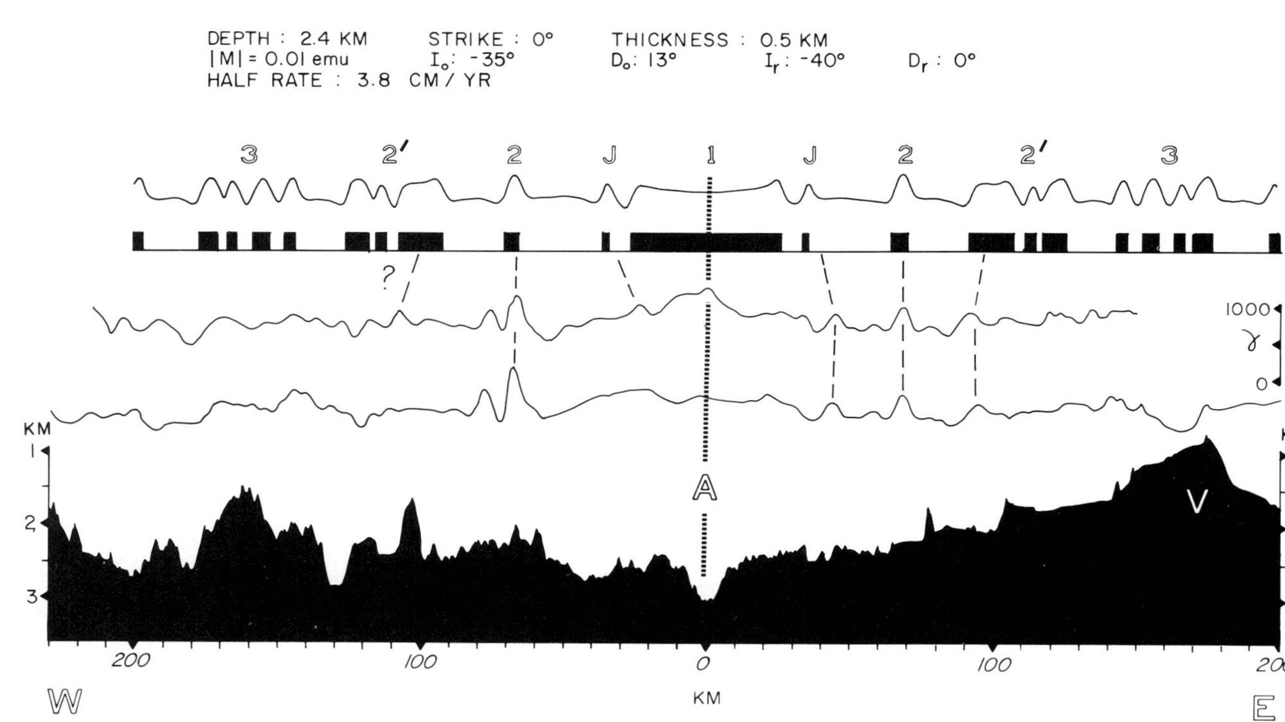

Figure 4. Comparison between two adjacent magnetic anomaly profiles from the Lau basin (Figure 2) and a synthetic profile calculated from a standard magnetic block model and the geomagnetic reversal time scale of Heirtzler et al., (1968). In the model parameters the 'o' and 'r' subscripts denote the ambient and remanent geomagnetic fields respectively. The morphology shown at the bottom is associated with the lower magnetic anomaly profile (Figure 3, profile 2). 'A' refers to the axial rift and 'V' refers to the volcanic axis on the Tonga island arc.

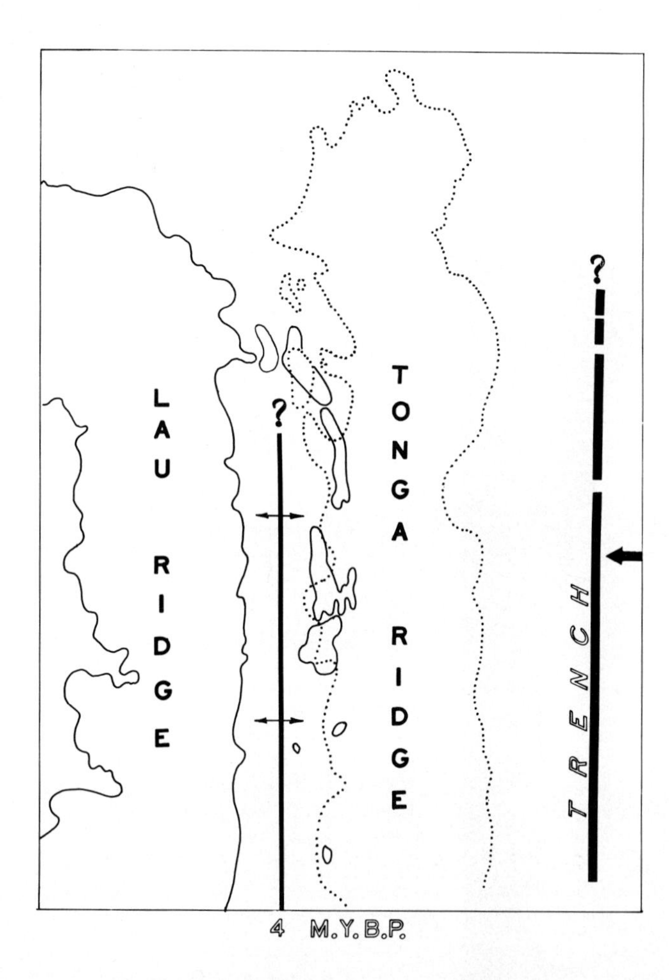

Figure 5. Schematic configuration of plate boundaries and the frontal and remnant arcs at 4 m.y.B.P.

Peggy ridge and the Roger fracture zone which lacks a distinctive morphologic and seismologic expression. Under the concepts of the interaction of rigid plates, a possible, although speculative, explanation for the differences between the characteristics of the two proposed strike-slip boundaries may be found in small changes in relative motion between the plates A, B, and C (Figure 6) during the last 3.5 m.y. From the magnetic lineations shown in Figures 2 and 8, the ridge axis at anomaly 2 time probably had an azimuth slightly different from the present strike. Assuming that spreading was perpendicular to the ridge, the direction of relative motion between plates A and B at anomaly 2 time (1.8 m.y. B.P.) was probably N 80°E as shown in Figure 7. Between the times of formation of anomalies 2 and J (1.8 and 0.9 m.y.B.P.), the segment of ridge crest south of the triple junction jumped to the west as discussed above. After the jump, the ridge assumed an almost north-south strike and east-west spreading became established (Figure 7). Assuming that the triple junction was stable before and after this slight change, the new relative velocity directions may provide an explanation for the differences between the two inferred strike-slip boundaries. Essentially, the directions of motion on the fracture zone boundaries adjusted to the change in spreading direction and the tectonic regimes at pre-existing parts of the fracture zones changed from purely strike-slip. The model predicts oblique compression and oblique extension across the older parts of the Peggy ridge and Roger fracture zone respectively during the past million years or so. Recent oblique compression explains why the Peggy ridge is more seismically active than the Roger fracture zone. Oblique compression could also account for the elevated, asymmetric morphology of the Peggy ridge [see Karig, 1970, Figure 7] which is reminiscent of the morphology of the Mendocino fracture zone [Silver, 1971]. The lack of distinctive morphology along the Roger fracture zone is consistent with a small amount of passive upwelling. I emphasize that this explanation for the differ-

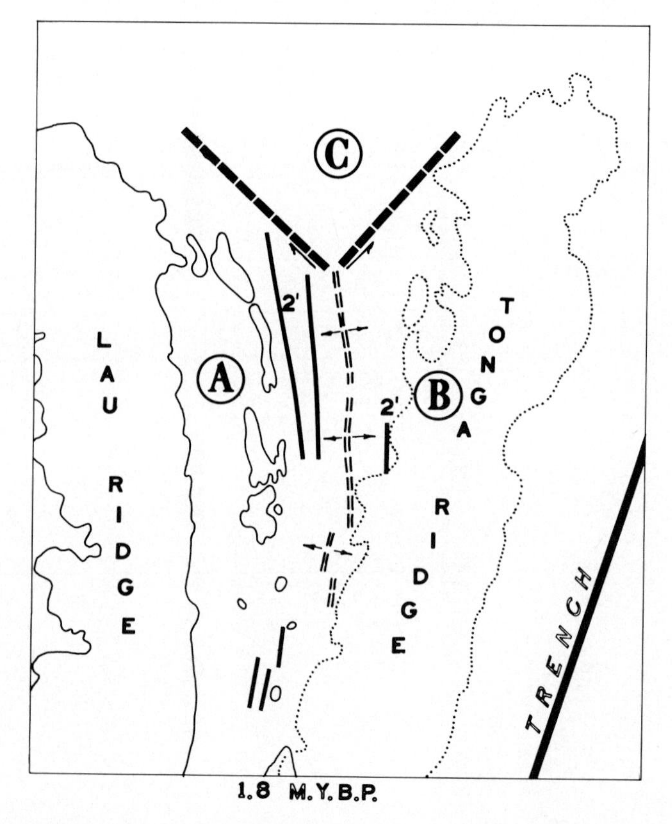

Figure 6. Tectonic fabric of the basin at the time of formation of anomaly 2 (1.8 m.y.B.P.). Strike-slip plate boundaries active at this time are denoted by heavy broken lines. Active accreting plate margin is indicated by broken parallel lines. Directions of relative motions between plates A, B, and C are shown by light arrows.

ences in seismicity and morphology between the two strike-slip plate boundaries is only one of many, equally plausible, explanations.

The possibility of an active triple junction in the Lau basin has been briefly explored in previous studies [Chase, 1971; Sclater et al., 1972]. This paper, however, presents much evidence to support the notion that the Lau basin has formed by the evolution of small rigid plates and that the boundaries between three of these plates met at a RFF triple junction during the last 3.5 m.y.

Conclusions

North-south trending magnetic lineations comprising anomalies 1 through 2' (0 to 3.5 m.y. B.P.) are present in the Lau basin south of the Peggy ridge. The magnetic pattern is obliquely bounded on the north by the Peggy ridge, a previously recognized strike-slip feature [Barazangi and Isacks, 1971], and a postulated northeast trending strike-slip boundary termed the Roger fracture zone. During the past 3.5 m.y. a ridge-fault-fault (RFF) triple junction has been active in the Lau basin. The magnetic lineations determine the history of the accreting plate boundary and the Peggy ridge and Roger fracture zone constitute the strike-slip boundaries. The part of the basin north of the strike-slip boundaries has been generated from still active ridge segments although associated magnetic lineations have not been mapped. Between about 6 and 3.5 m.y.B.P. an initial phase of crustal accretion occurred in a subsidiary basin which is presently preserved along the western margin of the Lau basin. Thus, the evolution of the Lau basin appears more complicated than a single episode of crustal accretion at a simple two-limb spreading system.

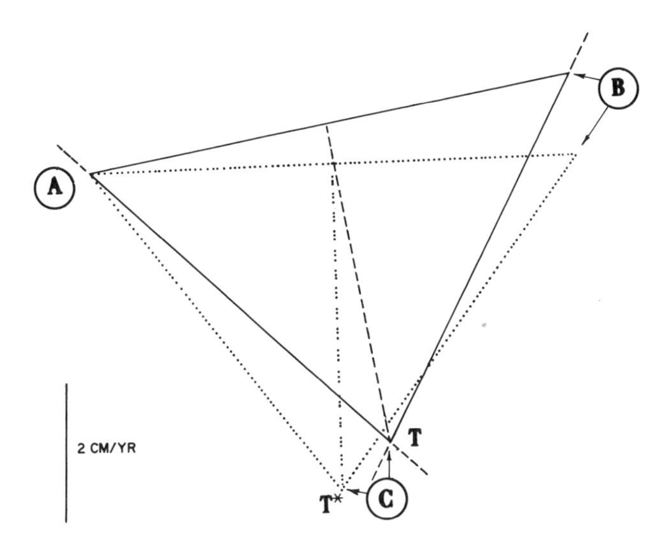

Figure 7. Relative motion vectors between plates A, B, and C of Figure 6 for 1.8 m.y.B.P. (solid lines) and for the present time (dotted lines). For stable conditions, velocity vector triangles ABC are isosceles and the triple junction (T and T*) is at rest with respect to plate C (see McKenzie and Morgan, 1969, for a more complete explanation). The dashed lines AC and CB indicate the azimuths of the purely strike-slip plate boundaries at 1.8 m.y.B.P. The angles between these lines and the new relative motion vectors (dotted lines AC and BC) indicate the oblique compression across the Peggy ridge and the oblique extension across the Roger fracture zone.

Figure 8. Present tectonic fabric. Active plate boundaries are denoted by heavy lines (broken where hypothetical) and directions of relative motion by light arrows. The bathymetric contours are modified after Hawkins [1974]. The 'V' symbols on the Tonga ridge represent active volcanoes of the island arc. 'T' denotes the location of the inferred RFF triple junction.

Lamont-Doherty Geological Observatory Contribution No. 2413

Acknowledgments. I would like to thank people who kindly provided unpublished magnetics data: David Handschumacher (University of Hawaii); James Hawkins (Scripps Institution of Oceanography); and Fred Davey (DSIR, New Zealand). This research was supported by National Science Foundation Grants DES 71-00214 and OPP 74-02238 and Office of Naval Research Contract N00014-75-C-0210.

References

Barazangi, M., and Isacks, B. L., Lateral variations of seismic-wave attenuation in the upper mantle above the inclined earthquake zone of the Tonga island arc: Deep anomaly in the upper mantle, J. Geophys. Res, 76, 8493-8516, 1971.

Burns, R. E., and Andrews, J. R., Regional aspects of deep sea drilling in the Southwest Pacific, In: Burns, R. E., Andrews, J. R., et al., Initial Reports of the Deep Sea Drilling Project, V. 21, Washington, D. C., 987-996, 1973.

Chase, C. G., Tectonic history of the Fiji Plateau, Geol. Soc. Am. Bull., 82, 3087-3110, 1971.

Chase, T. E., Smith, S. M., Newhouse, D. A., Crocker, W. L., Schoenbechler, M., Hydock, L., and Ritter, U., Track charts of SIO bathymetric data and track charts of SIO magnetic data in the Pacific Ocean, IMR TR-25 Sea Grant Publication No. 7, Scripps Inst. of Oceanography, 1972.

Foreman, J. A., A structural and tectonic study of the Lau-Havre-South Fiji Basin region, MS Thesis, Univ. of Hawaii, 1973.

Gill, J., and Gorton, M., A proposed geological and geochemical history of eastern Melanesia, In: The Western Pacific, P. J. Coleman, (editor), Crane, Russack and Co., Inc., New York, 543-566, 1973.

Hawkins, J. W., Geology of the Lau basin, a marginal sea behind the Tonga arc, In: Geology of continental margins, C. Burke and C. L. Drake (editors), Springer-Verlag, Berlin, 505-518, 1974.

Hawkins, J. W., Petrology and geochemistry of basaltic rocks of the Lau basin, Earth and Planet. Sci. Lett., 28, 283-297, 1976.

Hawkins, J. W., Petrologic and geochemical characteristics of marginal basin basalts, Maurice Ewing Series, A.G.U. Monograph 1977.

Heirtzler, J. R., Dickson, G. O., Herron, E. M., Pitman, W. C., and Le Pichon, X., Marine magnetic anomalies, geomagnetic field reversals and motions of the ocean floor and continents, J. Geophys. Res., 73, 2119-2136, 1968.

Hochstein, M. P., and Reilly, W. I., Magnetic measurements in the south-west Pacific Ocean, N. Z. Jour. Geol. Geophys., 10, 1527-1562, 1967.

Johnson, T., and Molnar, P., Focal mechanisms and plate tectonics of the southwest Pacific, J. Geophys. Res., 77, 5000-5032, 1972.

Karig, D. E., Ridges and basins of the Tonga-Kermadec island arc system, J. Geophys. Res., 75, 239-254, 1970.

Karig, D., Origin and development of marginal basins in the western Pacific, J. Geophys. Res., 76, 2542-2561, 1971.

Lawver, L. A., Hawkins, J. W., and Sclater, J. G., Magnetic anomalies and crustal dilation in the Lau basin, Earth and Planet. Sci. Lett., in press, 1976.

McKenzie, D. P., and Morgan, W. J., Evolution of triple junctions, Nature, 224, 125-133, 1969.

Sclater, J. G., and Francheteau, J., The implications of terrestrial heat flow observations on current tectonic and geochemical models of crust and upper mantle of the earth, Geophys. J. Roy. astr. Soc., 20, 509-542, 1970.

Sclater, J. G., Anderson, R. N., and Bell, M. L., The elevation of ridges and the evolution of the eastern Central Pacific, J. Geophys. Res., 76, 7888-7915, 1971.

Sclater, J. G., Hawkins, J. W. Jr., Mammerickx, J., and Chase, C. G., Crustal extension between the Tonga and Lau Ridges: Petrologic and geophysical evidence, Geol. Soc. Am. Bull., 83, 505-518, 1972.

Silver, E. A., Tectonics of the Mendocino triple junction, Geol. Soc. Am. Bull., 82, 2965-2978, 1971.

THE BERING SEA - A MULTIFARIOUS MARGINAL BASIN

Alan K. Cooper, Michael S. Marlow and David W. Scholl

U. S. Geological Survey, 345 Middlefield Road, Menlo Park, California 94025

Abstract. The Bering Sea Basin differs from other marginal basins of the Pacific Ocean in that a large section of its igneous crust is believed to be abandoned 'trapped' oceanic plate (eastern Bering Sea), a large negative magnetic anomaly exists over most of the central part of the Bering Sea Basin (Aleutian Basin), and a set of long-wavelength geophysical anomalies in the back-arc basins appears to correlate with different underthrusting regimes along the frontal Aleutian arc. Within the Bering Sea, a contrast in geophysical anomalies coincides with an apparent structural boundary, Shirshov Ridge, which separates the eastern (Aleutian and Bowers Basins) and western (Komandorsky Basin) abyssal basin areas of the Bering Sea Basin. At Shirshov Ridge the apparent regional trends in the heat flow, magnetic, and gravity anomalies change, from east-northeast in the Aleutian Basin to northwest in the Komandorsky Basin; amplitudes of the anomalies also change. Convergence between the Pacific and North American plates shifts along the Aleutian arc from oblique underthrusting in the eastern part to strike slip motion along the western segment. The Komandorsky Basin lies behind the strike slip segment of the arc while most of the Aleutian Basin lies behind the oblique underthrusting segment. Within the abyssal basins, the long-wavelength magnetic and heat flow anomalies may reflect regional variations in the depth to the Curie isotherm. Isotherm depths deduced from the heat flow data are generally shallowest in the central Aleutian Basin some 600 km behind the Aleutian arc; in the Komandorsky Basin, these depths are inferred to be shallowest directly adjacent to the western part of the Aleutian arc. Broad positive gravity anomalies in the same general areas suggest a thin oceanic layer 3 and shallow mantle depths. The cause of the geophysical disparity between the abyssal basins in the eastern and western parts of the Bering Sea Basin may be related to processes of heat transfer in the underlying lithospheric rocks. A model that includes convective flow within the asthenosphere and magmatic injection into the lithosphere is suggested for the Aleutian Basin; for the Komandorsky Basin, a model employing episodic mantle diapirism, currently adjacent to the Aleutian arc, is postulated.

Introduction

The marginal seas of the western and northern Pacific Ocean exhibit similarities in such geophysical characteristics as oceanic crustal sections, lineated magnetic anomalies, low free air gravity values, and locally high heat flow values. The basins of the western Pacific are generally thought to have formed by an extensional crustal process in which new basaltic sea floor is emplaced at spreading centers within the basin [Karig, 1971; Packham and Falvey, 1971; Sclater et al., 1972; Watts and Weissel, 1975; Weissel and Watts, 1976]. The variations in the ages of the basins, as inferred from geological and geophysical data, suggest that the extensional process has operated at different times. A unique and simple scheme for the evolutionary process cannot, however, be applied to all marginal basins, as evidenced by the large number of models proposed [Beloussov and Ruditch, 1961; Karig, 1971; Packham and Falvey, 1971; Sleep and Toksoz, 1971; Oxburg and Turcotte, 1971; Uyeda and Miyashiro, 1974].

The Bering Sea Basin, the northernmost marginal basin of the Pacific Ocean, is special in that crustal extension apparently is not the primary mechanism for the creation of the larger eastern part of the basin. Three basins lie within the Bering Sea Basin: the Aleutian, Bowers, and Komandorsky Basins (Fig. 1). The Aleutian and Bowers Basins appear to be underlain by remanent pieces of trapped Mesozoic sea floor [Scholl et al., 1975; Cooper et al., 1976a and 1976b]. However, the igneous crust of the Komandorsky Basin seems to be in part much younger than that of the Aleutian and Bowers Basins [Scholl et al., 1975; Cormier, 1975].

The tectonic setting of the Bering Sea Basin is not unlike that of other western Pacific basins, that is, it is bounded on one side by a seismically active island arc-trench system, and on the other side by a stable continental area. The interior of the basin is crossed by two aseismic ridges, Bowers Ridge and Shirshov Ridge.

These beveled ridges are in part composed of altered andesitic volcanic rocks, which suggests that they may be old island arcs which once stood near sea level [Ludwig et al., 1971; Scholl et al., 1975]. Others [Kienle, 1971; Karig, 1972] propose that Bowers Ridge is similar to the remanent arcs of the western Pacific marginal basins.

An examination of the limited geophysical data in the abyssal basins of the Bering Sea suggests that systematic internal variations in trends and amplitudes of the long-wavelength gravity, magnetic, and heat flow anomalies may exist between the eastern (Aleutian Basin) and western (Komandorsky Basin) parts of the Bering Sea Basin. The apparent change occurs at the north-south trending Shirshov Ridge, which appears to be an older structural boundary between the two abyssal basin areas. Similar long-wavelength variations have not been reported from the other Pacific marginal basins. Therefore, the presence of such variations in the Bering Sea, if confirmed, would be significant. The apparent change in the long-wavelength anomalies may be a consequence of two distinct processes beneath the eastern and western parts of the basin, which are related to differing plate convergence regimes along the eastern and western segments of the Aleutian arc [Minster et al., 1974; Cormier, 1975].

In the present paper, we wish to illustrate the apparent disparity in the Bering Sea Basin and to present hypothetical models relating the geophysical observations to regional tectonics.

Geologic Data

The apparent difference in the geologic ages of sediment layers in the Komandorsky and Aleutian Basins is shown in an interpretative structural section that crosses the southern part of Shirshov Ridge (Fig. 2). The illustration is modified from Scholl et al. [1975, Fig. 5], and includes the refraction data of Shor [1964], Ludwig et al. [1971], and Fornari and Shor [1976]. At DSDP site 191 in the Komandorsky Basin (Fig. 1), the cored section included rocks ranging in age from Pleistocene to late Miocene. The late Miocene sedimentary units were found directly overlying tholeiitic basalt of possible Oligocene age [29.6 m.y.; Creager, Scholl et al., 1973]. A similar section of late Cenozoic sedimentary rocks was recovered from the Aleutian Basin at DSDP site 190 near the base of Shirshov Ridge; however, the hole terminated above acoustic basement. The postulated age of basement in the Aleutian Basin, based on the identification of magnetic lineations in the vicinity of hole 190, is about 130 m.y. [Cooper et al., 1976a]. Altered andesitic tuffs, overlain by sediment of middle Miocene age or younger, have been sampled from the western flank of Shirshov Ridge along the profile in Figure 2; farther south, other andesitic tuffs that have been collected yield a K-Ar age date of 16.8 m.y., or early Miocene. The thick sediment cover, reaching a 2 km maximum thickness in some summit basins, and the concordance of the ridge with onshore late Mesozoic-early Cenozoic structural trends suggest that the ridge may have been initiated in early Tertiary time [Scholl et al., 1975]. The southern part of the ridge (56°N) and also the apparent junction with Bowers Ridge system (171° to 175°E) appear to be much younger late Cenozoic features [Scholl et al., 1975; Rabinowitz, 1974].

Seismic Refraction Data

The seismic reflection profiles show that the depth to acoustic basement (below sea level) is 4 to 5 km in the Komandorsky Basin, and 6 to 8 km in the central Aleutian Basin [Cooper et al., 1976a, Fig. 4]. Refraction data show that the deep crustal sections beneath the two areas are also appreciably different. The crustal sections in the Aleutian and Bowers Basins are characterized by thick sedimentary layers overlying igneous crustal sections that are similar, but not identical, to the average oceanic crustal section (Fig. 2). The depth to the M-discontinuity beneath both basins is about 15 km. The Komandorsky Basin has a more typical oceanic crustal section than does the adjacent Aleutian Basin. However, the thickness (3 km) of the 5.5 km/sec layer (oceanic layer 2) and the shallow depth of the M-discontinuity in the Komandorsky Basin are not typical characteristics of other marginal basins of the Pacific [Karig, 1971].

Gravity Data

A free-air gravity map of the Bering Sea Basin from Watts [1975] is shown in Figure 3. Gravity models made by Kienle [1971] suggest that the large positive and negative anomalies associated with the arcuate Bowers Ridge may be the signature of a high density (volcanic) ridge flanked on its north side by a low density (sediment-filled) trough. Although a thick section of low density sediments may be present along the northern part of Shirshov Ridge near the Kamchatka margin, a low density sediment trough is apparently not associated with other parts of the ridge [Kienle, 1971]. In those parts of the Aleutian and Komandorsky Basins where trackline control is good (west of longitude 175°E), the gravity anomalies are less than 40 mgal and appear to have wavelengths of 100 to 200 km (Fig. 3). Values tend to be slightly lower in the Aleutian Basin, averaging about zero; in contrast, the Komandorsky Basin has a higher average value, near 20 mgal. In this same area (west of 175°E), the trends of the long-wavelength anomalies appear to differ in the Aleutian and Komandorsky Basins. A relative gravity high (30 mgal) directly west of Shirshov Ridge trends northwest, nearly parallel to the western Aleutian arc (Fig. 3). Directly east of Shirshov Ridge, a broad 20 mgal high trends northeast, nearly parallel to parts of both Bowers and Shirshov Ridges. Because of poor trackline coverage in the center of the Aleutian

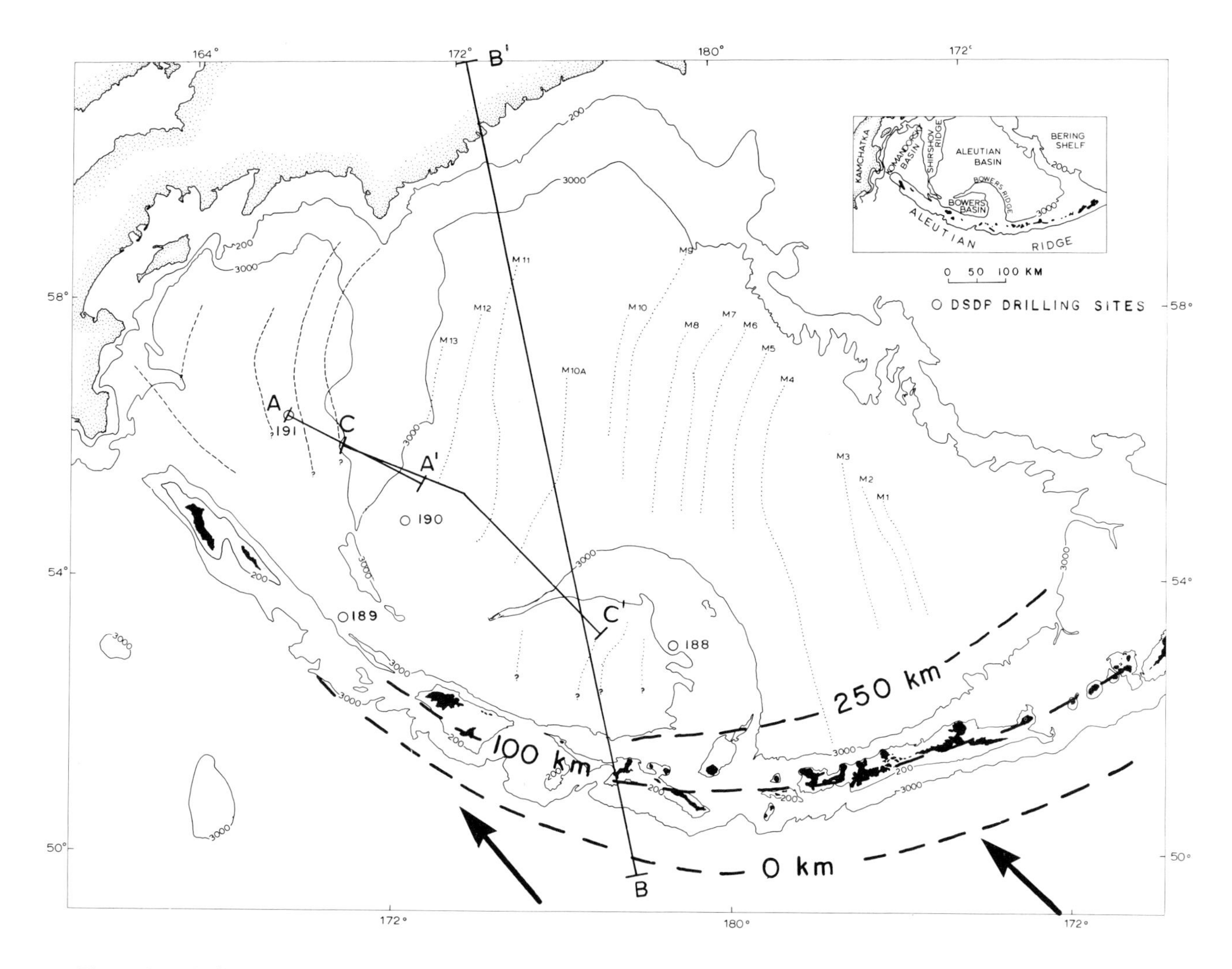

Fig. 1. Index map of the Bering Sea Basin showing magnetic lineations [Cooper et al., 1976a], depth to inclined seismic zone [Jacob, 1972], direction of relative Pacific-North American plate motion [Minster et al., 1974], and bathymetry in meters. Transverse mercator projection.

Basin and the presence of small amplitude gravity values along these tracklines, no long-wavelength trends in the data from this area may be reliably identified.

Magnetic Data

Two magnetic maps have been prepared from ship-track data for the Bering Sea Basin; the first, not shown here, is a compilation of residual (IGRF) anomalies [Kienle, 1971; Cooper et al., 1976a, c], and the second (Fig. 4) is a low-pass filtered version of the residual (IGRF) anomalies [Cooper et al., in prep.]. Figure 4 also includes the POGO satellite data [Regan, 1974, pers. comm.]. The low-pass map (Fig. 4) illustrates the approximate amplitude and shape of the magnetic anomalies which have wavelengths greater than 100 to 150 km. Cooper et al. [in prep.] note that small irregularities may be present in the map due to uncorrected diurnal variations and the filtering technique used to compile the map. However, they feel that the amplitudes and trends of the broader anomalies are essentially correct. Trackline control for Figure 4 is generally good except for some areas in the central and northern Aleutian Basin.

In the low-pass data (Fig. 4) the large submarine ridges (Bowers, Shirshov, and Aleutian) are associated with parallel belts of relative high and low magnetic values. These bipolar anomalies reflect the thick welt of volcanic rocks that are presumed to form the cores of these ridges [Scholl et al., 1975]. Within the abyssal basins, the anomalies east of Shirshov Ridge appear to be quite different from those west of the ridge. The Aleutian Basin is characterized by a mosaic of closed anomalies that generally trend northeast, while the Komandorsky Basin lacks closed anomalies and has magnetic contours that parallel the northwest trend of the western Aleutian arc.

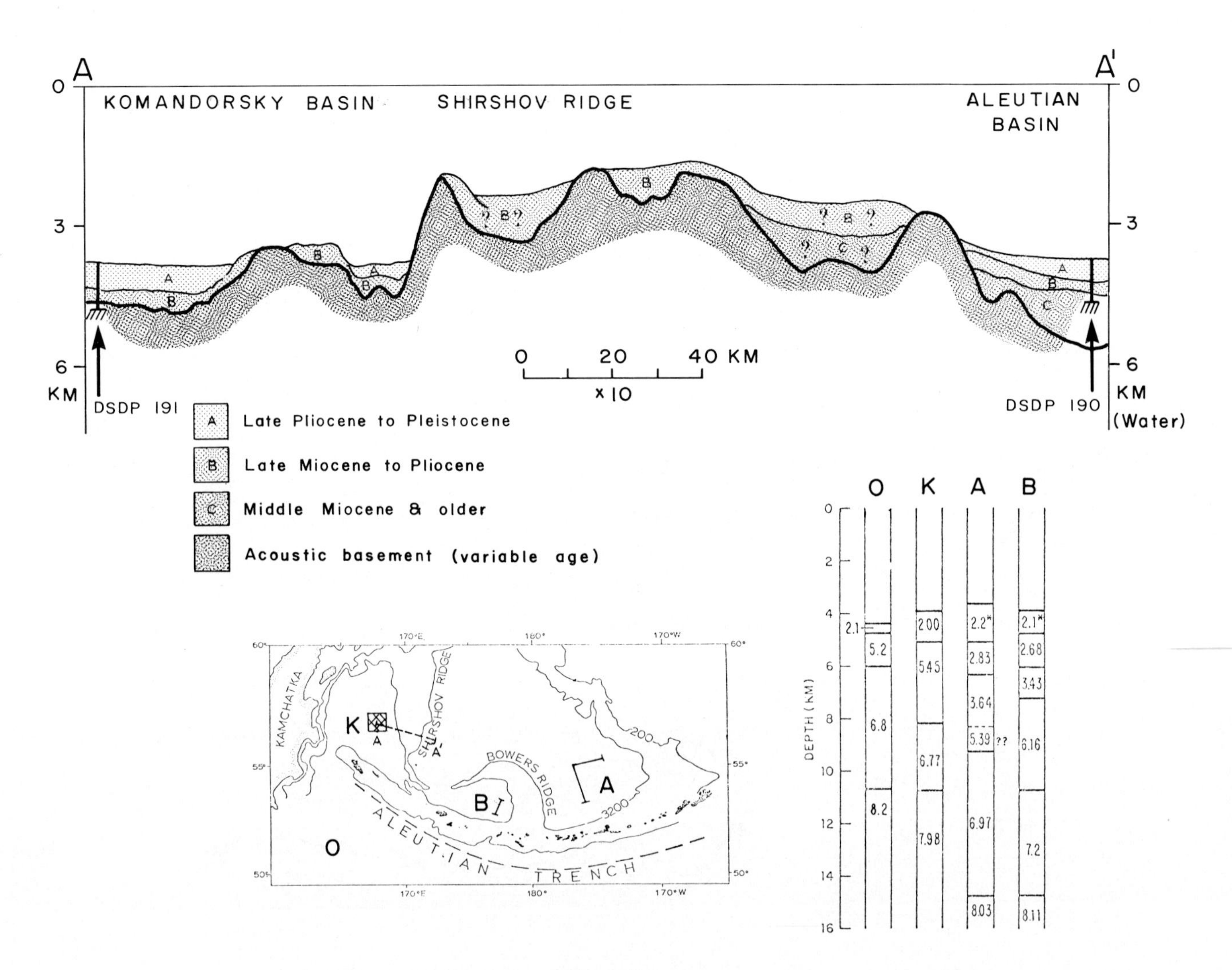

Fig. 2. Interpretative cross-section, modified from Scholl et al. [1975], across the eastern and western parts of the Bering Sea Basin. The refraction sections from Fornari and Shor [1976] are average sections for each of the areas shown; the "o" refers to a worldwide oceanic crustal section. Note the disparity in the sediment thickness (age), the depth to basaltic basement (v = 5.4 km/sec layer), and the mantle depths (v = 8.0 km/sec layer) between the Aleutian and Komandorsky Basins.

Satellite data from the Bering Sea (Fig. 4) show a relatively large negative anomaly (-4 gamma) over the Aleutian Basin. The zero contour of the anomaly generally follows the bathymetric edge of the abyssal basin areas, encompassing both the Aleutian and Komandorsky Basins. The center of the satellite low lies directly north of Bowers Ridge and roughly coincides with the elongate negative anomaly recorded in the long-wavelength surface data. In areal extent, the surface anomaly appears to be quite large, covering much of the central Aleutian Basin. This particular anomaly consists of an elongate low (-300 to -400 gamma) bordered on the north and south by relative highs (-50 to -175 gamma); it appears to extend nearly 600 km across the Aleutian Basin, from Shirshov Ridge to a point near the continental margin. A smaller set of relative highs and lows trending northeast is also found in the southeastern Aleutian Basin.

A comparison of the trends in the short-wavelength lineations (M-sequence, Fig. 1) and the apparent trends of the long-wavelength anomalies (Fig. 4) in the Aleutian Basin illustrates an interesting disparity in the data. The short-wavelength anomalies generally trend north-south, but the long-wavelength anomalies have different trends in different areas. This incongruity is most obvious in the Aleutian Basin. Here the narrow north-south lineations (M1 - M13 spreading anomalies) are superimposed upon the long-wavelength features that trend east-northeast.

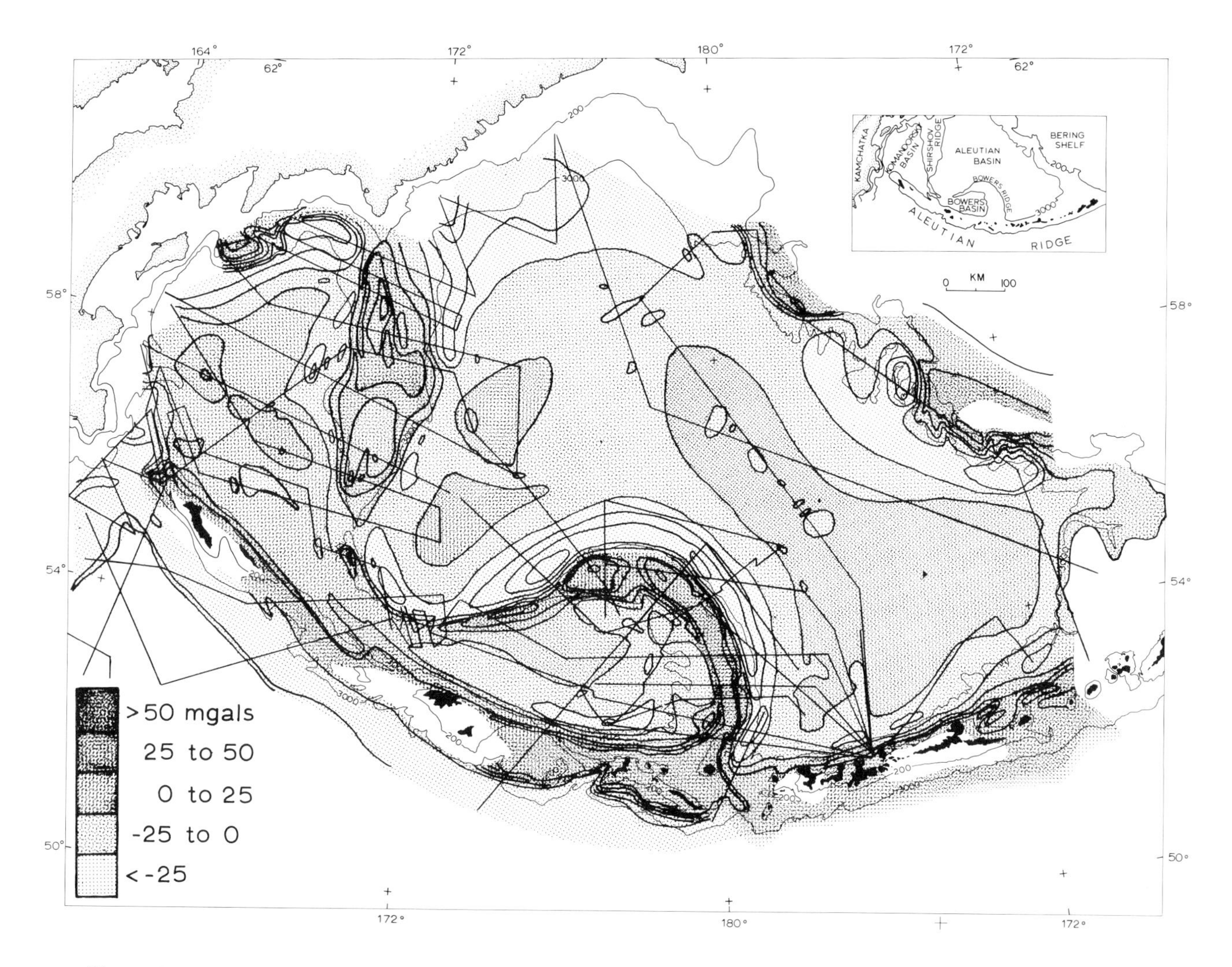

Fig. 3. Free-air gravity map of the Bering Sea Basin, from Watts [1975]. Note the change in the trend and the magnitude of the broad anomalies, which occurs at the north-south trending Shirshov Ridge. Transverse mercator projection.

Heat Flow

The measured heat flow values from the Bering Sea Basin [Foster, 1962; Sass and Munroe, 1970; Langseth and Horai, unpub. man.] are shown in Fig. 5. Most of the data (85%) are reported by Langseth and Horai [unpub. man.] and appear to be reasonably good measurements, with 95% of their values having a reliability factor of 7 or higher, of a possible factor of 10. Even so, many of the observed variations may reflect the uncertainties commonly associated with the heat flow measurements.

Pursuing the hypothesis that systematic long-wavelength variations may be present in the Bering Sea Basin, we have contoured the heat flow data to demonstrate the possibility of different regional trends in different parts of the basin. The contours in Figure 5 must be considered highly interpretive, due to the paucity of data; however, the proposed trends in the heat flow data do appear to be consistent with the trends in the long wavelength magnetic data from the Aleutian and Komandorsky Basins. Heat flow values in the Komandorsky Basin are high, exceeding 4.0 HFU adjacent to the Aleutian arc and decreasing to about 2.5 HFU in the center of the basin [Watanabe et al., this vol.]. The trend of these data appears to be to the northwest, parallel with the western segment of the Aleutian arc.

Langseth and Horai [unpub. man.] note that high sedimentation rates in the abyssal basins during the last 5 m.y. have suppressed the heat reaching the basin floors. They estimate that after adjustment for high sedimentation rates, the actual heat flow may be 20-25% larger than the observed measurements shown in Figure 5. These larger, corrected heat flow values indicate that the Bering Sea Basin has higher heat flow than was previously thought [Foster, 1962] and

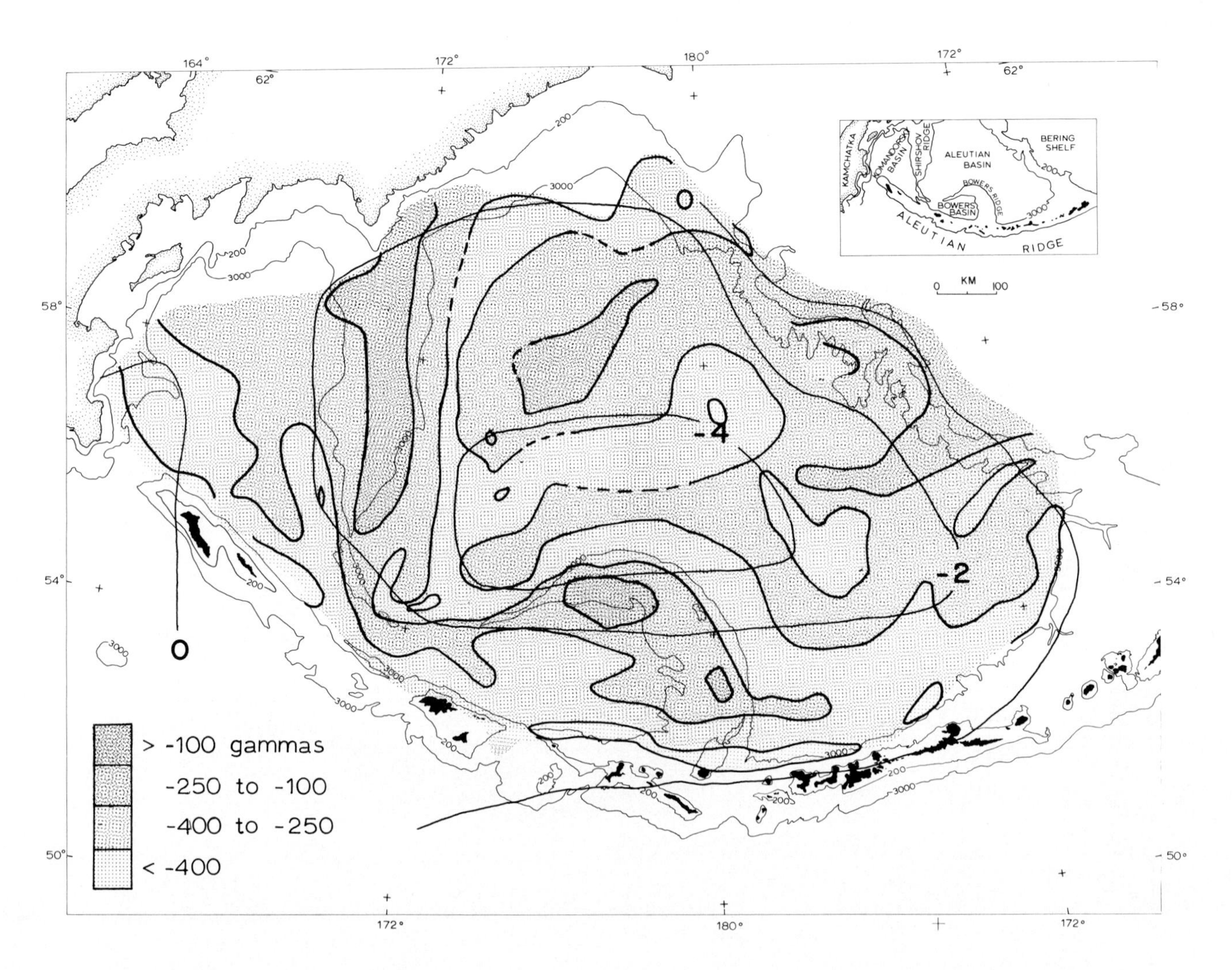

Fig. 4. Map of long wavelength magnetic anomalies (wavelengths greater than 100 to 150 km) in the Bering Sea Basin, from Cooper et al. [in prep.]. Contours of satellite magnetic data from Regan [1974, writ. comm.] are superimposed. Note the coincidence of the large amplitude negative anomaly seen in both the surface and satellite magnetic data. The broad anomalies in the abyssal basins, which may reflect regional variations in the depths to the Curie isotherm, appear to change in trend and magnitude at Shirshov Ridge, from east-northeast in the Aleutian Basin to northwest in the Komandorsky Basin. Transverse mercator projection.

that the heat flow in the Bering Sea Basin is close to values recorded in the other marginal basins of the Pacific [Sclater, 1972; Watanabe et al., this vol.].

Discussion

In gravity, magnetic, and heat flow data, a correlation can commonly be made between the wavelength of the observed anomaly and the depth to the source body causing the anomaly. Longer wavelength anomalies are generally associated with deeper sources. Although this does not apply in all situations, model studies using the geophysical data from the Bering Sea Basin suggest that deep-seated sources may be present. Supporting this contention, the refraction work of Shor [1964] in the eastern part of the Aleutian Basin indicates lateral variations in the compressional velocity of the upper mantle near the M-discontinuity. Cooper et al. [in prep.] note that this velocity anomaly coincides with anomalies in the gravity and heat flow data. Their depth solutions based on the gravity data suggest that the source of the gravity anomaly is also close to the M-discontinuity. The regional magnetic low observed in both satellite and surface data may also be attributed to a deep source, perhaps reflecting a rise in the Curie isotherm (580°C) at lower crust-upper mantle depths [Cooper et al., in prep.].

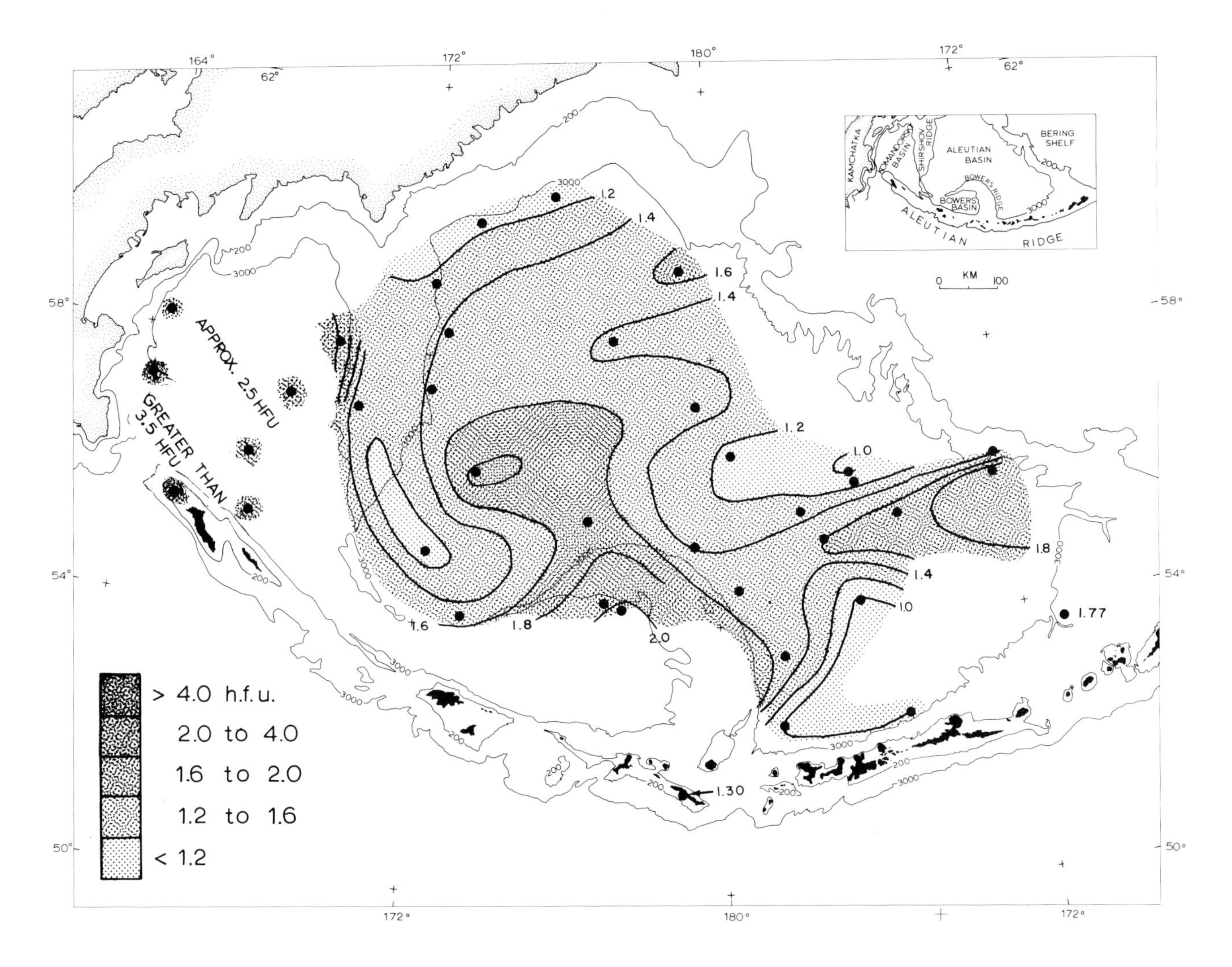

Fig. 5. Map of uncorrected heat flow anomalies in the Bering Sea Basin, contoured from heat flow data given by Langseth and Horai [unpub. man.]. The data are contoured to illustrate the similarity to the long wavelength magnetic anomalies in Figure 4. Values in the Komandorsky Basin exceed 4.0 HFU adjacent to the Aleutian arc and decrease uniformly to about 2.5 HFU in the center of the basin [Watanabe, et al., this vol.]. Transverse mercator projection.

Significance of Magnetic and Heat Flow Anomalies

If the apparent correlation of the low-pass magnetic anomalies (Fig. 4) with the observed heat flow anomalies (Fig. 5) in both the Aleutian and Komandorsky Basins is assumed to be valid, then one possible explanation for the regional anomalies is a variation in the depth to the Curie isotherm under the basins. Because of the complexity of both magnetic and heat flow anomalies, this simple explanation could probably not be applied throughout the Bering Sea Basin. However, Cooper et al. [in prep.] do provide support for the idea that some magnetic and heat flow anomalies may be closely related. Two of their magnetic models reproduce the large negative anomaly crossing the Aleutian Basin. These models were constructed using geometries that approximate the geometry of the Curie isotherm, which is obtained from observed heat flow values and one-dimensional heat flow models. One of their models [Cooper et al., in prep., Fig. 7] which crosses the Aleutian Basin between Bowers and Shirshov Ridges, is reproduced in Figure 6. As noted by Cooper et al. [in prep.], large relief on the boundaries of the deep-seated magnetic body (Curie isotherm?), as well as seemingly large induced crustal magnetizations, are necessary to duplicate the amplitude of the regional negative anomaly in the Aleutian Basin.

Despite the uncertainties in the heat flow measurements and in the interpretive contouring of the data, large variations may exist in the heat flow data from different areas of each basin. The largest recorded value of heat flow in each basin differs from the smallest value by nearly a factor

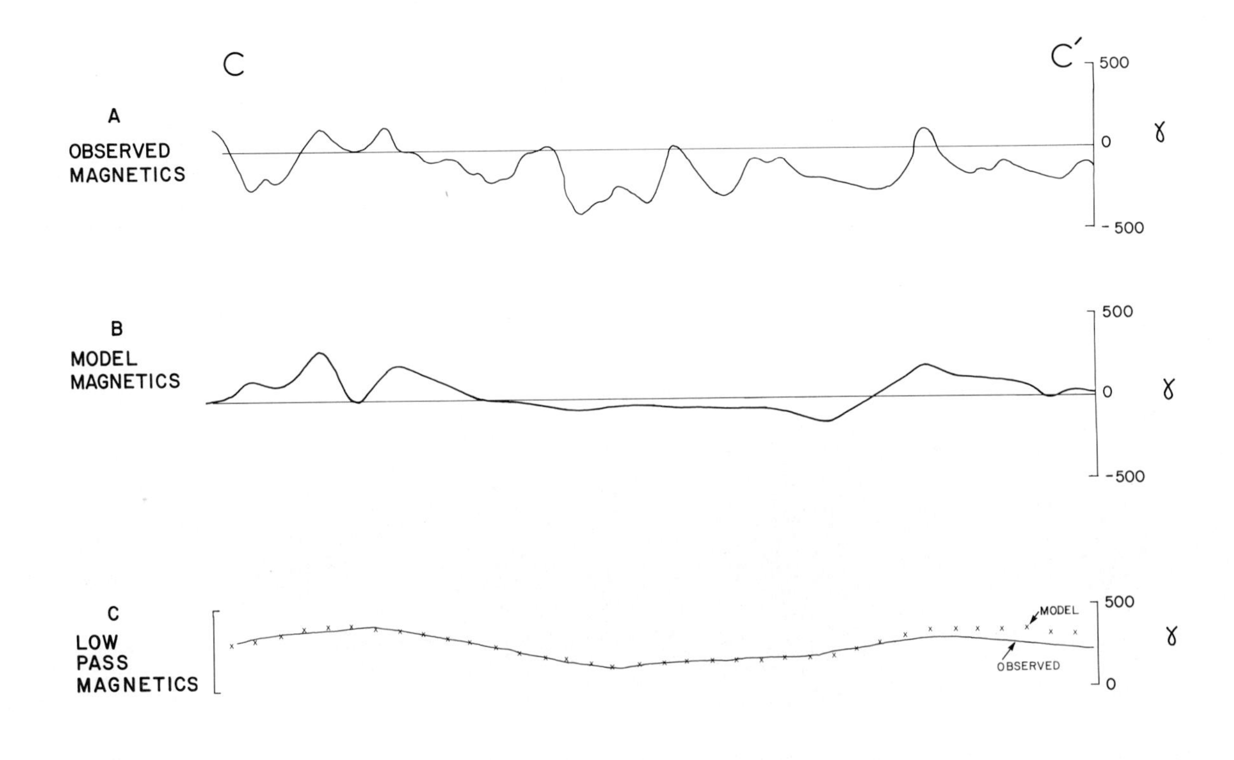

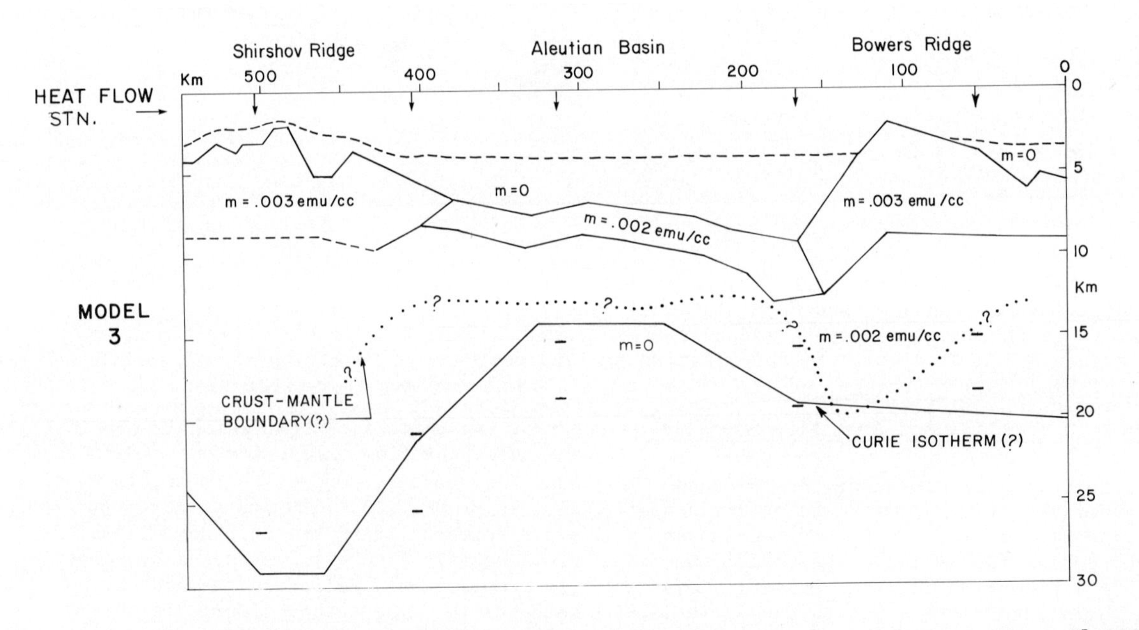

Fig. 6. Magnetic model for the Aleutian Basin [from Cooper et al., in prep., Fig. 7] illustrating the geometry of the Curie isotherm necessary to duplicate the long wavelength magnetic low over the central basin (see Curve C). Heavy dashes below each heat flow station are depth estimates for the 580°C isotherm (Curie isotherm) based on one-dimensional heat flow models. No attempt has been made to model the short wavelength anomalies in the center of the basin, as they do not contribute to the broad negative feature. Note that large vertical relief of the bottom boundary (Curie isotherm?) for the model is required to reproduce the amplitude of the observed anomaly. If lower values of deep crustal magnetizations are assumed, then larger relief of the bottom boundary is necessary to fit the anomaly.

of two. If a simple one-dimensional heat flow model is applied to areas in the Aleutian Basin where the sediment thickness and rock conductivities are assumed to be relatively uniform, then the two-fold difference in observed heat flow implies a two-fold increase in the depth to the Curie isotherm. Since the sediment thicknesses are only known approximately [unpublished reflection profiles; Cooper et al., 1976a, Fig. 4] and the uniformity of rock conductivity at depth is unknown, many of the observed heat flow variations may be related to changes in these parameters. Yet, a comparison of the magnetic (Fig. 4) and heat flow (Fig. 5) maps does suggest that, in most parts of the basin, variations in observed heat flow are generally comparable to amplitude changes in magnetic anomalies. This correspondence further suggests that the postulated association between the long-wavelength magnetic anomalies and the Curie isotherm depths [Cooper et al., in prep.] may be applicable over a large part of the Aleutian Basin.

The relative difference in the amplitude of broad magnetic anomalies (Fig. 4) is larger in the Aleutian Basin (-50 to -350 gamma) than in the Komandorsky Basin (-200 to -400); this may be related to the different geometry of the postulated magnetic source bodies (Curie isotherm depths). In the Aleutian Basin, the Curie isotherm is presumed to rise in the shape of an elongated prism beneath the center of the basin [Cooper et al., in prep., Fig. 7]. However, in the Komandorsky Basin the uniform decrease in the heat flow values from the Aleutian arc toward the center of the basin [Watanabe et al., this vol.] suggests that the isotherm may rise in the shape of a north-dipping ramp that is shallowest adjacent to the Aleutian arc and deepens toward the center of the basin. Simple magnetic models using these two isotherm geometries can readily explain the apparent two-fold difference between the maximum amplitudes of the long-wavelength anomalies observed in the two abyssal basin areas. In these Curie isotherm models a horizontal variation in the magnetization of the lower crust-upper mantle layers between the Aleutian and Komandorsky Basins is not required to explain this effect.

Significance of Gravity Anomalies

The long-wavelength gravity anomalies immediately east of Shirshov Ridge are quite different from those immediately west; the trends are almost at right angles to each other and the amplitudes differ by nearly a factor of two. Kienle's [1971, Fig. 9] gravity model in the Aleutian Basin implies that a rise of 1.5 to 2 km in the mantle depths or possibly an increase in the density of the mantle may explain the broad northeast-trending gravity high that lies between Bowers and Shirshov Ridges (Fig. 3). The anomaly does not appear to be a topographic edge effect of these ridges. A structure contour map on acoustic basement [Cooper et al., 1976a, Fig. 4] shows that a narrow (30 to 40 km wide) basement ridge with a relief of 2 to 3 km is associated with the broad gravity anomaly. This subsurface ridge, the Bartlett Ridge, is too narrow to cause the long-wavelength gravity anomaly; however, there is good correlation between the position of this ridge and the area where a thinner oceanic layer 3 and shallower mantle depths are suspected. Seismic reflection profiles over the Bartlett Ridge indicate relatively recent tectonic activity in this area. The ridge appears to be structurally deforming sedimentary layers of Pleistocene and possible Holocene age.

If locally higher temperatures are present beneath some areas of the central Aleutian Basin, as is suggested by several heat flow measurements, then slightly shallower sea floor depths and higher free-air gravity values might be anticipated in these areas; these effects may be similar to those observed over the mid-ocean spreading ridges [Talwani, 1971, Fig. 15]. The change in sea floor elevation, assuming a thermal expansion of $dV/Vdt = 4 \times 10^{-5}/^{o}C$ [Clark, 1966, p. 92] and the elevated Curie isotherm model (Fig. 6), should be small, on the order of 150 meters. A correspondingly small decrease in the rock densities (1 - 2%) in the higher temperature areas might also be expected with these parameters. The net change in the free-air gravity values, however, should be slightly positive, with the effects of topography being more important than the decreased densities at depth [Talwani, 1971].

Uniform water depths between Bowers and Shirshov Ridges indicate that the sea floor is flat and that there is not a 150-meter upward bulge. Buried sedimentary layers at sub-bottom depths exceeding 300 meters, however, may indeed be bowed up in the middle of the basin [Fig. 6; Ludwig et al., 1971, Fig. 5]. Because these layers are incorporated into Kienle's [1971, Fig. 9] gravity model for the same area, the gravity effect of the upbowed sediment layers is apparently not sufficient to explain the positive free air gravity values; a 1.5 to 2.0 km shallowing of the mantle is still required. Further refraction measurements are needed to establish the depths to sediment layers and to the mantle beneath other parts of the Aleutian Basin where locally higher temperatures are suspected.

Crustal refraction data from the Komandorsky Basin [Fornari and Shor, 1976] suggest that a thin oceanic layer 3 and a shallow mantle depth may also be associated with the broad positive gravity anomaly that trends northwest across the center of the basin. The refraction measurements (Fig. 2) were made over the southeast end of the broad gravity high. Although a buried basement ridge does not appear to be associated with the gravity high, as it is in the Aleutian Basin, the depths to the acoustic basement (basalt?) are generally shallower

(4 - 5 km below sea level) beneath the gravity high than they are throughout the northern and southern parts of the basin [Cooper et al., 1976a, Fig. 4]. Refraction measurements are not available in other parts of the Komandorsky Basin, so it is not possible to establish whether the thickness of oceanic layer 3 and the depth to the mantle both increase away from the center of the basin. By analogy to the Aleutian Basin, one may speculate that the gravity high in the Komandorsky Basin results from either a rise in the mantle depths or from an increase in the density of the mantle beneath the center of the basin.

Comparisons of observed free-air gravity values with theoretical ones which are based on refraction measurements (Fig. 2) and velocity-density curves [Bateman and Eaton, 1967; Ludwig et al., 1970] suggest that there may be a significant difference in crustal structure between the Aleutian and Komandorsky Basins. The observed gravity values, from areas where the refraction measurements were made, show a difference of 35 mgal between the basins; the theoretical gravity values indicate a difference of either 117 mgal [Bateman and Eaton, 1967] or 98 mgal [Ludwig et al., 1970], depending upon the velocity-density curves used. The Kormandorsky Basin has the higher observed and theoretical values. Although part of the 75 mgal discrepancy between the theoretical and observed values (110 - 35 = 75 mgal) is probably due to uncertainties in the velocity-density relations, the discrepancy is sufficiently large to suggest that variations in the mantle densities beneath the two basins may be present.

Crustal Models

Most of the models proposed for the development of the marginal basins peripheral to the Pacific Ocean rely on the tectonic process and magmatic consequence of lithospheric underthrusting at arc-trench systems. We believe that geophysical observations in the Bering Sea Basin can also be interpreted in terms of the likely consequences of subduction. The Bering Sea marginal basin may be unique in that the long-wavelength anomalies in the magnetic, gravity, and heat flow data are thought to reflect relatively recent tectonic processes - a set of processes that differ from those that formed the basin in early Cenozoic time. For example, crust beneath the Aleutian Basin is believed to have formed by the entrapment of a piece of Mesozoic sea floor that was imprinted with north-south magnetic lineations. Superimposed upon these sea floor-spreading lineations are broad northeast trending magnetic anomalies that may be the result of late Cenozoic processes acting beneath the basin.

The tectonic setting of the Aleutian arc is somewhat different from that of many other island arcs in that the relative motion of the Pacific/North America plates changes from oblique subduction to a strike slip mode along the Aleutian arc (Fig. 1). The transition is believed to occur near Buldir Island (176°E), the westernmost Quaternary volcano along the Aleutian arc [Marlow et al., 1973]; seismicity studies [Cormier, 1975] and plate reconstructions [Minster et al., 1974] also support this idea. A dredge sample of extrusive volcanic rocks (0.6 m.y. age) from west of Buldir Island [174.8°E; Scholl et al., 1976], however, suggests that the transition may actually occur farther west, between Buldir and Agattu Islands (173.5°E).

Shirshov Ridge, which forms the apparent structural boundary between the Aleutian and Komandorsky Basins, lies west of the area of transition in relative plate motions. The position (170.5°E) and early Tertiary (or older?) age of Shirshov Ridge suggest that it is not related to the current underthrusting regime on the frontal Aleutian arc. However, the apparent change in the trend of the long-wavelength anomalies at Shirshov Ridge suggests that it may be a major vertical boundary that effectively isolates the eastern and western parts of the Bering Sea Basin. Consequently, the apparent regional changes in geophysical data in this area could be related to subbasin processes generated by differing relative plate motions (oblique underthrusting and strike slip motion).

Eastern Bering Sea. In the eastern part of the Bering Sea Basin several relatively high heat flow values are found 700 to 800 km behind the Aleutian trench. Based on the prevalence of anomalously low magnetic values in the center of the Aleutian Basin (Fig. 4, 6) we postulate that the higher heat flow values here, although limited, may be characteristic of the central basin. If the heat flow is relatively high, then models that use mantle diapirs to explain the high heat flow in other marginal basins [Hasebe et al., 1970; Oxburgh and Turcotte, 1971; Karig, 1971; Packham and Falvey, 1971] would be difficult to apply in the Bering Sea because of the greater distances involved here (Fig. 1). A more likely model is one that incorporates convective flow within the asthenosphere [Sleep and Toksoz, 1971; Toksoz and Bird, 1976], as it can provide a source of heat at a large distance from the arc-trench system. If the Sleep and Toksoz [1971] model is applied to the Aleutian Basin (Fig. 7) and is scaled to a 350 km deep Benioff zone, the position of the upwelling convective cells would provide hotter material at the base of the lithosphere beneath the Aleutian Basin. The model in Figure 7 requires the assumption that the Benioff zone may extend 75 km deeper than the present zone of seismicity [Engdahl, 1973]; this assumption is based on plate tectonic models for the north Pacific [Larson and Pitman, 1972; Delong et al., 1975;

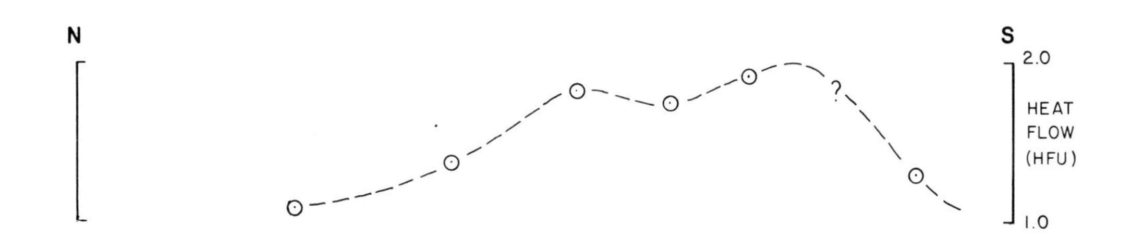

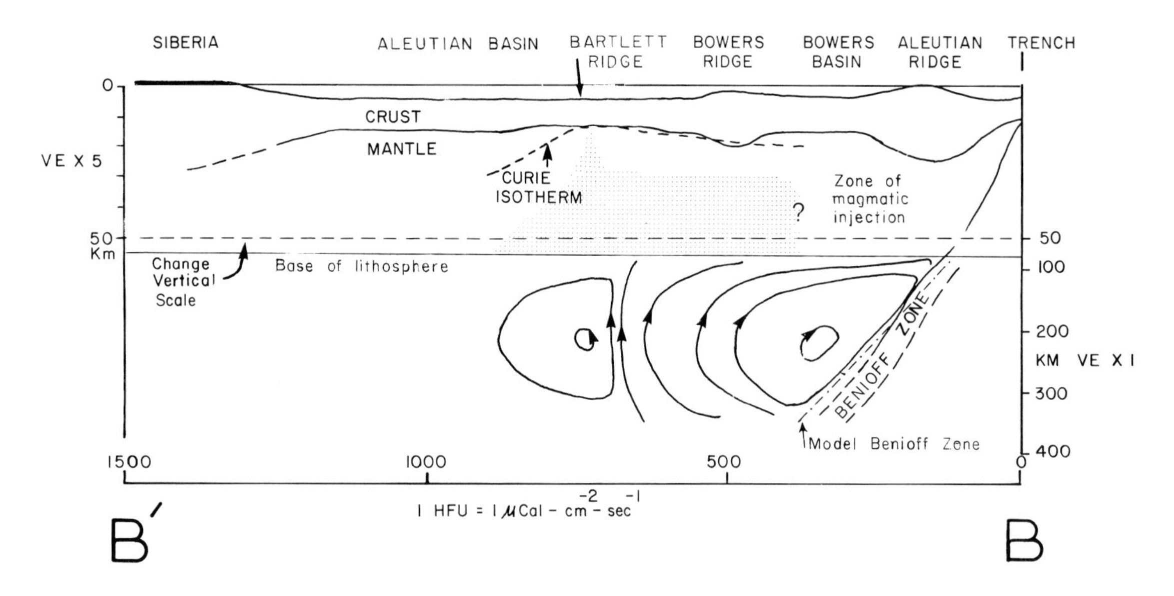

Fig. 7. Interpretative model for the eastern Bering Sea Basin, based on a generalized model for convective asthenospheric flow by Sleep and Toksoz [1971]. The Bering Sea model is scaled for asthenospheric depths ranging from 80 km to 350 km. The elevated depth to the Curie isotherm, based on Figure 6, may reflect the existence of magmatic injections within the upper part of the lithosphere. Heat flow values at top of figure are the uncorrected, observed values from Langseth and Horai [unpub. man.]. See Figure 1 for location of profile (B-B').

Cooper et al., 1976b] that imply that older oceanic plate has been subducted and may exist at greater depths than is reflected by the present seismicity. If a 275 km-deep Benioff zone is used, the divergence point for the convective cells shifts about 150 km to the east, to an area in the Aleutian Basin just north of Bowers Ridge. In either case, the area of maximum upwelling (Fig. 7) generally corresponds to the postulated rise in the Curie isotherm (Fig. 6) and with the area where some higher heat flow measurements are recorded (Fig. 5).

A purely convective model that does not include lithospheric thinning or magmatic injection into the lithosphere is difficult to apply to the Bering Sea. A conservative estimate of the time needed to develop a rise in the Curie isotherm, using a composite model which does not incorporate these processes, would include 10 to 15 m.y. for generation of the convective cells [Toksoz, pers. comm.], and an additional 30 m.y. for conductive heating of the upper part of the lithosphere [Lachenbruch et al., 1976, Fig. 10b]. If this composite process had operated in the Bering Sea for 40 to 45 m.y., we would also expect to see ample evidence for substantial northwest-southeast extension and for the presence of a seismic attenuation (low Q) zone beneath the Aleutian Basin. Because these phenomena have not been observed [Cooper et al., 1976a; Barazangi et al., 1975], then either the convective process has not operated, or convection has not developed sufficiently to manifest itself in a characteristic way. Another mechanism, such as injection of magma conduits into the upper parts of the lithosphere, could explain the presumed rise in the Curie isotherm, and would also provide a magma source for the late Cenozoic intrusive activity along the buried Bartlett Ridge and, incidentally, in the vicinity of the Bowers-Shirshov Ridge connection [Rabinowitz, 1974; Scholl et al., 1975].

An alternative explanation, similar to the one proposed for the Sea of Japan [Uyeda and Miyashiro, 1974] is that the present heat flow pattern in the Aleutian Basin may be a remnant of the thermal processes caused by the subduction of an active spreading ridge (Kula-Pacific ridge). Subduction of the Kula-Pacific ridge occurred prior to 35 m.y. ago [Delong et al., 1975] and probably about 45-50 m.y. ago [Marlow et al., 1973; Cooper et al., 1976b, Fig. 4]. These early Tertiary age estimates require the thermal effects of the subducted hot ridge to be present for at least 35 m.y. However, no evidence has been found for the large amount of back-arc extension (north-south (?)) that would presumably be associated with such a very long period of high heat flow, as exemplified by the Sea of Japan model of Uyeda and Miyashiro [1974, Fig. 7].

In fact, the most severe problem associated with the long-term convective model and the subducted ridge model is that they both predict formation of new igneous crust within the Bering Sea but, as noted, back-arc spreading is contradicted by both regional geologic [Scholl et al., 1975] and magnetic data [Cooper et al., 1976a]. Some evidence exists for minor subduction along the north side of Bowers Ridge, possibly due to readjustments of North America-Eurasian plate motions rather than to major internal basin spreading [Kienle, 1971; Ludwig et al., 1971; Pitman and Talwani, 1972]. This southeast directed, intrabasin subduction ceased in early Tertiary time, and therefore does not seem important to the present thermal regime. We do not know why the Aleutian Basin has apparently not been subjected to major episodes of extensional growth since its latest Mesozoic or earliest Tertiary formation, especially in view of the several thousand kilometers of oceanic lithosphere that have underthrust the Aleutian arc [Larson and Pitman, 1972].

We feel that a convective flow model for the Aleutian Basin (Fig. 7) is a plausible, although not definitive, mechanism for explaining the observed and postulated regional variations in magnetic, heat flow, and gravity data. The process may have started relatively recently, beginning 15 to 20 m.y. ago, with the initiation of sub-basin convective cells that triggered magma injection into the upper part of the lithosphere in late Cenozoic time. The first evidence of possible basin extension in the vicinity of the Shirshov-Bowers-Bartlett Ridge area is just now being seen.

Western Bering Sea. The tectonic setting of the Komandorsky Basin differs from that of the Aleutian Basin in that there is limited seismic evidence for subduction beneath the western strike-slip segment of the Aleutian arc [Fig. 1; Cormier, 1975]. Consequently, subduction-related processes that may explain the regional geophysical observations in the Aleutian Basin presumably cannot be applied to the Komandorsky Basin. Cormier [1975] has proposed that the rough basement topography, reduced sediment thickness, and high heat flow characteristic of the basin are the results of a mid-Cenozoic emplacement of basaltic magma. The magmatism was initiated when, along the western Aleutian arc, subduction of Kula lithosphere ceased (30 to 40 m.y. ago); release of compressive stress then allowed a mantle diapir to form beneath the basin. Since this time, the diapir has migrated toward the southwest part of the basin.

The high heat flow values and the associated regional magnetic lows now found directly behind the Aleutian arc are difficult to reconcile with Cormier's interesting model. These observations, mostly from areas of thin sediment (1 to 2 km thick), indicate that more heat is being liberated than could be provided by a mantle diapir that has been cooling for 30 m.y. This relation implies that either another magmatic episode has occurred within the past several million years, or that small amounts of underthrusting, sufficient to maintain the heat balance through diapiric intrusion, are occurring beneath the western sector of the Aleutian arc. The position of the regional geophysical anomalies, especially the heat flow, adjacent and parallel to the backside of the western Aleutian arc favors the idea of limited underthrusting, or possibly a 'bleeding' plate boundary [Scholl et al., 1973; Scholl et al., 1976].

Cormier's model is especially appealing because it explains why widespread Cenozoic magmatism occurred in the western and not the eastern part of the Bering Sea. An important corollary is that a deepseated mantle process, such as a diapir rising from a depth of several hundred kilometers, can apparently be confined to a laterally small area without 'leaking' into adjacent sub-basin areas.

Although a diapir may have formed beneath the Komandorsky Basin, there is little evidence for systematic spreading and subduction within the basin since the mid-Miocene [15 m.y., Scholl et al., 1975]. Weak north-south magnetic lineations [Cooper et al., 1976a, c] and a sediment-filled depression along the western basin margin may have resulted from asymmetric spreading during the mid-Tertiary; however, the data are presently too limited to unequivocally prove or disprove a spreading hypothesis. If the north-south trends of the short-wavelength anomalies are the result of diapiric injection and spreading, then the apparent discordance in the trends of the long and short wavelength anomalies suggests that Cormier's mantle diapir has altered its shape and position since its inception.

Summary and Concluding Remarks

The Aleutian Basin and the Komandorsky Basin differ in several important aspects; contrasts exist in the estimated ages of the basement rocks, in the total thicknesses (and possibly in the ages) of the sediments, and in the magnitudes and trends of the long-wavelength geophysical anomalies. This distinction may be due to differing sub-basin processes in the two areas, which may in turn be causally related to differing directions of plate convergence along the western segment (where the relative plate motions along the Aleutian arc are strike slip) and eastern segment (where the predominant motion is oblique subduction beneath most of the Aleutian Basin) of the Aleutian arc. The transition in the present relative plate motion occurs east of Shirshov Ridge. However, Shirshov Ridge, which is an early Tertiary feature, forms the structural boundary between the Aleutian and Komandorsky Basins and does not appear to be related to the present underthrusting regime along the Aleutian arc. Instead, the ridge appears to act as the structural barrier that isolates the different sub-basin processes postulated for the Aleutian Basin and Komandorsky Basin.

In the Aleutian Basin, a broad negative magnetic anomaly trends northeast, perpendicular to the underthrusting direction of the Pacific plate beneath the eastern Aleutian arc. The apparent association of this anomaly with similar northeast-trending heat flow and gravity anomalies 600 to 700 km behind the Aleutian arc may be explained using a model for convective flow in the asthenosphere (Fig. 7). The absence of a high-attenuation (low Q) seismic zone and also of an east-west fossil spreading center within the Aleutian Basin suggests that a convective system, if present, may be in a juvenile stage.

The Komandorsky Basin is characterized by higher heat flow values and by gravity and magnetic anomalies that generally parallel the direction of relative North America-Pacific plate motions along the western part of the Aleutian arc. The proximity of the highest heat flow and lowest magnetic values to the arc may be construed as evidence for some form of mantle diapirism in the Komandorsky Basin that may have been initiated in mid-Cenozoic time and that is currently being renewed by small amounts of episodic underthrusting.

Acknowledgments. We wish to thank fellow participants of the Ewing Symposium for numerous stimulating discussions. We also appreciate the cooperation of M. Langseth and K. Horai in giving us access to their unpublished heat flow data, and of G. Shor and D. Fornari in providing a preprint of their refraction data in the Komandorsky Basin. Critical reviews of the manuscript were made by A. Griscom, J. Case, J. Kienle, and S. Cande. Invaluable assistance in the preparation of the manuscript was provided by K. Bailey and J. Childs.

References

Barazangi, M., W. Pennington, and B. Isacks, Global study of seismic wave attenuation in the upper mantle behind island arcs using pP waves, J. Geophys. Res., 80, 1079-1092, 1975.

Bateman, P. C., and J. P. Eaton, Sierra Nevada Batholith, Science, 158, 1047-1417, 1967.

Beloussov, V. V., and E. M. Ruditch, Island arcs in the development of the Earth's structure, J. Geol., 69, 647-657, 1961.

Clark, S. P., Jr., Ed., Handbook of physical constants, Geol. Soc. Am. Bull., 97, 587 p., 1966.

Cooper, A. K., D. W. Scholl, and M. S. Marlow, Mesozoic magnetic lineations in the Bering Sea marginal basin, J. Geophys. Res., 81, 11, 1916, 1976a.

Cooper, A. K., D. W. Scholl, and M. S. Marlow, A plate tectonic model for evolution of the eastern Bering Sea Basin, Geol. Soc. Am. Bull., 87, 1119-1126, 1976b.

Cooper, A. K., K. Bailey, J. Howell, M. S. Marlow, and D. W. Scholl, Preliminary residual magnetic map of the Bering Sea Basin and Kamchatka Peninsula, U. S. Geological Survey Misc. Field Studies Map MF 715, scale 1:250,000, 1976c.

Cooper, A. K., M. S. Marlow, D. W. Scholl, and J. Howell, Evidence for thermal inhomogeneity beneath the Aleutian Basin, Bering Sea, in prep.

Cormier, V. F., Tectonics near the junction of the Aleutian and Küril-Kamchatka arcs and a mechanism for middle Tertiary magmatism in the Kamchatka Basin, Geol. Soc. Am. Bull., 86, 443-453, 1975.

Creager, J. S., D. W. Scholl, and others, Initial Report of the Deep Sea Drilling Project, 19, U. S. Government Printing Office, Washington, D. C., 1973.

Delong, S. E., P. J. Fox, and F. W. McDowell, Subduction of the Kula Ridge at the Aleutian Trench, EOS Transactions, Am. Geophys. Union, 56, 1066, 1975.

Engdahl, E. R., Relocation of intermediate depth earthquakes in the central Aleutians by seismic ray tracing: Nature, 245, 23-26, 1973.

Fornari, D. J., and G. G. Shor. Jr., Crustal structure of the Kamchatka Basin, western Bering Sea, EOS Transactions, Am. Geophy. Union, 57, 264, 1976.

Foster, T. D., Heat flow measurements in the northeast Pacific and in the Bering Sea, J. Geophys. Res., 67, 2991-2993, 1962.

Hasebe, K. N., N. Fujii, and S. Uyeda, Thermal processes under island arcs, Tectonophysics, 10, 335-355, 1970.

Jacob, K. H., Global tectonic implications of anomalous seismic p-travel times from the nuclear explosion Longshot, J. Geophys. Res., 77, 2556-2573, 1972.

Karig, D. E., Origin and development of marginal basins in western Pacific, J. Geophys. Res., 76, 2542-2561, 1971.

Karig, D. E., Remanent arcs, Geol. Soc. Am. Bull., 83, 1057-1068, 1972.

Kienle, J., Gravity and magnetic measurements over Bowers Ridge and Shirshov Ridge, Bering Sea, J. Geophys. Res., 76, 7138-7153, 1971.
Lachenbruch, A. H., J. H. Sass, R. J. Munroe, and T. H. Moses, Jr., Geothermal setting and simple heat conduction models for the Long Valley Caldera, J. Geophys. Res., 81, 769-784, 1976.
Langseth, M. G., and K. Horai, Heat flow in the marginal seas of the northern and western Pacific - part 1: Heat flow in the Bering Sea, unpublished manuscript.
Larson, R. L., and W. C. Pitman, III, World-wide correlation of Mesozoic magnetic anomalies and its implications, Geol. Soc. Amer. Bull., 83, 3645-3662, 1972.
Ludwig, W. J., Nafe, J. E., and C. L. Drake, Seismic refraction, in The Sea, 4, pt. 1, ed. A. E. Maxwell, 53-84, Wiley-Interscience, N. Y., 1970.
Ludwig, W. F., S. Murauchi, N. Den, M. Ewing, H. Hotta, R. E. Houtz, T. Yoshii, T. Asanuma, K. Hagiwara, T. Saito, and S. Ando, Structure of Bowers Ridge, Bering Sea, J. Geophys. Res., 76, 6350-6366, 1971.
Marlow, M. S., D. W. Scholl, E. C. Buffington, and T. R. Alpha, Tectonic history of the central Aleutian arc, Geol. Soc. Am. Bull., 84, 1555-1574, 1973.
Minster, J. B., T. H. Jordan, P. Molnar, E. Haines, Numerical modeling of instantaneous plate tectonics, Geophysics J. Roy. Astr. Soc., 36, 541-576, 1974.
Oxburgh, E. R., and D. H. Turcotte, Origin of paired metamorphic belts and crustal dilation in island arc regions, J. Geophys. Res., 76, 1315-1327, 1971.
Packham, G. H., and D. A. Falvey, An hypothesis for the formation of marginal seas in the western Pacific, Tectonophysics, 11, 79-109, 1971.
Pitman, W. C., III, and M. Talwani, Sea-floor spreading in the north Atlantic, Geol. Soc. Am. Bull., 83, 619-646,1972.
Rabinowitz, P. D., Seismic profiling between Bowers Ridge and Shirshov Ridge in the Bering Sea, J. Geophys. Res., 79, 4977-4979, 1974.
Regan, R. D., J. E. Cain, and W. M. David, A global magnetic anomaly map, J. Geophys. Res., 80, 794-802, 1975.
Sass, J. H. and R. J. Munroe, Heat flow from deep boreholes on two island arcs, J. Geophys. Res., 75, 4387-4395, 1970.
Scholl, D. W., M. S. Marlow and E. C. Buffington, Pliocene and Pleistocene eruptive rocks of the western Aleutian Ridge - a bleeding strike-slip plate boundary (?), Geol. Soc. Am. Abs. with programs, 5, No. 1, 101, 1973.
Scholl, D. W., E. C. Buffington, and M. S. Marlow, Plate tectonics and the structural evolution of the Aleutian-Bering Sea region, in Forbes, R. B., ed., The Geophysics and Geology of the Bering Sea Region, Geol. Soc. Amer. Mem., 151, 1975.
Scholl, D. W., M. S. Marlow, N. S. Macleod, E. C. Buffington, Episodic Aleutian Ridge igneous activity: Implications of Miocene and younger submarine volcanism west of Buldir Is., Geol. Soc. Am. Bull., 87, 547-554, 1976.
Sclater, J. G., J. W. Hawkins, J. Mammerickx, and C. Chase, Crustal extension between the Tonga and Lau ridges: Petrologic and geophysical evidence, Geol. Soc. Amer. Bull., 83, 505-518, 1972.
Sclater, J. G., Heat flow and elevation of the marginal basins of the western Pacific, J. Geophys. Res., 77, 5705-5720, 1972.
Shor, G. G., Jr., Structure of the Bering Sea and the Aleutian Ridge, Marine Geology, 1, 213-219, 1964.
Sleep, N., and M. N. Toksoz, Evolution of marginal basins, Nature, 23, 548-550, 1971.
Talwani, M., Gravity, in The Sea, 4, ed. A. E. Maxwell, 251-297, J. Wiley and Sons, New York, 1971.
Toksoz, M. N., and P. Bird, Effects of subducting lithosphere and marginal basins, in The Ewing Symposium, 24, 1976.
Uyeda, S., and A. Miyashiro, Plate tectonics and the Japanese Islands: A synthesis, Geol. Soc. Am. Bull., 85, 1159-1170, 1974.
Watanabe, T., M. G. Langseth, and R. N. Anderson, Heat flow in back-arc basins of the western Pacific, this volume.
Watts, A. B., Gravity field of the northwest Pacific Ocean basin and its margin: Aleutian arc-trench system, Geol. Soc. Am. Map and Chart Services, MC-10, 1975.
Watts, A. B., and J. L. Weissel, Tectonic history of the Shikoku marginal basin, Earth and Plan. Sci. Lett., 25, 239-250, 1975.
Weissel, J. K., and A. B. Watts, Middle to late Cenozoic evolution of the South Fiji and Lau marginal basins, in The Ewing Symposium, 25, 1976.

THE STRUCTURE AND AGE OF ACOUSTIC BASEMENT IN THE OKHOTSK SEA

C. A. Burk

Marine Science Institute, The University of Texas, P.O. Box 7999, Austin, Texas 78712

H. S. Gnibidenko

Sakhalin Complex Scientific Research Institute, Novoalexandrovsk, Sakhalin 694050, USSR

Abstract. This paper is the result of the authors' cooperative work during the 13th cruise of the *R/V Mendeleev* in the Okhotsk Sea in August-October 1974. During this cruise the authors dredged apparent outcrops of acoustic basement in the central part of the Okhotsk Sea, associated with normal faults and structural arches. The rocks dredged are related to the andesitic assemblages, consisting of igneous and volcanogenic sedimentary rocks. The compressional wave velocities of individual samples range from 4.3 to 7.6 km/sec. These rocks have locally been metamorphosed and strongly deformed. Ages based on K/Ar measurements indicate late Cretaceous metamorphic and igneous events (73 to 95 m.y.), with one anomalous middle Triassic age (206-209 m.y.). These rocks may represent ancient island arcs now covered by waters of the Okhotsk Sea.

Basic Geological Structure

The continental crust of the Okhotsk Sea (Fig. 1, 2) is distinct from the oceanic crust underlying the Kurile Basin. The regional geology of the Okhotsk Sea has been described recently by Burk (1975). According to deep acoustic-refraction structure, the continental crust is subdivided here into the northern part with a thickness of about 30 km and a southern part with thickness ranging from 10-25 km (Yanshin, 1976). These regions also differ in the upper-crustal structure, where a system of fault-blocks can be recognized in the southern part. In regional marine surveys, the basement complex (the acoustic basement) of the Okhotsk Sea averages 4.5-6.2 km/sec and the sedimentary cover averages 2.0+ km/sec (Fig. 3). In the Kurile Basin, the sedimentary cover is distinctly subdivided into stratified and transparent parts, the lower of which (transparent) has a velocity of about 2.4-3.0 km/sec. The Kurile Basin has been interpreted as a relict basin with suboceanic crust (Snegovskoi, 1974) or as a trapped bit of ancient oceanic crust (Burk, 1975).

The sedimentary cover of the shallower Okhotsk Sea overlaps and levels the rough surface of the basement complex. The sedimentary cover ranges from 0.25-0.75 in thickness, and in the basement troughs it increases to 1.5-2 km, rarely reaching 3-4 km, and is characterized by well-reflected horizons of wide distribution. The sedimentary cover can be subdivided into two strata, the lower of which seems to have a velocity of about 3 km/sec.

According to seismic profiling data, the relationships between the sedimentary cover and the acoustic basement near uplifts are nonconformable, but in a few cases, these relationships appear to be conformable. It is likely that some of the rounded material dredged from the basement outcrops are the result of conglomerates occurring at the foot of these uplifts, entirely within the Sea of Okhotsk. On the whole, this acoustic unconformity becomes rather more pronounced towards Sakhalin, Kamchatka and the northern coast of the Okhotsk Sea, where it can be correlated in some places with the boundary between the Cenozoic sedimentary-volcanogeneous deposits and more ancient folded and metamorphosed complexes.

Academy of Sciences Rise

Two samples of possible acoustic basement were dredged approximately 75 km north of the Academy of Sciences Rise (Fig. 1). The north and south flanks of this local exposure were both dredged (samples 961-1, 2) yielding cobbles and pebbles to a maximum size of about 15 cm. The rocks consisted of coarse-grained quartz diorite, various andesites (commonly porphyritic), rhyolite porphyry, sedimentary clastic rocks (graywackes) of apparent volcanogenic origin, and some bedded chert; metamorphism was only very slight. Overall composition of all rocks is compatible with the "andesite suite." No glacial

TABLE 2. Chemical Analyses of Dredged Rocks

	962-1 Rhyolite Porphyry	962-2 Granodiorite	967-1 Granodiorite
SiO_2	68.60	62.98	61.82
TiO_2	0.28	0.47	0.47
Al_2O_3	16.20	15.78	15.98
Fe_2O_3	1.75	3.02	2.48
FeO	1.98	3.77	3.69
MnO	0.10	0.14	0.16
MgO	0.50	2.51	2.76
CaO	3.12	6.28	5.25
Na_2O	3.78	1.85	2.70
K_2O	3.05	2.44	2.75
H_2O	0.92	0.71	1.69
SO_3	--	--	--
P_2O_5	0.04	0.04	0.05
Lost remainder	1.23	0.63	2.20
TOTAL	100.32	99.80	99.80

Analyst Sergeeva, Far East Geological Institute, Vladivostok, USSR

pressional-wave velocities ranging from 5.0-5.2 km/sec for rhyolite porphyry (density of 2.49 g/cm^3) and 5.4-6.5 km/sec for quartz-diorite (density of 2.68 g/cm^3) to 7.6 km/sec for andesite-porphyry (density of 2.68 g/cm^3). The densities of five samples of igneous rocks and one sample of phyllite range from 2.49-2.69 g/cm^3 and the seismic compressional-wave velocities of these samples are from 4.3 to 7.6 km/sec.

K-Ar Ages of the Dredged Rocks

K-Ar ages of the dredged samples from the Okhotsk Sea (Fig. 3, Tables 3 and 4) are largely Upper Cretaceous. According to A. Y. Illyov (Sakhalin Complex Scientific Research Institute, personal communication), the plutonic rocks dredged in the central part of the Academy of Sciences Rise have a whole-rock K-Ar age of 113-122 m.y. (late Lower Cretaceous), and the grandiorite from St. Iona Islet in the northwestern part of the Sea is 45-53 m.y. by whole-rock K-Ar determination (V. I. Nariyzhny, personal communication).

The Middle Triassic age of the granodiorite from the northern part of the Academy of Sciences Rise (206-209 m.y.) probably testifies to the Lower Triassic and possibly Upper Paleozoic age of the folded geosynclinal sequence which is cut by this granodiorite. The isopach scheme (Fig. 3) and magnetic (Kochergin, et al., 1970) and gravity (Gainanov, et al., 1974) anomaly patterns indicate a structural connection of the Academy of Sciences Rise and the Sredinny horst-anticlinal zone of the Kamchatka Peninsula through the Bolsheretsky Uplift (Fig. 3). The metamorphic complex in Sredinny is formed from Triassic and probably older geosynclinal volcanic and sedimentary rocks (Gnibidenko, et al., 1972). It is possible that in the west, the Academy of Sciences Rise is structurally connected with Lower Mesozoic geosynclinal complexes (Gnibidenko, 1971) in the central part of Sakhalin. On this basis it is reasonable to propose the existence of an Upper Paleozoic-Mesozoic island-arc geosynclinal system along the edge of the Okhotsk Sea continental margin, submerged in Neogene time.

The dredged samples were also dated and examined in detail by Dr. R. E. Dennison of Mobil Oil Corporation by separating mineral fractions that were present in sufficient amounts. This was in some cases less than ideal since the most suitable mineral often was not present in sufficient amounts. We would also liked to have dated some mineral pairs (e.g., hornblende-biotite) in order to evaluate the possibility of later heating, but again these pairs were not present in sufficient quantities from the limited specimens. Nonethe-

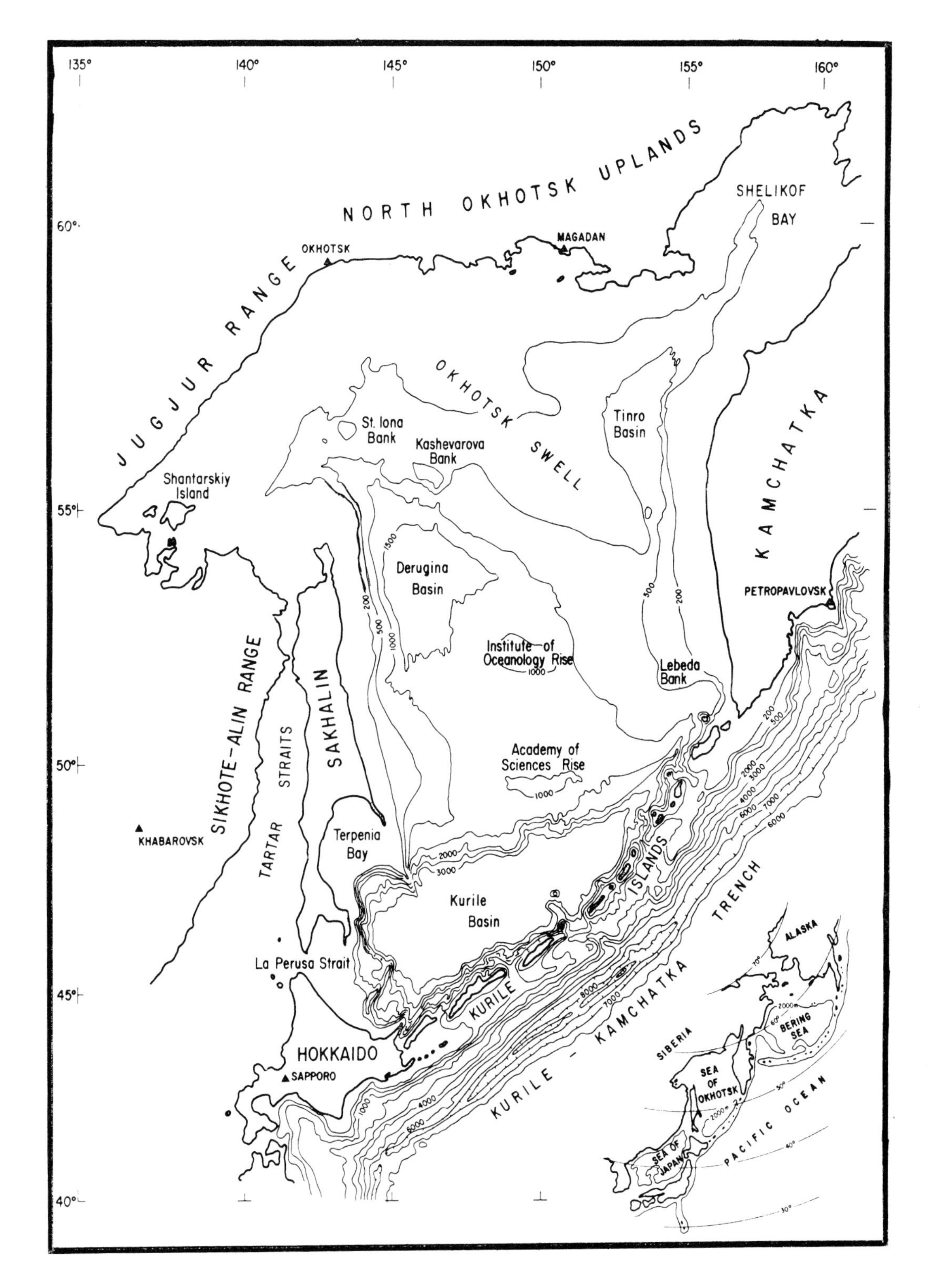

Fig. 1. Isobath map of the Okhotsk Sea showing major topographic features (after Burk, 1975).

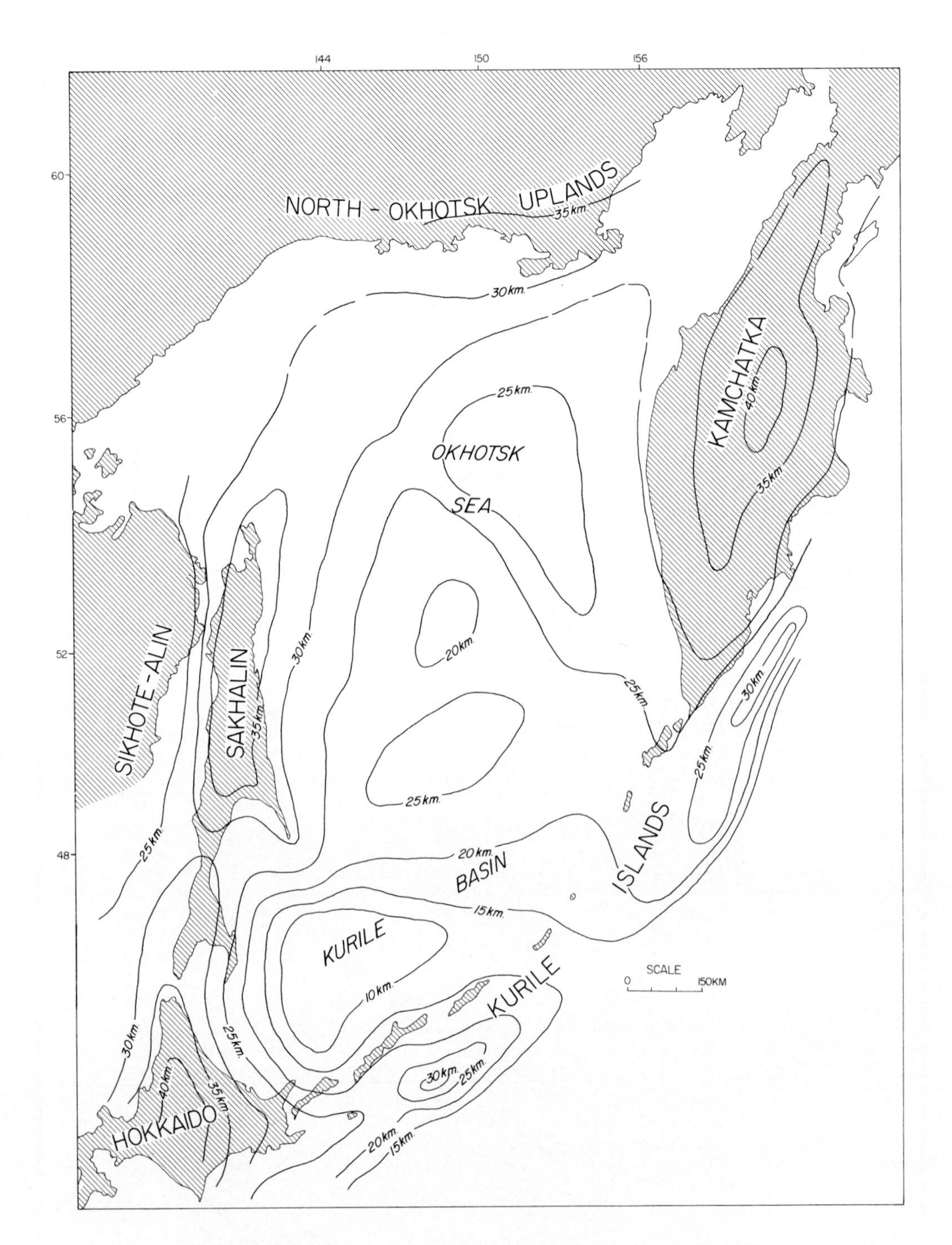

Fig. 2. Crustal thickness in the Okhotsk Sea and the surrounding area in isopachs (km) according to Gnibidenko (1971).

striae or faceting was observed on any of the rock samples (Tables 1 and 2).

A second site on the Academy of Sciences Rise (962), approximately 10 miles farther north, yielded a very large sample of cobbles and boulders (to approximately 60 cm), plus many angular fragments. The two largest boulders were quartz diorite and quartz rhyolite. Some granodiorite and granite also were present. Volcanic rocks were andesites of various types, and the sedimentary rocks were volcanogenic graywackes and tuffs, with some possible bedded and recrystallized chert. Some samples were slightly more metamorphosed than at the previous site, but none

TABLE 1. Bulk Mineral Composition of Plutonic Rocks

(R. E. Dennison, Mobil Oil Corporation)

Sample	Quartz	Plagioclase	K feldspar	Biotite	Hornblende	Chlorite	Iron Oxides	Others	Color Index
962-2	21.5	46.4	9.5	9.2	11.0	0.5	1.0	0.9	22.1
962-3	23.4	57.0	4.5	9.5	3.6	0.6	1.3	0.1	15.0
962-4	22.5	42.5	26.2	3.3	--	3.9	0.8	0.8	9.0
967-1	19.3	50.3	13.5	7.2	0.3	6.8	1.7	0.9	16.0

exceeded hornfels facies (some slaty cleavage). In general, the rocks again are compatible with each other and related to the "andesite suite"; but the samples from this site are more closely related to each other than to those of the first site in that all rock types are markedly quartz-rich. The quartz-diorite contains very abundant visual quartz, some of the andesites can be described only as "quartz-andesite", and the graywackes contain noticeable quartz. No glacial striae or faceting were noted.

It is impossible to establish definitely whether such rocks represent local erosion or were brought great distances in floating ice. However, several factors suggest that these rocks are locally derived and represent a fair sample of adjacent acoustic basement: all are of compatible lithology of the "andesite suite"; significant and consistent differences in quartz content distinguish the two localities; no entirely incompatible rocks are present to suggest distinctly different provenances; no indications of glacial scars, striae, or faceting were found.

Institute of Oceanology Rise

One outcrop of apparent acoustic basement was dredged about 75 km north of the Institute of Oceanology Rise (967), recovering a large volume of boulders, cobbles and pebbles up to about 40 cm. The largest was an angular block of quartz diorite, which was also common in smaller fragments. Also present was granodiorite, abundant quartz rhyolite, altered dacite and andesite; abundant blocks and stones of gray phyllite (commonly with slaty cleavage) and fragments of greenschist; one price of mica schist; numerous coarse-to-fine graywacke sandstones and siltstones, and some gray-to-brown bedded chert.

As with earlier dredges, it is impossible to establish that these fragments represent local erosion of acoustic basement, but several factors suggest that most were not transported far: (1) the rock assemblage is largely compatible with a limited outcrop area, and exotic kinds are rare; (2) the assemblage is distinct from all others in abundance of greenschist, phyllite and graywacke, and limited in abundance of andesite; and (3) no glacial striae or faceting were found.

Physical Properties of Dredged Rocks

Seismic compressional-wave velocities, as determined in the laboratory, of plutonic rocks dredged about 75 km north of the Institute of Oceanology Rise range from 4.3-5.5 km/sec (density of 2.60-2.68 g/cm^3) and in the phyllite from the same place, 4.7-6.6 km/sec (density of 2.69 g/cm^3). Bedrocks dredged from the north edge of the Academy of Sciences Rise have com-

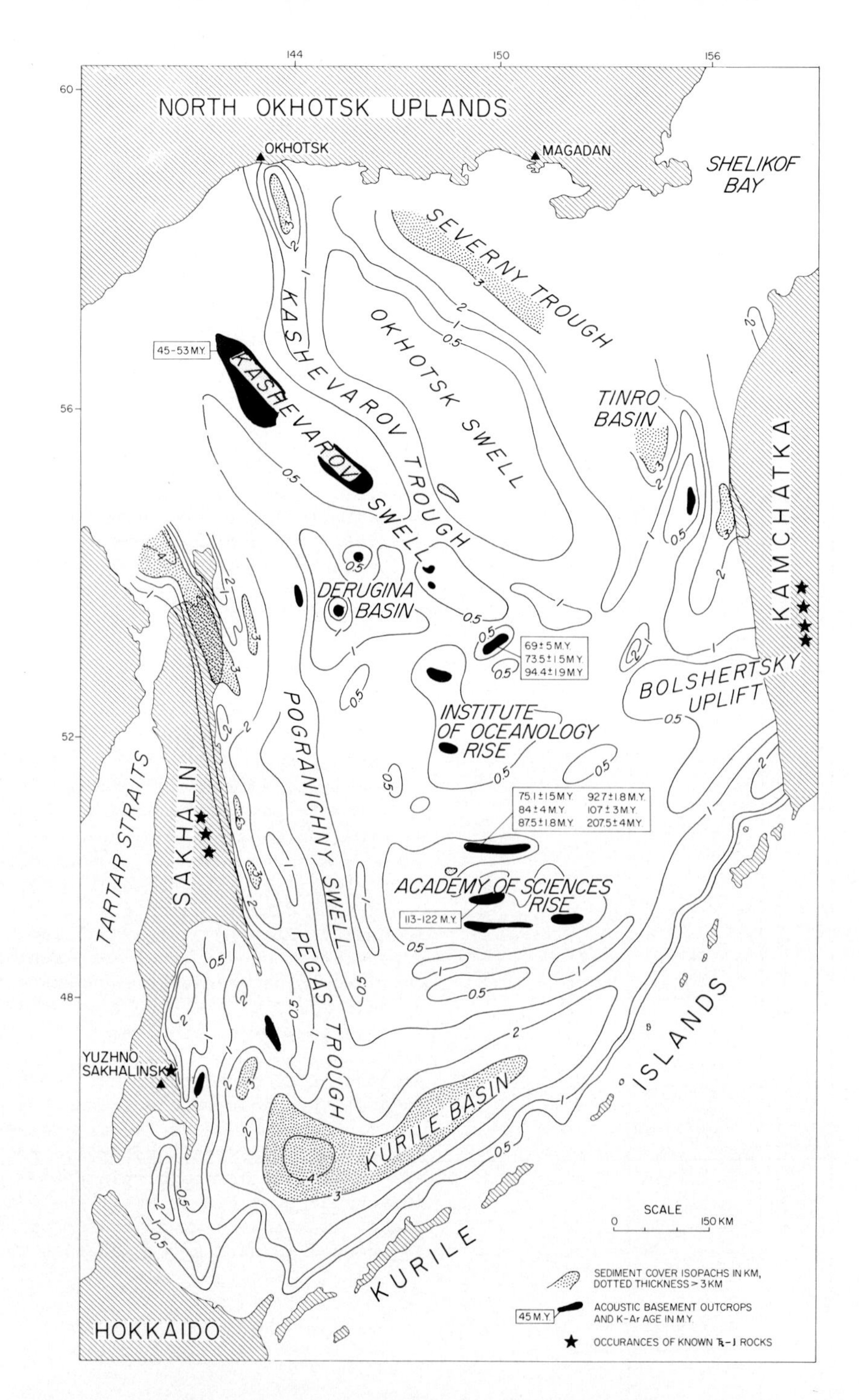

Fig. 3. Thickness of the sedimentary cover in the Okhotsk Sea having apparent average velocity of 2 km/sec: (1) isopach; (2) outcrops of the acoustic basement on the sea-bottom and radiometric age determinations; (3) occurrences of known and suspected geosynclinal Jurassic-Triassic and Upper Paleozoic rocks.

TABLE 3. K-Ar Age of the Rock Samples*

Sample N	Location	Rock	Age in m.y.		Time of the event
			By separating mineral fractions	Whole rock	
962-1	50° 30.2' N 149° 43.05' E Northern Academy of Sciences Rise	Rhyolite porphyry	87.7 ± 1.8 87.3 ± 1.8 (feldspar)	84 ± 4	Middle part of Upper Cretaceous
962-2		Granodiorite	91.0 ± 1.8 94.7 ± 1.9 (biotite)	107 ± 3	Early Upper Cretaceous
962-3		Granodiorite (mildly sheared)	206 ± 4 209 ± 4 (biotite)		Middle Triassic
962-4		Granite	75.8 ± 1.5 74.6 ± 1.5 (feldspar)		Late Upper Cretaceous
967-1	53° 76.55' N 149° 01.1' E Northern Institute of Oceanology Rise	Granodiorite	93.5 ± 1.9 95.4 ± 1.9 (feldspar)	69 ± 5	Early Upper Cretaceous
967-2		Biotite schist	74.0 ± 1.5 73.0 ± 1.5 (biotite)		Late Upper Cretaceous

*Radiometric K-Ar dating by separating mineral fractions was made by Dr. R. E. Dennison of Mobil Research and and Development Corporation and whole rock K-Ar dating was made by the Far East Geological Institute (Vladivostok).

TABLE 4. Analytical K-Ar Results

Sample # Mineral	Sample Wt. Grams	% K	Ar^{40*} x 10^{-10} moles	Ar^{40*} / Ar^{40} total	m.y.
962-1	0.762	3.990	4.846	92.2	87.7 ± 1.8
	0.701		4.443	89.1	87.3 ± 1.8
962-2	0.523	6.296	5.460	87.3	91.0 ± 1.8
	0.511		5.558	87.6	94.7 ± 1.9
962-3	0.361	6.129	8.575	94.1	206.0 ± 4.0
	0.366		8.799	93.7	209.0 ± 4.0
962-4	0.515	5.784	4.091	52.3	75.8 ± 1.5
	0.421		3.296	50.3	74.6 ± 1.5
967-1	0.411	7.474	5.235	80.5	93.5 ± 1.9
	0.434		5.637	81.8	95.4 ± 1.9
967-2	0.351	6.462	3.045	84.2	74.0 ± 1.5
	0.341		2.918	82.0	73.0 ± 1.5

Ar^{40*} = radiogenic argon

(Data provided by R.E. Dennison, Mobil Oil Corporation)

less the following results, summarized from Dennison, are reasonably consistent, indicating one period of intrusion-metamorphism, and suggesting another.

962-1 Rhyolite porphyry, Northern Academy of Sciences Rise
87.7 ± 1.8 million years K-Ar feldspar
87.3 ± 1.8 million years K-Ar feldspar

Phenocrysts of quartz, plagioclase, potash feldspar, and altered biotite are set in a finely crystalline groundmass of quartz-feldspar. The quartz phenocrysts are large and show cracks and broken-to-embayed outlines. The plagioclase is generally fresh but contains irregular masses of clay-sericite and has numerous cracks along which iron-stained clay minerals have invaded. It appears to be in the sodic oligoclase-calcic albite composition range. The biotite was originally red-brown, but is now much replaced by iron oxides and clay-like alteration material. The groundmass shows faint but well-defined eutaxitic structures that are particularly well-defined between large phenocrysts. Most of the iron oxides are now converted to hematite. Traces of zircon and apatite are accessory minerals.

This is the least suitable of all the rocks for dating. The fact that it yielded an age intermediate between the other Cretaceous ages is encouraging. It is almost certainly a minimum age and may actually have been extruded in mid-Cretaceous and early Late Cretaceous. The complete preservation of delicate original crystallization structures is excellent evidence of a simple thermal history.

962-2 Granodiorite, Northern Academy of Sciences Rise
91.0 ± 1.8 million years K-Ar biotite
94.7 ± 1.9 million years K-Ar biotite

Plagioclase, quartz, and perthite are the main rock-forming minerals with substantial amounts of both biotite and hornblende. The plagioclase is sodic andesine near the cores and is intermediate oliogoclase at the zoned margins. The plagioclase is mostly fresh, but contains locally well-developed sericite-clay mats and cracks along which iron oxides have been introduced. Some of these larger crystals contain small, round poikilitic inclusions of iron oxides, hornblende, and pyroxenes. Perthite is fresh and lightly dusted with hematite and vacuoles. Quartz is mildly strained to unstrained. The femic minerals are in clots in association with iron oxides. These include a pale green common hornblende, red-brown biotite and lesser amounts of pyroxene. Epidote and well-crystallized anhedral sphene are concentrated in the femic mineral clots. Minor amounts of the biotite have been replaced by chlorite. Traces of zircon and more numerous apatite needles are found as accessory minerals. Minor, sparse blue-green tourmaline is found as inclusions in some quartz. The perthite occupies a late intergranular position essentially between, and occasionally surrounding, plagioclase. Quartz tends to occupy this same general position.

The determined age is early Late Cretaceous and should be close to the time of crystallization of the intrusion. The hornblende is fresh and could have been used to check this age if a sufficient quantity could have been separated from the sample.

962-3 Granodiorite, Northern Academy of Sciences Rise
206 ± 4 million years K-Ar biotite
209 ± 4 million years K-Ar biotite

The rock shows a general mild shearing which is marked by straining and some recrystallization of quartz, and bent plagioclase laths and biotite books. Most of the rock is composed of plagioclase, quartz, and perthite with both biotite and hornblende as prominent accessory minerals. The plagioclase is irregularly zoned with cores in the calcic andesine range and the margins calcic oligoclase. The plagioclase carries minor irregularly-distributed sericitic alteration and is mostly fresh. The hornblende is a moderately pleochroic, common type. The biotite is a reddish brown color, contains minor chloritic and other alterations and, as previously noted, shows some bending and development of kink bands. Perthite is fresh and is found in large intergranular crystals. Iron oxides have an irregular shape and are found in association with relatively large and abundant apatite crystals. Minor amounts of zircon and sphene are found as primary minerals and epidote occurs as a secondary mineral.

The Triassic age was not expected. The rock has a composition well within the range of those yielding Cretaceous ages. The only difference is that this granodiorite has undergone shearing after crystallization, something that cannot be demonstrated for the other rocks of this composition. The age is tentatively interpreted as the time of crystallization of the granodiorite but may have been lowered during the shearing episode. An attempt to separate the hornblende to check the age did not meet with success due to the small sample.

962-4 Granite, Northern Academy of Sciences Rise
75.8 ± 1.5 million years K-Ar feldspar
74.6 ± 1.5 million years K-Ar feldspar

Discrete plagioclase crystals 3 to 4 mm in length are set in a matrix composed mostly of quartz-perthite micrographic intergrowth and smaller plagioclase crystals. In addition, amphibole, chlorite, biotite, iron oxides, epidote, apatite, zircon, feldspar alterations, tourmaline, and sphene are present. Red-brown biotite has nearly all been replaced by chlorite with attendant sphene. The amphibole is a common type in well-formed crystals and actinolite in secondary sheaves of crystals. The large plagioclase crystals are lightly and irregularly dusted with alterations, but generally clear, and are in the calcic oligoclase-sodic andesine composition range and are irregularly zoned. The smaller matrix crystals are clouded with uniform dust and vacuoles as well as some discrete sericite flakes. These plagioclases are in the sodic oligoclase-calcic albite composition range. Perthite is fresh and contains the same clouding dust-vacuoles of the small plagioclase. Quartz is very mildly strained to unstrained, and for the most part is in well-defined uniform intergrowths with perthite. Epidote is mostly associated with chlorite, but is also found as well-formed discrete crystals.

The determined age is in the younger part of Late Cretaceous and should be close to the time of crystallization. Other ages determined on biotite support this age.

967-1 Granodiorite, Northern Institute of Oceanology Rise
93.5 ± 1.9 million years K-Ar feldspar
95.4 ± 1.9 million years K-Ar feldspar

Most of the rock is quartz, plagioclase, and perthite with lesser amounts of hornblendes, pyroxene, chlorite, biotite, iron oxides, apatite, epidote, sphene, zircon, and calcite. The plagioclase is generally fresh, is strongly zoned only at the margins, and is in the sodic andesine-calcic oligoclase composition range. Some plagioclase crystals have extensive sericite development. Quartz is mildly strained and contains dust-like particles in linears. The perthite is fresh and occupies a late, intergranular position. The original biotite is a red-brown color, but most is replaced by chlorite with the development of finely granular sphene along cleavage of the pseudomorphs. Amphibole is found as a mildly pleochroic common hornblende as well as deuteric actinolite. The hornblende occurs as discrete crystals and as partial coronas around pyroxenes and pyroxene replacements. The latter are mostly chlorite, calcite, and actinolite. Iron oxides are found as discrete crystals commonly in clots of accessory minerals. Apatite is typically found as partial or total inclusions in the magnetite. Epidote is found in association with chlorite.

The determined age is early Late Cretaceous and is probably close to the time of intrusion. We would have preferred to date the hornblende, but a clear separation of sufficient quantity could not be made.

967-2 Biotite schist, Northern Institute of Oceanology Rise
74.0 ± 1.5 million years K-Ar biotite
73.0 ± 1.5 million years K-Ar biotite

The schist is composed mostly of biotite, quartz, and plagioclase with lesser amounts of hornblende, epidote, microcline, sphene, hematite, and chlorite. The preferred orientation of the biotite is well-marked, but that of the sparse hornblende is poorly, if at all, defined. The biotite is red-brown in color and is very fresh. Epidote occurs as scattered crystals most often associated with biotite. The plagioclase is poorly twinned, in the calcic albite-sodic oligoclase composition range, and contains locally well-developed fractures along which hematite has invaded. The quartz is mildly strained to unstrained. Microcline is sparse, well-twinned, and very fresh. Virtually the only alterations in the rock are in the form of hematite introduced between grains and along cleavage planes.

The schist is of uncertain origin although it has clearly been recrystallized to middle greenschist facies. The apparent age is in the younger Late Cretaceous and is interpreted as being close to the age of metamorphism. The granitic rocks yielding similar ages support this as, if not the age of metamorphism, the age of a strong thermal event which has reset the radiometric clock.

Discussion

The acoustic basement in the central part of the Okhotsk Sea thus involves folded, geosynclinal formations of Lower Paleogene and mainly Cretaceous ages and, probably, more ancient ages (Lower Triassic-Upper Paleozoic, as indicated by the age of about 207 m.y. for granodiorite from the north side of the Academy of Sciences Rise). The sediment cover involves a quasi-platform sedimentary complex having an age from Paleogene to Holocene. Within the Okhotsk Swell and the northwestern part of the Sea, the basement may be represented by Paleozoic and partly Precambrian folded formations, as indicated by apparent onshore connections.

The basins in the Okhotsk Sea are caused and deformed by Cenozoic movements over deep fracture zones, mainly normal faults of northwestern strike. The basin system (Fig. 3) is buried beneath the Neogene-Quaternary and partly Paleogene quasi-platform cover, and according to the seismic profiling data, the Quaternary deposits of the Okhotsk Sea platform have also been faulted. These data, as well as the earthquake occurrence in the northwestern part of the Okhotsk Sea (Tarakanov *et al*., 1975), testify to the very young and recent deformation over basement fractures, particularly along the flanks of the Okhotsk Swell.

The Kashevarov Trough, extending southeastward about 1,000 km, with a width of about 50-100 km, is a particularly prominent zone, with numerous earthquakes in its northern part (Fig. 3). There are no registered earthquakes in the southeast part of the zone, but this may result from an insufficient network of seismic stations. To the south of the Kashevarov Trough is the Kashevarov Swell, which is a system of horsts of Paleogene-Mesozoic and older folded basement, including Paleozoic and possibly Precambrian rocks in the northern part.

The Severny Trough is located along the northeastern margin of the Okhotsk Swell and extends more than 300 km, with a width of about 150 km. The northwestern part of this graben-synclinal basin is also characterized by relatively high seismicity.

In the central part of the Okhotsk Sea there is a system of grabens, oriented generally northwest. The Derugina Basin begins in the northwest, at the opening of the St. Iona graben. The graben-synclinal trough of the Schmidt Basin (at the northern end of Sakhalin Island) is an analogous and deeper zone. The en-echelon trough along the eastern Sakhalin Island is limited by very young bordering faults. The Institute of Oceanology and the Academy of Sciences Rises are separated from each other by the Makarov graben-synclinal zone. The Colygin graben zone at the southern tip of Kamchatka is the northeast extension of the Kurile Basin. The seismic profiling and earthquake data thus seem to establish the movements over fractures along grabens of the central part of the Okhotsk Sea, which have been continued to the present.

Rather intensive seismicity is established only along the zones of the East-Sakhalin fracture, while deeper earthquake foci are generally characteristic of the central part of the Sea. Active tectonism within this region is also indicated by a high and sharply differentiated heat flow within the region (T. Exara and Mr. Langseth, 1975, personal communication), considerable differences in thermodynamic conditions within the earth's crust and the upper mantle (Tikhomirov, 1970), and also by the positive and negative isostatic anomalies for the central and northern regions of the Okhotsk Sea (Gainanov *et al*, 1974).

If we suppose that the formation of the structural basins of the Okhotsk Sea resulted from normal faulting, then horizontal tensional stresses would be largely perpendicular to the general trends of the fractures of these active and restricted grabens. This suggests the coexistence of differently oriented tensional stresses for this region, leading to faulting, basement subsidence, and basin formation. This is difficult to reconcile with the unidirectional stress system for apparent subduction at the margin of the Okhotsk Sea, along the Kurile Trench and Kurile Island Arc.

An important related question is the evolution of the oceanic crust underlying the Kurile Basin. Was this originally continental crust, now oceanized, or was it very ancient oceanic crust trapped behind the young Kurile Volcanic Arc. Only deep-ocean drilling may answer this problem, but our new knowledge of the age and structure of the shallower parts of the Okhotsk Sea provide major new constraints on the interpretation of the geological evolution of this important part of the northwestern Pacific margin.

Acknowledgments. We are particularly obliged to Dr. R. E. Dennison of Mobil Oil Corporation and Mr. E. S. Ovcharek of the Soviet Far East Geological Institute for age determinations and petrographic analyses, and also to the scientific staff and crew of the *R/V Mendeleev* during our cruise in the Okhotsk Sea.

---

[1]University of Texas Marine Science Institute Contribution No. 98

References

Burk, C.A., Sea of Okhotsk--Thirteenth Cruise of the Dmitry Mendeleev, Geology, p. 141-144, 1975.

Gainanov, A.G., Pavlov, Ya A., Stroev, P.A., Sychev, P.M., Tuyezov, I.K., Anomalous gravity fields of the Far East marginal seas and the adjacent part of the Pacific Ocean, Novosibirsk, Nauka, 108 pp., (in Russian), 1974.

Galperin, E.I. and Kosminskaya, I.P. (eds.), Structure of the Earth's crust within the Asia-to-Pacific transition zone from DSS data, Moskva, Nauka, 307 pp., (in Russian), 1964.

Gnibidenko, H.S., Geology and deep structure of Sakhalin, Kuril Islands and Kamchatka, in Island Arc and Marginal Sea, Ed. S. Asano and G. Udintsev, Tokai University Press, 5-23, (in Japanese with English Abstract), 1971.

Gnibidenko, H.S., Gorbachev, S.Z., Lebedev, M.M., Marakhanov, V.I., Geology and deep structure of Kamchatka Peninsula, Pacific Geology, 7, 1-32, 1972.

Kochergin, E.V., Krasny, M.L., Sychev, P.M., and Tuyezov, I.K., Anomalous geomagnetic field of northwestern part of the Pacific mobile belt and its relation to the tectonic structure, Geology and Geophysics, 12, 77-79, (in Russian), 1970.

Kulakov, A.P., Quaternary coastal lines of the Okhotsk and Japan Seas, Novosibirsk, Nauka, 188 pp., (in Russian), 1973.

Snegovskoi, S.S., Reflection surveys and tectonics of the Okhotsk Sea southern part and the adjacent Pacific areas, Novosibirsk, Nauka, 88 pp., (in Russian), 1974.

Tarakanov, R.Z., Kim Chun Un, Sukhomlinova, R.I., The structure peculiarities of the Kuril-Kamchatka focal zone and the Japanese regions, in Seismicity and deep structure of Siberia and Far East, Ed. S.L. Soloviev, Trudi SakhNKII, v. 39, (in Russian), 1975.

Tikhomirov, V.M., Thermodynamic conditions in the Earth's crust and upper mantle of the Okhotsk Sea, Kurile Islands and the Near-Kurile part of the Pacific Ocean, in Geologiya i geofizika Tihookeanskogo poyasa, Trudi SakhKNII, v. 25, Ed. H.S. Gnibidenko, 23-33, (in Russian), 1970.

Yanshin, A.L. (ed.), The structure of the Earth's crust and upper mantle of the transition zone from Asian continent to NW Pacific, Nauka, Novosibirsk, (in Russian), 1976.

VOLCANOES AS POSSIBLE INDICATORS OF TECTONIC STRESS ORIENTATION -- ALEUTIANS AND ALASKA

Kazuaki Nakamura*, Klaus Jacob and John Davies

Lamont-Doherty Geological Observatory, Palisades, New York 10964

A new method for obtaining tectonic stress orientation from volcanic structures, proposed by Nakamura (1969, and in press), was applied to the Aleutian and Alaskan volcanoes and volcanic fields. The method is essentially the recognition of the preferred orientation of radial and parallel dike swarm development by means of their probable surface manifestations, such as flank crater distribution. By the method one obtains for individual volcanoes and volcanic fields primarily the trend of the maximum compression of the horizontal components of the tectonic stress. When the method is applied on a regional scale, one can further identify the trend, either as the maximum or as the intermediate axis of the tectonic stress.

The trends of the volcanic features which indicate the maximum compression of the horizontal components were obtained from fifteen volcanoes, including Buldir, the westernmost Aleutian volcano, and Iliamna Volcano, near the eastern end of the Aleutian volcanic belt. Trends of the volcanic features generally coincide well with the azimuths of slip vectors for the relative motion between the Pacific and North American Plates (Minster et al., 1974), thus providing evidence that the obtained horizontal directions at these volcanoes coincide with the maximum compressional axis of the tectonic stress.

General east-west trends were obtained from nine volcanoes and volcanic fields on islands and at the mainland coast of the Alaska-Bering Sea shelf. These volcanoes and volcanic fields are mostly of alkali basalt (Smith et al., 1973) and are generally associated with normal faults of similar trends. Therefore, the obtained directions are most probably those of the intermediate axis, with the maximum compression axis in the vertical direction.

These results give strong support to the idea that the flank eruption on polygenetic volcanoes can be regarded as large-scale natural experiments for magmatic hydrofracturing. Moreover, they have important implications for the tectonics of island arcs and back-arc regions: (1) Volcanic belts of some island arcs, at least of the Aleutian arc, are under compressional stress. For such arcs, it is improbable to split and form a marginal basin; (2) The compressional stress at the arc, probably generated by the underthrusting, appears to be transmitted across the entire arc structure, but is apparently replaced within several hundred kilometers by a different, tensional stress system in the back-arc region; (3) Both stress systems, the compressional and the tensional ones, are associated with volcanoes and volcanic fields, which, however, differ in the chemistry of their magma. This difference supports the idea that the back-arc stress system has its own source at considerable depth beneath the crust.

*On leave from Earthquake Research Institute, University of Tokyo

References

Minster, J. B., Jordan, T. H., Molnar, P. and Hains, E., Numerical modelling of instantaneous plate tectonics, Geophys. Jour. Roy. Astr. Soc., 36, 541-576, 1974.

Nakamura, K., Arrangement of parasitic cones as a possible key to regional stress field, Bull. Volc. Soc. Japan, 17, 8-20, 1974, (in Japanese with English abstract)

Nakamura, K., Volcanoes as possible indicators of tectonic stress orientation - principle and proposal, Jour. Volc. Geoth. Res., (in press)

Smith, R. L. and Soule, C. E., Western Alaska and Bering Sea Island; Alaska Peninsula and the Aleutian Islands and Range, in Data sheets of the Post-Miocene volcanoes of the world with index maps, IAVCEI, Roma, 1973.

DEVELOPMENT OF SEDIMENTARY BASINS ON THE LOWER TRENCH SLOPE

G. F. Moore and D. E. Karig

Department of Geological Sciences, Cornell University, Ithaca, New York 14853

In arc systems where an oceanic plate with a thick sedimentary cover is subducted, these sediments are scraped off and accreted to the inner trench slope. The accreted sediments form ridges behind which sedimentary basins may form. The width of these basins increases from 2-3 km near the base of the inner slope to 10 km near the trench slope break. Sediments within the basins increase in thickness from a few meters in the basins now at greatest depths to several kilometers in the basins on the shallowest part of the slope.

Slope basins begin to form at the base of the lower slope where sediments accumulate between adjacent thrust faults. Addition of more accreted material at the trench causes uplift and rotation of the thrust slices and overlying slope sediments. As deformation proceeds, displacement along some of the thrusts dies out, and the inactive thrusts become buried by slope sediments, increasing the size of the slope basins. Continuing, but decreasing, movement on the remaining thrusts deforms the overlying slope sediments.

In orogenic belts, sediments that were deposited on a "basement" of mélange and are now tectonically enclosed by mélange are hypothesized to be ancient slope basin deposits. Examples are the Late Cretaceous-Early Tertiary sediments that now lie tectonically within belts of Franciscan mélange in northern California.

THE UYAK COMPLEX, KODIAK ISLANDS, ALASKA: A SUBDUCTION COMPLEX OF EARLY MESOZOIC AGE

William Connelly, Malcolm Hill, Betsy Beyer Hill, and J. Casey Moore

Earth Sciences Board, University of California, Santa Cruz, CA 95064

Abstract. The Uyak Complex of the northwest side of the Kodiak Islands is a lithologically chaotic assemblage of deep-sea sedimentary and igneous rocks and blueschist-bearing crystalline schists which we interpret as a subduction complex that was emplaced in the early Mesozoic. Thinly interlayered and highly deformed gray chert and argillite with minor green tuff comprise about 52% of this tectonic melange; the gray chert is highly recrystallized but scarce ghosts of radiolaria suggest its biogenic origin. Various-sized blocks of internally sheared arkosic wacke (Q:F:L = 29:61:10, P/F = 1.0, V/L = 1.0, C/Q = 0.3) account for about 20% of the Complex. Non-vesicular pillowed and massive greenstones comprise about 20% of the Uyak; the greenstones petrographically and geochemically (ca. 50.0% SiO_2, 16.2% Al_2O_3, 3.2% Na_2O, 0.5% K_2O, and 1.4% TiO_2) resemble ocean floor basalts. Small bodies of contorted radiolarian chert are scattered throughout the melange (ca. 2% by volume) and locally occur in sedimentary contact with underlying pillowed greenstone. Tectonically bound slabs containing layered gabbro, clinopyroxenite, dunite, and plagioclase peridotite occur along the northwest margin of the Uyak and account for about 6% of the Complex. The simplest interpretation of these lithologies is that the gabbroic and ultramafic rocks and greenstone represent basal oceanic crust upon which the radiolarian chert, gray chert and argillite, and wacke were deposited respectively at a mid-ocean rise, abyssal ocean floor, and oceanic trench. Because the green tuff is interbedded with gray chert and argillite but not with radiolarian chert, it most likely is an air-fall tuff blown seaward from a volcanic arc rather than an aquagene tuff from a spreading center. Two blocks of limestone have yielded mid-Permian fusulinids of Tethyan affinities and therefore may have been conveyed to the Uyak plate margin from warmer latitudes. Radiolaria extracted from a Uyak chert are Paleozoic in age (Upper Jurassic or Lower Cretaceous radiolaria occur in a chert body not clearly in the Uyak, but closely associated with it). A limestone block containing an Upper Triassic hydrozoan quite similar to a genus found on the Alaska Peninsula suggests derivation from the structurally overlying plate.

During underthrusting, brittle rock-types were broken into phacoids of all sizes and suspended in the less competent gray chert and argillite matrix with their longest dimensions aligned subparallel to the cataclastic foliation. Prehnite and pumpellyite developed extensively in lithologies of suitable composition. In addition, the Complex includes two crystalline schist bodies containing quartz-mica schist, greenschist, blueschist, and epidote amphibolite.

Similarities in lithology, style of deformation, degree of metamorphism, and structural position suggest a correlation between the Uyak Complex and rocks on southern Kenai Peninsula presumed to be an extension of the McHugh Complex of the Anchorage area. Together the Uyak-McHugh belt defines a probable subduction complex trending northeast for at least 600 km along the SW Alaska margin. Early Jurassic K-Ar ages from blueschist terranes on the Kodiak Islands and southern Kenai Peninsula provide one measure for the time of emplacement of this melange.

The Uyak-McHugh melange underthrusts Upper Triassic vesicular pillow lava (which geochemically resembles volcanic-arc tholeiite) and volcaniclastic turbidites in the Kodiak Islands and on southern Kenai Peninsula. These volcanogenic sections are biostratigraphically equivalent and lithologically similar to bedded rocks on the Alaska Peninsula, and together outline a forearc basin. Primary volcanic deposits of the forearc basin were derived from a magmatic arc (Alaska-Aleutian Range batholith) to the northwest and range in age from Norian through Callovian (ca. 210 to 165 myBP). This stratigraphic measure of the duration of arc volcanism is consistent with K-Ar ages of 176 to 154 myBP from the presently exposed granitic rocks of the Alaska-Aleutian Range batholith. We interpret this volcanoplutonic arc and forearc basin to be associated with Uyak subduction and therefore to provide an estimate for its time of emplacement; this estimate is consistent with fossil and radiometric ages from the melange.

Note Added in Proof

Recently collected chert from the Uyak Complex has yielded Lower Cretaceous (upper Valanginian to Hauterivian) radiolaria: see "note added in proof" at the end of our paper, Mesozoic Tectonics of the Southern Alaska Margin, for a discussion of how this modifies some of our interpretations.

THIN ELASTIC PLATE ANALYSIS OF OUTER RISES

J.G. Caldwell, D.L. Turcotte, W.F. Haxby, and D.E. Karig

Department of Geological Sciences, Cornell University, Ithaca, New York 14853

The existence of the outer rise can be explained in terms of the flexure of an elastic lithosphere. The assumption that the oceanic lithosphere behaves like a thin elastic plate allows the calculation of a universal trench profile which agrees quite well in detail with several bathymetric profiles (e.g., profiles across the Middle America, Aleutian, Kuril, Bonin, and Mariana trenches) which have been corrected for variations in sediment thickness and lithospheric age. No horizontal force is required to obtain good agreement between theory and observation.

This analysis allows the determination of the flexural rigidity from the geometry of the outer rise; values obtained are near 1×10^{30} dyne cm. This value corresponds to an effective elastic thickness of the lithosphere of about 25 km. Maximum bending stresses of about 6-9 kb are predicted by the analysis for the upper surface of the plate, which is the region of the outer trench wall where near-surface block faulting is known to occur.

It is found that the outer rise is very small and may even be non-existent on very young ocean oceanic lithosphere; an example is the lithosphere being subducted off of Mexico. This implies that the effective elastic thickness becomes very small for hot young lithosphere and suggests a strong dependence of lithospheric rheology on temperature.

CENOZOIC TECTONICS OF EAST ASIA

Maurice J. Terman

U. S. Geological Survey, Reston, Virginia 22092

A synthesis of data on Cenozoic sedimentation, volcanism, faulting, and seismicity in East Asia supports the following plate-tectonic interpretations (see fig. 1).

The pre-Cenozoic motion of Eurasia appears to have been generally to the southeast, and, in the Late Cretaceous, the continental plate is envisioned as fronted by northeast Siberia, Primorye, southwest Japan, Ryukyus, Taiwan, Philippines, Borneo, and Sumatra. An inherited spreading-center system is postulated to have been quite active in Marine East Asia during the Late Cretaceous and into the Cenozoic; it fed subduction zones that generally dipped west under most of the Asian front and also possibly dipped east under the western edge of the Pacific oceanic plate, envisioned as fronted by Koryak-Sakhalin, northeast Japan, and West Marianas Ridge. During the Cenozoic, the Marine East Asia spreading-center system was itself subducted to the east progressively from north to south until it disappeared west of the Marianas during the Miocene. Since the early Cenozoic, the Pacific plate motion has been taken up along the trenches extending from Kamchatka to Yap and along the Caroline Basin margins. While these events were taking place, the motion of Eurasia was considerably affected by the Pacific plate collision in the northeast during the Eocene, by the Indian continent in the southwest since the Oligocene, and by the Luzon arc and Australian continent in the southeast since the Miocene. These collisions, particularly that with the Indian continent, have effectively interlocked several cratonal nuclei of the agglomerated Asian continent, but the relative motion between Eurasia and North America persisted throughout the Cenozoic; the rotation pole shifted to northeast Siberia in the late Cenozoic.

Northern Asia has only minor Cenozoic geologic features, but an active seismic zone indicates tensional separation and compressive juncture of the northern parts of the Eurasian (Northwest Asia) and American (Northeast Asia) plates. North-Central East Asia is characterized by many graben, widespread alkaline volcanism, and scattered shallow seismicity. All these indicate crustal extension in a region between the central constrained cratons and the continental margin from Primorye to Taiwan, a margin that has had no spreading pressures against it since at least the middle Miocene; the extensional phenomena appear to have a rotation pole near the Kamchatka-Aleutian juncture. South-Central East Asia is another region that has only minor Cenozoic features except those associated with the continent-arc (Taiwan-Luzon) collision at its leading edge since the Miocene. Southeast Asia is characterized by extensional graben and alkaline volcanism in the northwest and compressional subduction and collision features in the east; the region appears to be rotating counterclockwise as a unit about a pole in the Philippine Sea.

Thus, major plate interactions in East Asia can be grouped into 1) simplistic subduction patterns for the Wharton Basin part of the Indian plate (Andaman Islands to Sumba) underthrusting to the north and the Pacific plate (Kamchatka to Caroline Basin) underthrusting to the west; and 2) complex subduction and collision patterns for the Asian continental plate front (Primorye to Taiwan to Banda arcs), generally overriding a relatively stationary Marine East Asia region. The Indian continent collision has caused much seismicity and faulting within Southwest Asia along Paleozoic plate sutures, and considerable local uplift and subsidence in reactivated fold systems.

This plate-tectonic scenario for Asia prompts the following explanatory speculations: Pre-Cenozoic motion of the composite Eurasian lithospheric plate was dependent on asthenospheric convection currents flowing horizontally from thermal rises below midoceanic spreading ridges (Arctic and North Atlantic) to thermal sinks below subduction zones (along the Asian front from northeast Siberia to Sumatra). Collisions with other lithospheric plates caused variable reactions in the Eurasian plate east of the central region of constrained cratons. Crustal stability, as in South-Central East Asia, suggests that plate rigidity is maintained as long as there are opposing lithospheric pressures at plate margins. Crustal extension, as in North-Central East Asia, suggests a stretching of the lithosphere above the continued flow of horizon-

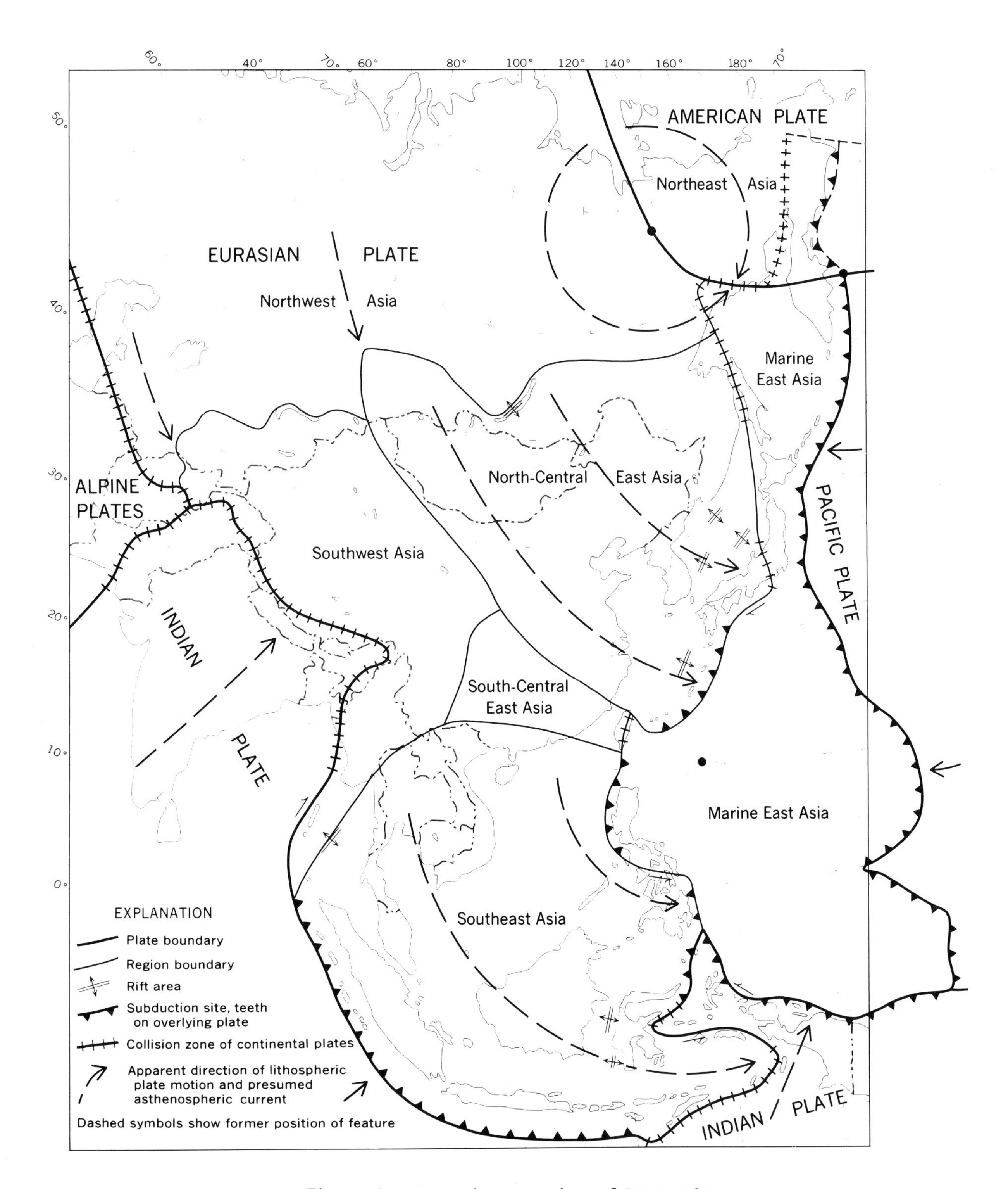

Figure 1. Cenozoic tectonics of East Asia.

tal asthenospheric currents away from a constrained cratonal nucleus and toward a plate margin against which there are no opposing currents or consequent lithospheric spreading pressures.

Although extensional phenomena take place more than 3,000 km from the Asian Continental edge (into Ozero Baykal), possibly the most important event tectonically might be the formation of two marginal seas or so-called back-arc basins (Sea of Japan and Okinawa Trough), as peripheral fragments of continental crust (respectively, southwest Honshu and Ryukyu Islands) rotate or migrate away from Asia, and new oceanic crust forms in their wake. The active overriding of oceanic crust by these continental fragments is speculated to be in response to the newly extended flow pattern of asthenospheric currents, and thus the principal period of formation of these marginal seas was not synchronous with the underthrusting of oceanic crust, but rather took place only after cessation of active spreading pressures against the Asian continent.